Ultra Low Power Bioelectronics

This book provides, for the first time, a broad and deep treatment of the fields of both ultra low power electronics and bioelectronics. It discusses fundamental principles and circuits for ultra low power electronic design and their applications in biomedical systems. It also discusses how ultra energy-efficient cellular and neural systems in biology can inspire revolutionary low power architectures in mixed-signal and RF electronics.

The book presents a unique, unifying view of ultra low power analog and digital electronics and emphasizes the use of the ultra energy-efficient subthreshold regime of transistor operation in both. Chapters on batteries, energy harvesting, and the future of energy provide an understanding of fundamental relationships between energy use and energy generation at small scales and at large scales. A wealth of insights and examples from brain implants, cochlear implants, biomolecular sensing, cardiac devices, and bio-inspired systems make the book useful and engaging for students and practicing engineers.

Rahul Sarpeshkar leads a research group on Bioelectronics at the Massachusetts Institute of Technology (MIT), where he has been a professor since 1999. This book is based on material from a course that Professor Sarpeshkar has taught at MIT for 10 years, where he has received both the Junior Bose Award and the Ruth and Joel Spira Award for excellence in teaching. He has won several awards for his interdisciplinary bioengineering research including the Packard Fellow Award given to outstanding faculty.

Ultra Low Power Bioelectronics

Fundamentals, Biomedical Applications, and Bio-inspired Systems

RAHUL SARPESHKAR

Massachusetts Institute of Technology

CAMBRIDGE UNIVERSITY PRESS
Cambridge, New York, Melbourne, Madrid, Cape Town, Singapore,
São Paulo, Delhi, Dubai, Tokyo

Cambridge University Press
The Edinburgh Building, Cambridge CB2 8RU, UK

Published in the United States of America by Cambridge University Press, New York

www.cambridge.org
Information on this title: www.cambridge.org/9780521857277

First published 2010

Printed in the United States of America

A catalogue record for this publication is available from the British Library

Library of Congress Cataloging-in-Publication Data

Sarpeshkar, Rahul
Ultra low power bioelectronics : fundamentals, biomedical applications, and bio-inspired systems /
Rahul Sarpeshkar.
    p.   cm.
  ISBN 978-0-521-85727-7 (hardback)
  1. Low power systems. 2. Biomimicry. 3. Bioelectronics. 4. Bionics. 5. Medical
electronics. I. Title.
  TK7881.15.S27 2010
  621.381–dc22

                                                                    2009047509

ISBN 978-0-521-85727-7 Hardback

Additional resources for this publication: www.cambridge.org/9780521857277

To my father, Pandi
who infected me with an enthusiasm and love for science.

To my mother, Nalini
whose boundless love and belief in me form the core of my being.

To my wife, Morgen
whose beauty, sweetness, wisdom, and infinite love empower me to be the man that I am.

Contents

Acknowledgements

I have many people to thank from whom I have learned a tremendous amount, with whom I have worked, and who have generously helped me in many ways. Without doubt, my biggest thanks go to my past and present graduate students and researchers whose joint work with me over the past decade has informed nearly every chapter of this book. These students and researchers include Scott Arfin, Michael Baker, Bruno do Valle, George Efthivoulidis, Daniel Kumar, Timothy Lu, Soumyajit Mandal, Alex Mevay, Micah O'Halloran, Benjamin Rapoport, Christopher Salthouse, Ji Jon Sit, Maziar Tavakoli, Lorenzo Turicchia, Woradorn Wattanapanitch, Keng Hoong Wee, Heemin Yang, and Serhii Zhak. Several students have served as teaching assistants and have helped improve the course that led to this book.

I would especially like to thank Soumyajit Mandal. While the work of most students has significantly influenced a portion of a chapter or a chapter, his joint work with me has significantly influenced no less than four chapters of this book, and his help in teaching versions of the class on which this book is based has been tremendous. He has speedily and generously conducted experiments for me on many occasions, provided useful feedback on many chapters, helped create online problems, and I have learned a tremendous amount about RF design from him.

Scott Arfin's assistance with redrawing and organizing many figures in this book has been critical and his help in gathering data has been very useful as well. Lorenzo Turicchia's generous help with MATLAB and insight into algorithms has been invaluable. Benjamin Rapoport's wonderful artwork forms the cover of this book and his medical perspective has been enriching. Woradorn Wattanapanitch has provided reviews of numerous chapters in the book, has caught various bugs, which have been very helpful in improving the book, and helped create online problems. Daniel Kumar's help with locating references and with simulations has been extremely useful, and he has helped create online slides. Bruno do Valle has provided valuable feedback on numerous chapters in the book, and helped create online problems.

I owe a great debt to my PhD adviser, Carver Mead. His powerful scientific intuition, broad perspective, intellectual courage, and pioneering vision have helped shape my own intellectual outlook in science and engineering. John Wyatt first turned my attention to bioelectronics as an undergraduate through his retinal implant project for the blind. Since that time, he has been my constant friend and mentor. I will always be obliged to him for his unshakable belief in me and his unfailing support of me. Jim Roberge's work on operational amplifiers and feedback has influenced my thinking on circuits since my undergraduate days at MIT. Yannis Tsividis's pioneering work on the MOS transistor, Eric Vittoz's pioneering work on subthreshold design, and the EKV model developed by Enz, Krummenacher, and Vittoz have strongly influenced the chapters on device physics in the book. The god-like mastery of both theory and experiment of my middle- and high-school chemistry teacher, David Chatterjee, taught me that it is the combination of theory and experiment that makes science and engineering so mysterious, powerful, interesting, magical, and divine.

Several colleagues and researchers in biology and engineering have been a pleasure to collaborate with in the past or in the present. Work that I have or am jointly performing with a few of them is featured in parts of the book; all have enriched my knowledge. I would like to thank Richard Andersen, Steve Beck, Emilio Bizzi, Edward Boyden, Peter Dallos, Rodney Douglas, Ananth Dodabalapur, Emad Eskandar, Michael Faltys, Michale Fee, Frank Guenther, Neil Gershenfeld, Richard Hahnloser, Jongyoon Han, Van Harrison, Christof Koch, Philip Loizou, Derek Lovley, Richard Lyon, Scott Manalis, Andrew Oxenham, Flavio Pardo, John Pezaris, Bhiksha Raj, Sridevi Sarma, Bent Schmidt-Nielsen, Hugh Secker-Walker, Sebastian Seung, and Bruce Tidor.

The following colleagues in engineering, medicine, biology, and physics have been an especial pleasure to interact with. Their perspectives, insights, and work have shaped my own knowledge and understanding. Some have helped review chapters and sections in the book. I would like to thank Jont Allen, John Allman, Andreas Andreou, Alyssa Apsel, Arvind, Friedrich Barth, Sangeeta Bhatia, Bernhard Boser, Raj Bridgelall, Bill Brownell, Gyorgy Buszaki, Gert Cauwenberghs, Ralph Cavin, Dmitiri Chklovskii, Isaac Chuang, James Collins, Eugene Culurciello, Munther Dahleh, Suman Das, Tobias Delbrück, John Doyle, Christian Enz, Ralph Etienne-Cummings, Dennis Freeman, Julius Georgiou, Maysam Ghovanloo, Barrie Gilbert, Ann Graybiel, Glen Griffith, John Guttag, Ali Hajimiri, Reid Harrison, Paul Hasler, John Hopfield, Timothy Horiuchi, Allyn Hubbard, Krzysztof Iniewski, Joe Jacobson, Charles Jennings, David Kewley, Peter Kinget, Jan Korvink, Abhi Kulkarni, Tor Sverre Lande, Robert Langer, Doug Lauffenburger, Simon Laughlin, Shih-Chih Liu, Timothy Lu, Hideo Mabuchi, Sanjoy Mahajan, Sunit Mahajan, Rajit Manohar, Robert Metcalfe, Bradley Minch, Lakshmi Misra, Partha Mitra, Sanjoy Mitter, Pedram Mohseni, Joe Paradiso, Dharmasena Peramunge, David Perrault, Lena Petersen, Tomaso Poggio, Behzad Razavi, Joseph D. Rizzo III, Tamás Roska, Mario Ruggero, Mohamad Sawan, Terry Sejnowski, Christopher Shera, Jean-Jacques Slotine, Shihab Shamma, Kenneth N. Stevens, Christofer Toumazou, Luke Theogarajan, Alexander van Oudenarden, George Verghese, Eric Vittoz, Lloyd Watts, Kensall Wise, Guang Zong Yang, Yuan Ting Zhang, and George Zweig.

I would like to thank those students who have taken my class, and helped me improve it through critical feedback. Some have done work in their class project, that have led to publications, including Jose Bohorquez, Leon Fay, Vinith Misra, and William Sanchez.

I would like to thank lab and department heads at MIT, who have encouraged my work over the years and who have ensured that politics and bureaucracy never slowed its progress. I would especially like to thank Jeffrey Shapiro, Rafael Reif, Duane Boning, John Guttag, Srinivas Devadas, Eric Grimson, and Robert Desimone. I would also like to thank all the present and former staff at the Research Laboratory of Electronics at MIT, particularly William Smith, Mary Young, and Nan Lin for their prompt and helpful responses to my administrative needs.

I would like to thank the sponsors that have supported research in my lab over the years, and without whose support a large fraction of the knowledge and systems described in the book would never have been discovered or built. They include the Office of Naval Research, the Advanced Bionics Corporation, the National Institutes of Health, the David and Lucille Packard Foundation, the Defense Advanced Research Projects Agency, Mitsubishi Electric Research Laboratories, Symbol Technologies, BAE Systems, the Center for Integration of Medicine and Innovative Technology, the National Science Foundation, the McGovern Institute NeuroTechnology program, and the Swartz Foundation.

I would like to thank the team at Cambridge University Press, who contributed to help produce this book. In particular, I would like to thank Philip Meyler, whose brief conversation with me one afternoon initiated this whole project. I would also like to thank Julie Lancashire who seeded me with ideas for how to organize and title the book. Finally, I would like to thank Sarah Matthews for her prompt and efficient management of the book's production and Sabine Koch for her assistance and patience.

I would especially like to thank Jay Nungesser of Big Blur Design Inc., whose incredibly fast skill and talent with figures was vital to producing this book. He has been a pleasure to work with.

Finally and most importantly, I would like to thank my wife, Morgen. She has given selflessly of her time and has been my anchor and support throughout the incredibly intense years during which this book was written. She has been infinitely patient, sweet, and supportive, and has contributed insight, clarity, and wisdom to much of my writing. Her attention to detail, administrative assistance, proof-reading, and organization propelled the book into its finished state. But most of all, without her abiding love and encouragement, I would not have been able to write this book. This book is primarily dedicated to her.

Rahul Sarpeshkar
Cambridge, Massachusetts

Section I

Foundations

1 The big picture

It is the harmony of the diverse parts, their symmetry, their happy balance; in a word it is all that introduces order, all that gives unity, that permits us to see clearly and to comprehend at once both the ensemble and the details.

It is through science that we prove, but through intuition that we discover.

Henri Poincaré

This book, *Ultra Low Power Bioelectronics*, is about ultra-low-power electronics, bioelectronics, and the synergy between these two fields. On the one hand it discusses how to architect robust ultra-low-power electronics with applications in implantable, noninvasive, wireless, sensing, and stimulating biomedical systems. On the other hand, it discusses how bio-inspired architectures from neurobiology and cell biology can revolutionize low-power, mixed-signal, and radio-frequency (RF) electronics design. The first ten chapters span feedback systems, transistor device physics, noise, and circuit-analysis techniques to provide a foundation upon which the book builds. Chapters that describe ultra-low-power building-block circuits that are useful in biomedical electronics expand on this foundational material, followed by chapters that describe the utilization of these circuits in implantable (invasive) and noninvasive medical systems. Some of these systems include cochlear implants for the deaf, brain implants for the blind and paralyzed, cardiac devices for non-invasive medical monitoring, and biomolecular sensing systems. Chapters that discuss fundamental principles for ultra-low-power digital, analog, and mixed-signal design unify and integrate common themes woven throughout the book. These principles for ultra-low-power design naturally progress to a discussion of systems that exemplify these principles most strongly, namely biological systems. Biological architectures contain many noisy, imprecise, and unreliable analog devices that collectively interact through analog and digital signals to solve complex tasks in real time, with precision, and with astoundingly low power. We provide examples of how bio-inspired systems, which mimic architectures in neurobiology and cell biology, lead to novel systems that operate at high speed and with low power. Finally, chapters on batteries, energy harvesting, and the future of energy discuss tradeoffs between energy density and power density, which are essential in architecting an overall low-power system, both at small scales and at large scales.

The book can serve as a text for senior or graduate students or as a reference for practicing engineers in the fields of

- Ultra-low-power Electronics: Chapters 1 through 22, 25, and 26.
- Biomedical Electronics: Chapters 1 through 22, 25, and 26.
- Bio-inspired Electronics: Chapters 1 through 18, 21 through 26.
- Analog and Mixed-Signal Electronics: Chapters 1 through 24.

In this busy day and age, many people with an interest in these fields may not have the time to read a whole book, especially one of this size. Therefore, the book has been written so that a reader interested in only a chapter or two can read the chapter and delve deeper if he/she would like. There is a slight amount of redundancy in each chapter to enable such sampling, with interconnections among the various chapters outlined throughout every chapter. The index should also be useful in this regard. Every reader should read Chapter 1 (this chapter). Chapter 2 on the fundamentals of feedback is also essential for a deeper understanding of many chapters. Chapters 1 through 10 provide a firm foundation, necessary for a deep understanding of the whole book.

Throughout this book, intuitive, geometric, and physical thinking are emphasized over formal, algebraic, and symbolic thinking. Physical intuition is extremely important in getting systems to work in the world since they do not always behave like they do in simulations or as the mathematical idealizations suggest they do. When the mathematics becomes intractable, usually the case in all but the simplest linear and idealized systems, intuitive and physical thinking can still yield powerful insights about a problem, insights that allow one to build high-performance circuits. Practice in physical thinking can lead to a lightning-fast understanding of a new circuit that lots of tedious algebra simply can never provide. Nevertheless, one must attempt to be as quantitative as possible for a deep understanding of any system and for theory and experiment to agree well. Thus, the book does not aim to substitute qualitative understanding for quantitative understanding; rather it attempts to maximize insight and minimize algebraic manipulations. We will always aim to look at problems in a physically insightful and original way such that the answer is intuitive and can be obtained exactly and quickly because the picture in our heads is clear.

Feedback is so fundamental to a deep understanding of how circuits work and how biology works that we shall begin this book with a review of feedback systems in Chapter 2. We shall see in this chapter that feedback is ubiquitous in physical, chemical, biological, and engineering systems even though the importance of feedback has been largely unappreciated. Throughout the book, we shall draw on our knowledge of feedback systems to derive or interpret results in a simple way that would not be possible without the use of this knowledge. For example, our discussion of physics in an MOS transistor will often use feedback analogies to understand the physics of their operation intuitively in Chapters 3 and 4. The equations of electron velocity saturation in an MOS transistor will be represented as a feedback loop in Chapter 6. We shall often avoid tedious Kirchoff's current law algebraic equations by simply drawing a feedback loop to provide all the answers for any transfer function, noise or offset analysis, robustness analysis, or dynamic analysis that we may need. We shall use feedback interpretations to understand how the noise in a transistor is affected by internal feedback within it. A deep understanding of feedback and circuits can enable a unified understanding of several systems in the world.

In both biomedical and bio-inspired electronics, it is important to deeply understand the biology. To understand and mimic biological systems in this book, we shall use circuits as a primary language rather than mathematics. Several nonlinear partial differential equations and structures in biology then translate into simple intuitive, lumped or distributed circuits. For example, we use such circuits to mimic the inner ear or cochlea, to understand the retina in the eye, to understand and mimic the heart, to mimic the vocal tract, to mimic spiking (pulsatile) neurons in the brain, and to understand and mimic biochemical gene–protein and protein–protein molecular networks within cells. Such circuits can help make engineers and physicists more comfortable with biology because it is described in a familiar language. Distributed circuits will help us understand Maxwell's equations and antennas intuitively. Circuits will help us quickly understand chemical reactions. Circuits will even help us understand the energy efficiency of cars.

In the rest of this chapter, we shall summarize some themes, ideas, principles, and biomedical and bio-inspired system examples that are discussed in depth elsewhere in the book. In this introductory chapter, the aim is to provide an intuitive 'big picture' without getting caught up in details, citations, proofs, mathematical equations and definitions, subtleties, and exceptions, which are addressed in the remaining chapters of the book. We shall start by discussing the importance of ultra-low-power electronics. We shall describe a power-efficient regime of transistor operation known as the subthreshold regime, which is enabling in low-power design. We shall then discuss important connections between information, energy, and power. We shall highlight some key themes for designing ultra-low-power mixed-signal systems that have analog and digital parts. We shall discuss examples of biomedical application contexts for low-power design, and fundamental principles of low-power design that are applicable to all systems, analog or digital, electronic or biological. After providing some numbers for the amazing energy efficiency of biological systems, we shall briefly discuss examples of systems inspired by neurobiology and by cell biology. Then, we provide a discussion of batteries and other energy sources, highly important components of low-power systems at small scales and at large scales. Finally, we shall conclude with a summary of the book's sections and some notes on conventions followed in the book.

1.1   Importance of ultra-low-power electronics

Ultra-low-power electronics in this book usually refers to systems that operate anywhere from a pico- to a milliwatt. However, the principles of ultra-low-power design are useful in all kinds of systems, even in low-power microprocessors, that dissipate 1 W, say, rather than 30 W, without compromising performance. In general, ultra-low-power electronic design is important in five different kinds of systems:

1. Systems that need to be portable or mobile and therefore operate with a battery or other energy source of reasonable size with a long lifetime or long time between recharges. The more miniature the system, the smaller the energy source, and the more stringent is the power constraint.
2. Systems that function by harvesting relatively small amounts of energy from their environment.
3. Systems that need to minimize heat dissipation.
4. Complex systems with many devices whose complexity simply will not scale unless the power and consequently heat dissipation of each of their components is kept within check.
5. Systems where the overall cost is a strong function of the size of the system or the cost of the battery in the system.

Biomedical systems are examples of systems where ultra-low-power electronics is paramount for multiple reasons. For example, biomedical systems that are implanted within the body need to be small and lightweight with minimal heat dissipation in the tissue that surrounds them. In some systems like cardiac pace-makers, the implanted units are often powered by a non-rechargeable battery. In others, like cochlear implants, the implants are traditionally constantly powered by rectified wireless energy provided by a unit outside the body. In either case, power dissipation dictates the size of the needed receiving coil, antenna, or battery, and therefore sets a minimal size constraint on the system. Size is important to ensure that there is space within the body for the implant and that surgical procedures are viable. Implant-grade batteries need conservative short-circuit-protection mechanisms to mitigate concerns about battery shorting and resultant tissue heating within the body. They are relatively expensive and the cost of implanted systems is strongly impacted by the costs of the battery. The costs of hermetic sealing and bio-compatibility of electronic implants are size dependent as well. A fully implanted system with a battery that has a limited number of wireless recharges must operate under stringent low-power constraints such that constant surgery is not needed to change the battery in a patient. The system must ideally function for 10 to 30 years without the need for a battery replacement. Implanted systems with ultra capacitors are capable of more recharges than batteries, but they have low energy densities such that more frequent recharging is necessary. Thus, ultra-low-power operation will always be paramount in implantable biomedical systems.

Noninvasive biomedical systems like cardiac medical tags are attached to the skin or to clothing for patient monitoring. They are powered by received RF energy or by a battery and also need to operate in an ultra-low-power fashion. In certain lab-on-a-chip or biomolecular-sensing instrumentation, ultra-low-noise and precision electronics is often more important than ultra-low-power operation. Fortunately, a good ultra-low-power designer is also a good ultra-low-noise designer, since both kinds of designers need to be skilled in the management of circuit noise. Therefore, many of the techniques for ultra-low-power design that we discuss in this book will also enable the reader to become a skilled ultra-low-noise designer. For example, we will discuss how to architect electronics for a micro-electro-mechanical-system (MEMS) vibration sensor with 0.125 parts-per-million sensitivity (23-bit precision), and for ultra-low-noise and micropower neural and cardiac amplifiers in the book.

This book will, in large part, use biomedical systems as examples. However, the principles, techniques, and circuits that we describe in this book are general and are applicable in several other portable systems such as in cell phones, next-generation radios, sensor networks, space, and military hardware. In fact, many of the principles of ultra-low-power design that we shall explicitly and implicitly discuss, e.g., energy recycling, apply to non-electrical systems and at much larger power scales as well. They can and already are being exploited in the design of low-power mechanical systems such as next-generation cars. Cars, which operate on average at nearly 42 kW when going at 30 mph today, cannot afford to operate with such power consumption in our planet's future.

1.2   The power-efficient subthreshold regime of transistor operation

Since power is the product of voltage and current, low-power systems must necessarily operate at low voltages and/or low currents[1]. Electronic systems that run on 0.5 V power-supply voltages already exist. It is hard to scale the power-supply voltage of electronic circuits below 0.25 V and preserve reliable perfor-mance in digital circuits, as we discuss in Chapter 21. The performance of analog circuits significantly degrades at such low power-supply voltages. Thus, power savings via voltage reduction is inherently limited in both the digital and the analog domains. For ultra-low-power systems, it is more promising to focus on currents, which can be scaled from pA to mA levels in modern-day MOS transistors with appropriate use of transistor geometries and bias voltages in the subthreshold regime of transistor operation.

The subthreshold region of operation is present in a transistor when it is operated below its *threshold voltage*, a region where digital circuits are sometimes approximated as being 'turned off'. In reality, the threshold voltage is merely a useful approximation to describe transistor operation. Just as the current in a diode decreases exponentially as the voltage across it is decreased, the current in a subthreshold transistor decreases exponentially as the magnitude of the transistor's gate-to-source voltage is decreased. Nevertheless, just as it is sometimes useful to

---

[1] In non-dissipative devices like ideal inductors or capacitors, current and voltage variables are always orthogonal to each other, such that power is never dissipated even at high voltages and/or high currents. However, even in such devices, in practice, finite dissipative losses are minimized with low voltages and/or low currents.

view a diode as having a 'turn-on threshold' of 0.6 V due to the steep nature of its exponential current–voltage characteristics, it is sometimes useful to view a transistor as having an abrupt threshold voltage at which it turns on.

The 'leakage current' in a transistor that is turned off in a digital system is dominated by the transistor's subthreshold current. Such leakage current can be considerable in large digital systems. For example, 100 million transistors × 1 nA of leakage current per transistor yields 100 mA of standby leakage current. Due to the lowering of the threshold voltage of the transistor in advanced transistor processes, the absolute value of the subthreshold leakage current increases as MOS technologies progress towards ever-smaller dimensions. Subthreshold operation also occupies an increasingly larger fraction of the range of power-supply operation as transistor sizes get progressively smaller. For all of these reasons, there has been a great renewal of interest in the subthreshold regime of operation. Subthreshold operation in both analog and digital circuits has almost become a necessity.

The maximal frequency of operation possible in diffusion-current-determined subthreshold operation scales inversely with the square of the transistor's channel length. In contrast, the maximal speed of velocity-saturated above-threshold operation only scales inversely with the channel length. Thus, subthreshold operation is rapidly allowing faster speeds of operation and may no longer be viewed as a 'slow regime' of operation of the transistor. For example, 1 GHz analog preamplifiers with all-subthreshold operation can now be built in a 0.18 μm process and digital circuits can be made to operate at such speeds as well.

The subthreshold region of operation is a region where the bandwidth available per ampere of current consumed in the transistor is maximal. The power-supply voltage needed for subthreshold operation is also minimal since the saturation voltage of a transistor in subthreshold is only 0.1 V. Due to the high bandwidth-per-ampere ratio and the ability to use small power-supply voltages, the bandwidth per watt of power consumed in the transistor is maximized in its subthreshold regime. Consequently, subthreshold operation is the most power-efficient regime of operation in a transistor.

For all these reasons, this book focuses heavily on the use of subthreshold circuits for ultra-low-power electronic design in the analog and the digital domains. In the subthreshold region of operation, often also referred to as the weak-inversion region of operation, it is important to ensure that systems are robust to transistor mismatch, power-supply-voltage noise, and temperature variations. Hence, circuit biasing and feedback techniques for ensuring robustness are important. We shall discuss them in various contexts in the book, but particularly in Chapter 19, where we discuss the design of ultra-low-power biomedical system chips for implantable applications, and in Chapter 22 on ultra-low-power digital design. Furthermore, the subthreshold regime is characterized by relatively high levels of noise, since there are few electrons per unit time to average over. Thus, throughout the book, we shall discuss device noise, how to mitigate it, how to analyze it, how to design around it, and, in some cases, how to even exploit it.

Since ultra-energy-efficient biological systems also operate with Boltzmann exponential devices, subthreshold operation is highly useful in mimicking their operation. Thus, subthreshold operation is enabling in bio-inspired systems as well.

1.3   Information, energy, and power

Information is always represented by the states of variables in a physical system, whether that system is a sensing, actuating, communicating, controlling, or computing system or a combination of all types. It costs energy to change or to maintain the states of physical variables. These states can be in the voltage of a piezoelectric sensor, in the mechanical displacement of a robot arm, in the current of an antenna, in the chemical concentration of a regulating enzyme in a cell, or in the voltage on a capacitor in a digital processor. Hence, it costs energy to process information, whether that energy is used by enzymes in biology to copy a strand of DNA or in electronics to filter an input.[2] To save energy, one must then reduce the amount of information that one wants to process. The higher the output precision and the higher the temporal bandwidth or speed at which the information needs to be processed, the higher is the rate of energy consumption, i.e., power. To save power, one must then reduce the rate of information processing. The information may be represented by analog state variables, digital state variables, or by both. The information processing can use analog processing, digital processing, or both.

The art of low-power design consists of decomposing the task to be solved in an intelligent fashion such that the rate of information processing is reduced as far as is possible without compromising the performance of the system. Intelligent decomposition of the task involves good architectural system decomposition, a good choice of topological circuits needed to implement various functions in the architecture, and a good choice of technological devices for implementing the circuits. Thus, low-power design requires a deep knowledge of devices, circuits, and systems. This book shall discuss principles and examples of low-power design at all of these levels. Figure 1.1 shows the "low-power hand". The low-power hand reminds us that the power consumption of a system is always defined by five considerations, which are represented by the five fingers of the hand: 1) the task that it performs; 2) the technology (or technologies) that it is implemented in; 3) the topology or architecture used to solve the task; 4) the speed or temporal bandwidth of the task; and, 5) the output precision of the task. As the complexity, speed, and output precision of a task increase, the rate of information processing is increased, and the power consumption of the devices implementing that task increases.

---

[2] In Chapter 22, we shall see that, technically, if one operates infinitely slowly and in a manner that allows the states of physical variables to be recovered even after they have been transformed, energy need not be dissipated. In practice, in both natural and artificial systems, which cannot compute infinitely slowly, and which always have finite losses, there is always an energy cost to changing or maintaining the states of physical variables.

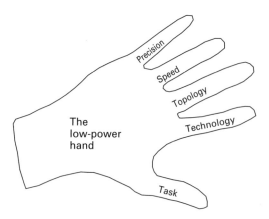

Figure 1.1. The low-power hand.

1.4   The optimum point for digitization in a mixed-signal system

The problem of low-power design may be formulated as follows: Suppose we are given an input $\mathbf{X}$, an output function $\mathbf{Y}(t) = f(\mathbf{X}, t)$, basis functions $\{i_{out1} = f_1(\mathbf{v}_{in}), i_{out2} = f_2(d\mathbf{v}_{in}/dt), i_{out3} = f_3(\int \mathbf{v}_{in}), ..\}$ formed by the current-voltage curves of a set of technological devices, and noise-resource equations for devices in a technology that describe how their noise or error is reduced by an increase in their power dissipation for a given bandwidth; such noise-resource equations are described in Equation (22.4) in Chapter 22. Then, find a topological implementation of the desired function in terms of these devices that maximizes the mutual information between the actual output $\mathbf{Y}(t)$ and the desired output $f(\mathbf{X},t)$ for a fixed power-consumption constraint or per unit system power consumption. Area may or may not be a simultaneous constraint in this optimization problem. A high value of mutual information, measured in units of bits per second, implies that the output encodes a significant amount of desired information about the input, with higher mutual information values typically requiring higher amounts of power consumption.

Hence, low-power design is in essence an information-encoding problem. How do you encode the function you want to compute, whether it is just a simple linear amplification of a sensed signal or a complex function of its input, into transistors and other devices that have particular basis functions given by their current-voltage curves? Note that this formulation is also true if one is trying the minimize the power of an actuator or sensor, since information is represented by physical state variables in both, and we would like to sense or transform these state variables in a fashion that extracts or conveys information at a given speed and precision. In non-electrical systems, through (current) and across (voltage) variables play the roles of current and voltage, respectively. For example, in a fluid-mechanical system, pressure is analogous to voltage while volume velocity of fluid flow is analogous to current.

A default encoding for many sensory computations in man-made systems is a high-speed high-precision analog-to-digital conversion followed by digital signal processing that computes $f(\ )$ with lots of calculations. This solution is highly flexible and robust but is rarely the most power-efficient encoding. It does not exploit the fact that the actual meaningful output information that we are after is often orders-of-magnitude less than the raw information in the numbers of the signal, and the fact that the technology's basis functions are more powerful than just switches. It may be better to preprocess the information in an analog fashion before digitization, and then digitize and sample higher-level information at significantly lower speed and/or precision.

A familiar example in the field of radio engineering illustrates why analog preprocessing before digitization is important in lowering overall system power. Suppose a radio with a 1 GHz carrier frequency is built with a 16-bit analog-to-digital converter (ADC) such that all operations are done directly in the digital domain. Such a design affords maximum flexibility and robustness and allows one to build a 'software radio' that can be programmed to work at any carrier frequency up to 1 GHz as long as a broadband antenna is available and to function over a nearly 96 dB input dynamic range. The power consumption of the ADC in the software radio alone would be at least 1 pJ/(quantization level) $\times 2^{16}$ quantization levels $\times 4 \times 10^9$ Hz $= 256$ W, an unacceptable number for the few hundreds of milliwatt power budget of a cell phone! The figure of 1 pJ/(quantization level) represents a very optimistic estimate for the energy efficiencies of ADCs working at such simultaneously high speeds and precisions, thus far never reported. This figure also assumes that the precision of the ADC is not thermal noise limited, an optimistic assumption, and that the signal is sampled at four times the Nyquist rate, typical in many applications. The digital processor has to process 16-bit numbers at 10^9 Hz, which makes its power consumption high as well. It is no wonder that most radios are built with analog preprocessing to amplify and mix the high-bandwidth signal down to baseband frequencies where it can be digitized by a much lower-speed ADC and processed by a digital signal processor (DSP) operating at a low information rate. The power of the ADC and the power of the DSP are then significantly reduced. The analog circuits are built with power-efficient passive filters, power-efficient tuned low-noise amplifiers, power-efficient oscillators, and power-efficient active mixers. The extra degrees of freedom inherent in analog processing, i.e., every signal is continuous and not just a '1' or a '0', and the fact that every device has a technological input-output curve that can be exploited in the computation and is not just a switch, saves energy. Function approximation is not merely done with quantization and Boolean logic functions but with basis functions which are fewer and more efficient for the task. In radios, the basis functions inherent in inductors and capacitors allow for very energy-efficient, relatively noise-free filtering.

One of the themes of this book is that the delayed-digitization example in the preceding paragraph generalizes to other designs where power consumption needs to be optimized. In general, Figures 1.2 (a) and 1.2 (b) illustrate how the overall

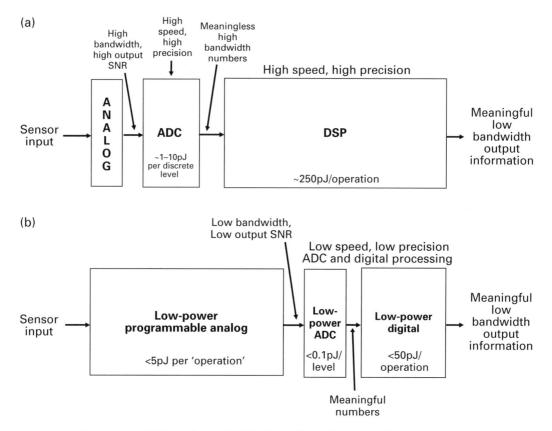

Figure 1.2a, b. (a) A traditional ADC-then-DSP architecture; (b) A more energy-efficient mixed-signal architecture. The numbers shown represent bounds that are constantly improving for both analog and digital technologies and that vary with implementation details.

power of an electronic system is lowered by delaying digitization. Figure 1.2 (a) shows a method of processing information where digitization is immediate, while Figure 1.2 (b) shows a method for processing information where digitization is delayed. The delaying of digitization leads to a significant lowering of power in the ADC and in the digital portions of the system but an increase in the analog power consumption. There is an optimal point for digitization where the overall power is minimized that varies depending on what is being computed. The lowering of power arises not only because speed and/or precision are lowered in the ADC or in the digital processing but also because it is easier to design energy-efficient ADCs and low-power digital systems at lower speeds and/or precisions. We shall discuss ultra-low-power ADCs in Chapter 15 and principles for ultra-low-power digital design in Chapter 21.

Several biomedical applications can exploit the architecture of Figure 1.2 (b) to lower power consumption. In these applications, the meaningful information bandwidth at the output is far less than the information bandwidth at the input determined by the raw input signal dynamic range and the maximum input frequency. For example, in cochlear-implant applications, 16-bit speech information sampled at 44 kHz, needed for wide-input-dynamic-range and audio bandwidth operation, yields 704 kb/s of raw input information; the 16-electrode output information after processing corresponds to 5-bit current-amplitude information at a 200 Hz sample rate, i.e., 16 kb/s of actual discriminable information by a subject. In a paralysis prosthetic, 100 neuronal electrodes in the brain may provide 8-bit, 30 kHz data or 24 Mb/s of raw input information; the output information rate corresponds to 3 output motor-control parameters, updated at 100 Hz and 8-bit precision, i.e., 2.4 kb/s, at best. In a pulse-oximetry application, the input photo-plethysmographic information is about 16 kb/s while the meaningful output information about oxygen saturation may be a few bits per second. All three of these applications will be discussed in Chapters 19 and 20. In such applications, the noise and mismatch inherent in analog devices serve to reduce the information content of the computation during analog preprocessing. However, if sufficient degrees of freedom are allocated to combat these, the output information content, typically significantly less than the raw input information content, is still robustly preserved. In essence, degrees of freedom in analog processing are not wasted in making every device and every signal robust but in making the overall output robust. The book provides concrete examples of systems that are architected in such a fashion to be robust to thermal noise, transistor mismatch, $1/f$ noise, temperature variations, power-supply noise, and mixed-signal cross talk. Chapter 19 discusses an ultra-low-power analog cochlear-implant or bionic-ear processor in depth to illustrate how such principles are concretely applied and the various system tradeoffs that are involved in any practical engineering design. In this processor, 373 digitally programmable bits allow 86 patient parameters to be changed. Thus, although the architecture of Figure 1.2 (b) is not as flexible as that of Figure 1.2 (a), enough flexibility is preserved such that the flexibility-efficiency tradeoff is at a more optimal point [1].

Since the analog preprocessor and the ADC are digitally calibrated or programmed and output digital bits, they can together appear as an energy-efficient 'coprocessor' from the point of view of the digital system. Other designs in the medical-electronics industry have also used low-power analog preprocessing to reduce power, for example, in electroencephalogram (EEG) systems [2]. The reliability of an analog system can be made to be as high as that of any ADC, and analog systems typically degrade gracefully rather than catastrophically. Thus, their use in a medical context where graceful degradation is important can actually be an advantage.

The traditional general-purpose architecture of ADC-then-DSP processing in Figure 1.2 (a) is highly useful for initial exploration of a system when one desires maximum flexibility because one does not know exactly what one must do. After this exploratory stage, when one knows what one must do, the task must be crafted with a more specialized mixed-signal architecture as in Figure 1.2 (b) if power consumption is to be minimized; otherwise, the extra degrees of freedom inherent in the general-purpose architecture will waste energy without performing any useful function in the special-purpose task. Even a small amount of analog preprocessing such as an analog automatic-gain-control system before digitization can be advantageous. An option that preserves both flexibility and efficiency is to have the high-power flexible system only periodically turn on to calibrate and reprogram parameters in the low-power less-flexible-but-highly-efficient system.

To architect the low-power system in Figure 1.2 (b), the output after analog preprocessing must be digitized and then processed digitally. Therefore, low-power ADC and low-power digital design are also essential. Hence, this book will discuss ultra-low-power ADC converter design in Chapter 15 and principles for ultra-low-power digital design in Chapter 21. Low-power electronic systems today incorporate analog sensors and amplifiers, ADCs, digital processors, digital-to-analog converters (DACs), radio frequency (RF) transceivers, actuator electronics, battery recharging and power-electronics circuits, and the battery all in one package. While much has been written about how the power of digital systems can be lowered, little attention has been paid to the analog portions of such systems. In many miniature biomedical and portable systems, the analog and RF portions can easily dominate the power of the overall system especially if programmable analog preprocessing reduces digital processing power to insignificant levels compared with sensor, actuator, or RF power. In this book, ultra-low-power analog and digital design are both discussed, though more emphasis is placed on analog and mixed-signal operation. Principles for low-power analog design do not seem to be commonly known and are extremely important in lowering the power of an overall system. Hence, several chapters in the book discuss low-power analog circuits useful for biomedical applications such as filters, amplifiers, RF inductive recharging circuits, RF energy-harvesting antenna circuits, RF telemetry circuits, electrode-stimulation circuits, MEMS sensor circuits, imager circuits, microphone circuits, automatic-gain-control circuits, and energy-extraction circuits. Some bio-inspired circuits that we discuss are useful in non-biological applications such as in wireless communication, in speech recognition, or in image processing.

1.5   Examples of biomedical application contexts

Figure 1.3 shows the mechanical topology of a cochlear implant or bionic ear whose configuration bears similarity to that in many biomedical implants today. Such cochlear implants enable profoundly deaf people, for whom a hearing aid is not beneficial, to hear. In this particular case, a microphone (1) transduces sound into an electrical signal, which is conveyed via a cable within the 'hook' to a speech processor located behind the ear (2). The speech processor relays the results of its spectral-analysis and gain-control processing to an RF coil (3). The RF coil

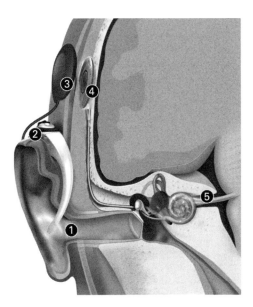

Figure 1.3. An example of a low-power cochlear implant or bionic ear. (Adapted and reprinted with kind permission of the Advanced Bionics Corporation, Valencia, California. The picture shown corresponds to their Harmony cochlear-implant processor.)

relays these results via a transcutaneous wireless transformer-like inductive link to an RF receiver that has been implanted within the body (4). The wireless link is important for ensuring that information is transmitted through the body without the need for wires, which are prone to infection. The RF receiver rectifies its received energy to create a dc voltage source that powers the implant. The RF receiver also demodulates the received RF energy to create signals suitable for stimulation of electrodes near the auditory nerve. These electrodes use charge-balanced stimulating ac currents to excite the nerve (5). A loud high-frequency sound will cause the basal electrodes near the beginning of the cochlea to output high values of charge-balanced ac current. A soft low-frequency sound will cause apical electrodes near the end of the cochlea to output low values of charge-balanced ac current. A cochlear implant mimics the frequency-to-place stimulation of the healthy auditory nerve and the compression or gain-control properties of the healthy ear. Cochlear implants have the speech processor in a behind-the-ear (BTE) unit as shown in Figure 1.3 today. They are evolving towards fully implanted systems, where all components will be implanted inside the body.

Figure 1.4 shows a mechanical topology in a brain implant or in a brain-machine interface (BMI) that could be used to treat blindness in the future. Electrodes that are implanted in the brain stimulate one of its visual regions, in this case, say, the deep-brain lateral geniculate nucleus (LGN) region. The electrodes connect to an implanted unit between the skin and the scalp via a flexible cable that can take up the slack caused by brain motion. The internal unit can

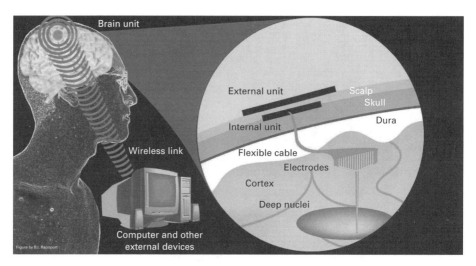

Figure by B.I. Rapoport

Figure 1.4. Configuration of a brain implant or brain-machine interface (BMI).

receive power and data wirelessly from an external unit as in cochlear implants. The received power can periodically recharge an implanted battery or constantly power the internal unit wirelessly. The external unit connects in a wired (or wireless) fashion with an imager and a processor, which can be worn on a set of 'glasses'. The external unit is capable of wireless communication to a computer via a standard Ultra Wide Band (UWB), Zigbee, or Bluetooth interface for programming, monitoring, and debugging functions. The external unit and internal unit are mechanically aligned via electronic magnets. The overall configuration is one of several possibilities but presents several advantages including stable coil coupling, a relatively benign bio-environment between skin and scalp for the implanted unit, minimization of tissue heating within the skull, the ability to be relatively stable during brain and head motion, and good RF link efficiency.

Figure 1.4 also illustrates how the same BMI is useful for the treatment of paralysis. In this case, neural electrical-recording (sensing) electrodes from the surface of a motor region of the brain can be used to decode the intention of a paralyzed patient to move their limb. The decoded information can then be used to stimulate a muscle or robot arm in the patient. The decoding can be fully or partially done in the external unit based on data conveyed by the internal unit. The external unit then wirelessly relays motor commands to the arm. Brain implants for epilepsy require simultaneous recording and stimulation circuitry since the detection of a seizure must trigger electrical stimulation to suppress it.

The examples of Figure 1.3 and Figure 1.4 represent two biomedical application contexts where several ultra-low-power RF, sensor, analog processing, electrode-recording, and electrode-stimulation building-block circuits, which are described in the book, are useful. Implants for several other biomedical applications, e.g., cardiac pacemakers, spinal cord stimulators, deep-brain stimulators for the treatment of Parkinson's disease, vagal-nerve stimulators, etc., utilize several of these same building-block circuits, to architect slightly different systems that all operate by and large with the same technology base. In fact, biocompatibility design, hermetic design, mechanical design, and electrode design share several similarities in all of these applications as well. We shall also discuss how to architect far-field energy-harvesting RF circuits for non-invasive cardiac medical applications, which will make the reader familiar with the basic principles of antenna design. Antenna-based RF communication systems transmit information through a relatively thick portion of the body, e.g., for deep implants such as electronic pills, used for diagnosis of gastrointestinal disorders. Chapters 10 through 18 discuss several circuits useful in low-power biomedical applications.

Chapter 19 focuses on implantable electronics, with an emphasis on cochlear implants and brain implants. It provides a concrete example of the power savings possible when analog preprocessing is used to delay digitization. Chapter 20 focuses on noninvasive medical electronics with an emphasis on cardiac devices and biomolecular sensing.

1.6   Principles for ultra-low-power design

The principles for low-power design discussed in the book apply to all systems that aim to use power efficiently, independent of their absolute power consumption. These principles are true for both biological systems and electronic systems. They include:

1. Encoding the task in the computational basis functions of technology in an efficient fashion to save power, e.g., the use of exponentials to efficiently compute logarithms in electronics or the use of chemical binding in biology to multiply.
2. Using energy-efficient regions of operation of technological devices, e.g., the use of the exponential subthreshold regime of transistor operation in analog and digital electronics and the use of exponential Boltzmann relations in biology.
3. Delaying digitization via analog preprocessing to reduce the information bandwidth in the computation, e.g., as in the filtering and mixing operations in a radio today or in the image preprocessing done by the retina in the eye.
4. Decomposing tasks into more energy-efficient slow-and-parallel architectures rather than fast-and-serial ones, e.g., as in several low-power digital architectures today or in slow-and-parallel computation in the brain.
5. Balancing the costs of computation and communication, e.g., as in a wireless biomedical system transmitting high-level versus low-level information or in the division of energy for computation versus communication in the brain's neuronal computing cells.
6. Reducing the amount of information that needs to be processed, e.g., as in automatic-gain-control circuits in analog electronics, in gated-clock circuits in digital electronics, in event-driven asynchronous analog and digital systems,

in calcium gain-control circuits in the photoreceptor of the eye, and in gated bio synthesis cell-regulation circuits.

7. Using feedback, knowledge, and learning to improve the efficiency of the computation via error-correction, compression, and optimization mechanisms, e.g., as in digitally calibrated analog-to-digital converters, auto-zeroing analog systems, negative auto-regulation feedback loops within cells, or in adaptive power-supply circuits in digital design.

8. Architecting the task such that its circuits do not need to be simultaneously fast and precise, e.g., as in comparators in electronics or in neuronal comparators in biology.

9. Operating slowly or 'adiabatically' with passive and active components that consume little power, e.g., as in adiabatically clocked digital circuits in electronics, in high-quality-factor circuits in electronics, or in high-quality-factor mechanical transmission lines in the biological inner ear or cochlea. The quality factor is a measure of the ratio of the reactive energy to dissipative energy in a system.

These low-power principles are discussed in detail in Chapter 22. Many of the low-power principles that we have listed apply to both low-power analog and low-power digital design although they are often manifested in seemingly different ways. Low-power digital design is discussed in Chapter 21. Chapter 22 discusses several similarities between low-power analog and low-power digital design. In both analog and digital systems, natural or artificial systems, robustness and flexibility trade off against the efficiency of the architecture. Extra degrees of freedom are always needed to attain robustness and flexibility, which compromise its efficiency. A good architecture must be designed to be efficient and robust without being needlessly flexible. We show in Chapter 22 that biological systems obey *all* of the energy-saving principles listed above in an exemplary fashion. They also obey another important principle that we term *collective analog* or *hybrid computation* and that we explain in Chapter 22 [3], [4]. In addition, they provide us with clues for building ultra-low-power systems that force us to think outside the box of traditional engineering designs.

1.7   Ultra-low-power information processing in biology

Biology has designed architectures where many noisy, imprecise, unreliable analog devices collectively interact through analog and digital signals to solve a task in real time, precisely, and with astoundingly low power. For example, a single neuron in the brain performs highly complex real-time pattern recognition, spatio-temporal filtering, and learning tasks with $\sim 0.66\,\text{nW}$ of power. The ~ 22 billion neurons of the brain consume only $\sim 14.6\,\text{W}$ of power in an average 65 kg male. Neurons collectively interact to perform sensorimotor tasks in noisy environments in real time that no computer can yet match. A single cell in the body performs ~ 10 million energy-consuming biochemical operations per second on its noisy molecular inputs with $\sim$1 pW of average power. Every cell implements a $\sim$30,000 node gene-protein molecular interaction network within its confines. All the $\sim$100 trillion cells of the human body consume $\sim$80 W of power at rest. The average energy for an elementary energy-consuming operation in a cell is about 20 kT, where kT is a unit of thermal energy. In deep submicron processes today, switching energies are nearly $10^4 - 10^5$ kT for just an elementary 0$\rightarrow$1 digital switching operation. Even at 10 nm, the likely end of business-as-usual transistor scaling in the future, it is unlikely that we will be able to match such energy efficiency. Unlike traditional digital computation, biological computation is tolerant to error in elementary devices and signals. Nature illustrates that it is significantly more energy efficient to compute with error-prone devices and signals and then correct for these errors through feedback-and-learning architectures than to make every device and every signal in a system robust, as in traditional digital paradigms thus far.

We can learn from Nature, for she has had 1 billion years of experimentation to evolve magnificent nanotechnological, low-power devices, circuits, topologies, and architectures in environments where food, and consequently energy, was scarce. What we learn can help inspire the design of novel engineering systems termed *bio-inspired systems*. Such inspiration from nature must always be combined with perspiration and rational design from engineering if the final result is to be useful. In this book, we shall always take an engineering approach and show that bio-inspired designs can and do result in impressive engineering architectures. Birds are not airplanes and airplanes are not birds, but one can shed insight into the operation of the other. A humble, open, and curious mindset toward ideas in nature is all that is needed for appreciating this field.

We shall find on several occasions that bio-inspired algorithms, architectures, and circuits frequently have biomedical applications. Not surprisingly, it helps to mimic how the biology works if one is attempting to fix it. For example, we shall discuss an asynchronous stochastic sampling strategy inspired by how the auditory-nerve works in Chapter 19 that enables an approximately 6$\times$ reduction in electrode stimulation power while maintaining a high effective rate of sampling. Bio-inspired systems, which have just scratched the surface of what will likely be possible in the future, already show that there are several clever ideas in biology that are useful in engineering, and not just for low-power design. Now, we shall highlight one example of a neuromorphic electronic system, i.e., an electronic system inspired by neurobiology. The term *neuromorphic electronics* was coined by the founder of the field, Carver Mead. We shall describe an *RF cochlea*, a fast, power-efficient radio-frequency spectrum analyzer useful in broadband wireless communication systems.

1.8   Neuromorphic system example: the RF cochlea

The biological inner ear, or cochlea, performs spectrum analysis on its incoming sound input through the use of a mechanical transmission-line architecture.

The transmission-line architecture of the cochlea has exponentially tapered time constants from its beginning or high-frequency base location to its ending or low-frequency apex location. The net result is that the cochlea separates sounds based on their frequencies, with high-frequency sounds stimulating the base or beginning of the cochlea and low-frequency sounds stimulating the apex or end of the cochlea. Thus, the cochlea functions as a spectrum analyzer that performs a frequency-to-space or frequency-to-location transformation on its sound input. It operates over an ultra-broadband 100:1 sound-carrier-frequency range of $\sim$100 Hz to 10 kHz with good sensitivity. The cochlea also performs gain control and compression on its sound input via a distributed amplification system within the transmission line that is highly energy efficient. Therefore, we can hear over a 120 dB input dynamic range with a minimum detectable signal of 0.05 Å at our ear drums at our most-sensitive frequency of $\sim$3 kHz. The power consumption of the cochlea is $\sim$14 μW even though it implements at least $\sim$1 billion floating-point operations per second. Nonlinearity and gain-control in the cochlea are important for enhancing signals in noisy environments and for our ability to hear speech in noise. An appendix in Chapter 23 discusses how we can estimate power and/or information processing rates in the ear, the eye, the brain, and the body.

The ear operates remarkably like an ultra-broadband super-radio for sound waves with 3,500 output spectral channels operating in parallel. The outer ear or pinna is a directional antenna. The middle ear is an impedance-matching transformer that matches the impedance of the antenna to the impedance of the inner ear or cochlea. The piezoelectric outer hair cells in the cochlea function like amplifiers that enhance the passive resonant gain of its membrane-and-fluid transmission-line structure. The inner hair cells in the cochlea function like rectifying demodulators that detect modulations of the carrier sound waves propagating through the cochlea. Finally, the auditory-nerve output spikes or pulses sample and partially quantize the inner-hair-cell cochlear output for eventual communication to the brain. Figure 19.1 in Chapter 19 reveals some of the anatomy of the ear.

We show in Chapter 23 that the exponentially tapered time-constant architecture of the cochlear transmission-line allows the cochlea to implement the fastest and most-efficient spectrum-analysis architecture that is currently known. For a spectrum analyzer with N output bins operating over a given input-frequency range, the cochlear architecture only consumes $O(N)$ resources in time and $O(N)$ resources in hardware to perform spectrum analysis. In contrast, a constant fractional-bandwidth analog filter bank spectrum analyzer consumes $O(N)$ resources in time and $O(N^2)$ resources in hardware. A fast-fourier-transform (FFT) fixed-bandwidth spectrum-analyzer consumes $O(N\log_2 N)$ resources in time and $O(N\log_2 N)$ resources in hardware. A traditional swept-sine spectrum analyzer consumes $O(N^2)$ resources in time and $O(1)$ resources in hardware. Thus, the cochlear architecture has a good tradeoff between the use of temporal resources and hardware resources, and for a given amount of hardware resources its architecture for spectrum analysis leads to the quickest performance.

Calibration
circuits   Single stage   Signal propagation

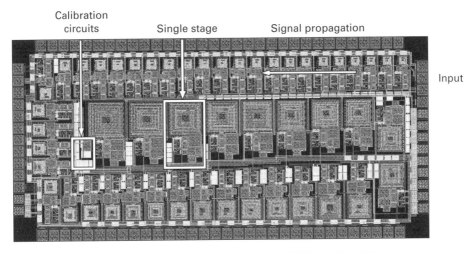

Input

Figure 1.5. Layout of a bidirectional 0.13 μm bio-inspired RF cochlea chip.

Figure 1.5 shows a 0.13 μm silicon RF cochlea, a bio-inspired radio-frequency spectrum-analysis chip that operates at million-fold higher carrier frequencies than the cochlea to perform ultra-broadband spectrum analysis of electromagnetic radio waves in a fast and power-efficient fashion. The chip mimics the architecture of the cochlea in fairly precise mathematical detail, solving almost the same partial differential equations for wave propagation as in the cochlea except at million-fold faster time scales. In the RF cochlea, inductors correspond to fluid masses in the biological cochlea, capacitors correspond to membrane stiffness, amplifiers correspond to amplifying outer hair cells, and rectifiers correspond to the inner hair cells. The physical layout of the RF cochlea has an exponential taper and spiral that mimics the biological cochlea as well. The RF cochlea inherits the fast spectrum-analysis and ultra-broadband algorithmic advantages of the biological cochlea. Therefore, it is able to perform very fast spectrum analysis over an experimentally demonstrated 600 MHz to 8 GHz carrier-frequency range (at present), with low power, and over a wide input dynamic range. Direct digitization of the input spectrum over an equivalent input dynamic range would consume 100× more power while an analog filter bank with the same number of output taps (50) would consume 20× more hardware.

The RF cochlea is useful in applications such as cognitive radio, where a fast survey of the input spectrum is needed for adapting communication strategies. In general, in the increasingly crowded RF spectrum of today, it is useful to know what one's RF environment is such that one can efficiently pick transmission and reception frequencies to minimize interference, improve bandwidth, and help enable the functioning of software and universal radios. Radio engineers have found that, statistically, a large fraction of the radio spectrum is unutilized and wasted such that an adaptive strategy for communication can maximize bandwidth and robustness in next-generation wireless systems.

The RF cochlea is an extremely recent neuromorphic architecture, which will undoubtedly evolve over time [5]. However, it has already excited interest and enthusiasm in the radio-engineering community. In general, inspiration from the ear and auditory system may lead to revolutionary RF architectures in the future.

1.9   Cytomorphic electronics

Circuits in cell biology and circuits in electronics may be viewed as being highly similar with biology using molecules, ions, proteins, and DNA rather than electrons and transistors. Just as neural circuits have led to biologically inspired neuromorphic electronics, cellular circuits can lead to a novel biologically inspired field that we introduce in this book and term *cytomorphic electronics*. We will show that there are many similarities between spiking-neuron computation and cellular computation in Chapter 24.

Figure 1.6 illustrates that there are striking similarities between chemical reaction dynamics (Figure 1.6 (a)) and electronic current flow in the subthreshold regime of transistor operation (Figure 1.6 (b)). Electron concentration at the source is analogous to reactant concentration; electron concentration at the drain is analogous to product concentration; forward and reverse current flows in the transistor are analogous to forward and reverse reaction rates in a chemical reaction; the forward and reverse currents in a transistor being exponential in voltage differences at its terminals are analogous to reaction rates being exponential in the free energy differences in a chemical reaction; increases in gate voltage lower energy barriers in a transistor increasing current flow are analogous to the effects of enzymes or catalysts in chemical reactions that increase reaction rates; and, the stochastics of Poisson shot noise in subthreshold transistors are

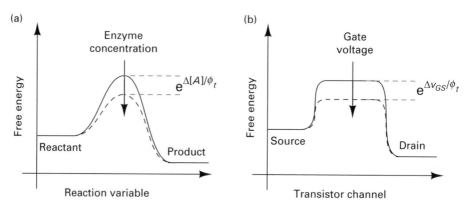

Figure 1.6a, b. Similarities between chemical reaction dynamics and subthreshold transistor electronic flow. Reprinted with permission from [6] (© 2009 IEEE).

analogous to the stochastics of molecular shot noise in reactions. These analogies suggest that one can mimic and model large-scale chemical-processing systems in biological and artificial networks very efficiently on an electronic chip at time scales that could potentially be a million times faster. No one thus far appears to have exploited the detailed similarity behind the equations of chemistry and the equations of electronics to build such networks. The single-transistor analogy of Figure 1.6 is already an exact representation of the chemical reaction $A \rightleftharpoons B$ including stochastics, with forward electron flow from source to drain corresponding to the $A \rightarrow B$ molecular flow and backward electron flow from drain to source corresponding to the $B \rightarrow A$ molecular flow. In Chapter 24, we shall build on the key idea of Figure 1.6 to show how to create current-mode subthreshold transistor circuits for modeling arbitrary chemical reactions. We can then create large-scale biochemical reaction networks from such circuits for modeling computation within and amongst cells.

Since extracellular cell-cell networks also rely on molecular binding and chemical reactions, networks such as hormonal networks or neuronal networks can be efficiently modeled using such circuits. Thus, in the future, we can potentially attempt to simulate cells, organs, and tissues with ultra-fast highly parallel analog and hybrid analog-digital circuits including molecular stochastics and cell-to-cell variability on large-scale electronic chips. Such molecular-dynamics simulations are extremely computationally intensive, especially when the effects of noise, nonlinearity, network-feedback effects, and cell-to-cell variability are included. Stochastics and cell-to-cell variability are highly important factors for predicting a cell's response to drug treatment, e.g., the response of tumor cells to chemotherapy treatments. We will show in Chapter 24 that circuit, feedback, and noise-analysis techniques described in the rest of this book can shed insight into the systems biology of the cell. For example, flux balance analysis is frequently used to reduce the search space of parameters in a cell. It is automatically implemented as Kirchhoff's current law in circuits since molecular fluxes map to circuit currents. Similarly, Kirchhoff's voltage law automatically implements the laws of thermo-dynamic energy balance in chemical-reaction loops. Robustness analysis of the circuit using feedback techniques can shed insight in the future into which genes, when mutated, will lead to disease in a network, and which will not. Circuit-design techniques can also be mapped to create synthetic-biology circuits that will perform useful functions in the future.

1.10 Energy sources

A highly important component of any low-power electronic system is the battery or energy source. Therefore, the last two chapters of the book are devoted to this topic. Chapter 25 discusses batteries and electrochemistry in some depth. It shows how a simple physical interpretation of the equations of electrochemistry lead to electrical large-signal and small-signal equivalent circuits that make the operation of batteries intuitive. These equivalent circuits are also useful for modeling recording and stimulation electrodes in biomedical systems. A new and simple formula for battery operation that characterizes the loss in battery capacity with increasing usage and increasing current draw is described. A discussion of how battery–ultra-capacitor and fuel-cell–battery hybrids can enable advantageous operation in many systems is presented. We shall see that, in low-power systems, a battery does not have a longer lifetime only because its power draw is low. Its lifetime is longer also because it becomes capable of more charge-recharge cycles and because its geometry within a fixed volume can be architected such that it has more charge capacity. Furthermore, when the battery supplies current, its output voltage is higher, increasing its energy efficiency. Thus, the rate of use of energy is intimately linked to its storage and generation.

Chapter 26 reviews prior work on energy harvesting, including the use of body motion, body heat, and solar energy in self-powered battery-free biomedical and portable systems. The fundamental principles of piezoelectric motion-energy harvesting, thermoelectric body-heat harvesting, and photovoltaic electricity generation are discussed. Circuit models for energy harvesting are found to be similar to circuit models for RF energy harvesting discussed in depth in Chapters 16 and 17. We shall find that the principles of low-power electronic design described in this book apply not just to electronic systems and at small scales, but are also useful for non-electrical systems and at large scales, e.g., in low-power cars. An equivalent circuit model of a car can help us understand issues related to the energy efficiency of transportation. Thus, the transport energy efficiency of, say, a cheetah versus an electric car or a bicycle, or a bird versus an airplane can be compared. The current power consumption of the world is largely dominated by transport, heating, and electricity costs and is a staggering 15 TW today. We shall summarize ideas actively being researched for architecting low-power systems that function with renewable sources of energy like solar power or biofuels, a highly likely necessity in our planet's future.

We shall now conclude our top-down view of some of the book's contents. In the next section, we will present a brief bottom-up view of the book's organization.

1.11   An overview of the book's chapters and organization

The book is organized into seven sections, each with several chapters. We shall discuss each section of the book briefly.

1. The **Foundations** section of the book contains ten chapters including this overview chapter, Chapter 1. This section contains a review of feedback

systems in Chapter 2, an in-depth discussion of the device physics of the MOS transistor in Chapters 3 through 6, and a discussion of noise in devices and circuits in Chapters 7 and 8, respectively. The section concludes with more material on feedback systems and feedback-circuit-analysis techniques in Chapters 9 and 10, respectively.

2. The **Low-Power Analog and Biomedical Circuits** section of the book is formed by Chapters 11 through 15. This section contains various circuits that are useful for low-power biomedical electronics and analog electronic systems in general. The foundational material from the first section enables design and analysis of these circuits.

3. The **Low-Power RF and Energy-Harvesting Circuits for Biomedical Systems** section of the book is formed by Chapters 16 through 18. It contains a description of energy-efficient power and data radio-frequency (RF) links that are uniquely suited to biomedical systems.

4. The **Biomedical Electronic Systems** section of the book contains a chapter on ultra-low-power implantable electronics, Chapter 19, and a chapter on ultra-low-power noninvasive medical electronics, Chapter 20. In Chapter 19, exemplary systems for cochlear implants for the deaf, brain implants for the blind and paralyzed, and other implantable systems are discussed. Building-block low-power circuits from the previous chapters and new circuits unique to implantable electronics show how large systems can be architected. In Chapter 20, cardiac devices for noninvasive medical monitoring are discussed. Principles for biomolecular sensing, such as in bioMEMS and microfluidic systems, are also discussed.

5. The **Principles for Ultra-low-power Analog and Digital Design** section of the book contains one chapter on principles for ultra-low-power digital design, Chapter 21, and one chapter on principles for ultra-low-power analog and mixed-signal design, Chapter 22. Similarities in principles for low-power analog and digital design are discussed. Ten principles for ultra-low-power design are discussed, all of which are obeyed by ultra-energy-efficient biological systems.

6. The **Bio-inspired Systems** section of the book comprises two chapters. The first chapter, Chapter 23, on *neuromorphic electronics* discusses electronics inspired by neurobiology. The second chapter, Chapter 24, discusses the novel form of electronics that we have termed *cytomorphic electronics*, i.e., electronics inspired by cell biology. Applications of neuromorphic and cytomorphic electronics to engineering and medicine are discussed.

7. The **Energy Sources** section of the book comprises Chapters 25 and 26. Chapter 25 on batteries and electrochemistry discusses how batteries work from a unique circuit viewpoint and presents important tradeoffs between

energy density and power density. Chapter 26 discusses energy harvesting in portable and biomedical systems at small scales and at larger scales. We show how some of the principles of low-power design that we have studied apply not only at small scales and in electronics but also at large scales and in non-electrical systems.

1.12   Some final notes

Problem sets, errata, and instructor material for the book will be available at the Cambridge University Press website (http://www.cambridge.org/). The author's full name will serve as an index term for navigating this site.

The book attempts as far as possible to follow IEEE convention for algebraic symbols: A dc variable is denoted by I_A, a small-signal variable is denoted by i_a, a total-signal variable is denoted by i_A, and a frequency-response or Laplace variable is denoted by $I_a(f)$ or $I_a(s)$, respectively. Thus, $i_A = I_A + i_a$. However, there are situations in the book where this convention has not been followed to improve clarity, or where doing so does not matter much in the discussion.

I have tried to write this book such that it is clear, accessible, and rewards the reader with a broad and deep knowledge in the fields of both ultra-low-power electronics and bioelectronics. As I have discussed, physically intuitive ways of thinking are emphasized throughout this book. It is worth noting that it was Einstein's impressive physical intuition that enabled him to see that the equations of special relativity, which had already been stated by Lorenz, had a transparently simple derivation if one accepted that the speed of light is a constant in nature. It is no surprise, then, that Einstein said, "The only real valuable thing is intuition." While most of us, including the author, do not have Einstein's physical intuition, we can take inspiration from Einstein to always strive for his simple intuitive way of understanding phenomena. It is by developing our scientific intuition that we can begin to see connections that illuminate the unity of all nature.

References

[1] R. Sarpeshkar, C. D. Salthouse, J. J. Sit, M. W. Baker, S. M. Zhak, T. K. T. Lu, L. Turicchia and S. Balster. An ultra-low-power programmable analog bionic ear processor. *IEEE Transactions on Biomedical Engineering*, **52** (2005), 711–727.

[2] A. T. Avestruz, W. Santa, D. Carlson, R. Jensen, S. Stanslaski, A. Helfenstine and T. Denison. A 5 μW/channel Spectral Analysis IC for Chronic Bidirectional Brain-Machine Interfaces. *IEEE Journal of Solid-State Circuits*, **43** (2008), 3006–3024.

[3] R. Sarpeshkar. Analog versus digital: extrapolating from electronics to neurobiology. *Neural Computation*, **10** (1998), 1601–1638.

[4] R. Sarpeshkar and M. O'Halloran. Scalable hybrid computation with spikes. *Neural Computation*, **14** (2002), 2003–2038.

[5] S. Mandal, S. M. Zhak and R. Sarpeshkar. A bio-inspired active radio-frequency silicon cochlea. *IEEE Journal of Solid-State Circuits*, **44** (2009), 1814–1828.

[6] S. Mandal and R. Sarpeshkar, Log-domain circuit models of chemical reactions. *Proceedings of the IEEE Symposium on Circuits and Systems (ISCAS)*, Taipei, Taiwan, 2009.

2 Feedback systems: fundamentals, benefits, and root-locus analysis

It is presumed that there exists a great unity in nature, in respect of the adequacy of a single cause to account for many different kinds of consequences.

Immanuel Kant

Devices that are hooked to each other at various terminals create a circuit. Almost all nontrivial circuits comprise topologies where output terminal(s) are directly or indirectly coupled back to input terminal(s), thus forming a *feedback circuit*. When the output feeds back to reduce the effects of the input, the feedback is termed *negative feedback*, and when the output feeds back to increase the effects of the input, the feedback is termed *positive feedback*. Purely feed-forward circuits are usually simple to build and easy to analyze, even when nonlinear, such that most of the complexity and richness in circuits arises from the feedback embedded within them.

Negative-feedback circuits function by creating forces within the circuit that attempt to restore its signals to a desired equilibrium point if these signals deviate away from this point. Negative-feedback circuits often serve regulatory functions improving the precision of the output to that provided by a precise equilibrium-setting reference input and/or that of a precise sensor or feedback network. Such precision is achieved in spite of imprecision in an actuator or feed-forward network in the circuit and/or disturbances present at the output of the circuit.

Positive-feedback circuits function by creating forces within the circuit that attempt to move its signals further away from a point if these signals deviate from that point. Positive-feedback circuits often serve to enhance the gain of circuits when the positive feedback is strong enough to enhance the gain to the input but weak enough to prevent instability, e.g., to improve the quality factor in resonators. When the positive feedback is strong enough to induce instability, positive-feedback circuits serve to create discrete states of operation that are stabilized by nonlinearities at limiting signal levels of operation, e.g., in latches and Schmitt-triggers.

Both negative-feedback and positive-feedback circuits can create desirable or undesirable oscillations and instabilities in circuits. Both forms of feedback alter the dynamics of circuit behavior as the parameters of the *feedback loop* are changed. The feedback loop expresses mathematical relationships amongst circuit signal variables that are chained together within it in a cascade-like fashion to create a circular dependence on one another. Changes in dynamics caused by feedback loops in circuits are often intentionally exploited to create circuits that perform faster, that respond only to changing inputs, or that reduce leakage rates on a capacitor, etc.

Feedback is at the heart of how *all* circuits work, whether electrical or non-electrical in nature. All nontrivial circuits arise from feedback interactions between devices in physics, chemistry, or biology that create a circuit, sometimes called a network or system. Thus, a deep understanding of feedback is important to an understanding of all systems, since feedback principles are at the heart of how simple devices, when hooked together, exhibit complex behavior in a circuit or system. We will begin with a few examples that illustrate how amazingly broad the reach of feedback circuits is in all of science and engineering.

2.1   Feedback is universal

Feedback circuits ensure that we are all alive and well! Several homeostatic systems use molecular negative-feedback circuits to ensure that our temperature, the pH of our blood and intracellular fluids, our weight, and the concentrations of various nutrients and molecules in our cells and bodies are regulated to be within an acceptable range. Gene-protein feedback networks within cells ensure normal functioning of the cell under various environmental conditions and developmental cycles. Cell growth is regulated to ensure normal growth and functioning of the body's organs and to prevent cancer. Positive and negative feedback circuits in the immune system help combat infection and implement self healing and repair. A malfunction in any of these feedback systems in our cells or bodies can lead to debilitating diseases like diabetes, hyperthyroidism, cancer, or auto-immune disease.

The sensory and motor systems of our bodies make extensive use of feedback circuits to improve their performance and dynamic range of operation. For example, pupil accommodation in the eye and calcium-based feedback circuits in our photoreceptors allow us to see over a wide range of light levels. Auditory-gain-control systems in our ears exploit nonlinearity and feedback in our outer hair cells to enable us to hear over a wide range of sound intensity levels. Our ability to continually gaze at moving objects is made precise via eye-movement pursuit and tracking feedback circuits. Our motor systems use sensory feedback to improve their performance, e.g., visual feedback for locomotion and auditory feedback for speech production.

The brain is a massive feedback circuit with extensive excitatory and inhibitory connections amongst its cells, termed *neurons*. There are significantly more feedback connections formed by the cells of the brain among themselves than feedforward connections received from the inputs to the brain. The pulsatile signal variables of neurons, known as *spikes*, are created by transient positive-feedback action in sodium conductances in neurons much as in a relaxation oscillator in electrical circuits. Learning may be viewed as feedback information from errors,

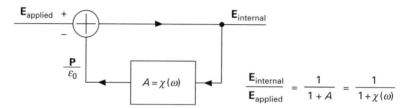

Figure 2.1. Block diagram of the negative-feedback loop that creates the dielectric constant.

failure, or from the structure of the input space that is ported into the adaptive architecture of the brain to improve its performance over time. Evolution comprises slower feedback from the environment to the genes.

All chemical reactions are bidirectional and have two embedded feedback loops within them that help establish their equilibrium: an increase in the rate of the backward reaction if the concentration of products becomes excessive and a decrease in the rate of the forward reaction as the concentration of reactants is used up. Several biochemical reaction pathways in the body have enzyme-mediated feedback loops that function by products enhancing or decreasing the rates of the reactions that created them. Bonds between atoms in a molecule or between molecules lead to an equilibrium separation established by forces that effectively implement a feedback mechanism that pushes the constituents apart if they get too close and that pulls them back together if they get too far apart.

In physics, feedback mechanisms are ubiquitously present, although they are not often described as such. For example, the electric field emanating from a charge in water is reduced by a factor of 81 from the electric field emanating from the same charge in air or in vacuum. Therefore, we say that water has a dielectric constant that is 81 times larger than that of air. Such an effect arises because polar water molecules strongly oppose any electric field within them by creating their own anti-aligned polarization electric field that subtracts from the input field and thus attempts to neutralize it via negative-feedback action. The small residual output electric field that remains after near-perfect neutralization is 81 times weaker because the negative-feedback loop gain A of this feedback loop is 80, and, as we shall discuss later in the chapter, the effect of such negative-feedback action is to weaken the response to the input by a factor of $1/(1+A)$. Figure 2.1 shows the operation of the feedback loop and the basic negative-feedback equations that explain this dielectric effect. As a consequence, electrostatic forces between charges in water are highly attenuated compared with forces in air. Therefore, ionic substances like salt easily dissolve in water due to the weakening of the forces which hold them together. The variable $\chi(\omega)$ in Figure 2.1 is called the *susceptibility*, which in our viewpoint is simply the frequency-dependent loop gain $A(\omega)$.

Phase transitions that cause a substance to change from one form, e.g., solid ice, to another form, e.g., liquid water, at a particular magical temperature known as the *melting point* are due to a positive-feedback loop. At the magical temperature, the thermal energy is sufficiently large such that the breaking of a few bonds between ice molecules causes bonds between other molecules to more easily break since they are now collectively more loosely held. Thus, the breakage of a few bonds causes easier breakage of more bonds through positive-feedback action until the entire substance melts into a new stable form, water. The whole process is similar to the positive-feedback action seen when a slight tear in a garment can lead to a larger tear and finally tearing of the entire garment. Presumably, at the magical melting point, the collective positive-feedback loop gain of the system is greater than 1, the critical value at which a positive-feedback loop just goes unstable.

Maxwell's equations inherently have feedback embedded into them because the distributions of charges and motions of charges that create electric and magnetic fields are themselves affected by these same fields. Thus, fields and charged matter form a reciprocal feedback network with each affecting the other.

A stable downward-hanging pendulum has forces at equilibrium that are an example of negative-feedback action that restore it to its lowest point. An unstable upward-hanging inverted pendulum has forces at equilibrium that are an example of positive-feedback action that move it away from its highest point.

Feedback circuits are so ubiquitous in control systems in engineering that the terms *feedback* and *control* are almost used synonymously even though control systems could be purely feed-forward in principle. Industrial-process control, aircraft control, mechanical system control, robotic systems, space-flight systems, guidance and navigation systems all exploit feedback circuits to function.

Distributed computing systems such as those present in the internet are composed of multiple interacting feedback loops within the communication network. All communication systems that have the possibility of sending messages from a transmitter to a receiver and then receiving one in turn via one or more hops in the network may be viewed as feedback networks. Social communication networks are an example of such networks as well.

Feedback is the essential ingredient that makes circuits work. Without feedback, analog circuits would be simple, imprecise, feed-forward, impractical curiosities. With feedback, analog systems become rich, complex, precise, practical systems that can make our hearts pound with excitement. Without a deep understanding of feedback, every analog circuit appears like a special case. With a deep understanding of feedback, almost all analog circuits begin to look like examples of a general pattern that has been studied before. That is why we tackle feedback systems before we tackle circuits in this book. Digital circuits incorporate feedback into a discrete dynamical system termed a *finite state machine*. Here, the output of the system feeds back and affects the processing of its input after a delay of one clock cycle. The discrete state of the system is concomitantly updated after this clock cycle as well. Digital systems also use positive feedback to create static-memory circuits, which store digital bits of information.

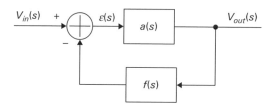

Figure 2.2. Basic block diagram used to describe negative-feedback loops.

2.2   The basic linear feedback loop

Much insight into the essence of feedback systems can be gained by studying linear feedback systems that are well modeled by the feedback loop shown in Figure 2.2. In negative-feedback loops, the output feeds back to decrease the effects of the input at dc; i.e., the adder in Figure 2.2 has a negative input feeding back at the output of $f(s)$ and $a(0)f(0)$ is positive. In positive-feedback loops, the output feeds back to increase the effects of the input at dc; i.e., the adder in Figure 2.2 has a positive input feeding back at the output of $f(s)$ and $a(0)f(0)$ is positive. Both forms are important. Without loss of generality, we shall focus mainly on negative-feedback systems; i.e., we shall assume that the adder shown in Figure 2.2 has a negative input with $a(0)f(0)$ being positive. We can then generalize to positive-feedback systems in a straightforward fashion as needed. *In this chapter, unless otherwise mentioned, all feedback systems will be assumed to be negative-feedback systems.* Chapter 9 has a brief discussion of positive-feedback systems.

In Figure 2.2, $V_{in}(s)$ represents the Laplace transform of the reference or tracking input, $a(s)$ represents the feed-forward system transfer function, typically that of a motor pathway or that of a controller and 'plant', $f(s)$ represents the feedback transfer function, typically that of a sensory pathway or the inverse of the desired input-output function, and $\varepsilon(s)$ is the transfer function of the error between the reference input and $f(s)V_{out}(s)$. From Figure 2.2, we can derive that

$$
\begin{aligned}
\varepsilon(s) &= V_{in}(s) - f(s)V_{out}(s) \\
V_{out}(s) &= a(s)\varepsilon(s) = a(s)[V_{in}(s) - f(s)V_{out}(s)] \\
\frac{V_{out}(s)}{V_{in}(s)} &= \frac{a(s)}{1 + a(s)f(s)} = \frac{\text{Feed-forward transfer function}}{1 + \text{Loop transmission}}
\end{aligned}
\tag{2.1}
$$

The latter formula in Equation (2.1), which describes the input-output transfer function of the feedback system, is called *Black's formula*, in honor of Harold Black, the inventor of the negative-feedback amplifier [1]. Black, just six years out of college, invented the negative-feedback amplifier in a brilliant flash of insight as he was commuting to work at Bell Labs on the Lackawanna Ferry [2]. Black's formula is one of the most important formulas in all of electrical engineering, perhaps *the* most important, and the reader is strongly urged to memorize it. It is one of the few formulas in engineering that is actually worth memorizing because of its seminal and ubiquitous presence.

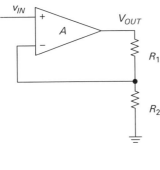

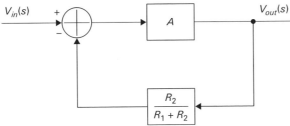

Figure 2.3. Inverting amplifier circuit (top) and associated block diagram (bottom).

From Equation (2.1), we can conclude that if the magnitude of the *loop transmission* $a(s)f(s)$, i.e., $|a(s)f(s)| \gg 1$,

$$\frac{a(s)}{1 + a(s)f(s)} \approx \frac{a(s)}{a(s)f(s)} = \frac{1}{f(s)} \tag{2.2}$$

Thus, the closed-loop transfer function $V_{out}(s)/V_{in}(s)$ is determined by $1/f(s)$ independent of $a(s)$. Thus, if we can control $f(s)$ well, independent of how poorly controlled $a(s)$ is, as long as the *loop gain* $|a(s)f(s)| \gg 1$, the closed-loop transfer function is well controlled. Figure 2.3, which illustrates a non-inverting amplifier configuration, provides a concrete example of the principle of Equation (2.2):

$$V_{in} = V_+$$
$$V_{out} = A(V_+ - V_-)$$
$$V_- = \frac{R_2}{R_1 + R_2} V_{out}$$
$$\frac{V_{out}}{V_{in}} = \frac{A}{1 + A\dfrac{R_2}{R_1 + R_2}} \tag{2.3}$$
$$\text{If } A\frac{R_2}{R_1 + R_2} \gg 1$$
$$\frac{V_{out}}{V_{in}} = \frac{1}{\left(\dfrac{R_2}{R_1 + R_2}\right)} = \frac{R_1 + R_2}{R_2}$$

As long as operational amplifier's gain A is sufficiently high, the closed-loop gain of the feedback loop is determined by the resistors independent of A.

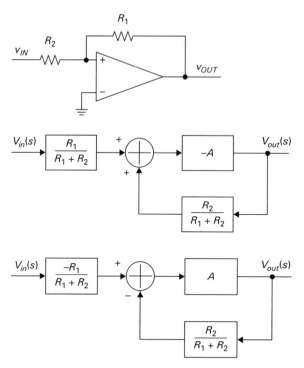

Figure 2.4. Non-inverting amplifier circuit (top) and associated block diagrams before (middle) and after (bottom) simplification into the standard negative-feedback form.

From Figure 2.2, we can also quickly derive that the error transfer function $\varepsilon(s)/V_{in}(s)$ is given by

$$\frac{\varepsilon(s)}{V_{in}(s)} = \frac{1}{1 + a(s)f(s)} \tag{2.4}$$

and is small if $|a(s)f(s)| \gg 1$. In fact, it is the fundamental principle of Equation (2.4) that leads to the rule of thumb that $v_- \approx v_+$ in an operational amplifier circuit. If the error voltage $v_+ - v_-$ builds in magnitude, a negative-feedback loop with a sufficiently high loop gain, which is due to sufficiently high operational-amplifier gain, will attenuate this error voltage until it is nearly zero. Figure 2.4 shows an inverting amplifier configuration. Simple manipulations of the block diagram describing its operation convert it to the standard feedback block diagram as shown. The ideal closed-loop transfer function between the input and output if the loop gain or equivalently the operational-amplifier gain is sufficiently large is then given by

$$\frac{V_{out}}{V_{in}} = \frac{-R_1}{R_1 + R_2} \times \frac{R_1 + R_2}{R_2} = \frac{-R_1}{R_2} \tag{2.5}$$

The non-inverting amplifier of Figure 2.3 and the inverting amplifier of Figure 2.4 both have the same value of loop transmission $a(s)f(s)$ but different closed-loop transfer functions due to the feed-forward block that is present in the block diagram of Figure 2.4 but not in the block diagram of Figure 2.3.

In general, the loop transmission of a feedback loop is denoted by $L(s)$ such that the loop gain $|L(s)| = |a(s)f(s)|$ and the loop phase $\angle L(s) = \angle a(s)f(s)$. Often, if we can express $a(s)f(s) = a_0f_0g(s)$ with $g(0) = 1$, then a_0f_0 is termed the *dc loop gain*. Note that a_0 and f_0 are frequency independent (i.e., do not depend on s).

The conclusions that we have drawn about the robustness of feedback loops to changes in $a(s)$ when the magnitude of the loop transmission $L(s)$ is large have been independent of the phase of the loop transmission $\angle L(s)$. Thus, it may appear that they are equally true of positive and negative feedback loops, i.e., independent of whether $a(0)f(0)$ has a positive sign corresponding to negative feedback or a negative sign corresponding to positive feedback. However, as we discuss in Chapter 9, if $|L(s)|$ is larger than 1, most positive-feedback loops are typically unstable such that our conclusions are invalid. Thus, the approximations of Equation (2.2) are implicitly valid only for stable negative-feedback loops.

2.3   Connections between feedback loops and circuits

There is a deep relationship between feedback loops and circuits. Figure 2.5 (a) shows a feedback loop and then shows that the feedback loop models equations exactly equivalent to those seen in a parallel-impedance circuit with a current input $I_{in}(s)$, a voltage output $V_{out}(s)$, a complex impedance of value $a(s)$ in parallel with a complex impedance of value $1/f(s)$, and a closed-loop impedance transfer function of value $V_{out}(s)/I_{in}(s)$. The error current corresponds to that which flows

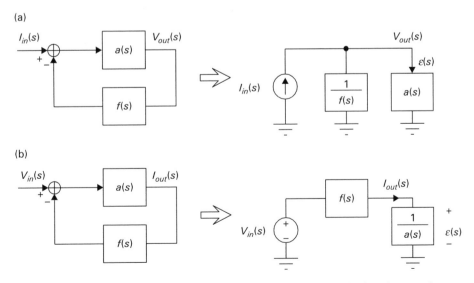

Figure 2.5a, b. Feedback loops can be modeled as circuits with transfer functions analogous to impedances: (a) the parallel form and (b) the series form.

into $a(s)$. If $|a(s)| \gg |1/f(s)|$, the closed-loop impedance is determined primarily by $1/f(s)$ and is very insensitive to changes in $a(s)$, as we might expect for the impedance of two parallel resistors when one resistor is much larger than the other. Figure 2.5 (b) shows a dual series version of the circuit of Figure 2.5 (a) with a voltage input $V_{in}(s)$, a current output $I_{out}(s)$, a complex impedance of value $1/a(s)$ in series with a complex impedance of value $f(s)$, and a closed-loop admittance transfer function of value $I_{out}(s)/V_{in}(s)$. The error voltage corresponds to that dropped across the impedance $1/a(s)$. In the series case of Figure 2.5 (b),

$$\frac{I_{out}(s)}{V_{in}(s)} = \frac{1}{\dfrac{1}{a(s)} + f(s)} = \frac{a(s)}{1 + a(s)f(s)} \tag{2.6}$$

We can again conclude that the admittance transfer function is very insensitive to changes in the load admittance $a(s)$ if the magnitude of the load admittance $a(s)$ is much greater than the magnitude of the coupling admittance $1/f(s)$.

In general, the nesting of feedback loops within each other and the cascading of feedback loops leads to circuits with passive impedances and active dependent generators.

2.4 The seven benefits of feedback

There are seven common benefits of feedback which make it ubiquitous. These benefits are all implicitly captured by Equation (2.1). However, like Maxwell's equations, Black's formula requires an understanding of several situations where it applies before its depth can truly be appreciated. These benefits primarily arise from negative feedback. We shall defer discussing positive feedback until Chapter 9.

2.4.1 The feedback advantage of reducing sensitivity

If we do the simple mathematical manipulations below,

$$c = \frac{a}{1 + af}$$
$$\ln(c) = \ln(a) - \ln(1 + af) \tag{2.7}$$
$$\frac{\partial c}{c} = \frac{\partial a}{a} - \frac{f \cdot \partial a}{1 + af} = \frac{\partial a}{a} \left(\frac{1}{1 + af} \right)$$

we note that the percentage change in closed-loop gain $\partial c/c$ is a small fraction $1/(1 + af)$ of the percentage change in the feed-forward transfer function $\partial a/a$ if the loop transmission $|af| \gg 1$. In contrast, the percentage change in c is almost exactly the percentage change in f if $|af| \gg 1$:

$$\frac{\partial c}{c} = \frac{\partial f}{f} \left(\frac{af}{1 + af} \right) \tag{2.8}$$

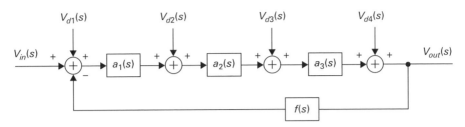

Figure 2.6. Block diagram showing how feedback loops attenuate disturbance signals present at various inputs in the loop.

Thus, if we implement f with well-controlled elements like resistors, which are relatively cheap and easy to make precise, and implement a with poorly controlled elements like amplifiers, which are relatively expensive and hard to make precise, our closed-loop gain is set precisely by f and is highly insensitive to a when $|af| \gg 1$. The insensitivity of the closed-loop gain to variations in a is one of the major benefits of feedback.

2.4.2   The effect of feedback on noise, disturbances, or distortion

Figure 2.6 shows a feedback loop composed of a feedback stage $f(s)$ and three feed-forward stages in cascade $a_1(s)$, $a_2(s)$, and $a_3(s)$, all of which add some noise or distortion at their output due to noise or nonlinearities present in the devices that comprise them. These terms are represented as $V_{d1}(s)$, $V_{d2}(s)$, $V_{d3}(s)$, and $V_{d4}(s)$ respectively and may also represent disturbances from external sources at these outputs due to unwanted coupling. The term $V_{d1}(s)$ also captures errors in the reference or tracking input $V_{in}(s)$. Applying Black's formula to the input and each disturbance input and keeping in mind that the feed-forward transfer function to the output is different for the various inputs but that the loop transmission is the same for all inputs, we get

$$\frac{V_{out}(s)}{V_{in}(s)} = V_{in}(s) \left(\frac{a_1(s)a_2(s)a_3(s)}{1 + a_1(s)a_2(s)a_3(s)f(s)} \right) + V_{d1}(s) \left(\frac{a_1(s)a_2(s)a_3(s)}{1 + a_1(s)a_2(s)a_3(s)f(s)} \right) +$$

$$V_{d2}(s) \left(\frac{a_2(s)a_3(s)}{1 + a_1(s)a_2(s)a_3(s)f(s)} \right) + V_{d3}(s) \left(\frac{a_3(s)}{1 + a_1(s)a_2(s)a_3(s)f(s)} \right) + \quad (2.9)$$

$$V_{d4}(s) \left(\frac{1}{1 + a_1(s)a_2(s)a_3(s)f(s)} \right)$$

Thus, compared with the gain of $V_{in}(s)$, the gains of $V_{d1}(s)$, $V_{d2}(s)$, $V_{d3}(s)$, and $V_{d4}(s)$ to the output are attenuated by 1, $1/a_1(s)$, $1/(a_1(s)a_2(s))$, and $1/(a_1(s)a_2(s)a_3(s))$, respectively. Thus, noise at the output is attenuated the most while noise at the input is unaffected.

This benefit of feedback is primarily due to the benefit of amplification and would also be seen in a feed-forward amplifier cascade with no feedback at all. However, in practice, feedback loops can have a large gain $a_1(s)a_2(s)a_3(s)$ and a

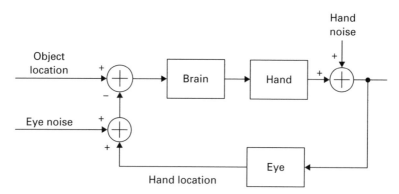

Figure 2.7. Block diagram of the brain-hand-eye feedback loop.

modest $1/f(s)$ such that the output is not saturated for most inputs due to a modest closed-loop gain; however, there is still significant attenuation of disturbances at the output due to the large loop gain of the feedback loop. We are effectively trading away the large gain we could achieve in a feed-forward system in return for more robustness in the feedback system.

Figure 2.7 reveals an example that illustrates the ability of feedback to attenuate noise at the output but not at the input. The brain, hand, and eye are in a feedback loop to allow accurate hand-reaching movements to grasp an object with visual feedback. If one's hand is noisy in its movements but one has good eyes and a good brain (high loop transmission in the motor pathway), the hand noise at the output can be vastly attenuated and one can reach for an object accurately. If the eye reports noisy, blurred, and incorrect information however, a good brain and good hand can't help one reach for an object accurately since the noise at the input cannot be attenuated. The ability of a high loop gain to attenuate noise in hand movements is dramatically revealed when one tries to perform fine motor movements, e.g., threading a needle under a microscope with $1500\times$ amplification. Due to the presence of high loop gain, very precise hand movements can be made that allow the task to be done easily.

2.4.3   The feedback advantage of linearization

The insensitivity of the closed-loop feedback transfer function to a as revealed by Equation (2.7) also makes it insensitive to nonlinearities in a. Thus, for example, in the non-inverting amplifier circuit of Figure 2.3, if the operational amplifier's gain changes from 10^5 to 10^4 as its differential input voltage v_{DIFF} changes from to 1 mV to 10 mV, the effects of such nonlinearities or kinks in a are extremely gentle in the closed-loop transfer function and almost unobservable. Further, since the negative-feedback loop equilibrates at a point such that v_{DIFF} is always very small, they are present only for very large input signals and/or large closed-loop gains. Both of the latter conditions tend to increase v_{DIFF} because v_{DIFF} scales with the

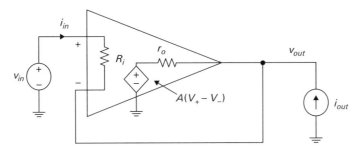

Figure 2.8. Operational-amplifier-based buffer or follower circuit used to illustrate the ability of feedback to change input and output impedances.

input and inversely with (1+ the loop gain), and the loop gain decreases for small f or equivalently for large closed-loop gains.

This insensitivity to nonlinearities in a is only present if the loop gain remains high in spite of the nonlinearity. If there is a dead zone in an operational amplifier's transfer function, e.g., a region where the gain is zero if $|v_{diff}| < \varepsilon$, the loop gain is zero in this region, and such nonlinearities do manifest in the closed-loop transfer function.

Nonlinearities at the output of a feedback loop that manifest *after* there has been amplification are greatly attenuated in the same manner that distortion is greatly attenuated at the output of a feedback loop. For example, if an operational-amplifier buffer or follower is modified by interposing a diode element between the output of the operational amplifier and the follower output, the diode dead zone of 0.6 V is attenuated to a closed-loop dead zone of 0.6 V $/ (1+A)$, where A is the gain of the operational amplifier.

2.4.4 The effect of feedback on resistance and impedance

Figure 2.8 shows an operational-amplifier follower or buffer circuit with an operational amplifier gain of A where the input and output impedances of the operational amplifier have been explicitly shown as R_i and r_o respectively. The input voltage source v_{in} and an output current source i_{out} are also shown. If we first set i_{out} to zero, and neglect feedthrough transmission of v_{in} to v_{out} via R_i (since good operational amplifiers always have $r_o \ll R_i$), then we can show from Black's formula that

$$v_{out} = \frac{A}{1+A} v_{in} \tag{2.10}$$

Thus, we can compute that

$$i_{in} = v_{in}\left(1 - \frac{A}{1+A}\right)/R_i = \frac{v_{in}}{R_i(1+A)}$$

$$\frac{v_{in}}{i_{in}} = \boxed{R_i(1+A)} \tag{2.11}$$

Hence, feedback increases the high input impedance of the operational amplifier by a factor of $(1+A)$ when the operational amplifier is configured as a follower.

To compute the output impedance of this topology, we short v_{in} and compute v_{out} when i_{out} is active. Assuming that all of i_{out} flows through r_o (an excellent approximation since $r_o \ll R_i$ and since the other end of r_o is at $-Av_{out}$), we find that

$$\frac{v_{out}(1-(-A))}{r_o} = i_{out} \tag{2.12}$$

Thus, the output impedance of the configuration is given by

$$\frac{v_{out}}{i_{out}} = \boxed{\frac{r_o}{1+A}} \tag{2.13}$$

Hence, feedback lowers the low output impedance of the operational amplifier by a factor of $(1+A)$ when the operational amplifier is configured as a buffer.

Figures 2.9 (a), (b), (c), and (d) and the associated caption show four basic two-port feedback configurations where resistance/impedance transformations due to feedback at the input or output all show multiplications or divisions by $(1+A)$, respectively. In Figure 2.9, we sense a voltage/current and feedback a voltage/current such that four possible configurations are created.

Shunt configurations at the input correspond to feeding back current while series configurations at the input correspond to feeding back voltage. Since the corrective action of a negative-feedback current is to try to keep the input voltage constant if the input current changes, shunt configurations at the input lower the input impedance. Since the corrective action of a feedback voltage is to try to keep the input current constant if the input voltage changes, series configurations at the input will raise the input impedance. Shunt configurations at the output correspond to sensing the output voltage while series configurations at the output correspond to sensing the output current. The action of negative feedback is to reject output disturbances and maintain the sensed variable to that determined by an input reference. Shunt output configurations sense output voltage and try to maintain this voltage to be constant independent of an output current disturbance and thus lower the output impedance. Series output configurations sense output current and try to maintain this current to be constant independent of an output voltage disturbance and thus raise the output impedance.

In summary, shunting, whether at the input or output, always attenuates impedances by a factor of $(1+A)$ in negative-feedback loops. Series connecting always raises impedances by a factor of $(1+A)$ in negative-feedback loops. In practice, the loading of the feedback network on the feed-forward network causes real circuits not to be directly mapped to the classic canonical topologies of Figure 2.9. Thus, the topologies of Figure 2.9 are not worth remembering, and it is far better to simply draw feedback block diagrams that take loading into account and that compute impedances from first principles. In Chapter 10, we will discuss how the techniques of return-ratio analysis, pioneered by Bode, can be used to analyze feedback circuits efficiently taking all loading effects into account [3].

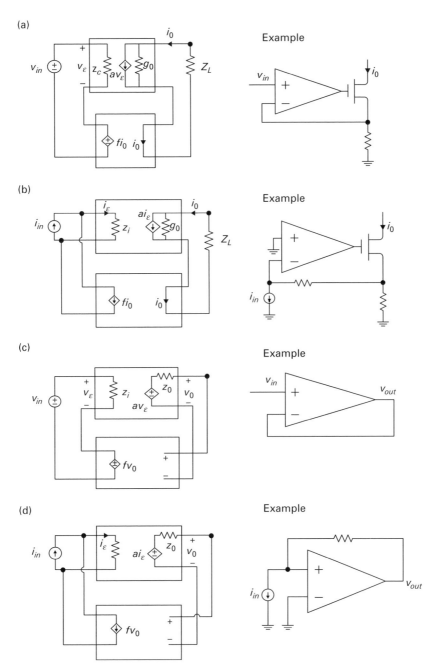

Figure 2.9a, b, c, d. Basic two-port feedback configurations: (a) series-series, (b) shunt-series, (c) series-shunt, and (d) shunt-shunt.

Figure 2.10. Block diagram showing that feedback can speed up slow time constants.

2.4.5   The feedback benefit of speedup

Figure 2.10 shows a simple feedback loop with a fast amplifier, modeled by A, and a significantly slower system, modeled by a low-pass transfer function. From Black's formula, we can derive that

$$\frac{V_{out}(s)}{V_{in}(s)} = \frac{\dfrac{A}{\tau s + 1}}{1 + \dfrac{A}{\tau s + 1}} = \frac{\left(\dfrac{A}{1+A}\right)}{\dfrac{\tau s}{1+A} + 1} \approx \frac{1}{\dfrac{\tau s}{1+A} + 1} \tag{2.14}$$

if A is large compared with 1. We see that the net effect of feedback has been to lower the open-loop time constant τ to a closed-loop value of $\tau/(1+A)$. This benefit is frequently used to speed up the response times of slow mechanical systems τ by combining them with fast electrical amplifiers A in a feedback loop. It is important to note that the speedup benefit of feedback is still limited by the smallest time constants available in the feedback loop, such that we are not magically increasing the speed of a system with feedback. For example, speedup schemes that attempt to decrease mechanical response times to electrical response times will necessarily result in more complicated dynamics since A is now really $A(s)$; the ultimate settling time achievable in a well-behaved system then turns out to be near that determined by the amplifier alone.

2.4.6   Feedback can implement inverse functions

In linear feedback loops with high loop gain, we achieve a closed-loop transfer function of $1/f(s)$, the inverse of the transfer function of the feedback system $f(s)$. More generally, if the feedback path contains an element with a nonlinear function and the overall feedback loop has high loop gain, then the closed-loop system implements the inverse of this nonlinear function. If there is high loop gain, the output of the feedback system tracks the input of the closed-loop system while the input of the feedback system is the output of the closed-loop system; so, for the closed-loop system, we have $f(output) = input$, i.e., $output = f^{-1}(input)$.

Figure 2.11 illustrates an example where the use of an exponential element in a feedback loop results in an inverse exponential, i.e., logarithmic input-output function. The feedback loop ensures that the current through the bipolar transistor is equal to the input current. Thus, from

$$i_{BIP} = I_S e^{v_{out}/\phi_t}$$
$$i_{IN} = i_{BIP} \tag{2.15}$$

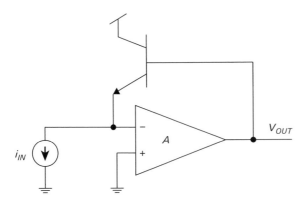

Figure 2.11. Feedback can create inverse transfer functions. An operational amplifier circuit with an exponential element in the feedback path realizes a logarithmic input-output transfer function.

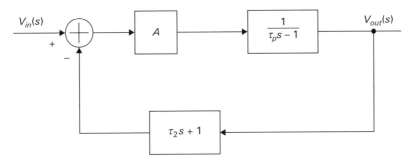

Figure 2.12. Block diagram showing that feedback can stabilize systems that are open-loop unstable.

we deduce that

$$v_{OUT} = +\frac{kT}{q}\ln\left(\frac{i_{IN}}{I_s}\right) \tag{2.16}$$

2.4.7 Feedback can stabilize systems

Figure 2.12 illustrates an unstable feed-forward system with $a(s) = 1/(\tau_p s - 1)$, i.e., an unstable system with a right half plane (RHP) pole at $1/\tau_p$. The system is stabilized by negative feedback with $f(s) = (\tau_2 s + 1)/(\tau_{high} s + 1)$ and with A being greater than 1. We shall assume that $\tau_{high} \ll \tau_2$ such that $f(s)$ can be well approximated by $(\tau_2 s + 1)$ alone. Even if this approximation is not valid, the stabilization conclusions are still valid; however, the algebra is simpler and the basic stabilization mechanism is similar if we just use $f(s) = (\tau_2 s + 1)$, so we shall do so. From Black's formula, we can show that

$$\frac{v_{out}(s)}{v_{in}(s)} = \frac{\dfrac{A}{\tau_p s - 1}}{1 + \dfrac{A(\tau_2 s + 1)}{(\tau_p s - 1)}} = \frac{A}{(\tau_p + A\tau_2)s + A - 1} = \frac{(A/(A-1))}{\dfrac{(\tau_p + A\tau_2)s}{A - 1} + 1} \tag{2.17}$$

Thus, if $A > 1$, the RHP pole of the open-loop unstable system is converted into a stable closed-loop left half plane (LHP) pole.

The seven benefits of feedback can be remembered by the mnemonic **Squirrels Do Love Running Swiftly In Snow**. The underlined letters help one remember the benefits of Sensitivity reduction to $a(s)$, Disturbance, distortion and noise rejection at the output, Linearization, Resistance and impedance transformations, Speedup, Inverse-Function implementation, and Stabilization.

2.5   Root-locus techniques

The previous discussion has shown that a high feedback loop gain is beneficial in achieving robustness to changes in $a(s)$, in sculpting precise closed-loop transfer functions, in transforming impedances, and in altering the dynamics of a system in a beneficial way, e.g., for speedup or for stabilization. Many of these benefits increase with increasing loop gain $|L(s)|$. However, if $|L(s)|$ is large, especially at high frequencies, feedback systems invariably begin to exhibit ringing and overshoot in response to changing inputs. If $|L(s)|$ is sufficiently large, feedback systems often go unstable. The physical cause of such instability is that delays and phase shifts in the feedback loop cause it to take action based on sensed information that reflects a past state of the output rather than on its current state, and/or to take actions which, because of their delay in manifestation, are inappropriate at the present time. The actions of the loop can consequently actually exacerbate its tracking error rather than reducing it. The larger the loop gain, the larger is the reactive action of the loop to error information, and the larger is the likelihood that strong-and-incorrect reactions to delayed and/or incorrect error information will lead to instability. Hence, in a feedback loop, robustness-and-precision considerations usually force one to use a loop gain that is as large as possible while stability considerations usually force one to use a loop gain that is not ambitiously large. Practical feedback system design is always a compromise between these two considerations. To understand how such tradeoffs work in detail, it is necessary to understand how feedback alters the dynamics of a closed-loop system. The brilliant control engineer Walter Evans devised a geometric and intuitive technique called *root locus* that allows one to rapidly understand how the effects of increasing loop gain alter the dynamics of a feedback system [4]. We shall now begin our study of these root-locus techniques. A good understanding of root locus is essential to the mastery of feedback systems.

Figure 2.13 shows a feedback loop wherein K, a scalar, is varied to alter the loop gain of the feedback loop. As K varies, the transfer function of the closed-loop system varies according to Black's formula and is given by

$$\frac{V_{out}(s)}{V_{in}(s)} = \frac{Ka(s)}{1 + Ka(s)f(s)} \qquad (2.18)$$

Thus, the roots of the denominator vary with K. We would like to understand how the locus of these roots, i.e., the values of the poles of the closed-loop system, vary as K is varied from 0 to ∞. Negative values of K are forbidden for now,

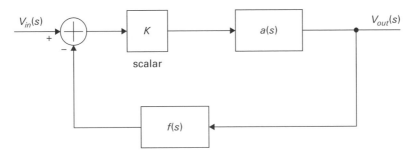

Figure 2.13. The basic context for root locus: the value of the scalar K is varied and the closed-loop pole locations are plotted on the complex plane.

since such values correspond to the case of positive feedback. If s' is a root of the denominator, then we must have

$$Ka(s')f(s') = -1 \qquad (2.19)$$

Thus,

$$\frac{1}{|a(s')f(s')|} = |K| \qquad (2.20)$$
$$\angle a(s')f(s') = (2n+1)\pi$$

The first condition in Equation (2.20) is called the root-locus magnitude condition while the second condition is called the root-locus angle condition. The angle condition is significantly more informative than the magnitude condition. If we can find a value of s' that satisfies the angle condition in the s plane, then we can always evaluate $|a(s')f(s')|$ at this point, and then find a $|K|$ that satisfies the magnitude condition by simply setting $|K|$ to the reciprocal of $|a(s')f(s')|$. Figure 2.14 illustrates how we attempt to locate such points s' in the s plane. We try to find all locations on the s plane where the sum of negative angles from the poles of $a(s)f(s)$ and the positive angles from the zeros of $a(s)f(s)$ will sum to an odd multiple of π, i.e., $(2n+1)\pi$, where n is an integer[1]. How many such points are there?

If we assume that

$$a(s)f(s) = \frac{\prod_{i=1}^{i=z}(s - z_i)}{\prod_{j=1}^{j=p}(s - p_j)}, \quad p \geq z \qquad (2.21)$$

i.e., the loop transmission has p poles and z zeros with $p \geq z$, then

$$1 + K\frac{\prod_{i=1}^{i=z}(s - z_i)}{\prod_{j=1}^{j=p}(s - p_i)} = 0 \qquad (2.22)$$

---

[1] In Figure 2.14, the poles are drawn as crosses and zeros are drawn as circles with positive angles defined in an anticlockwise sense starting from the positive real axis.

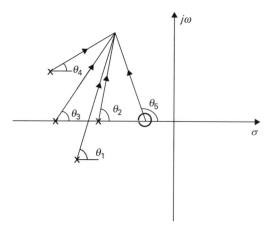

Figure 2.14. Any point on the root locus must satisfy the angle condition: the sum of the signed angles from the singularities in the loop transmission must be an odd multiple of 180°.

leads to p solutions for the pth-order polynomial or to p roots for every K. *Thus, as K varies there are p solutions for each K or equivalently p branches of the root locus.* The eight root-locus rules that we list below tell us where the p branches of the root locus lie on the s plane and how they vary with K. They are derived from either the magnitude condition or the angle condition. Taken together, they allow us to quickly see how the locations of the poles of a closed-loop system vary as the loop gain of the system is scaled from 0 to ∞.

2.6   Eight root-locus rules

2.6.1   The beginning rule

As K goes from $0 \rightarrow \infty$, branches of the root locus begin at the poles of $a(s)f(s)$ and end at its zeros. The $p-z$ branches that have nowhere to go end up at ∞.

Figure 2.15 reveals an example that illustrates this rule. The loop transmission has 3 poles and 1 zero. Thus, there are 3 branches of the root locus. Each branch starts at a pole and ends at a zero. Since there is only one zero in this example, only one branch heads towards this zero while the other two branches head towards ∞.

The reasoning behind this rule is straightforward. If $|Ka(s')f(s')| = 1$, and $K = 0$, then $|a(s)f(s)| = \infty$, i.e., we're at a pole of the loop transmission. If $|Ka(s')f(s')| = 1$ and $K = \infty$, then $|a(s)f(s)| = 0$, i.e., we're at a zero of the loop transmission, which occurs either at an explicit zero location or at ∞ where $a(s)f(s)$ goes to zero. Alternatively, a simpler proof is evident from inspection of Equation (2.22).

A zero effectively functions like an attractor negative charge on which one and only one root-locus field line ends. A pole effectively functions like a repeller positive charge from which one and only one root-locus field line begins.

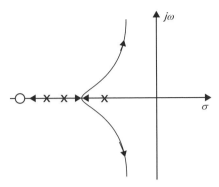

Figure 2.15. The beginning rule for root loci.

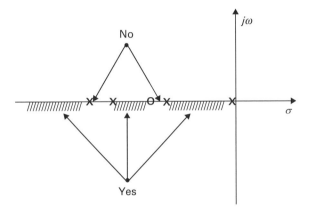

Figure 2.16. The real-axis rule for root loci.

2.6.2 The real-axis rule

All branches of the root locus that lie on the real axis must lie to the *left* of an odd number of singularities present on the real axis, whether these singularities are poles or zeros.

Figure 2.16 reveals an example for a loop transmission with four poles and one zero. The regions marked Yes are regions on the real axis to the left of an odd number of singularities while those marked No are regions to the left of an even number of singularities. For the application of this rule, it does not matter whether the singularities are poles or zeros, just that there is an odd number of them.

The rule is a direct consequence of the angle condition illustrated in Figure 2.14. On the real axis, any complex-conjugate poles or complex-conjugate zero pairs contribute a net zero phase and thus do not matter. All real-axis singularities for which a root-locus branch is on their right contribute zero angles to any point on this root-locus branch and thus also do not matter. Thus, the only singularities that can contribute to creating a net angle of π on a root-locus branch must have the root-locus branch on their left. Each such singularity, whether pole or zero,

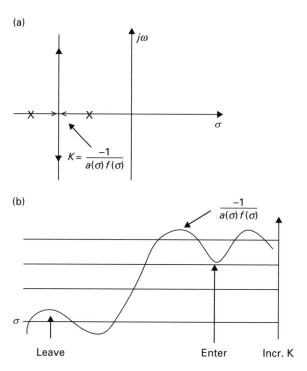

Figure 2.17a, b. The departure rule for root loci, (a) in the s-plane, and (b) in a K versus σ plot.

contributes an angle of π to any point on this branch (a pole contributes $-\pi$ and a zero contributes $+\pi$ but these angles are equal). Thus, to obtain a net contribution of π, the branch of interest must be to the left of an odd number of singularities such that the net contribution has an odd number of π's, i.e., is π. If the branch is to the left of an even number of singularities, the net angle would correspond to an even number of π's, i.e., would correspond to 0, and violate the angle condition.

2.6.3   The departure rule

Branches of the root locus depart the real axis at maxima of $-1/(a(\sigma)f(\sigma))$ and re-enter the real axis at minima of $-1/(a(\sigma)f(\sigma))$ where the complex variable $s = \sigma + j\omega$.

Figure 2.17 (a) illustrates how two branches of a root locus come together and depart the real axis as K is increased. When these branches just depart the real axis, the two solutions on the real axis corresponding to each branch have just merged into one at the critical value of K at which departure just occurs. For values of K greater than this critical value, there is no solution on the real axis and the branches depart the real axis and acquire a $j\omega$ component. As K is increased further, these same branches may re-enter the real axis at a different location

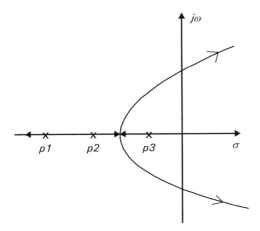

Figure 2.18. The mean rule (or Grant's rule) for root loci.

remote from the one that they departed from. At an entry point on the real axis, a sequence of events that is like a reversal of the events seen during departure occurs. The two solutions start as one at the entry point on the real axis and then split apart into two solutions on the real axis and diverge from each other as K is increased.

Figure 2.17 (b) illustrates the reasoning behind the rule. The root-locus magnitude condition implies that the intersection between a line of constant K and $-1/(a(\sigma)f(\sigma))$ yields points on the real axis that each correspond to one branch of a root-locus plot and that are the closed-loop pole locations at that value of K. As K increases, these roots move as the intersections change, leading to a continuous root-locus branch. At maxima of $-1/(a(\sigma)f(\sigma))$ and at a particular K two roots just merge together into one solution and then disappear as K is increased further, corresponding to departure of roots from the real axis. At minima of $-1/(a(\sigma)f(\sigma))$ and at a particular K one solution just manifests on the real axis and then diverges into two solutions as K is increased further.

2.6.4 The mean rule (Grant's rule)

If $p \geq z + 2$, then the arithmetic mean of all p roots is invariant with K.

The reasoning for this rule is evident by examining Equation (2.22). If $p \geq z + 2$, then the coefficient of the s^{p-1} term, which depends on the sum of the roots, is invariant with K. Figure 2.18 reveals an example of a 3-pole root-locus plot where $p = 3$ and $z = 0$ so the conditions of Grant's rule apply. In this example, since two poles move to the right while one pole moves to the left, to keep the arithmetic mean constant the leftward motion of the leftmost pole must be twice the rightward motion of either rightward and symmetrically moving pole.

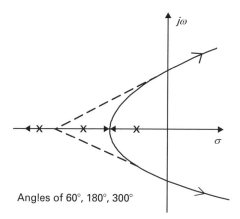

Angles of 60°, 180°, 300°

Figure 2.19. The asymptote rule for root loci.

2.6.5   The asymptote rule

The $p - z$ branches of the root locus that head off to ∞ all asymptotically appear to originate from a point on the real axis given by

$$\frac{\Sigma(\mathrm{Re}\{Poles\} - \mathrm{Re}\{Zeros\})}{p - z} \tag{2.23}$$

and make asymptotic angles with the real axis of

$$\frac{(2n + 1)\pi}{p - z}, n \in \{0, 1, 2, \ldots p - z - 1\} \tag{2.24}$$

The reasoning for Equation (2.23) is that, from a far enough distance, all the poles and zeros of $a(s)f(s)$ will appear to be one pole of multiplicity $p - z$ situated at the arithmetic mean of all the pole and zero locations with poles weighted positively and zeros weighted negatively (or vice versa). The arithmetic mean of all pole and zero locations is always on the real axis due to complex-conjugate symmetry. To satisfy the angle condition, the identical angles from all poles that constitute the pole of multiplicity $p - z$ must add up to $(2n + 1)\pi$ such that the angle from each and any pole at this location is given by Equation (2.24).

Figure 2.19 reveals an example that illustrates the asymptote rule for the case of $p = 3$. The three asymptotes make angles of 60°, 180°, and 300° with the real axis in accordance with the rule.

2.6.6   The complex-singularity rule

Near a complex pole the root-locus branch angle θ is evaluated as

$$\sum \angle Zeros - \sum \angle Poles - \theta = (2n + 1)\pi \tag{2.25}$$

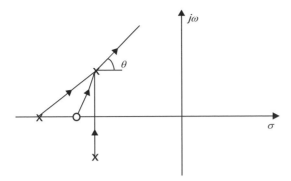

Figure 2.20. The complex-singularity rule (or departure angle rule) for root loci.

where the angles refer to angles from all other poles and zeros to the complex pole of interest; near a complex zero, the root-locus branch angle θ is evaluated as

$$\sum \angle Zeros - \sum \angle Poles + \theta = (2n+1)\pi \tag{2.26}$$

where the angles refer to angles from all other poles and zeros to the complex zero of interest.

Figure 2.20 illustrates the reasoning behind this rule for a complex pole. It is a straightforward application of the angle condition depicted in Figure 2.14 except that we approximate the angles from all other poles and zeros to the point s' by angles to the pole instead since s' is extremely close to the pole of interest. As Equations (2.25) and (2.26) reveal, the rule for a complex pole is similar to that for a complex zero except for the change in the sign of the angle θ.

2.6.7 The remote rule

To construct branches of the root locus near the origin ignore poles and zeros that are far from the origin.

Figure 2.21 illustrates the reasoning behind this rule. Essentially we are approximating the angle contributions of the remote poles and zeros as all being zero since they are so far away.

A corollary to this rule is that the contribution of remote poles and zeros in an RHP region may contribute a net angle of π or 0 at the origin, which can be treated as a constant in the root-locus angle condition of Equation (2.20). Similarly, the net angle contributions of poles and zeros at the origin can be treated as a constant when computing the root locus at a location remote from the origin.

2.6.8 The value rule

The value of K at which a particular location on a branch s_1 is attained is given by

$$K = \frac{1}{|a(s_1)f(s_1)|} \tag{2.27}$$

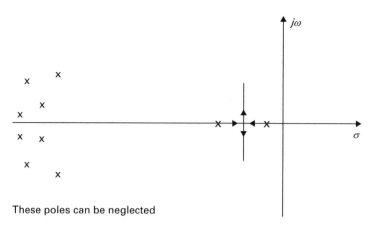

These poles can be neglected

Figure 2.21. The remote rule for root loci.

This rule is a straightforward restatement of the magnitude condition. There are some useful short cuts involved in applying this rule in common situations which are worth discussing. In physical systems, *every* pole and zero in $a(s)f(s)$ is often expressed in a normalized form within $a(s)f(s)$ such that its dc gain (gain at $s = 0$) is 1, i.e., as $1/(\tau_p s + 1)$, as $1/(\tau^2 s^2 + \tau s/Q + 1)$, or as $(\tau_z s + 1)$, such that K becomes the root-locus dc-loop-gain parameter and $a(s)f(s)$ captures all the normalized dynamics. The contribution in magnitude of a pole or zero to a given point s_1 is then determined by dividing the distance from the pole or zero to s_1 by the distance from the pole or zero to the origin to compute a normalized distance, which we shall term d_n; from Equation (2.27), the value of K that corresponds to the magnitude contribution of such a pole then scales with d_n while the value of K that corresponds to the magnitude contribution of such a zero scales with $1/d_n$. Thus, one can get a rapid estimate of the K that corresponds to a given point s_1 on the root-locus plot as

$$K = \frac{\prod_{i=1}^{i=p} d_{ni}^{poles}}{\prod_{j=1}^{j=z} d_{nj}^{zeros}} \qquad (2.28)$$

Equation (2.28) is useful in estimating the value of K that will create a closed-loop pole at a given location on a root-locus plot when all poles and zeros in $a(s)f(s)$ contribute unity gain at dc. Sometimes, the presence of a pole or zero at the origin, e.g., $1/\tau s$ or τs, violates this dc normalization condition. In the latter case, we simply compute d_n as the distance from the origin to the point s_1 divided by $1/\tau$; d_n still appears in the denominator for zeros and in the numerator for poles in Equation (2.28) as in the unity-gain normalized case. In general, the normalization constant for computing distances is always $1/\tau$, where τ is the parameter used to characterize poles and zeros in the normalized forms τs, $\tau s + 1$, or $\tau^2 s^2 + \tau s/Q + 1$.

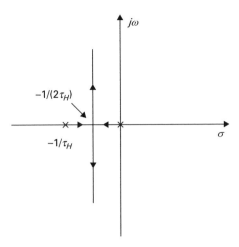

Figure 2.22. An example of the value rule for root loci.

Figure 2.22 illustrates a simple application of Equation (2.28) for $a(s)f(s) = 1/(\tau_L s(\tau_H s + 1))$. We would like to compute the value of K at which the pole at the origin and the pole at the $-1/\tau_H$ just meet before they depart from the real axis in the root-locus plot. This point is at $-1/(2\tau_H)$ by Grant's rule. Thus, the contribution of the normalized distance of the integrator pole at the origin is $\tau_L/(2\tau_H)$, the contribution of the normalized distance of the $-1/\tau_H$ pole is $\tau_H/(2\tau_H) = 1/2$. Thus, the value of K at which the two poles just meet before they depart from the real axis is given from Equation (2.28) to be the product of these normalized distances, i.e., $\tau_L/(4\tau_H)$. This leads to the well-known result that, if an integrator and a lowpass filter are placed together in a negative-feedback loop with $K = 1$, then the closed-loop poles are at 'critical damping' when the integrator's time constant is four times that of the lowpass filter's time constant.

The root-locus rules can be remembered by the mnemonic **Beaches Really Do Make A Calm Relaxing Vacation** where the underlined terms signify the Beginning, Real-axis, Departure, Mean, Asymptote, Complex-singularity, Remote, and Value rules respectively.

2.7   Example of a root-locus plot

Figure 2.23 (a) shows a feedback system composed of a motor with a lowpass transfer function $1/(\tau_m s + 1)$ between its input voltage and output angular velocity $\Omega_{out}(s)$ and an electrical controller transfer function $K(\tau_z s + 1)/(\tau_e s)$ composed of an integrator pole and a high-frequency zero. Figure 2.23 (b) plots the root-locus plot for this system as K varies from 0 to ∞. The root-locus plot of Figure 2.23 (b) is important in several feedback systems, including position and velocity control of motors, in transimpedance-amplifier feedback loops,

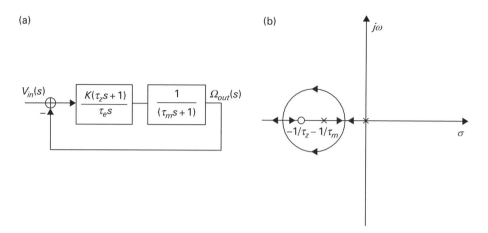

Figure 2.23a, b. An example of a 2-pole and 1-zero motor-velocity control system, (a) block diagram and (b) root-locus plot.

which are discussed in Chapter 11, and in phase-locked loops, which are discussed in Chapter 18.

To draw this plot, we apply the following root-locus rules: 1) The beginning rule states that there will be two branches of the root locus, each of which starts at one of the poles, one of which ends at one of the zeros, and another of which ends at ∞. 2) The real-axis rule states that any root-loci branches can only be present between the two poles (to the left of one pole) and to the left of the zero (to the left of three singularities). 3) The departure rule and value rule (as expressed in Equation (2.28)) cause the two root-loci branches to meet up and depart the real axis nearly midway between the two poles. The product of the two normalized distances from the integrator pole and the motor pole is maximal when these lengths are equal; at the departure point, the normalized distance contribution of the zero is nearly 1 and is not significant. After departing the real axis, these branches must re-enter the real axis at a high-enough value of K to the left of the zero and then diverge from each other with one branch heading towards the zero and the other branch heading towards ∞; we may deduce the latter scenario since one of the branches must end at the zero and both branches must remain complex conjugate while not on the real axis. 4) We have $p - z = 1$ such that the conditions of the mean rule do not apply implying that the net center-of-mass of the closed-loop system can change with increasing K and, in fact, does. 5) The asymptote rule requires one branch asymptotically heading to ∞ at large K and making an angle of 180° with the real axis which is consistent with the findings of the departure/re-entry rule. 6) The complex-singularity rule does not apply. 7) The remote rule helped us ignore the effect of the high-frequency zero when computing the root locus between the two poles. 8) The value rule can help us compute how high K needs to be for one of the poles to be right at the zero, say at $-1.01/\tau_z$. In this case, in Equation (2.28), the normalized distance contribution of the integrator pole is τ_e/τ_z, the normalized distance contribution of the motor pole is

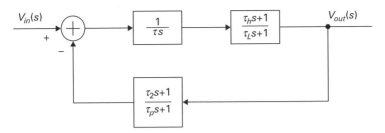

Figure 2.24. An example of a feedback block diagram that illustrates the location of zeros in a closed-loop transfer function.

$(1/\tau_z - 1/\tau_m)/(1/\tau_m) = (\tau_m/\tau_z - 1)$, and the normalized distance contribution of the zero is $1.01 - 1.0 = 0.01$. Thus, the value of K at which we are almost right at the zero is given by

$$K = \frac{\tau_e}{\tau_z}\left(\frac{\tau_m}{\tau_z} - 1\right)/0.01 \approx 100\frac{\tau_e\tau_m}{\tau_z^2} \qquad (2.29)$$

Equation (2.29) expresses the intuition that if the electrical and mechanical time constants are large and the zero is located at a very high frequency value, then a large K will be needed to move these poles towards the zero. It also reveals that getting closer and closer to the zero will require higher and higher gains, e.g., to locate the pole at $-1.001/\tau_z$ versus at $-1.01/\tau_z$ implies that the needed K is 10 times higher. This root-locus example reveals that perfect pole-zero cancellations are very hard to achieve.

2.8 The zeros of a closed-loop system

Thus far, we have focused on determining the poles of a closed-loop system. Where do the zeros of a closed-loop system lie? The answer is evident from Black's formula in Equation (2.1): at the zeros of $a(s)$ since this is the only term in the numerator of Black's formula and at the poles of $f(s)$ since the denominator of Black's formula then goes to ∞. Note that the poles of $a(s)$ are present in the numerator and denominator of Black's formula and thus do not cause the closed-loop system transfer function to go to zero at these values. Equation (2.21) along with Black's formula may be used to arrive at the same conclusion. Thus, **the zeros of a closed-loop system are at the zeros of $a(s)$ and at the poles of $f(s)$**.

Figure 2.24 reveals an example of a feedback loop. We may immediately conclude that the zeros of the closed-loop system are at the zeros of $a(s)$, i.e. at $-1/\tau_h$, and at the poles of $f(s)$, i.e., at $-1/\tau_p$.

2.9 Farewell to feedback systems

We shall bid farewell to feedback systems for now. However, we will draw on the amazingly deep concepts of feedback again and again throughout this book.

To start, we shall see that it affords simple and intuitive views of the device physics of an MOS transistor, particularly in Chapters 3, 4, and 6. In Chapter 8, we shall use feedback techniques to construct ultra-low-noise lock-in amplifiers for sensors in micro-electro-mechanical systems (MEMS). In Chapter 9, we shall discuss the analysis and design of feedback systems based on techniques primarily developed by Nyquist. In Chapter 10, feedback will appear in the form of return-ratio analysis, a brilliant circuit-analysis technique pioneered by Bode [3]. Feedback systems, in one form or another, will then manifest indirectly or directly in every chapter throughout the book.

References

[1] Harold S. Black, inventor. *Wave Translation System*, U.S. Patent Number 2,102,671.
[2] R. Kline. Harold Black and the negative-feedback amplifier. *IEEE Control Systems Magazine*, **13** (1993), 82–85.
[3] Hendrik W. Bode. *Network Analysis and Feedback Amplifier Design* (New York, NY: Van Nostrand, 1945).
[4] Walter R. Evans. *Control-System Dynamics* (New York, NY: McGraw-Hill, 1954).

3 MOS device physics: general treatment

Intuition will tell the thinking mind where to look next.

Jonas Salk

To deeply understand any electronic circuit, whether it is low power or not, it is essential to have a good mastery of the devices from which that circuit is made. In this chapter, we will begin our study of device physics with the metal oxide semiconductor (MOS) transistor, the most important active device in electronics today. The MOS transistor is a field effect transistor (FET) and MOSFETs are abbreviated as nFETs if their current is due to electron flow and as pFETs if their current is due to hole flow. In this chapter, we shall focus on fundamental principles and on exact mathematical descriptions that are applicable to transistors built in technologies with relatively long dimensions. In later chapters, we shall study practical approximations needed to simplify these exact mathematical descriptions (Chapter 4), study small-signal dynamic models of the MOS transistor (Chapter 5), and discuss effects observed in deep submicron transistors with relatively short dimensions (Chapter 6).

Figure 3.1 shows a zoomed-in view of an n-channel FET or nFET built in a standard bulk complementary metal oxide semiconductor (CMOS) process [1]. There are four terminals referred to as the gate (G), source (S), drain (D), and bulk (B), respectively. The control terminal, the metal-like polysilicon gate, is insulated from the silicon bulk via a silicon dioxide insulator; the source and drain terminals are created with n+ regions in the p-type silicon bulk. The voltage v_{GS} refers to the voltage between the gate and source terminals and v_{DS} refers to the voltage between the drain and source terminals. It is conventional to either reference all voltages to the source, i.e., make the source voltage ground, to create a source-referenced description or to reference all voltages to the bulk, i.e., make the bulk voltage ground, to create a bulk-referenced description. By convention, the drain terminal always has a higher voltage than the source terminal in nFETs. Conventional positively charged current, i_{DS} or i_D, flows from drain to source in the nFET while negatively charged electron current flows from source to drain in an nFET. In the particular example of Figure 3.1, the bulk is tied to the source such that v_{BS} is zero and all voltages are referenced to the source. Due to the symmetry of the source and drain terminals, bulk-referenced descriptions are more symmetric than source-referenced models but source-referenced descriptions are more widespread in use.

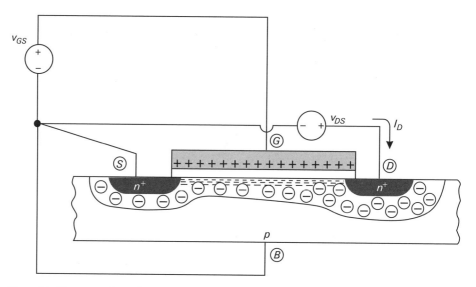

Figure 3.1. Zoomup of an nFET. Reprinted with kind permission from [3] by permission of Oxford University Press, Inc., and with kind permission from Professor Yannis Tsividis.

The gate terminal and bulk terminal of a transistor consume almost no dc current. The transistor operates such that increasing values of v_{GS} and v_{DS} increase i_{DS}. The current i_{DS} barely changes with v_{DS} after v_{DS} becomes larger than a saturation voltage, v_{DSAT}, and assumes a saturation value i_{DSAT} that is only dependent on v_{GS} in this saturation regime. For sufficiently low values of v_{GS}, i_{DSAT} is exponentially dependent on v_{GS}, a regime of operation termed *subthreshold* or *weak inversion*. In subthreshold, a 100 mV-per-decade increase in i_{DSAT} with v_{GS} is typical. For sufficiently large values of v_{GS}, i_{DSAT} changes with v_{GS} via a quadratic polynomial, a regime of operation termed *above-threshold* or *strong inversion*. The saturation voltage v_{DSAT} is constant at nearly 100 mV in subthreshold operation and varies linearly with v_{GS} in above-threshold operation. The transition from the linear region of operation, where $v_{DS} \ll v_{DSAT}$, to the saturation region follows an exponential function in subthreshold operation and a quadratic function in above-threshold operation. The currents in subthreshold operation are relatively small compared to those in above-threshold operation.

Figures 3.2 (a) and (b) show experimental data collected from a transistor that reveal subthreshold and above-threshold i_{DS} versus v_{DS} characteristics for various different v_{GS} values in subthreshold and above-threshold operation respectively. Figure 3.3 (a) shows why the approximation of a threshold appears to be a reasonable one: If the square root of the saturation current is plotted versus v_{GS} it appears to abruptly go to zero from above-threshold current values such that, from an above-threshold viewpoint, the approximation of a threshold is reasonable. In reality, the experimental data shown in the semi-logarithmic plot of Figure 3.3 (b) reveal that the saturation current smoothly transitions from exponential behavior to square-law behavior. Table 3.1 summarizes several differences

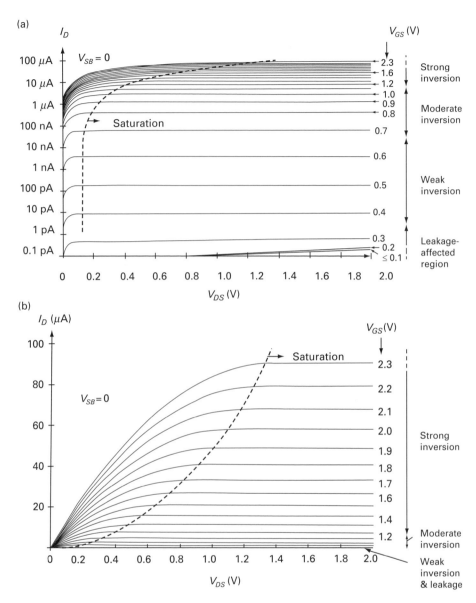

Figure 3.2a, b. Experimental (a) subthreshold and (b) above-threshold $i_D - v_{GS}$ characteristics. Reprinted with kind permission from [3] and [1], respectively, and with kind permission of Professor Yannis Tsividis.

between subthreshold operation and above-threshold operation. Before delving into the exact mathematics of MOS transistor operation that explains and describes why these curves have the shapes that they do, it is important to develop some intuition for MOS transistor operation. Intuitive descriptions and models, even if incomplete or inexact, are important in developing insight and in obtaining a big-picture view that is not obscured by details.

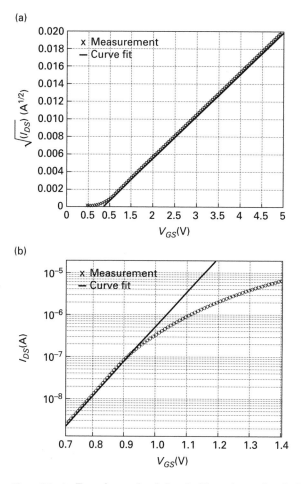

Figure 3.3a, b. Experimental subthreshold to above-threshold saturation characteristics from (a) square-root and (b) log plots.

3.1   Intuitive description of MOS transistor operation

Figure 3.1 also illustrates the basic principle of operation of the MOS transistor. Increasing values of gate voltage are associated with the gate terminal developing increasingly more positive charge. The positive charge on the gate repels the positively charged majority carrier holes in the p-type silicon bulk through capacitive action implemented by the gate, SiO_2 (silicon dioxide) insulator, and the bulk. Underlying negatively charged immobile dopant acceptor atoms in the bulk are then exposed. Since positive charges raise the potential of regions near them, the gate charge also increases the local *surface potential*, the potential beneath the gate in the bulk at the surface between the silicon bulk and the SiO_2. Since negatively charged electrons like to concentrate in low-energy positive potential regions, the rise in surface potential attracts minority carrier electrons from the source and

Table 3.1 Subthreshold versus above-threshold MOSFET operation

Subthreshold	Above-threshold
Saturation current is exponential in v_{GS}	Saturation current is square law in v_{GS}
v_{DSAT} is constant at approximately $100\,\mathrm{mV}$	v_{DSAT} varies linearly with gate voltage
Current flows by diffusion	Current flows mainly by drift
Charge concentrations are small	Charge concentrations are large
Currents are small	Currents are large
Good for ultra-low-power operation	Good for high-power operation
Power efficiency is constant with current	Power efficiency is lower and degrades with larger currents
High noise and offset	Low noise and offset
Can work on low power supply voltages	Needs higher power supply voltages
Linearity is hard to achieve	Linearity is easy to achieve
Math is usually beautiful and has closed-form solutions	Math is often ugly and lacks closed-form solutions
Increasingly important today and in the future	Traditional use of MOSFETs from the past
Suited for slow-and-parallel architectures	Suited for fast-and-serial architectures

drain n+ terminal regions of the transistor into the bulk. The n+ source and drain regions have lots of majority-carrier electrons that gravitate to the channel, a thin sheet-like region at and below the surface where the bulk potentials are highest, and therefore where most of the electrons in the bulk are exponentially more likely to congregate. Since the drain terminal is at a more positive potential than the source terminal, the energy required for electrons to leave the source and enter the channel is lower than that needed at the drain. Consequently, there is a higher concentration of electrons at the source end of the channel than at the drain end. This concentration gradient causes an electron current to flow from source to drain due to diffusion. This diffusion current is the current observed in *subthreshold* transistor operation, a region where the gate voltage is sufficiently small such that the electron charge concentration at the source and drain ends is significantly less than the dopant bulk charge concentration. The surface potential at the source and drain ends is then independent of the electron concentrations, is determined only by the positive charges on the gate and the negative charges on the dopant atoms, and is constant throughout the channel.

If the gate voltage becomes sufficiently large, the electron concentration at the source channel end and/or drain channel end can become larger than the dopant bulk charge concentration and begin to significantly affect the surface potential. Since there are more electrons at the source end of the channel than at the drain end, and negatively charged electrons lower the potential of the region that they are at, the surface potential at the source end of the channel becomes lower than the surface potential at the drain end. In this *above-threshold* regime, the surface-potential gradient establishes a lateral electric field between source and drain and a drift current begins to flow. If the drain voltage is made sufficiently high, it becomes energetically very expensive for electrons at the drain to enter the channel, such that the concentration of electrons at the drain end begins to approach zero and the surface potential at the drain end is determined only by the gate and bulk just as it is in the subthreshold regime. At this point, increasing the drain voltage any further will have little effect on either the charge concentration at the drain end of the channel, which continues to remain zero, or the surface potential at this end, which continues to remain constant. Thus, the lateral electric field in the channel and charge concentration in the transistor become independent of the drain voltage and dependent only on conditions at the source end of the channel, making i_{DS} nearly independent of v_{DS}, and causing the transistor to saturate. Saturation occurs for a similar reason in the subthreshold regime as well. With increasing drain voltage, if the charge concentration at the drain end of the channel continues to remain negligible compared with the charge concentration at the source end of the channel, the diffusion current becomes independent of the drain voltage.

In subthreshold operation, the charge concentrations are too small to cause a significant surface-potential gradient or lateral electric field. Thus, the current is dominated by diffusion. In above-threshold operation, the charge concentrations are large enough to cause a surface-potential gradient and lateral electric field, such that the dominant flow of current is via drift.

In subthreshold operation, the gate voltage is sufficiently small such that the concentration of electrons in the channel is significantly less than the concentration of dopant atoms in the channel and these electrons have little impact on the surface potential, which is primarily determined by the positive charges on the gate and the negatively charged dopant atoms. In above-threshold operation, the gate voltage is sufficiently large such that the concentration of electrons in the channel, especially at the source end, becomes significantly larger than the concentration of dopant atoms in the channel, and these electrons begin to significantly affect the surface potential. In fact, in the above-threshold regime, there are so many electrons in the channel that they have a regulatory effect on the surface potential in the channel through a negative-feedback loop: negatively charged electrons lower the potential of the region that they are at. Thus, if the surface potential tries to rise, for example, because the gate voltage is increased, electrons rush in and lower the surface potential; if the surface potential tries to fall, for example, because the gate voltage is decreased, electrons leave and raise the surface potential. This feedback action is stronger near the source end of the channel than near the drain end of the channel because there are more electrons at the source end than the drain end. When the transistor is saturated and the electron concentration at the drain channel end is zero, there is only feedback action at the source end of the channel. Due to the exponential dependence of charge carrier concentration on the surface potential, this feedback action is extremely strong at the source end of the channel and tends to keep the surface potential at this end nearly constant in above-threshold operation and independent of the gate voltage, functioning like a diode clamp voltage regulator in a circuit. The surface potential at the source channel end effectively behaves like a bottom-plate constant-voltage terminal of an oxide capacitance with the gate being the other terminal. The bulk is then screened from the effects of the gate near the source end of the channel such that the electron charge at this end changes with changes in gate voltage but the bulk charge and surface potential there remain nearly constant. The change in electron charge concentration at the source end of the channel is then well modeled by the change in charge across a linear C_{ox} capacitor with a constant-voltage bottom plate and a changing top-plate gate voltage, and therefore is linear with changes in gate voltage.

The above-threshold square-law dependence of the transistor saturation current on the gate voltage arises because increasing gate voltage increases the amount of mobile charge at the source linearly (as would be expected of a capacitor with a fixed voltage on its bottom plate), and the maximum lateral electric field in the channel also increases linearly with this increased charge yielding net square-law behavior for the drift current (drift current increases proportionately with more charge and with more electric field).

3.2   Intuitive model of MOS transistor operation

Figure 3.4 shows an intuitive model of the MOS transistor that is extremely useful in understanding its operation in both the subthreshold and above-threshold regimes. The transistor has a distributed oxide capacitance, modeled by the C_{ox} capacitors in the figure, and a distributed depletion capacitance due to bulk dopant charge, modeled by the C_{dep} capacitors in the figure. The surface potential varies along the length of the transistor and is comprised of the set of voltages at the junctions where capacitors meet in a distributed capacitive-divider-like configuration. At the source and drain ends of the channel, two diode-like elements represent pn junctions

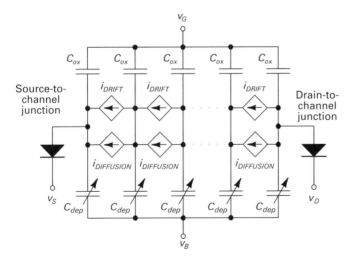

Figure 3.4. Intuitive model of an MOS transistor.

from source to channel and drain to channel, respectively. The diode-like elements intuitively represent the fact that the channel charge concentration at the source or drain channel boundaries increases exponentially with the channel-to-terminal voltage across the element. After a diode-like element turns on, it clamps the channel voltage near each terminal to an approximately constant diode drop above that terminal's voltage, and above-threshold operation is established. The current in the transistor is determined by the drift and diffusion generators responding to the derivative of the local channel potential or local charge concentration, respectively. The channel potential and charge concentrations equilibrate at values along the channel such that the net sum of the drift and diffusion current at any point in the transistor is constant. The diode-like elements and capacitors determine the boundary conditions for channel potential and charge concentration at the source and drain ends of the channel. The drift and diffusion generators then adjust their values to determine the current in the channel and ensure that it is constant throughout the channel. Diffusion current dominates in subthreshold operation while drift current dominates in above-threshold operation.

In subthreshold operation, both diode-like elements are off, such that the boundary conditions are determined only by the capacitive divider formed by the C_{ox} and C_{dep} at each terminal end; the surface potential is consequently the same at each end of the channel, and, therefore, throughout the channel; the charge concentration is determined by the exponential source-to-channel and drain-to-channel voltages across each junction at the source and drain ends of the channel respectively. In above-threshold operation, the source diode-like element is always on such that the surface potential is one diode-clamp voltage above the source terminal voltage at the source channel end; in linear above-threshold operation ($v_{DS} < v_{DSAT}$) the drain diode-like element sets the surface potential at the drain channel end to be one diode-clamp voltage above the drain terminal voltage; in saturated above-threshold operation ($v_{DS} \geq v_{DSAT}$), the drain diode-like element is off, and the surface potential at the drain end of the channel is determined by the capacitive-divider formed by the C_{ox} and C_{dep} capacitances, just as it is in subthreshold operation. The electron charge concentrations at the boundaries of the channel in linear above-threshold operation are well approximated by the gate charge across the C_{ox} capacitors at each end of the channel minus the fixed dopant charge across the C_{dep} capacitors when the diode-like elements just turn on; in saturated above-threshold operation, the source charge concentration is approximated as it is in above-threshold linear operation but the drain charge concentration is approximated as zero. The distributed C_{dep} capacitances, whose values are a square-root function of the local surface potential, are constant throughout the channel in subthreshold operation. They decrease from the source end to the drain end of the channel in above-threshold operation as the depletion region thickens due to the increasing surface potential (as shown in Figure 3.1).

It is extremely important to note that, as in a bipolar transistor, the drain-to-channel junction of Figure 3.4 can be reverse biased and still conduct lots of current. In fact, when the transistor is saturated, the source-to-channel junction injects all the electrons into the channel like the emitter-to-base junction in an NPN bipolar transistor. These electrons are then swept into the reverse-biased drain-to-channel junction just as they are swept into the reverse-biased collector-to-base junction in an NPN bipolar transistor. In a bipolar transistor, the current flow is always by diffusion, while in an MOS transistor, the current flow can occur via drift or via diffusion. In the linear regime of operation, both junctions in Figure 3.4 inject electrons into the channel of the transistor. Thus, the junctions should NOT be viewed as having independent I–V curves that exist without regard to the rest of the transistor but as junctions that determine boundary conditions for charge and/or potential within the transistor. The current is then determined by the drift and diffusion generators in the channel in a manner that is consistent with these boundary conditions and with an equilibrium charge and potential profile within the channel. A completely rigorous way of describing the situation would incorporate dependent current generators in parallel with each of the junctions in Figure 3.4 as in bipolar-transistor models. These generators would have a value that is determined by current injection from the opposite junction only. We have avoided such complexity in the intuitive model of Figure 3.4 to keep it simple. Its main purpose is to provide a picture in our heads that we will frequently use when deriving rigorous analytic results.

3.3   Intuitive energy viewpoint for MOS transistor operation

Figures 3.5 and 3.6 show an approximate electrical potential energy landscape for mobile electrons in a transistor as we move from the source to the channel and then to the drain. For readers with prior knowledge of device physics, this energy is the energy of the bottom edge of the conduction band. Figure 3.5 shows the energy landscape for subthreshold operation and Figure 3.6 shows the energy landscape for above-threshold operation. The landscapes are shown, first with the drain voltage being zero (equal to source and bulk voltage) and changing gate voltages, and then for changing drain voltages with a fixed gate voltage. The pn junctions at the channel-to-source end and channel-to-drain end have built-in energy barriers, which are lowered by increasing gate voltage that effectively attempts to forward bias these junctions. Note that a rising surface potential in the channel corresponds to a lowering in energy and vice versa as electrons are negatively charged.

Figure 3.5 shows that, in subthreshold operation, increasing gate voltages raise the surface potential and lower the energy level in the channel while increasing source and drain voltages lower the energy levels in the source and drain respectively. The surface potential is constant across the channel and unaffected by the source or drain voltages.

The surface potential in subthreshold operation changes proportionately with changes in the gate voltage but not one-for-one due to the presence of the distributed C_{ox} and C_{dep} capacitive divider shown in Figure 3.4. The surface potential changes the channel-to-source and channel-to-drain pn junction potential differences to exponentially modulate the charge concentrations at the source and drain

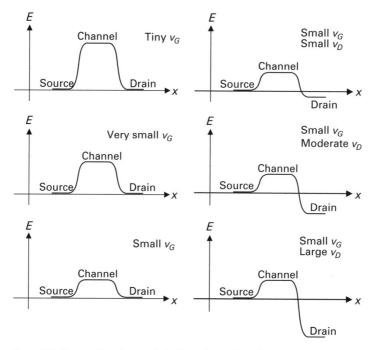

Figure 3.5. Energy-barrier modulation viewpoint of subthreshold operation.

ends of the channel, which then yields a diffusion current proportional to the difference between these two concentrations. The exponential dependence of the subthreshold saturation current on the gate voltage arises from the proportionate change in the surface potential with gate voltage and on the exponential dependence of the channel charge concentration on the surface potential [2].

Figure 3.6 shows that, in above-threshold operation, the surface potential is pinned by the source and/or drain diode-like elements to be nearly constant and is therefore unaffected by the gate voltage. The surface potential in above-threshold operation varies along the channel and is determined by the drain and source potentials. However, after saturation has been reached, the drain has little effect on the surface potential since the drain diode-like element turns off.

It is worth noting that a MOSFET in its linear regime of operation behaves like a bipolar transistor in its saturation region with both pn junctions being forward biased and on and the MOSFET channel region being analogous to the base region in a bipolar transistor. The MOSFET in its saturated regime of operation behaves like a bipolar transistor in its forward active region where the source-to-channel junction is on, or forward biased, but the drain-to-channel junction is off, or reverse biased. In the latter regime, the drain just collects the current coming from the source across its reverse-biased depletion region. The analogy to bipolar operation is almost exact in the subthreshold regime because, as in a bipolar transistor, each junction regulates its minority-carrier concentration in accord with an exponential relationship. We shall discuss more similarities (and differences) between MOS and bipolar transistor operation in Chapter 5.

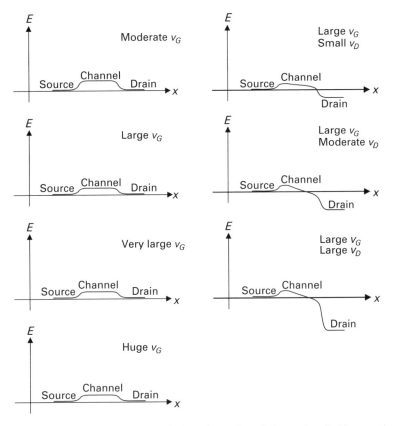

Figure 3.6. Energy-barrier modulation viewpoint of above-threshold operation.

How does the energy change as we move away from the surface and into the bulk? Since the potential in the bulk region is highest near the surface, and the bulk has negatively charged dopant ions in the depletion region that cause the potential to fall as we move deeper into the bulk, the energy level for electrons in the bulk is a minimum at the surface and rises as we move deeper into the bulk. Thus, any thermally generated electrons in the depletion region or bulk are attracted downhill towards the surface and contribute to the equilibrium carrier concentration in the channel. The C_{dep} capacitors marked in Figure 3.4 have properties similar to that of the depletion capacitance in a reverse-biased pn junction with the p terminal being the bulk outside the depletion region and the n terminal being the surface.

How are the electrons distributed in the depletion region in the bulk and what is the effective depth of the channel? To answer such questions, we will need to understand the transistor in quantitative detail. To do so, it is very useful to first understand a simpler device, the MOSCAP or MOSFET capacitor, quantitatively. The MOSCAP is a MOSFET with only two terminals, the gate and the bulk, and no source and drain. Ideas of subthreshold and above-threshold operation or equivalently those of weak and strong inversion originate in the MOSCAP and transfer to the MOSFET. Once we understand the MOSCAP, it is easy to understand the MOSFET by progressively moving from a two-terminal MOSCAP to a three-terminal half MOSFET, which has a source region imposing boundary conditions on one edge of the MOSCAP but no current flow; then to a four-terminal MOSFET with source and drain regions imposing boundary conditions on both edges of the MOSCAP and with current flow caused by drift and diffusion. This method of studying the MOSFET is adapted from the treatment by Tsividis in [3]. Our study uses graphical techniques and is more condensed than the treatment in [3].

3.4　The MOS capacitor (MOSCAP)

Figure 3.7 shows a MOSCAP, a capacitor formed by a polysilicon gate, SiO_2 insulator, and the silicon bulk. The connection to the p bulk is established through a p+ region, which is not shown in the figure, and which without loss of generality is set at ground. The potential of the positively charged gate, the surface potential, and the potential deep in the bulk are marked in the figure as v_G, ψ_s, and 0 respectively. The negative charge in the bulk region, needed to balance the positive charge on the gate, is composed partly of mobile electrons and partly of immobile dopant atoms in the depletion region, a region formed by the repulsion of majority-carrier holes in the bulk by the positively charged gate. At low values of v_G, almost all the electric field lines that begin at the gate end on dopant atoms since there are very few electrons per unit area; this scenario corresponds to the subthreshold or weak inversion region of operation of an MOS transistor. At high values of v_G almost all the electric field lines that begin at the gate end on mobile electrons. The electrons are so many in number that they shield the field lines from penetrating through to the dopant atoms by establishing an almost constant

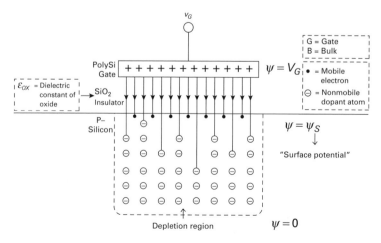

Figure 3.7. Basic idea of the MOS capacitor (MOSCAP).

surface potential in the bulk via a negative-feedback loop; the dopant charge stays fixed, almost independent of v_G; this scenario corresponds to the above-threshold or strong-inversion region of operation of an MOS transistor.

There is a magical gate voltage with respect to the bulk, referred to as the *flatband voltage*, at which there are no electric field lines in the MOS capacitor, either in the SiO_2 insulator, or in the bulk. At this magical voltage, the externally imposed charge on the capacitor cancels inbuilt charge in the capacitor present due to traps in the oxide, charge implanted during fabrication, and charge present at surface boundaries between SiO_2 and Si and between SiO_2 and polysilicon. For gate voltages above the flatband voltage, the net positive charge on the gate repels holes in the bulk creating a depletion region and the MOSCAP operates in a mode referred to as *depletion mode*. This is the normal mode of operation that corresponds to both the subthreshold and above-threshold regime of operation of transistors. For gate voltages below the flatband voltage, the net negative charge on the gate attracts holes in the bulk which accumulate near the surface between Si and SiO_2 forming a capacitor with a positively charged bottom plate in the bulk and the MOSCAP operates in a mode referred to as *accumulation mode*. This accumulation mode is normally not exercised during normal transistor action, and typically needs the gate voltage to be very near or below ground. In this book, we shall assume depletion-mode operation for all devices unless otherwise mentioned.

Figures 3.8 (a), 3.8 (b), 3.8 (c), 3.8 (d), and 3.8 (e) show how field lines, mobile charge, and dopant charge change as v_G is increased from the flatband voltage through weak inversion and strong inversion. Increasing values of v_G lead to increasing positive charge on the gate and increasing numbers of electric field lines that begin at the gate. For relatively low values of v_G, the effect of increasing positive charge on the gate is to deepen the extent of the depletion region and provide more dopant charge to absorb the increase in the number of field lines; a few field lines also end at mobile electrons near the surface, which have been attracted there by the increase in surface potential. As the value of v_G increases, the number of field lines, the depth of the depletion region, and the number of mobile electrons increase. There is an almost linear change in the surface potential with v_G and an exponential increase in the number of mobile electrons. Such behavior corresponds to that seen in the weak inversion or subthreshold region of operation of an MOS transistor. At a value of v_G that corresponds to the onset of strong inversion, the number of mobile electrons is so high that the electron concentration per unit volume becomes greater than the dopant atom concentration per unit volume. For voltages on the gate beyond this value of v_G the mobile electrons begin to have a large say in determining the surface potential, and the exponential increase in the number of mobile electrons with surface potential results in a powerful diode-clamp-like negative-feedback loop. This feedback loop keeps the surface potential nearly constant and independent of further increases in v_G; the constant surface potential shields additional field lines created beyond the onset of strong inversion from penetrating through to the dopant atoms and these field lines all almost end on mobile electrons. Thus, in strong inversion, although the concentration of the

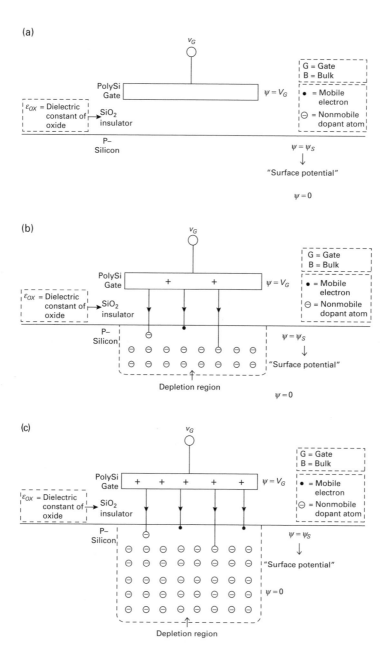

Figure 3.8a, b, c, d, e. Progression of field lines from flatband through strong inversion as the gate voltage v_G is increased from a small value in (a) to a large value in (e) in a MOSCAP.

mobile electrons increases with v_G, the depth of the depletion region or the surface potential barely increase with further increases in v_G.

The electrons in a MOSCAP do not arise from the source or drain, which provide a plentiful supply in an MOS transistor, but from a relatively scarce

(d)

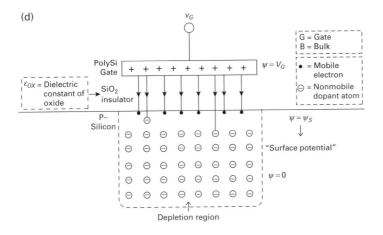

(e)

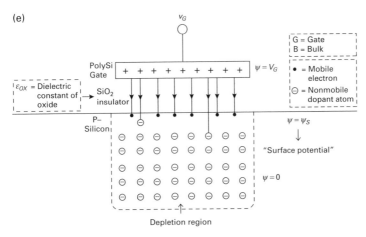

Figure 3.8. (*cont.*)

supply of thermally generated electron-hole pairs in the deep bulk and depletion regions. The electrons gravitate downhill towards the low-energy surface. Thus, if v_G changes, the time necessary for the electron concentration to reach its new equilibrium value is significantly longer in a MOSCAP than in an MOS transistor. In imagers built with charge-coupled devices (CCDs), the MOSCAP is used to collect electrons that have been generated by light, with higher light levels resulting in more electrons collected over a time window during which v_G is pulsed high.

3.5   Quantitative discussion of the MOSCAP

Thus far, we have been casual in our use of the words "charge" or "charge concentration", sometimes meaning total charge, sometimes meaning charge concentration per unit volume, and sometimes meaning charge concentration

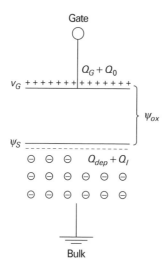

Figure 3.9. Charge balance in a MOSCAP.

per unit area. In intuitive discussions, it is usually clear from context as to what is meant, and such distinctions are not usually important in understanding ideas. For a quantitative development, however, it is important to be consistent, clear, and explicit in what is meant. Therefore, from now on, we adopt the convention that all symbols that denote charges in equations will be assumed to mean charge per unit area unless otherwise stated. The use of charge per unit area is natural in MOSFETs and MOSCAPs since most of the mobile electrons in the bulk exist in a thin two-dimensional charge sheet or channel at the Si/SiO$_2$ surface. The charge per unit length is then the charge per unit area times the width of the channel as the charge per unit area is usually uniform across the width of the channel. The total charge in the channel is found by integrating the charge per unit length, which is usually different in different regions of the channel, across the channel's length.

Figure 3.9 shows the charge distribution on the MOSCAP. The charges Q_{dep} and Q_I are the depletion charge (per unit area) due to fixed dopant atoms and the mobile charge (per unit area) due to electrons, respectively. Both these charges are negative. The charge (per unit area) $Q_G = C_{ox}\psi_{ox}$, where ψ_{ox} is the potential difference across the oxide, is the charge on the gate that varies with voltage ψ_{ox} and is positive; C_{ox}, the oxide capacitance per unit area, is given by ε_{ox}/t_{ox}, where ε_{ox} is the dielectric permittivity of SiO$_2$ and t_{ox} is the oxide thickness. The charge (per unit area) Q_0 is the fixed charge on the gate necessary to balance any fixed surface charge, implanted charge, impurity charge, or defect charge that is present in the MOSCAP from the time of fabrication. The MOSCAP must be charge balanced such that field lines from positive charges on the gate terminate on an equal number of negative charges on dopant atoms or on mobile electrons. Thus, we must have

$$C_{ox}\psi_{ox} + Q_0 + (Q_I + Q_{dep}) = 0 \tag{3.1}$$

The junction between any two dissimilar materials develops a *contact potential* due to the differing energies of free electrons in the two materials. Thus, the Si/SiO$_2$ junctions and the SiO$_2$/polysilicon surface junctions develop contact potentials. Each such contact potential has a value given by the difference in work functions of the two materials at the junction; the work function of a material is a measure of the free energy of electrons in the material with respect to electrons in vacuum, a good reference energy for all materials. The net potential difference between the polysilicon and the silicon due to these contact potentials, given by the sum of the contact potentials at the Si/SiO$_2$ and SiO$_2$/polysilicon junctions, is independent of the work function of SiO$_2$ and only dependent on the difference in work functions between polysilicon and silicon: the work function of SiO$_2$ contributes with opposite signs to the Si/SiO$_2$ and SiO$_2$/polysilicon contact potentials and cancels out. The net contact potential between polysilicon and silicon is denoted by ϕ_{MS}, and given by the difference in work functions between polysilicon and silicon.

Kirchhoff's voltage law applied to the voltages of Figure 3.9 yields

$$v_G = \psi_s + \psi_{ox} + \phi_{MS} \tag{3.2}$$

From Equations (3.1) and (3.2), we get

$$C_{ox}(v_G - \psi_S - \phi_{MS}) + Q_0 = -(Q_I + Q_{dep})$$

$$C_{ox}\left(v_G - \psi_S - \left(\phi_{MS} - \frac{Q_0}{C_{ox}}\right)\right) = -(Q_I + Q_{dep}) \tag{3.3}$$

$$C_{ox}(v_G - \psi_S - V_{FB}) = -(Q_I + Q_{dep})$$

where the flatband voltage V_{FB} is defined to be

$$V_{FB} = \phi_{MS} - \frac{Q_0}{C_{ox}} \tag{3.4}$$

From Equations (3.3) and (3.4), we then see that the flatband voltage acts as an effective offset voltage that must be subtracted from the voltage across the oxide capacitance to yield the net positive charge on the gate that balances the negative charge in the bulk due to dopant atoms and mobile electrons. We must now determine $(Q_I + Q_{dep})$ in the bulk when the surface potential is at a given value ψ_S.

3.6   Determining $(Q_I + Q_{dep})$ in a MOSCAP for a given ψ_S

Figure 3.10 shows the electrostatics problem that we must solve to determine $(Q_I + Q_{dep})$ in the bulk region of a MOSCAP. The potential is ψ_S at the surface where $y = 0$ and is zero deep in the bulk where the depletion region ends at $y = \infty$. It changes from ψ_S to 0 as we traverse the depletion region. We shall assume from the symmetry of the problem that the variation in potential is only a function of the y dimension and independent of other geometric dimensions, a good approximation in our long-channel transistor. Thus, the electric field lines are vertically

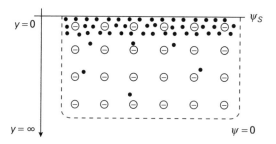

Figure 3.10. The hard physics: solving for the exact charge distribution in an MOS transistor or MOSCAP via the Poisson-Boltzmann equation.

oriented and directed towards the deep bulk from the surface; they have a finite value at $y = 0$ and go to zero in the deep bulk at $y = \infty$.

Electrons are exponentially more likely to congregate in regions where the potential is highest while holes are exponentially more likely to congregate in regions where the potential is lowest. Thus, electrons, shown as black dots in the figure, are most concentrated near the surface while holes are most concentrated outside the depletion region. Electrons and holes themselves affect the potential creating a feedback loop: the potential determines the electron and hole concentrations but the electron and hole concentrations in turn affect the potential.

To describe this problem mathematically, we first define some standard symbols. The intrinsic carrier concentration of silicon is denoted as n_i, the dopant atom density is denoted as N_A, the dielectric constant of silicon is denoted as ε_{si}, the electron charge is denoted as q, the thermal voltage is denoted as $\phi_t = kT/q$ where k is Boltzmann's constant and T is the absolute temperature, the electron concentration is denoted as $n_o(y)$, and the hole concentration is denoted as $p_o(y)$. It is also useful to define a potential ϕ_F given by

$$\phi_F = \phi_t \ln\left(\frac{N_A}{n_i}\right) \tag{3.5}$$

such that

$$n_i = N_A e^{-\phi_F/\phi_t} \tag{3.6}$$

Then the formal statement of our problem can be expressed as
Solve

$$\frac{d^2\psi(y)}{dy^2} = -\frac{q(p_0(y) - n_0(y) - N_A)}{\varepsilon_{si}}$$
$$p_0(y) = N_A e^{-\psi(y)/\phi_t}$$
$$n_0(y)p_0(y) = n_i^2 \tag{3.7}$$
$$n_0(y) = \frac{n_i^2}{p_0(y)} = N_A e^{-2\phi_F/\phi_t} e^{+\psi(y)/\phi_t}$$

subject to the boundary conditions

$$\psi(0) = \psi_S$$

$$-\frac{d\psi}{dy}\bigg|_{y=0} = E(0) = -\frac{(Q_I + Q_{dep})}{\varepsilon_{si}}$$

$$\psi(\infty) = 0 \tag{3.8}$$

$$-\frac{d\psi}{dy}\bigg|_{y=\infty} = E(\infty) = 0$$

where $\psi(y)$ and $E(y)$ are functions that describe the potential and electric field distribution. Note that the boundary condition for the electric field or derivative of the potential at $y = 0$ is obtained by a simple application of Gauss's law of electromagnetism to a pillbox formed with the surface as its upper plate and a surface outside the depletion region as its lower plate; $Q_I + Q_{dep}$ is the entire charge per unit area enclosed within this pillbox.

Equation (3.7) is the Poisson-Boltzmann equation and is almost identical to a similar equation seen in electrochemistry. In response to a potential on a metal electrode, ions in solution, rather than electrons and holes, distribute themselves according to Poisson electrostatics and Boltzmann statistics, just as in Equation (3.7). The depth of the inversion layer in a MOSFET is analogous to the depth of the Debye layer in electrochemistry. Electrochemistry is discussed in Chapter 25. Not surprisingly, electrical circuit models of electrodes and batteries in Chapter 25 have some similarities to MOS device physics in this chapter.

The solution to Equations (3.7) and (3.8) is a simple exercise in the solution of differential equations with boundary conditions and presented in Appendix F of the well-known text on the MOS transistor [3]. It is given by

$$(Q_I + Q_{dep}) = -\gamma C_{ox}\sqrt{\psi_S + \phi_t e^{(\psi_s - 2\phi_F)/\phi_t}} \tag{3.9}$$

with γ, often termed the *body-effect coefficient*, given by

$$\gamma = \frac{\sqrt{2q\varepsilon_{si}N_A}}{C_{ox}} \tag{3.10}$$

Note that although we have the total charge per unit area $(Q_I + Q_{dep})$ in the MOSCAP, we don't yet individually know Q_I and Q_{dep}. To get Q_I and Q_{dep} individually, we make an approximation that all of Q_I resides in a two-dimensional charge sheet only at the surface and is zero everywhere else. This charge-sheet approximation for the mobile charge is a good one since electrons are exponentially more likely to be at the surface than at a point anywhere else in the bulk and is well borne out by experiment. With this approximation, the charge Q_{dep} can be computed to be that present in a standard depletion capacitance with uniform doping, e.g., in a reverse-biased diode, with a top-plate voltage of ψ_S and a bottom-plate voltage of 0. It is given by

$$Q_{dep} = -\gamma C_{ox}\sqrt{\psi_s}$$
$$C_{dep} = -\frac{dQ_{dep}}{d\psi_s} = \frac{\gamma C_{ox}}{2\sqrt{\psi_s}} \tag{3.11}$$

From Equations (3.9) and (3.11), we can then compute that

$$Q_I = -\gamma C_{ox}\left(\sqrt{\psi_s + \phi_t e^{(\psi_s - 2\phi_F)/\phi_t}} - \sqrt{\psi_s}\right) \tag{3.12}$$

If $\phi_t e^{(\psi_s - 2\phi_F)/\phi_t}$ is small compared to ψ_s in Equation (3.12), then $Q_I \approx 0$ and we are in a region corresponding to subthreshold operation or weak inversion. If $\phi_t e^{(\psi_s - 2\phi_F)/\phi_t}$ is large compared to ψ_s, then Q_I is large and we are in a region corresponding to above-threshold operation or strong inversion. Note that when $\psi_s = 2\phi_F$, from Equation (3.9), we see that the electron concentration $n_0(y) = N_A$ and the charge concentration per unit volume of mobile electrons equals the charge concentration of dopant atoms at the surface. When ψ_s starts to exceed $2\phi_F$ by even a few ϕ_t, typically $6\phi_t$, Equation (3.12) shows that there is a steep exponential rise in Q_I with ψ_s.

The notion of 'inversion' describes the fact that, when $\psi_S = 2\phi_F$, the surface has a concentration per unit volume of mobile minority carrier electrons that is the same as the concentration of mobile majority carrier holes in the deep bulk. Thus it appears as though the p-type characteristics of the bulk semiconductor have been inverted to the characteristics of an n-type semiconductor at the surface. In weak inversion $\psi_S \leq 2\phi_F$ and in strong inversion $\psi_S > 2\phi_F$ by a few ϕ_t. In moderate inversion, ψ_S transitions between these two regimes. The reason that ψ_S does not increase beyond $2\phi_F$ by a few ϕ_t is because the exponential diode-clamp-like negative-feedback loop kicks in and prevents further increases in ψ_S with increases in v_G.

3.7   Equating the gate charge and bulk charge

From Equation (3.3), we know that the gate charge is $C_{ox}(v_G - V_{FB} - \psi_S)$. From Equation (3.9), we know how the bulk charge $Q_I + Q_{dep}$ depends on ψ_S. Since these two charges must be equal in magnitude and opposite in sign, we can derive that

$$C_{ox}(v_G - V_{FB} - \psi_S) = \gamma C_{ox}\sqrt{\psi_S + \phi_t e^{(\psi_s - 2\phi_F)/\phi_t}} \tag{3.13}$$

or, eliminating C_{ox} from both sides of Equation (3.13), we can compute the relationship between v_G and ψ_S to be

$$v_G = V_{FB} + \psi_S + \gamma\sqrt{\psi_S + \phi_t e^{(\psi_s - 2\phi_F)/\phi_t}} \tag{3.14}$$

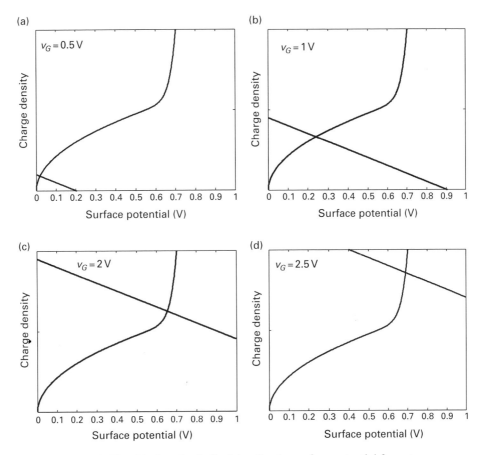

Figure 3.11a, b, c, d. Graphical method of solving for the surface potential for gate voltages of $v_G = 0.5\,\text{V}$, $1\,\text{V}$, $2\,\text{V}$ and $2.5\,\text{V}$. The straight line is the positive gate charge density while the curved line is the negative charge density due to mobile electrons and fixed dopant charge. At the intersection of the two curves the two charge densities are equal, which is required by the laws of charge conservation.

Equation (3.14) tells us how v_G depends on ψ_S. What we would really like to know is how ψ_S, an unknown internal potential in the MOSCAP, depends on v_G, a known external terminal voltage. Unfortunately, Equation (3.14) is not invertible in closed form. Nevertheless, we can obtain some graphical insight into the problem by plotting the left and right hand sides of Equation (3.13) as a function of ψ_S and noting that the intersection of the two graphs yields the solution for the value of ψ_S in the MOSCAP for a given value of v_G. Figures 3.11 (a) and (b) show how this intersection point varies for two values of v_G in subthreshold or weak-inversion operation and Figures 3.11 (c) and (d) show how it varies for two values of v_G in above-threshold or strong-inversion operation.

In weak inversion, the ψ_S term in Equation (3.13) on the right hand side is the only significant term such that Figures 3.11 (a) and (b) reveal a significant change in ψ_S with v_G. In fact, if we approximate the square-root function of Equation (3.13) by a Taylor-series linear approximation, the change in ψ_S with v_G is described by the varying intersection of two straight lines, one fixed, and one with varying y-intercept as v_G changes, such that the change in ψ_S with v_G is linear. The straight-line approximation is equivalent to modeling the depletion charge by a linear capacitance C_{dep} such that the change in ψ_S with v_G is determined by a capacitive divider formed by C_{ox} and C_{dep}, and ψ_S is proportional to $\left(C_{ox}/(C_{ox}+C_{dep})\right)v_G$. We shall use this straight-line approximation extensively in the following chapter, referring to it as the κ approximation with κ given by $\left(C_{ox}/(C_{ox}+C_{dep})\right)$. The parameter κ is also equal to $1/n$, where n is termed the *subthreshold slope coefficient* in some texts.

In strong inversion, the exponential term in Equation (3.13) on the right hand side is the only significant term such that Figures 3.11 (c) and (d) reveal a small change in ψ_S with v_G. In fact, if we approximate the exponential function of Equation (3.13) by an infinitely steep vertical line, the change in ψ_S with v_G is described by the varying intersection of two straight lines, one fixed and vertical, and one with varying y-intercept as v_G changes, such that there is no change in ψ_S with v_G. This approximation in the MOSCAP corresponds to the charge-diode-clamp approximation that we have discussed before for the diode-like elements of Figure 3.4 in the MOSFET. Since the steep exponential rise is extremely abrupt a few ϕ_t beyond $2\phi_F$ in Figures 3.11 (c) and (d), we may approximate $\psi_S \approx 2\phi_F + 6\phi_t = \phi_o$ in all of strong-inversion operation. Of course, in reality, the change in ψ_S with v_G is not constant but logarithmic. We shall use the charge-diode-clamp approximation extensively in the following chapter in conjunction with the diode-like elements of Figure 3.4.

Figure 3.12 summarizes the physical intuition behind the operation of the MOSCAP in subthreshold and above-threshold operation respectively. In subthreshold, capacitive-divider operation causes the capacitance per unit area from gate to bulk to be $(C_{ox}C_{dep})/(C_{ox}+C_{dep})$. In above-threshold operation, the charge-diode-clamped fixed bottom plate causes the capacitance per unit area

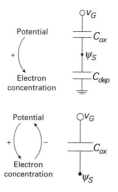

Figure 3.12. MOSCAP physical intuition in weak (top) and strong inversion (bottom). The feedback loop between potential and electron concentration is only active in strong inversion.

from gate to bulk to be C_{ox}. The figure also reminds us that the reason for the charge-diode-clamp behavior in strong inversion is the presence of a strong negative-feedback loop between electron concentration and surface potential.

3.8   Quantitative discussion of the MOSFET

Figure 3.13 shows a half MOSFET, i.e., an n+ region bordering a MOSCAP. The n+ region now presents an alternate region of low energy for electrons to reside and also provides an abundant supply of electrons. Increasing the potential v_{CB} of the n+ region with respect to the bulk lowers the energy of the n+ region with respect to the bulk. It also decreases the concentration of minority-carrier electrons in the bulk region near the n+ border by $\exp(-v_{CB}/\phi_t)$ just as in a reverse-biased diode. Across any diode-like junction, a balance between drift currents (due to electric fields in depletion regions of the diode) and diffusion currents (due to electron concentration differences across the diode junction) leads to an equilibrium minority carrier concentration at the edge of the junction that is exponentially dependent on the voltage across the junction. In effect, a positive value of v_{CB} functions to contribute a reverse-bias voltage to the diode-like element of Figure 3.4, which in a half MOSFET only exists on one side. The lack of current flow in the half MOSFET means that the electron concentration and surface potential are the same all along the horizontal x dimension of the channel at any given value of y. The variation in electron concentration $n_0(y)$ along the vertical y dimension is still described by the exponential relationship of Equation (3.7) in the MOSCAP in a long-channel scenario except that we now have an extra factor of $\exp(-v_{CB}/\phi_t)$ in the equation:

$$n_0(y) = N_A e^{-v_{CB}/\phi_t} e^{-2\phi_F} e^{+\psi(y)} \qquad (3.15)$$

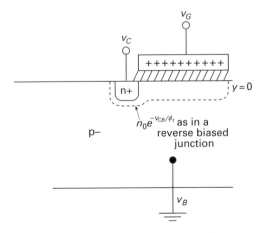

Figure 3.13. MOSCAP + 1 control terminal.

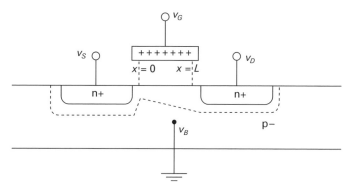

Figure 3.14. MOSCAP + 2 control terminals.

The global depression of electron concentration by $\exp(-v_{CB}/\phi_t)$ along the y dimension in the half MOSFET preserves equilibrium minority carrier concentrations such that drift currents (due to the vertical downward electric field in the depletion region of the bulk) and diffusion currents (due to variations in electron concentration along the y dimension) at any y remain in balance.

There are no other changes in Equations (3.7) or (3.8) such that the relationship between the gate voltage and surface potential in the half MOSFET is computed by simply replacing $2\phi_F$ with $(2\phi_F + v_{CB})$ in Equation (3.14) for the MOSCAP. Thus, we get

$$v_G = V_{FB} + \psi_S + \gamma\sqrt{\psi_S + \phi_t e^{(\psi_S - (2\phi_F + v_{CB}))/\phi_t}} \tag{3.16}$$

Note that, in the half MOSFET, all drift and diffusion generators in Figure 3.4 are nonexistent since two terminals between which current can flow do not exist and all concentrations and potentials are uniform in the x dimension at equilibrium. In a MOSCAP, the drift and diffusion generators and the diode-like elements do not exist, just the distributed C_{ox} and C_{dep} capacitances of Figure 3.4.

Figure 3.14 shows that the MOS transistor is two half MOSFETs, i.e., two n+ regions bordering a MOSCAP with potentials with respect to the bulk at the source and drain terminals of v_{SB} and v_{DB} respectively. Each n+ region will attempt to depress the minority carrier concentration at the channel borders by $\exp(-v_{SB}/\phi_t)$ and $\exp(-v_{DB}/\phi_t)$ at the source and drain edges of channel respectively, which are assumed to have surface potentials of ψ_{S0} and ψ_{SL} respectively. At each channel edge, the C_{ox}, C_{dep}, and diode-like elements of Figure 3.4 will make the edge appear like a half MOSFET such that Equation (3.16) applies and

$$v_G = V_{FB} + \psi_{S0} + \gamma\sqrt{\psi_{S0} + \phi_t e^{(\psi_{S0} - (2\phi_F + v_{SB}))/\phi_t}}$$
$$v_G = V_{FB} + \psi_{SL} + \gamma\sqrt{\psi_{SL} + \phi_t e^{(\psi_{SL} - (2\phi_F + v_{DB}))/\phi_t}} \tag{3.17}$$

From the charge sheet approximation, we can also compute the mobile charge at the source end of the channel, Q_S, and at the drain end of the channel, Q_D, by subtracting the depletion charge from the gate charge.

$$\begin{aligned} Q_S &= -C_{ox}(v_G - V_{FB} - \psi_{S0} - \gamma\sqrt{\psi_{S0}}) \\ Q_D &= -C_{ox}(v_G - V_{FB} - \psi_{SL} - \gamma\sqrt{\psi_{SL}}) \end{aligned} \tag{3.18}$$

Equations (3.17) and (3.18) establish the boundary conditions for surface potential and charge at the source and drain ends of the channel. To understand what happens in between these ends in the channel, we must finally invoke the drift and diffusion generators of Figure 3.4. To ensure that the surface potential and charge concentration at all points along the channel are at equilibrium, we must require that the sum of the drift and diffusion currents flowing towards any point in the channel is balanced by the sum of drift and diffusion currents flowing away from this point; to ensure current continuity all along the channel, the sum of the drift and diffusion currents flowing towards or away from any point must also be the same throughout the channel.

The drain current, i_{DS}, along any point in the channel is composed of a drift current, proportional to the electric field or derivative of the surface potential, and a diffusion current, proportional to the derivative of the charge density, such that

$$\begin{aligned} i_{DS} &= W\left[-\mu Q_I \frac{d\psi_S(x)}{dx} + D\frac{dQ_I}{dx}\right] \\ i_{DS} &= W\left[-\mu Q_I \frac{d\psi_S(x)}{dx} + \mu\phi_t\frac{dQ_I}{dx}\right] \end{aligned} \tag{3.19}$$

where the Einstein relation, $D = \mu\phi_t$, has been used in the latter equation, and μ is the mobility of electrons in the channel. Some simple algebra then yields

$$\begin{aligned} \int_0^L i_{DS}dx &= W\left[\int_{\Psi_{S0}}^{\Psi_{SL}} -\mu Q_I d\psi_S(x) + \int_{Q_{I0}}^{Q_{IL}} \mu\phi_t dQ_I\right] \\ i_{DS} &= \frac{\mu W}{L}\left[\int_{\Psi_{S0}}^{\Psi_{SL}} -Q_I d\psi_S(x) + \phi_t\int_{Q_{I0}}^{Q_{IL}} dQ_I\right] \end{aligned} \tag{3.20}$$

Substituting

$$Q_I(x) = -C_{ox}\left(v_G - V_{FB} - \psi_S(x) - \gamma\sqrt{\psi_S(x)}\right) \tag{3.21}$$

in Equation (3.20), we get

$$i_{DS} = \frac{\mu W}{L}C_{ox}\int_{\psi_{S0}}^{\psi_{SL}} \left(v_G - V_{FB} - \psi_S - \gamma\sqrt{\psi_S}\right)d\psi_S + \frac{\mu W\phi_t}{L}\int_{Q_{I0}}^{Q_{IL}} dQ_I \tag{3.22}$$

Simple integration and substitution of the integration limits then yields

$$
i_{DS} = \mu C_{ox} \frac{W}{L} \left[(v_G - V_{FB})(\psi_{SL} - \psi_{S0}) - \frac{1}{2}(\psi_{SL}^2 - \psi_{S0}^2) - \frac{2\gamma}{3}\left(\psi_{SL}^{3/2} - \psi_{S0}^{3/2}\right) \right]
$$
$$
+ \mu C_{ox} \frac{W}{L} \phi_t \left[(\psi_{SL} - \psi_{S0}) + \gamma\left(\psi_{SL}^{1/2} - \psi_{S0}^{1/2}\right) \right]
$$
(3.23)

Equation (3.23) along with the boundary conditions for the surface potential in Equation (3.17) is the complete solution for the current in the MOS transistor valid in all regions of operation, linear or saturation, weak inversion or strong inversion. Note that Equation (3.17) leads to implicit solutions for ψ_{SL} and ψ_{S0} as functions of the terminal voltages v_G, v_{SB}, and v_{DB}, which may then be used in Equation (3.23) to compute the current.

Equation (3.23) is a body-referenced model, with currents antisymmetric in the source and drain voltages, i.e., $i_{DS} = f(\psi_{SL}) - f(\psi_{S0})$, and the current changes sign if the source and drain voltages, or equivalently ψ_{SL} and ψ_{S0} in Equation (3.23), are interchanged. The square-root terms in Equations (3.21) and (3.22) lead to the messy half-power and three-half-power terms. Clearly Equation (3.23) is useful for simulation but too complex for hand analysis and practical circuit design.

In the next chapter, we will make three approximations, the κ approximation, the charge-diode-clamp approximation, and a Taylor-series-square-root approximation that will allow us to vastly simplify Equation (3.23) into more useful and insightful forms. However, it is important to understand the fundamental physics of the MOS transistor in depth such that a complete solution with no approximations like Equation (3.23) can be obtained from first principles.

3.9   Summary of physical insights

Since this chapter involved a lot of equations, to ensure that the key physical insights were not drowned in Greek letters, we review them below:

1. An nFET MOSCAP goes from a capacitive divider in weak inversion or subthreshold operation to a capacitor in strong inversion or above-threshold operation as its gate-to-bulk voltage is increased. The bottom plate of the MOSCAP has its potential nearly fixed in strong inversion due to a negative-feedback charge-diode-clamp effect.
2. The solution to the Poisson-Boltzmann equation with appropriate boundary conditions yields the MOSCAP's bulk charge dependence on surface potential and gate-to-bulk voltage.
3. A MOSFET is a MOSCAP with differing boundary conditions for the surface potential (and hence the mobile charge) at its source and drain ends.
4. The differing boundary conditions cause current flow through drift due to differing source and drain surface potentials, and through diffusion due to differing source and drain electron charge concentrations.

5. Energy diagrams represent useful and intuitive ways for thinking about transistor operation. In particular, the source and drain ends may be viewed as channel-to-source and channel-to-drain pn junctions whose minority carrier concentrations are controlled by the gate modulation of the energy barrier between the source/drain and the channel. The modulation acts via changes in the nearby surface potential thereby modulating the bias of the channel-to-source or channel-to-drain junctions.

6. Figure 3.4 presents an intuitive picture of the MOS transistor that is valid in all regions of operation.

References

[1] Yannis Tsividis. *Mixed Analog-Digital VLSI Devices and Technology* (Singapore: World Scientific Publishing Company, 2002).

[2] Carver Mead. *Analog VLSI and Neural Systems* (Reading, Mass.: Addison-Wesley, 1989).

[3] Yannis Tsividis. *Operation and Modeling of the MOS Transistor*, 3rd ed. (New York: Oxford University Press, 2008).

4 MOS device physics: practical treatment

Although this may seem a paradox, all exact science is dominated by the idea of approximation.

Bertrand Russell

In Chapter 3, we discussed an intuitive model to describe transistor operation shown in Figure 3.4. We will now use this intuitive model to simplify Equation (3.23) into more practical and insightful forms in subthreshold and above-threshold operation. To do so, we will use three approximations listed below.

1. **The κ approximation.** We approximate $\gamma\sqrt{\psi_S}$ by a constant term $\gamma\sqrt{\psi_{se}}$ around an operating point ψ_{se} plus a Taylor-series linear term $(\gamma/\sqrt{(2\psi_{se})})(\psi_S - \psi_{se})$ to describe deviations from the operating point. This approximation is equivalent to modeling the distributed nonlinear depletion capacitance C_{dep}, shown in Figure 3.4 as a linear capacitance with some fixed value around a given ψ_{se}. The surface potential ψ_s varies with x in above-threshold operation, such that C_{dep} varies along the channel in Figure 3.4. We ignore this variation and assume one value for C_{dep} throughout the channel given by $\gamma C_{ox}/(2\sqrt{\psi_{se}})$ from Equation (3.11). The chosen operating point ψ_{se} is usually at the subthreshold to above-threshold transition where we have good knowledge of the surface potential and, in addition, is at the source channel end in source-referenced models. The value of $\kappa = C_{ox}/(C_{ox} + C_{dep})$ is always between 0 and 1 and given by a capacitive-divider ratio. The κ approximation is useful in weak inversion and strong inversion. **The parameter $\kappa = 1/n$ by definition, where n is termed the subthreshold slope coefficient in some texts.**

2. **The Taylor-series square-root approximation.** In weak inversion, we assume that the $\phi_t\exp(\dots)$ term in $\gamma C_{ox}\sqrt{(\psi_S + \phi_t\exp((\psi_S - 2\phi_F - v_{CB})/\phi_t)))}$, an expression that describes the total charge at the source channel end or drain channel end, is much smaller than the ψ_S term in the expression. Therefore, we use a Taylor-series approximation for the square root again to evaluate the total charge in the expression. This approximation is useful in weak inversion only. Note that $v_{CB} = v_{SB}$ at the source end and $v_{CB} = v_{DB}$ at the drain end.

3. **The diode-clamp approximation.** In strong inversion, we will assume that the diode-like element of Figure 3.4 clamps the surface potential at the source end of the channel to $(2\phi_F + 6\phi_t) + v_{SB} = \phi_0^s + v_{SB}$ due to the operation of a strong negative-feedback loop. The surface potential at the drain end of the channel

is also clamped by a diode-like element to $\phi_0^s + v_{DB}$ if the transistor is not in saturation such that the diode-like element at the drain end is on. If the transistor is saturated, the diode-like element at the drain end is off and the drain surface potential is at a value determined by the capacitive divider formed by C_{ox} and C_{dep} in weak inversion. The mobile charge in strong inversion decreases along the channel as the bottom-plate surface potential on the distributed C_{ox} capacitors changes from a low value at the source end to a high value at the drain end, thus increasing the dopant charge and decreasing the total gate charge as we move from the source to the drain. In saturation, the mobile charge is assumed to abruptly go to zero at a pinchoff point along the channel very near the drain end where the mobile charge concentration is so low that the strong-inversion approximation no longer holds.

The diode-clamp approximation is only useful in strong inversion and is invalid in weak inversion. Weak-inversion operation occurs when the surface potential is less than or equal to $\phi_0^w + v_{SB} = 2\phi_F + v_{SB}$ at the source end of the channel. Note that we always use the w superscript for weak-inversion operation and the s superscript for strong-inversion operation. Thus, ϕ_0^w is the voltage across the diode-like elements above which weak-inversion operation is no longer valid while ϕ_0^s is the diode-clamp voltage across the diode-like elements in strong inversion, with $\phi_0^s - \phi_0^w = 6\phi_t$. In between the two extremes, where $\phi_0^w < \psi_S < \phi_0^s$ or equivalently, $2\phi_F < \psi_S < 2\phi_F + 6\phi_t$, we have moderate-inversion operation and neither the subthreshold exponential-junction nor the above-threshold diode-clamp approximation are valid.

We shall begin by discussing how the κ approximation is useful in multiple ways in subthreshold and above-threshold operation, especially when combined with the diode-clamp approximation. **The κ approximation could also be termed the n approximation since κ is defined to be $1/n$.**

4.1   The κ approximation

The κ approximation is useful in six different aspects of transistor operation, all of which may be derived from the intuitive model of Figure 3.4.

4.1.1   The capacitive divider of weak inversion

In weak inversion, κ is defined to be

$$\kappa = \frac{\partial \psi_S}{\partial v_G} \tag{4.1}$$

In weak inversion, the source and drain terminal voltages have little effect on the surface potential, which is constant throughout the channel and has a value denoted by $\psi_{sa}[1]$. To compute ψ_{sa} we can use Equation (3.16) with the exponential term in the square root set to zero such that

$$v_G = V_{FB} + \psi_{sa} + \gamma\sqrt{\psi_{sa}} \qquad (4.2)$$

Then, by differentiating Equation (4.2) and inverting, we may derive that

$$\kappa = \frac{1}{1 + \dfrac{\gamma}{2\sqrt{\psi_{sa}}}} \qquad (4.3)$$

However, from Figure 3.4, we can see that ψ_S will change with v_G in accordance with a capacitive-divider relationship if the diode-like elements are off. Thus,

$$\kappa = \frac{C_{ox}}{C_{ox} + C_{dep}}, \qquad (4.4)$$

where C_{dep} is $\gamma C_{ox}/(2\sqrt{\psi_{sa}})$ from Equation (3.11). Equations (4.3) and (4.4) are then seen to be identical but Equation (4.4) is intuitively obvious unlike Equation (4.3). Since κ is defined to be $1/n$, we can also write

$$n = \frac{C_{ox} + C_{dep}}{C_{ox}} \qquad (4.5)$$

4.1.2   Approximating the surface potential in weak inversion

Equation (4.2) is a quadratic equation in $\sqrt{\psi_{sa}}$ and may thus be solved for $\sqrt{\psi_{sa}}$ as a function of v_G. Solving this equation and squaring, we obtain that

$$\psi_{sa} = \left(-\frac{\gamma}{2} + \sqrt{\frac{\gamma^2}{4} + (v_G - V_{FB})}\right)^2 \qquad (4.6)$$

an exact expression that is rarely used in practice. Note that v_G means v_{GB} since our discussion has implicitly grounded the bulk and all voltages are thus bulk referenced.

An approximate expression for ψ_{sa} is more useful. When we are just at the edge of weak inversion, and $v_{SB} = 0$, the diode-like element at the source end of Figure 3.4 is just starting to turn on with $\psi_{sa} = \phi_0^w = 2\phi_F$. We define the threshold voltage of weak inversion V_{T0}^w to be that gate voltage $v_G = v_{GB}$ at which $\psi_{sa} = \phi_0^w$. Then, from Equation (4.2), we must have

$$V_{T0}^w = V_{FB} + \phi_0^w + \gamma\sqrt{\phi_0^w} \qquad (4.7)$$

Now, if v_G is slightly less than V_{T0}^w, we may use Equation (4.1), which is valid only in weak inversion, to create a Taylor-series approximation for ψ_{sa} such that

$$\psi_{sa} \approx \phi_0^w + \kappa_0^w(v_G - V_{T0}^w) \qquad (4.8)$$

where

$$\kappa_0^w = \frac{1}{\left(1 + \dfrac{\gamma}{2\sqrt{\phi_0^w}}\right)} \tag{4.9}$$

Note that Equation (4.8) is a good approximation only when $v_G < V_{T0}^w$, i.e., in weak inversion. In strong inversion, the diode-clamp approximation would predict from Equation (4.1) that $\kappa = 0$, a false prediction.

We define the threshold voltage of strong inversion, V_{T0}^s, to be that gate voltage $v_G = v_{GB}$ at which the surface potential $\psi_S = \phi_0^s$. We can compute V_{T0}^s to be

$$V_{T0}^s = V_{FB} + \phi_0^s + \gamma\sqrt{\phi_0^s} \tag{4.10}$$

We can similarly define a κ_0^s, a capacitive-divider ratio corresponding to the value of C_{dep} when $\psi_S = \phi_0^s$ for strong inversion, given by

$$\kappa_0^s = \frac{1}{\left(1 + \dfrac{\gamma}{2\sqrt{\phi_0^s}}\right)} \tag{4.11}$$

An approximation for ψ_{sa} at the border of strong inversion is then given by

$$\psi_{sa} \approx \phi_0^s + \kappa_0^s\left(v_G - V_{T0}^s\right) \tag{4.12}$$

The reason that ϕ_0, κ, and V_{T0} differ in weak and strong inversion is that the diode-clamp voltage in strong inversion is larger than the voltage at which we just leave weak-inversion operation and start to turn on by about $6\phi_t$. Thus, the value of C_{dep} is slightly different at the upper limit of weak inversion ($\phi_0^w = 2\phi_F$) and the lower limit of strong inversion ($\phi_0^s = 2\phi_F + 6\phi_t$). The threshold voltage and capacitive-divider ratio in strong inversion are then slightly larger than the corresponding parameters in weak inversion. Essentially, we are assuming that, from the above-threshold point of view, the diode-like elements turn on and clamp when the voltage across them is ϕ_0^s. From the subthreshold point of view, the diode-like elements must have a voltage across them that is less than ϕ_0^w for exponential behavior to be preserved. When the voltage across the diode-like element is between ϕ_0^w and ϕ_0^s, we are operating in moderate inversion and neither weak-inversion nor strong-inversion approximations of transistor behavior are correct.

Hereafter, we shall drop the w and s superscripts on ϕ_0, V_{T0}, and κ_0 to avoid infusing algebraic expressions with tedious notation. *It should be understood that expressions in weak inversion use values corresponding to the w superscript and equations in strong inversion use values corresponding to the s superscript.* If it is unclear from the context as to which parameter is relevant, we will explicitly use the superscripts.

4.1.3   Approximating the body effect

We may ask: how does the threshold voltage V_{T0} change when $v_{SB} \neq 0$? That is, at what new value of $v_{GB} = V'_T$ or equivalently at what new value of $v_{GS} = V'_T - v_{SB}$ with $v_{SB} = V_{SB}$ will the diode-like element at the source end just turn on or clamp? We know from Figure 3.4 that at this point, $\psi_{sa} = \phi_0 + V_{SB}$. Thus, from Equation (4.2) we must have

$$
\begin{aligned}
V'_T &= V_{FB} + (\phi_0 + V_{SB}) + \gamma\sqrt{(\phi_0 + V_{SB})} \\
V_{TS} &= V'_T - V_{SB} = V_{FB} + \phi_0 + \gamma\sqrt{(\phi_0 + V_{SB})} \\
V_{TS} &= \left(V_{FB} + \phi_0 + \gamma\sqrt{\phi_0}\right) + \left(\gamma\sqrt{(\phi_0 + V_{SB})} - \gamma\sqrt{\phi_0}\right) \\
V_{TS} &= V_{T0} + \gamma\left(\sqrt{(\phi_0 + V_{SB})} - \sqrt{\phi_0}\right)
\end{aligned}
\tag{4.13}
$$

The expression on the last line of Equation (4.13) is often referred to as the body effect equation in MOS transistor theory since a voltage between the source and body (bulk) V_{SB} affects the threshold at which the transistor just turns on, with a higher threshold at higher values of V_{SB} in accordance with Equation (4.13). Thus, the MOSFET is not like a bipolar transistor whose I–V characteristics are a function of v_{BE}, the voltage between the base (analogous to gate in the MOSFET), and emitter (analogous to source in the MOSFET) and independent of the absolute value of v_E with respect to a grounded bulk.

For small values of V_{SB} the last line of Equation (4.13) with the usual Taylor series approximation to a square root and simple algebraic manipulations of Equation (4.9) yields

$$
V_{TS} \approx V_{T0} + V_{SB}\left(\frac{1}{\kappa_0} - 1\right)
\tag{4.14}
$$

However, Figure 3.4 illustrates intuitively why Equation (4.14) must be true. If we are operating right at threshold with $v_G = V_{T0}$ and $V_{SB} = 0$ and we change V_{SB} by a small amount, to continue to operate right at threshold, the surface potential must change from ϕ_0 to $\phi_0 + V_{SB}$. Referred back to the gate, the change in surface potential of V_{SB} can only have occurred due to a change in gate voltage of V_{SB}/κ_0 since the capacitive divider attenuation from the gate to the surface potential is κ_0 near $V_{SB} = 0$. Thus, we must add a change of V_{SB}/κ_0 to the threshold voltage V_{T0} to get the new gate voltage at threshold. The new value of v_{GS} at threshold, i.e., V_{TS}, is then obtained by subtracting an additional V_{SB}, which then yields Equation (4.14). Note that C_{dep} behaves like a linear capacitance if V_{SB} is a small change such that the κ (or n) approximation is valid.

If C_{dep} were a linear capacitance there would still be a body effect and Equation (4.14) would be an exact representation of Equation (4.13) instead of just an approximation. Furthermore, Equation (4.8) would be exact as well. However, since C_{dep} is not a linear capacitance, at values of $V_{SB} \neq 0$, it is wise to not use the value of κ_0 computed in Equation (4.9) with the value of V_{T0}

in Equation (4.7) to approximate the surface potential in weak inversion, ψ_{sa}, with Equation (4.8). Rather, it is better to change the operating point about which one performs the Taylor expansion from ϕ_0 to $\phi_0 + V_{SB}$ to approximate the surface potential. Thus, at non-zero values of V_{SB}, a significantly better approximation for the surface potential in weak inversion is

$$\psi_{sa} \approx (\phi_0 + V_{SB}) + \kappa_S(v_{GS} - V_{TS})$$

$$\kappa_S = \cfrac{1}{1 + \cfrac{\gamma}{2\sqrt{\phi_0 + V_{SB}}}} \tag{4.15}$$

rather than the approximation used in Equation (4.8). Note that Equation (4.15) is a source-referenced rather than body-referenced model where both the gate voltage and the threshold gate voltage have been referenced to the source, and $\psi_{sa} - V_{SB}$ is the surface potential referenced to the source. We shall use Equation (4.15) for approximating the surface potential in weak inversion in source-referenced models and Equation (4.8) for approximating the surface potential in weak inversion in body-referenced models. Source-referenced models with the κ approximation are significantly more accurate than body-referenced models if V_{SB} is nonzero in weak inversion. Alternatively, we may just choose to use the exact expression for ψ_{sa} in Equation (4.6) in weak inversion but it adds little insight.

4.1.4 An estimate for the pinchoff voltage in strong inversion

The pinchoff voltage is that voltage, v_{CB}, which, when applied to the source terminal or the drain terminal, causes the source end or drain end of the channel to be right at the border between strong inversion and weak inversion operation, respectively. At the pinchoff voltage, the diode-like elements at the source end or drain end are just turned on. For voltages higher than the pinchoff voltage, the diode-like elements are off. As the gate voltage v_{GB} increases, the pinchoff voltage increases since the diode-like elements are turned off at a higher value of v_{CB} where the v_{GS} or v_{GD} are no longer sufficient to maintain the respective source end or drain end in above-threshold operation. The pinchoff voltage is a useful concept for understanding how the surface potential in the channel varies when source or drain terminal voltages vary and the gate voltage is fixed. It is the dual of the concept for understanding how the surface potential in the channel varies when the gate terminal voltage varies and the source and drain voltages are fixed. The pinchoff voltage is useful in understanding how the saturation voltage v_{DSAT} of a transistor varies with v_{GS} since v_{DSAT} corresponds to a value of v_{DS} at which the drain end of the channel is just pinched off.

Figure 4.1 shows how the surface potential ψ_S at the source end or drain end of the channel varies when we increase v_{CB} with v_{GB} being held constant. Thus, $v_{CB} = v_{SB}$ if we are discussing the source end and $v_{CB} = v_{DB}$ if we are discussing the drain end. Assume that v_{GB} is sufficiently larger than V_{T0} such that strong inversion operation is maintained for a range of positive v_{CB} because v_{GC}

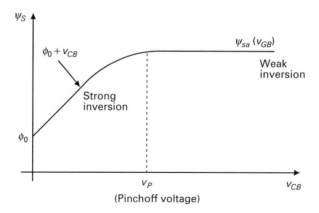

Figure 4.1. The pinchoff voltage.

is sufficiently large to continue to maintain above-threshold operation for this range. From Figure 3.4, as we increase v_{CB} from zero at either the source or drain terminals, at first the diode-like element at the respective terminal will raise the surface potential one-for-one at its respective channel end to maintain a constant diode-clamp voltage ϕ_0 across it. Thus, ψ_S initially increases linearly with v_{CB} in Figure 4.1 to be $v_{CB} + \phi_0$. However, when v_{CB} gets to a value equal to the pinchoff voltage v_P, the surface potential ψ_S rises to a value equal to ψ_{sa}, the surface potential corresponding to weak inversion operation. If we raise v_{CB} beyond v_P the diode-like element is turned off because the capacitive divider of weak inversion maintains ψ_S at ψ_{sa} and the voltage across the diode falls below the voltage ϕ_0 at which it just turns on. Further increases in v_{CB} have no effect on the surface potential ψ_S, which remains constant at ψ_{sa} and independent of the source or drain voltages as the diode-like elements begin to enter weak inversion more deeply and the voltages across them decrease further. Thus ψ_S is well approximated as the minimum of ψ_{sa} and $v_{CB} + \phi_0$.

The surface potential in weak inversion ψ_{sa} is a function of v_{GB} and given by Equation (4.6). From Figure 3.4 and the discussion of the previous paragraph, we know that when the source diode-like element or the drain diode-like element are just turned on at $v_{CB} = v_P$, we must have a voltage of ϕ_0 across the diodes. Thus, the pinchoff voltage v_P is given by

$$v_P = \psi_{sa} - \phi_0 \tag{4.16}$$

If we use the approximation for ψ_{sa} given by Equation (4.12), we find that the pinchoff voltage is given by

$$v_P \approx \kappa_0(v_G - V_{T0}) \tag{4.17}$$

Thus, as v_G increases beyond V_{T0} the pinchoff voltage increases. Equation (4.17) is not valid for voltages below V_{T0}. Equation (4.16) is significantly more accurate than Equation (4.17) if v_G is significantly above V_{T0}, a likely occurrence when v_{SB} is significantly above zero and v_G needs to be sufficiently high to just maintain the transistor in strong inversion. Thus, in source-referenced models, it is preferable to use Equation (4.16) with an approximation for ψ_{sa} given by Equation (4.15). The pinchoff voltage in such models is then given by

$$v_P = v_{SB} + \kappa_S(v_{GS} - V_{TS}) \tag{4.18}$$

If v_{DB} exceeds the pinchoff voltage in Equation (4.18), the charge at the drain end may be approximated as zero in strong inversion.

4.1.5 Approximating the mobile charge at the source end or drain end in strong inversion

The mobile charge in strong inversion at the source end ($v_{CB} = v_{SB}$) or drain end ($v_{CB} = v_{DB}$) is given by

$$Q_I = -C_{ox}\left(v_G - V_{FB} - (\phi_0 + v_{CB}) - \gamma\sqrt{\phi_0 + v_{CB}}\right) \tag{4.19}$$

where we have subtracted the dopant charge from the gate charge to get the mobile charge. Simple differentiation and algebraic manipulation reveal that

$$\frac{dQ_I}{dv_{CB}} = C_{ox}\left(1 + \frac{\gamma}{2\sqrt{\phi_0 + v_{CB}}}\right) = \frac{C_{ox}}{\kappa_C} \tag{4.20}$$

where we have evaluated κ at v_{CB} like our evaluation in Equation (4.15). At the pinchoff voltage, v_P, Q_I may be approximated as zero in strong inversion. Thus for voltages less than the pinchoff voltage and near it, we may use a Taylor series approximation with a slope given by Equation (4.20) to estimate the charge in strong inversion as

$$Q_I = -\frac{C_{ox}}{\kappa_C}(v_P - v_{CB}), \quad 0 < v_{CB} < v_P \tag{4.21}$$

Equation (4.21) is intuitive. A decrease in v_{CB} from the pinchoff voltage translates to a one-for-one decrease in surface potential via the diode-like element, which in turn translates to an increase in mobile charge due to an increase in gate charge across the C_{ox} capacitor and a reduction in dopant charge across the C_{dep} capacitor; or, equivalently, the surface potential decrease causes an increase in negative mobile charge because this potential is expressed across a C_{ox} capacitor looking upwards from the channel in parallel with a C_{dep} capacitor looking downwards from the channel. These charges add, and, from Equation (4.4), we see that $(C_{ox} + C_{dep}) = C_{ox}/\kappa$, consistent with Equation (4.21).

4.1.6 Relating strong-inversion mobile charge variation along the channel to surface-potential variation along the channel

The mobile charge anywhere along the channel is given by

$$Q_I = -C_{ox}\left(v_G - V_{FB} - \psi_s(x) - \gamma\sqrt{\psi_s(x)}\right) \tag{4.22}$$

in both strong inversion and weak inversion. Note that Equation (4.22) is identical to Equation (4.19) except that we're evaluating charge anywhere along the channel not just at the source-end or drain-end channel boundaries. Thus, just like in Equation (4.20), we may derive that

$$\frac{dQ_I}{d\psi_S} = C_{ox}\left(1 + \frac{\gamma}{2\sqrt{\psi_S}}\right) = \frac{C_{ox}}{\kappa_\psi} \tag{4.23}$$

where κ_ψ changes with x as C_{dep} varies along the channel. Typically, we shall ignore this variation and approximate κ_ψ all along the channel as κ_S, the κ given by Equation (4.15) in source-referenced models or the κ_0 given by Equation (4.9) in body-referenced models. Such an approximation is equivalent to using a value for C_{dep} all along the channel to be that of C_{dep} at the source or that of C_{dep} when $v_{SB} = 0$ and $v_{GB} = V_{T0}$, respectively. The notion of looking upwards into C_{ox} and downwards into C_{dep} that intuitively explained Equation (4.21) is also valid in Equation (4.23).

4.2   Charge-based current models with the κ approximation

Equation (3.20), which describes current flow in the MOSFET, is given by

$$\int_0^L i_{DS}dx = W\left[\int_{\Psi_{S0}}^{\Psi_{SL}} -\mu Q_I d\psi_S(x) + \int_{Q_{I0}}^{Q_{IL}} \mu\phi_t dQ_I\right]$$
$$i_{DS} = \frac{\mu W}{L}\left[\int_{\Psi_{S0}}^{\Psi_{SL}} -Q_I d\psi_S(x) + \phi_t \int_{Q_{I0}}^{Q_{IL}} dQ_I\right] \tag{4.24}$$

Instead of substituting the exact relationship between charge and surface potential, i.e., Equation (4.22), and proceeding with the derivation described in Chapter 3, we now use the κ approximation of Equation (4.23) to obtain

$$i_{DS} = \frac{\mu W}{L}\left[\int_{\Psi_{S0}}^{\Psi_{SL}} -Q_I \frac{dQ_I}{(C_{ox}/\kappa)} + \phi_t \int_{Q_{I0}}^{Q_{IL}} dQ_I\right], \tag{4.25}$$

where κ is approximated at a constant value throughout the channel. In body-referenced models κ is chosen to be κ_0, and in source-referenced models κ is chosen to be κ_S. Some simple integration then yields

$$i_{DS} = \frac{\kappa\mu}{2}\frac{W}{L}\frac{1}{C_{ox}}(Q_{I0}^2 - Q_{IL}^2) + \mu\phi_t\frac{W}{L}(Q_{IL} - Q_{I0}) \tag{4.26}$$

Equation (4.26) changes sign when Q_{I0} and Q_{IL}, the charge at the source and drain respectively, are interchanged indicating the symmetry of the MOSFET equations with respect to the source and drain. The first term, due to drift, dominates when Q_{I0} and Q_{IL} are large while the second term, due to diffusion, dominates when Q_{I0} and Q_{IL} are small. When the transistor is saturated, Q_{IL} may be approximated as being negligible compared with Q_{I0}. The first term dominates in above-threshold operation while the second term dominates in subthreshold operation. It is interesting to also group the terms of Equation (4.26) in the form below

$$i_{DS} = \mu \frac{W}{L}\left(\frac{\kappa}{2C_{ox}}Q_{I0}^2 - \phi_t Q_{I0}\right) - \mu \frac{W}{L}\left(\frac{\kappa}{2C_{ox}}Q_{IL}^2 - \phi_t Q_{IL}\right) \tag{4.27}$$

such that the first term represents the current when $Q_{IL} = 0$ while the second term represents the current when $Q_{I0} = 0$. Hence the MOSFET current may be viewed as being composed of the difference of a forward saturation current from source to drain and a reverse saturation current from drain to source. It is also interesting to note that in each of these saturation currents, the quadratic term becomes comparable to the linear term only if

$$Q_{I0,L} \geq \frac{2C_{ox}\phi_t}{\kappa} \tag{4.28}$$

4.3   Derivation of current in weak inversion

In weak inversion, the mobile charges at the source and drain are given by

$$\begin{aligned}Q_{I0} &= -\gamma C_{ox}\left(\sqrt{\psi_{sa} + \phi_t e^{(\psi_{sa} - (2\phi_F + v_{SB}))/\phi_t}} - \sqrt{\psi_{sa}}\right) \\ Q_{IL} &= -\gamma C_{ox}\left(\sqrt{\psi_{sa} + \phi_t e^{(\psi_{sa} - (2\phi_F + v_{DB}))/\phi_t}} - \sqrt{\psi_{sa}}\right)\end{aligned} \tag{4.29}$$

Since the exponential term is small, we may approximate

$$\sqrt{C + \varepsilon} - \sqrt{C} = \frac{\varepsilon}{2\sqrt{C}} \tag{4.30}$$

the Taylor-series-square-root approximation such that

$$i_{DS} = \mu \phi_t \frac{W}{L}(Q_{IL} - Q_{I0}) = \mu \phi_t^2 \frac{W}{L}\frac{\gamma C_{ox}}{2\sqrt{\psi_{sa}}}e^{(\psi_{sa} - 2\phi_F)/\phi_t}\left(e^{-v_{SB}/\phi_t} - e^{-v_{DB}/\phi_t}\right) \tag{4.31}$$

Substituting for ψ_{sa} from Equation (4.8) and using an equation analogous to Equation (4.9) to substitute for $\gamma/(2\sqrt{\psi_{sa}})$, we get

$$i_{DS} = \left[\mu C_{ox}\phi_t^2 \frac{W}{L}\left(\frac{1 - \kappa_{sa}}{\kappa_{sa}}\right)e^{(\phi_0^w - 2\phi_F)/\phi_t}e^{-\kappa_0 V_{T0}^w/\phi_t}\right]e^{\kappa_0^w v_{GB}/\phi_t}\left(e^{-v_{SB}/\phi_t} - e^{-v_{DB}/\phi_t}\right) \tag{4.32}$$

or

$$i_{DS} = I_0 e^{\kappa_0^w v_{GB}/\phi_t}\left(e^{-v_{SB}/\phi_t} - e^{-v_{DB}/\phi_t}\right) \tag{4.33}$$

where

$$I_0 = \mu C_{ox}\phi_t^2 \frac{W}{L}\left(\frac{1 - \kappa_{sa}}{\kappa_{sa}}\right)e^{-\kappa_0^w V_{T0}^w/\phi_t} \tag{4.34}$$

and $\phi_0^w = 2\phi_F$. Typically κ_{sa} is approximated as κ_0^w.

Equation (4.33) is the classic body-referenced equation for subthreshold operation, revealing the exponential relationship between the current and v_{GB}, and the symmetric exponential dependence of the current on v_{SB} and v_{DB}. If κ_0 is approximately 0.6, Equation (4.33) predicts a 100 mV/decade rise in current with increasing v_{GB}. Typically κ_0 is in the 0.5–0.8 range for common semiconductor processes.

When v_{SB} is greater than zero by a few hundred mV, Equations (4.33) and (4.34) are not as accurate a predictor of current as the prediction obtained via a source-referenced model. In source-referenced models, Equation (4.31) is approximated with a more accurate expression for ψ_{sa}, given by Equation (4.15) rather than by the expressions for ψ_{sa} given by Equations (4.8) and (4.9). Simple algebraic manipulations on Equation (4.31) with these approximations then yield

$$i_{DS} = I_{0S} e^{\kappa_S^w v_{GS}/\phi_t}(1 - e^{-v_{DS}/\phi_t})$$

$$I_{0S} = \mu C_{ox} \phi_t^2 \frac{W}{L} \left(\frac{1 - \kappa_{sa}}{\kappa_{sa}}\right) e^{-\kappa_S V_{TS}^w/\phi_t} \tag{4.35}$$

where V_{TS} and κ_S are defined as in Equations (4.13) and (4.15) with $\phi_0 = \phi_0^w = 2\phi_F$ and κ_{sa} often approximated as κ_S. Note that $I_{0S} = I_0$ and $\kappa_S = \kappa_0$ when $v_{SB} = 0$ so that the body-referenced Equation (4.33) and the source-referenced Equation (4.35) are identical in this case.

From Equations (4.34) and (4.35) I_0 and I_{0S} exponentially increase with falling threshold voltages V_{T0} and V_{TS}, respectively, and both rise with increasing C_{ox}. Thus, in advanced submicron processes where the threshold voltage decreases and C_{ox} increases, the subthreshold current for the same terminal voltages is considerably larger. The leakage current in digital circuits, the subthreshold current when $v_{GS} = 0$, i.e., I_{0S} (or I_0), then increases. Equation (4.35) predicts a saturation voltage v_{DSAT} of $4\phi_t \approx 100$ mV since $\exp(-v_{DS}/\phi_t)$ is nearly zero for v_{DS} greater than $4\phi_t$.

We shall now show experimental measurements and simulations taken from a range of processes. Figure 4.2 shows the i_D versus v_{DS} characteristics for a $8\,\mu$m $\times 8\,\mu$m transistor in subthreshold at $v_{GS} = 0.79$ V, $v_{SB} = 0.0$ V obtained from a $1.2\,\mu$m process with a V_{T0} near 0.95 V. The experimental data were fit to an equation of the form $i_{DS} = 7.5 \times 10^{-9}(1 - \exp(-v_{DS}/0.0291))$ with nonlinear regression. The fit is in accord with Equation (4.35) except that ϕ_t was found to be 29.1 mV instead of 25.9 mV. Figure 4.3 shows the i_{DS} versus v_{GS} characteristics for several different v_{GS} values in subthreshold. We observe that v_{DSAT} is indeed constant near $4\phi_t$. Figure 4.4 shows that the saturation current, i_{DSAT}, versus v_{GS} at $v_{SB} = 0$ is fit by an exponential of the form $I_0 \exp(\kappa_0 v_{GS}/\phi_t)$ with $\kappa_0 = 0.67$, $I_0 = 1.6 \times 10^{-13}$ A for v_{GS} below 0.46 V in a $0.18\,\mu$m process. The low-current measurements were obtained via the use of a Keithley electrometer. The simulation curves in Figure 4.5 show that, when v_{SB} changes from 0.0 V to 1.0 V, I_0 changes to a lower value I_{0S} shifting the $i_D - v_{GS}$ saturation-current curve to the right, and κ_0 changes to a higher value κ_S increasing the slope of the $i_D - v_{GS}$

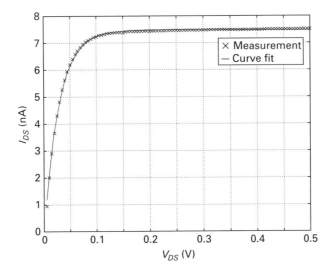

Figure 4.2. Subthreshold current equation fitted to experimental data.

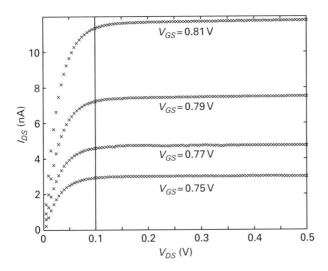

Figure 4.3. Subthreshold v_{DSAT} fitted to experimental data.

saturation-current curve. Such behavior is in accord with Equations (4.33), (4.34), (4.35), (4.9), and (4.15). In the latter example, the slope changed from a value corresponding to $\kappa_0 = 0.614$ to a value corresponding to $\kappa_S = 0.684$.

4.4 Derivation of current in strong inversion

If we assume that only the drift-current quadratic terms from Equation (4.26) are significant in strong inversion, and set $\kappa = \kappa_0$, an approximation suitable for a

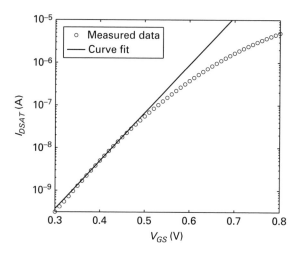

Figure 4.4. Experimentally measured exponential current characteristics in the subthreshold regime.

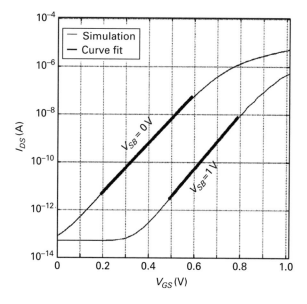

Figure 4.5. Body effect in the subthreshold regime.

small v_{SB} body-referenced model that we shall first investigate, we obtain the current in strong inversion to be

$$i_{DS} = \frac{\mu \kappa_0}{2 C_{ox}} \frac{W}{L} \left(Q_{I0}^2 - Q_{IL}^2 \right) \tag{4.36}$$

From Equations (4.17) and (4.21) with $\kappa_C = \kappa_0$, $v_{CB} = v_{SB}$ at the source end, and $v_{CB} = v_{DB}$ at the drain end, we obtain

$$Q_{I0} = -\frac{C_{ox}}{\kappa_0}[\kappa_0(v_{GB} - V_{T0}) - v_{SB}]$$
$$Q_{IL} = -\frac{C_{ox}}{\kappa_0}[\kappa_0(v_{GB} - V_{T0}) - v_{DB}] \tag{4.37}$$

where we have explicitly substituted v_{GB} for v_G to emphasize the body-referenced nature of our model. Then, from simple algebraic manipulations of Equations (4.36) and (4.37), we can derive that

$$i_{DS} = \frac{\kappa_0 \mu C_{ox}}{2}\frac{W}{L}\left[\left(v_G - V_{T0} - \frac{v_{SB}}{\kappa_0}\right)^2 - \left(v_G - V_{T0} - \frac{v_{DB}}{\kappa_0}\right)^2\right] \tag{4.38}$$

Equation (4.38) is anti-symmetric in the source and drain voltages. When the drain charge, i.e., the second term in Equation (4.38), goes to zero, the transistor is saturated. If $v_{SB} = 0$, the place where such a body-referenced model is most useful and likely to be valid, we have $v_{DB} = v_{DS}$, and

$$v_{DSAT} = \kappa_0(v_{GB} - V_{T0}) \tag{4.39}$$

The saturation current i_{DSAT} in this case is given from Equation (4.38) with $v_{SB} = 0$ and the second term at zero to be

$$i_{DSAT} = \frac{\kappa_0 \mu C_{ox}}{2}\frac{W}{L}(v_{GB} - V_{T0})^2 \tag{4.40}$$

4.5 Source-referenced model for strong inversion

If $v_{SB} \neq 0$, it is significantly more accurate to use Equation (4.18) and Equation (4.21) with $\kappa_C = \kappa_S$ and $\kappa = \kappa_S$ in the quadratic terms of Equation (4.26) to estimate Q_{I0} and Q_{IL} and thus obtain the transistor current instead of Equation (4.17) and Equation (4.21) as we did in our previous discussion for body-referenced models. With these substitutions, and noting that $v_{CB} = v_{SB}$ when we are evaluating Q_{I0} at the source and $v_{CB} = v_{DB}$ when we are evaluating Q_{IL} at the drain, we get

$$Q_{I0} = -\frac{C_{ox}}{\kappa_S}((v_{SB} + \kappa_S(v_{GS} - V_{TS})) - v_{SB})$$
$$Q_{IL} = -\frac{C_{ox}}{\kappa_S}((v_{SB} + \kappa_S(v_{GS} - V_{TS})) - v_{DB}) \tag{4.41}$$

or

$$Q_{I0} = -C_{ox}(v_{GS} - V_{TS})$$
$$Q_{IL} = -C_{ox}\left((v_{GS} - V_{TS}) - \frac{v_{DS}}{\kappa_S}\right) \tag{4.42}$$

Substituting for Q_{I0} and Q_{IL} in Equation (4.36) with κ_0 replaced by κ_S, we then obtain

$$i_{DS} = \frac{\kappa_S \mu C_{ox}}{2} \frac{W}{L} \left[(v_{GS} - V_{TS})^2 - \left(v_G - V_{TS} - \frac{v_{DS}}{\kappa_S} \right)^2 \right] \qquad (4.43)$$

Some simple algebraic simplification then yields the result for the source-referenced current in strong inversion,

$$i_{DS} = \mu C_{ox} \frac{W}{L} \left[(v_{GS} - V_{TS})v_{DS} - \frac{v_{DS}^2}{2\kappa_S} \right] \qquad (4.44)$$

The mobile charge at the drain end, Q_{IL}, goes to zero when the second quadratic term of Equation (4.43) is zero. At this point $v_{DS} = v_{DSAT}$ such that

$$v_{DSAT} = \kappa_S(v_{GS} - V_{TS}) \qquad (4.45)$$

The saturation current i_{DSAT} is then just computed from the first term of Equation (4.43) to be

$$i_{DSAT} = \frac{\kappa_S \mu C_{ox}}{2} \frac{W}{L} (v_{GS} - V_{TS})^2 \qquad (4.46)$$

Equations (4.43) and (4.44) are only valid for $v_{DS} \le v_{DSAT}$ beyond which the current is at its saturated value given by Equation (4.46).

Above-threshold equations in many textbooks ignore κ_S altogether, effectively pretending that it is always 1.0. Such an approximation is equivalent to pretending that the depletion capacitance C_{dep} is zero in strong inversion. The predicted curves correspond to $\kappa_S = 1$ in Equations (4.44), (4.45), and (4.46) and do not match experimentally observed data. In particular, the saturation voltage is significantly larger than is experimentally observed.

Figure 4.6 shows above-threshold i_{DS} versus v_{DS} current measurements for a $8\,\mu m \times 8\,\mu m$ transistor in a $1.2\,\mu m$ process with $v_{SB} = 0$ and v_{GS} fixed at 1.8 V. Nonlinear regression reveals that the measurements were well fit by Equation (4.44) with $\mu C_{ox} \frac{W}{L} = 71.2\,\mu A$, $\kappa_S = 0.7284$, and $V_{TS} = 0.95$V. We observe that the saturation voltage is slightly less than what we would predict from theory. If we use $\kappa_S = 1.0$, the fit is significantly worse and drastically under-predicts experimental values especially near the transition from the linear region to the saturated region of operation.

Figure 4.7 shows experimental measurements of the $i_{DS} - v_{DS}$ curves for the same transistor for several different v_{GS} values in the above-threshold region of operation with $v_{SB} = 0$. The saturation voltage is well predicted by Equation (4.45). Figure 4.8 shows that the square-root of the saturation current as predicted from Equation (4.46) is also in good accord with observations with $\mu C_{ox} \frac{W}{L} = 71.2\,\mu A$, $\kappa_S = 0.7$, and $V_{TS} = 0.9$ V providing a good fit to the data. Figure 4.9 shows that the threshold of a (different) transistor shifts from $V_{T0} = 0.544$ V to $V_{TS} = 0.834$ V when v_{SB} is changed from 0.0 V to 1.0 V in good accord with the predictions of Equation (4.13) with $\gamma = 0.84$ and $\phi_0 = 0.95$ V.

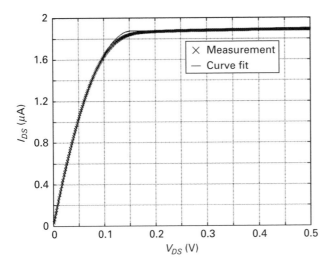

Figure 4.6. Above-threshold current equation fitted to experimental data.

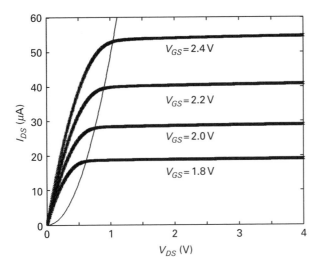

Figure 4.7. Above-threshold v_{DSAT} fitted to experimental data.

Table 4.1 summarizes equations in subthreshold and above-threshold operation in body-referenced and source-referenced models respectively. Note that I_0 and I_{0S} are often inaccurately predicted because of their exponential sensitivity to the threshold voltages V_{T0} or V_{TS}, respectively, which are not meaningfully defined to extreme accuracy. Nevertheless, the forms of the equations for I_0 and I_{0S} are useful for determining how I_0 and I_{0S} vary with process parameters. Extrapolating a square-root straight line down to zero yields a value for V_{T0}^s from an above-threshold viewpoint, e.g., as in Figure 4.8; however, fitting an exponential to subthreshold data, e.g., as in Figure 4.4, produces a strict exponential with little

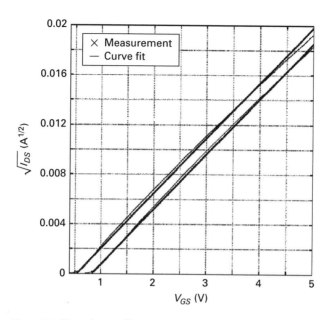

Figure 4.8. Experimentally measured square-root current characteristics in the above-threshold regime.

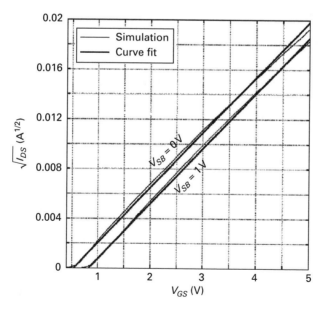

Figure 4.9. Simulated body effect in the above-threshold regime.

Table 4.1 Table summarizing source- and body-referenced equations

	Weak inversion $i_{DS} = \mu\phi_t \frac{W}{L}(Q_{IL} - Q_{I0})$	Strong inversion $i_{DS} = \frac{\mu\kappa_s}{2C_{ox}}\frac{W}{L}(Q_{I0}^2 - Q_{IL}^2)$
$\kappa_x \equiv 1/n_x$		
Body-referenced	$i_{DS} = I_0 e^{\kappa_0 v_{GB}/\phi_t}\left(e^{-v_{SB}/\phi_t} - e^{-v_{DB}/\phi_t}\right)$	$i_{DS} = \frac{\kappa_0\mu C_{ox}}{2}\frac{W}{L}\left[\left(v_G - V_{T0} - \frac{v_{SB}}{\kappa_0}\right)^2\right.$
	$I_0 = \mu C_{ox}\phi_t^2\frac{W}{L}\left(\frac{1-\kappa_{sa}}{\kappa_{sa}}\right)e^{-\kappa_0 V_{T0}/\phi_t}$	$\left. - \left(v_G - V_{T0} - \frac{v_{DB}}{\kappa_0}\right)^2\right]$
	$V_{T0} = V_{FB} + \phi_0 + \gamma\sqrt{\phi_0},\ \boxed{\phi_0 = 2\phi_F}$	
	$\kappa_{sa} = \dfrac{1}{1+\dfrac{\gamma}{2\sqrt{\psi_{sa}}}},\quad \kappa_0 = \dfrac{1}{1+\dfrac{\gamma}{2\sqrt{\phi_0}}} = \dfrac{1}{n_0}$	$V_{T0} = V_{FB} + \phi_0 + \gamma\sqrt{\phi_0},\ \boxed{\phi_0 = 2\phi_F + 6\phi_t}$
		$v_{DSAT} = \kappa_0(v_{GB} - V_{T0}),\quad v_{SB} = 0$
	$\psi_{sa} = \left(-\dfrac{\gamma}{2} + \sqrt{\dfrac{\gamma^2}{4} + (v_G - V_{FB})}\right)^2$	$i_{DSAT} = \dfrac{\kappa_0\mu C_{ox}}{2}\dfrac{W}{L}(v_{GB} - V_{T0})^2$
Source-referenced	$i_{DS} = I_{0S}e^{\kappa_s v_{GS}/\phi_t}\left(1 - e^{-v_{DS}/\phi_t}\right)$	$i_{DS} = \mu C_{ox}\dfrac{W}{L}\left[(v_{GS} - V_{TS})v_{DS} - \dfrac{v_{DS}^2}{2\kappa_s}\right]$
	$I_{0S} = \mu C_{ox}\phi_t^2\dfrac{W}{L}\left(\dfrac{1-\kappa_{sa}}{\kappa_{sa}}\right)e^{-\kappa_s V_{TS}/\phi_t}$	$V_{TS} = V_{T0} + \gamma\left(\sqrt{(\phi_0 + V_{SB})} - \sqrt{\phi_0}\right)$
	$i_{DS} = i_{DSAT}(1 - e^{-v_{DS}/\phi_t}),\ V_{DSAT} \approx 4\phi_t$	$v_{DSAT} = \kappa_s(v_{GS} - V_{TS})$
	$V_{TS} = V_{T0} + \gamma\left(\sqrt{(\phi_0 + V_{SB})} - \sqrt{\phi_0}\right)$	$i_{DSAT} = \dfrac{\kappa_s\mu C_{ox}}{2}\dfrac{W}{L}(v_{GS} - V_{TS})^2$

curvature only if v_{GS} is below V_{T0}^w. The threshold voltage V_{T0}^s is greater than V_{T0}^w by approximately $6\phi_t/\kappa_0$.

The saturation current, I_T, at the edge of subthreshold operation at $v_{GB} = V_{T0}^w$, $v_{SB} = 0$, may be approximated by substituting an overdrive voltage of $(v_{GS} - V_{TS}) \approx \phi_t/\kappa_0$ in the above-threshold square-law relation given by Equation (4.46) with κ_0 assumed near 0.5. The form of Equation (4.34) then suggests that I_0, the leakage current in digital circuits when $v_{GS} = 0$, may be estimated by attenuating I_T exponentially with one decade of attenuation every $(\phi_t/\kappa_0)\ln(10)$ V from $v_{GS} = V_{T0}^w$ down to $v_{GS} = 0$. For example, if κ_0 is near 0.6, we have nearly 100 mV per decade of attenuation; if $V_{T0}^s = 0.5$, $V_{T0}^w \approx 0.25$ V, and I_T is computed to be 1 μA, then the leakage current when $v_{GS} = 0$ is estimated by attenuating 1 μA over two-and-a-half 100 mV-decades to approximately 3.2 nA; if V_{T0}^s were 0.4 V, I_0 would be 32 nA. This example illustrates why leakage currents in modern semiconductor processes have been rising dramatically with reductions in threshold voltages. The speeds available in subthreshold operation have been rising dramatically as well.

4.6   Moderate inversion

An empirical equation for describing the current that is valid in weak inversion, moderate inversion, and strong inversion has been proposed and is often called the empirical EKV model after its authors, Enz, Krummenacher, and Vittoz [2]. The body-referenced model is described by

$$i_{DS} = \frac{2\mu C_{ox}}{\kappa_0}\frac{W}{L}\phi_t^2\left(\left[\ln\left(1+e^{\frac{\kappa_0(v_{GB}-V_{T0})-v_{SB}}{2\phi_t}}\right)\right]^2 - \left[\ln\left(1+e^{\frac{\kappa_0(v_{GB}-V_{T0})-v_{DB}}{2\phi_t}}\right)\right]^2\right) \quad (4.47)$$

The source-referenced model is described by

$$i_{DS} = \frac{2\mu C_{ox}}{\kappa_S}\frac{W}{L}\phi_t^2\left(\left[\ln\left(1+e^{\frac{\kappa_S(v_{GS}-V_{TS})}{2\phi_t}}\right)\right]^2 - \left[\ln\left(1+e^{\frac{\kappa_S(v_{GS}-V_{TS})-v_{DS}}{2\phi_t}}\right)\right]^2\right) \quad (4.48)$$

with V_{T0} and V_{TS} in these equations corresponding to the point where the surface potential is at $2\phi_F$ above the source, i.e., the surface potential corresponding to the upper limit of weak inversion is used as a reference. Note that the model actually uses $n_0 = 1/\kappa_0$ and $n_S = 1/\kappa_S$ for its formulation, but we have used κ_0 and κ_S.

Since $\ln(1+e^x) \approx e^x$ for x near and below 0, $\ln(1+e^x) \approx x$ for x significantly above 0, and $\ln(x^2) = 2\ln(x)$, the expressions above reduce to the equations listed in Table 4.1 for subthreshold and above-threshold operation except for slight differences in pre-constants in the subthreshold equations due to the exact manner in which the threshold is defined. The empirical EKV model is an elegant way of interpolating between the two regimes of operation in an intuitive fashion and is useful for analytical purposes. However, it should be remembered that it does not fit the actual experimental data in moderate inversion too closely. The equations of Chapter 3 fit the data more closely although they are far less analytically tractable. In Chapter 6, we shall discuss a more accurate EKV model that is not empirical. This EKV model does not result in closed-form expressions for the current as the empirical Equations (4.47) or (4.48) do. Nevertheless, they are tremendously useful in describing the transistor in all regions of operation and in modeling deep submicron effects such as velocity saturation.

References

[1] Carver Mead. *Analog VLSI and Neural Systems* (Reading, Mass: Addison-Wesley, 1989).
[2] Christian Enz and Eric A. Vittoz. *Charge-based MOS Transistor Modeling: The EKV Model for Low-Power and RF IC Design* (Chichester, England; Hoboken, NJ: John Wiley, 2006).

5 MOS device physics: small-signal operation

All difficult things have their origin in that which is easy, and great things in that which is small.

Lao Tzu

In many systems, certain important transistors in an architecture that determine most of its performance are operated such that the current through them has a large-signal dc bias component around which there are small-signal ac deviations. The voltages of these transistors correspondingly also have a dc large-signal operating point and small-signal ac deviations. If the ac deviations are sufficiently small, the transistor may be characterized as a linear system in its ac small-signal variables with the parameters of the linear system being determined by the dc large-signal variables. In this chapter, we will focus on the small-signal properties of the transistor.

Given that the transistor is a highly nonlinear device, it may be surprising that we are interested in its linear small-signal behavior. However, the transistor's linear small-signal behavior is most important in determining its behavior in analog feedback loops that are intentionally designed to have a linear input-output relationship in spite of nonlinear devices in the architecture. For example, most operational amplifier (opamp) circuits are architected such that negative feedback inherent in the topology ensures that the input terminal voltages of the opamp, v^+ and v^-, will be very near each other and, therefore, that the small-signal properties of the transistors in the opamp determine its stability and convergence in most situations.[1] Even in nonlinear analog circuits, the stability and dynamics of the circuit around its equilibrium values are determined by the small-signal linear properties of its devices around these equilibrium values. The noise inherent in transistors is well modeled as a random ac small-signal deviation in the transistor's current, with the value and properties of the small-signal noise deviations once again determined by large-signal dc operating parameters. In digital systems, which operate almost exclusively in highly nonlinear switching modes, the linear small-signal properties of the transistor provide insight into the maximal possible switching speed, the robustness of the digital circuit to transistor mismatch and power-supply

---

[1] In situations where v^+ and v^- are not near each other, the opamp's nonlinear behavior such as its maximum charging or discharging rate, termed its slew rate, determine the opamp's behavior.

variation, and the gain and signal-restorative properties of the digital circuit (see Chapter 21). Hence, small-signal models of transistors play a key role in all circuits, which is why we shall be devoting a whole chapter to them.

To avoid confusion about whether a variable is a small-signal variable or a large-signal variable, we shall adopt IEEE convention in this chapter such that

$$
\begin{aligned}
i_{DS} &= I_{DS} + i_{ds} \\
v_{GS} &= V_{GS} + v_{gs} \\
v_{DS} &= V_{DS} + v_{ds} \\
v_{BS} &= V_{BS} + v_{bs}
\end{aligned}
\tag{5.1}
$$

As Equation (5.1) illustrates, the total variable x_Y is the sum of a large-signal dc variable X_Y and an ac small-signal variable x_y. The variable $X_y(s)$ is reserved for the Laplace transform of x_y. Equation (5.1) lists the small-signal variables for a source-referenced description of a transistor. Throughout this chapter, we will use source-referenced descriptions unless otherwise noted. The generalization from source-referenced descriptions to body-referenced descriptions or vice versa is straightforward.

The small-signal equation that describes a transistor's operation is then given by

$$
\begin{aligned}
i_{ds} &= \frac{\partial i_{DS}}{\partial v_{GS}} v_{gs} + \frac{\partial i_{DS}}{\partial v_{BS}} v_{bs} + \frac{\partial i_{DS}}{\partial v_{DS}} v_{ds} \\
i_{ds} &= g_m v_{gs} + g_{mb} v_{bs} + g_{ds} v_{ds}
\end{aligned}
\tag{5.2}
$$

where g_m, g_{mb}, and g_{ds} are defined as the corresponding partial derivatives in Equation (5.2). These derivatives are functions of the values of the dc operating point (I_{DS}, V_{GS}, V_{BS}, V_{DS}). We shall evaluate these derivatives first in weak inversion and then in strong inversion. The parameter g_m is referred to as the transconductance, g_{mb} is referred to as the back-gate or bulk transconductance, and g_{ds} is referred to as the drain-to-source conductance.

5.1   Weak-inversion small-signal models

First, we review the basic equations for body-referenced and source-referenced subthreshold models. The body-referenced model is described by

$$
i_{DS} = I_0 e^{\kappa_0 v_{GB}/\phi_t} \left(e^{-v_{SB}/\phi_t} - e^{-v_{DB}/\phi_t} \right)
\tag{5.3}
$$

where

$$
\begin{aligned}
I_0 &= \mu C_{ox} \phi_t^2 \frac{W}{L} \left(\frac{1 - \kappa_{sa}}{\kappa_{sa}} \right) e^{-\kappa_0 V_{T0}/\phi_t} \\
\kappa_0 &= \frac{1}{1 + \dfrac{\gamma}{2\sqrt{\phi_0}}} = \frac{1}{n_0} \\
V_{T0} &= V_{FB} + \phi_0 + \gamma \sqrt{\phi_0}
\end{aligned}
\tag{5.4}
$$

The source-referenced model is described by

$$i_{DS} = I_{0S} e^{\kappa_S v_{GS}/\phi_t} (1 - e^{-v_{DS}/\phi_t}) \tag{5.5}$$

where

$$I_{0S} = \mu C_{ox} \phi_t^2 \frac{W}{L} \left(\frac{1 - \kappa_{sa}}{\kappa_{sa}} \right) e^{-\kappa_S V_{TS}/\phi_t}$$

$$\kappa_S = \frac{1}{1 + \dfrac{\gamma}{2\sqrt{\phi_0 + V_{SB}}}} = \frac{1}{n_S} \tag{5.6}$$

$$V_{TS} = V_{T0} + \gamma \left(\sqrt{(\phi_0 + V_{SB})} - \sqrt{\phi_0} \right)$$

By differentiating Equation (5.5) with respect to v_{GS}, we can see that

$$g_m = \frac{\kappa_S I_{DS}}{\phi_t} \tag{5.7}$$

Thus, the g_m of a subthreshold MOS transistor is very similar to the g_m of a bipolar transistor, I_C/ϕ_t, where I_C is the collector current except for the κ_S term.

We evaluate g_{mb} from Equations (5.5) and (5.6) to be

$$
\begin{aligned}
g_{mb} &= \frac{\partial I_{0S}}{\partial V_{TS}} \times \frac{\partial V_{TS}}{\partial v_{BS}} = \left(-\frac{\kappa_S}{\phi_t} I_{DS} \right) \times \left(-\frac{1 - \kappa_S}{\kappa_S} \right) \\
&= \frac{(1 - \kappa_S) I_{DS}}{\phi_t}
\end{aligned}
\tag{5.8}
$$

It is interesting to derive the same result from a body-referenced viewpoint as well. Equation (5.3) can be rewritten as follows:

$$
\begin{aligned}
i_{DS} &= I_0 e^{(\kappa_0 v_{GB} - v_{SB})/\phi_t} (1 - e^{-v_{DS}/\phi_t}) \\
&= I_0 e^{(\kappa_0 (v_{GB} - v_{SB}) - (1 - \kappa_0) v_{SB})/\phi_t} (1 - e^{-v_{DS}/\phi_t}) \\
&= I_0 e^{\kappa_0 v_{GS}/\phi_t} e^{(1 - \kappa_0) v_{BS}/\phi_t} (1 - e^{-v_{DS}/\phi_t})
\end{aligned}
\tag{5.9}
$$

from which we can see that $g_m = \kappa_0 I_{DS}/\phi_t$ and $g_{mb} = (1 - \kappa_0) I_{DS}/\phi_t$.

Equation (5.9) highlights the symmetry of the gate and bulk terminals in controlling the transistor. In fact, an even simpler way to derive g_{mb} is to note, from the discussion in Chapter 4, that the transistor current is given by

$$i_{DS} = (\ldots) e^{\psi_{sa}/\phi_t} \left(1 - e^{-v_{DS}/\phi_t} \right) \tag{5.10}$$

where the $(\ldots)$ refers to terms independent of v_{SB}. Then, from the capacitive divider relationship obvious from Figure 3.4,

$$
\begin{aligned}
g_{mb} &= \frac{\partial i_{DS}}{\partial v_{BS}} = \frac{I_{DS}}{\phi_t} \frac{\partial \psi_{sa}}{\partial v_{SB}} \\
&= \frac{I_{DS}}{\phi_t} \left(\frac{C_{dep}}{C_{dep} + C_{ox}} \right) \\
&= \frac{I_{DS}}{\phi_t} (1 - \kappa)
\end{aligned}
\tag{5.11}
$$

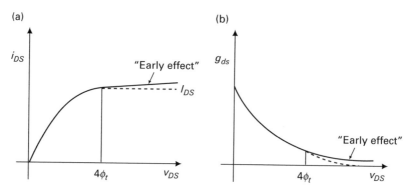

Figure 5.1a, b. Early voltage curves in the subthreshold regime.

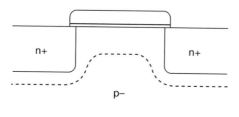

Figure 5.2. Depletion regions at the source and drain illustrating the origin of the Early effect.

Equations (5.7) and (5.11) illustrate that both the gate and bulk behave like good control terminals with a gate and bulk transconductance respectively. In Chapter 12, we will exploit circuits that use the bulk (the well terminal in a pFET) as a legitimate control terminal instead of merely tying it to the power supply, as is often done.

By differentiating Equation (5.5), we can derive that,

$$\frac{\partial i_{DS}}{\partial v_{DS}} = \frac{I_{OS}e^{\kappa V_{GS}/\phi_t}e^{-V_{DS}/\phi_t}}{\phi_t}$$

$$g_{ds} = \frac{I_{DSAT}}{\phi_t}e^{-V_{DS}/\phi_t} \tag{5.12}$$

Equation (5.12) predicts that g_{ds} asymptotically approaches zero after $V_{DS} \geq 4\phi_t$. However, in practice, an effect called the Early effect causes g_{ds} to remain finite after the saturation voltage as illustrated by the $i_{DS} - v_{DS}$ curve and $g_{ds} - v_{DS}$ curve of Figure 5.1.

Figure 5.2 shows that the length l_p, the sum of the depletion region length at the source end and drain end of the channel, reduces the effective channel length of the transistor from L to $L - l_p$: The edges of the depletion regions at the source end and drain end of the channel are the locations where the boundary conditions for the surface potential and charge are valid, and between which integration of the drift and diffusion flow equations yields the current in the transistor. Therefore, as v_{SB} or v_{DB} increase, increasing l_p, the effective channel length of the transistor reduces, increasing the current. When v_{SB} is fixed at V_{SB} for a particular source-referenced model, we may parameterize all of the change in l_p as being that due to a change in v_{DS} at the drain end of the channel. That is,

$$i_{DS} = (\dots\dots) \frac{W}{L - l_p}$$

$$\frac{\partial i_{DS}}{\partial v_{DS}} = (\dots) \frac{W}{(L - l_p)^2} \frac{\partial l_p}{\partial v_{DS}} \tag{5.13}$$

$$g_{ds} = \frac{I_{DS}}{L - l_p} \frac{\partial l_p}{\partial v_{DS}}$$

We define an Early voltage, V_A, such that

$$g_{ds} = \frac{I_{DS}}{V_A}$$

$$\frac{1}{V_A} = \frac{1}{L - l_p} \frac{\partial l_p}{\partial v_{DS}} \approx \frac{1}{L} \frac{\partial l_p}{\partial v_{DS}} \tag{5.14}$$

A large Early voltage signifies a weak dependence of the transistor current on the drain-to-source voltage. The Early voltage is hard to compute since the electric field lines right near the drain end of the channel depend on both the gate voltage and the drain voltage in a complex 3D electrostatic geometry. At higher gate voltages, the gate exerts a relatively larger influence on the field lines weakening the influence of the drain. Thus, the Early voltage is higher in strong inversion than in weak inversion. Empirically, it is found that

$$V_A = \frac{L}{l} [V_E + (V_{DS} - V_{DSAT})], \tag{5.15}$$

$$l = \sqrt{3 t_{ox} d_j},$$

where d_j is the depletion region thickness, and V_E is a process-dependent parameter near 1 V [1]. The increase in V_A with drawn channel length L is intuitive since a change in l_p will have less influence on the overall channel length of the transistor if l_p is a small fraction of L. The increase in V_A with V_{DS} is also intuitive since the rate of change of a depletion region with changing reverse bias across it slows as the reverse bias gets larger. Frequently, V_A is assumed to be constant for all $V_{DS} \geq V_{DSAT}$. The transistor equation in weak inversion is then modified to read

$$i_{DS} = I_{0S} e^{\kappa v_{GS}} \frac{W}{L_{eff}} (1 - e^{-v_{DS}/\phi_t}) \left(1 + \frac{v_{DS}}{V_A(L)}\right) \tag{5.16}$$

where $L_{eff} = L - l_p(0)$ is the effective channel length at zero drain-to-source voltage, and $1/V_A(L)$ is often denoted as λ, a parameter in SPICE, the commonly used circuit simulation program.

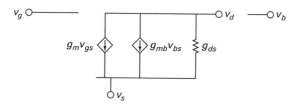

Figure 5.3. Small-signal model of an MOS transistor.

The small-signal equivalent circuit of Figure 5.3 and the relations below

$$g_m = \frac{\kappa_S I_{DS}}{\phi_t}$$

$$g_{mb} = \frac{(1 - \kappa_S) I_{DS}}{\phi_t}$$

$$g_{ds} = \frac{I_{DSAT}}{\phi_t} e^{-v_{DS}/\phi_t}, \quad V_{DS} \leq V_{DSAT}$$

$$g_{ds} = \frac{1}{r_0} = \frac{I_{DSAT}}{V_A}, \quad V_{DS} > V_{DSAT}$$

(5.17)

provide a summary of small-signal operation in weak inversion.

5.2   Strong-inversion small-signal models

The strong-inversion relationship between drain-to-source current and terminal voltages is given by

$$i_{DS} = \mu C_{ox} \frac{W}{L} \left[(v_{GS} - V_{TS}) v_{DS} - \frac{v_{DS}^2}{2\kappa_S} \right]$$

$$i_{DSAT} = \frac{\kappa_S}{2} \mu C_{ox} \frac{W}{L} (v_{GS} - V_{TS})^2$$

$$V_{TS} = V_{T0} + \gamma \left(\sqrt{\phi_0 + V_{SB}} - \sqrt{\phi_0} \right)$$

(5.18)

with the i_{DS} equation applying if $v_{DS} \leq v_{DSAT} = \kappa_S (v_{GS} - V_{TS})$ and the i_{DSAT} equation applying if $v_{DS} > v_{DSAT}$. The κ_S, ϕ_0, and V_{T0} parameters assume above-threshold values rather than subthreshold values as we discussed in Chapter 4 and itemized in Table 4.1. Then, we can show that

$$g_m = \frac{\partial i_{DS}}{\partial v_{GS}}$$

$$= \mu C_{ox} \frac{W}{L} V_{DS}, \quad V_{DS} \leq \kappa_S (V_{GS} - V_{TS})$$

(5.19)

$$= \mu C_{ox} \frac{W}{L} V_{DSAT}, \quad V_{DS} > \kappa_S (V_{GS} - V_{TS})$$

Thus, g_m is independent of V_{GS} and linearly dependent on V_{DS} in the linear regime of operation where $V_{DS} \leq V_{DSAT}$ and independent of V_{DS} and linearly dependent on V_{GS} in the saturation regime where $v_{DS} > v_{DSAT}$ (since $V_{DSAT} = \kappa_S(V_{GS} - V_{TS})$).

In the saturation regime, there are three equivalent forms for g_m and they are given by

$$g_m = \kappa_S \mu C_{ox} \frac{W}{L} (V_{GS} - V_{TS})$$

$$= \sqrt{2 I_{DSAT} \left(\kappa_S \mu C_{ox} \frac{W}{L} \right)} \qquad (5.20)$$

$$= \frac{I_{DSAT}}{\left(\dfrac{V_{GS} - V_{TS}}{2} \right)}$$

They are derived directly from Equation (5.19), by eliminating $(V_{GS} - V_{TS})$ using Equations (5.19) and (5.18), and by eliminating $\left(\kappa_S \mu C_{ox} \frac{W}{L} \right)$ using Equations (5.19) and (5.18) respectively. They emphasize the dependence of g_m on operating overdrive voltage $(V_{GS} - V_{TS})$ and geometry, on operating current and geometry, or on operating current and overdrive voltage, respectively.

The latter form of Equation (5.20) is particularly useful since it indicates that instead of being $I_{DS}/(\phi_t/\kappa_S)$ in weak inversion, g_m is $I_{DS}/((V_{GS} - V_{TS})/2)$ in strong inversion. Thus, the characteristic voltage ϕ_t/κ_S in weak inversion corresponds to $(V_{GS} - V_{TS})/2$ in strong inversion. This correspondence is frequently useful in generalizing intuitively and rapidly from a circuit operating in weak inversion to the identical circuit operating in strong inversion. For example, the range of linearity of a simple differential-pair circuit changes from $2\phi_t/\kappa_S$ in weak inversion to $2((V_{GS} - V_{TS})/2)$ of a bias-current-setting transistor in strong inversion; the gain of a current-source-loaded common-source amplifier changes from $V_A/(\phi_t/\kappa_S)$ in weak inversion to $V_A/((V_{GS} - V_{TS})/2)$ in strong inversion; the expected percentage error in the currents of transistors across a chip due to varying threshold voltages with a standard deviation σ_{VT0} across the chip changes from $\sigma_{VT0}/(\phi_t/\kappa_S)$ to $\sigma_{VT0}/((V_{GS} - V_{TS})/2)$. Often, the analysis of circuits is easier in weak inversion due to the pretty mathematics of exponentials. The results of the weak-inversion analysis can then be used to quickly generate a prediction of the behavior of the same circuit in strong inversion without having to solve intractable and messy square-and-square-root equations. Such intuitive generalizations work because the small-signal circuits in both regimes of operation are identical and can be described by the same algebraic parameters. Thus, an answer obtained in terms of these parameters is identical in both regimes. It is just the dependence of these parameters on the dc operating point variables that is different between the regimes, a dependence that is easily substituted at the very end of the small-signal analysis.

For a given I_{DS}, one would like to get as much g_m as possible since the maximal speed (bandwidth) of operation is often limited by the value of g_m, and it is advantageous to consume as little power as is needed in order to operate at this speed. The current-and-geometry form corresponding to the middle of Equation (5.20) tells us that we can increase g_m for a given I_{DS} by increasing the W/L of the transistor. The current-and-voltage form corresponding to the last relationship of Equation (5.20) tells us that when we increase the W/L of a transistor at constant I_{DS}, its operating $(V_{GS} - V_{TS})/2$ decreases, increasing its g_m. If the W/L is increased indefinitely, the $(V_{GS} - V_{TS})/2$ continues to decrease until it asymptotically approaches ϕ_t/κ_S and the transistor enters weak inversion. Further increases in W/L of the transistor once it enters weak inversion do not improve the g_m/I_{DS} ratio of the transistor, since the g_m of a weakly inverted transistor is independent of its geometry and only dependent on I_{DS}.

The g_m/I_{DS} ratio is an important metric for determining the speed available for a given level of power consumption. The g_m/I_{DS} ratio is derived from Equations (5.7) and (5.20) to be

$$\frac{g_m}{I_{DS}}\Big|_{subthreshold} = \frac{1}{(\phi_t/\kappa_S)}$$
$$\frac{g_m}{I_{DS}}\Big|_{above-threshold} = \sqrt{\frac{2\kappa_S \mu C_{ox}(W/L)}{I_{DS}}}$$
(5.21)

Figure 5.4 and Equation (5.21) show that the g_m/I_{DS} ratio is maximal and constant in weak inversion and degrades like an inverse square-root with increasing I_{DS} in a fixed-geometry transistor. I_{T0} is the current in the transistor near the border between weak and strong inversion. In power-efficient circuits, it is best to always operate in weak inversion if the W/L needed to do so is not too large such that area consumption is significantly compromised, degrading circuit compactness, or the input capacitance presented by the transistor to its driving source is significantly compromised, degrading circuit speed.

Often, operating a circuit in moderate inversion yields a satisfactory tradeoff between area efficiency and power efficiency. Subthreshold operation is also less robust to transistor mismatches, power-supply noise, and temperature, issues that must be addressed via appropriate architectural and biasing techniques, which are discussed within a system context in Chapter 19.

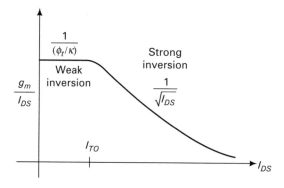

Figure 5.4. g_m/I for a MOSFET in the subthreshold and above-threshold regimes.

The bulk transconductance, g_{mb}, is derived from Equation (5.18) to be

$$\frac{\partial i_{DS}}{\partial v_{BS}} = \left(\frac{\partial i_{DS}}{\partial v_{GS}}\right)\left(\frac{-\partial V_{TS}}{\partial v_{BS}}\right)$$

$$= g_m\left(\frac{\gamma}{2\sqrt{\phi_0 + V_{SB}}}\right) \tag{5.22}$$

$$g_{mb} = g_m\left(\frac{1 - \kappa_S}{\kappa_S}\right)$$

The latter relationship in Equation (5.22) is also true in subthreshold as is seen from Equation (5.17) and true whether the transistor is saturated or not. We can get some insight into why this simple relationship holds in subthreshold and above threshold by examining Figure 3.4. In subthreshold, the mobile charge is exponential in the surface potential. A bulk potential change affects the surface potential via C_{dep} while a gate potential change affects the surface potential via C_{ox} such that the mobile charge scales proportionally with C_{ox} or C_{dep} if the gate or bulk potential change respectively. In above-threshold, the diode-like element(s) clamp(s) the surface potential such that the change in mobile charge is once again due to charge change across the C_{ox} capacitor when the gate potential changes and due to charge change across the C_{dep} capacitor when the bulk potential changes.[2] Thus, in both subthreshold and above-threshold the charge change due to varying v_G or v_B scales with C_{ox} or C_{dep} respectively. The current dependence on the charge change is linear in the subthreshold regime and quadratic in the above-threshold regime but this dependence is irrelevant to the proportionality between g_m and g_{mb} since it is common to both parameters. Thus $g_{mb}/g_m = C_{dep}/C_{ox}$ always, making the latter relationship of Equation (5.22) true in both subthreshold and above-threshold operation. The relationship is true whether the transistor is linear or saturated because the current change for a given charge change at the source or drain is identical and independent of whether that charge change is caused by v_B changes or v_G changes scaling g_m and g_{mb} identically.

The small-signal parameter g_{ds} is evaluated from Equation (5.18) to be

$$g_{ds} = \frac{\partial i_{DS}}{\partial v_{DS}} = \mu C_{ox}\frac{W}{L}(V_{GS} - V_{TS}) = \frac{g_m^{sat}}{\kappa_S}, \quad V_{DS} = 0$$

$$= \mu C_{ox}\frac{W}{L}\left[(V_{GS} - V_{TS}) - \frac{V_{DS}}{\kappa_S}\right], \quad V_{DS} \leq V_{DSAT} \tag{5.23}$$

$$= \mu C_{ox}\frac{W}{L}[(V_{GS} - V_{TS}) - (V_{GS} - V_{TS})] = 0, \quad V_{DS} > V_{DSAT}$$

However, the Early effect is also seen in strong inversion as illustrated in Figure 5.5 such that a more accurate model for the MOS transistor in strong inversion is given by

---

[2] When the charge across C_{dep} changes due to a varying v_{BS} and a fixed surface potential, the mobile charge, which is the difference between the gate charge and the bulk charge, changes one for one with the bulk charge since the gate charge is constant as the gate potential and surface potential are constant.

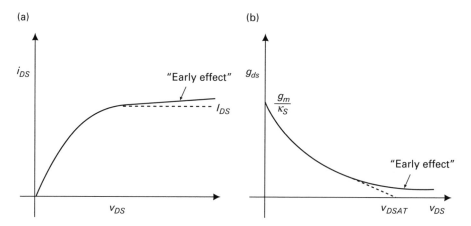

Figure 5.5a, b. Early voltage curves in strong inversion.

$$i_{DS} = \mu C_{ox} \frac{W}{L_{eff}} \left[(v_{GS} - V_{TS})v_{DS} - \frac{v_{DS}^2}{2\kappa_S} \right] \left(1 + \frac{v_{DS}}{V_A(L)} \right) \tag{5.24}$$

similar to that in Equation (5.16) in weak inversion. The Early voltage is typically higher in strong inversion than in weak inversion as previously discussed. In the square-bracketed term in Equation (5.24) we replace v_{DS} with $v_{DSAT} = \kappa_S(v_{GS} - V_{TS})$ for all $v_{DS} \geq v_{DSAT}$.

To summarize, the small-signal equivalent circuit in strong inversion is identical to that shown in Figure 5.3 for weak inversion except that the small-signal parameters are given by

$$
\begin{aligned}
g_m &= \mu C_{ox} \frac{W}{L} V_{DS}, \quad V_{DS} \leq V_{DSAT} \\
&= \mu C_{ox} \frac{W}{L} V_{DSAT}, \quad V_{DS} > V_{DSAT} \\
g_{mb} &= \left(\frac{1 - \kappa_S}{\kappa_S} \right) g_m \\
g_{ds} &= \mu C_{ox} \frac{W}{L} \left[(V_{GS} - V_{TS}) - \frac{V_{DS}}{\kappa_S} \right], \quad V_{DS} \leq V_{DSAT} \\
&= \frac{I_{DSAT}}{V_A}, \quad V_{DS} > V_{DSAT}
\end{aligned}
\tag{5.25}
$$

The small-signal drain resistance $r_0 = 1/g_{ds}$ is frequently used in small-signal notation especially in the saturation regime.

Thus far, we have kept the source potential constant in all our source-referenced models. The small-signal parameter g_{ss} describes the change in current when the source voltage changes and is given by

$$\frac{\partial i_{DS}}{\partial v_S} = g_{ss} = g_m + g_{mb} + g_{ds} \tag{5.26}$$

In saturation, g_{ds} is approximately zero such that

$$g_{ss}^{sat} = g_m^{sat} + g_{mb}^{sat}$$
$$= g_m^{sat} + g_m^{sat}\left(\frac{1 - \kappa_S}{\kappa_S}\right) \tag{5.27}$$
$$= \frac{g_m^{sat}}{\kappa_S}$$

Equation (5.27) is true in both strong inversion and weak inversion in the saturation regime because the proportionality between g_{mb} and g_m is maintained in both regimes of operation.

5.3 Small-signal capacitance models in strong inversion

In an electrical device, it is not just currents that are a function of its terminal voltages, but internal charges in the device as well. In the MOS transistor, the gate, dopant (bulk), and inversion charges are a function of its terminal voltages. Just as $\partial i/\partial v$ gives rise to small-signal conductances or transconductances in a device, $\partial q/\partial v$ gives rise to small-signal capacitances or transcapacitance currents in the device. We shall now evaluate these small-signal capacitances in the MOS transistor. We begin by assuming that the quasistatic approximation holds: at frequencies of operation that are not too fast, the internal charges in the device are instantaneous functions of the terminal voltages. In practice, the inversion charge Q_I, which must ultimately be supplied by the drain/source terminals, will take time to equilibrate because of finite drift or diffusion transport times. We shall assume that such equilibration times are extremely fast with respect to the time scales over which the quasistatic approximation holds. *We shall begin with capacitance models in strong inversion, which are then generalized to weak inversion.*

If Q_G^{TOT} and Q_B^{TOT} are the total gate and bulk charge in the transistor, obtained by integrating the charge per unit area over the entire area of the transistor, then the five small-signal capacitances in the transistor are defined as

$$C_{gs} = \frac{-\partial Q_G^{TOT}}{\partial v_S}; \; C_{gd} = \frac{-\partial Q_G^{TOT}}{\partial v_D}; \; C_{gb} = \frac{-\partial Q_G^{TOT}}{\partial v_B}$$
$$C_{bs} = \frac{-\partial Q_B^{TOT}}{\partial v_S}; \; C_{bd} = \frac{-\partial Q_B^{TOT}}{\partial v_D} \tag{5.28}$$

Note that these capacitances are actual capacitances, not capacitances per unit area. As we have discussed previously in Chapter 3, the sum of the inversion charge and bulk charge, both negative, must equal the positive gate charge in each infinitesimal area of the channel due to the necessity for charge balance. Thus,

$$Q_I(x) + Q_B(x) + Q_G(x) = 0$$
$$\left(Q_I^{TOT} + Q_B^{TOT}\right) + Q_G^{TOT} = 0 \tag{5.29}$$

Hence, from Equation (5.29), the inversion charge can always be derived from the gate and bulk charges.

Since the gate charge Q_G^{TOT} is positive, the bulk charge Q_B^{TOT} is negative, and the gate charge reduces when v_S, v_D, or v_B increase, while the magnitude of the bulk charge increases when v_S or v_D increase, all defined capacitances are always positive. The intuitive model of Figure 3.4 is useful in understanding the nature of the charge changes. With a fixed v_G and v_B, an increase in v_S or v_D changes the surface potential at the source or drain end of the channel in a one-for-one fashion via the diode-clamp mechanism in strong inversion. These boundary-condition changes result in increases in surface potential throughout the channel as well via the drift generators of Figure 3.4. Thus, the gate charge all along the channel decreases due to falling voltage differences across the C_{ox} capacitances, while the magnitude of the bulk charge all along the channel increases due to the increasing voltage differences across the C_{dep} capacitances making C_{gs}, C_{gd}, C_{bs}, and C_{bd} all positive. When the drain end is in saturation, the drain diode-like element is off, there is no effect of v_D on the charge or surface potential in the channel, hence no effect on the gate or bulk charges, and C_{gd} and C_{bd} are zero in this case. The capacitance C_{gb} is due to the capacitive divider formed by C_{ox} and C_{dep} and is positive. However, the presence of inversion charge partially screens the bulk from the gate by maintaining a bulk-independent surface potential at the diode-clamped boundaries of the channel but not throughout the interior of the channel. Thus, the gate charge changes weakly with changes in the bulk potential, making C_{gb} small. The capacitance C_{gb} is maximum when the diode-like element at the drain end is off and there is no screening at the drain end, i.e., when the transistor is saturated. The capacitance C_{gb} is 0 when $V_{DS} = 0$, and the surface potential is independent of the bulk and only determined by the source or drain-diode-clamped voltage everywhere in the channel.

To derive the small-signal capacitances in the transistor, we use the following relations listed below:

$$Q_I(x) = -\sqrt{Q_{I0}^2\left(1 - \frac{x}{L}\right) + Q_{IL}^2\frac{x}{L}}, \quad Q_I^{TOT} = W\int_0^L Q_I(x)dx$$

$$Q_{I0} = -C_{ox}(v_{GS} - V_{TS}), \quad Q_{IL} = -C_{ox}\left(v_{GS} - V_{TS} - \frac{v_{DS}}{\kappa_S}\right)$$

$$\psi_S(x) = (\phi_0 + V_{SB}) + \kappa_S\left[(v_{GS} - V_{TS}) + \frac{Q_I(x)}{C_{ox}}\right]$$

$$Q_G(x) = C_{ox}(V_{GB} - V_{FB} - \psi_S(x)), \quad Q_G^{TOT} = W\int_0^L Q_G(x)dx \tag{5.30}$$

$$Q_B(x) = -C_{ox}\left[\gamma\sqrt{\phi_0 + V_{SB}} + (1 - \kappa_S)\left(v_{GS} - V_{TS} + \frac{Q_I(x)}{C_{ox}}\right)\right]$$

$$Q_B^{TOT} = W\int_0^L Q_B(x)dx$$

$$V_{TS} = V_{T0} + \gamma\left(\sqrt{\phi_0 + V_{SB}} - \sqrt{\phi_0}\right)$$

$$\frac{dV_{TS}}{dV_{SB}} = \left(\frac{1 - \kappa_S}{\kappa_S}\right)$$

These relations yield the inversion charge, surface potential, gate charge, and bulk charge along the channel with Q_{I0} representing the inversion charge at the source end and Q_{IL} representing the inversion charge at the drain end. The various charges per unit area, assumed uniform over the width of the transistor, are integrated over its width, and then integrated along its length to yield the total charge. The relations are only valid for $v_{DS} \leq v_{DSAT} = \kappa_S(v_{GS} - V_{TS})$, after which the various functions are held constant at the values assumed at v_{DSAT}. All the charge relations are obtained by using known values for the charge at the boundary ends of the channel and a linear depletion capacitance approximation (the κ_S approximation).

The charge relation for $Q_I(x)$ on the first line of Equation (5.30) is derived by a simple current-constancy constraint. It states that the current in the transistor, being constant throughout the channel, can be computed by integrating the drift transport equation from the source to any point, x, along the channel instead of integrating from the source to the drain as we formerly did, i.e,

$$\frac{\frac{\kappa\mu W}{2C_{ox}}\left(Q_{I0}^2 - Q_I^2(x)\right)/x}{\frac{\kappa\mu W}{2C_{ox}}\left(Q_{I0}^2 - Q_{IL}^2\right)/L} = \frac{I_{DS}}{I_{DS}} = 1, \quad 0 < x < L \tag{5.31}$$

In effect we are equating currents for a transistor of length x and of length L to obtain

$$\frac{Q_{I0}^2 - Q_I^2(x)}{Q_{I0}^2 - Q_{IL}^2} = \frac{x}{L}$$

$$Q_I^2(x) = Q_{I0}^2\left(1 - \frac{x}{L}\right) + Q_{IL}^2\left(\frac{x}{L}\right) \tag{5.32}$$

$$Q_I(x) = -\sqrt{Q_{I0}^2\left(1 - \frac{x}{L}\right) + Q_{IL}^2\frac{x}{L}}$$

Hence, the transistor's inversion charge at x is a nonlinear weighted average of the charge at the source end and drain end of the channel with the weights decreasing with increasing distance from the respective boundary end. If the transistor is saturated, $Q_{IL} = 0$, such that the inversion charge has a square-root shape along the channel given by

$$Q_I(x) = -Q_{I0}\sqrt{1 - \frac{x}{L}} \tag{5.33}$$

Figure 5.6 plots the inversion charge in a saturated transistor given by Equation (5.33). The infinite slope at the drain end where $x = L$ corresponds to the strong-inversion need to have the product of the derivative of $Q_I(x)$, which is proportional to the electric field at x, and $Q_I(x)$ be a constant throughout the channel to maintain a constant drift current. Since the charge at the drain end is assumed zero in saturation, the slope of $Q_I(x)$ becomes infinity. In practice, the charge at the drain end is nonzero and corresponds to that seen in a weakly inverted transistor. Thus, the infinite slope is actually finite and inversely proportional to the weakly inverted nonzero charge at the drain end. Chapter 6 will discuss such issues in more depth, but we shall continue to use simple approximations in this chapter.

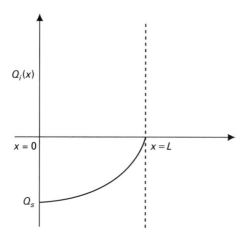

Figure 5.6. Square-root charge density profile within the channel of a strongly inverted MOSFET.

We note from Figure 3.4 that looking upwards from the channel we see C_{ox} towards the gate, and looking downwards from the channel we see C_{dep} towards the bulk. The inversion charge is expressed across $C_{ox} + C_{dep}$, the gate charge is expressed across C_{ox} and the bulk charge is expressed across C_{dep} with the surface potential representing the common terminal of the capacitive divider. Therefore, for a given change in surface potential at any point in the channel, the inversion charge, gate charge and bulk charge changes are in the ratio $(C_{ox} + C_{dep}) : C_{ox} : C_{dep}$ respectively. Since this relationship is true at all points in the channel, we may integrate over the entire channel to conclude that

$$C_{gs} = \kappa_S \frac{\partial Q_I^{TOT}}{\partial v_S}, \quad C_{gd} = \kappa_S \frac{\partial Q_I^{TOT}}{\partial v_D}$$

$$C_{bs} = (1 - \kappa_S) \frac{\partial Q_I^{TOT}}{\partial v_S}, \quad C_{bd} = (1 - \kappa_S) \frac{\partial Q_I^{TOT}}{\partial v_D} \tag{5.34}$$

Hence, we may immediately derive that

$$C_{bs} = \left(\frac{1 - \kappa_S}{\kappa_S}\right) C_{gs}, \quad C_{bd} = \left(\frac{1 - \kappa_S}{\kappa_S}\right) C_{gd} \tag{5.35}$$

a relationship that can also be formally derived from Equations (5.28) and (5.30) with some algebra.

From Figure 3.4, we note that the change in gate charge, ΔQ_G, for a given change in the bulk voltage, Δv_B, is given by

$$\Delta Q_G(x) = -C_{ox}\left(\frac{C_{dep}}{C_{dep} + C_{ox}} + \frac{1}{C_{ox} + C_{dep}} \frac{\Delta Q_I(x)}{\Delta v_B}\right) \Delta v_B \tag{5.36}$$

such that by integrating along the channel we obtain

$$C_{gb} = WLC_{ox}(1 - \kappa_S) + \kappa_S \frac{\partial Q_I^{TOT}}{\partial v_B} \tag{5.37}$$

The relations of Equations (5.34) and (5.37) along with Equation (5.30) can be used to derive capacitance results for each V_{DS} operating point in strong inversion. The derivations are simple mathematical exercises in differentiation and integration. Since all capacitances in Equations (5.34) and (5.37) are expressed in terms of Q_I^{TOT}, it is extremely useful to evaluate it explicitly from simple integration of $Q_I(x)$ in Equation (5.32) to obtain

$$Q_I^{TOT} = W \int_0^L Q_I(x)dx$$
$$Q_I^{TOT} = \frac{2}{3}WL\left[\frac{Q_{I0}^2 + Q_{IL}^2 + Q_{I0}Q_{IL}}{(Q_{I0} + Q_{IL})}\right] \qquad (5.38)$$
$$Q_I^{TOT} = \frac{2}{3}WL\left[(Q_{I0} + Q_{IL}) - \frac{Q_{I0}Q_{IL}}{(Q_{I0} + Q_{IL})}\right]$$

If we define

$$\eta = 1 - \frac{V_{DS}}{V_{DSAT}} = \frac{Q_{IL}}{Q_{I0}} \qquad (5.39)$$

such that η goes from 1 to 0 as V_{DS} goes from 0 to V_{DSAT}, then it can be shown that

$$C_{gs} = \frac{2}{3}WLC_{ox}\left(\frac{1 + 2\eta}{(1 + \eta)^2}\right), \quad C_{gd} = \frac{2}{3}WLC_{ox}\left(\frac{\eta^2 + 2\eta}{(1 + \eta)^2}\right)$$
$$C_{bs} = \left(\frac{1 - \kappa_S}{\kappa_S}\right)C_{gs}, \quad C_{bd} = \left(\frac{1 - \kappa_S}{\kappa_S}\right)C_{gd} \qquad (5.40)$$
$$C_{gb} = WLC_{ox}\left(\frac{1 - \kappa_S}{3}\right)\left(\frac{1 - \eta}{1 + \eta}\right)^2 = (1 - \kappa_S)\left(C_{ox} - \left(C_{gs} + C_{gd}\right)\right)$$

with

$$i_{DS} = i_{DSAT}\left(1 - \eta^2\right) \qquad (5.41)$$

Figure 5.7 plots Equation (5.40) and reveals how the various small-signal capacitances in the MOS transistor change as V_{DS} changes. The most significant point to note is that C_{gs} changes from 1/2 to 2/3 of WLC_{ox} while C_{gd} changes from 1/2 to 0 of WLC_{ox} as the transistor moves from the linear to saturation regime. The next significant point to note is that C_{bs} and C_{bd} are proportional to C_{gs} and C_{gd} with the same proportionality factor of $g_{mb}/g_m = (1 - \kappa_S)/\kappa_S$. Finally, C_{gb} rises from 0 to a small but finite value as we transition from the linear to saturation regime in the transistor. Figure 5.8 (a) shows the 'capacitance diamond', the five capacitances incorporated into a small-signal equivalent circuit. Figure 5.8 (b) shows the capacitances, conductance, and transconductances of the transistor all in one complete small-signal model. Often, v_B is a fixed voltage. Then, C_{gb}, C_{bs}, and C_{bd} all form grounded capacitances, and C_{gs} and C_{gd} are analogous to C_π and C_μ in bipolar transistors, respectively.

It is interesting to derive the general results of Equation (5.40) for certain special cases that provide insight, easily obscured by algebra. If the transistor is saturated,

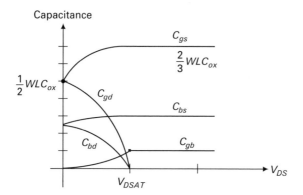

Figure 5.7. Dependence of various intrinsic capacitances on V_{DS} in an above-threshold MOSFET.

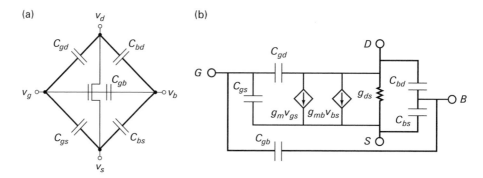

Figure 5.8a, b. The MOSFET "capacitance diamond" in (a) and complete small-signal model in (b).

$Q_{IL} = 0$, and the shape of the inversion charge distribution along the channel is a square-root function as revealed by Equation (5.33) and Figure 5.6. If we increase the source terminal voltage by an infinitesimal amount Δv_S, the diode-clamp mechanism raises the surface potential at the source end of the channel by Δv_S, which lowers the gate charge by $C_{ox}\Delta v_S$, and lowers the magnitude of the inversion charge by $(C_{ox} + C_{dep})\Delta v_S$ at this end. This lowering of charge at the source end lowers the total charge in the channel. The *total* charge change along the channel is then given by

$$-\frac{\Delta Q_G^{TOT}}{\Delta v_s} = \kappa_S \frac{\Delta Q_I^{TOT}}{\Delta v_s}$$

$$= \kappa_S \left(\frac{C_{ox}}{\kappa_S} W \int_0^L \sqrt{1 - \frac{x}{L}} dx \right)$$

$$= C_{ox} WL \int_0^1 \sqrt{z} dz$$

$$C_{gs} = \frac{2}{3} WL C_{ox}$$

(5.42)

The capacitance C_{gd} is zero in the saturation regime since $Q_{IL} = 0$, such that changes in v_D do not affect $Q_I(x)$ or any other charges in the transistor. In practice, slight changes in the length of the transistor with changes in v_D (Early effect changes in the length) do reduce the total inversion charge such that C_{gd} is non zero.

If the transistor is in its linear regime with $V_{DS} = 0$, the square-root distribution for $Q_I(x)$ changes to a flat distribution with $Q_{I0} = Q_{IL}$. An infinitesimally small change in charge at the source end or at the drain end caused by a Δv_S or Δv_D respectively will result in the flat distribution $Q_I(x)$ acquiring a small triangular increase in area due to the change in Q_{I0} with Q_{IL} fixed or vice versa. Since the area of a triangle is $(1/2)$ base $\times$ height, only half of the charge increase at the source end or drain end gets integrated into the total charge increase in the channel. Thus, we can show by a similar derivation as in Equation (5.42) that, when $V_{DS} = 0$,

$$C_{gs} = C_{gd} = \frac{1}{2}WLC_{ox} \qquad (5.43)$$

The capacitances C_{bs} and C_{bd} remain proportional to C_{gs} and C_{gd} with $(1 - \kappa_S)/\kappa_S$ as the constant of proportionality as C_{gs} and C_{gd} vary with V_{DS}.

The capacitance C_{gb} may be understood intuitively as follows. An increase in v_B results in no boundary-condition change in the surface potential at the diode-clamped source end of the channel and thus no change in the gate charge at this end. It does however result in a decrease in the magnitude of the bulk charge at the source end due to the falling voltage difference across C_{dep}. There is then a concomitant increase in the magnitude of the inversion charge at the source end to preserve charge balance. The increase in inversion charge at the source end results in an increase in inversion charge throughout the channel due to the drift generators in the channel. At the drain end of the channel, an increase in v_B increases the surface potential via a capacitive-divider mechanism if and only if the drain end is in saturation since the diode-like element at that end is then off. In general, v_B increases will attempt to increase the surface potential via a capacitive-divider mechanism throughout the channel. However, the increase in inversion charge due to the v_B increase at the source (and/or drain end of the channel if it is not in saturation) reduces the surface potential, thus attenuating the capacitive-divider increase via a negative-feedback loop. The loop effectively operates via the diode clamps at the boundary ends of the channel and propagates via the drift generators into the interior of the channel. The net effect of the v_B increase is then to leave the surface potential and gate charge unaffected at the source end of the channel and increase the surface potential in the interior of the channel slightly with stronger increases towards the drain end of the channel and a maximal increase at the drain end after it enters saturation. The total gate charge is reduced due to falling voltages across the C_{ox} capacitances when v_B is increased, thus making C_{gb} positive. The capacitance C_{gb} is zero if and only if $V_{DS} = 0$, because, in this case, the increase in inversion charge at the boundary ends of the channel does not alter the flat-and-constant surface potential in the channel, keeping the voltage differences across all C_{ox} capacitances fixed. In this case, the charge at the gate is effectively perfectly screened by the flat-and-constant surface potential from the bulk. In general, however, $C_{gb} \neq 0$ in strong inversion except when $V_{DS} = 0$, as is sometimes erroneously stated.

The above intuition is captured in Equation (5.36) and Equation (5.37) and may be used to derive C_{gb} from these equations:

$$
\begin{aligned}
C_{gb} &= WLC_{ox}(1 - \kappa_S) + \kappa_S \frac{\partial Q_I^{TOT}}{\partial v_B} \\
&= WLC_{ox}(1 - \kappa_S) + \kappa_S \left(\frac{\partial Q_I^{TOT}}{\partial Q_{I0}} \frac{\partial Q_{I0}}{\partial V_{TS}} \frac{\partial V_{TS}}{\partial v_B} + \frac{\partial Q_I^{TOT}}{\partial Q_{IL}} \frac{\partial Q_{IL}}{\partial V_{TS}} \frac{\partial V_{TS}}{\partial v_B} \right) \\
&= WLC_{ox}(1 - \kappa_S) + \kappa_S \left(\frac{\partial Q_I^{TOT}}{\partial Q_{I0}} C_{ox} \frac{-(1 - \kappa_S)}{\kappa_S} + \frac{\partial Q_I^{TOT}}{\partial Q_{IL}} C_{ox} \frac{-(1 - \kappa_S)}{\kappa_S} \right) \\
&= (1 - \kappa_S) \left[WLC_{ox} - \left(\frac{\partial Q_I^{TOT}}{\partial Q_{I0}} C_{ox} + \frac{\partial Q_I^{TOT}}{\partial Q_{IL}} C_{ox} \right) \right] \\
&= (1 - \kappa_S) \left[WLC_{ox} - \left(\kappa_S \frac{\partial Q_I^{TOT}}{\partial Q_{I0}} \frac{C_{ox}}{\kappa_S} + \kappa_S \frac{\partial Q_I^{TOT}}{\partial Q_{IL}} \frac{C_{ox}}{\kappa_S} \right) \right] \\
&= (1 - \kappa_S) \left[WLC_{ox} - \left(\kappa_S \frac{\partial Q_I^{TOT}}{\partial Q_{I0}} \frac{\partial Q_{I0}}{\partial v_S} + \kappa_S \frac{\partial Q_I^{TOT}}{\partial Q_{IL}} \frac{\partial Q_{I0}}{\partial v_D} \right) \right] \qquad (5.44) \\
&= (1 - \kappa_S) \left[WLC_{ox} - \left(\kappa_S \frac{\partial Q_I^{TOT}}{\partial v_S} + \kappa_S \frac{\partial Q_I^{TOT}}{\partial v_D} \right) \right] \\
&= (1 - \kappa_S) \left[WLC_{ox} - \left(\frac{\partial Q_G^{TOT}}{\partial v_S} + \frac{\partial Q_G^{TOT}}{\partial v_D} \right) \right] \\
&= (1 - \kappa_S) \left[WLC_{ox} - (C_{gs} + C_{gd}) \right] \\
&= (1 - \kappa_S) WLC_{ox} \left[1 - \frac{2}{3} \left(\frac{1 + 2\eta}{(1 + \eta)^2} \right) - \frac{2}{3} \left(\frac{\eta^2 + 2\eta}{(1 + \eta)^2} \right) \right] \\
C_{gb} &= (1 - \kappa_S) \frac{WLC_{ox}}{3} \left[\frac{(1 - \eta)^2}{(1 + \eta)^2} \right]
\end{aligned}
$$

5.4   Extrinsic or parasitic capacitances

The capacitances that we have discussed so far all involve charges that are fundamental to the operation of the MOS transistor and are termed *intrinsic capacitances*. In addition to these capacitances, the fabrication of the MOS transistor results in parasitic capacitances termed *extrinsic capacitances* that are not a direct consequence of how the transistor operates but that are significant in practice.

Figures 5.9 (a), (b), and (c) show how the extrinsic capacitances arise. The extrinsic capacitances C_{gse} and C_{gde} are due to the overlap of the gate with source or drain diffusion regions outside the channel respectively as shown in Figure 5.9 (a). During modern fabrication processes, the source and drain diffusion regions

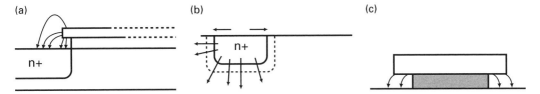

Figure 5.9a, b, c. Extrinsic capacitances: (a) C_{gse} and C_{gde}; (b) C_{bse} and C_{bde}; (c) C_{gbe}.

inevitably creep under the gate a little such that the finite overlap of the gate with the diffusion regions results in some field lines from the gate ending on source or drain diffusions, thus adding to C_{gs} and C_{gd} respectively. Since this overlap occurs all along the width of the transistor they can be parametrized as

$$C_{gse} = W(\alpha)C_{gso}$$
$$C_{gde} = W(\alpha)C_{gdo} \tag{5.45}$$

where α is a constant parameter reflecting the overlap length and C_{gso} and C_{gdo} represent capacitances per unit area. Often simulators simply multiply the width of the transistor by an empirical parameter that represents αC_{gso} or αC_{gdo} as a single parameter, the overlap capacitance per unit length at the source or drain, respectively.

The extrinsic capacitances C_{bse} and C_{bde} arise due to field lines from the source and drain diffusion regions terminating in charges in the transistor bulk outside the channel as shown in Figure 5.9 (b). They are depletion capacitances seen in reverse-biased pn junctions. They may be parametrized as

$$C_{bse} = AS.C_1 + (W + 2L_D)C_2$$
$$C_{bde} = AD.C_1 + (W + 2L_D)C_2 \tag{5.46}$$

where AS and AD represent source and drain diffusion areas, C_1 represents the capacitance per unit area from the bottom of the source/drain diffusion region pointing vertically into the bulk, L_D represents the length of the diffusion regions, $W + 2L_D$ represents the perimeter of the side wall, the three side surfaces of the source/drain diffusion region pointing horizontally towards the bulk and not facing the channel, and C_2 represents the capacitance per unit length of the sidewall regions, a parameter which implicitly factors in the depth of the sidewall. To minimize C_{bse} and C_{bde}, we must try to minimize the dimensions of the source and drain diffusion regions. Note that the sidewall capacitance does *not* include the capacitance of the diffusion sidewall facing the channel because the bulk charge facing the diffusion sidewall mostly changes in response to changes in the source or drain voltage in a manner already modeled by the intrinsic capacitances C_{bs} and C_{bd}.

The extrinsic capacitance C_{gbe} shown in Figure 5.9 (c) arises because field lines from the gate terminate in the bulk outside the transistor's channel. The poly-silicon gate has a width that is larger than the width of the transistor's channel since its width is required to exceed the diffusion-region width for correct fabrication and operation of the transistor. The value of C_{gbe} is then primarily determined by the area capacitance of the overhanging gate region, whose insulator SiO_2

transitions gradually in a 'bird's-beak-like' fashion from a thick-oxide overhanging region to a thin-oxide channel region.

The actual capacitances of the transistor are composed of a portion due to intrinsic capacitances and a portion due to extrinsic capacitances. If all intrinsic portions are subscripted with an i and all extrinsic portions are subscripted with an e, we get

$$
\begin{aligned}
C_{gs} &= C_{gsi} + C_{gse} \\
C_{ds} &= C_{dsi} + C_{dse} \\
C_{bs} &= C_{bsi} + C_{bse} \\
C_{bd} &= C_{bdi} + C_{bde} \\
C_{gb} &= C_{gbi} + C_{gbe}
\end{aligned}
\tag{5.47}
$$

In addition to these capacitances, a well-to-substrate depletion capacitance, C_{bbe}, is present in the transistor. This capacitance is important in circuits where the well voltage changes, e.g., when the well is used as an input. The extrinsic capacitances C_{bse}, C_{bde}, C_{gbe}, and C_{bbe} are nonlinear depletion capacitances whose values change with voltage, typically like an inverse-square-root function in abrupt junctions.

5.5   Small-signal capacitance models in weak inversion

In weak inversion, to first order, the bulk charge and gate charge are equal in magnitude and the inversion charge is zero. The source and drain terminal voltages have no effect on the surface potential and thus no effect on the gate charge or bulk charge. Thus, C_{gsi}, C_{gdi}, C_{bsi}, and C_{bdi} are zero. The only intrinsic capacitance that is significant in weak inversion is

$$
\begin{aligned}
C_{gbi} &= \frac{C_{ox}C_{dep}}{C_{ox} + C_{dep}} \\
&= \frac{C_{ox}C_{ox}\left(\dfrac{1 - \kappa_S}{\kappa_S}\right)}{C_{ox} + C_{ox}\left(\dfrac{1 - \kappa_S}{\kappa_S}\right)} \\
C_{gbi} &= C_{ox}(1 - \kappa_S)
\end{aligned}
\tag{5.48}
$$

The extrinsic capacitances in weak inversion are all identical to those in strong inversion if the voltage dependencies of these capacitances are taken into account. Thus, the capacitances in weak inversion are given by

$$
\begin{aligned}
C_{gs} &= C_{gse}, & C_{gd} &= C_{gde} \\
C_{bs} &= C_{bse}, & C_{bd} &= C_{bde} \\
C_{gb} &= C_{ox}(1 - \kappa_S) + C_{gbe}
\end{aligned}
\tag{5.49}
$$

The overall small-signal circuit in weak inversion is identical to that depicted in Figure 5.8 for strong inversion except that the values of the elements in the circuit are those corresponding to weak inversion.

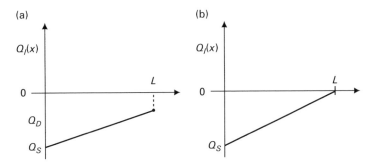

Figure 5.10a, b. Linear charge density profiles within the channel of a subthreshold MOSFET.

It is also interesting to compute the distribution of inversion charge across the channel in weak inversion. The diffusion current in weak inversion is given by

$$i_{DS} = \mu \phi_t W \frac{\partial Q_I(x)}{\partial x} \tag{5.50}$$

Thus, to ensure current continuity all along the channel, for all x, we must have $\frac{\partial Q_I(x)}{\partial x}$ be a constant. Thus, $Q_I(x)$ is a straight line with a constant slope proportional to the current and with boundary values of Q_{I0} and Q_{IL}. Hence

$$
\begin{aligned}
Q_I(x) &= Q_{I0} + x\left(\frac{Q_{IL} - Q_{I0}}{L}\right) \\
&= Q_{I0}\left(1 - \frac{x}{L}\right) + Q_{IL}\frac{x}{L}
\end{aligned}
\tag{5.51}
$$

Figure 5.10 plots the weak-inversion $Q_I(x)$ when the transistor is operating in its linear regime and when it is operating in saturation.

5.6   The transit time

The quasi-static approximation assumes that charges in the transistor change instantaneously when the terminal voltages are changed. However, due to finite diffusion and drift transport times, the inversion charge in the transistor cannot change instantaneously. To estimate the limits of the quasi-static approximation, we compute the *transit time*, the average time taken by an electron to travel the length of the channel. The capacitance results that we have derived are then valid for all frequencies of operation that are significantly less than the reciprocal of the transit time. The transit time is an important measure of the maximum speed of operation of the transistor.

Since the current I_{DS} is given by the total amount of charge arriving within a transit time, the transit time τ can be estimated as

$$\tau = \frac{|Q_I^{TOT}|}{I_{DS}} \tag{5.52}$$

In saturated weak-inversion operation, Q_I^{TOT} is $(Q_{I0}WL)/2$ and I_{DS} is $\mu\phi_t Q_{I0}W/L$ such that

$$
\begin{aligned}
\tau_{sub} &= \frac{Q_{I0}WL}{2\mu\phi_t Q_{I0}\left(\frac{W}{L}\right)} \\
&= \frac{L^2}{2\mu\phi_t} = \frac{L^2}{2D}
\end{aligned}
\tag{5.53}
$$

Not surprisingly, the transit time is just the diffusion time of an electron.

In saturated strong-inversion operation, we may similarly compute the transit time τ_{above} to be

$$
\begin{aligned}
\tau_{above} &= \frac{\frac{2}{3}Q_{I0}WL}{\frac{\mu\kappa_S}{2C_{ox}}\frac{W}{L}Q_{I0}^2} \\
&= \frac{4C_{ox}L^2}{3\kappa_S\mu Q_{I0}} \\
&= \frac{4C_{ox}L^2}{3\kappa_S\mu C_{ox}(V_{GS}-V_{TS})} \\
&= \frac{4L^2}{3\kappa_S\mu(V_{GS}-V_{TS})}
\end{aligned}
\tag{5.54}
$$

Not surprisingly, the transit time is the distance traveled, L, divided by the average drift velocity, v_{drift}, given by

$$
\begin{aligned}
v_{drift} &= \mu\frac{(3/4)\kappa_S(V_{GS}-V_{TS})}{L} \\
&= \mu\frac{(3/4)V_{DSAT}}{L}
\end{aligned}
\tag{5.55}
$$

In practice, due to velocity-saturation effects described in Chapter 6, v_{drift} reaches a maximum value v_{max} when V_{GS} is sufficiently large. The transit time in strong inversion operation then reaches a minimum given by

$$
\tau_{min} = \frac{L}{v_{max}}
\tag{5.56}
$$

Figure 5.11 plots the transit time versus V_{GS} in the transistor as we move from weak-inversion to strong-inversion to the velocity-saturated region of operation. It is worth noting that as L gets ever smaller in modern fabrication processes, the diffusion transit time, τ_{sub}, which scales like L^2, gets nearer and nearer to the minimum drift transit time, τ_{min}, which only scales like L. Thus, subthreshold operation becomes faster and faster relative to strong-inversion operation. In fact, it is interesting to compute the length L_c at which the two become equal, at least as per the predictions of Equations (5.53) and (5.56):

$$
\begin{aligned}
\frac{L_c^2}{2\mu\phi_t} &= \frac{L_c}{v_{max}} \\
L_c &= \frac{2\mu\phi_t}{v_{max}}
\end{aligned}
\tag{5.57}
$$

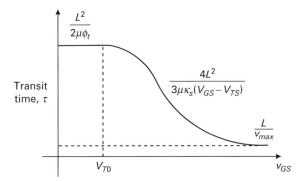

Figure 5.11. Transit time for a MOSFET in strong and weak inversion.

For v_{max} given by 10^5 m/s and $\mu = 500\,\text{cm}^2/\text{Vs}$ we find that L_c is 25 nm. The thermal velocity of electrons is given approximately by $\sqrt{kT/m_e}$, where m_e is the mass of the electron, and is about 0.67×10^5 m/s at 300 K. This discussion is highly oversimplified but provides an intuitive way of understanding why diffusive transport and subthreshold operation are getting increasingly faster. Chapter 6 will discuss, in a more rigorous fashion, interesting regimes of transistor operation in very-short-channel transistors where the charge density is relatively large but the mechanisms of transport are well described by thermal and scattering parameters in the transistor rather than by drift.

5.7   The 'beta' of an MOS transistor

Figure 5.12 (a) shows the small-signal equivalent of a source-and-bulk-grounded MOS transistor. The base (b) is analogous to the gate, the emitter (e) is analogous to the source, and the collector (c) is analogous to the drain. The input current $i_{gs}(s)$ through C_{gs} in Figure 5.12 (a) generates a current through the MOS g_m generator of $i_{gmm} = [g_m/(C_{gs}s)]i_{gs}(s)$ if we assume that C_{gb} may be neglected. The capacitances C_{gb} and C_{bd} are tied to the bulk terminal and effectively appear as grounded capacitances to the gate and drain respectively. Figure 5.12 (b) shows the small-signal equivalent of a bipolar transistor (BJT). The input current $i_b(s)$ through the r_π and $1/(C_\pi s)$ impedances generates a current through the bipolar g_m generator of $i_{gmb} = [g_m r_\pi/(r_\pi C s + 1)]i_b(s)$. If we define $\beta_m(s) = i_{gmm}(s)/i_{gs}(s)$ to be the beta of an MOS transistor, analogous to $\beta_b(s) = i_{gmb}(s)/i_b(s)$, then we get

$$
\begin{aligned}
\beta_m(s) &= \frac{g_m}{C_{gs}s} \\
\beta_b(s) &= \frac{g_m r_\pi}{r_\pi C_\pi s + 1} = \frac{\beta_0}{r_\pi C_\pi s + 1}
\end{aligned}
\tag{5.58}
$$

Figure 5.13 plots the 'beta' $\beta_m(s)$ of an MOS transistor and contrasts it with the beta $\beta_b(s)$ of a bipolar transistor with $C_\pi = C_{gs}$ and the g_m at the same value

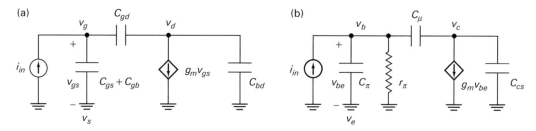

Figure 5.12a, b. The small-signal model of (a) an MOS transistor, and (b) a bipolar transistor.

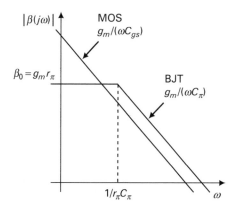

Figure 5.13. The current gain of an MOS and a bipolar transistor.

(we have ignored C_{gb} for simplicity). We see that the bipolar beta has a finite value of β_0 at dc while the MOS beta is infinite at dc. For most frequencies, however, they both behave like integrating functions $g_m/(C_{gs}s)$ or $g_m/(C_\pi s)$, respectively. The use of $\beta_m(s)$ is frequently useful in making analogies between bipolar and MOS circuits, and also in computing impedance transformations in circuits, often used in bipolar circuits. For example, an impedance from the source terminal to ground is multiplied by $(\beta_m(s)+1)$ when seen at the gate; an impedance from the gate terminal to ground is divided by $(\beta_m(s)+1)$ when seen at the source. Chapter 10 will compute such driving-point impedances from a feedback perspective.

The bipolar transistor is of course not symmetric in collector and emitter terminals, unlike a MOSFET with symmetric drain and source terminals, and $\kappa = 1$ in the bipolar transistor. Nevertheless, the Ebers-Moll models of a bipolar transistor may be viewed as two asymmetric MOS subthreshold transistors, each in saturation, and arranged in parallel; one transistor creates a forward current from emitter to collector and the other transistor creates a reverse current from collector to emitter with the net current from emitter to collector being the difference of the two, and the base current, defined by a forward beta and reverse beta respectively, being the sum of the base currents in each direction. Apart from the finite beta effects, this operation is very similar to the forward and reverse currents defined for MOS operation in Chapter 4. The forward and reverse currents are significant when the collector-to-emitter voltage is small, and only the emitter-to-collector current is significant when the bipolar transistor is in its forward active region, a region corresponding to the saturation regime of MOS transistor operation. Unfortunately, due to a poor choice of historical terminology, the saturation regime of bipolar operation corresponds to the linear regime of MOS transistor operation.

The short-circuit current gain of an MOS transistor $i_{ds}(s)/i_g(s)$, computed with the drain-to-source voltage, bulk voltage, and source voltage at ac ground, can be shown from Figure 5.12 (a) to be

$$\beta(s) = \frac{g_m}{(C_{gs} + C_{gd} + C_{gb})s} \tag{5.59}$$

The frequency where this gain is 1 is called the f_T of the transistor and is a measure of the highest frequency of operation where the transistor still has current gain with respect to its gate input. It is given by

$$f_T = \frac{g_m}{2\pi(C_{gs} + C_{gd} + C_{gb})} \tag{5.60}$$

By analogy, we may define an f_{TB} with respect to the bulk input to be

$$f_{TB} = \frac{g_{mb}}{2\pi(C_{bs} + C_{bd} + C_{gb})} \tag{5.61}$$

If we ignore the value of C_{gb}, usually small, then $f_T \approx f_{TB}$ since $g_{mb}/g_m = C_{bs}/C_{gs} = C_{bd}/C_{gd}$.

The angular frequency $\omega_T = 2\pi f_T$ may be shown to be $2/\tau_{above}$ in above-threshold saturated operation. In saturated subthreshold operation, $\omega_T = 2/\tau_{sub}$, when $V_{GS} = V_{TS}^w$, i.e, at the limit of weak-inversion operation where exponential operation is just maintained. For V_{GS} below V_{TS}^w, the f_T in subthreshold operation falls exponentially with V_{GS}. These results may be derived by substituting for g_m and the capacitances in Equation (5.60) with C_{gb} approximated as the only nonzero capacitance in weak inversion, C_{gs} approximated as the only nonzero capacitance in strong inversion, and relating the results to Equations (5.53) and (5.54), respectively.

$$f_{Tsub} = \frac{1}{\pi \tau_{sub}} e^{\kappa_s (V_{GS} - V_{TS}^w)/\phi_t}, \quad V_{GS} \le V_{TS}^w$$
$$f_{Tabove} = \frac{1}{\pi \tau_{above}}, \quad V_{GS} \ge V_{TS}^s \tag{5.62}$$

In Equation (5.62), we have explicitly reminded the reader that the threshold in strong inversion, V_{TS}^s, is greater than that in weak inversion, V_{TS}^w, as we discussed in Chapter 4.

We now have an in-depth background in device physics for long-channel transistors. Since subthreshold behavior in short-channel and in long-channel transistors is quite similar, this background will be of great utility in low-power design in all technologies. In the next chapter, we will finish our tour of MOS transistor device physics by studying the behavior of deep submicron transistors with very short channel lengths.

Reference

[1] Yannis Tsividis. *Operation and Modeling of the MOS Transistor*, 2nd ed. (Boston: WCB/ McGraw-Hill, 1999).

6 Deep submicron effects in MOS transistors

It [Moore's law] can't continue forever. The nature of exponentials is that you push them out and eventually disaster happens.

Gordon Moore, 2005

In this chapter, we discuss how to model effects that arise in transistors with short channel lengths. Such short-channel transistors have correspondingly thin gate oxides, shallow source/drain junctions, high doping, and small operational supply and threshold voltages. The models are accurate for transistor lengths that range from $1\,\mu\text{m}$ to $0.01\,\mu\text{m}$ ($10\,\text{nm}$), where semi-classical approximations of transistor function are still valid [1]. We shall allude to some quantum phenomena as well. We shall begin by first describing the EKV model in a long-channel transistor [2]. The EKV model is an insightful charge-based model that captures subthreshold and above-threshold operation in a simple analytic fashion. It is named after its originators, Enz, Krummenacher, and Vittoz. Then, based on a short-channel model in [2], we shall modify the long-channel EKV model to describe a short-channel effect that is important in above-threshold operation, namely velocity saturation. As the lateral electric field in a transistor increases with increasing drain-to-source voltage, the drift velocity of electrons in short-channel transistors begins to saturate and limit at a maximum velocity v_{sat}; the drift velocity is then no longer related to the lateral electric field via the mobility proportionality constant as it is in long-channel transistors. As a consequence, the saturation current i_{DSAT}, the limiting frequency of operation f_T, the transconductance g_m, and the saturation voltage v_{DSAT} of the short-channel transistor are all lower than that predicted from long-channel transistor equations. We shall show how a simple feedback loop can quantitatively represent the effects of velocity saturation in a compact block diagram. Then we describe a short-channel effect that is significant in subthreshold operation, namely drain induced barrier lowering (DIBL). DIBL causes the transistor's threshold voltage to decrease as its drain-to-source voltage increases; consequently, drain-to-source current increases in a weakly exponential fashion with increasing drain-to-source voltage and the output conductance of the transistor is severely degraded. In addition to these effects, we will describe three vertical-field effects, namely mobility degradation, polysilicon gate depletion, and bandgap widening, which worsen with high vertical fields in the transistor, and are consequently more severe in the above-threshold regime than in the subthreshold regime. In practice, all these vertical-field effects are empirically modeled by simulators such that an intuitive understanding of the physics is more valuable to a circuit designer than an exact mathematical expression with empirically fit parameters.

When transistors are operated at very high frequencies, extrinsic parasitics such as the substrate and terminal resistances and diode-junction terminal capacitances can affect operation. At very high frequencies, finite transit-time effects caused by distributed charge within the transistor can cause its effective small-signal trans-conductance and capacitance values to become a function of frequency rather than being constant. We shall summarize such effects within the context of an EKV model, which is described in more detail in [3]. Since many RF systems operate well below the f_T of their devices, such distributed-charge effects are usually unimportant and may be neglected to a first approximation.

When transistor channel-length dimensions are comparable to a mean free length (10–15 nm typically), an electron shooting out of the source may suffer no collision with any impurity or lattice wave such that it is not scattered. The electron simply makes its way to the drain in a ballistic or scatter-free fashion. The limits of transistor transport when there is no scattering in the channel are referred to as ballistic transport [4], [5]. We shall discuss models for transistor operation that are ballistic and nearly ballistic, i.e., there are very few scattering events in the channel [6].

At low nanometer channel lengths, tunneling from the source to the drain can dominate the current flow. In fact, many advanced carbon nano tube (CNT) and single-electron transistors at nanometer lengths function via bidirectional quantum mechanical tunneling from source to drain and from drain to source [7], [8]. In conventional nanoscale transistors of the future, quantum-mechanical tunneling from the source to the drain will increase subthreshold current but reduce above-threshold current. Source-to-drain tunneling represents an ultimate limit on transistor performance [9].

Quantum-mechanical tunneling effects from the channel to the gate become significant when the gate-oxide thickness is lower than 3 nm. Gate-oxide leakage via quantum-mechanical tunneling is already significant in the transistors of today that do not have high-dielectric-constant insulators and therefore require a thin insulator to function. In fact, limiting gate-oxide tunneling is an important constraint for ensuring transistor scaling into the future [10].

We conclude by discussing the scaling of transistors in the future as their channel length decreases. We discuss four scaling laws that have been found to empirically represent the scaling of transistor bulk doping, supply voltage, oxide thickness, and threshold voltage with channel length [11]. It is expected that semi-classical transistor operation as described in this chapter will represent transistor characteristics well till channel lengths reach 10 nm [6].

6.1   The dimensionless EKV model

As in several fields of physics, the EKV model represents all transistor variables such as charge Q, current I, voltage V, and channel distance from source x in dimensionless forms q, i, v, and ξ by dividing them by normalizing constants Q_n, I_n, V_n, and L_n respectively. The charge-based Maher-Mead transistor model [12], [13] with which the EKV model shares some similarities also uses dimensionless variables. Thus,

$$
\begin{aligned}
v &= \frac{V}{V_n}; \ V_n = \phi_t \\[2mm]
q &= \frac{Q}{Q_n}; \ Q_n = -2C_{ox}\frac{\phi_t}{\kappa} \\[2mm]
i &= \frac{I}{I_n}; \ I_n = 2\mu C_{ox}\frac{W}{L_{eff}}\frac{\phi_t^2}{\kappa} \\[2mm]
\xi &= \frac{x}{L_n}; \ L_n = L_{eff}
\end{aligned}
\tag{6.1}
$$

The normalization charge Q_n and normalization current I_n correspond to values of the mobile charge and drain current at the boundary between weak inversion and strong inversion. Thus, for q or i greater than 1, we enter strong inversion, while for q or i less than 1, we operate in weak inversion. To avoid confusion in this section, we shall adopt the convention of using capital letters for actual variables and small letters for dimensionless variables in the EKV model. Since we will sometimes discuss small-signal parameters with regard to the EKV model, the overlap of this convention with the usual IEEE small-signal-vs.-large-signal convention will be clear from context: two sides of a dimensionless equation must both be dimensionless or the dimensions must agree on both sides in an equation with dimensions.

We define a voltage V_i that is a quasi-Fermi level or electrochemical potential that represents electron concentration at any point in the channel according to

$$
Q_i = N_A e^{(\psi_s - (\phi_0 + V_i))/\phi_t},
\tag{6.2}
$$

where N_A is the acceptor concentration and $\phi_0 = 2\phi_F$. Thus, V_i varies from V_S to V_D as we move from the source to the drain. An important advantage of using V_i rather than the surface potential is that the usual drift-and-diffusion transport equation can then be expressed in a particularly simple form as derived in Equation (6.3) below; a second advantage is that V_i *is* the source or drain boundary voltage when one is at the source or drain such that, unlike ψ_S, it is directly related to terminal voltages. From the usual drift and diffusion equations of Chapter 3 and 4, and Equation (6.2), we can then derive that

$$
\begin{aligned}
I_{DS} &= \mu W\left(-Q_i\frac{d\psi_S}{dx} + \phi_t\frac{dQ_i}{dx}\right) \\[2mm]
\frac{dQ_i}{dx} &= \frac{Q_i}{\phi_t}\left(\frac{d\psi_s}{dx} - \frac{dV_i}{dx}\right) \\[2mm]
I_{DS} &= \mu W(-Q_i)\frac{dV_i}{dx}
\end{aligned}
\tag{6.3}
$$

$$
\boxed{i_{DS} = q_i\frac{dv_i}{d\xi}}
$$

The inversion charge Q_i given by the charge-sheet approximation $-C_{ox}(V_G - V_{FB} - \psi_S - \gamma\sqrt{(\psi_S)})$ may be further approximated with the usual κ or linear-depletion-capacitance Taylor-series expansion for the square-root term as discussed in Chapter 3. Thus, the inversion charge and surface potential are approximately linearly related to each other all along the channel. However, as Figure 3.11 and the extensive discussion in Chapter 3 illustrate, the inversion charge must also be consistent with the square-root-and-exponential relation given by Poisson-Boltzmann electrostatics in the channel, which implies that the inversion charge and surface potential are actually related to each other in a more complex nonlinear fashion. This dilemma is resolved by using the linear κ approximation within the nonlinear Poisson-Boltzmann equation to solve for a self-consistent solution [14]. The solution, obtained after some simple but tedious algebra, leads to Equation (6.4), the key equation of the EKV model [2]:

$$\boxed{2q_i + \ln q_i = v_p - v_i = \kappa(\nu_G - \nu_{T0}) - v_i} \tag{6.4}$$

In Equation (6.4) v_i varies from V_S/ϕ_t to V_D/ϕ_t as one moves from source to drain, and q_i varies from Q_S/Q_n to Q_D/Q_n as one moves from source to drain. Note that $\kappa(\nu_G - \nu_{T0})$ is the dimensionless pinchoff voltage v_p of strong inversion. Equation (6.4) represents the self-consistent electrostatics of the MOSFET all along the channel in weak, moderate, and strong inversion. The $2q_i$ term dominates in strong inversion, the $\ln q_i$ term dominates in weak inversion, and both terms are important in moderate inversion. From Equations (6.3) and (6.4), we can derive that

$$i_{DS}\int_0^1 d\xi = \int_{v_S}^{v_D} q_i(2 + \frac{1}{q_i})(-dq_i)$$

$$i_{DS} = (q_S^2 - q_D^2) + (q_S - q_D) \tag{6.5}$$

The square terms dominate when q_S and q_D are greater than 1, i.e., in strong inversion, and the linear terms dominate when q_S and q_D are less than 1, i.e., in weak inversion. Note that Q_n in Equation (6.1) is a measure of the threshold charge, right at the border of strong inversion and weak inversion, such that $Q_S = Q_n q_S$, $Q_D = Q_n q_D$. Thus, if q_S or q_D exceed 1, the charge at the respective terminal boundary is in moderate or strong inversion. If only the q_i term or the $\ln(q_i)$ term dominate in Equation (6.4), it is easily inverted to solve for q_S and q_D as an explicit function of the terminal voltages, which when substituted in Equation (6.5), leads to the usual body-referenced dimensionless forms of the weak-inversion or strong-inversion current equations of Chapter 3:

$$i_{DS}^{weak} = e^{\kappa(\nu_G - \nu_{T0})}(e^{-v_S} - e^{-v_D})$$

$$i_{DS}^{strong} = \frac{\kappa^2}{4}\left[\left(\nu_G - \nu_{T0} - \frac{v_S}{\kappa}\right)^2 - \left(\nu_G - \nu_{T0} - \frac{v_D}{\kappa}\right)^2\right] \tag{6.6}$$

We have used $\kappa = 1/n$ as a parameter in our formulation of the EKV model, but the published EKV model uses n.

In moderate inversion, Equation (6.4) cannot be inverted for an exact closed-form solution for q_S or q_D as a function of the source or drain terminal voltages. However, by setting $q_D = 0$ and $i_{DS} = i_{DSAT}$ in Equation (6.5), differentiating Equations (6.4) and (6.5) using the chain rule and solving a quadratic equation for q_s as a function of i_{DSAT}, we can get a closed-form solution for the dimensionless saturation transconductance g_{msat} that is valid in all regions of inversion:

$$
\begin{aligned}
g_{msat} &= \frac{\partial i_{DSAT}}{\partial q_S} \frac{\partial q_S}{\partial v_G} \\
&= \kappa q_S \\
&= \kappa \frac{\sqrt{1 + 4 i_{DSAT}} - 1}{2}.
\end{aligned}
\tag{6.7}
$$

Not surprisingly, $g_{msat} = \kappa \sqrt{(i_{DSAT})}$ when i_{DSAT} is significantly above 1 $(i_{DSAT} \gg I_n)$, i.e., in strong inversion, and $g_{msat} = \kappa i_{DSAT}$ when i_{DSAT} is significantly below 1 $(i_{DSAT} \ll I_n)$, i.e., in weak inversion. Similarly, we can show in all regions of inversion that

$$
\begin{aligned}
g_m &= \kappa(q_S - q_D) \\
&= \kappa(g_S - g_D).
\end{aligned}
\tag{6.8}
$$

The normalization constant for all conductances and transconductances is I_n/V_n.

6.2 Velocity saturation

As the lateral electric field E in a transistor increases and the mean drift velocity v of electrons in the channel increases, a negative-feedback loop begins to limit this velocity: the faster the mean velocity, the shorter is the mean time between scattering collisions, and the more likely is the electron to lose energy and momentum to the lattice. As the mean electron velocity nears the electron thermal velocity, the loop gain of this feedback loop becomes large enough such that the mean electron velocity may be assumed to be nearly invariant at a saturation value v_{sat} (see Chapter 2 for a review of feedback). One model of velocity saturation shown in Figure 6.1 captures this phenomenon according to the following equations

$$
\begin{aligned}
v &= v_{sat} \frac{\dfrac{E}{E_c}}{1 + \dfrac{E}{2E_c}}; \quad E \leq 2E_c \\
v &= v_{sat}; \quad E > 2E_c \\
v_{sat} &= \mu_0 E_c,
\end{aligned}
\tag{6.9}
$$

with $\mu_0 = 500 \, \text{cm}^2/\text{Vs}$ being the low-field intrinsic electron mobility, $E_c \approx 2 \times 10^6 \, \text{V/m}$, and $v_{sat} = 10^5 \, \text{m/s}$ [2]. At values of $E \ll E_c$, Equation (6.9) reduces to the $v = \mu_0 E$ relation assumed in the previous chapters. Holes have nearly the same saturation velocity as electrons but a value of E_c that is nearly

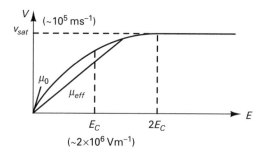

Figure 6.1. The basic phenomenon of velocity saturation.

2.5× as large, consistent with their lower intrinsic mobility. As the slope of the chord in Figure 6.1 shows, the effective electron mobility μ_{eff} is given by

$$\mu_{eff} = \frac{\mu_0}{1 + \frac{E}{2E_c}}; \quad E < 2E \qquad (6.10)$$

Since velocity saturation is an electric-field-based effect on drift transport, it is important only in above-threshold operation: the lateral electric field between source and drain is small in subthreshold operation. Therefore, we shall only discuss its impact on modifying transistor properties in above-threshold operation. We shall begin by analyzing the consequences of Equations (6.9) and (6.10) when $E \leq 2E_c$ throughout the channel and then discuss saturation effects that occur when $E > 2E_c$ at some point in the channel. When $E \leq 2E_c$, we may write

$$I_{DS} = W(-Q_i)\mu_{eff}\frac{d\psi_S}{dx}$$

$$I_{DS} = W(-Q_i)\mu_0 \frac{\frac{d\psi_S}{dx}}{1 + \frac{1}{2E_c}\frac{d\psi_S}{dx}}$$

$$I_{DS} = W(q_i\mu_0 Q_n)\frac{\phi_t}{L_{eff}} \frac{-\frac{dv_i}{d\xi}}{1 + \frac{1}{2E_c}\frac{\phi_t}{L_{eff}}\frac{dv_i}{d\xi}} \qquad (6.11)$$

$$I_{DS} = -I_n q_i \frac{-\frac{dv_i}{d\xi}}{1 + \frac{1}{2E_c}\frac{\phi_t}{L_{eff}}\frac{dv_i}{d\xi}}$$

$$i_{DS} = q_i \frac{\frac{dv_i}{d\xi}}{1 + \frac{1}{2E_c}\frac{\phi_t}{L_{eff}}\frac{dv_i}{d\xi}}$$

If we differentiate Equation (6.4) while ignoring the ln( ) term, the dimensionless form of the final line of Equation (6.11) can be written as

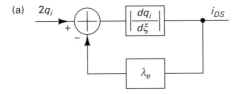

(a)

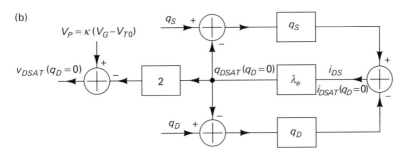

(b)

Figure 6.2a, b. (a) The feedback loop of velocity saturation present at every point in the channel. (b) The feedback loop of velocity saturation with q_S and q_D boundary charges.

$$i_{DS} = q_i \frac{-2\dfrac{dq_i}{d\xi}}{1 - \dfrac{2}{2E_c L_{eff}} \phi_t \dfrac{dq_i}{d\xi}}$$

$$i_{DS} = 2q_i \frac{\left\|\dfrac{dq_i}{d\xi}\right\|}{1 + \lambda_e \left\|\dfrac{dq_i}{d\xi}\right\|} \tag{6.12}$$

$$\lambda_e = \frac{\phi_t}{E_c L_{eff}}$$

Note that $dq_i/d\xi$ is negative since the unsigned dimensionless charge q_i decreases as ξ changes from 0 to 1, which is why we have used the absolute value in the second line of Equation (6.12). The parameter λ_e is a dimensionless electric field and is a measure of the strength of the short-channel effect. When L_{eff} is large and λ_e approaches 0, the current i_{ds} approaches its above-threshold value. Figure 6.2 (a) shows a feedback-loop interpretation of Equation (6.12). As λ_e increases, the feedback gain in Figure 6.2 (a) increases, attenuating the current from the above-threshold value given by just the feed-forward path to that given by Equation (6.12). Since i_{DS} is constant throughout the channel, the top line of Equation (6.12) is easily integrated as ξ ranges from 0 to 1 as we move from source to drain. We then get

$$i_{DS} = \frac{q_S^2 - q_D^2}{1 + \lambda_e(q_S - q_D)} \tag{6.13}$$

We notice that the above-threshold quadratic terms of Equation (6.5) have been modified by short-channel effects with a negative-feedback loop gain of

$\lambda_e(q_S - q_D)$ such that the current is attenuated by a factor of $1/(1 + \lambda_e(q_S - q_D))$ in Equation (6.13). Equation (6.13) is conveniently visualized in the negative-feedback block diagram of Figure 6.2(b). Since Equation (6.13) is merely the integrated form of Equation (6.12), we notice that the common feedback path in Figure 6.2 (b) has a gain proportional to λ_e just as it does in Figure 6.2 (a), and that there are forward (q_S) and reverse (q_D) paths that get subtracted from each other to yield the net current i_{DS}. Equation (6.13) is only valid for v_{DS} less than a certain saturation value, v_{DSAT}, when $E \leq 2E_c$ at all points along the channel. We now discuss the saturation limits of Equation (6.13) that arise when $E > 2E_c$.

The current at any point in the channel, whether in velocity saturation or not, is given by

$$I_{DS} = W(-Q_i)V^i_{drft} \tag{6.14}$$

where V^i_{drft} is the mean drift velocity at that point in the channel and Q_i is the areal charge concentration at that point in the channel. From Equations (6.1) and (6.12), if we define a dimensionless velocity, $v^i_{drft} = V^i_{drft}/v_{sat}$, Equation (6.14) transforms to a dimensionless equivalent given by

$$i_{DS} = (W \frac{Q_n}{I_n} v_{sat}) q_i v^i_{drft}$$

$$= \frac{v_{sat}}{(E_c \mu_0) \dfrac{\phi_t}{L_{eff} E_c}} q_i v^i_{drft} \tag{6.15}$$

$$\boxed{i_{DS} = \frac{q_i v^i_{drft}}{\lambda_e}}$$

Since $V^i_{drft} \leq v_{sat}$, v^i_{drft} has a maximum value of 1 in short-channel operation and cannot go to ∞ as we previously assumed in long-channel above-threshold operation. Thus, the maximal value of the current at any point in a short channel is given by

$$i^{max}_{DS} = \frac{q_i}{\lambda_e} \tag{6.16}$$

The product of the charge and velocity maintains a constant current all along the channel in long-channel or in short-channel operation. At the source, the charge is high and the electric field and velocity are low, while at the drain, the charge is low, and the electric field and velocity are high, but the product of the charge and velocity is the same all along the channel. In long-channel or short-channel operation, the charge q_i is at its lowest value and the electric field and velocity v^i_{drft} are at their highest value at the drain end of the channel. However, unlike in long-channel above-threshold operation, in short-channel above-threshold operation, since v_{drft} is limited to a maximally high value of 1, q_D is limited to a minimally low value of q_{DSAT} and cannot go to 0. Thus, when v_{DS} is sufficiently large such that the electric field at the drain end of the channel just reaches $2E_c$ and the velocity at the drain end of the channel correspondingly just reaches v_{sat}, from Equation (6.16) we get

$$i_{DSAT} = \frac{q_{DSAT}}{\lambda_e} \tag{6.17}$$

In Equation (6.13), by setting $q_D = q_{DSAT}$ and $i_{DSAT} = q_{DSAT}/\lambda_e$, we can solve for q_{DSAT} to find that

$$q_{DSAT} = q_S \left(\frac{\lambda_e q_S}{1 + \lambda_e q_S} \right) \qquad (6.18)$$

Correspondingly, from Equation (6.17) i_{DSAT} is then given by

$$i_{DSAT} = \frac{q_S^2}{1 + \lambda_e q_S} \qquad (6.19)$$

If we assume strong above-threshold operation and ignore the $\ln q$ term in Equation (6.4), the normalized saturation voltage v_{DSAT} is lower than the pinchoff voltage $v_p = \kappa(v_G - v_{T0})$ and given by

$$v_{DSAT} = \kappa(v_G - v_{T0}) - 2q_{DSAT} \qquad (6.20)$$

Not surprisingly, saturation occurs at a smaller saturation voltage in short-channel operation than in long-channel operation since it occurs when $q_D = q_{DSAT}$ rather than when $q_D = 0$.

In the feedback block diagram of Figure 6.2 (b), when saturation occurs, $q_D = q_{DSAT}$ and $i_D = i_{DSAT}$. Since $\lambda_e i_{DSAT}$ is q_{DSAT}, the positive input to the bottom-most adder block of Figure 6.2 (b) is q_{DSAT}. Since the negative input to the bottom-most adder block is also q_{DSAT}, the entire reverse (q_D) pathway is zeroed in Figure 6.2 (b) since it receives no net differential output from the bottom-most adder. Zeroing of the q_D pathway can also be more simply accomplished by setting $q_D = 0$ throughout the block diagram. Indeed, setting $q_D = 0$ in Equation (6.13) yields the same answer for i_{DSAT} in Equation (6.19) as we previously obtained. Furthermore, setting $q_D = 0$ makes the positive input to the bottom-most adder of Figure 6.2 (b) easily computable by Black's feedback formula for a classic negative-feedback loop (see Chapter 2). We find that Black's feedback formula yields the same value of q_{DSAT} as that predicted by Equation (6.18). Hence the block diagram of Figure 6.2 (b) is a very succinct way of yielding *all* the dimensionless equations of short-channel and long-channel above-threshold operation (λ_e is set to 0 in long-channel operation). In fact, we can derive the saturation voltage v_{DSAT} from the q_{DSAT} output as indicated in the block diagram as well. Thus, we assume that after q_D falls below q_{DSAT} (or equivalently, v_{DS} exceeds v_{DSAT}) the bottom pathway of the block diagram is zeroed out and the current is constant at a saturation value of i_{DSAT} set by q_S.

What really happens when the voltage v_{DS} is larger than v_{DSAT}? In practice, the current increases slightly via an Early-effect-like channel-length modulation (CLM) effect rather than staying constant as we have assumed. The value of $q_D = q_{DSAT}$ is established at a point in the channel that is located at a distance ΔL_{eff} from the drain with ΔL_{eff} dependent on v_{DS}. In the channel region from $L_{eff} - \Delta L_{eff}$ to L_{eff}, the voltage v_i (as defined by Equation (6.2)) changes from v_{DSAT} to v_{DS} creating a high electric field that is determined by gate *and* drain 2D electrostatics and that increases strongly from $2E_c$ at $L_{eff} - \Delta L_{eff}$ to an even higher value at L_{eff}. The velocity remains constant at v_{sat} throughout this region. The charge is approximated as being constant at q_{DSAT} throughout the region as well. However, the value of q_{DSAT} is slightly higher than that predicted by Equation (6.18) because λ_e, as defined by Equation (6.12), increases when $L_{eff} \rightarrow L_{eff} - \Delta L_{eff}$. Similarly, the value of the saturation current is slightly higher because q_{DSAT} is slightly higher. Note that Equation (6.17) might suggest that the saturation current could increase or decrease since both q_{DSAT} and λ_e increase with CLM. However, the actual un-normalized current, $I_{DSAT} = i_{DSAT}I_n$; the normalization constant I_n has an L_{eff} dependence computed in Equation (6.1) that cancels out the L_{eff} dependence in λ_e in i_{DSAT}; thus, the increase in q_{DSAT} increases the actual current I_{DSAT}.

What is ΔL_{eff}? The 2D electrostatics calculation in [15] and [2] predicts that, if we define a characteristic length l determined by such electrostatics to be

$$l = \sqrt{\frac{\varepsilon_{si}}{\varepsilon_{ox}} x_j t_{ox}}, \tag{6.21}$$

where the ε parameters refer to dielectric constants for silicon and the gate oxide, x_j is the junction depth at the drain, and t_{ox} is the oxide thickness, and we define a dimensionless electric field

$$u = \frac{V_{DS} - V_{DSAT}}{2E_c l}, \tag{6.22}$$

then ΔL_{eff} is given by

$$\Delta L_{eff} = l \ln(u + \sqrt{1 + u^2}) \tag{6.23}$$

The logarithmic dependence predicted by Equation (6.23) implies that the CLM effect is relatively gentle with changes in V_{DS}. The derivation of Equations (6.21), (6.22), and (6.23) is conceptually simple and involves the use of two-dimensional Gaussian pillboxes and some algebra [2].

We can derive the saturation transconductance g_{msat} in a manner similar to that used in Equation (6.7), with Equations (6.19) and (6.4) used to compute the partial derivatives (the ln term in Equation (6.4) is ignored since we are in above-threshold operation). We find that

$$\begin{aligned} g_{msat} &= \kappa \frac{q_S + \frac{\lambda_e q_S^2}{2}}{(1 + \lambda_e q_S)^2} \\ &\approx \frac{\kappa}{2\lambda_e}; \text{ at large } q_S \end{aligned} \tag{6.24}$$

Thus, due to velocity saturation, if the current is sufficiently large, g_{msat} becomes constant and independent of current level in contrast with the long-channel expression of Equation (6.7). The derivative procedure used to derive Equation (6.7) is identical such that Equation (6.8) is still valid in velocity saturation.

While dimensionless equations represent an elegant theoretical way to derive transistor behavior, to make them experimentally useful, normalization constants must be reincorporated into them. The reincorporation of these constants can sometimes shed insight that is obscured by dimensionless addition of charges to voltages to get currents, which is not possible in equations with dimensions. The re-incorporation of the constants of Equation (6.1) together with expressions for Q_S and Q_D derived in the previous chapters yields the following set of body-referenced and source-referenced equations. The body-referenced equations correspond to our discussion thus far and the source-referenced equations are derived by straightforward extensions of them as in Chapters 3 and 4.

Body-referenced equations:

$$I_{DS} = \frac{\kappa \mu_0 C_{ox}}{2} \frac{W}{L_{eff}} \frac{(V_G - V_{T0} - \frac{V_S}{\kappa})^2 - (V_G - V_{T0} - \frac{V_D}{\kappa})^2}{1 + \frac{V_{DS}}{2E_c L_{eff}}}$$

$$I_{DSAT} = \frac{\kappa \mu_0 C_{ox}}{2} \frac{W}{L_{eff}} \frac{(V_G - V_{T0} - \frac{V_S}{\kappa})^2}{1 + \frac{\kappa(V_G - V_{T0}) - V_S}{2E_c L_{eff}}}$$

$$\boxed{\begin{array}{l} I_{DSAT} = W C_{ox}\left(V_G - V_{T0} - \frac{V_S}{\kappa}\right) v_{sat}; E_c L_{eff} \to 0 \\[2mm] g_{mSAT} = W C_{ox} v_{sat}; \qquad\qquad E_c L_{eff} \to 0 \end{array}}$$

(6.25)

$$Q_{DSAT} = -C_{ox}\left(V_G - V_{T0} - \frac{V_S}{\kappa}\right) \frac{\frac{\kappa(V_G - V_{T0}) - V_S}{2E_c L_{eff}}}{1 + \frac{\kappa(V_G - V_{T0}) - V_S}{2E_c L_{eff}}}$$

$$V_{DSAT} - V_S = \frac{\kappa(V_G - V_{T0}) - V_S}{1 + \frac{\kappa(V_G - V_{T0}) - V_S}{2E_c L_{eff}}}$$

Source-referenced equations:

$$I_{DS} = \mu_0 C_{ox} \frac{W}{L_{eff}} \frac{(V_{GS} - V_{TS} - \frac{V_{DS}}{2\kappa_S}) V_{DS}}{1 + \frac{V_{DS}}{2E_c L_{eff}}}$$

$$I_{DSAT} = \frac{\kappa_S \mu_0 C_{ox}}{2} \frac{W}{L_{eff}} \frac{(V_{GS} - V_{TS})^2}{1 + \frac{\kappa_S(V_{GS} - V_{TS})}{2E_c L_{eff}}}$$

$$\boxed{\begin{array}{l} I_{DSAT} = W C_{ox}(V_{GS} - V_{TS}) v_{sat}; E_c L_{eff} \to 0 \\[2mm] g_{mSAT} = W C_{ox} v_{sat}; \qquad\qquad E_c L_{eff} \to 0 \end{array}}$$

(6.26)

$$Q_{DSAT} = -C_{ox}(V_{GS} - V_{TS}) \frac{\frac{\kappa_S(V_{GS} - V_{TS})}{2E_c L_{eff}}}{1 + \frac{\kappa_S(V_{GS} - V_{TS})}{2E_c L_{eff}}}$$

$$V_{DSAT} = \frac{\kappa_S(V_{GS} - V_{TS})}{1 + \frac{\kappa_S(V_{GS} - V_{TS})}{2E_c L_{eff}}}$$

We note from either the body-referenced or the source-referenced equations that we can define a voltage ratio, $(\kappa(V_G - V_{T0}) - V_S)/(2E_cL_{eff})$ or $\kappa_S(V_{GS} - V_{TS})/(2E_cL_{eff})$ in saturation, that is analogous to the loop gain $\lambda_e q_S$ in Figure 6.2 (b) in saturation. Similarly, the voltage ratio $(V_{DS}/2E_cL_{eff})$ is analogous to the loop gain $\lambda_e(q_S - q_D)$ prior to saturation in Figure 6.2 (b). Thus, as in Equations (6.13), (6.18), (6.19), and (6.20), this loop gain serves to attenuate current, attenuate saturation current, make the source and drain saturation charge more nearly equal, and attenuate saturation voltage in Equation (6.25) or Equation (6.26) respectively. Two limits are especially of note, and we discuss them with respect to the source-referenced Equation (6.26): when $E_cL \to \infty$, all equations reduce to their long-channel forms; when $E_cL \to 0$, saturation occurs at an extremely small V_{DSAT} of $2E_cL_{eff}$, the current saturates at a length-independent value of $C_{ox}W(V_{GS} - V_{TS})v_{sat}$, the transconductance becomes invariant with current or gate voltage, and the source and drain charge become almost equal. *In the latter limit, the current is determined by the uniform channel charge $C_{ox}W(V_{GS} - V_{TS})$ flowing at a velocity of v_{sat}.*

High electric fields can accelerate electrons to drift velocities that are comparable to their thermal velocities. These relatively high-energy electrons only lose a small average amount of energy, $\sim kT$, per scattering collision to the lattice. Electrons at and near the drain can get hot and effectively be at a higher average temperature than that of the lattice. Models of carrier heating that correspond to the velocity-saturation model of Equation (6.9) posit that the effective carrier temperature T_C and lattice temperature T_L are related by

$$\begin{aligned}
\frac{T_C}{T_L} &= \left(\frac{\mu_0}{\mu_{eff}}\right)^2 \\
&= \left(1 + \frac{|E|}{2E_c}\right)^2; \, |E| \leq 2E_c \\
&= \left(\frac{|E|}{E_c}\right)^2; \, |E| > 2E_c
\end{aligned} \tag{6.27}$$

Thus, the effective electron temperature varies along the channel, and the higher temperature leads to higher thermal noise. Thermal noise is discussed in Chapter 7. Hot electrons are intentionally exploited in floating-gate circuits to inject high-energy carriers that surmount the silicon–silicon-dioxide energy barrier, enter the gate oxide, and then lower the threshold voltage of the transistor [16].

6.3   Drain induced barrier lowering (DIBL)

Figure 6.3 (a) shows the subthreshold potential in a long-channel transistor from $x = 0$, the edge of the source, to $x = L$, the edge of the drain. The potential profile changes from $\phi_B + V_S$ at $x = 0$ to ψ_{sa} in the channel center to $\phi_B + V_D$ at $x = L$. The potential ϕ_B is the built-in potential of the source-to-bulk or drain-to-bulk junctions given by

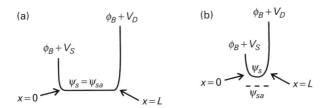

Figure 6.3a, b. (a) Potential profile in a long-channel subthreshold transistor. (b) Potential profile in a short-channel transistor that illustrates drain-induced barrier lowering (DIBL).

$$\phi_B = \phi_t \ln \frac{N_A N_D}{n_i^2} \tag{6.28}$$

From our discussion in Chapter 3, the potential ψ_{sa} in subthreshold is well approximated by

$$\psi_{sa} = \phi_0 + \kappa(V_G - V_{T0}). \tag{6.29}$$

From the solution of Poisson's equation, taking the 2D electrostatics of the gate and source/drain terminals into account [2], we may derive that the transition from the potential at the boundaries of the channel to ψ_{sa} in the channel center is well approximated by an exponential settling profile with a characteristic space-constant length given by

$$l_{DBL} = \sqrt{\frac{\varepsilon_{si}}{\varepsilon_{ox}} t_{ox} t_{dep}}, \tag{6.30}$$

where t_{dep} is the depth of the depletion region of C_{dep}. Thus,

$$\begin{aligned}\psi_S &= \psi_{sa} + (V_S + \phi_B - \psi_{sa})e^{-x/l_{DBL}} + (V_D + \phi_B - \psi_{sa})e^{-(L-x)/l_{DBL}} \\ &= \psi_{sa} + \Delta\psi_s(x).\end{aligned} \tag{6.31}$$

In a long-channel transistor, $L \gg l_{DBL}$ such that it is a good approximation to assume that the surface potential $\psi_S = \psi_{sa}$ everywhere from $x=0$ to $x=L$. In a short-channel transistor, L is comparable to l_{DBL} such that the source and drain terminal voltages can influence the surface potential in the channel through the exponential settling tails of Equation (6.31), thus elevating it above ψ_{sa} by an amount $\Delta\psi_s$ as shown in Figure 6.3 (b). The minimum elevation occurs near the middle of the channel and is given by

$$\Delta\psi_{S,min} = 2e^{-L/2l_{DBL}}\sqrt{(V_S + \phi_B - \psi_{sa})(V_D + \phi_B - \psi_{sa})} \tag{6.32}$$

The DIBL or drain induced barrier lowering effect is so called because the increase in surface potential described by Equation (6.32) causes increasing drain-to-source voltage to lower the energy barrier from source to channel and effectively serves to increase the charge concentration in the channel. Since ψ_{sa} is always less than $\phi_0 + V_S$ or $\phi_0 + V_D$ in subthreshold and $\phi_0 < \phi_B$, both terms in the square-root expression of Equation (6.32) are positive but get smaller as we near moderate inversion. As we near and enter strong inversion, the surface potential begins to be clamped to $\phi_0 + V_S$ near the source and Equation (6.32) is no longer

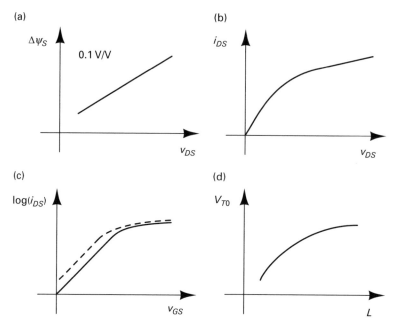

Figure 6.4a, b, c, d. (a) Linear approximation to DIBL. (b) Manifestation of DIBL in transistor $i_{DS}-v_D$ characteristic. (c) DIBL effect on saturation current at a fixed V_{DS}. (d) DIBL effect on transistor threshold.

valid. The DIBL effect can typically be neglected in strong inversion. The decrease in t_{dep} in Equation (6.30) with decreases in ψ_{sa} or V_G makes l_{DBL} decrease at very low values of V_G and weakens DIBL at these very low gate voltages as well.

The change in surface potential given by Equation (6.32) leads to an exponential increase in current proportional to $\exp(\Delta\psi_{S,min}/\phi_t)$. As in the κ approximation, where linear-Taylor-series expansions are used to approximate slowly changing square-root terms, DIBL is often approximated by source-referenced or body-referenced equations of the form

$$
\begin{aligned}
I_{DS} &= I_{0S}e^{\kappa_S(V_{GS}-V_{TS})/\phi_t}e^{\eta V_{DS}/\phi_t}\left(1 - e^{-V_{DS}/\phi_t}\right) \\
I_{DS} &= I_0 e^{\kappa(V_G-V_{T0})/\phi_t}e^{\eta(V_S+V_D)/\phi_t}\left(e^{-V_S/\phi_t} - e^{-V_D/\phi_t}\right),
\end{aligned}
\tag{6.33}
$$

where η is typically less than 0.1 V/V. Figure 6.4 (a) illustrates the linear approximation that characterizes surface-potential change versus drain-to-source voltage. Figure 6.4 (b) illustrates the effect of this approximation on the drain-current characteristic. Equation (6.33) and Figure 6.4 (c) show that at a given V_{DS}, the net effect of DIBL is to shift the subthreshold saturation-current characteristic as though there were a negative threshold-voltage change with V_{DS} given by

$$
\Delta V_{TS}^{DIBL} \approx -\frac{\eta}{\kappa}V_{DS}
\tag{6.34}
$$

Since increasing drain voltage helps the surface potential reach $\phi_0 + V_S$ at a smaller gate voltage, this threshold-voltage shift is manifested in the above-threshold characteristics as well. As channel length decreases, L/l_{DBL} falls such that the surface-potential change predicted by Equation (6.32) increases, which increases the threshold-voltage change due to DIBL. Thus, Figure 6.4 (d) illustrates that the threshold voltage is smaller at smaller channel lengths.

6.4 Vertical-field effects

Velocity saturation and DIBL are lateral-electric-field effects in short-channel transistors. In addition to these effects, we now describe three vertical-field effects, namely mobility degradation, polysilicon gate depletion, and bandgap widening. All of these effects increase with increasing vertical fields in the transistor channel or across its gate oxide such that they are more significant in above-threshold operation rather than in subthreshold operation. Mobility degradation is a reduction in the mobility (and diffusion constant) of electrons in short-channel transistors due to increased scattering with increased vertical field in the channel. Polysilicon gate depletion is a shift in the threshold voltage and/or the subthreshold exponential constant κ (or n) of the transistor due to voltage drops across the nonlinear depletion capacitance formed *within* its polysilicon gate; such drops, which are usually negligible in long-channel transistors, can be significant in short-channel transistors where doping of the gate is no longer orders-of-magnitude greater than the doping of the transistor bulk. Bandgap widening is a quantum-mechanical phenomenon that widens the effective bandgap seen by electrons as their vertical confinement in the channel increases at high vertical fields. Since bandgap widening lowers the intrinsic carrier concentration of electrons, its net effect is to lower the number of electrons present in the channel at a given surface potential, which can be mimicked by lowering the oxide capacitance. The effects of bandgap widening are manifest in a similar fashion to those of polysilicon gate depletion although the causes are very different.

6.4.1 Vertical-field mobility reduction (VMR)

Just as increased lateral fields reduce the mobility of electrons in a transistor, increased vertical fields at its surface reduce mobility as well. The mobility of electrons in the inversion layer of the transistor can be 2–3 times smaller than that in the silicon bulk. The reduction in mobility is described by an equation similar to that of Equation (6.10):

$$\mu_{0v} = \frac{\mu_0}{1 + \dfrac{E_v}{E_{cv}}}. \tag{6.35}$$

The value of $E_{cv} = 4 \times 10^7$ V/m is typically an order-of-magnitude greater than that of E_c in Equation (6.10). The vertical field E_v for electrons in the inversion layer can be approximated as being the average at the top and bottom of the inversion layer. Thus, it varies with the inversion charge Q_i along the channel and is given by

$$E_v = -\frac{(Q_i + Q_{dep}) + Q_{dep}}{2\varepsilon_{si}} \tag{6.36}$$

The dimensionless forms of Equations (6.35) and (6.36) can be combined with a transport equation as in Equation (6.5) to get an expression for the current [2]. The exact analytic answer is rarely of consequence to a designer with a simulator at hand. Therefore, we shall focus on the physical intuition and results. In subthreshold operation, the vertical field is invariant along the channel such that there is just a simple reduction in the mobility (and diffusivity), current, and transconductance without any change in the nature of the transistor I–V curve; due to the increase in $|Q_{dep}|$ with V_G, the reduction is larger as one nears moderate inversion. In above-threshold operation, a negative-feedback loop reduces current by reducing the effective mobility whenever $|Q_i|$ increases such that the current changes more weakly with V_G than a square law, especially for relatively large V_G. Often, a simple reduction in the overall current by a constant factor is a good-enough approximation in subthreshold and above-threshold operation.

VMR reduces the value of μ_0 to μ_{0v} in accord with Equation (6.35). In Equation (6.9), if μ_0 decreases and v_{sat} is constant, the critical field E_c that characterizes the onset of velocity saturation is increased to v_{sat}/μ_{0v}. Thus, an important consequence of VMR is that the onset of velocity saturation is delayed to a larger lateral electric field or to a larger V_{DS} value. Short-channel transistors in deep submicron processes typically have higher vertical fields due to increased doping and reduced oxide thicknesses. Thus, even though short-channel transistors are more prone to the effects of velocity saturation, they are also more prone to being protected from it by simple mobility degradation due to vertical-field effects.

6.4.2   Poly-gate depletion

We have thus far pretended that the polysilicon gate effectively behaves like a metal. In practice, due to its finite doping, depletion and accumulation regions occur within the polysilicon gate to create the charge that we term the gate charge. For example, if the polysilicon gate is an n + gate and the substrate is the p-type silicon of an nMOS transistor, a depletion region composed of positively charged dopant atoms in the polysilicon forms the positive gate charge. This charge exists at the interface of the polysilicon and silicon dioxide. Like C_{dep} in the substrate, this dopant charge then creates a nonlinear depletion capacitance C_{plydep} in series with C_{ox}. The presence of this depletion capacitance causes a small voltage drop across it that reduces the gate drive of the transistor and thus effectively increases its threshold voltage. It also reduces the gate transconductance of the transistor since C_{ox} and κ are now effectively reduced to values given by

$$C_{ox}^{eff} = \frac{C_{ox}C_{plydep}}{C_{ox} + C_{plydep}}$$

$$\kappa^{eff} = \frac{C_{ox}^{eff}}{C_{ox}^{eff} + C_{dep}}.$$

$$(6.37)$$

The reduction of C_{ox} reduces the specific values of Q_n and I_n in Equation (6.1) and thus the values of all currents and source/drain transconductances. The reduction of κ then reduces the gate transconductance even further. While such effects are not significant in long-channel devices, they are becoming increasingly important in deep submicron devices.

6.4.3   Bandgap widening

Electrons in a transistor are confined by an electric potential-energy well. The lowest energy portion of the well is at the silicon/silicon-dioxide border where the potential is highest. The highest energy portion of the well lies deep in the bulk where the potential is at its lowest. The rest of the electric potential well is formed by the voltage drop across the depletion region as we move from the surface to the bulk. Quantum mechanics requires that electrons have wave-like properties. Wave confinement in the vertical dimension results in constraints on the lowest energy that an electron can have.

The net result of these constraints is that the energy minimum of the conduction-band electron energy rises and the energy maximum of the valence-band electron energy falls; the bandgap, the difference between these energy extrema, then increases. The increase in the bandgap energy causes the intrinsic electron carrier concentration, n_i, to fall and $2\phi_F$, the inversion potential, to rise. Hence, the threshold voltage of the transistor rises. More importantly, the increase in surface potential with rising gate voltage causes stronger electron confinement such that n_i is a weak but monotonically decreasing function of ψ_S; the net effect is that the effective channel charge, compared with its nonquantum value, reduces when ψ_S increases. If the latter effect is modeled as being approximately linear with surface potential, it is effectively mimicked by simply reducing C_{ox} to C_{ox}^{eff}. Hence, the effects of bandgap widening are almost identical to those of poly-gate depletion: the threshold voltage rises, C_{ox} falls, and κ reduces. Not surprisingly, poly-gate depletion and bandgap widening are often combined into one empirical effect.

6.5   Effect on the intuitive model

Short-channel effects alter the intuitive model of Figure 3.4 as follows. Velocity saturation is modeled with more complex drift and diffusion generators. Note that the drift and diffusion generators are intimately related via an effective local temperature. Since DIBL is primarily important in subthreshold operation, the simplest approximation to model DIBL is to introduce one capacitance from the source and one capacitance from the drain to a central equipotential node in the channel. The phenomenon of VMR can be modeled by requiring the drift (and diffusion) generators to be surface-potential dependent. Poly-gate depletion and bandgap widening can be modeled by interposing a capacitance between the gate and C_{ox}.

6.6   High-frequency transistor models

At very high frequencies, various parasitics in the transistor begin to become important. Small resistances, which are a few ohms and normally negligible, become significant compared with other RF resistances, which may be near 50 Ω. Figure 6.5 reveals that these parasitics are primarily composed of extrinsic overlap and fringing capacitances, terminal resistances determined by salicide, contact, and sheet resistances, substrate resistances to power-supply contacts caused by finite-doping effects, and the reverse-biased source-to-bulk and drain-to-bulk junctions. The use of salicide on source/drain regions and copper interconnect helps reduce contact and interconnect resistances.

At extremely high frequencies, the quasi-static approximation for transistor operation, i.e., that the charge within it instantaneously equilibrates on the time scale of terminal-voltage changes, is no longer valid. The transistor is a device with distributed charge as Figure 3.4 indicates. In [3], if there is no velocity saturation, it has been derived that the dimensionless frequency limits for quasi-static approximation are given by

$$\Omega_{qs} = \frac{\omega_{qs}}{\omega_n} = \frac{1/\tau_{qs}}{\omega_n}$$

$$\omega_n = \frac{\mu_0 \phi_t}{L^2}$$

$$\Omega_{qs} = 30 \frac{(q_S + q_D + 1)^3}{4q_S^2 + 4q_D^2 + 12q_S q_D + 10q_S + 10q_D + 5}$$

(6.38)

In saturation, where $q_S \gg q_D$,

$$\Omega_{qs} \approx 6; \text{subthreshold}$$
$$\Omega_{qs} \approx 7.5q_S; \text{above-threshold}$$

(6.39)

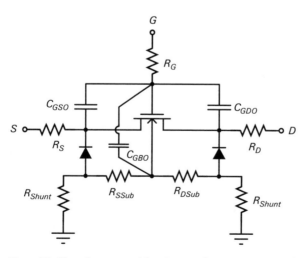

Figure 6.5. Transistor parasitics that are important at RF frequencies.

The input frequency must be 6–7 times less than Ω_{qs} for the quasi-static approximation to be valid. If dimensions are added back to Equation (6.39), the input frequency where quasi-static approximations hold is then near the reciprocal of the channel transit time that we computed in Chapter 5, i.e., near the maximal f_T. What occurs if we are near or greater than f_T? It has been found that a good approximation is to assume that all intrinsic transconductances and capacitances become frequency dependent in the small-signal model [2]. For example,

$$
\begin{aligned}
G_m(j\omega) &\approx \frac{G_m(0)}{1 + j\omega\tau_{qs}} \\
C_{GS}(j\omega) &\approx \frac{C_{GS}(0)}{1 + j\omega\frac{\tau_{qs}}{2}}
\end{aligned}
\tag{6.40}
$$

Note that the phase changes in $G_m(j\omega)$ and $C_{GS}(j\omega)$ in Equation (6.40) imply that these are no longer simple transconductances and capacitances but complex admittances. Nevertheless, we have used the form in Equation (6.40) because it is a simple and easily remembered modification on the classic low-frequency small-signal model.

6.7   Ballistic transport

When transistor channel length is comparable to or less than 100 nm, the thermal velocity of electron carriers, v_{th}, ejected at the transistor's source becomes important in limiting the maximal on current and maximal speed of operation of the transistor even in above-threshold operation [4], [5], [6]. Although the charge boundary condition at the source end of the channel is largely unchanged in above-threshold operation, the current flow in the above-threshold regime can be viewed as being more like that in a bipolar transistor [1]; it consists of a current-limiting initial diffusion component in the low-field channel region near the source (analogous to that in the base of a bipolar transistor) followed by a rapid drift component in the high-field channel regions near the drain (analogous to that across the depletion region in the collector of a bipolar transistor). As in a bipolar transistor, it is the initial diffusion component that limits current. High velocities are attained near the drain region of the transistor, but the charge and transport characteristics at the transistor's source establish limits on its flow. In ultra-short-channel regimes where the channel length of the transistor is comparable to or less than the mean free length for a scattering collision event for an electron (typically 10–15 nm) the transport can be almost ballistic: there is no scattering event for an electron in the channel as it just shoots out of the source and makes its way to the drain, losing little energy to the lattice in the channel. Therefore, the velocity near the drain can exceed v_{sat} significantly, a phenomenon termed velocity overshoot, but the current is still limited by the charge and thermal velocity at the source. In transistors with channel lengths between 20 nm and 100 nm, where there can be a handful of or no scattering events in the channel, it is more useful to think directly in terms of electron ballistics and scattering events rather than in terms of mobility, drift, and diffusion transport parameters: the latter transport parameters represent macroscopic concepts invented for averaging over many microscopic scattering events rather than just a few. We shall therefore discuss models for ballistic transistors and transport models based on scattering [4], [5], [6]. In such models, above-threshold equations look like the hybrid of a normal above-threshold charge factor multiplied by a diffusion-like subthreshold transport factor.

In the ballistic regime, parameters that are based on scattering phenomena such as diffusion or mobility constants are not relevant. What then sets the limits to transistor current? At any point in the channel, current is the product of the local charge per unit length and the local velocity. If any point in the channel sets limits on these quantities, the current in the overall transistor will be limited. As we discussed in the section on velocity saturation, the charge in the channel redistributes itself such that the current is constant throughout the channel. Thus far, we have discussed limitations on transistor current imposed by the maximal velocity attainable near the drain end of the channel. In ballistic transport, limitations in current that occur at the source end of the channel become important.

The mean speed of electrons emerging from the source, v_{th}, which have crossed the source-to-channel barrier and that have a positive drain-directed velocity is related to their thermal energy and given by

$$v_{th} = \sqrt{\frac{2kT}{\pi m^*}}; m^* = 0.19\, m_0$$
$$v_{th} \approx 1.2 \times 10^5\, \text{m/s}, \tag{6.41}$$

where m^* is the effective rest mass of the electron when it moves in the lattice and m_0 is its rest mass [1]. Note that v_{th} is near v_{sat}, the maximal drift saturation velocity near the drain. Thus, the maximal current possible at the source is given by $Q(0)Wv_{th}$ where $Q(0)$ is the areal charge concentration at the source. When $V_{DS} = 0$, the charge at the source end has a $Q(0)/2$ component that arises from positive-velocity carriers in the source and a $Q(0)/2$ component that arises from negative-velocity carriers in the drain. When V_{DS} is a few kT/q, the contribution of carriers from the drain end drops to zero and $Q(0)$ is made up mostly of carriers from the source. At all V_{DS}, however, the total charge at the source end is a constant $Q(0)$ determined by electrostatic boundary considerations and given by

$$Q(0) = C_{ox}(V_{GS} - V_{TS})$$
$$Q(0) = Q_S(1 + e^{-qV_{DS}/kT}), \tag{6.42}$$

where Q_S is the component of charge arising from the positive-velocity source carriers and $Q_S e^{-qV_{DS}/kT}$ is the component of charge arising from the negative-velocity drain carriers. We have assumed that the charges arising from the source and drain are in thermal equilibrium with their terminals and follow non-degenerate exponential Maxwell-Boltzmann statistics. The current in the transistor is given by the difference between the positive velocity carriers from the source and the negative velocity carriers from the drain and given by

$$I_{DS} = WC_{ox}(V_{GS} - V_{TS})v_{th}\left(\frac{1 - e^{-qV_{DS}/kT}}{1 + e^{-qV_{DS}/kT}}\right). \tag{6.43}$$

Equation (6.43) was first postulated by Natori [4], [5]. It is an interesting hybrid of the product of a subthreshold and an above-threshold expression: the charge is determined by an above-threshold $C_{ox}W(V_{GS} - V_{TS})$ expression but the transport is given by a diffusion-like subthreshold expression with v_{th} and the exponential terms. In subthreshold operation, we can postulate that Equation (6.43) has its charge portion replaced by $WC_{ox}(\phi_t/\kappa)e^{-q\kappa(V_{GS}-V_{TS})/kT}$ but that the transport v_{th} and exponential terms are unchanged.

6.8 Transport in nanoscale MOSFETs

What happens if we have a few scattering collisions in the channel and the transport is not purely ballistic? Lundstrom has derived a simple ballistic-like model that describes this case with a scattering parameter based on the mean free length λ [6]. The key physical insight behind this model is illustrated in the intuitive energy diagram of Figure 6.6 shown for a transistor in saturation. Electrons that leave the source for the channel are highly unlikely to return to the source from a scattering collision unless this collision occurs within a length l from the source. The length l corresponds to a kT drop in the electron's potential energy. Each collision imparts on the order of kT of energy to the electron. Electrons outside l have gathered sufficient kinetic energy as they move down the energy hill such that several probabilistically unlikely collisions are needed to reverse their downhill course to the drain and return via an uphill course to the source. The probability that they will scatter and return to the source is represented by a reflection coefficient R computed to be

$$R = \frac{l}{l + \lambda}. \tag{6.44}$$

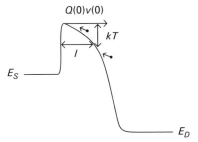

Figure 6.6. Energy diagram for computing scattering in a nanoscale MOSFET based on Lundstrom's elementary scattering theory.

As in the discussion on ballistic transport, we assume that the positive-velocity-directed electrons and negative-velocity-directed scattered electrons jointly form the charge concentration at the source, which is set by an electrostatic boundary condition. That is,

$$C_{ox}(V_{GS} - V_{TS}) = Q_S(1 + R),\tag{6.45}$$

where Q_S represents the areal charge concentration of positive-velocity-directed electrons leaving the source and RQ_S represents the areal charge concentration of negative-velocity-directed electrons that return. The current is given by the difference between the fluxes of the positive-velocity-directed and negative-velocity-directed carriers and is given by

$$I_{DS} = WC_{ox}(V_{GS} - V_{TS})v_{th}\left(\frac{1 - R}{1 + R}\right)\tag{6.46}$$

The similarity of Equations (6.46) and (6.43) is striking: the $e^{-qV_{DS}/kT}$ term in Equation (6.43) is replaced by the R term in Equation (6.46) with R given by Equation (6.44).

The distance l can be estimated by approximating the distance needed to drop a voltage of kT/q with the electric field just past the source, $E(0+)$. The latter electric field is computed from self-consistent MOS charge distributions in the channel. Thus, we may write

$$l \approx \frac{kT/q}{E(0+)}\tag{6.47}$$

Thus, if V_{DS} and consequently $E(0+)$ are large, l can be less than λ such that $R < 0.5$ in Equation (6.44); an R near 0 corresponds to completely ballistic transport and yields the same current in saturation as Equation (6.43). At small V_{DS}, Equation (6.47) predicts that $l \to \infty$. The R in Equation (6.44) then reaches a finite zero-field value R_0 given by

$$R_0 = \left(\frac{L}{L + \lambda}\right)\tag{6.48}$$

Hence, the R in Equation (6.44) is more accurately given by

$$\begin{aligned} R &= R_0\left(\frac{l}{l + \lambda}\right) \\ &= \left(\frac{L}{L + \lambda}\right)\left(\frac{l}{l + \lambda}\right) \end{aligned}\tag{6.49}$$

The value of λ can be related to the low-field mobility via the following expression and the use of the Einstein relation:

$$\begin{aligned} D_n &= \frac{\lambda v_{th}}{2} \\ \lambda &= \frac{2\mu_0 \frac{kT}{q}}{v_{th}} \end{aligned}\tag{6.50}$$

Thus, improvements in low-field mobility such as in strained silicon do translate to longer mean-field lengths, less scattering, and larger currents in nanoscale transistors, but the current will eventually asymptote at the ballistic limit given by Equation (6.43).

6.9   Tunneling

Since electrons have wave-like properties, they have the ability to tunnel through energy barriers from one side to another like evanescent waves with a propagation function proportional to e^{-kx}, where k is the wave number, and x is the tunneling distance. For those readers unfamiliar with wave physics, a brief introduction to waves and wave propagation can be found in the initial portion of Chapter 17 on antennas. Thus, if the gate-oxide thickness is sufficiently small (x is small) and/or the voltage across the gate oxide is sufficiently large (k is increased), electrons will tunnel through the built-in energy barrier from silicon to silicon dioxide and increase the negative charge on the polysilicon gate. The tunneling effect is intentionally exploited in programming E^2PROM nonvolatile memories and in floating-gate circuits.

When transistor oxide thicknesses in deep submicron transistors are less than 3 nm, voltages of even a few volts can cause tunneling to create gate-leakage currents. Gate-leakage current can increase by a factor of 10 for every 0.2 nm reduction in gate-oxide thickness. A detailed derivation of such currents would take us deep into quantum mechanics and is beyond the scope of this book. Fortunately, simple equations allow us to summarize the dependence of this current on various parameters [2]. The tunneling gate current is given by

$$J_{gate} = K_G \left(\frac{V_{ox}}{t_{ox}}\right)\left(\frac{-Q_i}{\varepsilon_{ox}}\right) p_{tun}; K_G = 30\,\mu A/V^2$$

$$p_{tun} = e^{-\frac{E_B t_{ox}}{V_{ox}}(1-(1-\min(1,\frac{V_{ox}}{X_B})))}; E_B = 29\,V/nm; X_B = 3.1\,V \qquad (6.51)$$

$$V_{ox} = E_{ox}t_{ox} = -\frac{Q_{dep}+Q_i}{C_{ox}}$$

The values of K_G, E_B, and X_B are 40 $\mu A/V^2$, 43 V/nm, and 4.5 V for holes, respectively. The tunneling current of Equation (6.51) varies as the oxide field and inversion charge vary along the channel and must be integrated along the channel to get the total gate current. The use of high-K dielectrics allows oxide thickness to be reduced, significantly reducing gate leakage, while still maintaining a high value of C_{ox}.

When transistor channel lengths are in the nm range, tunneling of electrons through the source-to-drain barrier can occur [9]. Such tunneling sets one ultimate limit on transistor scaling to nanoscale dimensions. Tunneling from the drain to the bulk can also cause drain-leakage currents [11]. Tunneling has been exploited in carbon-nanotube transistors with the gate controlling the tunneling barrier [7]. Such transistors are still at an early stage of evolution today but, due to their high mobility and relative ease of manufacture, may represent an important direction for the future.

6.10   Scaling of transistors in the future

Transistor channel length, L, is decreasing by $\sim 1.4\times$ in every 18-month technology revision today. On a semilog plot, transistor dimensions appear to obey an approximately constant geometric or exponential scaling law with time, as first observed by Moore [17] and now termed Moore's law. Moore's law is slowing as we reach fundamental physical limits that make well-controlled fabrication at small dimensions increasingly more expensive and difficult. The doping of the bulk N_A, supply voltage V_{DD}, oxide thickness t_{ox}, and threshold voltage V_{T0} of transistors appear to obey some simple empirical trends with L. These trends as postulated by Mead [11] are summarized below:

$$
\begin{aligned}
N_A &= (4 \times 10^{16})L^{-1.6} \\
V_{DD} &= 5L^{0.75} \\
t_{ox} &= \max(21L^{0.77}, 14L^{0.55}) \\
V_{T0} &= 0.55L^{0.23},
\end{aligned}
\tag{6.52}
$$

where L is in μm units, N_A is in cm^{-3} units, V_{DD} is in V units, t_{ox} is in nm units, and V_{T0} is in V units. The scaling of N_A and V_{DD} with L are necessary to limit punchthrough, DIBL, drain-junction breakdown, drain tunneling, and drain-conductance degradation; to ensure that κ is not too low and that the gate continues to have an effect on the surface potential, t_{ox} must then be lowered as well; the scaling of V_{T0} is based on the practicalities of doping implants and materials issues and to limit 'subthreshold leakage' in the off-state of digital transistors. Empirically, it is often found that

$$
\sigma_{vth} = \frac{A_{vth}(L)}{\sqrt{WL}},
\tag{6.53}
$$

with σ_{vth} being the standard deviation of random threshold-voltage mismatch across transistors in a process with width and length dimensions W and L respectively. A regression fit by the author of A_{vth} versus L data compiled in [18] reveals that

$$
A_{vth}(L) = 17L^{0.71},
\tag{6.54}
$$

where A_{vth} is in mV. The scaling of the threshold-voltage mismatch in Equation (6.53) with transistor dimensions is similar to the scaling of $1/f$ noise in transistors. As we discuss in Chapter 7, $1/f$ noise may be viewed as a dynamically varying threshold voltage caused predominantly by the occurrence of random trap-filling processes in the oxide of the transistor. Equation (6.54) can be viewed as an addition to the four scaling laws of Equation (6.52).

Some ultimate limits of scaling based on the Shannon limit of energy dissipation per bit ($kT \ln 2$ as explained in Chapter 22), tunneling from source to drain (~ 1 nm), and quantum channel transit times ($L/v_{quantum}$) are discussed in [9] and [1]. They predict a power dissipation of $3.7\,\text{MW/cm}^2$ if we continue scaling as usual. However, as is correctly pointed out, it is unlikely that more than $100\,\text{W/cm}^2$

will be tolerable based on heat-sink concerns. System parameters such as clock speed are no longer mostly limited by transistor f_T's, L/v_{sat}, or L/v_{th}, but mostly by parasitics and interconnect. A website presents frequent updates on the expected scaling of transistors in the future [10].

The significantly slower scaling of V_{T0} than V_{DD} implies that subthreshold operation has become and will become an even increasingly larger fraction of the voltage operating range in the future. Fortunately, subthreshold or moderate-inversion operation implies four benefits:

1. The improved g_m/I_{DS} ratio allows more power-efficient operation.
2. Velocity saturation is avoided due to lower drain-to-source voltages, which allows $1/L^2$ scaling rather than $1/L$ scaling until the ballistic limit is reached. It is relatively easy to attain f_T's in excess of a few GHz in moderate-inversion operation today.
3. Carrier heating effects that lead to excess noise and transistor-current degradation are avoided.
4. Lower supply-voltage and lower-current operation imply that ultra-low-power operation is possible.

Hence a large part of this book will focus on subthreshold operation for ultra-low-power operation since it is timely, advantageous, and necessary. Moderate inversion is sometimes a good compromise between degraded linearity and speed in subthreshold operation versus good power efficiency. Subthreshold operation is also more subject to extrinsic parasitics limiting the speed of operation (usually only to within a factor of 2), to transistor mismatch, to temperature variations, and to power-supply noise. This book will discuss through several circuits-and-systems examples and principles how robust and efficient operation can be achieved without compromising performance, especially in biomedical applications. In particular, Chapters 19, 21, and 22 will outline how large low-power systems that are robust and efficient can be built by exploiting the best of the analog and the digital worlds. Subthreshold operation is also uniquely suited to mimicking highly energy-efficient architectures in neurobiology (Chapter 23) and in cell biology (Chapter 24). Biology's significant accomplishment has been to architect ultra-energy-efficient, precise, complex systems with noisy mismatched components.

References

[1] Mark Lundstrom and Jing Guo. *Nanoscale Transistors: Device Physics, Modeling and Simulation* (New York: Springer, 2006).
[2] Christian Enz and Eric A. Vittoz. *Charge-based MOS Transistor Modeling: The EKV Model for Low-Power and RF IC Design* (Chichester, England; Hoboken, NJ: John Wiley, 2006).
[3] A.S. Porret, J.M. Sallese and C.C. Enz. A compact non-quasi-static extension of a charge-based MOS model. *IEEE Transactions on Electron Devices*, **48** (2001), 1647–1654.

[4] K. Natori. Ballistic metal-oxide-semiconductor field effect transistor. *Journal of Applied Physics*, **76** (1994), 4879–4890.

[5] K. Natori. Scaling Limit of the MOS Transistor–A Ballistic MOSFET. *IEICE Transactions on Electronics*, **E84-C** (2001), 10291036.

[6] M. Lundstrom. Elementary scattering theory of the Si MOSFET. *IEEE Electron Device Letters*, **18** (1997), 361–363.

[7] A. Javey, R. Tu, D.B. Farmer, J. Guo, R.G. Gordon and H. Dai. High Performance n-Type Carbon Nanotube Field-Effect Transistors with Chemically Doped Contacts. *Nano Letters*, **5** (2005), 345–348.

[8] K.K. Likharev. Single-electron devices and their applications. *Proceedings of the IEEE*, **87** (1999), 606–632.

[9] V.V. Zhirnov, R.K. Cavin III, J.A. Hutchby and G.I. Bourianoff. Limits to binary logic switch scaling – a gedanken model. *Proceedings of the IEEE*, **91** (2003), 1934–1939.

[10] International Technology Roadmap for Semiconductors. Available from: http://public.itrs.net/.

[11] C.A. Mead. Scaling of MOS technology to submicrometer feature sizes. *The Journal of VLSI Signal Processing*, **8** (1994), 9–25.

[12] M.A. Maher and C.A. Mead, A physical charge-controlled model for the MOS transistor. *Proceedings of the Advanced Research in VLSI Conference*, Stanford, CA, 1987.

[13] M.A. Maher and C.A. Mead. Fine Points of Transistor Physics. In *Analog VLSI and Neural Systems,* ed. C Mead. (Reading, MA: Addison-Wesley; 1989), pp. 319–338.

[14] J.M. Sallese, M. Bucher, F. Krummenacher and P. Fazan. Inversion charge linearization in MOSFET modeling and rigorous derivation of the EKV compact model. *Solid State Electronics*, **47** (2003), 677–683.

[15] P.K. Ko. Approaches to Scaling. In *Advanced MOS Device Physics*, ed. N. G. Einspruch, Gildenblat G. S. (San Diego: Academic Press; 1989), pp. 1–37.

[16] C. Diorio, P. Hasler, A. Minch and C.A. Mead. A single-transistor silicon synapse. *IEEE Transactions on Electron Devices*, **43** (1996), 1972–1980.

[17] G.E. Moore. Cramming more components onto integrated circuits. *Electronics*, **38** (1965), 114–117.

[18] K. Bult. Analog design in deep sub-micron CMOS. *Proceedings of the 26th European Solid-State Circuits Conference* (*ESSCIRC*), Stockholm, Sweden, 126–132, 2000.

7 Noise in devices

But the real glory of science is that we can find a way of thinking such that the law is evident.
Richard P. Feynman

Noise ultimately limits the performance of all systems. For example, the maximum gain of an amplifier is limited to V_{DD}/v_n where V_{DD} is the power-supply voltage and v_n is the noise floor at the input of the amplifier. Gains higher than this limiting value will simply amplify noise to saturating power-supply values and leave no output dynamic range available for discerning input signals.

Since power is the product of voltage and current, low-power systems have low voltages and/or low current signal levels. Hence, they are more prone to the effects of small signals such as noise. A deep understanding of noise is essential in order to design architectures that are immune to it, in order to efficiently allocate power, area, and averaging-time resources to reduce it, and in order to exploit it. We will begin our study of noise in physical devices from a first-principles view of some of the fundamental concepts and mathematics behind it.

7.1 The mathematics of noise

We pretend that macroscopic current is the flow of a smooth continuous fluid. However, the current is actually made up of tiny microscopic discrete charged particles that flow in a semi-orderly fashion. The random disorderly portion of the charged-particle motion manifests itself in the macroscopic current as noise. Figure 7.1 reveals how macroscopic current is actually made up of tiny fluctuations around its mean value which constitute current noise. Thus,

$$\begin{aligned} \overline{I(t) - \overline{I(t)}} &= 0 \\ \overline{\left(I(t) - \overline{I(t)}\right)^2} &= \sigma_I^2 \neq 0, \end{aligned} \qquad (7.1)$$

where $\overline{I(t)}$ is the mean value of the current, and σ_I^2 is the variance of the current around this mean, i.e., the current-noise power.

Imagine that the current flow is Poisson and has an event arrival rate of charged particles at the terminal of a device of λ: the probability that a charged particle arrives between time τ and time $\tau + d\tau$ is $\lambda d\tau$ and is uncorrelated with any past or

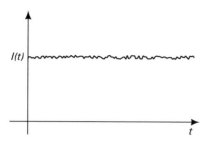

Figure 7.1. The current $I(t)$ shows fluctuations around a mean value because electronic charge is discrete.

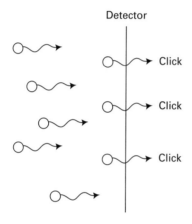

Figure 7.2. The random arrival of electrons at a detector leads to impulsive events, which are denoted by 'Clicks' in the detector.

future arrivals of charged particles, i.e., there is no memory in the arrivals. A good analogy is that of raindrops arriving at a detector. There is an average current of water that results from a rate of arrival of raindrops but the arrival of each raindrop is random and uncorrelated with that of any other. Figure 7.2 illustrates event arrivals at a detector by discrete particles. Now imagine that each event arrival generates a microscopic current waveform $f(t)$ at the detector whose net integral is the charge of the particle. The waveform $f(t)$ could correspond to the dynamics of the detector's response to impulsive arrival events and/or to the microscopic induced current generated at the terminal of a device when a charged particle arrives at it. Any arrival event must generate some $f(t)$ even if it is just an impulse. Figure 7.3 illustrates a single microscopic $f(t)$ current waveform and several of these microscopic current waveforms, all starting at random times but with some mean rate of occurrence. These microscopic current waveforms all add together in the detector to generate a macroscopic current waveform such as that of Figure 7.1 with some mean and variance. We shall now compute

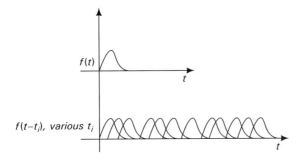

Figure 7.3. The waveform $f(t)$ is the response of the detector to a single electron that arrives at $t = 0$. Reponses due to electrons arriving at different times overlap with each other and add to produce the total current.

the mean and variance of the macroscopic waveform from the nature of the underlying microscopic behavior.

We are interested in computing the mean of the macroscopic quantity $F(t)$ given by

$$F(t) = \sum_{t_i} f(t - t_i) \tag{7.2}$$

in the limit that t_i is described by a continuous Poisson process with an event arrival rate λ. In this continuous limit, we may replace sums by integrals and weight each $f(t - t_i)$ by its arrival probability to obtain

$$\overline{F(t)} = \int_{-\infty}^{t} f(t - t_i) \lambda dt_i = \int_{\infty}^{0} -f(u) \lambda du = \int_{0}^{\infty} \lambda f(u) du$$

$$\overline{F(t)} = \int_{0}^{\infty} \lambda f(\tau) d\tau \tag{7.3}$$

Not surprisingly, a larger event arrival rate and a larger integral of the microscopic current waveform, i.e., larger charge, generate a larger mean current.

The variance of $F(t)$ can be computed from the mean $\overline{F(t)}$ if we can compute $\overline{F^2(t)}$ using the classic relation

$$\overline{\left(F(t) - \overline{F(t)}\right)^2} = \overline{F^2(t)} - \left(\overline{F(t)}\right)^2 \tag{7.4}$$

To compute $\overline{F^2(t)}$, we need to compute

$$F^2(t) = [f(t - t_1) + f(t - t_2) + \ldots + f(t - t_n) + \ldots]^2$$

$$= \sum_{t_i} f^2(t - t_i) + \sum_{t_i} \sum_{t_j} f(t - t_i) f(t - t_j) \tag{7.5}$$

in the continuous limit that t_i and t_j are described by the same continuous Poisson process with event arrival rate λ. Once again, in this continuous limit, we replace sums by integrals and weight each $f(t - t_i)$ or $f(t - t_j)$ by its arrival probability to obtain

$$\overline{F^2(t)} = \int_{-\infty}^{t} f^2(t - t_i)\lambda dt_i + \int_{-\infty}^{t} \int_{-\infty}^{t} f(t - t_i)f(t - t_j)\lambda dt_i \lambda dt_j$$

$$= \int_{0}^{\infty} \lambda f^2(u)du + \left(\int_{0}^{\infty} \lambda f(u)du\right)\left(\int_{0}^{\infty} \lambda f(u)du\right) \tag{7.6}$$

$$= \int_{0}^{\infty} \lambda f^2(u)du + \left(\overline{F(t)}\right)^2$$

Thus, we finally obtain the variance of $F(t)$ to be

$$\overline{F^2(t)} - \left(\overline{F(t)}\right)^2 = \int_{0}^{\infty} \lambda f^2(\tau)d\tau \equiv \sigma_{F(t)}^2 \tag{7.7}$$

Equation (7.7) was first discovered by Carson and Campbell [1] and is a highly important result in the mathematics of noise.

Now, let $\mathcal{F}(f)$ be the Fourier transform of $f(t)$. Then, by Parseval's theorem and the even symmetry of $|\mathcal{F}(f)|$ for real-valued $f(t)$, we obtain that

$$\sigma_{F(t)}^2 = \lambda \int_{0}^{\infty} f^2(\tau)d\tau = \lambda \int_{-\infty}^{\infty} |\mathcal{F}(f)|^2 df$$

$$= 2\lambda \int_{0}^{\infty} |\mathcal{F}(f)|^2 df$$

$$= \int_{0}^{\infty} \underbrace{\left(2\lambda|\mathcal{F}(f)|^2\right)}_{P(f)} df \tag{7.8}$$

$$= \int_{0}^{\infty} P(f)df,$$

where $P(f)$ is called the power spectral density since the variance is computed by integrating the power spectral density over all frequencies

$$\sigma_{F(t)}^2 = \int_{0}^{\infty} P(f)df \text{ with } P(f) = 2\lambda|\mathcal{F}(f)|^2 \tag{7.9}$$

The diffusion current of diodes, bipolar transistors, tunneling junctions, and in subthreshold MOSFETs is Poisson and is therefore described by Equation (7.3) with the integral $q = \int_{0}^{\infty} f(\tau)d\tau$ being the charge of the electron. Thus,

$$\bar{I} = q\lambda, \quad q = \text{charge on electron.} \tag{7.10}$$

Thus, by direct application of Carson and Campbell's theorem of Equation (7.7), we find that

$$\overline{\Delta I^2} = \overline{(I - \bar{I})^2} = q^2 \int_{0}^{\infty} 2\lambda|\mathcal{F}(f)|^2 df. \tag{7.11}$$

If we assume that $f(t)$ is an impulse because electron arrivals at terminals cause impulses of charge, then, $|\mathcal{F}(f)|^2 = 1$, $\forall f$ such that

$$\overline{\Delta I^2} = \overline{(I - \bar{I})^2} = 2q^2\lambda \int_0^\infty df$$

$$= 2q\bar{I} \int_0^\infty df.$$

(7.12)

Thus, in a certain bandwidth Δf, the shot noise of the Poisson electron current is given by

$$\overline{\Delta I^2} = 2q\bar{I}\Delta f.$$

(7.13)

The bandwidth Δf is established by a lowpass filter that is implicit or explicit in the detector or in an external system that the detector is coupled to. The time window over which the current is averaged in the Δf filter is reciprocally related to Δf such that a small Δf implies a long averaging window and a large Δf implies a short averaging window. As Δf goes to zero, we average for increasingly longer times, increasingly fewer frequency components in the current contribute to the noise variance, and the noise variance goes toward zero. As $\Delta f \to \infty$, we average for increasingly shorter times, increasingly more frequency components in the current contribute to the noise variance, and the noise variance goes toward ∞. It is misleading to talk about the total noise of a system without specifying the noise bandwidth since noise may always be lowered by lowering bandwidth.

The shot-noise result of Equation (7.13) is so widely useful in the study of noise that it is worth memorizing. Note that $q\Delta f$ has units of current and $\bar{I}$ has units of current such that their product has units of square current, consistent with the units of variance. The power spectral density of the current noise $2q\bar{I}$ is quoted in units of A^2/Hz. The 'amplitude' of the power spectral density is often quoted in units of $A/\sqrt{Hz}$. Similarly voltage noise is quoted in units of V^2/Hz or $V/\sqrt{Hz}$. The power spectral density of shot noise is flat or white because the impulsive nature of electron arrivals results in $f(t)$ being impulse like, which results in $|\mathcal{F}(f)|^2$ being flat. White noise may be viewed as being composed of a set of sine waves of all frequencies but whose relative phases are random with respect to each other. The power spectral density $P(f)$ is a measure of the power of the sinusoidal components between frequency f and frequency $f + df$.

An important insight into the nature of noise emerges from computing the relative power of the noise with respect to the mean signal

$$\overline{\Delta I^2} = 2q\bar{I}\Delta f$$

$$\left(\frac{\overline{\Delta I^2}}{\bar{I}^2}\right) = 2\left(\frac{q\Delta f}{\bar{I}}\right) = \left(\frac{q}{\bar{I}T_{avg}}\right) = \left(\frac{q}{Q_{TOT}}\right) = \left(\frac{1}{N_{TOT}}\right),$$

(7.14)

where $T_{avg} = 1/(2\Delta f)$ is the effective averaging window length, Q_{TOT} is the total charge collected within the averaging window, and N_{TOT} is the total number of electrons collected within the averaging window. We notice that the relative

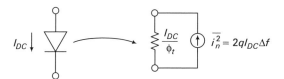

Figure 7.4. Small-signal noise model of a diode.

strength of the noise power compared with the mean signal power falls as we collect more electrons within the averaging window, an intuitively pleasing result from the laws of probabilities and large numbers. As N_{TOT} increases, the standard deviation of the collected charge scales like $\sqrt{N_{TOT}}$ while the total charge scales like N_{TOT} such that the ratio of their powers scales like $\left(\sqrt{N_{TOT}}/N_{TOT}\right)^2$, i.e., $1/N_{TOT}$, as expressed by Equation (7.14). Thus, we obtain more precision in the mean simply by collecting more electrons, either by increasing our mean current $\bar{I}$, which increases power consumption, or by increasing the averaging window length T_{avg}, or both.

Noise is typically small enough to be well modeled as a small random ac deviation that exists on top of a large dc mean signal. An example of a noise model is shown in Figure 7.4. The small-signal circuit of a diode is modified to include a shot-noise generator of mean square value $2qI_{DC}\Delta f$ in parallel with its small-signal resistance of ϕ_t/I_{DC}. The noise generator only has a well-defined mean square value with the direction of the current being irrelevant. The bandwidth Δf is typically set by other resistances and capacitances in the circuit that the diode is a part of. The diode's own intrinsic capacitance and resistance can set the value of Δf as well.

When estimating the variances of uncorrelated random variables, e.g., if i_1 and i_2 are independent noise generators in a small-signal circuit with a small-signal output i_{out} such that

$$i_{out} = \alpha_1 i_1 \pm \alpha_2 i_2, \tag{7.15}$$

then the estimated noise of i_{out} is given by

$$\sigma_{iout}^2 = \alpha_1^2 \sigma_{i_1}^2 + \alpha_2^2 \sigma_{i_2}^2 \tag{7.16}$$

Note that the variance of i_{out} in Equation (7.16) is independent of whether the sign of the $\pm$ term of Equation (7.15) is positive or negative. If each of the noise generators can be modeled with a shot-noise term given by their dc operating currents, Equation (7.16) leads to

$$\sigma_{iout}^2 = \alpha_1^2 2qI_1 \Delta f_1 + \alpha_2^2 2qI_2 \Delta f_2 \tag{7.17}$$

If the bandwidth is not explicit and the transfer functions are frequency dependent and described by $\alpha_1(f)$ and $\alpha_2(f)$, we would need to evaluate σ_{iout}^2 according to

$$i_{out}(f) = \alpha_1(f)i_1 \pm \alpha_2(f)i_2$$
$$\sigma_{iout}^2 = 2qI_1 \int_0^\infty |\alpha_1(f)|^2 df + 2qI_2 \int_0^\infty |\alpha_2(f)|^2 df \tag{7.18}$$

Examples of such calculations are described in more detail in the following chapter. The above discussion, however, implicitly assumed that the two noise generators were independent and uncorrelated noise sources. If correlations are present, then they need to be taken into account and lead to an extra cross-correlation term in Equation (7.16):

$$i_{out} = \alpha_1 i_1 \pm \alpha_2 i_2$$
$$\sigma_{iout}^2 = \alpha_1^2 \sigma_{i_1}^2 + \alpha_2^2 \sigma_{i_2}^2 \pm 2\alpha_1\alpha_2 \overline{i_1 i_2}$$

$$(7.19)$$

7.2   Noise in subthreshold MOS transistors

Electrons in an MOS transistor have large thermal velocities and are constantly suffering collisions that randomize their direction of motion. If the lateral electric field in the transistor is small, as it is in subthreshold MOS operation, electrons in the channel are equally likely to move in a direction towards the drain or move in a direction towards the source. The net measurable electronic current in the external voltage source that maintains a constant drain-to-source potential difference is the difference between anticlockwise currents flowing from source to drain and then back through the voltage source to the source and clockwise currents flowing from drain to source and then back through the voltage source from source to drain. Figure 7.5 (a) shows an anticlockwise current flow and Figure 7.5 (b) shows a clockwise current flow. Since there are more electrons at the source boundary end of the channel than at the drain boundary end of the channel, on average, more electrons flow from the source to the drain than vice versa and generate a net diffusion current from source to drain, i.e., the externally measured anticlockwise electronic current exceeds the clockwise electronic current. Both clockwise and anticlockwise diffusion currents exhibit electronic shot noise even though it is their difference that manifests as a net external current, i.e., Equation (7.15) and Equation (7.16) with $\alpha_1 = 1$ and $\alpha_2 = 1$ apply. Thus, at zero drain-to-source voltage even when the external transistor current is zero due to the equality of clockwise and anticlockwise diffusion currents, the noise current is finite, and is in fact maximal as we discuss below.

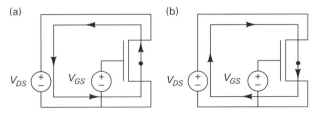

Figure 7.5a, b. Electrons must flow in closed paths (loops). In the figure, (a) anticlockwise and (b) clockwise current flow paths through a transistor are shown.

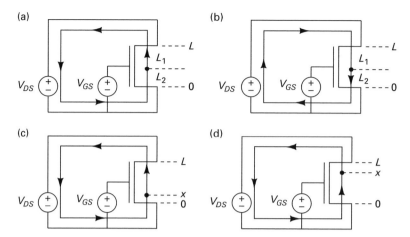

Figure 7.6a, b, c, d. The total distance that an electron must diffuse before it completes a loop and return to its starting point is independent of the sense of the loop (parts (a) and (b)). Parts (c) and (d) show that it is also independent of the location of the starting point.

Thermally generated diffusion currents exhibit shot noise because the variance in the round-trip clockwise or anticlockwise circuit travel times is well modeled by a random Poisson process. Larger temperatures, which cause higher thermal velocities, and larger mean free times between collisions yield a larger value for the diffusion constant and larger diffusion currents. If L is the effective channel length between the edge of the source depletion region and the edge of the drain depletion region, the transit time $\tau = L^2/2D_n$ where D_n is the diffusion constant for electrons in silicon. If we assume that the depletion-region travel times and voltage-source travel times are negligible, a good approximation in long-channel transistors, the transit time is a measure of the average time taken by an electron in the channel to complete a round-trip clockwise or anticlockwise journey through the full circuit and return to its starting point. Note also that, in our method of accounting for Poisson events, a full circuit must be completed for the external voltage source to register an electronic current. Incomplete motions of electrons within the channel that do not result in external current flow do not register as current events in the external circuit. Such motions may involve lots of path retracing within the channel even after a long time interval has passed but will eventually result in a round-trip motion. These motions are all part of the normal randomness that causes variance in event arrival times in a Poisson process. The external voltage source may be viewed as a small-signal short that couples electrons that come in at one end directly to the other end with an almost-zero delay. Thus, the diffusive motions may be viewed as occurring on a loop such that the point of origin of an electron in the channel does not matter. Figure 7.6 (a) and Figure 7.6 (b) illustrate that the mean travel time for an anticlockwise journey and a clockwise journey are the same since $L_1 + L_2 = L_2 + L_1 = L$, respectively. Figure 7.6 (c) and Figure 7.6 (d) illustrate that the mean travel time is also independent of the point of origin of the electron in the channel since it is the total round-trip time that matters.

From our prior discussion, the rate of round-trip events in our Poisson process, whether clockwise or anticlockwise, has an event arrival rate of N_{TOT}/τ where N_{TOT} is the total number of electrons in the channel and τ is the transit time across the channel. The event rate of the Poisson process is N_{TOT}/τ because each electron's motion is completely uncorrelated with that of any others, there are N_{TOT} of them, they are indistinguishable from any other, and the statistics of the motion of any electron is well modeled as being identical to that of any other in a large population of electrons. Thus, the net electronic current noise is given by the shot-noise formula of Equation (7.13) to be

$$\overline{\Delta I_{ds}^2} = 2q\left(\frac{qN_{TOT}}{\tau}\right)\Delta f; \quad \tau = \frac{L^2}{2D_n}$$

$$\overline{\Delta I_{ds}^2} = 4qD_n\left(\frac{Q_I^{TOT}}{L^2}\right)\Delta f; \quad D_n = \mu_n\frac{kT}{q} \tag{7.20}$$

$$\overline{\Delta I_{ds}^2} = 4kT\mu_n\left(\frac{Q_I^{TOT}}{L^2}\right)\Delta f,$$

where $Q_I^{TOT} = qN_{TOT}$ is the total electronic charge in the channel and the Einstein relation $D_n = \mu_n\frac{kT}{q}$ has been used. Since the charge concentration in a subthreshold MOS transistor varies linearly from Q_{I0} at the source to Q_{IL} at the drain, we can compute that

$$Q_I^{TOT} = WL\left(\frac{Q_{I0} + Q_{IL}}{2}\right). \tag{7.21}$$

Since

$$I_{DSAT} = \mu_n\phi_t\frac{W}{L}Q_{I0} \text{ with } Q_{IL} = Q_{I0}e^{-V_{DS}/\phi_t}, \tag{7.22}$$

the noise of Equation (7.20) can be computed to be

$$\overline{\Delta I_{ds}^2} = 4kT\frac{\mu_n}{L^2}WL\left(\frac{I_{DSAT}}{2\mu_n\phi_t(W/L)}\right)\left(1 + e^{-V_{DS}/\phi_t}\right)\Delta f$$

$$= 2qI_{DSAT}\left(1 + e^{-V_{DS}/\phi_t}\right)\Delta f. \tag{7.23}$$

Since

$$i_{DS} = I_{DSAT}(1 - e^{-v_{DS}/\phi_t}), \tag{7.24}$$

the current in a subthreshold MOS transistor increases as we move from the linear region to the saturation region and eventually asymptotes. In contrast, the noise in a subthreshold MOS transistor, as computed from Equation (7.23), decreases as we move from the linear region to the saturation region and eventually asymptotes. The noise spectral density in saturation, $2qI_{DSAT}$, is half the noise spectral

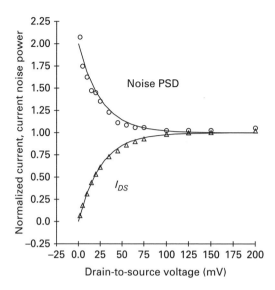

Figure 7.7. Experimentally measured power spectral density of drain-current noise in a subthreshold MOS transistor as the region of operation of the device is varied from linear to saturation. Reprinted with kind permission from [2] (©1993 IEEE).

density in the linear region, $4qI_{DSAT}$. Figure 7.7 shows the first experimental measurements of the electronic current and power spectral density of noise in subthreshold MOS transistors [2]. The measurements are in good accord with Equations (7.24) and (7.23), respectively. The power spectral density of the noise in the subthreshold transistor decreases by a half because the total charge in the device is reduced by a half as we move from the linear region to the saturation region. Alternatively, if we view the external current as being composed of the difference between a forward anticlockwise current of I_F and a reverse clockwise current of I_R, Equations (7.23) and (7.24) are analogous to stating that

$$I_{ds} = I_F - I_R$$
$$I_F = I_{DSAT}$$
$$I_R = I_F e^{-V_{DS}/\phi_t}$$
$$\overline{\Delta I_{ds}^2} = 2q(I_F + I_R)\Delta f.$$
(7.25)

Thus, in this interpretation, the noise spectral density is due to the sum of the shot noise of a forward diffusion current and a reverse diffusion current. It reduces by a factor of two as we move from the linear region to the saturation region because the shot noise of the reverse current, which is equal to that of the forward current when $V_{DS} = 0$, goes to zero as the reverse current goes to zero. Equation (7.25) reveals that the variance of the difference of two currents is maximal when both are present even though the mean is minimal at this point, a straightforward application of Equations (7.15) and (7.16), with $\alpha_1 = 1$ and $a_2 = 1$. Figure 7.8 also reveals that

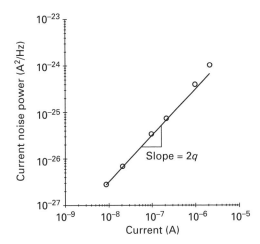

Figure 7.8. Experimentally measured power spectral density of drain-current noise in a subthreshold MOS transistor in saturation as a function of its saturation current, I_{DSAT}. The slope of the curve is $2q$, where q is the charge of an electron. Reprinted with kind permission from [2] (©1993 IEEE).

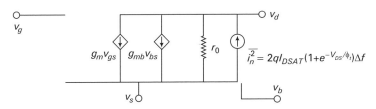

Figure 7.9. Small-signal model of a subthreshold MOSFET, with the drain-current noise generator included.

the current noise spectral density in saturation in a subthreshold MOS transistor is experimentally well modeled as $2qI_{DSAT}$ as predicted from Equation (7.23) [2]. Figure 7.9 shows how we incorporate a noise generator into the subthreshold MOS small-signal circuit. The capacitances of the small-signal model are not shown for clarity in the figure.

7.3   Noise in resistors

In a resistor, current flow occurs mainly by drift; that is, in addition to random thermal velocities with no preferred directions, all the electrons in a resistor also have a mean drift velocity with a preferred direction set by the direction of the electric field in the resistor. The mean value of the current is then determined primarily by the drift velocity and results in a predictable travel time for the whole population of electrons in the resistor. Typical drift velocities are extremely small for modest electric fields (on the order of $1/150\,\text{m/s}$) compared with thermal velocities ($10^5\,\text{m/s}$ at room temperature), so we can imagine that the electrons in the resistor are moving like a swarm of bees on a slowly moving train with tremendously random motions within the swarm but the entire swarm has a slowly moving velocity. Thus, even though the mean current in a resistor is mainly due to drift, the variance in this current will still primarily be due to diffusion. Effectively, we are assuming that the probability for an electron to move toward the right is only slightly greater than that to move to the left, such that the variance of the overall random process is still well modeled by a diffusive flow with equal probabilities for the electron moving to the right or to the left. If this assumption is true, then the *thermally generated internal diffusion currents in the resistor will exhibit shot noise just like they do in other electronic devices*. Thus, the noise in the resistor can be modeled exactly by Equation (7.20) as we did for subthreshold MOS transistors, i.e.,

$$\overline{\Delta I_{res}^2} = 4kT\mu_n\left(\frac{Q_I^{TOT}}{L^2}\right)\Delta f$$

$$= 4kT\mu_n\left(\frac{q(nAL)}{L^2}\right)\Delta f$$

$$= 4kT\left(\frac{(\mu_n qn)A}{L}\right)\Delta f \tag{7.26}$$

$$= 4kT\left(\frac{\sigma A}{L}\right)\Delta f$$

$$= 4kTG\Delta f,$$

where n is the charge concentration per unit volume, A is the area of cross-section of the resistor, L is its length, σ is the specific conductivity constant of the resistive material, and $G = 1/R$ is the conductance of the resistor. The final expression of Equation (7.26) is the famous Nyquist-Johnson result for the current noise of a resistor [3]. In fact, if we re-examine Equation (7.20) for a subthreshold MOS transistor, we find that

$$\overline{\Delta I_{ds}^2} = 4kT\mu_n\left(\frac{Q_I^{TOT}}{L^2}\right)\Delta f$$

$$\overline{\Delta I_{ds}^2} = 4kT\mu_n\frac{W}{L}\left(\frac{Q_I^{TOT}}{WL}\right)\Delta f$$

$$= 4kT\underbrace{\left(\mu_n\frac{W}{L}\overline{Q_I}\right)}_{\substack{\text{Conductance of a}\\\text{sheet resistor with}\\\text{charge concentration}\\\text{per unit area}=\overline{Q_I}}}\Delta f \tag{7.27}$$

$$\equiv 4kTG\Delta f,$$

where Q_I^{TOT} is the total electron charge, and $\overline{Q_I}$ is the mean charge concentration per unit area across the whole channel. That is, the noise within a subthreshold MOS transistor may be viewed as being the same as that of an equivalent sheet resistor with charge concentration per unit area of $\overline{Q_I}$.

7.4 Unity between thermal noise and shot noise

Thermal noise and shot noise are often mistakenly viewed to be different forms of white noise, i.e., noise with a flat power spectrum. However, our discussion and results suggest that

1. If there is no increase in variance due to large-electric-field effects, a good approximation is to assume that only diffusion currents cause noise in all electronic devices, and that drift currents are noiseless.
2. Shot noise is fundamental and due to the mathematics of Poisson processes.
3. *Thermal noise is shot noise due to internal thermally generated diffusion currents in physical devices.* The latter statement is true, to good approximation even if the dominant external current in the device is due to drift as long as the electric fields in the devices are such that the drift velocity is well below the thermal velocity.

This method of viewing white noise in electronic devices helps one avoid double counting shot noise and thermal noise sources and views them as different ways of expressing the same white-noise source. Consider the example shown in Figure 7.10 where we have a modest dc voltage V_{DC} across a resistor and we would like to know the noise current variance in this situation. Is the current noise given by

$$\left[2q\left(\frac{V_{DC}}{R}\right) + 4kTG\right]\Delta f?$$ (7.28)

Such a view might arise because we get shot noise from the dc current flow and thermal noise from the fact that it is a resistor and exhibits Nyquist-Johnson noise. In fact, measurements show that, if V_{DC} is a reasonable commonly used voltage of a few volts such that the electric fields in the resistor are not so large as to cause hot electron effects or drift velocities comparable to thermal velocities, then the noise in the resistor is simply $4kTG\Delta f$ and is independent of V_{DC}. The internal diffusion currents which cause noise in the resistor are unchanged by V_{DC} and

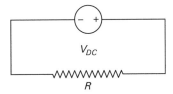

Figure 7.10. Finding the noise of a resistor R with a dc voltage source V_{DC} connected across its terminals.

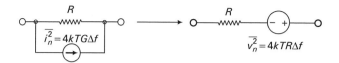

Figure 7.11. Noise models of resistors can be represented as equivalent Norton (left) and Thevenin (right) forms.

therefore yield the same noise. The $2q(V_{DC}/R)\Delta f$ term is meaningless and erroneous, since to first approximation the drift current V_{DC}/R is noiseless and affects the mean of the current but not its variance. If the variance of the velocity distribution of electrons is largely unchanged from the zero-field equilibrium value, the Einstein relation $D_n/\mu_n = kT/q$, which is strictly only valid in a zero-field thermal-equilibrium situation, is still valid. If the temperature goes to 0, the diffusion constant goes to 0 via the Einstein relation such that there is no thermal noise at zero temperature, as we would expect.

Figure 7.11 shows that we may Thevenize the Norton $4kTG\Delta f$ current-noise generator of a resistor into a voltage-noise generator of mean square value $4kTGR^2\Delta f$, i.e., of value $4kTR\Delta f$ as shown in the figure. Both forms of determining the noise in the resistor are equivalent and both are frequently used.

7.5   Noise in above-threshold MOS transistors

Our discussion thus far has revealed that white noise in an electronic device is primarily due to the shot noise of internal diffusion currents in the device even if the dominant external current is due to drift as long as electric fields in the device are moderate. Thus, even in an above-threshold long-channel MOS transistor, the noise is well described by Equation (7.20). Therefore, we obtain the noise in an above-threshold MOS transistor from the relationships below and some algebra:

$$\overline{\Delta I_{ds}^2} = 4kT\mu_n\left(\frac{Q_I^{TOT}}{L^2}\right)\Delta f$$

$$Q_I(x) = -\sqrt{Q_{I0}^2\left(1 - \frac{x}{L}\right) + Q_{IL}^2\left(\frac{x}{L}\right)}; \quad Q_I^{TOT} = W\int_0^L Q_I(x)dx \equiv WL\overline{Q_I}$$

$$Q_{I0} = -C_{ox}(V_{GS} - V_{TS}), \quad Q_{IL} = -C_{ox}\left(V_{GS} - V_{TS} - \frac{V_{DS}}{\kappa_S}\right)$$ (7.29)

$$\text{Define } \eta = 1 - \frac{V_{DS}}{V_{DSAT}} = \frac{Q_{IL}}{Q_{I0}},$$

$$\overline{\Delta I_{ds}^2} = 4kT\mu_n\frac{W}{L}C_{ox}(V_{GS} - V_{TS})\frac{2}{3}\left(\frac{1 + \eta + \eta^2}{1 + \eta}\right)\Delta f$$

We note that as we move from the linear region to saturation (η goes from 1 to 0 in Equation (7.29)), the power spectral density of the current noise reduces by a factor of 2/3 since the total charge in the MOSFET reduces by a factor of 2/3. In comparison, the power spectral density of the current noise reduces by a factor of 1/2 as we move from the linear region to the saturation region in the subthreshold regime. In both cases, the reduction is simply because of the presence of less charge in the saturation region.

7.6 Input-referred gate noise

From Equation (7.27) and the relationships for the transconductance of a transistor in Chapter 5, we can derive by substitution that the current noise in subthreshold and above-threshold operation is given by

$$\overline{\Delta I_{ds}^2} = 4kT(\gamma g_{ds0})\Delta f$$
$$\overline{\Delta I_{ds}^2} = 4kT\left(\gamma \frac{g_{msat}}{\kappa_S}\right)\Delta f \tag{7.30}$$

where g_{ds0} is the channel conductance at zero electric field and g_{msat} is the transconductance of the transistor in its saturation region. The parameter γ goes from 1 to 1/2 as we move from the linear region to the saturation region in the subthreshold regime. It goes from 1 to 2/3 as we move from the linear region to the saturation region in the above-threshold regime. The input-referred voltage noise $\overline{v_{gn}^2}$ is the equivalent noise on the gate needed to produce the output current noise of the transistor assuming that all the noise of the transistor is referred back to its input. Thus, it is given by

$$\overline{v_{gn}^2} = \frac{4kT(\gamma g_{ds0})}{g_m^2}\Delta f$$
$$\overline{v_{gn}^2} = \frac{4kT\left(\gamma \frac{g_{msat}}{\kappa_S}\right)}{g_m^2}\Delta f \tag{7.31}$$

At $V_{DS} = 0$ in the linear region where γ is at its highest ($\gamma = 1$) and g_{msat}/g_m^2 is infinite, $\overline{v_{gn}^2}$ is infinite. At $V_{DS} \geq V_{DSAT}$ in the saturation region where γ is at its lowest ($\gamma = 1/2$ in subthreshold and 2/3 in above-threshold operation) and g_{msat}/g_m^2 is at its highest value of $1/g_{msat}$, we find that

$$\overline{v_{gn}^2} = 4kT\left(\frac{\gamma}{\kappa_S g_{msat}}\right)\Delta f \tag{7.32}$$

in both subthreshold and above-threshold operation. Thus, we find that we need to have a high g_{msat} to reduce input-referred white noise per unit bandwidth, i.e., we need to burn power by operating at a high value of I_{DS}. Since the g_m/I_{DS} ratio is maximized in subthreshold operation, the input-referred noise per unit bandwidth obtained for a given power consumption is the least in subthreshold operation.

For the same reason, the bandwidth obtained for a given power consumption is also maximal in subthreshold operation.

7.7   1/*f* or flicker noise in MOS transistors

Thus far, we have only discussed white noise with a flat power spectrum as shown in Figure 7.12. White noise is fundamental, unavoidable, and due to the basic laws of random process and thermodynamics. However, $1/f$ noise, flicker noise, or pink noise is also a significant source of noise in the MOSFET at low frequencies. It is believed to be due to fluctuations in the number of electrons shuttling between traps in the gate oxide and the channel in the McWorther model [4] and fluctuations in the mobility of transistors in the Hooge model [5]. The quantification of $1/f$ noise is still largely based on empirical considerations.

Figure 7.13 shows a power spectral density plot of the current noise in an MOS transistor that includes white noise and $1/f$ noise. The $1/f$ noise has a power spectral density of $1/f^n$ with n typically near 1. The corner frequency where the white noise and $1/f$ noise yield the same value for power spectral density, f_c, is also shown in Figure 7.13. It can vary from a few 100 Hz in high-quality clean processes with few impurities during fabrication to a few MHz or a few tens of MHz in lower-quality dirty processes with more impurities during fabrication.

Figure 7.12. White noise has a power spectral density $P(f)$ that is flat (independent of f).

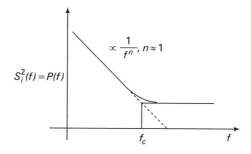

Figure 7.13. The actual current noise produced by a transistor has a power spectral density $S_i^2(f)$ that behaves like $1/f^n$ at low frequencies less than f_c. It becomes independent of f (white) for frequencies greater than f_c.

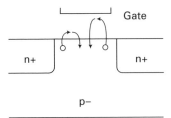

Figure 7.14. Intuitive explanation of flicker noise as the result of traps within the gate oxide capturing and releasing electrons from the channel.

The power spectral density of $1/f$ current noise is found to be described by

$$\overline{\Delta I_{ds}^2} = \frac{KI_{ds}^2}{f^{n_1}}\Delta f, \quad \text{in subthreshold } n_1 \approx 1$$
$$= \frac{KI_{ds}}{f^{n_2}}\Delta f, \quad \text{above threshold } n_2 \approx 1 \tag{7.33}$$
$$\propto g_m^2, \quad \text{both above and below threshold}$$

Figure 7.14 illustrates the trapping theory for $1/f$ noise, which is successful in explaining many observations about $1/f$ noise. These include explaining why the current noise is proportional to g_m^2 in both subthreshold and above-threshold operation and why input-referred $1/f$ voltage noise is reduced if transistors are made larger. The trapping theory states that electrons move into and out of traps or low-energy states in the gate oxide. This motion modulates Q_0, the fixed charge in the oxide, and consequently varies the flatband voltage V_{FB} such that there is a dynamic modulation of the threshold voltage of the transistor. These random time-varying threshold-voltage changes manifest as $1/f$ noise or flicker noise.

If the noise is caused by changes in the threshold voltage described by a voltage power spectral density given by

$$\overline{\Delta V_{T0}^2} \equiv S_v^2(f)\Delta f = \frac{K}{f^n}\Delta f, \tag{7.34}$$

then the current-noise power spectral density is given by

$$S_i^2(f) = g_m^2 S_v^2(f)$$
$$= \frac{KI_{ds}^2}{f^n}\Delta f, \quad \text{in subthreshold}$$
$$= \frac{KI_{ds}}{f^n}\Delta f, \quad \text{above threshold} \tag{7.35}$$

consistent with Equation (7.33).

If we model the traps or impurities as being randomly scattered throughout the wafer with a Poisson arrival rate per unit area of N_{imp}, the impurity charge density per unit area $B_{imp} = qN_{imp}$. From the mathematics of Poisson processes, we may then conclude that

$$\overline{\Delta V_{T0}^2} \equiv S_v^2(f)\Delta f = \left(\frac{\overline{\Delta Q_0^2}}{C_{ox}^2(WL)^2}\right)\frac{\Delta f}{f} \approx \left(\frac{qQ_0}{C_{ox}^2(WL)^2}\right)\frac{\Delta f}{f}$$
$$= \left(\frac{q(qN_{imp}(WL))}{C_{ox}^2(WL)^2}\right)\frac{\Delta f}{f} = \left(\frac{qB_{imp}}{fC_{ox}^2 WL}\right)\Delta f$$

(7.36)

Thus, we note that the input-referred voltage power spectral density of $1/f$ noise, $S_v^2(f)$, is reduced by having transistors of large area WL. The $1/f$ noise is larger in processes with a higher value of B_{imp}. The result of Equation (7.36) is analogous to the result of Equation (7.14). To lower white noise, we need to collect more electrons such that the variance relative to the mean is lowered, which means that we need to average over larger times; to lower $1/f$ noise, we need to have more traps such that the variance in the threshold voltage relative to the mean threshold voltage is lowered, which means that we need to average over larger areas of the transistor. Note that although larger areas imply more traps, they also imply more capacitance such that the mean threshold voltage is unchanged when the transistor size is increased.

Why is the power spectral density proportional to $1/f$? One hypothesis is as follows. There is a uniform distribution of traps in the gate oxide with shallow traps in the oxide near the channel having a high mean frequency of fluctuation, deep traps in the oxide near the gate having a low mean frequency of fluctuation, and traps at intermediate depths in the oxide having an intermediate mean frequency of fluctuation. The power spectrum of all the traps at different depths is then described by summing the power spectra of a set of bandpass filters with uniformly distributed center frequencies (first-order lowpass filters can be used instead; the results that follow remain essentially the same). If the bandpass power spectra are well approximated as being of constant Q (constant center-frequency-over-bandwidth ratio) because the behavior of the various traps at the different time scales is scale invariant or similar at all time scales, the power contained between any constant ratio of frequencies, such as an octave, obtained via summation will be flat or white. Thus the power spectral density of the noise will appear to be $1/f$. Figure 7.15 (a) shows the individual bandpass power spectra on a logarithmic frequency scale. Figure 7.15 (b) shows that the power per octave produced by summing together the bandpass spectra is indeed constant over the frequency range where the bandpass filters are present. Figure 7.15 (c) illustrates the summed power spectral density and shows that it is proportional to $1/f$ over this same frequency range. In general, if we find the overall power spectrum to be described by $1/f^n$ with n larger than 1, say, such a finding may imply a greater density of deep low-frequency traps than shallow high-frequency traps.

Figure 7.16 (a) shows the $S_i^2(f)$ power spectrum of $1/f$ noise extracted from measurements of current in an MOS transistor for over 6 months [6]. Figure 7.16 (b) shows the amplitude distribution of the $1/f$ noise, which is nearly Gaussian.

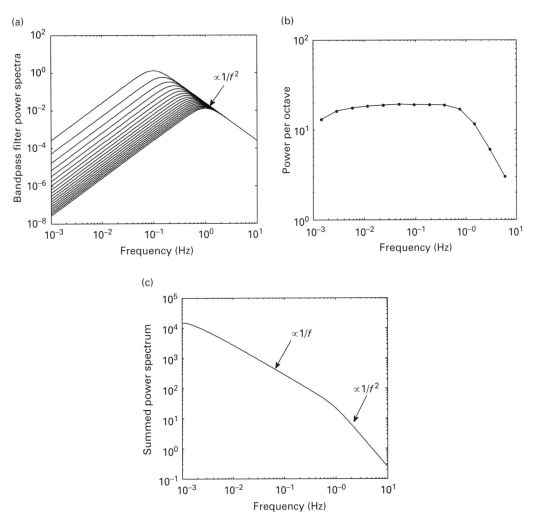

Figure 7.15a, b, c. Summing up the response of many constant-Q bandpass filters (with uniformly distributed center frequencies) produces a $1/f$ power spectrum. Individual bandpass filter responses are shown in part (a). Part (b) shows that the sum of these responses has constant power per octave (or any other constant frequency ratio). Part (c) plots the summed response and shows that it has a $1/f$ power spectrum.

7.8   Some notes on 1/*f* noise

We shall now itemize some points of note regarding $1/f$ noise:

1. Transistors that are pFETs usually have much less $1/f$ noise than nFETs presumably because electrons need much more energy to enter the oxide in pFETs than in nFETs: the mean energy of electrons in the channel of a pFET is near the bottom of the valence band rather than near the top of the conduction band in nFETs. Thus, many low-noise circuits are built with pFET transistors in their very first input-sensing gain stage.

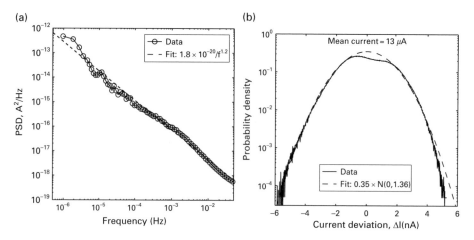

Figure 7.16a, b. Experimentally measured $1/f$ noise from an NMOS transistor: (a) power spectrum and (b) amplitude distribution. Reprinted with kind permission from [6].

2. JFETs and buried-channel MOSFETs usually have much lower $1/f$ noise because conduction is away from the surface where most of the traps lie near the surface. Many low-noise operational amplifiers, e.g., TL074, are built with JFET input transistors.

3. Bipolar transistors usually have significantly lower noise than MOSFETs since there are fewer traps at base-emitter junctions than at the silicon/silicon-dioxide interface.

4. It is likely that both mobility and trap-based number fluctuations are involved in causing $1/f$ noise. The $1/f$ noise in pFETs is likely significantly due to mobility fluctuations while that in nFETs may primarily be due to trap-based number fluctuations. However, even in nFETs, it has been shown that mobility fluctuations dominate in very weak inversion because of their inverse dependence on channel charge [7], [8]. Mobility fluctuations also have an inverse oxide capacitance dependence rather than a squared-inverse oxide capacitance dependence in Equation (7.36). In very strong inversion, where terminal resistances in the source exceed $1/g_m$, the $1/f$ noise in these terminal resistances can determine the overall $1/f$ noise.

5. The presence of $1/f$ noise is ubiquitous: nerve membranes, heart beats, and earthquake vibrations all exhibit $1/f$ noise. No successful universal theory of $1/f$ noise has been found.

6. The noise power in any octave or in any frequency interval with a constant frequency ratio is constant for $1/f$ noise. In contrast, the noise power in any frequency interval with a constant bandwidth is constant for white noise.

$$\int_{f_L}^{f_H} \frac{df}{f} = \ln\left(\frac{f_H}{f_L}\right)$$

$$\int_{f_L}^{f_H} df = f_H - f_L \tag{7.37}$$

The logarithmic scaling of $1/f$ noise explains why its 'infinite value' at 0 is not a practical issue: the increase in integrated $1/f$ noise as we decrease f_L scales extremely slowly. The contribution of the frequency band from $1/(300 \text{ years})$ until the present day is nearly the same as that of the contribution of the frequency band from $1/(1 \text{ day})$ until 1 Hz which is nearly the same as the contribution of the frequency band from $1/(1 \text{ s})$ until $1/(10^{-5} \text{ s})$. The latter frequency of $1/(10^{-5} \text{ s}) = 10^5 \text{ radians/s}$ is a reasonable flicker-noise corner frequency beyond which white noise is dominant. Thus, even if our electronics was manufactured 300 years ago (not possible since the MOSFET was not even invented until 1934 and most electronics today is less than 50 years old), the integrated $1/f$ noise from a start time at its time of manufacture till the present day would only be 3 times as large as the integrated $1/f$ noise from $1/(1 \text{ s})$ until $1/(10^{-5} \text{ s})$. In practice, the start time for integrating noise is frequently reset since we switch off our electronics and start again. Furthermore, there are only a finite number of traps and the slowest trap fluctuation frequency is limited by the gate-oxide thickness. Thus, at very low frequencies, environmental fluctuations due to temperature or the weather may determine the noise of the transistor rather than its own intrinsic $1/f$ noise. Recent measurements [6] that subtract out such temperature variations via a simple and inexpensive common-mode noise-cancellation technique have shown $1/f^{-1.2}$ power spectra down to $0.9 \, \mu\text{Hz}$ with a Gaussian amplitude distribution as revealed in Figures 7.16 (a) and 7.16 (b), respectively.

7. The approximately scale-invariant behavior of $1/f$ noise has led to fractal theories of $1/f$ noise.

8. The input-referred voltage noise per unit bandwidth for white noise, $v_{gn}^2(f)$, implicitly stated in Equation (7.32), and the input-referred voltage noise per unit bandwidth for flicker noise, $S_v^2(f)$, implicitly stated in Equations (7.34) and (7.36), are explicitly restated below.

$$v_{gn}^2(f) = 4kT\left(\frac{\gamma}{\kappa_S g_{msat}}\right)$$
$$S_v^2(f) = \left(\frac{qB_{imp}}{C_{ox}^2 WL}\right)\frac{1}{f} \tag{7.38}$$

The two right-hand sides of the expressions of Equation (7.38) are equal at the flicker-noise corner frequency f_c. Thus, by equating these two expressions, we may conclude that

$$f_c = \left(\frac{qB_{imp}}{C_{ox}^2 WL}\right)\left(\frac{\kappa_S g_{msat}}{4kT\gamma}\right)$$
$$f_c = \left(\frac{qB_{imp}}{4kT\gamma C_{ox}}\right)\left(\frac{\kappa_S g_{msat}}{C_{ox}WL}\right) \tag{7.39}$$
$$f_c \approx \left(\frac{qB_{imp}}{4kT\gamma C_{ox}}\right)f_T$$

Since the f_T of transistors is lower in the subthreshold regime rather than in the above-threshold regime, the flicker-noise corner frequency is lower in the subthreshold regime than in the above-threshold regime. Therefore, it is common to see increasingly more $1/f$ noise in a circuit as its mode of operation is changed from the subthreshold regime to the above-threshold regime. Figure 12.15 in Chapter 12 reveals experimental measurements of this phenomenon.

9. The complete small-signal noise model that incorporates white noise and $1/f$ noise into the noise generator of Figure 7.9 and that is valid in both the linear and saturation regions of operation is listed below.

$$\overline{i_n^2} = \left[2qI_{sat}\left(1 + e^{-V_{DS}/\phi_t}\right) + \frac{qB_{imp}}{C_{ox}^2 WL} \frac{I_{sat}^2 \kappa_s^2 \left(1 - e^{-V_{DS}/\phi_t}\right)}{\phi_t^2} \frac{1}{f} \right] \Delta f \quad \leftarrow \quad \text{Subthreshold}$$

$$\overline{i_n^2} = \left[4kT\mu\frac{W}{L}C_{ox}(V_{GS} - V_{TS})\frac{2}{3}\left(\frac{1 + \eta + \eta^2}{1 + \eta}\right) \right. \tag{7.40}$$

$$\left. + \frac{qB_{imp}}{C_{ox}^2} \frac{2\kappa\mu C_{ox}I_{sat}}{L^2} (1 - \eta^2)^2 \frac{1}{f} \right] \Delta f \quad \leftarrow \quad \text{Above threshold}$$

10. In deep submicron devices where oxide capacitances are small, it is possible to see individual filling and emptying of traps such that $1/f$ noise manifests as discrete jumps in current noise or input-referred voltage noise in a random telegraph fashion.

11. It is possible to reduce $1/f$ noise by cycling the gate of a MOSFET between accumulation, which empties the trap since it is energetically unfavorable for the electron to remain in it, and regular-inversion operation, which is not exercised long enough for the trap to refill [9]. Other techniques for reducing $1/f$ noise and transistor threshold-voltage mismatch effects, which are the dc manifestation of $1/f$ noise from extremely slow-time-constant traps, have been reviewed in [10]. These include autozeroing, correlated double sampling, and chopper stabilization. In Chapter 8, we shall discuss the technique of lock-in amplification, upon which chopper stabilization is based, which also mitigates the effects of $1/f$ noise.

7.9   Thermal noise in short-channel devices

In short-channel devices that have high lateral and vertical electric fields, the phenomena of velocity saturation, vertical mobility reduction, carrier heating, and channel-length modulation all conspire to affect thermal noise in above-threshold operation. These phenomena are described in Chapter 6 and a review of this chapter is helpful for understanding this section. The thermal noise in subthreshold operation is largely unaffected so we shall focus only on above-threshold operation. In the velocity-saturated above-threshold regime, electrons near the drain end of the channel are at a drift velocity v_{sat} that is near their thermal velocity and electrons near the source end of the channel have a drift velocity that is quite small. Thus, we cannot assume that the drift velocity is much less than the thermal velocity in all regions of the channel such that diffusive random processes are largely unaffected throughout the channel.

The shot noise due to internal thermally generated diffusion currents within the transistors is still the source of its white noise at each local point in the transistor channel. That is, the drift current still contributes no noise at any point in the channel since it is still the variance of a distribution that contributes noise, not its mean. However, the variance of the distribution can no longer be assumed to be independent of its mean such that the effective value of the diffusion constant is no longer D_n. In fact, the temperature T, mobility μ_n, diffusion constant D_n and conductance G have an effective value that is different along each portion of the channel. Thus, computations of the noise require care and attention to these variations all of which affect the macroscopic observed noise. A semi-empirical approach to noise generation with several infinitesimal Norton-like $4kT(x)G(x)$ white-noise-generation circuits like those in Figure 7.11 that are distributed throughout the channel, and with parameters $R(x)$ and $G(x)$ corresponding to the local areal sheet-charge density as in Equation (7.27), is often used to compute the noise. Each such noise-generation circuit within the channel sees the impedance of the rest of the channel to its right, which corresponds to looking into one terminal of a half transistor, and the impedance of the rest of the channel to its left, which corresponds to looking into the terminal of another half transistor. The two half transistors are hooked in series via an external V_{DS} source, which forms an ac short, and are effectively in parallel with each Norton noise-generation circuit. The macroscopic noise due to one noise-generation circuit in the channel then corresponds to the current propagated through from the noise generator to either half transistor. The macroscopic mean-squared noise due to all noise generators in the channel is computed by integrating all such mean-squared contributions from each generator over all x in the channel. Such a computation has been done in [8], but the mathematics is algebraically complicated. We shall focus on providing intuition and answers that are useful for design. The reader should keep in mind that noise in short-channel transistors is still an unsolved problem, as far as a deep understanding from first principles is concerned.

Figure 3.6 in Chapter 3 shows an energy diagram for large-v_G and large-v_D above-threshold operation that is helpful for visualizing a typical electron's journey within a transistor. The electrons behave like a set of skiers who start out in a crowded and thermally unruly pack at the high-energy source-region summit peak. They then quickly thin out and speed up to a nearly constant drift velocity v_{sat} near the low-energy drain-region slope bottom. The electrons get increasingly hot as they accelerate down the ski slope from the source end of the slope towards the drain end. The hot electrons finally fall off the ski slope into the drain pit, cool off to normal thermal velocities, and then take the speed-of-light V_{DS} transport lift back to source base camp. They then wait for thermal energy to lift them out of source base camp and into the high-energy source-region summit peak to begin their ski journey all over again. Many millions of skier electrons do this over and over again to create what we know as a macroscopic steady-state electronic current. The variations in the round-trip times of the electrons leads to a stochastic arrival process whose macroscopic consequence is current noise.

To compute the value of the $4kT(x)G(x)$ white-noise generator at a point x along the ski channel, we assume that the carrier temperature $T(x)$ varies according to Equation (6.27) in Chapter 6, and that $G(x)$ corresponds to the differential tangential mobility at the given local value of electric field. Thus, there is no $4kT(x)G(x)$ noise generation at portions of the channel where v is constant at v_{sat} and the differential mobility is zero (so $G(x)$ is 0) even though the carrier temperature there is large. That is, the skiers near the bottom of the ski slope are an ordered bunch. In contrast, $4kT(x)G(x)$ is larger near the source regions where the differential mobility and temperature are near their low-electric-field non-velocity-saturated values. That is, the skiers at the source-region summit peak are an unruly bunch. Since voltage variations at a local point in the channel caused by local skier current fluctuations affect the slope that skiers arriving to that point see and skiers departing from that point see, these variations lead to changes in the fluxes of electrons behind and ahead of the skiers. Thus, the noise generation of each bunch of skiers along different portions of the ski slope propagates throughout the ski slope behind them and ahead of them such that it causes macroscopic measurable fluctuations in skier fluxes at the V_{DS} transport lift.

Velocity saturation tries to make the charge at the source, Q_S, and the charge at the drain, Q_D, equal via the negative-feedback loop of Figure 6.2 (b) such that Q_D is increased compared to its non-velocity-saturated value. It also increases the noise generation at several portions in the channel where the electrons are hot. The increase in the size of Q_D increases the number of carriers in the channel and the increase in average channel temperature both increase the current noise in velocity-saturated devices. However, the decrease in channel conductance reduces noise-propagation transfer functions and the presence of noise-generator-free regions in the channel near the drain serve to reduce current noise in velocity-saturated devices. The net overall effect of velocity saturation is still an increase in the overall current noise. Vertical mobility reduction (VMR) delays the onset of velocity saturation (see Chapter 6) and serves to reduce current noise in velocity-saturated devices. Channel length modulation reduces the effective length of the transistor and serves to increase current noise via the thermal-noise-current dependence on transistor length in Equation (7.20), and also by increasing the average electric field throughout the channel making the noise increase due to velocity-saturation more potent. When all of these effects are combined, we find that the overall noise in a velocity-saturated transistor can still be described by Equations (7.30) and (7.31) except that γ, the excess noise factor, is no longer 2/3 in above-threshold operation (or 1/2 in subthreshold operation) but has an empirical value that is near 1 or 2 in many processes [11]. These values of γ occur at high values of V_{GS} where vertical mobility reduction is effective. At lower values of V_{GS},

γ can almost double and in some reported measurements can be even as high as 10. High values of V_{GS} also help attenuate noise by making Q_S and Q_D more equal, thus reducing lateral electric fields and all the associated effects of velocity saturation.

7.10 Thermal noise in moderate inversion

Equation (7.27) for the thermal noise is valid in all regions of operation as long as electric fields are not high enough to create hot electrons. Thus, it applies in weak inversion, moderate inversion, and strong inversion. In weak inversion and saturated operation, it can be evaluated as

$$\overline{\Delta I_{ds}^2} = 2qI_{DSAT}\Delta f \tag{7.41}$$

In strong inversion and saturated operation, it can be evaluated as

$$\overline{\Delta I_{ds}^2} = 4kT\left(\frac{2}{3}\frac{g_{msat}}{\kappa_S}\right)\Delta f \tag{7.42}$$

In moderate inversion, there is no exact closed-form expression for Equation (7.27) in terms of the terminal voltages of the transistor. Nevertheless, in the spirit of the empirical EKV equation of Chapter 4 (Equation 4.48), we can invent a formula that interpolates between the two regimes:

$$\boxed{\overline{\Delta I_{ds}^2} = \frac{2qI_{DSAT}}{1 + \dfrac{(3/4)\kappa_s(V_{GS} - V_{TS})}{2\phi_t}}\Delta f} \tag{7.43}$$

Figure 7.17 reveals that simulations of the thermal noise in a transistor of $200\,\mu$m width and $2\,\mu$m length in a UMC $0.13\,\mu$m process are consistent with

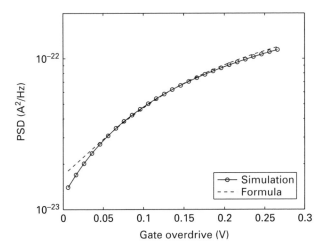

Figure 7.17. Thermal noise in moderate inversion.

the formula of Equation (7.43). In the simulation data, I_{DSAT} is nearly $52\,\mu A$ when $(V_{GS} - V_{TS}) \simeq 0$ and $I_{DSAT} = 1.03$ mA when $(V_{GS} - V_{TS}) \simeq 0.2$ V with $V_{TS} = 0.234$ V. The overdrive voltage, $(V_{GS} - V_{TS})$, is directly plotted on the horizontal axis with $\kappa_S = g_m/(g_m + g_{mb}) \approx 0.88$ used in the fit. Note that Equation (7.43) is only valid for positive values of $(V_{GS} - V_{TS})$. The simulation data were graciously provided to the author by Dr. Soumyajit Mandal.

The ratio

$$A = \frac{(3/4)\kappa_S(V_{GS} - V_{TS})}{2\phi_t} \tag{7.44}$$

is interestingly the ratio of the diffusion-to-drift transit time in a channel from Equations (5.53) and (5.54) in Chapter 5. It is a measure of how much stronger the drift current is than the diffusion current. The white-noise power in the case that the current is entirely diffusive, $2qI_{DSAT}\Delta f$, is reduced by a factor of $1/(1 + A)$ to a lower value given by Equation (7.43). This value corresponds to the current having a drift component in it that is A times stronger than its diffusive component. Equation (7.43) manifests a different way of expressing the notion that the noise is entirely due to diffusion: if $2qI_{diffusion}$ describes the noise PSD, but the current flow is mainly by drift, then there will appear to be a reduction in the $2qI_{drift}$ noise of $I_{diffusion}/(I_{diffusion} + I_{drift})$. Figure 12.16 (b) in Chapter 12 provides an experimental example that illustrates the concepts behind Equation (7.43).

The reduction of diffusion-based shot noise in Equation (7.43) is reminiscent of a feedback loop in vacuum tubes that serves to reduce the shot noise in the current as the intensity of current increases due to an effect termed space charge smoothing. If a excessively large number of electrons are emitted from the cathode due to stochastic fluctuations, they lower the potential in a space-charge layer near the cathode preventing future electrons from being emitted as easily [12], [13]. Thus, there is feedback regulation in the numbers of electrons that are emitted from the cathode. This feedback regulation serves to reduce the noise variation in the current to a value that is below that expected from purely Poisson statistics $(2qI\Delta f)$, and we say that space charge smoothing of the noise has occurred.

Electrostatic feedback interactions occur at the channel-to-source junction, channel-to-drain junction, and throughout the transistor channel in Figure 3.4 (in Chapter 3) as well. If too many skier electrons get bunched together at a point in the channel for stochastic reasons, they lower the voltage to a value that is below the equilibrium value at that point or equivalently raise the energy at that point; the flux of skiers arriving behind them will then decrease in number since the slope of the energy landscape decreases for them; the flux of skiers departing from the point will increase in number since the slope of the energy landscape increases for them. Thus, a stochastic increase in the number of skiers at a point leads to more skiers departing away from that point than arriving at it, which serves to reduce the original stochastic fluctuation. Thus, the number fluctuations at any local point in a transistor channel are constantly being reduced through electrostatic feedback. Such feedback is created by a drift-and-diffusion inter-action as the slope of the ski slope governs drift flows. Drift-and-diffusion feedback interactions are similarly important in setting the equilibrium value of the minority carrier concentration at the edge of a pn junction. In fact, Equation (7.43) may be viewed as a reduction in the shot-noise flux of source-to-channel junction emissions due to local electrostatic feedback that regulates it. Since the electrostatic feedback is non-existent in subthreshold operation (see Figure 3.12), we experience the full shot noise in this regime.

7.11 Induced gate noise

In certain RF circuits, it may be necessary to include an induced gate-noise generator in an MOS device. At high frequencies, some of the diffusive currents in the channel may be shunted by the oxide capacitance and thus appear at the gate terminal rather than at the drain or source terminals. This generator is correlated with the normal drain-to-source current noise generator since both noise currents arise from the same diffusive currents in the channel. The induced gate-noise generator is only of importance when the operating RF frequencies are very near the f_T of the device, a bad strategy in ultra-low-power design as we discuss in Chapter 22. Thus, we shall not focus much on the latter noise generator in this book. It is discussed extensively in other texts, e.g., [14]. Nevertheless, we summarize certain key points about it.

The induced gate noise $\overline{i_{ng}^2}$ behaves as though the drain-current noise were reflected back into the gate of the MOS transistor via a $1/\beta^2$ factor with $\beta = \omega C_{gs}/g_m$ and with a factor γ_{ng} that accounts for the distributed nature of the MOS capacitance in Figure 3.4:

$$\overline{i_{ng}^2} = 4kT\gamma_{ng}g_m\left(\frac{\omega C_{gs}}{g_m}\right)^2$$

$$\gamma_{ng} \approx \frac{4\kappa}{15}$$

(7.45)

However, it is correlated with the normal drain current noise $\overline{i_{nd}^2}$ according to [8]:

$$c = \frac{i_{ng}i_{nd}^*}{\sqrt{|i_{nd}|^2|i_{nd}|^2}}$$

$$c = +j\frac{1}{\sqrt{3}} \text{ in weak inversion}$$

(7.46)

$$c = +j\sqrt{\frac{5}{32}} \text{ in strong inversion}$$

7.12   Some caveats about noise

Purely reactive devices like inductors, capacitors, masses, springs, i.e., devices that store energy but do not dissipate it as heat, do not exhibit noise. They do not exhibit noise because of a fundamental theorem known as the fluctuation-dissipation theorem [15]. The theorem states that in order for a device to exhibit thermal fluctuations, it must dissipate energy. The physical insight behind the theorem is that dissipation couples a device to the many degrees of freedom of the heat reservoir that always surrounds it. The device dissipates energy into these degrees of freedom and these degrees of freedom, in turn, pour energy back into the device which manifests as noise. If a device is ideal and lossless, then it is not coupled to its surroundings, nothing can affect it, and therefore it does not exhibit noise. The fluctuation-dissipation theorem is a generalization of the Nyquist-Johnson formula for noise in an electrical resistor, i.e., $\overline{\Delta I_{res}^2} = 4kTG\Delta f$, to noise in any device that dissipates energy whether mechanical, chemical, etc. In Chapter 8, we shall use this theorem to compute noise in mechanical MEMS resonators. In Chapter 23, we shall use the concepts of noise discussed in this chapter and in Chapter 8 to compute noise fluxes in chemical reactions.

Small-signal resistances that are not physical but that merely arise from definitions, e.g. r_π and r_o, which are small-signal ac resistances (obtained by linearizing the I–V curve of the transistor about an operating point) do not exhibit noise. On the other hand, real resistances like r_S and r_G that may be present at the source or gate terminals do exhibit noise.

References

[1] Aldert Van der Ziel. *Noise: Sources, Characterization, Measurement* (Englewood Cliffs, NJ: Prentice-Hall, 1970).

[2] R. Sarpeshkar, T. Delbruck and C. A. Mead. White noise in MOS transistors and resistors. *IEEE Circuits and Devices Magazine*, **9** (1993), 23–29.

[3] H. Nyquist. Thermal agitation of electric charge in conductors. *Physical Review*, **32** (1928), 110–113.

[4] A. L. McWorther. $1/f$ noise and germanium surface properties. In *Semiconductor Surface Physics*, ed. RH Kingston (Philadelphia, Penn.: University of Pennsylvania Press; 1957), pp. xvi, 413 p.

[5] F. N. Hooge. $1/f$ noise. *Physica*, **83** (1976), 14–23.

[6] S. Mandal, S. K. Arfin and R. Sarpeshkar. Sub-μHz MOSFET $1/f$ noise measurements. *Electronics Letters*, **45** (2009).

[7] R. Kolarova, T. Skotnicki and J. A. Chroboczek. Low frequency noise in thin gate oxide MOSFETs. *Microelectronics Reliability*, **41** (2001), 579–585.

[8] Christian Enz and Eric A. Vittoz. *Charge-based MOS Transistor Modeling: The EKV Model for Low-Power and RF IC Design* (Chichester, England; Hoboken, NJ: John Wiley, 2006).

[9] E. A. M. Klumperink, S. L. J. Gierkink, A. P. Van der Wel and B. Nauta. Reducing MOSFET $1/f$ noise and power consumption by switched biasing. *IEEE Journal of Solid-State Circuits*, **35** (2000), 994–1001.

[10] C. C. Enz and G. C. Temes. Circuit techniques for reducing the effects of op-amp imperfections: autozeroing, correlated double sampling, and chopper stabilization. *Proceedings of the IEEE*, **84** (1996), 1584–1614.

[11] C. H. Chen and M. J. Deen. Channel noise modeling of deep submicron MOSFETs. *IEEE Transactions on Electron Devices*, **49** (2002), 1484–1487.

[12] G. L. Pearson. Shot Effect and Thermal Agitation in an Electron Current Limited by Space Charge. *Physics*, **6** (1935).

[13] B. J. Thompson, D. O. North and W. A. Harris. Fluctuations in space-charge-limited currents at moderately high frequencies. *RCA Review*, **4** (1940).

[14] Thomas H. Lee. *The Design of CMOS Radio-Frequency Integrated Circuits*. 2nd ed. (Cambridge, UK; New York: Cambridge University Press, 2004).

[15] H. B. Callen and T. A. Welton. Irreversibility and generalized noise. *Physical Review*, **83** (1951), 34–40.

8 Noise in electrical and non-electrical circuits

When we tug on a single thing in nature, we find it attached to everything else.

John Muir

Devices that dissipate energy, such as resistors and transistors, always generate noise. This noise can be modeled by the inclusion of current-noise generators in the small-signal models of these devices. When several such devices interact together in a circuit, the noise from each of these generators contributes to the total current or total voltage noise of a particular signal in a circuit. In this chapter, we will understand how to compute the total noise in a circuit signal due to noise contributions from several devices in it. We shall begin by discussing simple examples of an RC circuit and of a subthreshold photoreceptor circuit. We shall see that the noise of both of these circuits behaves in a similar way. We shall discuss the equipartition theorem, an important theorem from statistical mechanics, which sheds insight into the similar noise behavior of circuits in all physical systems. We shall then outline a general procedure for computing noise in circuits and apply it to the example of a simple transconductance amplifier and its use in a lowpass filter circuit. We will then be armed with the tools needed to understand and predict the noise of complicated circuits, and to design ultra-low-noise circuits.

We shall conclude by presenting an example of an ultra-low-noise micro-electro-mechanical system (MEMS), a capacitance-sensing system capable of sensing a 0.125 parts-per-million (ppm) change in a small MEMS capacitance (23-bit precision in sensing) [1]. This system will help us understand how mechanical noise and electrical circuit noise both determine the minimum detectable signal of the sensor. In fact, we will see that noise in non-electrical systems can be treated and understood with the same fundamental concepts that we have used in electrical systems. In Chapter 24, we will see that such noise concepts can be extended to understand noise in chemical and molecular systems in biology as well.

The example will also help us understand the important technique of lock-in detection, an ingenious technique for removing non-fundamental $1/f$ noise in circuits such that detection limits are only set by fundamental thermal noise. Lock-in detection techniques require the use of multiplier or mixer circuits. Lock-in amplifiers that implement these multipliers with passive switch-based mixers are often referred to as chopper-modulated amplifiers. The lock-in techniques will be presented in the context of a vibration sensor. However, such techniques can be adapted for several applications including DNA and bio-molecular detection in cantilever-based MEMS systems or in low-power Electroencephalogram (EEG) amplifiers [2], [3].

8.1   Noise in an RC lowpass-filter circuit

Figure 8.1 (a) shows a simple RC circuit. Figure 8.1 (b) shows the same circuit with the resistor replaced with a noise generator of $4kTRf$ in series with it. Figure 8.1 (c) is the circuit that we use for computing the noise, obtained by grounding v_{IN}. When computing noise, we ground all input voltage sources and open all input current sources such that we can focus on the noise of the circuit even when no input is present. We find that:

$$v_{out}(s) = \frac{v_{in}(s)}{\tau s + 1}, \quad \tau = RC = \frac{1}{2\pi f_c}, \quad s = j\omega$$

$$\overline{v_{out}^2}(\omega) = \frac{\overline{v_{in}^2}(\omega)}{|1 + j\omega\tau|^2} = \frac{\overline{v_{in}^2}(\omega)}{1 + \omega^2\tau^2}$$

$$\overline{v_{out}^2} = \int_0^\infty \frac{\overline{v_{in}^2}(f)df}{1 + (2\pi f)^2/(2\pi f_c)^2}$$

$$= \int_0^\infty \frac{4kTRdf}{1 + (f/f_c)^2} \tag{8.1}$$

$$= 4kTR \int_0^\infty \frac{f_c du}{1 + u^2}$$

$$= 4kTRf_c \tan^{-1}(u)\big|_0^\infty$$

$$= 4kTR \times \frac{1}{2\pi RC} \times \frac{\pi}{2}$$

$$\overline{v_{out}^2} = \frac{kT}{C}$$

The calculations of Equation (8.1) illustrate how we calculate the output voltage noise measured at the capacitor of the RC circuit. We first find the transfer function from the noise generator to the output as though it were a regular input. Then, we evaluate the mean-square output noise per unit bandwidth between frequencies f and $f + df$ by multiplying the mean-square input noise per unit bandwidth due to the noise generator between these frequencies and the square of the magnitude of the transfer function that we have just computed at frequency f. Finally, we integrate the mean-square output noise over all frequencies from 0 to ∞ to compute the total noise to be kT/C.

The surprisingly simple kT/C result arises from the fact that the noise per unit frequency bandwidth is $4kTR$, the bandwidth is $1/RC$ in ω units or $1/(2\pi RC)$ in

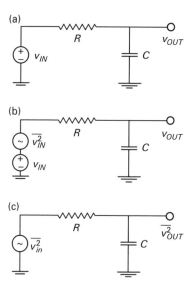

Figure 8.1a, b, c. A simple RC circuit. (a) The actual circuit, (b) the circuit with resistor noise added, and (c) the circuit used for making noise calculations.

frequency units, and the equivalent rectangular bandwidth of a single-pole low-pass filter is $\pi/2$ greater than $1/\tau$ such that the total output noise is given by

$$4kTR \times \frac{1}{RC} \times \frac{1}{2\pi} \times \frac{\pi}{2} = \frac{kT}{C}. \tag{8.2}$$

Figure 8.2 illustrates where the $\pi/2$ factor comes from: the gradual rolloff of a single-pole lowpass filter means that the total area enclosed by the filter's magnitude-squared transfer function can be computed by integrating the area under an equivalent rectangular filter with an infinitely sharp rolloff slightly past the single-pole filter's corner frequency. As Figure 8.2 shows, when we are a factor of $\pi/2$ past the single-pole filter's corner frequency $1/\tau$, the extra area contributed by the rectangular filter before $\pi/(2\tau)$ is exactly equal to the extra area contributed by the single-pole filter past $\pi/(2\tau)$. Double-pole and higher-order lowpass filters will have factors like $\pi/2$ as well except that they will be nearer 1.0 due to their sharper rolloffs.

Figure 8.3 shows experimental power spectra of an active resistor-like circuit with noise properties similar to that of a resistor [4]. We see that a large value of resistance corresponds to a large noise per unit bandwidth and a small bandwidth while a small value of resistance corresponds to a small noise per unit bandwidth and a large bandwidth. A moderate value of resistance has an intermediate noise per unit bandwidth and intermediate bandwidth. In all three cases, as expected from Equation (8.2), the total squared area of the power spectra integrated over all frequencies is identical.

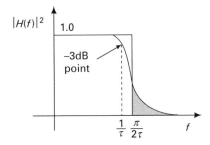

Figure 8.2. Effective noise bandwidth of a first-order lowpass filter.

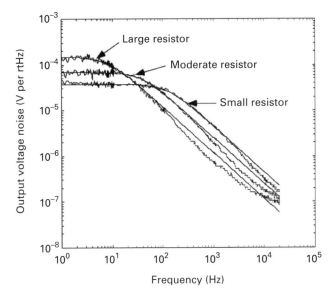

Figure 8.3. Experimentally measured output voltage noise of various resistor-like circuits. Reprinted with kind permission from [4], Springer Science and Business Media.

8.2   A subthreshold photoreceptor circuit

Figure 8.4 (a) reveals a subthreshold photoreceptor circuit. It is composed of a reverse-biased photodiode that generates a photocurrent proportional to the light level, an MOS transistor that operates in its subthreshold regime due to the small photocurrent that it equilibrates to, and an intentional or parasitic capacitance C. The photodiode is easily built with an np junction, which can be the same as the np junction formed between the source and the bulk of the MOS transistor in Figure 8.4 (a). Each photon generates an electron-hole pair with a quantum efficiency of α. The value of α is dependent on the wavelength of the photon but is typically near 0.3. The holes travel towards the p side of the junction, a region of low energy for holes, while the electrons travel toward the n side of the junction,

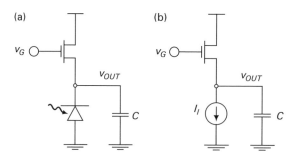

Figure 8.4a, b. A subthreshold photoreceptor using a photodiode. (a) Actual circuit and (b) an equivalent circuit with a light-dependent current source replacing the photodiode.

a region of low energy for electrons. The net resulting photocurrent may then be represented by the circuit of Figure 8.4 (b) where I_l is a light-dependent photocurrent such that

$$\alpha I_{photon} = \frac{I_l}{q} \tag{8.3}$$

The value of I_{photon} is the number of photons per second impinging on the light-collection area of the photodiode and is proportional to the intensity of the light shining on the photodiode. The photon arrivals exhibit Poisson statistics, the probability that an electron-hole pair is generated from a photon is a random process, and the flow of photoelectrons is governed by random collisions as well. Thus, the electronic photocurrent obeys Poisson statistics and its noise statistics are well modeled by a $2qI_l$ shot-noise term. Since the current through the sub-threshold MOS transistor is also I_l in equilibrium, and the transistor operates in saturation if V_{DD} is sufficiently high, its noise statistics are also well modeled by a $2qI_l$ shot-noise term. Thus, the small-signal noise circuit corresponding to Figure 8.4 is that of Figure 8.5 (a) with

$$g_m = \frac{\kappa_s I_l}{\phi_t}$$

$$g_{mb} = (1 - \kappa_s)\frac{I_l}{\phi_t} \tag{8.4}$$

$$g_m + g_{mb} = g_s = \frac{I_l}{\phi_t}$$

Some simplifications lead Figure 8.5 (a) to transform to Figure 8.5 (b) where the drawn resistor is a noiseless small-signal ac resistance of value $1/g_s$. The circuit of Figure 8.5 (b) then simplifies to that of Figure 8.5 (c) with the current source being the only noise generator. Simple analysis then reveals that

$$v_{out} = \frac{i_{in}R}{1 + sRC}, \quad R = \frac{1}{g_s}, \quad \tau = \frac{1}{2\pi f_c} = \frac{C}{g_s}$$

$$\overline{v_{out}^2} = \frac{\overline{i_{in}^2}R^2}{1 + \tau^2 \omega^2} = \int_0^\infty \frac{4qI_lR^2 df}{1 + (2\pi f)^2/(2\pi f_c)^2}$$

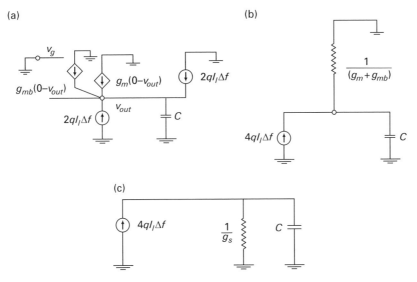

Figure 8.5a, b, c. Small-signal equivalent circuit of the subthreshold photoreceptor. (a) Actual circuit, (b) simplified circuit, and (c) simplified circuit with $(g_m + g_{mb})$ replaced with g_s.

$$= 4qI_l \times \frac{1}{g_s^2} \times f_c \times \int_0^\infty \frac{du}{1 + u^2}$$

$$= 4qI_l \times \frac{\phi_t^2}{I_l^2} \times \frac{I_l}{2\pi C \phi_t} \times \frac{\pi}{2}$$

$$= 4q \frac{\phi_t}{C} \times \frac{1}{2\pi} \times \frac{\pi}{2} \tag{8.5}$$

$$\overline{v_{out}^2} = \frac{kT}{C}$$

Once again, we find that, independent of the light level or equivalently g_s, the total output noise of the photoreceptor is given by

$$\frac{4kT}{g_s} \times \frac{1}{2\pi} \times \frac{g_s}{C} \times \frac{\pi}{2} = \frac{kT}{C}, \tag{8.6}$$

similar to the RC circuit in Equation (8.2). That is, the noise per unit bandwidth of $4kT/g_s$ is integrated over an equivalent rectangular bandwidth of $g_s/(2\pi C) \times (\pi/2)$ such that the total noise is again kT/C. Experimental measurements of this photoreceptor reveal that the predicted noise is in good accord with theory and is indeed invariant with light level [5].

It is natural to wonder if there is something more fundamental that caused the total output noise of the two circuits to magically come out equal to kT/C. It is time to discuss the equipartition theorem, which will shed some light on our findings.

8.3   The equipartition theorem

Suppose that in any physical system, the energy E, which is stored amongst a set of degrees of freedom, can be expressed as

$$E = \varepsilon(q_1) + H(q_2, q_3, \ldots), \tag{8.7}$$

where $\varepsilon(q_1)$ is the energy stored in the degree of freedom corresponding to q_1. The degree of freedom q_1 could represent the position of a spring, the mass of an object, the charge on a capacitor, the current through an inductor, or any degree of freedom that the energy of the system depends on. Then, if this degree of freedom is coupled to the thermal environment, the mean fluctuation energy coupled to this degree of freedom by the thermal environment is given by

$$\bar{\varepsilon} = \frac{\int_{-\infty}^{\infty} \varepsilon(q_1) e^{-\varepsilon(q_1)/kT} dq_1}{\int_{-\infty}^{\infty} e^{-\varepsilon(q_1)/kT} dq_1} = -\frac{\partial}{\partial \beta} \ln\left(\int_{-\infty}^{\infty} e^{-\varepsilon(q_1)/kT} dq_1\right),$$

$$\text{where} \quad \beta = \frac{1}{kT}, \tag{8.8}$$

and where we assume that the probability that the degree of freedom has an energy $\varepsilon(q_1)$ is proportional to $\exp(-\varepsilon(q_1)/kT)$. If $\varepsilon(q_1) = bq_1^n$, then

$$\begin{aligned} \ln\left(\int_{-\infty}^{\infty} e^{-bq_1^n \beta} dq_1\right) &= \ln\left(\frac{1}{\beta^{1/n}} \int_{-\infty}^{\infty} e^{-bu^n} du\right) \\ &= -\frac{1}{n}\ln(\beta) + \ln\left(\int_{-\infty}^{\infty} e^{-bu^n} du\right) \end{aligned} \tag{8.9}$$

$$\bar{\varepsilon} = -\frac{\partial}{\partial \beta} \ln\left(\int_{-\infty}^{\infty} e^{-bq_1^n \beta} dq_1\right) = \frac{1}{n\beta} = \frac{kT}{n}$$

Thus, if $b = (1/2)m$, $q_1 = v$, $n = 2$, and $\varepsilon(q_1) = (1/2)mv^2$ is the kinetic energy, then the mean value of the kinetic-energy degree of freedom is $kT/2$.

Hence if the energy is quadratic in some degree of freedom and the probability for being in a state with energy E in that degree of freedom is proportional to $\exp(-E/kT)$, then the average thermal energy for that degree of freedom is $(1/2)kT$. The latter statement is called the equipartition theorem. The equipartition theorem is so called because, if there are several degrees of freedom that satisfy the conditions of the theorem, then each has an equal share of the thermal energy given by $(1/2)kT$.

Figure 8.6 (a) shows an RC circuit. Applying the equipartition theorem to the capacitor, we have

$$\begin{aligned} \frac{1}{2}C\overline{V^2} &= \frac{1}{2}kT \\ \Rightarrow \overline{V^2} &= \frac{kT}{C} \end{aligned} \tag{8.10}$$

independent of R. The resistor R serves to couple the energy-storing degree of freedom in the capacitor, the voltage, to the thermal environment, such that it develops a mean-square noise voltage consistent with the equipartition theorem. The equipartition theorem is a much simpler way of getting the result of Equation (8.1).

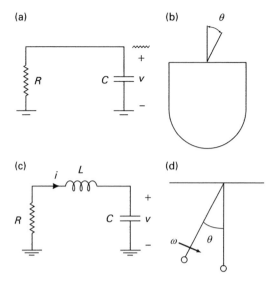

Figure 8.6a, b, c, d. Circuits and physical systems used to demonstrate the equipartition theorem: (a) an RC circuit, (b) a hair cell (stereocilium), (c) a series RCL circuit, and (d) a pendulum.

Figure 8.6 (b) reveals a stereocilium, a tiny hair, on a hair cell in our inner ears coupled to the thermal environment via damping in its fluid surround. If the torsional stiffness of the hair cell to angular deviations is κ, then by the equipartition theorem,

$$\frac{1}{2}\kappa\overline{\theta^2} = \frac{1}{2}kT$$
$$\overline{\theta^2} = \frac{kT}{\kappa} \tag{8.11}$$

is the mean-square fluctuation in its angular variable, independent of the damping.

Figure 8.6 (c) reveals a series LCR circuit with two energy-storing degrees of freedom, the inductor current i and the capacitor voltage v. Applying the equipartition theorem,

$$\frac{1}{2}L\overline{i^2} = \frac{1}{2}kT \Rightarrow \overline{i^2} = \frac{kT}{L}$$
$$\frac{1}{2}C\overline{v^2} = \frac{1}{2}kT \Rightarrow \overline{v^2} = \frac{kT}{C}. \tag{8.12}$$

Thus, the mean-square noise in the capacitor voltage and the inductor current can be rapidly computed and are both independent of the resistance R. The value of R does determine the quality factor or Q of the power spectrum with larger values of R corresponding to a smaller Q and vice versa. The Q is a measure of the resonant gain and narrowness of the frequency response of the capacitor or inductor voltage to a voltage source input in the series loop. High values of Q

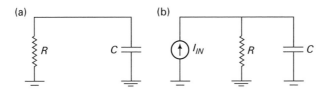

Figure 8.7a, b. Examples of a system (an RC circuit) that is in (a) thermal equilibrium and (b) a non-equilibrium steady-state system.

correspond to a large resonant gain and narrow frequency response; low values of Q correspond to a small resonant gain and broad frequency response. Small values of R lead to small voltage noise per unit bandwidth but a sharply tuned and large resonant gain and large values of R lead to large voltage noise per unit bandwidth but a broadly tuned and small resonant gain such that the total noise over all frequencies is constant, independent of R, and only determined by the equipartition theorem. In Chapter 13, we shall prove this result quantitatively in a slightly different context, but the equipartition theorem allows us not to have to!

Figure 8.6 (d) reveals a string pendulum with angular and angular-velocity degrees of freedom with a dangling mass m, angle θ, angular velocity ω, and with the acceleration due to gravity being g. Applying the equipartition theorem, we can show that, independent of the damping of the pendulum, the mean-square angular and angular-velocity fluctuations are given by

$$E = mgl(1 - \cos\theta) + \frac{1}{2}ml^2\omega^2 \approx mgl\frac{\theta^2}{2} + \frac{1}{2}ml^2\overline{\omega^2}$$

$$mgl\frac{\overline{\theta^2}}{2} = \frac{1}{2}kT \quad \Rightarrow \quad \overline{\theta^2} = \frac{kT}{mgl} \tag{8.13}$$

$$\frac{1}{2}ml^2\overline{\omega^2} = \frac{1}{2}kT \quad \Rightarrow \quad \overline{\omega^2} = \frac{kT}{ml^2}$$

The equipartition theorem strictly applies to systems in thermal equilibrium. The RC circuit of Figure 8.7 (a) is in thermal equilibrium if the capacitor voltage has settled to 0 and there are only thermal voltage fluctuations on it. In contrast, the RC circuit of Figure 8.7 (b) is in steady-state operation with a constant dc current flowing through the resistor and with the capacitor voltage being at a non-zero constant, but it is not in thermal equilibrium. Most circuits, unless they are switched off, do not operate in thermal equilibrium, so our discussion of the equipartition theorem is not strictly relevant to these cases, although it does add some insight into them. Nevertheless, the derivations of current in electronic devices, as in the MOSFET, assume that the charge concentrations are distributed according to a Boltzmann exponential energy distribution. For example, in Chapters 3, 4, 5, and 6, Boltzmann energy distributions determine charge concentrations throughout the channel of the MOSFET. The equilibrium between drift and diffusion currents yields a Boltzmann exponential distribution for the distribution of charge in the depletion region bordering a MOSFET channel.

So, non-equilibrium situations are often still modeled as being subject to exponential probability distributions. Not surprisingly, the answers that we compute for the total noise often come out to be $\gamma kT/C$, where $\gamma > 0$ and depends on the details of the circuit. Often, the non-equilibrium noise per unit bandwidth has a value $4kT\gamma R$ and an effective bandwidth of $1/(4RC)$ such that the total noise is $\gamma kT/C$:

$$\overline{v_n^2} = 4kT\gamma R \times \frac{1}{4RC} = \gamma \frac{kT}{C} \qquad (8.14)$$

We shall now show how to compute the effective value of γ from first principles in circuits using the non-equilibrium noise formulas derived in Chapter 7. We have actually already embarked on this path in the subthreshold photoreceptor example, where we used non-equilibrium shot-noise formulas for the transistor and photo-diode to calculate the total noise of the photoreceptor. Now, we shall focus on computing the total noise of a commonly used subthreshold transconductance amplifier. The latter example is a good tutorial circuit for illustrating how to compute the total noise in any general circuit.

8.4   Noise in a subthreshold transconductance amplifier

Figure 8.8 (a) reveals a symbolic representation of a transconductance amplifier whose output current i_{out} is a linear function of the differential input voltage $(v_+ - v_-)$ across its positive and negative terminals, i.e., $i_{out} = G(v_+ - v_-)$, where G is said to be the transconductance of the amplifier. A transconductance amplifier is similar to an operational amplifier except that its output current rather than its output voltage is linearly related to its input differential voltage. Transconductance amplifiers are often the first stage in many operational-amplifier topologies. In most transconductance amplifiers, the current i_{out} is relatively independent of the output voltage. We shall assume operation in this practical regime.

Figure 8.8 (b) reveals a common configuration for a differential amplifier. The transistors M_0, M_1, and M_2 form a differential pair with M_0's bias voltage determining the bias or tail current of the transconductance amplifier denoted I_B, and M_1 and M_2's gate voltages, v_+ and v_-, respectively, determining how much of I_B is steered through the M_1 vs. M_2 arm. If $v_+ > v_-$, more of I_B will flow through M_1 rather than M_2 and vice versa. The transistors M_3 and M_4 form a current mirror such that the input sink current from the M_1 arm is mirrored out as a source current from the positive V_{DD} rail by M_4. The difference between the latter source current and the current in the M_2 arm is then output by the transconductance amplifier as i_{OUT} and flows into the voltage source V_{OUT}. In this example, the voltage at the output node of the transconductance amplifier is set by the voltage source to be V_{OUT}, but it is usually implicitly determined by the circuit in which the transconductance amplifier is used. If V_{OUT} is not too close to either power-supply

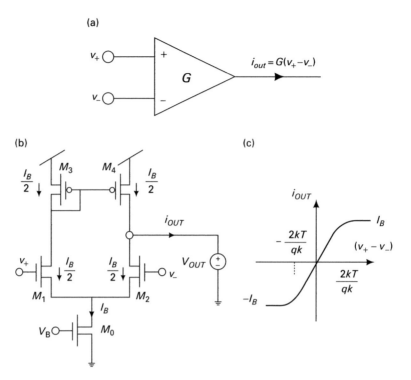

Figure 8.8a, b. The operational transconductance amplifier (OTA). (a) Block diagram, (b) circuit diagram, and (c) tanh (saturating) dc $I - V$ characteristic.

rail and if V_{DD} is not too small, all transistors in the amplifier operate in the saturation region, and we shall assume operation in this scenario.

Simple analysis of the transconductance amplifier using relationships for the subthreshold current through each transistor reveals that

$$i_{\text{out}} = I_B \tanh\left(\frac{(v_+ - v_-)}{(2kT/q\kappa)}\right), \tag{8.15}$$

where κ is the subthreshold exponential coefficient of the M_1 and M_2 transistors [6]. If

$$|v_+ - v_-| < \frac{2kT}{q\kappa}$$

$$i_{out} \approx \frac{I_B}{(2kT/q\kappa)}(v_+ - v_-) = G(v_+ - v_-) \tag{8.16}$$

$$G = \frac{I_B}{(2kT/q\kappa)} \equiv \frac{I_B}{V_L}$$

Figure 8.8 (c) illustrates the input-output transfer function of the transconductance amplifier as predicted by Equation (8.15). We shall assume that the transconductance amplifier operates with the magnitude of its differential voltage less

Figure 8.9. Transforming the output current noise of an OTA into an equivalent input-referred voltage noise source.

than its linear range $V_L = 2kT/q\kappa$ such that the linear equations of Equation (8.16) apply and the transconductance of the amplifier is $G = I_B/V_L$.

First, let us assume that the amplifier is near balance such that $v_+ \approx v_-$. In this situation, we would like to find the input-referred noise power per unit bandwidth due to noise contributions from all of its devices. As Figure 8.9 shows, doing so implies that we are going to transform the amplifier from one with noisy devices and a noiseless input to one with noiseless devices and a noisy input. Such a transformation is convenient because it allows us to directly perform calculations with one noisy signal at the input of the transconductance amplifier when it is used. It also allows us to directly compare and contrast the noise strength with the actual input strength at the same terminals. The procedure for doing so is as follows:

1. Draw a small-signal circuit including a noise generator for each device in the amplifier. Assume that all voltage/current inputs are ac shorts/opens, respectively.
2. Find the transfer function from each noise generator to the transconductance amplifier output, $\alpha_k(f)$, while zeroing out all other noise generators.
3. Find the total output noise from the various generators by superposition, but remember to add the squared magnitudes of the transfer functions, not the transfer functions themselves: $\overline{i_{out}^2(f)} = \sum_{\forall k} |\alpha_k(f)|^2 \overline{i_k^2(f)}$.
4. Find the input-referred noise per unit bandwidth, given by $\overline{v_{in}^2(f)} = \frac{\overline{i_{out}^2(f)}}{G^2}$. The value of $\overline{v_{in}^2(f)}$ may be a function of frequency due to $1/f$ noise generators and or filtering in the small-signal circuit.

Once the input-referred noise power per unit bandwidth is available, we can compute the output noise due to the transconductance amplifier between frequencies f and $f + df$ by multiplying its noise power per unit bandwidth at f by the square of the magnitude of the transfer function from the input to the output at f in the circuit it is used in, i.e., in a filter, in an operational amplifier, in an amplifier, etc. If we integrate over all frequencies from 0 to ∞, we then obtain the total output noise over all frequencies in the circuit due to the transconductance amplifier. The procedure that we have outlined is easily generalized to any circuit, not just a transconductance amplifier. We shall now illustrate the application of the procedure to the transconductance amplifier in detail.

Figure 8.10 illustrates the small-signal circuit of the transconductance amplifier with five noise generators added, one for each transistor. It is convenient to

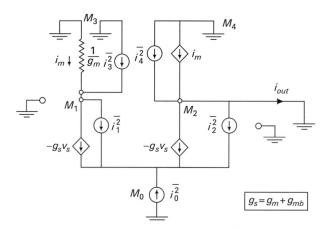

Figure 8.10. Complete small-signal noise circuit of the OTA.

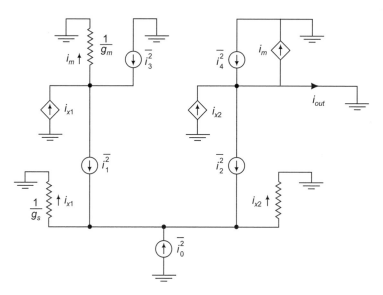

Figure 8.11. Simplified small-signal noise circuit of the OTA.

replace the g_s generators of M_1 and M_2 in Figure 8.10 with $1/g_s$ resistors and i_{x1} and i_{x2} generators as shown in Figure 8.11. We shall analyze the squared noise-transfer functions α_k^2 from each noise generator $\overline{i_k^2}$ to the output as k ranges from 1 to 5.

Computation of α_0^2: $\alpha_0^2 = (1/2 - 1/2)^2 = 0$. The current i_0 divides equally through both $1/g_s$ resistors so $i_{x1} = i_{x2} = i_0/2$. Each $i_0/2$ half current makes it to the output and cancels the other half out, i.e., $i_m = i_{x1}$; $i_{out} = i_{x2} - i_m = i_{x2} - i_{x1} = 0$. Thus M_0 contributes no noise to the output.

Computation of α_1^2: $\alpha_1^2 = (1/2 - (-1/2))^2 = 1$. The current i_1 divides equally amongst the two $1/g_s$ resistors, making $i_{x1} = i_{x2} = i_1/2$; i_{x1} gets subtracted from i_1 and makes it to the output as $i_m = i_{x1} - i_1 = -i_1/2$. Thus, at the output $i_{out} = i_{x2} - i_m = i_1/2 - (-i_1/2) = i_1$. Thus, M_1 contributes all its noise to the output.

Computation of α_2^2: $\alpha_2^2 = (-1/2 - 1/2)^2 = 1$. The current i_2 divides equally amongst the two $1/g_s$ resistors, making $i_{x1} = i_{x2} = i_2/2$; i_{x2} gets subtracted from i_2 and makes it to the output as $-i_2/2$; i_{x1} makes it to the output as i_m. Thus, at the output $i_{out} = -i_2/2 - i_m = -i_2$. Thus, M_2 contributes all of its noise to the output.

Computation of α_3^2: $\alpha_3^2 = (-1)^2 = 1$. The current i_3 makes it to the output directly as $-i_m$ such that $i_{out} = -i_3$. Thus, M_3 contributes all of its noise to the output.

Computation of α_4^2: $\alpha_4^2 = (1)^2 = 1$. The current i_{out} is simply equal to i_4 such that M_4 contributes all of its noise to the output.

Thus, the total output noise is given by

$$\overline{i_{out}^2(f)} = \sum_{k=0} |\alpha_k|^2 \overline{i_k^2(f)}, \quad \text{A}^2/\text{Hz}$$

$$= [0^2 + 1^2 + 1^2 + 1^2 + 1^2] 2q\left(\frac{I_B}{2}\right) \qquad (8.17)$$

$$= 4\left[2q\left(\frac{I_B}{2}\right)\right]$$

Hence,

$$\overline{v_{in}^2(f)} = \frac{\overline{i_{out}^2(f)}}{G^2} = \frac{4\left[2q\left(\frac{I_B}{2}\right)\right]}{\left(\frac{\kappa I_B}{2\phi_t}\right)^2} = \frac{16}{\kappa^2}\left(\frac{kT}{q}\right)^2 \frac{q}{I_B}, \quad \text{V}^2/\text{Hz}. \qquad (8.18)$$

We notice from Equation (8.18) that the input-referred voltage noise per unit bandwidth decreases with the bias current I_B.

Suppose the transconductance amplifier is used in a circuit to create a $G_m - C$ lowpass filter as shown in Figure 8.12 (a). We can show for this circuit that

$$G(v_{in} - v_{out}) = sCv_{out}$$

If we define $\quad \tau = \frac{C}{G}$

$$v_{in} - v_{out} = \tau s v_{out}$$

$$v_{in} = (1 + \tau s)v_{out} \quad \text{or} \qquad (8.19)$$

$$v_{out}(s) = \frac{1}{1 + \tau s} v_{in}(s).$$

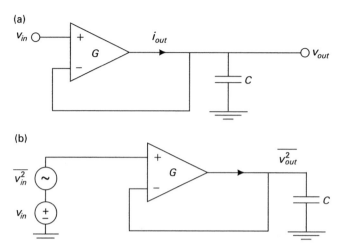

Figure 8.12a, b. A first-order $G_m - C$ lowpass filter, (a) the circuit and (b) the circuit and its output and input-referred noise voltages.

Figure 8.12 (b) includes the input-referred noise of the transconductance amplifier into the circuit such that we can compute the total output voltage noise of the filter to be:

$$
\begin{aligned}
\overline{v_{out}^2} &= \int_0^\infty \frac{\overline{v_{in}^2}(f)df}{1 + \tau^2(2\pi f)^2} \\
&= \frac{4\left[2q\left(\frac{I_B}{2}\right)\right]}{G^2} \times \underbrace{\frac{1}{2\pi}}_{\substack{\omega \to f \\ \text{conversion}}} \times \underbrace{\frac{\pi}{2}}_{\substack{\text{Effective} \\ \text{noise} \\ \text{bandwidth}}} \times \underbrace{\frac{G}{C}}_{\text{Like } \frac{1}{RC}}
\end{aligned}
\tag{8.20}
$$

$$
= \frac{2q\left(\frac{I_B}{2}\right)}{GC} = \frac{2q\left(\frac{I_B}{2}\right)}{\left(\frac{\kappa I_B}{2\phi_t}\right)C} = 2\frac{kT}{\kappa q} \times \frac{q}{C} = \frac{2}{\kappa}\left(\frac{kT}{C}\right)
$$

Once again we find that the total output voltage noise is independent of the bias current I_B and depends only on (kT/C). The excess noise factor, i.e., noise above kT/C, is $(2/\kappa)$ in this case.

A useful concept in circuits with noise is the concept of the minimum detectable signal. The minimum detectable signal at an input terminal of a circuit is defined to be the amplitude of the signal at which the output signal power is just equal to the output noise power, i.e., the signal-to-noise ratio is 1. If the output noise power over all frequencies is v_{noise}^2, and the gain from the input to the output at angular frequency ω is $|G(\omega)|$, then the minimum detectable signal at ω is given by

$$
\frac{|G(\omega)|^2 v_{mds}^2(\omega)}{v_{noise}^2} = 1 \Rightarrow v_{mds}(\omega) = \frac{\sqrt{v_{noise}^2}}{|G(\omega)|}
\tag{8.21}
$$

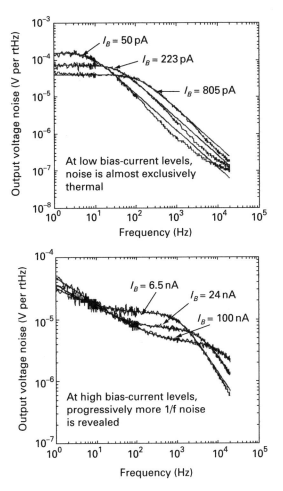

Figure 8.13. Experimental measurements of the output voltage noise PSD of a first-order $G_m - C$ filter, (a) for low bias current levels and (b) for higher bias current levels. Reprinted with kind permission from [4], Springer Science and Business Media.

If we apply this definition to the $G_m - C$ lowpass filter with $C = 1\text{pF}$ and $\kappa = 0.5$, then we find that the minimum detectable signal for low-frequency passband signals with a gain of 1 is given by

$$\frac{(1)^2 v_{mds}^2}{\dfrac{2}{\kappa}\left(\dfrac{kT}{C}\right)} = 1 \Rightarrow v_{mds} = \sqrt{\frac{2kT}{\kappa C}} = 130 \ \mu\text{V} \tag{8.22}$$

Signals outside the passband of this filter (where the gain is less than 1) will have a minimum detectable signal that scales with the reciprocal of their gain.

Experimental measurements of noise are generally in very good accord with theory in subthreshold circuits where the noise is predominantly thermal. Figure 8.13 (a) shows that the noise is predominantly thermal at low bias-current values in $G_m - C$ filters when the f_T of the operating devices is low as we discussed in Chapter 7 [4]. The data was taken for a $G_m - C$ filter different from the tutorial example in the text but the power spectra in both cases have similar shapes, and in both cases the total noise power is independent of the bias current I_B and only dependent on the capacitance C. Figure 8.13 (b) shows that progressively more $1/f$ noise is revealed at higher bias-current values as the operating f_T of the devices in the transconductance amplifiers increases. Even for the example of Figure 8.13 (b), however, the total noise power was found to be dominated by thermal noise. In Chapter 12, we will discuss examples that calculate the total noise power when $1/f$ noise terms are included.

8.5   Noise in general circuits

The procedure described for calculating the input-referred noise per unit bandwidth, the total output noise power, or the minimum detectable signal for the transconductance amplifier is easily generalized to other circuits. We compute the total noise power per unit bandwidth due to all noise generators in the circuit at the output by inactivating inputs and summing products of squared noise transfer functions and mean-square noise-generator values; we can refer this noise power back to the input of interest by dividing by the magnitude of the squared transfer function from that input to the output; we can find the total output noise power by integrating the output noise power per unit bandwidth over all frequencies; and we can find the minimum detectable signal at a particular angular frequency ω and at a particular input by dividing the square-root of the total output noise power by the gain from that input to the output at ω.

In some treatments of noise, it is customary to define an input-referred noise-voltage generator and an input-referred noise-current generator such that the noise may be calculated for varying source impedances at the input terminal. We have not found this strategy to be useful since it is hard to keep track of the correlations between these input-referred noise generators, which are caused by device noise in one transistor leading to terms in both input-referred noise generators. Rather, in our strategy, for a given parameterized source impedance and voltage input, we compute an input-referred voltage noise in series with the voltage input; or, for a given parameterized source admittance and a given current input, we compute an input-referred current noise in parallel with the current input. Thus, we always refer the noise back to the actual input such that it is automatically in whatever coordinates this input happens to be in (voltage or current), the effects of the source impedance automatically get taken into account when computing transfer functions from input to output, and any noise power due to the source impedance merely adds to the input-referred noise. The complication of correlations between input-referred voltage and current noise generators does not arise in our method since there is always only one input noise generator.

We have glibly assumed that the noise generators from different devices in a circuit are all independent and uncorrelated when making our calculations. This assumption may seem strange since each generator affects the circuit's voltages and currents. Therefore each noise generator alters the operating-point parameters of the whole circuit, thereby affecting the noise produced by all the other generators. Thus, the noise from various generators ought to be correlated. The key to this paradox is *small signal*. We assume that the net effect of all the noise generators is still small enough such that the operating-point parameters of the circuit are barely changed. To first order then, all the noise generators remain uncorrelated because the electrons in one device flow without regard to electrons in other devices and the noise from one device does not alter the operating-point parameters of another device.

In certain RF circuits it may be necessary to include induced gate noise in an MOS device with inherent correlation between a drain-to-source current noise generator and a gate-to-source current noise generator in the elementary noise model of the device itself (see Equations (7.45) and (7.46) in Chapter 7). The induced gate-noise generator is only of importance when the operating RF frequencies are very near the f_T of the device, a bad strategy for ultra-low-power design, as we discuss in Chapter 22. Thus, we shall not focus much on the latter noise generator in this book. It is discussed extensively in other texts, e.g. [7]. In such cases, one can use Equation (7.19) in Chapter 7 to account for correlations, and refer back to the actual input of the system at the very end.

8.6   An ultra-low-noise MEMS capacitance sensor

Figures 8.14 (a) and 8.14 (b) illustrate the principle of lock-in signal detection in the context of a MEMS capacitance sensor, similar to the one used in several commercial accelerometers and vibration sensors. The capacitances C_{s1} and C_{s2} are differential capacitances that differ from one another ever so slightly, with one increasing by ΔC when the other decreases by ΔC and vice versa. The differential nature of the capacitance arises because the V_x node is formed by a mechanically mobile plate common to both capacitances: in response to a mechanical linear acceleration in one direction, the mobile plate moves towards the static plate of C_{s1} and away from the static plate of C_{s2}; in response to a mechanical linear acceleration in the other direction, the mobile plate moves towards the static plate of C_{s2} and away from the static plate of C_{s1}. The mobile plate of mass m is tethered via a spring of stiffness k to a substrate, such that the mechanical equilibrium displacement of the mobile plate in response to an acceleration a is $(ma)/k$. For the relatively small accelerations that we shall be considering, the sensor may be assumed linear with the fractional capacitance change of either capacitance proportional to its fractional displacement. For a typical k/m of $2\pi \times 20$ kHz, a few mg of acceleration will cause displacements on the order of 0.01 Å, significantly less than the nominal $2\,\mu$m plate separation. Thus, we need to detect

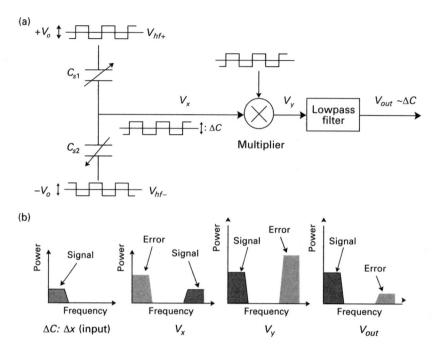

Figure 8.14a, b. Principle of lock-in detection in the context of capacitive sensing in (a) and and an explanation of its operation in the frequency domain in (b). Reprinted with kind permission from [1] (© 2003 IEEE), and the kind permission of Professor Bernhard Boser.

displacements, or equivalently capacitance changes, that are less than 1 part in a million! Therefore, the electronics needed to detect such motions needs to be at least 20-bit precise, which means that the noise in the system must be very low indeed. We must definitely not be at the mercy of non-fundamental low-frequency noise such as $1/f$ noise, substrate coupling, or offset. The lock-in topology of Figure 8.14 (a) can help make the influence of such low-frequency noise negligible.

The static plates of the capacitances are driven by two anti-phasic high-frequency carrier signals V_{hf}^+ and V_{hf}^- with $V_{hf}^+ = -V_{hf}^-$ and an amplitude of V_0 such that the voltage at the middle node V_x is given by

$$V_x = \left(\frac{C_{s1} V_{hf}^+(t) + C_{s2} V_{hf}^-(t)}{C_{s1} + C_{s2}} \right)$$

$$V_{hf}^+(t) = -V_{hf}^-(t)$$

$$C_{s1} = C_0 + \Delta C(x)$$

$$C_{s2} = C_0 - \Delta C(x)$$

$$V_x(t) = \left(V_{hf}^+(t) \right) \left(\frac{\Delta C(x)}{C_0} \right)$$

(8.23)

Hence, we have multiplied or modulated the desired signal of interest $(\Delta C(x)/C_0)$ by a high-frequency carrier $V_{hf}^+(t)$ in a relatively noiseless fashion (capacitors are noiseless). Therefore, analogous to AM radio, $V_x(t)$ is now a signal centered at the square-wave carrier frequency with a modulation signal bandwidth around this carrier frequency. The modulation bandwidth is the bandwidth of the baseband signal of interest ($\Delta C(x)/C_0$), which slowly varies as x alters in accordance with acceleration forces. We will ignore the harmonics of the square wave and pretend that the square wave effectively behaves like a pure sine wave frequency in all that follows: the nature of the lock-in technique is such that it will not matter much, as we shall see later. The key motivation for the modulation is that the signal information can be made to reside in frequencies that are greater than the corner frequency where the $1/f$ noise power spectrum and the white-noise power spectrum intersect. Thus, signal and low-frequency error now reside in different parts of the frequency space such that distinguishing the two becomes easier. In particular, effects of low-frequency errors due to any circuitry after the modulation can be made negligible. In this particular case, the modulation is accomplished by using the fact that the $Q = CV$ relation creates a multiplier in the sensor itself: we are trying to sense C and we multiply by V. In other cases, e.g., in chopper modulation, passive mixer switches are used to create the modulation in a relatively noiseless fashion [3].

Figure 8.14 (b) illustrates how the error and signal are now distinguished. The modulated signal V_x is demodulated by multiplying it with the square-wave carrier again in an amplifying mixer, i.e., an analog multiplier, to create V_y. In the signal V_y, the desired signal gets amplified and down-converted back to baseband while the error gets amplified and up-converted. The lowpass filter following V_y then filters out the error, which is now at high frequencies. The desired signal lies in the passband of the lowpass filter and remains unaltered. Thus, the signal V_{out} now contains an amplified version of the signal with a strongly attenuated error. Clearly, the higher the frequency of V_{hf} compared with the corner of the lowpass filter, the more attenuated is the error. Typically, lock-in amplifiers operate with at least a factor of 20 or more in the ratio of the carrier frequency to the lowpass corner frequency. Operating with a higher carrier frequency is advantageous for attenuating effects due to square-wave harmonics as well. However, operating at a very high carrier frequency can impose demands on the f_T of the amplifying mixer which then needs to operate at a higher bias current. A higher carrier frequency can also lead to an increased incidence of parasitic coupling effects. There is usually an optimum carrier frequency at which the minimum detectable signal is the lowest.

8.6.1 Practical implementation

Figure 8.15 shows the overall topology of a practical implementation that includes two improvements to the basic idea of lock-in detection. The first improvement is the use of a bandpass inverting preamplifier that is tuned around the carrier frequency and interposed between the sensing node and the amplifying mixer. It is implemented with a bandpass operational amplifier and inverting feedback.

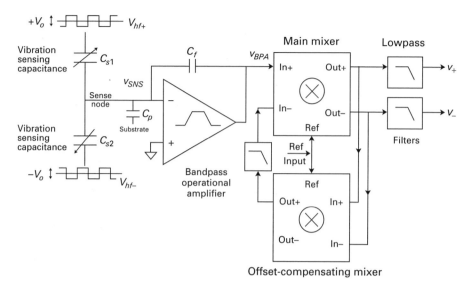

Figure 8.15. A low-noise lock-in architecture for capacitive sensing. Reprinted with kind permission from [1] (© 2003 IEEE).

The bandpass preamplifier ensures that the uncontrolled parasitic capacitance C_p does not affect the overall closed-loop gain of the amplifier and provides amplification that eases noise requirements on the mixer. The second improvement is the use of a mixer with feedback offset compensation. We shall describe the bandpass preamplifier first and then the offset-compensated mixer.

In Figure 8.15, if the bandpass preamplifier (BPA) were absent, the value of the magnitude of V_x would, from an analysis similar to that of Equation (8.23), be given by

$$V_x = V_0 \frac{\Delta C}{C_o} \times \frac{1}{\left(1 + \frac{C_p}{2C_o}\right)} \tag{8.24}$$

Thus, V_x would be attenuated by the parasitic C_p and the overall gain of the amplifier would be determined by a poorly controlled parasitic capacitance to substrate. By using the BPA we can ensure that node voltage v_{SNS} is at a virtual-ground value such that C_p is effectively 'shorted'. Hence, independent of C_p, the magnitude of the BPA output voltage, v_{BPA}, in Figure 8.15 is given by

$$v_{BPA} = \frac{\Delta C}{C_f} V_0 \tag{8.25}$$

The parasitic capacitance does lead to amplification of the input-referred noise, $\widetilde{v_{nBPA}}$, of the BPA, which must therefore be maintained at a low value. The diagrams of Figures 8.16 (a) and 8.16 (b) illustrate how amplification of v_{nBPA} occurs via a classic non-inverting topology. In these figures, $G_{adp} = 1/R_{adp}$ where

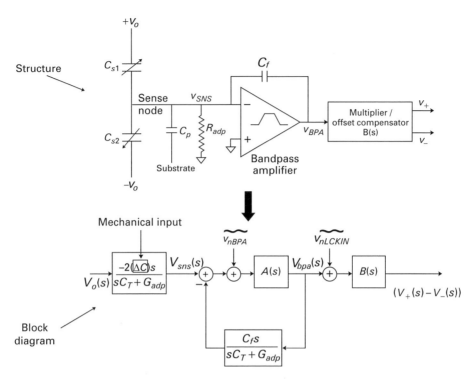

Figure 8.16. Topology of the sensing node and bandpass amplifier (top) and a block diagram representation (below). Reprinted with kind permission from [1] (© 2003 IEEE).

R_{adpt} is an extremely large resistance used to set the dc value of v_{SNS} at the sense node. The value of the total capacitance at the sense node, C_T, is given by

$$C_T = C_{s1} + C_{s2} + C_p + C_f$$
$$C_T = 2C_0 + C_p + C_f$$

(8.26)

The input-referred noise of the lock-in amplifier (mixer/multiplier/demodulator) $\widetilde{v_{nLCKIN}}$ is also shown. The BPA provides enough gain around the carrier frequency such that the input-referred noise of the lock-in amplifier at these frequencies is less critical. The $1/f$ noise of the BPA and the lock-in amplifier have negligible effects due to the inherent benefits of lock-in amplification. Nevertheless, any dc offset at the input of the lock-in amplifier, caused primarily by offsets in MOS transistors in the BPA, causes differential-pair inputs within it to become imbalanced. The imbalance reduces the effective g_m of differential-pair transistors, reduces the lock-in gain, and increases nonlinearity and distortion in its output signal. The topology of Figure 8.15 solves this problem by using another offset-compensating mixer in a feedback loop around the main lock-in mixer.

The v_{BPA} signal input to the main mixer in Figure 8.15 is down-converted or demodulated to base band by the reference carrier input of the mixer.

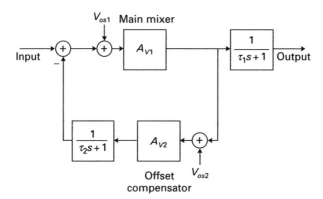

Figure 8.17. The feedback block diagram that illustrates how offset compensation works in the lock-in amplifier. Reprinted with kind permission from [1] (© 2003 IEEE).

Simultaneously, the input dc offset of the main mixer is up-converted to the carrier frequency. Thus, if we demodulate the output of the mixer with the offset-compensating mixer in Figure 8.15, and lowpass filter this demodulated output, the dc offset information of the main mixer can be extracted. If we then set the In^- input of the mixer to this extracted value, the dc offset of the mixer is strongly attenuated through negative-feedback action: negative feedback forces the input In^- to follow the dc value of In^+ such that the differential input $(In^+ - In^-)$ of the main mixer has zero dc value. Figure 8.17 shows the equivalent feedback loop at baseband with all modulation and demodulation operations removed for clarity. If V_{os1} is the input-referred offset of the main mixer with gain A_{v1}, V_{os2} is the input-referred offset of the offset-compensating mixer with gain A_{v2}, τ_1 and τ_2 are the lowpass filter time constants at their outputs as in Figure 8.15, and the loop gain of the feedback loop $A_{v1}A_{v2}$ is large, then the dc offset V_{os}^{out} at the output of the main mixer is given from Figure 8.17 and Black's formula to be

$$V_{os}^{out} = \frac{V_{os1}}{A_{v2}} + V_{os2} \tag{8.27}$$

The input-referred offset of the whole lock-in amplifier is therefore

$$V_{os}^{in} = \frac{V_{os1}}{A_{v1}A_{v2}} + \frac{V_{os2}}{A_{v1}} \tag{8.28}$$

The topology of Figure 8.15 does nothing for any dc offsets that may be present at the reference input of the main mixer. However, mixers, which are typically architected by cascading differential pairs in series [1], have saturating tail currents when a square wave of sufficient amplitude is applied to either differential reference input; thus, offsets in the reference inputs are typically of less concern as long as the input square-wave signal plus the offset is always large enough to fully saturate a differential pair.

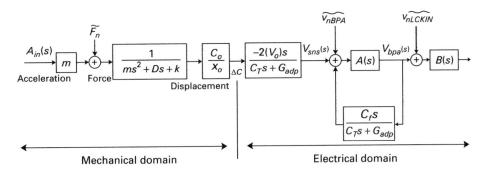

Figure 8.18. Illustration of a block diagram for noise analysis in both the electrical and mechanical domains.

8.6.2 Noise analysis

Figure 8.18 reveals the transfer function for the complete mechano-electrical system with mechanical and electrical noise sources. The noise input $\widetilde{F_n}$ represents the power spectral density of the mechanical force noise, $\widetilde{v_{nBPA}}$ represents the input-referred power spectral density of the bandpass preamplifier with transfer function $A(s)$, and $\widetilde{v_{nLCKIN}}$ represents the input-referred power spectral density of the lock-in amplifier with transfer function $B(s)$.

The mechanical force noise PSD $\widetilde{F_n}$ is related to the mechanical damping D just as the electrical current noise PSD is related to the electrical conductance, i.e.,

$$\widetilde{F_n^2} = 4kTD, \tag{8.29}$$

analogous to

$$\widetilde{i_n^2} = 4kTG \tag{8.30}$$

In fact, the transfer function from force to displacement shown in Figure 8.18 is identical to that of a parallel LCR circuit with L analogous to $1/k$, C analogous to m, D analogous to G, the force analogous to an input current, and the velocity analogous to a voltage. Thus, the mechanical quality factor or Q of the system is given by

$$Q = \frac{R}{\sqrt{L/C}} = \frac{RC}{\sqrt{LC}} = \frac{\omega_{res}m}{D} \tag{8.31}$$

Since ω_{res}, m, and Q are easily measured for a spring-mass system, we can measure D in terms of Q. From Figure 8.18, the input-referred mechanical acceleration noise PSD in units of g, the gravitational acceleration, are then given by dividing Equation (8.29) by m^2g^2. Thus, the input-referred acceleration noise PSD due to mechanical noise is given from Equations (8.29) and (8.31) to be

$$\widetilde{a_{mech}}^2 = \frac{4kT\omega_{res}}{Qm(9.8)^2} \ \text{g}^2/\text{Hz} \tag{8.32}$$

For typical values of $Q = 2$, $\omega_{res} = 20$ kHz, $m = 1.4$ μg, this value works out to 96 $\mu g/\sqrt{(\text{Hz})}$, which for a vibration-sensing bandwidth of 1 Hz – 100 Hz, yields a minimum detectable acceleration of 1 mg. The vibration-sensing bandwidth is set by the bandwidth of the lowpass filter of the main lock-in mixer of Figure 8.15. For electrical noise to be insignificant compared with mechanical noise, it must lead to an input-referred mechanical acceleration noise that is significantly below 96 $\mu g/\sqrt{(\text{Hz})}$.

Figure 8.18 shows how we map from the mechanical to electrical domain. Since fractional displacements lead to fractional changes in capacitance, we find that

$$\Delta C = \left(\frac{C_0}{x_0}\right)\Delta x_0, \tag{8.33}$$

with x_0 being the nominal capacitance plate separation. For the sensor being studied, various design/simulation/measurements revealed that $C_0 = 150$ fF, $x_0 = 2$ μm, $C_f = 50$ fF in the BPA, and $C_p = 680$ fF [1]. With all of these parameters, all that is left is to estimate $\widetilde{v_{nBPA}}$ and $\widetilde{v_{nLCKIN}}$ in Figure 8.18 and refer them back to the acceleration input. To do so, it is convenient to refer them back to the v_{SNS} input at the sense node first.

$$\widetilde{v_{nSNS}}^2 = \widetilde{v_{nBPA}}^2 + \left(\frac{C_f}{C_T}\right)^2 \widetilde{v_{nLCKIN}}^2 \quad \text{V}^2/\text{Hz} \tag{8.34}$$

From Figure 8.18, the transfer function v_{SNS}/a_{in} for low frequencies in the pass band of the vibration sensor is given by

$$\frac{V_{sns}(s)}{A_{in}(s)} = 9.8\left(\frac{1}{\omega_{res}^2}\right)\left(\frac{V_o}{x_o}\right)\left(\frac{2C_o}{C_f}\right)\text{V}/\text{g} \tag{8.35}$$

With V_0 at 2 Vrms, we now have all the information needed for estimating the electrical noise. We note from Equation (8.34), that because of the large value of $C_T = 2C_0 + C_f + C_p$ compared with C_f, the BPA's noise will be considerably more dominant than the noise of the lock-in amplifier. Thus, we can focus our attention and resources on minimizing the thermal noise of the BPA (1/f noise of both the BPA and lock-in amplifier are negligible due to our use of the lock-in technique).

Figure 8.19 (a) shows the transistor schematic of the BPA and Figure 8.19 (b) shows a block diagram equivalent. The transistors T_1, T_3, and T_{10} form a differential pair that is loaded by the resistors R to create the primary input stage with gain A_{v1} and input-referred noise PSD $\widetilde{v_{n1}}$. Similarly, the transistors T_2, T_4, and T_{11} and the resistors R create a secondary input stage with gain A_{v2} and input-referred noise PSD $\widetilde{v_{n2}}$. Transistors T_5, T_6, T_{12}, T_7, and T_8 and C_2 form a transconductor that performs differential-to-single-ended conversion. They effectively implement a stage with gain A_{v3}, time constant τ_{BPA2}, and input-referred noise PSD $\widetilde{v_{n3}}$. The output of the A_{v3} stage drives a source-follower stage. The transistors T_9 and T_{13}, form this source-follower stage, which together with C_1 creates a lowpass filter of time constant τ_{BPA1} and input-referred noise PSD $\widetilde{v_{nbuf1}}$. The lowpass filter's output is an input to the A_{v2} stage such that a negative-feedback loop with a

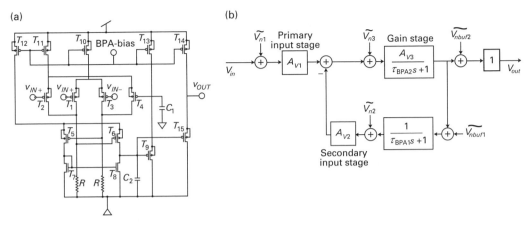

Figure 8.19a, b. Circuit schematic of the bandpass amplifier in (a) and its block-diagram equivalent in (b). Reprinted with kind permission from [1] (© 2003 IEEE).

lowpass filter in the feedback path is created, as shown in Figure 8.19 (b). The transistors T_{15} and T_{14} provide a low-impedance source-follower-buffered output for the BPA with input-referred noise PSD $\widetilde{v_{nbuf2}}$. A simple root-locus analysis (see Chapter 2) reveals that the two time constants in the negative-feedback loop create a bandpass amplifier. Furthermore, if we assume that all transistors operate in an above-threshold saturated regime and that all differential pairs are balanced, then we can show from noise analysis that the various PSDs of the most dominant noise sources are given by

$$\widetilde{v_{n1}}^2 = \frac{16kT}{3g_{mT1}}$$

$$\widetilde{v_{n2}}^2 = \frac{16kT}{3g_{mT2}}$$

$$\widetilde{v_{n3}}^2 = 8kTR + \frac{16kT}{3g_{mT5}}\left[1 + \frac{g_{mT7}}{g_{mT5}}\right] \tag{8.36}$$

$$\widetilde{v_{nBPA}}^2 = \widetilde{v_{n1}}^2 + \left(\frac{A_{V2}}{A_{V1}}\right)^2 \widetilde{v_{n2}}^2 + \left(\frac{1}{A_{V1}}\right)^2 \widetilde{v_{n3}}^2$$

Thus, the power spectral density of the BPA is dominated by that of $\widetilde{v_{n1}}$ and $\widetilde{v_{n2}}$. Like in most ultra-low-noise amplifiers, a resistively loaded first stage with a gain greater than 10 ensures that the input-referred noise power of the resistors in the first stage is an order of magnitude beneath that of the input-differential-pair transistors. Thus, these differential-input transistors dominate the overall noise performance in a well-designed amplifier and must, therefore, have large g_m values. All other noise in the amplifier is typically insignificant if the gain of the first stage is sufficiently large. Non-resistively loaded amplifiers with high gain are always noisier than amplifiers with resistive loads because the active-current-source loads contribute further noise, unlike resistors. However, as we can see in the expression for $\widetilde{v_{n3}}$ in Equation (8.36), the noise in such transistors or in mirror transistors can be reduced by architecting their g_ms to be small with regard to the g_ms of the input differential-pair transistors. Chapter 19 exploits some of the same concepts for low-noise neural-amplifier design.

8.6.3   Experimental performance

With the value of $\widetilde{v_{nBPA}}$ in Equation (8.36) and the transfer function from a_{in} to v_{SNS} in Equation (8.35), we can design the BPA and the overall topology of Figure 8.15 such that the input-referred acceleration noise PSD due to electrical noise is significantly smaller than that due to mechanical noise given in Equation (8.32). For example, the design described here and in more detail in [1] achieved $30\ \mu g/\sqrt{(\text{Hz})}$ of experimental electrical noise compared with $96\ \mu g/\sqrt{(\text{Hz})}$ of experimental mechanical noise. The electronics alone in a 1 Hz – 100 Hz bandwidth would thus have been able to sense 0.375 mg or 2.5 mÅ of motion, which reflects 0.125 ppm sensitivity or a 1-part-per-8-million change in capacitance; but the mechanical noise limits the minimum detectable signal to ~1 mg or 0.01 Å of motion or to a 1-part-per-2.5-million change in capacitance. The predicted minimum detectable signal was 1 mg, and 1.2 mg was measured; similarly, the predicted and measured electrical noise at the sense node were extremely close $(13.5\ \text{nV}/\sqrt{(\text{Hz})})$ versus $(16\ \text{nV}/\sqrt{(\text{Hz})})$. Once the principles of low-noise design have been mastered, good agreement between theory and experiment with ultra-low-noise circuits is not unusual.

It is worth noting that more than an Avogadro's number of atoms moves in concert on a mechanical structure such that sensing motions that are far below the diameter of a hydrogen atom is not unusual. The noise in the mean displacement of an Avogadro's number of atoms moving in concert is a factor of approximately $\sqrt{(6.02 \times 10^{23})}$ less than the noise in the displacement of a single atom. As another example, the human ear senses 0.05 angstroms of motion at the eardrum at the best threshold of hearing, which is near 3 kHz.

8.6.4   Electrostatic actuation

The electrostatic force of attraction between two capacitor plates is found by differentiating the stored $(1/2)C(x)V^2$ energy with respect to x. We find that its magnitude is given by

$$F_{attr} = \frac{1}{2}\frac{C_0}{x_0}V_0^2 \tag{8.37}$$

In our sensor, the electrostatic forces on the moving plate due to each capacitance are nominally balanced. However, a slight mismatch between the two capacitances can occur due to fabrication errors. The equilibrium position of the moving plate with no acceleration will then establish itself at a value such that stiffness forces balance any electrostatic force mismatch. One undesirable consequence of the capacitor mismatch is that a false dc acceleration is reported at the output of the lock-in amplifier (the output after demodulation and filtering) when none exists. Fortunately, in the vibration sensor that we have described such an effect is easily cancelled: the dc value of the lock-in amplifier's output is fed back to alter the dc value of one of the static plates in a negative-feedback loop such that electrostatic forces reestablish exact equality in the capacitances [1]. The closed-loop bandwidth of this negative-feedback loop is set around 1 Hz to create the highpass cutoff frequency of the vibration sensor.

Electrostatic force feedback is frequently used in BioMEMS applications to create oscillations in MEMS resonators. In these cases, positive feedback rather than negative feedback is utilized to excite the oscillation [2]. It is worth noting that the $1/x_0$ dependence of the force in Equation (8.37) results in a negative electrostatic spring stiffness due to positive feedback in any application: the closer two capacitive plates get, the stronger is the electrostatic force between them, which brings the plates even closer together. For overall stability in non-oscillatory applications, the normal positive spring stiffness must exceed the negative spring stiffness set by electrostatics. The MEMS example of this chapter provides useful background for the BioMEMS applications discussed in Chapter 20.

Chapters 7 and 8 conclude our treatment of noise. The material of these chapters is important for a deeper understanding of several chapters in the book, including Chapters 11, 12, 13, 14, 15, 18, 19, 20, 22, and 24. These chapters also provide further examples of low-noise low-power circuits.

References

[1] M. Tavakoli and R. Sarpeshkar. An offset-canceling low-noise lock-in architecture for capacitive sensing. *IEEE Journal of Solid-State Circuits*, **38** (2003), 244–253.

[2] T. P. Burg, M. Godin, S. M. Knudsen, W. Shen, G. Carlson, J. S. Foster, K. Babcock and S. R. Manalis. Weighing of biomolecules, single cells and single nanoparticles in fluid. *Nature*, **446** (2007), 1066–1069.

[3] T. Denison, K. Consoer, A. Kelly, A. Hachenburg, W. Santa, M. N. Technol and C. Heights. A 2.2 μW 94nV/Hz, Chopper-Stabilized Instrumentation Amplifier for EEG Detection in Chronic Implants. *Digest of Technical Papers of the IEEE International Solid-State Circuits Conference (ISSCC 2007)*, San Francisco, CA, 162–594, 2007.

[4] R. Sarpeshkar, R. F. Lyon and C. A. Mead. A low-power wide-linear-range transconductance amplifier. *Analog Integrated Circuits and Signal Processing*, **13** (1997), 123–151.

[5] T. Delbruck and C. A. Mead. Analog VLSI phototransduction by continuous-time, adaptive, logarithmic photoreceptor circuits. *CalTech CNS Memo*, **30** (1994).

[6] Carver Mead. *Analog VLSI and Neural Systems* (Reading, Mass.: Addison-Wesley, 1989).

[7] Thomas H. Lee. *The Design of CMOS Radio-Frequency Integrated Circuits*, 2nd ed. (Cambridge, UK; New York: Cambridge University Press, 2004).

9 Feedback systems

Nothing is more practical than a good theory.

Ludwig Boltzmann

In this chapter, we shall continue our study of feedback systems by understanding the Nyquist criterion for the stability of a feedback system [1]. Nyquist-based empirical techniques like gain margin and phase margin help us design feedback systems with acceptable levels of robustness. Compensation techniques allow us to sculpt the dynamics of a feedback loop to ensure its stability while preserving important aspects of its performance. We shall also briefly discuss second-order, conditionally stable, nonlinear, and positive-feedback systems, and outline how our study of negative feedback can help analyze such systems. We will conclude with a discussion of why feedback loops are ubiquitous in circuits and describe a technique for analyzing small-signal circuits with dependent generators through feedback. The latter technique illustrates how feedback loops are deeply embedded in circuits and allows one to leverage off the techniques of feedback analysis to understand how circuits work.

9.1 The Nyquist criterion for stability

For a linear feedback system described by Black's formula, i.e.,

$$\frac{V_{out}(s)}{V_{in}(s)} = \frac{a(s)}{1 + a(s)f(s)}, \tag{9.1}$$

to be stable all the roots of $1 + a(s)f(s)$ must have their real parts in the left half plane (LHP). Otherwise, tiny impulses to the system, e.g., caused by noise in its own devices, can cause the system to respond in a growing exponential fashion and thus beget instability. Nyquist described a technique for determining if all the roots of a feedback system were in the LHP without actually having to find those roots. His technique functions even if the singularities of $a(s)$ or $f(s)$ are unknown, and even if $a(s)$ and $f(s)$ are distributed systems that cannot be characterized by poles and zeros. All that is needed is the frequency response of the loop transmission, i.e., $L(j\omega) = a(j\omega)f(j\omega)$. The Nyquist criterion for stability is based on an interesting property of functions in the complex plane, which we now describe.

(a)

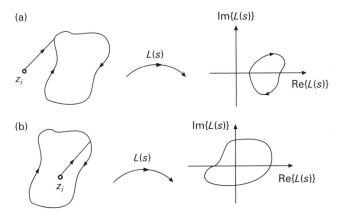

(b)

Figure 9.1a, b. Mapping the behavior of the complex function $L(s) = (s - z_i)$ on s-plane contours that (a) do not encircle z_i and (b) that encircle z_i.

Figures 9.1 (a) and 9.1 (b) reveal the difference in the behavior of the phase of $L(s)$ in the $L(s)$-plane if $L(s)$ contains a singularity (pole or zero) outside a closed contour in the s-plane (Figure 9.1 (a)) versus inside a closed contour in the s-plane (Figure 9.1 (b)). The $L(s)$ function in this example contains a simple zero and is described by $L(s) = s - z_i$ where z_i is the location of the zero. When the singularity is outside the contour as in Figure 9.1 (a), $\sphericalangle L(s)$ is non-monotonic on the contour, does not complete a 2π circuit, and thus does not encircle the origin in the $L(s)$-plane. When the singularity is inside the contour as in Figure 9.1 (b), $\sphericalangle L(s)$ is monotonic on the contour, completes a 2π circuit, and thus encircles the origin in the $L(s)$-plane.

If there are Z zeros and P poles of $L(s)$ located inside a closed s-plane contour and we evaluate $L(s)$ on this contour in the s-plane in a clockwise circuit then each of the Z zeros contributes one clockwise encirclement of the origin in the $L(s)$-plane and each of the P poles contributes one anticlockwise encirclement of the origin in the $L(s)$-plane. Thus, the net number of clockwise encirclements of the origin in the $L(s)$-plane is then $Z - P$. Note that the zeros and poles of $L(s)$ that lie outside the s-plane contour contribute no net phase in the $L(s)$-plane and thus do not affect encirclements around the origin in the $L(s)$-plane.

We are almost ready to understand the Nyquist criterion. Figure 9.2 illustrates another simple but important concept needed to understand the criterion. Since $1 + L(s)$ simply maps any closed contour in the $L(s)$-plane into the same contour in the $1 + L(s) = L'(s)$-plane, except that the contour gets translated by 1 along the real axis, an encirclement of the origin in the $L'(s)$-plane corresponds to an encirclement of -1 in the $L(s)$-plane.

Figure 9.3 illustrates the Nyquist criterion. If we evaluate $L(s)$ on an infinite closed D-contour, i.e., on the $j\omega$-axis from 0 to ∞, then clockwise on a semicircular contour in the right half plane (RHP) of infinite radius, and then close the contour on the $j\omega$-axis from $-\infty$ to 0, the presence of any zeros or poles of $1 + L(s)$

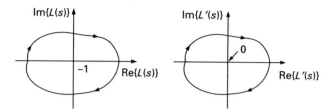

Figure 9.2. Mapping between $L(s)$ and $(1 + L(s))$ on the complex plane.

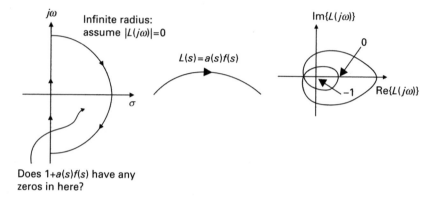

Figure 9.3. The right-half-plane D contour used to generate the Nyquist plot.

in the RHP will lead to encirclements of -1 in the $L(s)$-plane. If $1 + L(s)$ contains Z zeros and P poles in the RHP, then the net number of clockwise encirclements of -1 in the $L(s)$-plane, N, is given by

$$N = Z - P. \tag{9.2}$$

The Nyquist criterion states that if Z is zero, then there are no zeros in the RHP and therefore that the system is stable. We would like to find that Z is zero to ensure stability. The Nyquist plot in the $L(s)$-plane gives us information about N, the number of clockwise encirclements of -1. To evaluate Z, we also need to know the value of P, the number of RHP poles of $L(s) = a(s)f(s)$ from other prior information. The overall feedback system can still be stable with a nonzero P because the open-loop system may contain unstable RHP poles in $L(s)$ that are stabilized by feedback. For example, we could have $P = 2$ and find from the Nyquist plot that $N = -2$; thus, we can conclude that $Z = 0$, which corresponds to two unstable RHP poles of $L(s)$ stabilized by feedback such that $1 + L(s)$ has no RHP zeros; therefore, the closed-loop feedback system is stable. Most open-loop systems are stable to start with such that $P = 0$; therefore, typically, N and Z have the same value. Also, note that the value of $L(s)$ on the curved part of the infinite D contour is assumed to go to zero since all practical systems have a zero frequency response at ∞; thus, the Nyquist criterion actually only uses information about the frequency response $L(j\omega)$, i.e., $L(s)$ evaluated along the $s = j\omega$ axis, to evaluate stability.

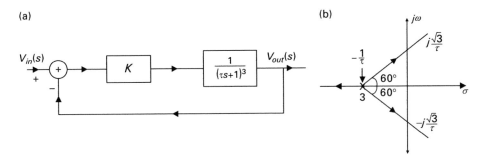

Figure 9.4a, b. A phase-shift oscillator with three delay stages, (a) its block diagram and (b) its root-locus plot.

Figure 9.4 (a) shows a feedback-system example composed of three identical low pass filters in a negative-feedback loop, the classic topology of a phase-shift oscillator. We would like to know for what values of K the system is unstable and for what critical value of K are any of the poles of the closed-loop system just on the verge of instability, i.e., on the $j\omega$-axis. We shall first solve this example by root-locus techniques and then via the Nyquist criterion. Figure 9.4 (b) shows a root-locus plot of the system under negative feedback. Since all three poles are at the same location, contributions of $60°$, $180°$, or $300°$ phase from each pole yields an overall phase shift of $180°$ by the asymptote rule and yields three root-loci branches as shown. Both of the rightward branches cross the $j\omega$-axis for sufficiently large K. The normalized distance to either of the $j\omega$-axis crossing points is $1/\cos(60°)$ from each of the three poles, such that by the root-locus value rule (Equation (2.27)) the critical K at which instability or oscillation just sets in is $(1/(1/2))^3 = 8$. The normalized root-locus value rule (Equation (2.28)) would have given us the answer directly as $2^3 = 8$. Can we obtain this result by the Nyquist criterion as well?

The straightforward application of the Nyquist criterion would evaluate $L(s) = K/(\tau s + 1)^3$ on the $j\omega$ D-contour in the s-plane for various K and generate Nyquist plots in the $L(s)$-plane for these K. We would then examine these Nyquist plots to deduce the critical K at which encirclements of the -1 point just begin to happen. Such a strategy is awkward since it needs us to generate many Nyquist plots. A simpler strategy is to plot $a(s)f(s) = 1/(\tau s + 1)^3$ on the $j\omega$ D-contour in the s-plane, generate one Nyquist plot in the $a(s)f(s)$ plane, and look for encirclements of the $-1/K$ point in the $a(s)f(s)$ plane. Then, as K varies, the $-1/K$ point moves with respect to the fixed Nyquist plot and at some critical value begins to just be encircled by the plot. Most practical applications of the Nyquist criterion use the latter strategy with a fixed plot and a changing $-1/K$ point rather than a fixed -1 point and a changing plot.

Figure 9.5 illustrates the application of the Nyquist criterion to the phase-shift oscillator case. Each pole contributes $-60°$ of phase at the $-180°$ frequency, $j\sqrt{3}/\tau$, such that the magnitude of $1/(\tau s + 1)^3$ at this point is $1/\left(\sqrt{((\sqrt{3})^2+1^2)}\right)^3 = 1/8$. Each pole contributes a total phase of $-90°$ such that at infinite frequency

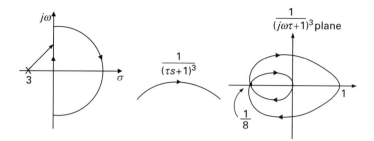

Figure 9.5. Nyquist plot of the phase-shift oscillator example.

the magnitude is 0 and the overall phase is $-270°$. For physical systems, the magnitude of $a(j\omega)f(j\omega)$ is an even function of ω while the phase is an odd function of ω. The Nyquist plot for negative frequencies is then an exact replica of that for positive frequencies except that it is mirrored about the real axis such that $a(-j\omega)f(-j\omega) = [a(j\omega)f(j\omega)]^*$, i.e., complex-conjugate symmetry is preserved.

Figure 9.5 shows that for $0 \le K \le 8$, the $-1/K$ point falls outside both encirclements of the Nyquist plot. For $K > 8$, the $-1/K$ point is encircled twice by two clockwise encirclements corresponding to $N = 2$. Since $P = 0$, $Z = N + P = 2$, i.e., we have two zeros of $1 + Ka(s)f(s)$ in the RHP for $K > 8$, indicating instability for these K. The Nyquist diagram is consistent with our root-locus answer. In fact, it provides additional information as well. We note that for $K < -1$, the $-1/K$ point has one encirclement corresponding to one unstable pole for these K. The latter case corresponds to the case of positive feedback since K is negative and corresponds to the finding that the phase-shift oscillator is unstable if the positive-feedback loop gain exceeds 1. Root-locus plots for positive feedback are discussed later in this chapter and yield the same answer. For the phase-shift oscillator, the root-locus plot for positive feedback has three asymptotes making angles of $0°$, $120°$, and $240°$ with the real axis; the $0°$ asymptote leads to the unstable RHP pole for $|K| > 1$, i.e., $K < -1$.

The example of Figure 9.5 reveals that the Nyquist plot and root-locus techniques yield similar information. However, the Nyquist method can use experimental information from Bode plots of $a(j\omega)f(j\omega)$ even if knowledge about the poles and zeros of $a(s)f(s)$ is not available. Nyquist techniques can therefore function when root-locus techniques fail and are consequently more general.

9.2   Nyquist-based criteria for robustness: Gain margin and phase margin

From the following algebraic manipulations on Black's formula

$$\frac{V_{out}(s)}{V_{in}(s)} = \frac{a(s)}{1 + a(s)f(s)} = \frac{1}{f(s)}\left[\frac{a(s)f(s)}{1 + a(s)f(s)}\right] = \frac{1}{f(s)}\left[\frac{L(s)}{1 + L(s)}\right], \quad (9.3)$$

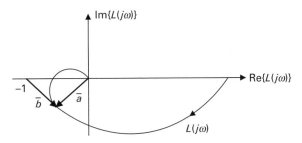

Figure 9.6. Simple performance criteria based on the Nyquist plot.

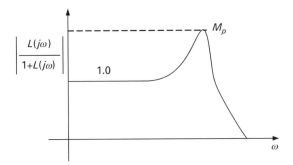

Figure 9.7. The peak value, M_p, occurs in the closed-loop frequency response when $L(j\omega)$ becomes close to -1.

we observe that if

$$\frac{L(s)}{1 + L(s)} \approx 1, \tag{9.4}$$

the closed-loop transfer function of Equation (9.3) is nearly the ideal $1/f(s)$ that we design the closed-loop system to be. If we assume that $1/f(s)$ only contains desirable dynamics that we intend, then all unintended dynamics in our system arise from $L(s)/(1 + L(s))$ not being nearly 1. Figure 9.6 is a Nyquist-like plot of a typical $L(j\omega)$ seen in many feedback systems. We must have

$$\frac{L(j\omega)}{1 + L(j\omega)} = \frac{\bar{a}}{\bar{b}}, \tag{9.5}$$

where the vectors or complex numbers $\bar{a}$ and $\bar{b}$ are marked as shown in the figure. When $L(j\omega)$ is near -1, the magnitude of $\bar{b}$ is small, which causes $\bar{a}/\bar{b}$ to become large. Figure 9.7 plots the frequency response of $|L(j\omega)/(1 + L(j\omega))|$, which reveals peaking when $L(j\omega)$ is near -1 or equivalently when $|L(j\omega)|$ is near 1. We can define M_p such that

$$M_p = \max_{\omega} \left| \frac{L(j\omega)}{1 + L(j\omega)} \right|. \tag{9.6}$$

The nearer $L(j\omega)$ gets to -1, the larger is M_p, the more non-ideal is our desired transfer function, and the larger is M_p. A large M_p is indicative of a badly behaved closed-loop system with lots of ringing and overshoot in its step or impulse response and portends that instability is near. A peaky frequency response is indicative of high-Q complex poles being present in a closed-loop system, which is highly undesirable in most feedback loops unless they are intended to create high-Q filters, resonators, or oscillators. High-Q systems have poles that make small angles with the $j\omega$-axis θ_q given by

$$\sin(\theta_q) = \frac{1}{2Q} = \zeta \qquad (9.7)$$

with Q known as the quality factor and ζ known as the damping. Figure 9.19 (a) shows an example of high-Q complex poles.

A large majority of, but not all, plots of $L(j\omega)$ look like that shown in Figure 9.6 with a smooth monotonic frequency response as ω goes from 0 to ∞. They have a low-frequency region where $L(j\omega)$ is large, a bandwidth-limiting region where $L(j\omega)$ is near -1 but does not encircle -1 to stay stable, and a region at high frequencies past the bandwidth-limiting region of the loop where $L(j\omega)$ goes to zero. For such loops, we can get information about the dynamics of closed-loop behavior without knowing any details of the loop transmission's poles or zeros if we simply characterize the behavior of $L(j\omega)$ near -1 with some empirical parameters that measure how near -1 $L(j\omega)$ gets. The gain margin and phase margin parameters shown in Figure 9.8 are empirical measures that characterize how well behaved a feedback loop is. The further away from -1 $L(j\omega)$ is, the lower is the M_p and the more likely is the loop to be well behaved, but the less likely it is to achieve aggressively high performance in terms of closed-loop bandwidth, disturbance rejection, or linearity. Thus, feedback-system designers constantly attempt to get near enough to -1 to achieve good-enough perform-ance without being too near to cause ringing and instability. That is, they try to have adequate gain margins and phase margins but not excessively conservative margins that detract from system performance.

Figure 9.8 (a) illustrates the definition of gain margin and phase margin. The gain margin is defined as $1/a$, i.e., at $-180°$ of loop transmission phase, we measure the reciprocal of the loop's gain and call it the gain margin (GM). A multiplicative increase in the loop gain by a factor equal to the gain margin is just small enough to avoid an encirclement of the -1 point and maintain loop stability. The phase margin is defined as θ, i.e., at the unity-gain or crossover frequency ω_1, we measure the difference between the loop's phase and $-180°$ and call it the phase margin (PM). The addition of extra negative phase to the loop of magnitude equal to the phase margin is just small enough to avoid an encirclement of the -1 point and maintain loop stability.

It is empirically true in many feedback loops that $\text{GM} \geq 3$ and $45° < \text{PM} < 60°$ yield settling dynamics with acceptable overshoot without compromising

(a)

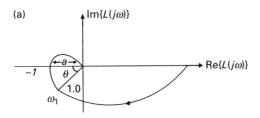

(b)

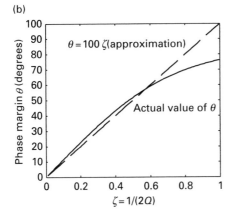

Figure 9.8a, b. (a) Evaluating gain and phase margin from a Nyquist plot. (b) Damping and phase margin are approximately proportional to each other for a second-order system.

bandwidth much. The PM measure is only useful if $GM \geq 3$. For such loops, if we denote $\theta = PM$ we can estimate that

$$M_p = \frac{1}{\sin \theta}$$
$$\zeta = \frac{1}{2Q} \approx \frac{\theta}{100}$$

(9.8)

where ζ is the damping parameter, Q is the quality factor of an equivalent second-order system that the loop's dynamics may be approximated by, and θ is measured in degrees. Second-order systems are discussed in depth in Chapter 13. For $45° < \theta < 60°$, such loops have $0.86 < Q < 1.1$ and thus exhibit little overshoot in their step responses. Figure 9.8 (b) shows that the second approximation in Equation (9.8) is a good one for a wide range of ζ values.

A high phase margin and conservative design may be necessary in an application like aircraft control while a lower phase margin and more aggressive design may be tolerable in a high-speed voltage buffer communicating internet data.

9.3   Compensation techniques

Feedback loops can be compensated to improve their dynamic behavior to transient inputs by improving the phase margin present at their unity-gain or

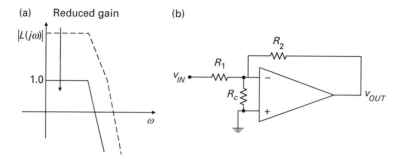

Figure 9.9a, b. Reduced-gain compensation: (a) Bode plot and (b) circuit example.

crossover frequency. There are five common compensation techniques, which we describe below [2].

9.3.1 Reduced-gain compensation

In this scheme, the loop gain of the feedback loop is simply lowered until there is acceptable phase margin. Figure 9.9 (a) reveals an example Bode plot of a loop transmission (a log-log plot of $|L(j\omega)|$ vs. ω) before and after compensation. By simply lowering the loop gain, the unity-gain crossover frequency is forcibly lowered such that phase contributions from only one of the two poles in $|L(j\omega)|$ are significant near crossover. Thus, the phase margin improves from nearly $0°$ in the uncompensated case where both poles contribute an almost full $-90°$ of phase each, to nearly $90°$ in the compensated case where only one pole contributes $-90°$. Figure 9.9 (b) reveals an inverting-amplifier circuit example which uses this form of compensation. The presence of R_c between the v_- and v_+ inputs of the operational amplifier alters the loop gain of the uncompensated system from

$$\frac{A_v(s)G_2}{G_2 + G_1} \quad \text{to} \quad \frac{A_v(s)G_2}{G_2 + G_1 + G_c} \tag{9.9}$$

in the compensated system without noticeably affecting the overall closed-loop gain at low frequencies where $|A_v(s)| \gg 1$. Here, $A_v(s)$ is the operational-amplifier gain, and $G_i = 1/R_i$. While the compensation improves the transient performance of the loop considerably, the loop's crossover frequency and consequently closed-loop bandwidth are considerably reduced. Equation (9.3) reveals that a closed-loop system's frequency response begins to significantly deviate from its ideal $1/f(j\omega)$ behavior near and after $|L(j\omega)| = 1$; thus, the closed-loop bandwidth, i.e., the range of frequencies over which ideal behavior is exhibited, scales with the loop's unity-gain or crossover frequency.

9.3.2 Dominant-pole compensation

In this scheme, a slow large-time-constant system is introduced into the loop such that, near the loop's crossover frequency, the dynamics of the loop transmission

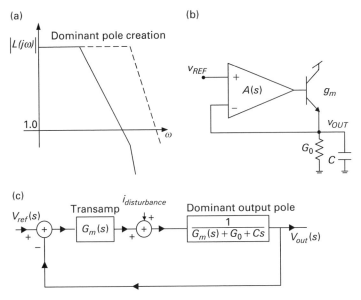

Figure 9.10a, b, c. Dominant-pole compensation: (a) Bode plot, (b) circuit example, and (c) block diagram.

are dominated by those of the slow system and the loop behaves largely like a single-time-constant system. Figure 9.10 (a) reveals an example Bode plot of a loop transmission before and after dominant-pole compensation. The dominant pole improves the phase margin from 0° before compensation to almost 90° after compensation since the poles of the uncompensated loop only exert their influence well beyond the crossover frequency and thus do not affect the loop's phase margin. As in reduced-gain compensation, this scheme achieves better phase margin at the price of lowered closed-loop bandwidth. However, unlike reduced-gain compensation, the benefits of high loop gain are still available at low frequencies since the loop gain has not been reduced everywhere.

In feedback systems like voltage regulators, slow closed-loop behavior is actually desirable since the output of the system is expected to always be constant and the input is an unchanging reference. Figure 9.10 (b) shows a linear voltage-regulator circuit where a dominant pole is created by intentionally adding a large load capacitor. Figure 9.10 (c) is a feedback block diagram that models the behavior of the circuit. For simplicity, the bipolar transistor and the transconductor have been replaced as having an effective transfer function $G_m(s) = (1 + A(s))g_m$. The large load capacitor creates a dominant pole which improves phase margin and improves disturbance rejection of high-frequency currents in the regulator. The regulator is mathematically described by

$$V_{out}(s) = \frac{G_m(s)}{G_m(s) + G_0 + Cs} V_{ref}(s) + \frac{1}{G_m(s) + G_0 + Cs} I_{disturbance}(s) \qquad (9.10)$$

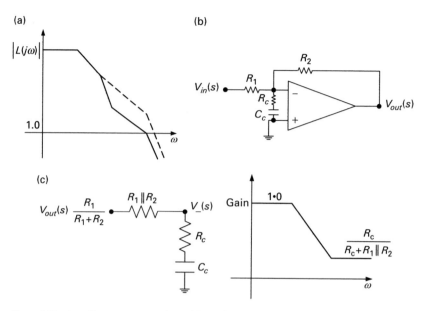

Figure 9.11a, b, c. Lag compensation: (a) Bode plot, (b) circuit example, and (c) typical feedback network.

Thus, the regulator's dc response to $V_{ref}(0)$ is left intact by the dominant pole while the disturbance rejection is improved.

9.3.3   Lag compensation

In this scheme, a pole-and-zero lag network sculpts the loop-transmission frequency response such that the loop gain and, consequently, crossover frequency are reduced to a value where good phase margin is achievable. However, the loop transmission is only reduced after the lag network's pole-frequency location such that the benefits of high loop gain are still available over a wide range of frequencies below crossover, not just at low frequencies. Figure 9.11 (a) reveals an example Bode plot of a loop transmission before and after lag compensation. The phase margin is improved from a few degrees before compensation to 45° after compensation because the loop crossover frequency is lowered to be *at* one of the loop-transmission's poles rather than *beyond* it. Figure 9.11 (b) illustrates an operational-amplifier circuit which implements lag compensation. The similarity of the circuit to reduced-gain compensation in Figure 9.9 should be noted; unlike reduced-gain compensation, however, the presence of the capacitor in the lag network ensures that the loop-gain reduction only manifests at high frequencies, not at all frequencies. Figure 9.11 (c) reveals the general topology and frequency response of a lag network; in Figure 9.11 (b), the Thevenin-equivalent voltage $V_{out}(R_1/(R_1 + R_2))$ of the resistive divider formed by the coupling resistance R_2, and the load resistance R_1 (with $V_{in}(s)$ grounded) and the Thevenin-equivalent resistance $R_1 \| R_2$ provide the input voltage and the input coupling resistance of the lag network of Figure 9.11 (c), respectively.

Root-locus plots of lag-compensated feedback loops reveal that the zero of the lag network attracts a pole towards it in the closed-loop system. The resultant pole-zero cancellation in the closed-loop system is not perfect since the loop gain required for near-perfect cancellation is extremely high. Thus, residual settling dynamics at the lag network's zero-frequency location manifest as long settling tails in the closed-loop system's transient response. The residual settling dynamics occur at time scales that are much slower than that expected from the crossover frequency of the loop, and constitute one of lag compensation's disadvantages.

9.3.4   Lead compensation

In this scheme, a zero-and-pole lead network with a zero-frequency location right near the crossover frequency adds positive phase to the loop transmission such that the phase margin of the loop is improved. The crossover frequency of the loop is not affected much since the magnitude consequences of the zero at crossover are mild while its positive phase contribution of approximately 45° can be significant. The pole-frequency location of the lead network occurs after the crossover frequency such that its negative-phase contribution to the loop transmission is slight.

Figure 9.12 (a) reveals an example Bode plot of a loop transmission before and after lead compensation. The magnitude of the loop transmission and crossover

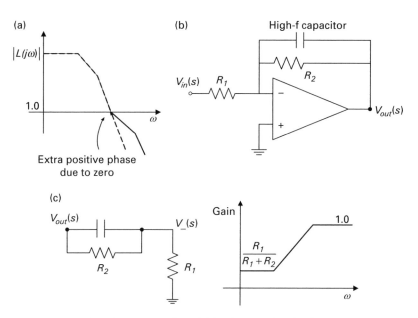

Figure 9.12a, b, c. Lead compensation: (a) Bode plot, (b) circuit example, and (c) typical feedback network.

frequency are nearly unchanged although the phase margin is improved. Figure 9.12 (b) reveals how a lead network sculpts the loop transmission in an operational-amplifier circuit. Figure 9.12 (c) reveals the general architecture of a lead network and its frequency response. Note that the closed-loop transfer function of the inverting amplifier of Figure 9.12 (b) is relatively unchanged over most of the closed-loop's operating range of frequency since the capacitor's admittance is only significant near the crossover frequency of the loop.

Lead compensation is the most aggressive form of compensation in terms of preserving the benefits of high loop gain over all frequencies. However, it may be inappropriate in feedback loops with a delay where a moderate positive-phase contribution at crossover is not sufficient to improve the phase margin of the loop or if the larger closed-loop bandwidth results in an increase in noise.

9.3.5   Minor-loop compensation

In this scheme, sometimes referred to as pole-splitting compensation, a minor feedback loop is created within the principal or major feedback loop transmission. The effect of the minor loop is to split poles that are relatively near to each other in the uncompensated major loop transmission into very widely separated poles in the compensated major loop transmission. One of these poles is a very-low-frequency dominant pole and the other of these poles is a very-high-frequency pole well beyond the crossover frequency of the major loop transmission. Thus, the phase margin of the major loop transmission begins to resemble that due to a single dominant pole and improves from a few degrees in the uncompensated case to nearly $90°$ in the compensated case. The minor feedback loop is created via a coupling capacitor that creates a zero in its feedback path; the net effect of this zero is to cause the closed-loop transfer function of the minor loop to develop widely separated poles. The closed-loop transfer function of the minor loop serves to transform the major loop transmission that it is embedded within into one with high phase margin.

One of the most common applications of minor-loop compensation is in the architecture of operational amplifiers. These amplifiers are created with a minor loop within their circuitry that effectively causes them to have a high-gain dominant-pole transfer function and one very-high-frequency pole. When the operational amplifiers are used, for example, to create a unity-gain buffer in an external circuit, the major loop transmission of the unity-gain buffer has nearly $90°$ of phase margin since the dominant pole is the only pole below crossover. Minor-loop compensation is perhaps the most widely used form of compensation in integrated-circuit design with the operational amplifier being a very common example. In contrast, reduced-gain, lag, and lead compensation are more common in discrete-circuit design. We shall begin our discussion of minor-loop compensation by describing the architecture of operational amplifiers.

Figure 9.13 illustrates a common operational-amplifier architecture with v_+ and v_- being the input terminals and v_{out} being the output terminal. The

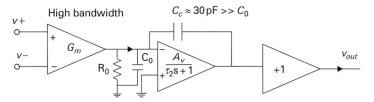

Figure 9.13. Typical two-stage integrated operational-amplifier architecture.

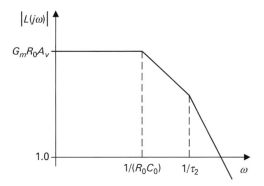

Figure 9.14. Forward-path of the minor loop in the operational-amplifier example.

transconductance amplifier G_m, R_0, and C_0 form one high-gain stage and create one pole and the $A_v/(\tau_2 s + 1)$ block forms another high-gain stage and creates another pole. The capacitor C_c implements minor-loop compensation. A major loop is created when the operational amplifier is used in a circuit, for example when v_{out} is hooked to v_- to create a unity-gain buffer. We shall begin by first discussing the uncompensated case when C_c is absent.

Figure 9.14 shows the uncompensated major loop transmission corresponding to a unity-gain buffer, i.e., the transfer function of the operational amplifier without C_c. We note that the high gain $G_m R_0 A_v$ is accompanied by two relatively closely spaced poles at $1/R_0 C_0$ and $1/\tau_2$. Thus, at crossover, which is well beyond both poles, each pole contributes nearly $-90°$ of phase resulting in a poor phase margin of a few degrees.

Figure 9.15 is a feedback block diagram that reveals the role of C_c in the minor loop. The primary effect of C_c is to feed back a current $sC_c v_{out}$ to the first high-gain stage's load and thus create a zero in the minor loop. A secondary effect of C_c is to load the first gain stage with its capacitance. Since $C_c \gg C_0$ typically, we have assumed that most of the capacitance at the load of the first high-gain stage is due to C_c. We have, for simplicity, also neglected the feedthrough transmission through C_c from the negative input of the A_v stage to the output of the A_v stage. In practice, such feedthrough transmission can cause a right-half-plane (RHP) zero, whose deleterious negative-phase effects are alleviated by adding a resistance in series with C_c [2].

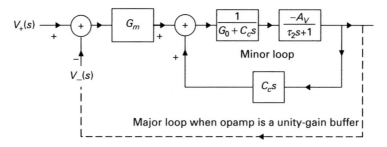

Figure 9.15. Block diagram of the minor-loop compensation scheme used in the operational amplifier example.

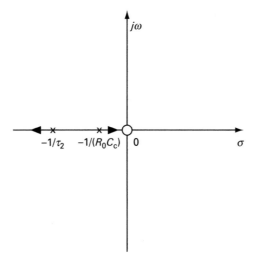

Figure 9.16. Root-locus plot of the minor loop used in the operational amplifier example.

Figure 9.16 shows a root-locus plot that illustrates how the zero in the minor loop causes poles to separate. The large A_v and large value of C_c/G_0 together function to create a large root-locus gain parameter ensuring a wide separation of poles.

A more insightful way of understanding minor-loop compensation is to understand that any feedback loop's closed-loop transfer function may be expressed as

$$\left| \frac{a(s)\frac{1}{f(s)}}{a(s) + \frac{1}{f(s)}} \right| \approx \min\left(|a(s)|, \frac{1}{|f(s)|} \right) \tag{9.11}$$

Thus, as we discussed in the circuit analogy of Chapter 2 (Figure 2.5), we are effectively computing the equivalent impedance of an $a(s)$ impedance in parallel with that of a $1/f(s)$ impedance such that the closed-loop transfer function is dominated by the impedance with the lower magnitude. Hence in Figure 9.15,

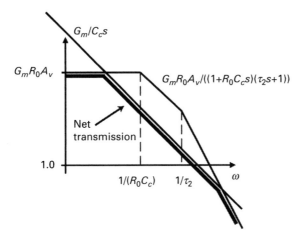

Figure 9.17. Operational amplifier example, showing that the closed-loop transmission of the minor loop closely follows the minimum of its feed-forward and feedback paths.

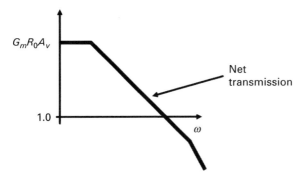

Figure 9.18. The net closed-loop transmission of the minor loop in the operational amplifier example.

the minor loop's $a(s)$ and $1/f(s)$ interact to create a transfer function that is approximately the minimum of the two. This transfer function is scaled by G_m to become the major loop transmission in a unity-gain buffer. Since G_m is just a scale factor multiplying both $a(s)$ and $1/f(s)$, the major loop transmission, which is plotted in Figure 9.17, is the minimum of the two intersecting curves as shown. The presence of a dominant pole well below crossover and a high-frequency pole above crossover are evident in the kinks of the net transmission. The locations of the poles are computed from Figure 9.17 to be at the locations of these kinks, which are very nearly at $1/(R_0(A_V + 1)C_c)$ and at $(A_V + 1)/\tau_2$.

Figure 9.18 reveals that the net transmission of the overall forward path of Figure 9.15 has almost 90° of phase margin even for unity-gain feedback, the most challenging case for the use of an operational amplifier. In the latter case, the feedback path gain and major loop gain are at their highest values such that a good phase margin for this case guarantees a good phase margin for other cases where the feedback path gain has an attenuating value, and the major loop gain and crossover frequency are consequently lower. The shape of the net transmission is controlled by a well-defined passive component C_c and the minor loop has about $90°$ of phase margin as well.

9.4   The closed-loop two-pole τ-and-Q rules for feedback systems

Frequently, well-behaved loop transmissions in circuit design can be approximated with a feedback loop that has two time constants and a dc loop gain A_{lp} especially near their crossover frequency:

$$L(s) = A_{lp}\left(\frac{1}{\tau_{big}s + 1}\right)\left(\frac{1}{\tau_{sml}s + 1}\right) \tag{9.12}$$

The closed-loop transfer function of a system with this loop transmission entirely in its feedforward path and with a unity-gain feedback path is then given by

$$H_{cl}(s) = \frac{L(s)}{1 + L(s)}$$

$$= \frac{\left(\dfrac{A_{lp}}{A_{lp} + 1}\right)}{\dfrac{\tau_{big}\tau_{sml}}{A_{lp} + 1}s^2 + (\tau_{big} + \tau_{sml})s + 1} \tag{9.13}$$

Equation (9.13) can be rewritten in a form that describes the closed-loop system as a second-order system with its transfer function in a canonical form

$$H_{cl}(s) = \frac{A_{cl}}{\tau_{cl}^2 s^2 + \dfrac{\tau_{cl}s}{Q_{cl}} + 1} \tag{9.14}$$

By performing some simple algebra on Equation (9.13), the values of A_{cl}, τ_{cl}, and Q_{cl} in Equation (9.14) can be found to be

$$A_{cl} = \frac{A_{lp}}{A_{lp} + 1}$$

$$\frac{1}{\tau_{cl}} = \omega_n = \sqrt{\frac{1 + A_{lp}}{\tau_{sml}\tau_{big}}} \tag{9.15}$$

$$Q_{cl} = \frac{\sqrt{(1 + A_{lp})}}{\sqrt{\dfrac{\tau_{big}}{\tau_{sml}}} + \sqrt{\dfrac{\tau_{sml}}{\tau_{big}}}}$$

In the case where the $1/(\tau_{big}s + 1)$ degenerates to an integrator pole at the origin, $1/(\tau_{big}s)$, then the expressions of Equation (9.15) get modified to those of Equation (9.16):

$$A_{cl} = 1$$

$$\frac{1}{\tau_{cl}} = \omega_n = \sqrt{\frac{A_{lp}}{\tau_{sml}\tau_{big}}}$$

$$Q_{cl} = \frac{\sqrt{A_{lp}}}{\sqrt{\frac{\tau_{big}}{\tau_{sml}}}}.$$

(9.16)

The characterization of the closed-loop second-order system by canonical parameters τ (or $\omega_n = 1/\tau$) and Q is useful because τ is a measure of the closed-loop system's 10% to 90% rise time and Q is a measure of its ringing in its step response or its peakiness in its frequency response. The poles of the closed-loop system in Figure 9.19 (a) and the frequency-response plots of Figures 9.19 (b) and 9.19 (c) reveal that higher Q's lead to more peakiness (and more ringing) and that higher ω_n's lead to a higher closed-loop bandwidth. Thus, in this second-order approximation, Q represents M_p and ω_n represents the cutoff, unity-gain, or crossover frequency in Figure 9.7. We notice from Equations (9.15) or (9.16) that a wide separation of time constants is needed to ensure a relatively low value of Q_{cl}, and that a high value of A_{lp} increases ω_n but also increases Q_{cl}. The root-locus plot of Figure 2.22 for a two-time-constant feedback loop combined with the definition of Q_{cl} in Figure 9.19 (a) provides geometric intuition for the algebraic predictions of Equations (9.15) or (9.16).

We shall refer to Equations (9.15) or (9.16) as the closed-loop two-pole τ-and-Q rules for feedback systems. The closed-loop two-pole τ-rule refers to formulas for τ_{cl} while the closed-loop two-pole Q-rule refers to formulas for Q_{cl}. We shall apply these rules in Chapter 11, where we will present them again in a more intuitive loop-transmission and root-locus context with $A_{lp} = 1$ in Equation (9.16). The rules are very useful for two reasons. First, almost all well-behaved negative feedback systems can be well approximated by an equivalent closed-loop second-order system. Second, because there is a wealth of knowledge on second-order systems, which we discuss in Chapter 13, characterizing feedback systems in terms of second-order parameters allows us to leverage off this knowledge.

9.5   Conditional stability

Feedback loops can sometimes go unstable when the loop gain is too low, rather than when it is too high. A common example of such a feedback loop occurs when there are three identical low-frequency poles near the origin designed to maximize dc gain and low-frequency performance in the loop transmission and two identical zeros in a mid-frequency region of the loop transmission designed to ensure that

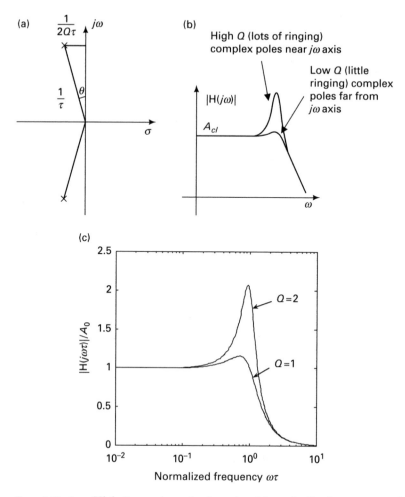

Figure 9.19a, b, c. High-Q complex poles in a closed-loop feedback system are shown in (a) and a corresponding frequency-response plot is shown in (b). The figure in (c) shows more quantitative detail in (b).

the feedback loop has 90° of phase margin at its crossover frequency: the two zeros add $+180°$ to the loop transmission to cancel the $-270°$ of phase from the three poles such that the net phase near the crossover frequency is $-90°$. However, if the loop gain is not high enough, the crossover frequency may occur at or before the location of the mid-frequency zeros such that the positive phase contribution of the zeros is nearer $+90°$, and the loop then has nearly zero degrees of phase margin. At some value of the loop gain that is sufficiently low such that we have zero degrees of phase margin, the loop will go unstable. The loop stays stable until very high loop gains again cause instability due to the occurrence of high-frequency poles that occur past the frequencies of the zeros. Such a feedback loop is said to be conditionally stable since it is only stable under certain conditions of loop gain, not just loop gains that are sufficiently low.

A common reason for the reduction of loop gain is the presence of saturation nonlinearities in a feedback system. Such nonlinearities can cause a feedback loop to be stable at small signals but go unstable at large signals where its effective loop gain is lower. For example, in a tanh amplifier such as the one we described in Chapter 8, as long as differential inputs are within the linear range of the tanh, the differential input-output gain is set by the slope of the tanh. Outside the linear range of the tanh, the input-output gain of the tanh lowers to a smaller and smaller effective value as its differential input is made larger and larger. If such an amplifier is part of a feedback loop with conditionally stable dynamics, instability can result. High-order $\Sigma\Delta$ analog-to-digital converters (see Chapter 15) can also exhibit such conditionally stable dynamics within their feedback loops. Amplifiers in such converters are intentionally designed to have larger linear ranges. A semi-empirical and semi-quantitative technique called describing-function analysis helps us analyze the effects of nonlinearities in such feedback systems.

9.6   Describing-function analysis of nonlinear feedback systems

A describing function $G_D(E,\omega)$ defines the input-output gain and phase of a nonlinear system to a sinusoidal input of amplitude E and angular frequency ω [2]. The gain and phase of $G_D(E,\omega)$ are computed by pretending that only the component of the output at frequency ω, i.e., only the fundamental component in a set of harmonic outputs, matters for the input-output function. In essence, we have a parametric linear system with a transfer function depending on the input amplitude E and the input angular frequency ω, rather than just on ω. Thus, for example, a function with a threshold at 0 and output E_0 that generates square waves of amplitude E_0 from sine waves has a describing function with a gain that is $4E_0/(\pi E)$ and identically zero phase. A Schmitt trigger nonlinear system with hysteretic thresholds of $\pm E_S$ has

$$G_D(E) = 0; E \le E_S$$

$$G_D(E) = \left(\frac{4E_0}{\pi E}\right) e^{-j\arcsin(E_S/E)}; E > E_S \tag{9.17}$$

Intuitively, a Schmitt trigger seems to respond with a delay that decreases for larger inputs such that it has a nonzero phase function. A simple thresholding element has no delay in its response, and therefore a zero phase function. In both cases, the gain decreases with the amplitude of the input E since the output remains saturated at E_0 even while the input E increases. Thus, the gain of the system effectively decreases with increasing E. In both of these cases, there is no frequency dependence in the static nonlinearity.

Describing functions can serve as transfer function blocks inside feedback loops and help us analyze issues such as stability. For example, we can apply the Nyquist criterion for stability with E as a parameter. Thus, they are most useful in analyzing the critical amplitude of oscillation at which an oscillator just stabilizes its amplitude of oscillation. Or, in the case of a conditionally unstable feedback loop, they can help us understand at what critically large input amplitude the loop gain of a feedback loop might become too low resulting in instability. The assumption that only the fundamental sinusoidal frequency matters is usually a good one in oscillators and in feedback loops where the harmonics are strongly attenuated via bandpass or lowpass filtering. If such filtering does not occur, the utility of describing functions is questionable and a full-blown nonlinear analysis of the system may be necessary.

9.7   Positive feedback

Figure 9.20 (a) shows a block diagram where the feedback is positive. The negative sign assumed in the block diagram till now becomes positive. Thus, Black's formula in this case would have the form

$$\frac{V_{out}(s)}{V_{in}(s)} = \frac{a(s)}{1 - a(s)f(s)}. \tag{9.18}$$

Hence the root-locus criteria have a magnitude and angle condition given by

$$|a(s_1)f(s_1)| = 1$$
$$\angle a(s_1)f(s_1) = 2n\pi \tag{9.19}$$

The root-locus rules for positive feedback are derived from the angle and magnitude conditions of Equation (9.19) with a root-locus parameter K changing from 0 to $+\infty$ as before but with a positive-feedback $2n\pi$ angle condition instead of the negative-feedback $(2n+1)\pi$ angle condition. All of the rules remain substantially unchanged except for the real-axis, asymptote, and complex singularity rules: the real-axis rule is changed to require any branch to be to the left of an *even* number of singularities; the asymptote rule is changed with $2n\pi$ replacing $(2n+1)\pi$ in its statement; the complex-singularity rule is changed with $2n\pi$ replacing $(2n+1)\pi$ in its statement. These rules may be applied to show that

(a)   (b)

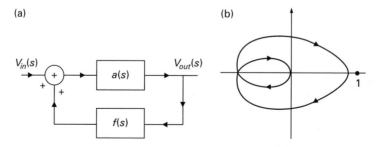

Figure 9.20a, b. Positive feedback: (a) block diagram and (b) Nyquist plot.

the negative-feedback root-locus plot of Figure 9.4 (b) is transformed to a positive-feedback root-locus plot where the asymptotes make angles with the real axis of 0°, 120°, and 240°; the 0° asymptote enters the right half plane when the positive-feedback loop gain exceeds 1 consistent with the findings of the Nyquist-criterion analysis performed for this example.

From Equation (9.18), we derive that the Nyquist criterion for positive-feedback systems requires zero clockwise encirclements of the 1 point in the $L(s)$-plane for stability as shown in Figure 9.20 (b). In contrast, zero clockwise encirclements of the -1 point are needed for stability in negative-feedback systems. The $L(j\omega)$ contour plots are identical for both systems if they have the same $L(s)$ but differing signs of feedback.

For positive-feedback systems to be stable, the dc loop gain usually needs to be less than 1 to prevent encirclements of the 1 point. Stability is thus typically decided by low-frequency behavior rather than by high-frequency behavior as in negative-feedback systems. Pathological cases, e.g., stable systems with a Nyquist contour plot for which $\text{Re}\{L(j\omega)\} > 1$ always such that they do not ever encircle the 1 point while still having a dc loop gain >1, can be imagined, but they are not typical.

Digital circuits like latches and flip-flops exploit positive feedback to create unstable systems that stabilize at a 1 or a 0 due to nonlinearities. Thus, they form the foundation of memory circuits.

Analog circuits use positive feedback to improve resonant gain Q and or amplification, e.g., $Q \rightarrow Q/(1 - \alpha)$ where $0 < \alpha < 1$ is the positive-feedback loop gain. Positive feedback is used to drive resonators unstable and to thus create electrical oscillators or optical lasers. Regenerative amplifiers invented by Armstrong exploit instabilities to sample and amplify initial conditions followed by a quenching of the instability in a periodic manner. If the input is not changing too quickly between samples, a lowpass filtered version of the output of a regenerative amplifier is an amplified version of the input. Amplifiers built with positive feedback are often prone to amplifying noise and are thus not common in high-fidelity systems.

9.8   Feedback in small-signal circuits

Active devices such as transistors in a circuit contain dependent current or voltage sources whose values depend on other control voltage or control current variables in the circuit. For example, the drain-to-source current in a transistor is a function of its gate-to-source voltage. In small-signal circuits, these dependent generators have a linear dependence on the control variable, for example, $i_{ds} = g_m v_{gs}$. Since the dependent sources are themselves part of a circuit involving the control signals, they affect the values of their own control signals. Thus, feedback loops are created within circuits due to the presence of dependent sources. Figure 9.21 (a) shows how feedback loops are created by dependent sources in general and Figure 9.21 (b) shows a simple small-signal example.

(a)

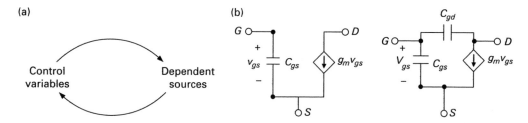

Figure 9.21a, b. Feedback dependence between control variables and small-signal dependent sources in circuits: (a) general concept and (b) a simple example with C_{gd} in the MOS transistor.

In this example, v_{gs} controls the value of the dependent current source via g_m. The current source in turn affects the value of v_{gs} through C_{gd}. Since small-signal circuits are linear, amenable to linear feedback analysis, and preserve the embedded feedback loops of the circuit, we shall discuss how to analyze small-signal active circuits with feedback analysis. Such analysis can shed insight into large-signal behavior as well.

The presence of feedback loops in active circuits often leads to difficulty in the quick analysis of circuits with dependent sources. For example, superposition cannot be used with dependent sources like independent sources, forcing one to solve Kirchhoff's current law equations to analyze the circuit. Such equations are useful for computer simulations but rarely provide any insight into what the circuit is doing and are often an algebraic mess. Armed with our knowledge of feedback systems, however, we can understand these circuits more insightfully by analyzing them via a circuit-analysis trick that preserves insight and transforms tedious algebra into simple feedback block-diagram manipulations. The final feedback block diagram and Black's formula can then be used to quickly write down answers to what the circuit is doing with little algebra. The block diagram also allows other questions to be asked and answered easily, e.g., the quick computation of arbitrary input-output transfer functions in the circuit by just reapplying Black's formula with different feed-forward transmissions and the same feedback transmission, the robustness of the circuit to parameter changes in the circuit, the noise performance of the circuit if noise generators are added at various locations in the circuit, analysis of the circuit via root-locus to see how its dynamics are changed if certain parameters in the circuit are varied etc. Ideas for how to alter the circuit or to compensate it to prevent oscillations are also more easily generated and analyzed. We shall explore the true depth of using feedback to analyze circuits in the next chapter (Chapter 10) on return-ratio analysis, a feedback-circuit-analysis technique pioneered by Bode, a great genius of electrical engineering [3]. However, the trick presented below will serve as a brief preview and foundation for return-ratio analysis.

9.9   The 'fake label' circuit-analysis trick

1. We first re-label all dependent sources with new and unique names, e.g., i_{x1}, v_{yz}, etc. We erase their dependencies on their control variables.

2. The circuit cannot tell the difference between a dependent source with a 'fake label' and an independent source with a 'real label'. A current source is a current source, independent of what it is called.

3. Superposition now applies to all sources! Find via superposition the values of all control variables as a function of the independent sources and the newly independent sources. Find via superposition the values of any desired output variables as a function of the independent sources and the newly independent sources. It's acceptable to short or open the newly independent sources as is customary during superposition. Form a block diagram.

4. Now reinsert the dependencies of the newly independent sources on their control variables to make them dependent again and thus complete a feedback block diagram.

5. Analyze the feedback block diagram via block-diagram simplifications and Black's formula and obtain insight into the circuit.

Figures 9.22 (a) and (b) illustrate why this trick works. We are essentially breaking feedback loops involving the control variables (Figure 9.22 (a)) and then reforming them (Figure 9.22 (b)) by removing dependencies and then reinserting them. Superposition applies at the adder block in Figure 9.22 (a). The figures show how control variables of dependent generators can lead to feedback loops. Once the values of the dependent generators have been computed by this trick, the independent and dependent generators values can be used to determine any output variables of interest, once again by superposition at the adder as in Figure 9.22 (c).

9.10   A circuit example

Figure 9.23 shows the small-signal circuit for a source-degenerated single-transistor amplifier. We'd like to find the input-output transfer function $v_{out}(s)/v_{in}(s)$ using the strategy described above. Since v_{in} sets the value of v_g, v_s is the only unknown control variable that determines the value of the g_m and g_{mb} generators. Thus, we label the g_m and g_{mb} generators with fake labels i_{x1} and i_{x2} respectively to make them independent sources as shown in Figure 9.24 (a) and compute via superposition the value of v_s as a function of v_{in}, i_{x1}, and i_{x2}. We also compute v_{out} as a function of the same three sources by superposition. Since i_{x1} and i_{x2} are in parallel with each other, their transfer functions to both v_s and v_{out} are identical.

The superposition subcircuit obtained by opening i_{x1} and i_{x2} is shown in Figure 9.24 (b) while that obtained by shorting v_{in} is shown in Figure 9.24 (c). From these subcircuits, we can see by inspection that

(a)

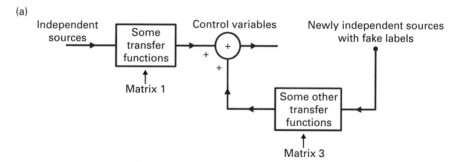

(b)

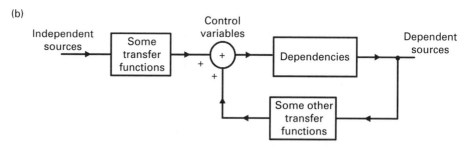

(c)

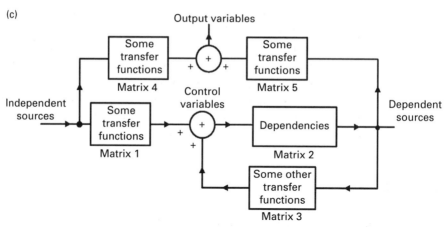

Figure 9.22a, b. Development of the fake-label circuit analysis technique: (a) remove dependencies of controlled sources and apply superposition, (b) put the dependencies back in and complete a feedback block diagram, and (c) combine independent and dependent sources to create new output variables.

$$v_{out} = \left(\frac{C_{gd}s}{G_L + C_{gd}s}\right)v_{in} - \left(\frac{1}{G_L + C_{gd}s}\right)i_{x1} - \left(\frac{1}{G_L + C_{gd}s}\right)i_{x2}. \qquad (9.20)$$

and

$$v_s = \left(\frac{C_{gs}s}{G_S + C_{gs}s}\right)v_{in} + \left(\frac{1}{G_S + C_{gs}s}\right)i_{x1} + \left(\frac{1}{G_S + C_{gs}s}\right)i_{x2} \qquad (9.21)$$

Frequently, in circuits such as this one, the weighted-conductance formula that describes the circuit of Figure 9.25 is a useful primitive with $G_i = 1/R_i$. It is the

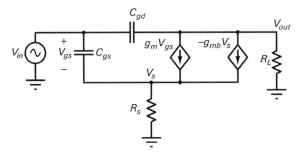

Figure 9.23. Small signal equivalent circuit of a source-degenerated common-source amplifier with a resistive load R_L.

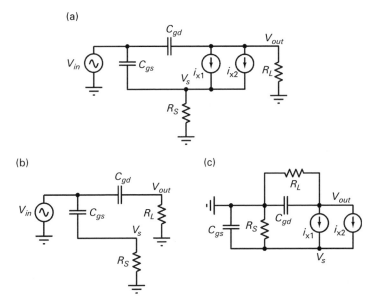

Figure 9.24a, b, c. Common-source amplifier example: (a) small-signal fake-label circuit, (b) and (c) two half-circuits that result from using superposition.

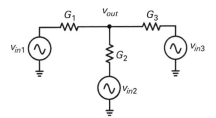

Figure 9.25. A useful circuit primitive (building-block) for the fake-label technique: the weighted conductance summer.

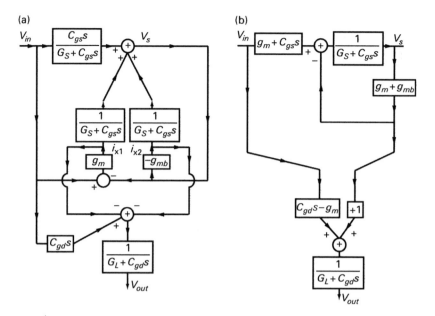

Figure 9.26. Common-source amplifier example: (a) original and (b) simplified block diagram.

multiple-input generalization (only three inputs are shown in Figure 9.25) of the usual resistive-divider formulas. The weighted-conductance formula for Figure 9.25 is given by

$$v_{out} = \frac{G_1 v_{in1} + G_2 v_{in2} + G_3 v_{in3}}{G_1 + G_2 + G_3} \tag{9.22}$$

Now we put the dependencies back in i_{x1} and i_{x2}:

$$i_{x1} = g_m(v_{in} - v_s) \tag{9.23}$$

$$i_{x2} = -g_{mb} v_s \tag{9.24}$$

and draw the feedback block diagram shown in Figure 9.26 (a). Each 3-input adder in the block diagram corresponds to Equation (9.20) for v_{out} or Equation (9.21) for v_s respectively. Some factorizations and grouping simplifications of the block diagram then result in the simplified block diagram of Figure 9.26 (b). In Figure 9.26 (b), we can see that there is one path to the output via $-g_m$ and another path to the output via $+ g_m(L(s)/1 + L(s))$, where $L(s) = (g_m + g_{mb})/(G_s + sC_{gs})$ is the loop transmission of the source-degeneration feedback loop in Figure 9.26 (b). The two contributions that are each proportional to g_m sum to give a net contribution equal to $- g_m(1/(1 + L(s)))$. The contribution to the output via sC_{gs} occurs via $L(s)/(1 + L(s))$. Thus,

$$\frac{v_{out}(s)}{v_{in}(s)} = \frac{\dfrac{-g_m + sC_{gs}\left(\dfrac{g_m + g_{mb}}{G_s + sC_{gs}}\right)}{1 + \left(\dfrac{g_m + g_{mb}}{G_s + sC_{gs}}\right)} + sC_{gd}}{(G_L + C_{gd}s)} \tag{9.25}$$

Physically, the source-degeneration feedback loop arises because any current increase through the $(sC_{gs} + G_s)$ conductance at the v_s node in Figure 9.23 causes an increase in v_s which feeds back to reduce this current increase via the $g_m + g_{mb}$ generators.

For the simple circuit of Figure 9.23, it may seem like it is simpler to just write KCL equations and not bother with a feedback block diagram analysis if all one wants is the transfer function from v_{in} to v_{out}. However, once the feedback block diagram is drawn, questions like, "What is the power-supply rejection?" are easily answered: we add a $G_L v_{dd}$ input to the output adder block in Figure 9.26 (b) and compare gain to v_{dd} versus gain to v_{in}. Or, "How does the dc gain of the amplifier change with changes in g_s?": we see that it is simply the sensitivity of the closed-loop gain in Figure 9.26 (b) to variations in the feed-forward path gain of its source-degeneration feedback loop. Or, "How does the dynamic response of the circuit change as we change I_{DS} or equivalently $g_m + g_{mb}$?": we draw a root-locus plot with gain parameter proportional to $g_s = g_m + g_{mb}$. Or, "What is the output noise of the circuit?": we introduce a $(2qI_{DS} + 4kTG_S)\Delta f$ noise input to an adder inserted at the output of $g_m + g_{mb}$ block in Figure 9.26 (b) and before it feeds back to the negative adder input; we also introduce a $(-2qI_{DS} + 4kTG_L)\Delta f$ input to the output adder block in Figure 9.26 (b). All of the above questions are answered by simple additions to the block diagram of Figure 9.26 (b) and the use of Black's formula. Thus, the insight/algebra ratio is maximized and we have a graphical way of simultaneously accounting for all aspects of the circuit in one unified diagram rather than performing multiple analyses of the same circuit.

References

[1] H. Nyquist. Regeneration theory. *Bell Systems Technical Journal*, **11** (1932), 126–147.
[2] James K. Roberge. *Operational Amplifiers: Theory and Practice* (New York: Wiley, 1975).
[3] Hendrik W. Bode. *Network Analysis and Feedback Amplifier Design* (New York, NY: Van Nostrand, 1945).

10 Return-ratio analysis

You don't understand anything until you learn it more than one way.

Marvin Minsky

In this chapter, we shall discuss a feedback technique for analyzing linear circuits, invented by Hendrik W. Bode in his landmark book, *Network Analysis and Feedback Amplifier Design* published in 1945 [1]. The technique, known as return-ratio analysis, allows one to compute the return ratio of an active dependent generator or passive impedance in a linear circuit as a function of its dependent gain or of its passive impedance, respectively. The return ratio is a quantity analogous to the loop transmission in a feedback loop: just as we can use the loop transmission in a feedback loop to analyze how the dynamics of the loop changes as we vary its dc gain, we can use the return ratio of an element to analyze how transfer functions in the circuit change as we vary the dependent gain or passive impedance of the element. The return ratio also gives us a measure of the robustness of the circuit to changes in the gain or impedance of the element in the same manner that the loop transmission gives us a measure of the robustness of a feedback loop to changes in its feedforward gain. The return ratio explicitly realizes that circuits are composed of bidirectional elements and loading such that the creation of unidirectional block diagrams with feedforward gain $a(s)$ and feedback gain $f(s)$ to analyze them is not unique and sometimes cumbersome. The return ratio $R(s)$ is a bidirectional loop-transmission quantity and is unique, independent of what $a(s)$ and $f(s)$ are used to construct the block diagram, and automatically takes all loading in the circuit into account. Return-ratio techniques lead to methods for computing impedances and transfer functions in circuits such as Blackman's impedance formula [2] and Middlebrook's extra element theorem [3]. These methods focus on how changes in one element or a new added extra element to a circuit alters its transfer function without requiring one to re-compute the entire transfer function of the circuit. In fact, Thevenin's theorem is itself a special case of return-ratio analysis. Return-ratio analysis allows us to write down transfer functions in circuits with feedback loops with little algebra and a few visual manipulations of the circuit taking all loading effects into account. Return-ratio analysis is an indispensable and fundamental tool for every circuit designer that reveals the deep connection between feedback loops and circuits. It allows one to understand circuits insightfully in terms of their underlying feedback loops.

We shall introduce return-ratio analysis by exploiting the fake-label concept described in Chapter 9. Bode himself did not describe return-ratio analysis with fake labels per se, although they are implicit in his description. We shall then compute the return ratio, R, for dependent generators and passive impedances and apply them to create formulas for computing transfer functions in circuits in terms of return ratios. We shall show that the robustness of a circuit to parameter variations in one of its elements is proportional to the reciprocal of the return difference, $1 - R$, of that element; thus, the return ratio of an element is a measure of the robustness of the circuit to variations in this element's parameters. After providing three examples of application of return-ratio analysis in an inverting operational-amplifier circuit, a resistive-bridge circuit, and a bridged-T network, we present Blackman's impedance theorem, a special case of the return-ratio formulas, useful for computing impedances in circuits. We show examples of application of Blackman's impedance formula to a cascode impedance, a cascode impedance with a resistive load, and driving-point impedances of single-transistor circuits. Then, we discuss Middlebrook's extra-element theorem and illustrate its application to an example. We then show that Thevenin's theorem is also a special case of the return-ratio formulas. We conclude the chapter with two examples that illustrate the application of return-ratio analysis. The first example shows how return-ratio techniques may be used in a hierarchical fashion to analyze ever-more complex circuits built up on simpler circuits that have themselves been analyzed by return-ratio techniques. The second example analyzes a super-buffer circuit via return-ratio techniques and shows the equivalence of return-ratio techniques to more conventional feedback-block-diagram techniques such that the two may be compared.

10.1 Return ratio for a dependent generator

Figure 10.1 illustrates a dependent generator in a big linear circuit that has been replaced with a fake label that allows us to pretend temporarily that it is an independent generator of value $i_{fakelabel}$. The transfer function from the independent generator to the dependent generator's control variable v_ε is given by $Z_{fakelabel}$, which is a transfer impedance due to the other elements of the circuit. Thus, we may write

$$v_\varepsilon = i_{fakelabel}Z_{fakelabel} \tag{10.1}$$

However, in reality, if we put the dependency back in

$$i_{fakelabel} = -g_m v_\varepsilon, \tag{10.2}$$

we can define a return ratio R_{dep} for the dependent generator to be

$$R_{dep} = -g_m Z_{fakelabel}, \tag{10.3}$$

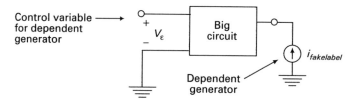

Figure 10.1. Fake-label circuit for a dependent generator.

i.e., as the product of the transfer function $Z_{fakelabel}$ from the independent generator to the control variable and the dependent generator's gain with regard to the control variable $-g_m$. The example shown is for a voltage-controlled current generator but the idea is easily generalized to any kind of dependent generator. *The return ratio of a dependent generator is the product of the transfer function from the fake-labeled generator to its control variable and the gain of the dependent generator with respect to this control variable.* In all cases R_{dep} is dimensionless.

The return ratio R is analogous to a feedback loop transmission caused by $i_{fakelabel}$ reducing the control variable in the big circuit that is responsible for generating it. The return difference is defined as $1 - R_{dep}$ and given by

$$1 - R_{dep} = 1 + g_m Z_{fakelabel} \qquad (10.4)$$

Most return ratios of dependent generators in circuits are negative as in Equation (10.3), which is why the return difference of Equation (10.4) is positive. When computing the return ratio for a dependent generator, the effects of all other elements of the big circuit that it is part of in Figure 10.1 have to be taken into account. That is, $Z_{fakelabel}$ may itself be hard to compute if the big circuit itself has several other dependent generators whose effects are hard to compute or if it has passive elements arranged in bridge-like or other configurations that prevent a simple series-parallel computation of $Z_{fakelabel}$. In this case, one may need to solve Kirchhoff's current law (KCL) equations, use impedance transformations, or use Thevenin's theorem to simplify the big circuit itself and obtain $Z_{fakelabel}$. So, the return ratio as defined here is most useful when

1. The big circuit is not hard to analyze or is small.
2. We want to understand how changing the value of, say, a particular dependent gain affects an overall transfer function in the circuit. Bode invented the concept of the return ratio to understand the robustness of a circuit to variations in a given element in it. Such analysis generalizes the robustness analysis of a simple feedback loop, which is robust to changes in its feedforward gain but is not robust to changes in its feedback gain, if its loop transmission is large.
3. We are adding an element to a circuit and would like to compute the resulting new transfer function amongst circuit variables and characterize how this transfer function varies as we make changes in the parameters of the added element.

10.2   Return ratio for a passive impedance

Figure 10.2 (a) illustrates a passive impedance in a big linear circuit that has been replaced with a fake label that allows us to pretend temporarily that it is an independent current source of value $i_{fakelabel}$. Figure 10.2 (a) shows that in the case of a passive impedance, the control variable that determines the current $i_{fakelabel}$ through that element, v_ε, is the voltage across that element itself rather than some other control voltage, as it is for the dependent current generator shown in Figure 10.1. Thus, if we temporarily ignore the dependency between voltage and current inherent in a passive impedance, we may write

$$v_\varepsilon = i_{fakelabel} Z_{fakelabel} \qquad (10.5)$$

Note that $Z_{fakelabel}$ is just the impedance looking into the big circuit across the element of interest. Now, we re-introduce the dependency of $i_{fakelabel}$ on v_ε to compute that

$$i_{fakelabel} = \frac{-v_\varepsilon}{Z_{element}} \qquad (10.6)$$

By analogy with the dependent generator, the return ratio R of a passive impedance in parallel with the big circuit is then computed to be

$$R_{psv\_prl} = -\frac{Z_{fakelabel}}{Z_{element}} \qquad (10.7)$$

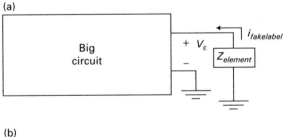

(a)

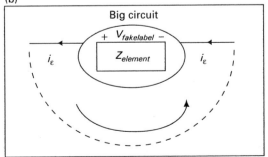

(b)

Figure 10.2a, b. Fake-label circuit for a passive impedance: (a) parallel case and (b) series case.

The return ratio R_{psv_prl} is the feedback loop transmission created by v_{ε} feeding back to reduce the value of $i_{fakelabel}$ that created it. Note that the return ratio of the passive impedance of Equation (10.7) is exactly analogous to the return ratio of the dependent generator of Equation (10.3) with $1/Z_{element}$ corresponding to g_m. The return difference is $1 - R_{psv\_prl}$. Figure 10.2 (a) illustrates a scenario where a passive impedance is in parallel with a big circuit. If a passive impedance is in series with a big circuit as shown in Figure 10.2 (b), we can pretend temporarily that is an independent voltage source of value $v_{fakelabel}$. Then, we may compute the return ratio by noting that

$$i_{\varepsilon} = \frac{v_{fakelabel}}{Z_{fakelabel}}$$

$$v_{fakelabel} = -Z_{element}i_{\varepsilon} \qquad (10.8)$$

$$R_{psv\_srs} = \frac{-Z_{element}}{Z_{fakelabel}}$$

The return ratio of the passive element is still given by an impedance ratio that is determined by the impedance looking across the element and the element's impedance. However, the return ratio for the series case in Equation (10.8) is the reciprocal of the return ratio for the parallel case in Equation (10.7). Thus, in the parallel case, the return ratio is high if $Z_{fakelabel} \gg Z_{element}$, while in the series case, the return ratio is high if $Z_{fakelabel} \ll Z_{element}$. Intuitively, an impedance that is small compared with that of the network in parallel with it creates a strong negative-feedback loop transmission that attenuates current disturbances that attempt to flow into the network by absorbing these disturbances itself. An impedance that is large compared with that of the network in series with it creates a strong negative-feedback loop transmission that attenuates voltage disturbances that attempt to change the voltage across the network by absorbing most of the effects of the voltage disturbance across itself. The strong negative-feedback loop transmission in both cases also ensures that the circuit is quite robust to small percentage changes in the value of $Z_{element}$ as long as the return ratio, a measure of the negative feedback loop transmission, is of sufficiently large magnitude. From Chapter 2, we know that, as long as the loop transmission has a sufficiently large magnitude, negative-feedback loops attenuate disturbances and are robust to their feedforward gains; return-ratio analysis expresses an analogous concept in the circuit domain.

10.3   Transfer function modification with the return ratio

In any linear circuit with an input where there is a dependent current generator, we may represent this generator by $i_{fakelabel}$ and characterize it by the superposition of transfer functions from the input i_{in} and $i_{fakelabel}$ to the control variable V_{ε} and the output variable V_{out} respectively. Thus,

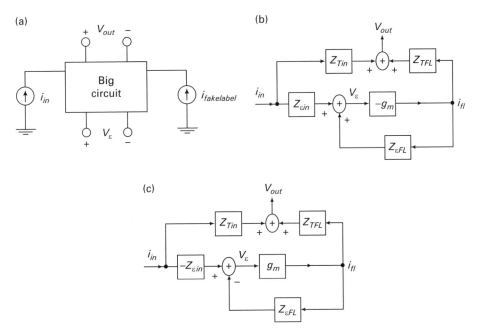

Figure 10.3a, b, c. Transfer-function modification using return ratios: (a) circuit, (b) block diagram, and (c) simplified block diagram.

$$V_{out} = Z_{Tin}i_{in} + Z_{TFL}i_{fakelabel}$$

$$V_\varepsilon = Z_{\varepsilon in}i_{in} + Z_{\varepsilon FL}i_{fakelabel} \qquad (10.9)$$

$$i_{fakelabel} = -g_m V_\varepsilon$$

Figure 10.3 (a) shows the linear circuit and Figure 10.3 (b) shows the block diagram that represents Equation (10.9). The current $i_{fakelabel}$ is denoted by i_{fl} in the block diagram to avoid lengthy subscripts. Although Equation (10.9) is written for a dependent current generator and current input, in general any linear circuit can be characterized by the expressions in Equation (10.9) with appropriate input, output, control, and dependent variables replacing the ones shown. Figure 10.3 (c) simplifies the block diagram of Figure 10.3 (b) into a more canonical negative-feedback-loop-like representation. We notice immediately that $-g_m Z_{\varepsilon FL}$ is a negative-feedback loop transmission and is what we computed R_{dep} in Equation (10.3) to be. From the block diagram of Figure 10.3 (c), Black's formula, and simple algebraic simplifications we get

$$V_{out} = i_{in}\left(Z_{Tin} - \left(\frac{Z_{\varepsilon in}g_m}{1 + g_m Z_{\varepsilon FL}} \right) Z_{TFL} \right)$$

$$V_{out} = i_{in}\left(\frac{Z_{Tin}(1 + g_m Z_{\varepsilon FL}) - Z_{\varepsilon in}g_m Z_{TFL}}{1 + g_m Z_{\varepsilon FL}} \right)$$

$$= i_{in}Z_{Tin}\left[\frac{1 + g_m\left(Z_{\varepsilon FL} - \frac{Z_{\varepsilon in}}{Z_{Tin}}Z_{TFL}\right)}{1 + g_m Z_{\varepsilon FL}}\right]$$

$$= i_{in}Z_{Tin}\left[\frac{1 - g_m\left(\frac{Z_{\varepsilon in}}{Z_{Tin}}Z_{TFL} - Z_{\varepsilon FL}\right)}{1 - (-g_m Z_{\varepsilon FL})}\right] \tag{10.10}$$

$$= i_{in}Z_{Tin}\left[\frac{1 - R_{outputnulled}}{1 - R_{inputnulled}}\right]$$

where $R_{outputnulled}$ can be shown to be the return ratio of the g_m generator with the input appropriately chosen such that the output is zero and $R_{inputnulled}$ is the return ratio with the input set at zero. The return ratio $R_{inputnulled}$ is the usual return ratio R_{dep} of Equation (10.3) and is seen by setting $i_{in} = 0$ in Figure 10.3 (c) to be given by

$$R_{inputnulled} = -g_m Z_{\varepsilon FL} \tag{10.11}$$

We will now prove that $R_{outputnulled}$ is given by

$$R_{outputnulled} = -g_m\left(\frac{Z_{\varepsilon in}}{Z_{Tin}}Z_{TFL} - Z_{\varepsilon FL}\right) \tag{10.12}$$

as Equation (10.10) implies.

If we choose the input for a given i_{fl} such that the output is at zero,

$$i_{fl}Z_{TFL} + i_{in}Z_{Tin} = 0 \Rightarrow i_{in} = \frac{-i_{FL}Z_{TFL}}{Z_{Tin}} \tag{10.13}$$

Then,

$$\begin{aligned}V_\varepsilon &= -Z_{\varepsilon in}i_{in} - Z_{\varepsilon FL}i_{fl}\\ &= +\frac{Z_{\varepsilon in}Z_{TFL}}{Z_{Tin}}i_{fl} - Z_{\varepsilon FL}i_{fl}\\ &= \left(\frac{Z_{\varepsilon in}Z_{TFL}}{Z_{Tin}} - Z_{\varepsilon FL}\right)i_{fl}\end{aligned} \tag{10.14}$$

Thus, from Figure 10.3 (b) we see that i_{FL} returns as

$$-g_m V_\varepsilon = -g_m\left(\frac{Z_{\varepsilon in}Z_{TFL}}{Z_{Tin}} - Z_{\varepsilon FL}\right)i_{fl}$$

$$\text{i.e., } R_{outputnulled} = -g_m\left(\frac{Z_{\varepsilon in}Z_{TFL}}{Z_{Tin}} - Z_{\varepsilon FL}\right) \tag{10.15}$$

consistent with Equations (10.10) and (10.12).

Most return ratios in circuits are negative at dc like most loop transmissions such that *we can assume that all return ratios are negative by default unless otherwise mentioned.* If we do so, Equation (10.10) can be rewritten as

$$\frac{V_{out}}{i_{in}} = Z_{Tin}\left[\frac{1 + R_{outputnulled}}{1 + R_{inputnulled}}\right] \tag{10.16}$$

The result of Equation (10.16) is true for *any* linear system whether we use a dependent generator or a passive impedance, current or voltage inputs, current or voltage outputs, or dependent generators with arbitrary control and output variables. Our formulation of Equation (10.9) is generalized to an identical mathematical description in any such case and described by the block diagram structure of Figure 10.3 (c). Thus, any linear system may be described as

$$\boxed{TF_{gm} = TF_0\left[\frac{1 + R_{outputnulled}}{1 + R_{inputnulled}}\right]} \tag{10.17}$$

where TF_{gm} is the transfer function with the element whose return-ratio is being computed that has a return-ratio gain parameter given by g_m, TF_0 is the transfer function with the gain parameter $g_m = 0$ such that TF_0 corresponds to Z_{Tin} in Equation (10.16), $R_{outputnulled}$ is the return ratio of the element computed with the output nulled, and $R_{inputnulled}$ is the return ratio of the element computed with the input nulled. Both $R_{inputnulled}$ and $R_{outputnulled}$ are proportional to a scaling parameter like g_m for a dependent generator, or $1/Z_{element}$ for a passive impedance in parallel with a circuit, or $Z_{element}$ for a passive impedance in series with a circuit.

The evaluation of $R_{inputnulled}$ is straightforward since one sets the input to zero and computes a loop gain as we have frequently done. The evaluation of $R_{outputnulled}$, however, appears to be cumbersome since it requires us to first compute the input that will null the fake-label input's contribution to the output, then evaluate the return ratio by computing the transfer function from the fake-label source to the control variable with this value of input, and finally multiply this output-nulled transfer function by the control gain. It may seem easier to just analyze the original circuit with KCL equations! Fortunately, there are two ways out of this quandary:

1. We back-propagate the constraint that the output is zero and do not solve for the input at all. With such back propagation, the value of the control variable can be implicitly obtained as a function of the fake-label input without needing to find the input. Computations of $R_{outputnulled}$ can then even be simpler than computations of $R_{inputnulled}$.

2. The formula can be modified to be written in alternate form by noting that $R_{inputnulled}$ and $R_{outputnulled}$ go to ∞ as g_m goes to ∞. Thus, from Equation (10.17),

$$TF_{gm} = TF_0\left[\frac{1 + R_{outputnulled}}{1 + R_{inputnulled}}\right]$$

$$= \frac{R_{outputnulled}TF_0}{R_{inputnulled}}\left[\frac{\dfrac{1}{R_{outputnulled}} + 1}{\dfrac{1}{R_{inputnulled}} + 1}\right] \tag{10.18}$$

$$TF_\infty = TF_{gm}_{\substack{gm \to \infty}} = \frac{TF_0 R_{outputnulled}}{R_{inputnulled}}$$

If we substitute for $R_{outputnulled}$ as a function of TF_∞ from Equation (10.18), Equation (10.17) can be rewritten as

$$TF_{gm} = TF_0 \frac{1 + R_{outputnulled}}{1 + R_{inputnulled}}$$

$$= \frac{TF_0}{1 + R_{inputnulled}} + TF_0 \frac{R_{outputnulled}}{1 + R_{inputnulled}}$$

$$R_{outputnulled} = \frac{TF_\infty}{TF_0} R_{inputnulled} \qquad (10.19)$$

$$TF_{gm} = \frac{TF_0}{1 + R_{inputnulled}} + TF_\infty \frac{R_{inputnulled}}{1 + R_{inputnulled}}$$

$$\boxed{TF_{gm} = TF_0 \left(\frac{1}{1 + R_{inputnulled}} \right) + TF_\infty \left(\frac{R_{inputnulled}}{1 + R_{inputnulled}} \right)}$$

The form of Equation (10.19) is in terms of $R_{inputnulled}$ only and has a pleasing form where the transfer function at any g_m is the weighted average of its value when $g_m = 0$ (TF_0) and when $g_m = \infty$ (TF_∞) with the weights being proportional to 1 and $R_{inputnulled}$ for TF_0 and TF_∞, respectively. The formula is only useful if TF_∞ is easy to compute in a circuit. Fortunately, it is! TF_∞ corresponds to the ideal transfer function in a circuit when the feedback loop has infinite loop transmission such that the control variable is zero and we may make assumptions like $v_+ - v_- = 0$ in an ideal operational amplifier. Thus, it is the easy case to evaluate when the g_m generator is so strong that we may make assumptions that zero its control variable. Zeroing the control variable usually leads to easy computation of TF_∞ via virtual-ground-like assumptions. In the block diagram of Figure 10.3 (c),

$$TF_\infty = Z_{Tin} - \frac{Z_{\varepsilon in}}{Z_{\varepsilon FL}} Z_{TFL} \qquad (10.20)$$

If the g_m generator really represents a passive impedance in parallel with a network, TF_0 corresponds to opening the impedance, i.e., making it non-existent, and TF_∞ corresponds to shorting it. If the g_m generator is really a passive impedance in series with a network, TF_0 corresponds to shorting the impedance, i.e., making it non-existent, and TF_∞ corresponds to opening it.

Any linear system may then be represented by a block diagram as in Figure 10.4 that represents the boxed expression in Equation (10.19) schematically. We see that TF_∞ is the ideal transfer function when $R_{inputnulled} = \infty$. The transfer function TF_0 serves to create a disturbance that is annulled by the feedback loop transmission $R_{inputnulled}$.

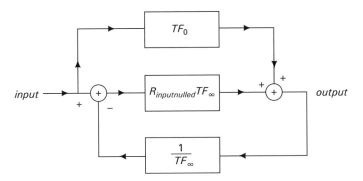

Figure 10.4. Block diagram showing relationships between TF_0 and TF_∞.

10.4 Robustness analysis with the return ratio

From Equation (10.17), through some simple differentiation and algebra, we can show that

$$TF_{gm} - TF_0 = TF_0 \left(\frac{R_{outputnulled} - R_{inputnulled}}{1 + R_{inputnulled}} \right)$$

$$\frac{d(TF_{gm})}{dg_m} = TF_0 \frac{\left[(1 + R_{inputnulled}) \left(\frac{dR_{outputnulled}}{dg_m} - \frac{dR_{inputnulled}}{dg_m} \right) - (R_{outputnulled} - R_{inputnulled}) \frac{dR_{inputnulled}}{dg_m} \right]}{(1 + R_{inputnulled})^2}$$

$$\frac{d(TF_{gm})}{dg_m} = TF_0 \frac{\left[(1 + R_{inputnulled}) \left(\frac{R_{outputnulled} - R_{inputnulled}}{g_m} \right) - (R_{outputnulled} - R_{inputnulled}) \frac{R_{inputnulled}}{g_m} \right]}{(1 + R_{inputnulled})^2}$$

$$\frac{d(TF_{gm})}{dg_m/g_m} = TF_0 \frac{\left[(1 + R_{inputnulled}) (R_{outputnulled} - R_{inputnulled}) - (R_{outputnulled} - R_{inputnulled}) R_{inputnulled} \right]}{(1 + R_{inputnulled})^2} \quad (10.21)$$

$$\frac{d(TF_{gm})}{dg_m/g_m} = TF_0 \frac{R_{outputnulled} - R_{inputnulled}}{1 + R_{inputnulled}} \frac{1}{(1 + R_{inputnulled})}$$

$$\frac{d(TF_{gm})}{dg_m/g_m} = (TF_{gm} - TF_0) \frac{1}{(1 + R_{inputnulled})}$$

$$\boxed{\frac{d(TF_{gm})}{(TF_{gm} - TF_0)} = \frac{d(TF_{gm} - TF_0)}{(TF_{gm} - TF_0)} = \frac{dg_m/g_m}{(1 + R_{inputnulled})}}$$

Thus, we see from the boxed expression in Equation (10.21) that the fractional change in the transfer function caused by the presence of the g_m generator, $TF_{gm} - TF_0$, is the product of the fractional change in the g_m generator's value and the reciprocal of the return difference of the generator $1/(1 + R_{inputnulled})$. Note that $d(TF_0) = 0$ since it does not depend on g_m.

The result of Equation (10.21) is similar to the sensitivity of the closed-loop gain of a feedback loop to changes in its feedforward gain $a(s)$ that we discussed in Chapter 2. In fact, Equation (10.21) describes the sensitivity of simple feedback loops to changes in $a(s)$ and $f(s)$ as well since these loops form a special case of the structure of Figure 10.3 (c) for which Equation (10.21) was derived. For simple feedback loops, $TF_0 = 0$ when we are computing the sensitivity to changes in $a(s)$ and $R_{inputnulled}$ corresponds to the loop transmission $a(s)f(s)$. We find by applying Equation (10.21) that the fractional change in the closed-loop transfer function is the fractional change in $a(s)$ attenuated by $1/(1 + a(s)f(s))$ making the loop robust to changes in $a(s)$ if $a(s)f(s)$ is large consistent with the findings of Chapter 2. In contrast, $TF_0 = a(s)$ when we are computing sensitivity to changes in the feedback gain $f(s)$ and $R_{inputnulled}$ still corresponds to the loop transmission $a(s)f(s)$; in this case, we find by applying Equation (10.21) that the fractional change in the closed-loop transfer function is the fractional change in $f(s)$ multiplied by $(a(s)f(s))/(1 + a(s)f(s))$ such that the loop is not robust to changes in $f(s)$ if the loop transmission is large. In general, we may write from Equation (10.21) that

$$\frac{d(TF_{gm})}{TF_{gm}} = \frac{TF_{gm} - TF_0}{TF_{gm}} \frac{dg_m/g_m}{1 + R_{inputnulled}}$$

$$\boxed{\frac{d(TF_{gm})}{TF_{gm}} = \left(1 - \frac{TF_0}{TF_{gm}}\right)\left(\frac{dg_m/g_m}{1 + R_{inputnulled}}\right)} \tag{10.22}$$

Thus, if TF_0/TF_{gm} is large, a circuit may not be robust to changes in the parameters of an element even if the return ratio of that element in the circuit is large. The topology and circuit in which an element resides is thus important to determining its robustness in that circuit.

The concept of the return ratio explains why the noise of a cascode transistor barely affects the noise of a common-source cascoded amplifier: the return ratio of the g_m generator of the cascode transistor is large as we discuss later such that variations in the cascode transistor's g_m, which may be viewed as causing its noise or offset, are highly attenuated. It also explains why the gain of the common-source amplifier is sensitive to the value of its input transistor whose g_m determines its gain: the return ratio of the g_m generator of the input gain-determining transistor is 0 since the input determines the control gate variable of this transistor and there is no feedback to this node; thus fractional changes in the input g_m directly manifest as fractional changes in the circuit's gain.

10.5   Examples of return-ratio analysis

10.5.1   An inverting-amplifier circuit

Figure 10.5 shows an inverting operational amplifier circuit depicted with its dependent generator with gain A, operational-amplifier output impedance R_{out},

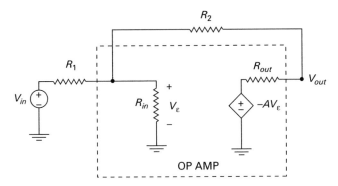

Figure 10.5. Inverting-gain operational amplifier circuit.

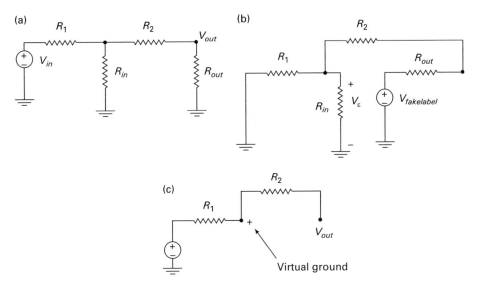

Figure 10.6a, b, c. Inverting-gain operational amplifier circuit: (a) feed-forward circuit, (b) return-ratio circuit, and (c) 'ideal' operational amplifier circuit.

operational-amplifier input impedance R_{in}, and gain-setting resistors R_1 and R_2. We would like to compute the transfer function from V_{in} to V_{out} taking all loading effects into account. A simple way to compute the transfer function is to compute the transfer function TF_0 when the operational amplifier gain $A = 0$, and then use the boxed expression of Equation (10.19) with the return ratio of the dependent voltage source to compute the transfer function.

Figure 10.6 (a) shows that when $A = 0$ TF_0 is computed from resistive-divider analysis to be

$$\frac{V_{out}}{V_{in}}\bigg|_{A=0} = TF_0 = \left[\frac{R_{in}\|(R_{out} + R_2)}{R_{in}\|(R_{out} + R_2) + R_1}\right]\left[\frac{R_{out}}{R_{out} + R_2}\right] \qquad (10.23)$$

Figure 10.6 (b) shows the sub circuit for computing the return ratio of the dependent generator. The return ratio with the input zeroed, $R_{inputnulled}$, is computed by multiplying the transfer function from the drawn fake-label voltage source to V_ε by A. Thus, we find that

$$R_{inputnulled} = A \frac{(R_{in} \| R_1)}{(R_{in} \| R_1) + (R_{out} + R_2)} \tag{10.24}$$

The transfer function TF_∞ is the transfer function when the operational amplifier gain A goes to ∞, at which point V_ε becomes a virtual ground as shown in Figure 10.6 (c). Thus,

$$TF_\infty = -\frac{R_2}{R_1} \tag{10.25}$$

i.e., this is the ideal gain of the inverting operational amplifier. Then, from Equation (10.19) we can compute that

$$TF_A = \left[\frac{\left[\dfrac{R_{in} \| (R_{out} + R_2)}{R_{in} \| (R_{out} + R_2) + R_1}\right]\left[\dfrac{R_{out}}{R_{out} + R_2}\right]}{1 + \dfrac{A(R_{in} \| R_1)}{R_{in} \| R_1 + (R_{out} + R_2)}} \right] + \left[\frac{\dfrac{A(R_{in} \| R_1)}{R_{in} \| R_1 + (R_{out} + R_2)}}{1 + \dfrac{A(R_{in} \| R_1)}{R_{in} \| R_1 + (R_{out} + R_2)}} \right] \left(-\frac{R_2}{R_1} \right)$$

$$\tag{10.26}$$

Equation (10.26) would have taken considerably more work to compute by solving KCL equations after which we would have no insight into the circuit anyway. Now we immediately realize how R_{in} and R_{out} affect the return ratio and why the feedthrough term, i.e., the numerator TF_0 of the first term of Equation (10.26), is usually negligible compared with the second $TF_\infty = (-R_2/R_1)$ term. We see that the circuit is a weighted combination of the feedthrough and ideal inverting-amplifier gain terms with the weights being 1 and $R_{inputnulled}$ as computed in Equation (10.24). The use of feedback block diagrams would have helped shed some insight but the use of return-ratio analysis allows us to not even need to draw these diagrams and just write down the answer by inspection: we visualize the three sub-circuits in Figure 10.6 in our heads from simple visual manipulations of Figure 10.5.

Return-ratio analysis allows us to use the highly evolved visual pattern recognition software of our brains instead of the highly unevolved, grungy algebraic equation-solving methods more suited to a computer. In this example, each piece of the three portions of the analysis, TF_0, TF_∞, and $R_{inputnulled}$, is simple because it involves simplifying abstractions that make analysis easier: we zero out the amplifier, virtual ground the control variable, or break the feedback loop inherent in the circuit through the use of a fake label in Figure 10.5, respectively. We compose the more complex answer out of three simpler pieces that can be analyzed quickly. While one could simplify Equation (10.26) further and expand out the parallel operator $x \| y = xy/(x + y)$ it is better not to do so and leave the answer in the transparent form of Equation (10.26) where all physical effects are evident.

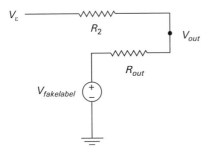

Figure 10.7. Inverting-gain operational amplifier circuit: output nulled.

It is possible to use the $R_{outputnulled}$ form of Equation (10.17) to compute the transfer function of Figure 10.5 as well. If we do so, with the output zeroed, Figure 10.7 and the analysis below reveal that

$$V_{out} = \frac{G_2 V_\varepsilon + G_{out} V_{fakelabel}}{G_2 + G_{out}}$$

$$V_{out} = 0$$

$$\Rightarrow V_\varepsilon = -\left(\frac{G_{out}}{G_2}\right) V_{fakelabel} \tag{10.27}$$

$$\Rightarrow R_{outputnulled} = -A\left(\frac{G_{out}}{G_2}\right)$$

Note that we have computed $R_{outputnulled}$ via a back-propagation constraint that evaluated V_ε without our needing to actually calculate the output-nulling input. Although the $R_{outputnulled}$ technique of Equation (10.17) is equivalent to the TF_∞ technique of Equation (10.19) in return-ratio analysis, the author typically uses Equation (10.19) since it allows one to obtain insight about the behavior of the circuit with $R_{inputnulled} = \infty$.

10.5.2   A resistive-bridge circuit

Figure 10.8 shows a resistive-bridge circuit. We would like to compute the impedance across the terminals indicated in the figure. If we use R_{bridge} as the element whose return ratio we compute, then its return ratio as shown in Figure 10.9 (a) is given from Equation (10.7) to be the ratio of the impedance across the bridge element to the bridge element's impedance.

$$R = \frac{Z_{FL}}{R_{bridge}} = \frac{(R_1 + R_3)||(R_2 + R_4)}{R_{bridge}} \tag{10.28}$$

Note that we have assumed that R is negative as per our convention. Figure 10.9 (b) shows that TF_0, the impedance when there is no bridge element, is

$$TF_0 = (R_1 + R_2)||(R_3 + R_4) \tag{10.29}$$

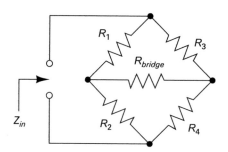

Figure 10.8. Resistive bridge circuit.

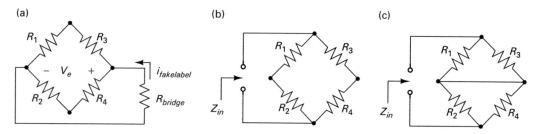

Figure 10.9a, b, c. Return-ratios for a resistive bridge circuit: (a) fake-label circuit, (b) node-open, and (c) node-shorted circuits.

The impedance when the return ratio is infinite corresponds to the bridge element being shorted. From Figure 10.9 (c), this impedance is given by

$$TF_\infty = (R_1\|R_3) + (R_2\|R_4) \tag{10.30}$$

Thus, from Equation (10.19), we get

$$Z_{in} = \frac{(R_1 + R_2)\|(R_3 + R_4)}{\left(1 + \dfrac{(R_1 + R_3)\|(R_2 + R_4)}{R_{bridge}}\right)} + \frac{\left(\dfrac{(R_1 + R_3)\|(R_2 + R_4)}{R_{bridge}}\right)}{\left(1 + \dfrac{(R_1 + R_3)\|(R_2 + R_4)}{R_{bridge}}\right)}[(R_1\|R_3) + (R_2\|R_4)]$$

$$\tag{10.31}$$

This formula would take considerable algebra to derive via KCL, and a feedback block diagram analysis would require work too. In contrast, Equation (10.31) tells us that the impedance of the bridge is the weighted sum of the impedances that would be present if the bridge element were open versus shorted with the weights being 1 and the ratio of the impedance across the bridge to the bridge element's impedance respectively. The entire answer could be written down by inspection in one line as skill in the use of return-ratio analysis develops.

10.5.3   A bridged-T network

Figure 10.10 shows a bridged-T network where we would like to compute the transfer function from V_{in} to V_{out}. Clearly, the problem would be simple if the

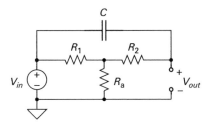

Figure 10.10. Bridged-T network.

bridged capacitor were not present. Therefore, we compute the bridged-capacitor's return ratio and use it to compute the transfer function that we want. We shall not even draw the TF_0, $R_{inputnulled}$ and TF_∞ subcircuits this time but just evaluate them visually to be the no-capacitor-present transfer function, the ratio of the impedance across the capacitor to the capacitor's impedance with the input grounded, and the transfer function when the capacitor is a short. Thus we find that

$$TF_0 = \frac{R_a}{R_a + R_1}$$

$$TF_\infty = 1 \tag{10.32}$$

$$R_{inputnulled} = sC((R_a\|R_1) + R_2) = sCR_p$$

Thus, from Equation (10.19), we find that

$$\frac{V_{out}(s)}{V_{in}(s)} = \left(\frac{R_a}{R_a + R_1}\right)\frac{1}{1 + R_p sC} + (1)\frac{R_p sC}{1 + R_p sC} \tag{10.33}$$

The examples that we have described should illustrate that return-ratio analysis can often simplify the problem of finding the transfer function of a complex feedback circuit into three simpler problems from whose solutions the transfer function is composed via Equation (10.19): the computation of TF_0, the computation of TF_∞, and, the computation of $R_{inputnulled}$. Computing TF_0 is often simple because it represents an idealized basis point of operation for the circuit where, for example, the impedance of the return-ratio element is an open or short, or the g_m generator is absent. Computing TF_∞ is often simple because it represents another idealized basis point of operation for the circuit, where, for example, the impedance of the return-ratio element is a short or open, or $g_m = \infty$ such that the control variable is assumed to be at zero and virtual-ground-like assumptions can be made. Computing $R_{inputnulled}$ is often simple because feedback loops in the circuit are broken via the introduction of fake-label sources. A judicious choice of an element in a circuit around which the return-ratio analysis is organized is often one in whose absence the circuit's analysis would be much simpler. That is why we performed a return-ratio analysis organized around the operational-amplifier dependent voltage source, the bridge resistance, and the capacitance in the bridged-network.

10.6   Blackman's impedance formula

Blackman's impedance formula is a useful special case of the more general Equation (10.17). Suppose we want to compute a driving-point impedance at a node in a certain circuit while we change the parameter of another variable in the circuit, e.g., the g_m of a voltage-dependent current source in the circuit. That is, we want to compute the transfer function $Z_{node} = V_{node}/I_{node}$ when we drive it with a current source I_{node} and measure the voltage at this same node, V_{node}, as a function of g_m. From Equation (10.17), we may then conclude that

$$Z_{node} = Z_0 \left(\frac{1 + R_{nodeshorted}}{1 + R_{nodeopen}} \right), \tag{10.34}$$

where $Z_0 = TF_0$ is the transfer function when $g_m = 0$, i.e., when the element is not present; $R_{nodeshorted}$ is the return ratio of the g_m generator with the output nulled, i.e., the output node voltage is zero or equivalently the node is shorted; and $R_{nodeopen}$ is the return ratio with the input nulled, i.e., the driving current source I_{node} is set to 0 or equivalently the node is opened. Equation (10.34) is called Blackman's impedance formula.

The dual formula for conductance is obtained when we drive with a voltage input V_{node} and attempt to measure the current supplied by the voltage source at the node, I_{node}, such that input nulling corresponds to a short at the node and output nulling corresponds to an open at the node. Then, we have

$$G_{node} = G_0 \left(\frac{1 + R_{node\ open}}{1 + R_{node\ shorted}} \right) \tag{10.35}$$

as indeed we should if we reciprocate Equation (10.34).

Blackman's impedance formula is easily remembered: the Ω symbol for impedance has two straight lines joined by a circular wavy line. The two straight lines would be open but a wavy short connects them and since the wavy short is above the open straight lines, we have a short over an open. Thus, in Equation (10.34), we must have $R_{nodeshorted}$ over $R_{nodeopened}$, i.e., $R_{nodeshorted}$ is in the numerator and $R_{nodeopen}$ is in the denominator.

Often one or the other of $R_{nodeshorted}$ or $R_{nodeopen}$ is 0 such that Equation (10.34) predicts a decrease or increase in impedance by a factor of $(1+R)$, respectively. The formula is thus a generalization of the observation that feedback increases or decreases the impedance by a factor of $1+L(s)$, where $L(s)$ is the loop transmission.

Figure 10.11 (a) shows a cascoded current source with v_{IN} setting the value of the series drain current and V_B being the cascode bias voltage. Figure 10.11 (b) shows a small-signal circuit with output resistances r_{01} and r_{02} and a g_s generator if V_B and v_{IN} are kept constant. We would like to compute the driving-point output impedance of the cascode current source between the terminal marked in Figure 10.11 (b) and ground. Figure 10.11 (c) shows the node-shorted subcircuit, and Figure 10.11 (d) shows the node-open subcircuit. From these circuits, we can compute the return ratio with the fake-label current source by computing

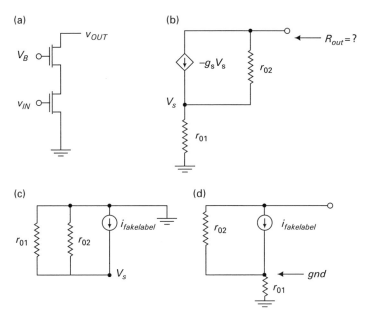

Figure 10.11a, b, c, d. Input impedance of a cascode circuit: (a) basic cascode circuit, (b) small-signal equivalent, (c) $R_{nodeshorted}$ circuit, and (d) $R_{nodeopen}$ circuit.

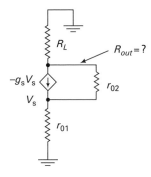

Figure 10.12. Cascode circuit with load impedance.

the transfer function from this source to the node V_s and multiplying by g_s. In Figure 10.11 (d), V_s is forced to ground since no current can flow through r_{01}. Thus, we find from straightforward application of Equation (10.34) that

$$Z_{cascode} = (r_{01} + r_{02}) \frac{1 + g_s(r_{01}||r_{02})}{1 + g_s.0}$$

$$= (r_{01} + r_{02} + g_s r_{01} r_{02})$$

(10.36)

Figure 10.12 shows a small-signal cascode circuit with a load impedance R_L. In this case, $R_{nodeopen}$ is no longer zero since the fake-label current has a path for current flow via the series combination of r_{01} and R_L. Applying Blackman's impedance formula of Equation (10.34), we get

$$Z_{cascRL} = (R_L||(r_{01}+r_{02})) \frac{1+g_s(r_{01}||r_{02})}{1+g_s\left(\dfrac{r_{02}}{r_{02}+r_{01}+R_L}\right)r_{01}} \qquad (10.37)$$

In computing Equation (10.37), we used a current-divider relationship to obtain the current flow through r_{01} and R_L in series but the V_s fake-label transfer voltage is only due to the voltage drop across the r_{01} resistor. We could have computed Equation (10.37) as $R_L||Z_{cascode}$, where $Z_{cascode}$ is described by Equation (10.36); however, Equation (10.37) is a good example that illustrates how $R_{nodeshorted}$ and $R_{nodeopen}$ can both be nonzero in a circuit and how Blackman's formula can be directly applied in one step to compute the whole impedance. The topology of Figure 10.12 also frequently occurs in driving-point impedance calculations where the effects of r_0 are important.

10.7   Driving-point transistor impedances with Blackman's formula

Figures 10.13 (a), 10.13 (b), and 10.13 (c) reveal three driving-point impedances at the source, gate, and drain of a transistor, respectively, with load resistances at one terminal and the other terminal being grounded. In Figure 10.13 (a), we may compute the impedance Z_{src} seen at the source due to the g_s generator, R_L, $Z_{gs} = 1/(C_{gs}s)$ and r_0 by using Blackman's impedance formula in Equation (10.34) with return-ratio computations for the g_s generator that are very similar to that performed for $R_{nodeopen}$ in Figure 10.12.

$$Z_{src} = [(R_L+r_0)||Z_{gs}]\left[\frac{1+g_s.0}{1+g_s\left(\dfrac{r_0}{r_0+R_L+Z_{gs}}\right)Z_{gs}}\right]$$

$$Z_{src} = \frac{[(R_L+r_0)||Z_{gs}]}{1+g_s\left(\dfrac{r_0}{r_0+R_L+Z_{gs}}\right)Z_{gs}}$$

$$\boxed{\begin{aligned} Z_{src}(\text{dc}) &= \frac{R_L+r_0}{1+g_sr_0} \approx \frac{1}{g_s} \\ r_0 \to \infty \Rightarrow Z_{src} &= \frac{Z_{gs}}{1+g_sZ_{gs}} = \frac{Z_{gs}}{1+\beta_{MOS}} \approx \frac{1}{g_s} \end{aligned}} \qquad (10.38)$$

Note that the impedance seen at the source at dc is affected by that at the drain contrary to popular belief unless $r_0 \gg R_L$. Also, the last expression in the box in Equation (10.38) is the MOS analog of the bipolar impedance Z_{gs} being reflected into the source with a $1/(1+\beta)$ transformation factor, where β corresponds to

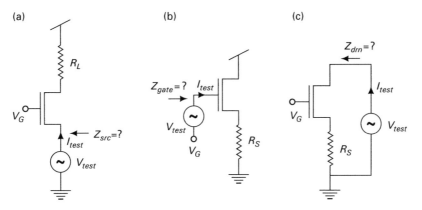

Figure 10.13a, b, c. Driving-point impedance circuits: (a) source, (b) gate, and (c) drain.

$g_s Z_{gs}$. In general, the presence of the body effect in MOS transistors ($g_s \neq g_m$) causes more care to be needed in the application of such reflection formulas because the current through the dependent generator is $(g_m v_g - g_s v_s) \neq g_m v_{gs}$. For example, if there were a grounded impedance in the gate Z_g rather than the gate being at ground in Figure 10.13 (a), the effective impedance seen in the source when $r_0 \to \infty$ is found from return-ratio analysis similar to that in Equation (10.38) to be

$$Z_{src} = \frac{Z_{gs} + Z_g}{1 + g_s(Z_{gs} + Z_g(1 - \kappa_S))}$$

$$\beta_{MOS} = g_s(Z_{gs} + Z_g(1 - \kappa_S)) \tag{10.39}$$

An exact analysis of the driving-point impedance with $Z_g \neq 0$ and r_0 not being infinite generalizes Equation (10.38) to

$$Z_{src} = \frac{[(R_L + r_0) \| (Z_{gs} + Z_g)]}{1 + g_s \left(\dfrac{r_0}{r_0 + R_L + Z_{gs} + Z_g} \right) (Z_{gs} + (1 - \kappa_S)Z_g)} \tag{10.40}$$

To compute the impedance at the gate in Figure 10.13 (b), we first define

$$Z_s = R_S \| \frac{1}{g_{mb}}, \tag{10.41}$$

such that the dependent current source from drain to source may be described by $g_m v_{gs}$ rather than by $g_m v_g - g_s v_s$, i.e., we absorb the effects of g_{mb} into Z_s in Equation (10.41). If we assume that $r_0 \gg R_S$, applying Blackman's impedance formula in Equation (10.34) for the g_m generator yields

$$Z_{gate} = (Z_{gs} + Z_s) \frac{1 + g_m(Z_{gs} \| Z_s)}{1 + g_m . 0}$$

$$Z_{gate} = Z_{gs} + Z_s + g_m Z_{gs} Z_s \tag{10.42}$$

$$\boxed{\begin{aligned} Z_{gate} &= Z_{gs} + (g_m Z_{gs} + 1)Z_s \\ Z_{gate} &= Z_{gs} + (\beta_{MOS} + 1)Z_s \end{aligned}}$$

In deriving Equation (10.42), we have used the fact that $R_{nodeopen}$ in Blackman's impedance formula is 0 if no current flows through Z_{gs} such that no voltage is developed across Z_{gs}. Equation (10.42) is the MOS version of the well-known β reflection formula for bipolar transistors, where impedances in the emitter get reflected into the base with a gain of $(\beta_{MOS} + 1)$. As we pointed out in Chapter 5, $g_m Z_{gs}$ behaves like β_{MOS} in MOS transistors. Note that β_{MOS} in Equation (10.38) is $g_s Z_{gs}$ rather than $g_m Z_{gs}$ as it is in Equation (10.42), and when there is an impedance in the gate β_{MOS} is still different as in Equation (10.39). An exact analysis of the input gate impedance including the effects of r_0 and with a nonzero drain resistance R_L can be found by Blackman's impedance formula to be

$$Z_{gate} = (Z_{gs} + (R_S || (R_L + r_0))) \frac{1 + g_s \left(\dfrac{r_0}{r_0 + R_L + R_S || Z_{gs}} \right) (R_S || Z_{gs})}{1 + g_s (1 - \kappa_S) \left(\dfrac{r_0}{r_0 + R_L + R_S} \right) R_S} \qquad (10.43)$$

To compute the impedance at the drain in Figure 10.13 (c), Blackman's impedance formula for the g_s generator may be applied to a circuit identical to the $R_{nodeshorted}$ small-signal cascode circuit of Figure 10.11 (c):

$$Z_s = R_S || Z_{gs}$$

$$Z_{drn} = (r_0 + Z_s) \frac{1 + g_s(r_0 || Z_s)}{1 + g_s.0} \qquad (10.44)$$

$$\boxed{Z_{drn} = r_0 + Z_s + g_s r_0 Z_s},$$

where $R_{nodeopen} = 0$ in this case just as in the cascode circuit of Figure 10.11 (b). The presence of a finite impedance at the gate Z_g rather than a grounded gate changes Equation (10.44) to

$$Z_s = R_S || (Z_{gs} + Z_g)$$

$$Z_{drn} = (r_0 + Z_s) \frac{1 + g_s(r_0 || Z_s) \left(1 - \kappa_s \left(\dfrac{Z_g}{Z_g + Z_{gs}} \right) \right)}{1 + g_s.0} \qquad (10.45)$$

$$Z_{drn} = r_0 + Z_s + g_s r_0 Z_s \left(1 - \kappa_s \frac{Z_g}{Z_g + Z_{gs}} \right)$$

Equations (10.38), (10.42), and (10.44) yield the driving-point impedances at the source, gate, and drain respectively with an impedance in one of non-driving-point terminals and the other non-driving-point terminal being grounded. Equations (10.40), (10.43), and (10.45) yield the driving-point impedances at the source, gate and drain, respectively, with impedances present in both non-driving-point terminals. In general, since the circuits attached to non-driving-point transistor terminals can always be converted into Thevenin equivalents with some Thevenin impedance, the impedances computed in Equations (10.40), (10.43), and (10.45) can be used to compute impedances seen at transistor terminals including effects of the entire rest of the network that the transistor is attached to.

10.8   Middlebrook's extra-element theorem

Figure 10.14 (a) shows a big circuit in which we want to insert an extra passive impedance element with impedance Z across two terminals of the big circuit, i.e., in parallel with the rest of the circuitry across these two terminals. We would like to study how a transfer function in the big circuit is changed by this new addition without having to redo all of the work that went into computing the original transfer function of the circuit. From Equation (10.17) and Equation (10.7), we can write

$$TF_z^{prl} = TF_{open}\left(\frac{1 + R_{outputnulled}}{1 + R_{inputnulled}}\right)$$

$$R_{outputnulled} = \frac{Z_n}{Z}$$

$$R_{inputnulled} = \frac{Z_d}{Z}$$ (10.46)

$$\boxed{TF_z^{prl} = TF_{open}\left(\frac{1 + Z_n/Z}{1 + Z_d/Z}\right)}$$

where return ratios have been assumed negative, Z_d is the impedance looking across the element with the input nulled, and Z_n is the impedance looking across the element with the output nulled. Equation (10.46) was first discovered by Professor Middlebrook at CalTech [3]. If we use Equations (10.19) and (10.7), we may derive an alternative version of the extra-element theorem which is sometimes more convenient to work with since nulling the output via double injection at the input and fake-label sources is not necessary. We see that

$$\boxed{TF_z^{prl} = \frac{TF_{open}}{1 + Z_d/Z} + TF_{short}\left(\frac{Z_d/Z}{Z_d/Z + 1}\right)}$$ (10.47)

We shall call Equations (10.46) or (10.47) the parallel extra-element theorem.

(a)
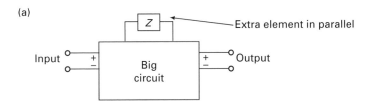

(b)
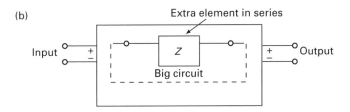

Figure 10.14a, b. Middlebrook's extra-element theorem: (a) parallel and (b) series.

Figure 10.14 (b) shows a big circuit where the extra element has been added in series with the big circuit rather than in parallel. From Equations (10.17) and (10.8), we have

$$R_{outputnulled} = \frac{Z}{Z_n}$$

$$R_{inputnulled} = \frac{Z}{Z_d} \tag{10.48}$$

$$\boxed{TF_z^{srs} = TF_{short}\left(\frac{1 + Z/Z_n}{1 + Z/Z_d}\right)}$$

If we use Equations (10.19) and (10.8), we can also write that

$$\boxed{TF_z^{srs} = TF_{short}\left(\frac{1}{1 + Z/Z_d}\right) + TF_{open}\left(\frac{Z/Z_d}{Z/Z_d + 1}\right)} \tag{10.49}$$

Equations (10.48) or (10.49) are called the series extra-element theorem. Note that in the series and parallel extra-element theorems, all computed functions whether TF_{short}, TF_{open}, Z_n, or Z_d are independent of the extra element and only a property of the network. The series and parallel cases differ in how the new transfer function is composed from these component functions. Furthermore, since

$$TF_z^{prl} = TF_{open}\left(\frac{1 + Z_n/Z}{1 + Z_d/Z}\right) = \left(TF_{open}\frac{Z_n}{Z_d}\right)\frac{\left(\frac{Z}{Z_n} + 1\right)}{\left(\frac{Z}{Z_d} + 1\right)} \tag{10.50}$$

when $Z = 0$, we must have $TF_z^{prl} = TF_{short}$ such that

$$TF_{short} = TF_{open}\frac{Z_n}{Z_d}$$

$$\frac{TF_{open}}{TF_{short}} = \frac{Z_d}{Z_n} \tag{10.51}$$

Equation (10.51) is useful in computing Z_d from Z_n and vice versa if the shorted or opened transfer functions are known. The output-nulled impedance Z_n is sometimes easier to compute than Z_d since one can back-propagate the output-nulled constraint without needing to find a nulling input.

Figure 10.15 (a) shows an emitter-degenerated bipolar amplifier with a load R_L and with parasitic capacitances in the bipolar transistor of C_π and C_μ analogous to C_{gs} and C_{gd} in the MOSFET, respectively. Figure 10.15 (b) shows the small-signal equivalent circuit of Figure 10.15 (a). Figure 10.15 (b) reveals a r_π resistance between the base of the bipolar transistor and a g_m generator with $g_m r_\pi = \beta_0$, the dc ratio of collector to base current. The capacitance C_μ will now be analyzed as if it were an extra element in the small-signal circuit for which we would like to compute the transfer function from $V_{in}(s)$ to $V_{out}(s)$.

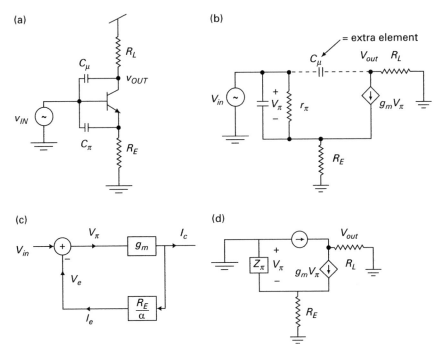

Figure 10.15a, b, c, d. Common emitter amplifier including C_μ: (a) basic circuit, (b) small-signal circuit, (c) source-degeneration feedback loop, and (d) small-signal circuit for computing the impedance across C_μ.

If we denote

$$Z_\pi = \frac{1}{C_\pi s} \| r_\pi$$

$$\alpha(s) = \frac{g_m Z_\pi}{g_m Z_\pi + 1}$$

(10.52)

then the simple circuit of Figure 10.15 (b) without the C_μ element has an effective transconductance $G_{meff}(s)$ from $V_{in}(s)$ to $I_c(s)$ given by Black's formula applied to the feedback loop of Figure 10.15 (c):

$$G_{meff}(s) = -\frac{g_m}{1 + g_m \dfrac{R_E}{\left(\dfrac{g_m Z_\pi}{g_m Z_\pi + 1}\right)}}$$

(10.53)

$$\frac{V_{out}(s)}{V_{in}(s)} = -G_{meff}(s) R_L$$

Equation (10.53) corresponds to TF_{open} in the extra-element theorem. The transfer function TF_{short} corresponds to C_μ going to ∞ such that Z_μ is a short. Therefore $TF_{short} = 1$.

The impedance across C_μ when the input is nulled is determined from the circuit of Figure 10.15 (d). The current through the g_m generator and Z_π is zero because $I_c(s) = G_{meff}(s).0 = 0$. Thus, the test current of Figure 10.15 (d) flows through R_L such that

$$R_{inputnulled} = \frac{Z_d}{Z} = \frac{R_L}{1/C_\mu s} = R_L C_\mu s \qquad (10.54)$$

Since we now know TF_{open}, TF_{short}, and Z_d/Z, we may apply Equation (10.47) to get

$$\begin{aligned}
\frac{V_{out}(s)}{V_{in}(s)} &= \frac{-G_m^{eff} R_L}{1 + R_L C_\mu s} + \frac{(1)R_L C_\mu s}{1 + R_L C_\mu s} \\
&= \frac{(C_\mu s - G_m^{eff})R_L}{1 + R_L C_\mu s}
\end{aligned} \qquad (10.55)$$

Thus, we see that the element C_μ yields a right half plane (RHP) zero and a left half plane (LHP) pole.

10.9   Thevenin's theorem as a special case of return-ratio analysis

Figure 10.16 (a) shows an input and an output voltage V_{out} across the terminals of a big circuit. Suppose we now attach an impedance Z across these output terminals and would like to compute the output voltage across these terminals due to this input and all other inputs in the big circuit. From return-ratio analysis, the return ratio of Z with the output voltage nulled is 0 for any input in the circuit since the control variable for Z is the output voltage. The return ratio of Z with all inputs nulled is Z_{across}/Z where Z_{across} is the impedance across the output voltage terminals in the big circuit. Thus, for the transfer function from any input to the output terminal voltage we may write

$$\begin{aligned}
\frac{V_{out}}{In_i} &= TF_i \frac{1 + 0}{1 + \dfrac{Z_{across}}{Z}} \\
&= \frac{TF_i}{1 + \dfrac{Z_{across}}{Z}}
\end{aligned} \qquad (10.56)$$

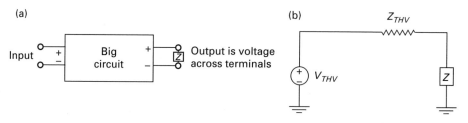

Figure 10.16a, b. Thevenin's theorem: (a) abstract circuit and (b) voltage divider.

where In_i is the i_{th} input in the circuit with transfer function TF_i to the open-circuit voltage V_{out} when Z is absent from the circuit, i.e., TF_i corresponds to TF_0 in Equation (10.17) for input In_i. Since $R_{outputnulled} = 0$ and $R_{inputnulled} = Z_{across}/Z$ for any input in the circuit, Equation (10.56) applies to all inputs of the circuit. The net output voltage may then be found by superposition of all transfer functions from all inputs with transfer functions described by Equation (10.56) to be

$$
\begin{aligned}
V_{out} &= \frac{1}{1 + \frac{Z_{across}}{Z}} \sum_i TF_i.In_i \\
&= \frac{V_{THV}}{1 + \frac{Z_{Th}}{Z}} \\
&= \left(\frac{Z}{Z_{THV} + Z} \right) V_{THV}
\end{aligned}
\tag{10.57}
$$

In Equation (10.57), Z_{across} has merely been relabeled as Z_{THV} and V_{THV} is $\sum_i TF_i.In_i$, the open-circuit voltage due to all input sources by superposition. Thus, any big circuit across two of whose terminals Z is hooked up in parallel to may be represented by the voltage-divider circuit of Figure 10.16 (b) since it may be described by Equation (10.57).

10.10 Two final examples of return-ratio analysis

10.10.1 A resistive-bridge circuit with a g_m generator

Figure 10.17 (a) shows a resistive-bridge circuit across whose indicated terminals, we would like to compute the impedance. This example and Figure 10.18 are adapted from a discussion in [4]. Although we have solved this problem through direct application of Equation (10.19) to Figures 10.8 and 10.9, now we shall re-compute the impedance across the bridge with Blackman's formula. Figure 10.17 (b) shows the node-shorted subcircuit and Figure 10.17 (c) shows the node-open subcircuit such that we may find the impedance to be

$$
\begin{aligned}
Z_0 &= Z_{in, R_{bridge} \to \infty} = (R_1 + R_2)||(R_3 + R_4) \\
R_{nodeshorted} &= \frac{Z_{nodeshorted}}{R_{bridge}} = \frac{(R_1||R_2) + (R_3||R_4)}{R_{bridge}} \\
R_{nodeopen} &= \frac{Z_{nodeopen}}{R_{bridge}} = \frac{(R_1 + R_3)||(R_2 + R_4)}{R_{bridge}} \\
Z_{in} &= (R_1 + R_2)||(R_3 + R_4) \frac{1 + \dfrac{(R_1||R_2) + (R_3||R_4)}{R_{bridge}}}{1 + \dfrac{(R_1 + R_3)||(R_2 + R_4)}{R_{bridge}}}
\end{aligned}
\tag{10.58}
$$

Figure 10.18 (a) shows a bridge circuit with a g_m generator. Now we shall treat the bridge impedance formula of Equation (10.58) as our base-case impedance

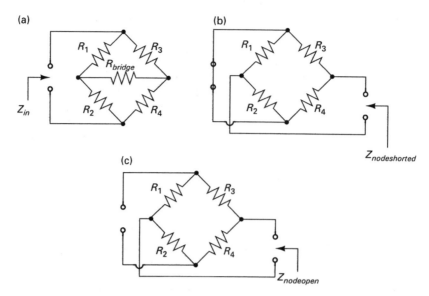

Figure 10.17a, b, c. Resistive bridge: (a) basic circuit, (b) node-shorted, and (c) node-open circuits.

and use Blackman's impedance formula with return ratios computed for the g_m generator to find the impedance across the indicated terminals in Figure 10.18 (a). Thus, we are applying Blackman's impedance formula, first to R_{bridge}, and then to the g_m generator, to understand the circuit of Figure 10.18 (a) in a hierarchical fashion. That is,

$$Z_{in} = Z_0 \left(\frac{1 + R_{nodeshorted}}{1 + R_{nodeopen}} \right)$$

$$Z_0 = Z_{in, g_m = 0} = (R_1 + R_2) || (R_3 + R_4) \left(\frac{1 + \dfrac{(R_1 || R_2) + (R_3 || R_4)}{R_{bridge}}}{1 + \dfrac{(R_1 + R_3) || (R_2 + R_4)}{R_{bridge}}} \right) \qquad (10.59)$$

The node-shorted and node-opened return-ratio subcircuits with a fake-labeled g_m generator are shown in Figures 10.18 (b) and 10.18 (c), respectively. From these subcircuits, we can derive that

$$R_{nodeshorted} = g_m \frac{R_{bridge}(R_1 || R_2)}{R_{bridge} + (R_3 || R_4) + (R_1 || R_2)}$$

$$R_{nodeopen} = g_m \frac{R_1 \left(R_{bridge} || (R_2 + R_4) \right)}{R_{bridge} || (R_2 + R_4) + R_1 + R_3} \qquad (10.60)$$

Combining Equations (10.59) and (10.60), the overall impedance of the bridge with a g_m generator from Blackman's impedance formula is given by

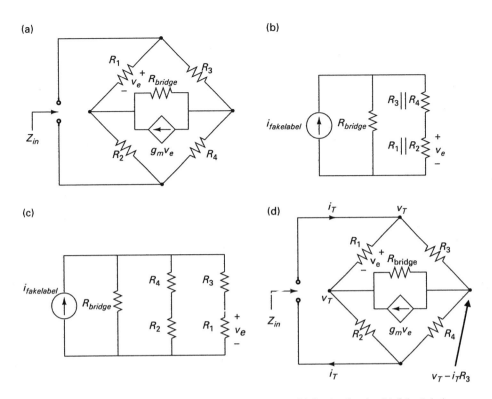

Figure 10.18a, b, c, d. Resistive bridge with g_m generator: (a) basic circuit, (b) fake-label, node-shorted (redrawn), (c) fake-label, node-open (redrawn), and (d) TF_∞ circuits. Adapted from [4]; reproduction with permission.

$$Z_{in} = (R_1 + R_2)||(R_3 + R_4)\left(\frac{1 + \dfrac{(R_1||R_2) + (R_3||R_4)}{R_{bridge}}}{1 + \dfrac{(R_1 + R_3)||(R_2 + R_4)}{R_{bridge}}}\right)$$

$$\cdot \left(\frac{1 + g_m \dfrac{R_{bridge}(R_1||R_2)}{R_{bridge} + (R_3||R_4) + (R_1||R_2)}}{1 + g_m \dfrac{R_1\left(R_{bridge}||(R_2 + R_4)\right)}{R_{bridge}||(R_2 + R_4) + R_1 + R_3}}\right)$$

(10.61)

The use of return-ratio analysis has allowed us to write down a complex expression for the impedance with almost no algebra and a few visual manipulations of the circuit. Furthermore, the answer transparently reveals the dependence of the impedance on g_m.

It is also interesting to compute the impedance of the circuit of Figure 10.18 (a) when $g_m \to \infty$. Figure 10.18 (d) shows a subcircuit drawn for this calculation along with the pertinent variables for the test voltage generator and resulting test current used to compute the impedance. We find by virtual zeroing of the control variable v_e and the application of KCL at the bottom node of the bridge that

$$i_T = \frac{v_T}{R_2} + \frac{v_T - i_T R_3}{R_4}$$
$$\frac{v_T}{i_T} = Z_{in} = (R_2 \| R_4)(1 + R_3/R_4) \tag{10.62}$$

The impedance is thus independent of R_1 if g_m is infinite.

10.10.2   A super-buffer circuit

Figure 10.19 (a) illustrates a super-buffer circuit, i.e., a circuit whose output v_{OUT} tracks its input v_{IN} and that has a very low output impedance. Intuitively, the output impedance of a traditional source follower formed by M_1 and I_2 is decreased because changes in node voltage caused by a current disturbance at v_{OUT} are sensed by the source-driven amplifier formed by M_1 and I_1 and strongly attenuated via a negative-feedback correction current provided by M_2. The equilibrium biasing conditions of the circuit are such that I_1 flows through M_1 and $I_2 - I_1$ flows through M_2. The constant current of I_1 through M_1 means that v_{OUT} tracks v_{IN} just as in a traditional source follower. Figure 10.19 (b) shows a small-signal circuit of the super buffer assuming that the output impedances of all current sources are sufficiently large such that they are negligible when compared

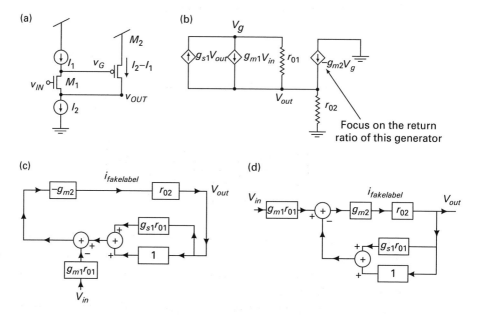

Figure 10.19a, b, c, d. Super-buffer circuit: (a) actual circuit, (b) small-signal circuit, (c) block diagram, and (d) simplified block diagram.

with the transistor output impedances. We shall analyze the circuit by focusing on the return-ratio of the g_{m2} generator due to the M_2 transistor.

The impedance at node V_{out} that the fake-label g_{m2} generator sees in Figure 10.19 (b) is simply r_{02}: the dependent generators in parallel with r_{01} form a floating loop that is incapable of accepting any net current coming from the V_{out} node since that current has nowhere to exit. Alternatively, we may view the driving-point impedance looking into the source of transistor M_1 from Equation (10.38) to be $(R_L + r_{01})/(1 + g_s r_{01}) = \infty$ since R_L is infinite. Thus, the small-signal block diagram of Figure 10.19 (c) which expresses the circuit relationships of Figure 10.19 (b) shows that the transfer function of $i_{fakelabel}$ back to itself, i.e., $R_{inputnulled}$, is given by

$$R_{inputnulled} = g_{m2}r_{02}(1 + g_{s1}r_{01}) \tag{10.63}$$

It is useful to simplify Figure 10.19 (c) into a canonical Black's-formula-type negative-feedback loop as shown in Figure 10.19 (d). From Figure 10.19 (d), we see that

$$TF_{gm_2=0} = 0$$
$$TF_{gm_2=\infty} = \frac{g_{m1}r_{01}}{1 + g_{s1}r_{01}} \cong \frac{g_{m1}}{g_{s1}} \tag{10.64}$$

From Equations (10.63), (10.64), and (10.19) we can then deduce that the transfer function of the circuit is given by

$$\frac{V_{out}}{V_{in}} = \left[\frac{g_{m1}r_{01}}{1 + g_{s1}r_{01}}\right]\left[\frac{g_{m2}r_{02}(1 + g_{s1}r_{01})}{1 + g_{m2}r_{02}(1 + g_{s1}r_{01})}\right]$$
$$\approx \frac{g_{m1}}{g_{s1}} \tag{10.65}$$

The output impedance of the circuit is given by

$$R_{out} = r_{02}\left(\frac{1 + R_{vout\ shorted}}{1 + R_{vout\ open}}\right) = r_{02}\left(\frac{1 + 0}{1 + g_{m2}r_{02}(1 + g_{s1}r_{01})}\right) \approx \frac{1}{g_{m2}(g_{s1}r_{01})} \tag{10.66}$$

A noise analysis of the circuit would add noise generators corresponding to the noise in M_1 and the noise in M_2 as shown in the block diagram of Figure 10.20.

Return-ratio analysis is in many ways just conventional feedback block diagram analysis in this example. However, it helped us organize our analysis around

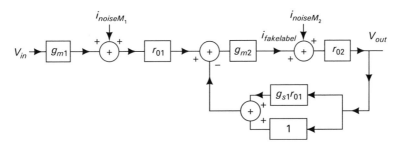

Figure 10.20. Super-buffer circuit block diagram with noise sources included.

the g_{m2} generator and rapidly compute the output impedance with Blackman's formula. As one gets more practiced, return-ratio techniques can be done mentally, without the need for an explicit feedback block diagram.

10.11   Summary of key results

Since several inter-related ideas were presented in this chapter, it is worth summarizing the five key ideas derived from them:

1. The return ratio of a dependent generator is $-g_m Z_{fakelabel}$. The return ratio of a parallel passive impedance Z is $-Z_{fakelabel}/Z$ and the return ratio of a series passive impedance Z is $-Z/Z_{fakelabel}$. Return ratios are assumed negative, hereafter.

2. The transfer function of an element with a varying parameter g_m such as the g_m of a dependent generator or the Z of a passive impedance is given by

$$TF_{gm} = TF_0 \left(\frac{1 + R_{outputnulled}}{1 + R_{inputnulled}} \right)$$

$$= \frac{TF_0}{1 + R_{inputnulled}} + TF_\infty \left(\frac{R_{inputnulled}}{1 + R_{inputnulled}} \right) \qquad (10.67)$$

$$TF_0 R_{outputnulled} = TF_\infty R_{inputnulled}$$

3. Middlebrook's extra-element theorem for a parallel passive impedance Z is given by

$$TF_Z = TF_{open} \left(\frac{1}{1 + \dfrac{Z_d}{Z}} \right) + TF_{short} \left(\frac{\dfrac{Z_d}{Z}}{\dfrac{Z_d}{Z} + 1} \right) \qquad (10.68)$$

where Z_d is the impedance across the element. The extra-element theorem for a series passive impedance Z is given by

$$TF_Z = TF_{short} \left(\frac{1}{1 + \dfrac{Z}{Z_d}} \right) + TF_{open} \left(\frac{\dfrac{Z}{Z_d}}{\dfrac{Z}{Z_d} + 1} \right) \qquad (10.69)$$

4. Blackman's formula for a driving-point impedance in a circuit is given by

$$Z_{gm} = Z_{gm=0} \left(\frac{1 + R_{nodeshorted}}{1 + R_{nodeopen}} \right), \qquad (10.70)$$

where $R_{nodeshorted}$ and $R_{nodeopen}$ are return ratios for an element with varying parameter g_m computed when the impedance node of interest is shorted or opened respectively.

5. Thevenin's theorem is a special case of return-ratio analysis.

References

[1] Hendrik W. Bode. *Network analysis and feedback amplifier design* (New York: Van Nostrand, 1945).

[2] R. B. Blackman. Effect of feedback on impedance. *Bell Systems Technical Journal*, **22** (1943).

[3] R. D. Middlebrook. Null double injection and the extra element theorem. *IEEE Transactions on Education*, **32** (1989), 167–180.

[4] Vatché Vorpérian. *Fast Analytical Techniques for Electrical and Electronic Circuits* (Cambridge; New York: Cambridge University Press, 2002).

Section II

Low-power analog and biomedical circuits

11 Low-power transimpedance amplifiers and photoreceptors

In the study of this membrane [the retina] ... I felt more profoundly than in any other subject of study the shuddering sensation of the unfathomable mystery of life.

Santiago Ramón y Cajal

With this chapter, we begin our study of circuits by understanding a classic feedback topology used to sense currents and convert them into voltage, i.e., transimpedance amplifiers. Transimpedance amplifiers are widely used in several sensor and communication applications including microphone preamplifiers, amperometric molecular and chemical sensors, patch-clamp amplifiers in biological experiments, photoreceptors for optical communication links, barcode scanners, medical pulse oximeters for oxygen-saturation measurements, and current-to-voltage conversion within and between chips. After a brief introduction to transimpedance amplifiers, we shall focus on a specific application, i.e., the creation of a photoreceptor, a transimpedance amplifier for sensing photocurrents on a chip. The photoreceptor example will serve as a good vehicle for concretely illuminating several issues that are typical in sensors and transimpedance-amplifier design. It will also serve as a good application for illustrating how the fundamentals of device physics, feedback systems, and noise affect the operation of real circuits and systems.

The photoreceptor discussed here was inspired by the operation of photoreceptors in turtle cones and was first described in [1]. A photoreceptor similar to the one described in this chapter is used to create the low-power pulse oximeter described in Chapter 20 and in [2]. The photodiode basics described here are also useful in understanding how low-power imagers (see Chapter 19) and solar cells (see Chapter 26) work. Transimpedance amplifiers are useful in understanding how biomolecular amperometric sensors and electrochemical sensors, which are briefly discussed in Chapter 20, work. Many of the feedback principles described in this chapter, such as the two-pole Q approximation and the two-pole time-constant rule, are useful in the design of other feedback systems as well.

11.1   Transimpedance amplifiers

Figure 11.1 (a) illustrates a classic transimpedance amplifier topology. A current i_{IN} at a node with capacitance C_{in} is sensed and converted to a voltage

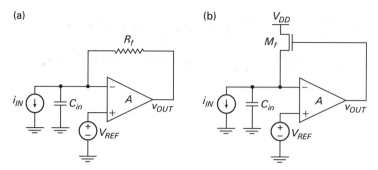

Figure 11.1a, b. Transimpedance amplifier topologies (a) linear and (b) logarithmic.

$v_{OUT} = V_{REF} + i_{IN}R_f$. If the amplifier block has a gain A and very fast dynamics, then the time constant for sensing currents is improved from $R_f C_{in}$ to $R_f C_{in}/(1+A)$ by the feedback loop as a root-locus plot with a single-real-axis-pole would predict. By drawing a feedback block diagram, we can show that the closed-loop transfer function from current to voltage is given by

$$\frac{V_{out}(s)}{I_{in}(s)} = R_f \left(\frac{A/(1+A)}{1 + \dfrac{sC_{in}R_f}{1+A}} \right) \tag{11.1}$$

Thus, the sense amplifier can sense currents and convert them to voltage while speeding up the time constant for sensing from $R_f C_{in}$ to $R_f C_{in}/(1+A)$ such that rapid changes in the input current are not filtered out. Such changes would be filtered out if we merely used R_f to directly convert i_{IN} to v_{OUT}. A further and related benefit is that the voltage v_{IN} at the input node is kept at a virtual reference value of V_{REF} by the feedback loop such that it does not change even if the input current changes. Thus, variations in current caused by variations in v_{IN}, e.g., due to Early effects or other effects in the sensor, are minimized. The dynamic range in output-voltage variation due to input-current variation can be maximized by allowing the amplifier to operate on a large power supply while the input voltage is maintained at V_{REF} such that the sensor's voltage-breakdown characteristics are not compromised.

Figure 11.1 (b) illustrates a logarithmic transimpedance amplifier topology built by having the feedback element of the transimpedance amplifier be a subthreshold transistor with an exponential parameter of κ_S and a pre-exponential constant of I_{OS}. Here, the input current is converted into a voltage output according to

$$v_{OUT} = V_{REF} + \frac{kT}{q\kappa_S} \ln\left(\frac{i_{IN}}{I_{OS}} \right) \tag{11.2}$$

The small-signal relationships between the input current and output voltage of this amplifier are similar to that of the linear amplifier in Figure 11.1 (a) and given by

$$\frac{v_{out}(s)}{i_{in}(s)} = \frac{1}{g_f}\left(\frac{A/(1+A)}{1 + \dfrac{s(C_{in}/g_f)}{1+A}}\right)$$

$$\text{where } g_f = \frac{I_{IN}}{(kT/q)} \qquad\qquad (11.3)$$

$$\Rightarrow v_{out}(s) = \frac{kT}{q}\left(\frac{i_{in}}{I_{IN}}\right)\left(\frac{A/(1+A)}{1 + \dfrac{s(C_{in}/g_f)}{1+A}}\right)$$

Since i_{in} is a small change in the operating-point current of I_{IN}, i.e., $i_{in} = \Delta I_{IN}$, Equation (11.3) may be re-expressed as

$$v_{out}(s) = \frac{kT}{q}\left(\frac{\Delta I_{IN}(s)}{I_{IN}}\right)\left(\frac{A/(1+A)}{1 + \dfrac{s(C_{in}/g_f)}{1+A}}\right) \qquad\qquad (11.4)$$

Thus, at any operating point I_{IN}, the logarithmic transimpedance amplifier transduces small fractional changes in its input $\Delta I_{IN}/I_{IN}$ into an output voltage while speeding up the input time constant by $(1 + A)$. Since g_f varies linearly with the input I_{IN} according to Equation (11.3), the small-signal time constant of the logarithmic transimpedance amplifier varies linearly with I_{IN} and is operating-point dependent unlike that of a linear transimpedance amplifier. The logarithmic transimpedance amplifier is capable of wide dynamic range operation with a modest power-supply voltage. Its scale-invariant fractional amplification is beneficial in several applications where percentage changes rather than absolute changes carry information, e.g., objects reflect or absorb a fixed fraction of the light incident on them independent of the light intensity; consequently, it is the relative $\Delta I/I$ contrast in time or space that carries information in a visual image rather than the absolute intensity. The minimum detectable contrast in a logarithmic transimpedance amplifier is input-current-intensity invariant because the bandwidth and consequently integration interval of the system scales with input current such that a constant number of electrons is always gathered during sensing. In a linear transimpedance amplifier, the bandwidth and consequently integration interval of the system are fixed such that the minimum detectable contrast is worsened at low input intensities due to the gathering of few electrons and improved at high input intensities due to the gathering of more electrons. A logarithmic transimpedance amplifier effectively functions like a linear transimpedance amplifier with built-in gain control where R_f is changed with I_{IN} such that $I_{IN}R_f$ is a constant. Operation in the logarithmic versus linear domains is analogous to floating-point versus fixed-point operation in digital number systems. Both domains of operation are useful in various applications, with logarithmic transduction being particularly beneficial in systems that need scale-invariant and wide-dynamic-range operation.

To understand the tutorial example of this chapter, i.e., a photoreceptor that senses photocurrents via a transimpedance amplifier topology, it is useful but not strictly necessary to understand how photocurrents are generated in silicon. Thus, we shall begin by reviewing some background material on the transduction of light to electrons. Readers unfamiliar with basic semiconductor device physics or readers only interested in transimpedance amplifiers may skip the following section. They will then need to abstract a photodiode as a device that generates a photocurrent proportional to the photon current or equivalently to the light intensity striking it.

11.2   Phototransduction in silicon

Figure 11.2 (a) shows a pn junction and Figure 11.2 (b) reveals an energy band diagram of the pn junction with zero voltage across it. The Fermi level, marked E_F, represents the average energy of an electron, which at zero voltage is the same all along the junction. The lowermost energy level of the conduction band (the conduction band edge) and the uppermost energy level of the valence band (the valence band edge) are drawn in bold wavy lines in the figure. They maintain a constant energy difference between each other, the bandgap energy, while the absolute values of these energies change due to the built-in depletion region or equivalently the built-in potential of the junction. In the n-type region, the Fermi level is near the bottom of the conduction band edge reflecting the fact that lots of donor dopant atoms have contributed to creating a large population of electrons in the conduction band. In the p-type region, the Fermi level is near the top of the valence band edge reflecting the fact that lots of acceptor dopant atoms have contributed to creating a large population of holes in the valence band.

An optical photon with an energy higher than the bandgap energy (1.12 eV in silicon) can create an electron-hole pair when it strikes a silicon atom: the photon promotes an electron from a valence-band energy level to a conduction-band

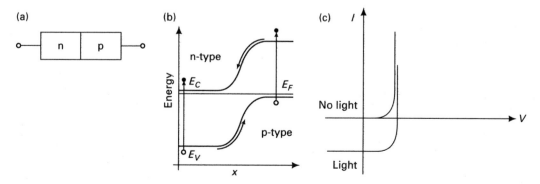

Figure 11.2a, b, c. An pn junction (a), its energy band diagram (b), and its $I - V$ curve with and without the presence of light (c).

energy level, thus creating a hole in the valence band and an electron in the conduction band. Figure 11.2 (b) illustrates that electron-hole pairs created in the p-type region are such that the holes remain in their region of low-energy, i.e., p-type, while electrons travel downhill towards their region of low-energy, i.e., n-type. Similarly, Figure 11.2 (b) also illustrates that electron-hole pairs created in the n-type region are such that the electrons remain in their region of low-energy, i.e., n-type, while holes travel uphill towards their region of low-energy, i.e., p-type. Thus, light tends to cause carriers to go home towards their native majority-carrier type. The net result is that light tends to forward-bias a junction since extra electrons in an n-type region will make the n-type region more negative and extra holes in a p-type region will make the p-type region more positive. Typically most pn junctions that are exposed to light, i.e., photodiodes, are reverse biased such that the energy difference due to the reverse bias makes it even more favorable for carriers to want to go home to their native majority-carrier type (remember that the n-type region is positively biased and the p-type region is negatively biased in a reverse-biased diode). The large reverse bias also reduces the depletion capacitance of the photodiode.

Figure 11.2 (c) shows that light alters the I–V curve of a diode by shifting it downward towards negative currents. In the presence of light, minority carriers are injected back home, thereby creating reverse currents in the diode. The short-circuit zero-voltage bias photocurrent i_{PHOTO} is given by

$$i_{PHOTO} = \alpha q N_{photon} A, \qquad (11.5)$$

where α is the quantum efficiency, the probability that a photon generates an electron-hole pair, typically between 0.3 and 1, N_{photon} is the number of photons per unit area per second arriving at the photodiode and A is the cross-sectional area of the junction exposed to incoming photons. Thus a photodiode of area A receiving light of power intensity per unit area of P_L (in W/m^2) of wavelength λ_{pk} has a photocurrent given by

$$i_{PHOTO} = \alpha q \underbrace{\frac{P_L A}{\left(\dfrac{hc}{\lambda_{pk}}\right)}}_{\text{quanta/s}}, \qquad (11.6)$$

where $h = 6.626 \times 10^{-34}$ J s is Planck's constant and $c = 3 \times 10^8$ m/s is the speed of light in vacuum. The spectrum of sunlight peaks at $\lambda_{pk} = 555$ nm. Moonlight corresponds to an intensity of $P_L \approx 1$ mW/m^2, office fluorescent lighting corresponds to $P_L \approx 1$ W/m^2, and bright midday cloud-free sunlight with no earth tilt corresponds to $P_L \approx 1$ kW/m^2. In practice, the average power due to solar radiation is often between 100 W/m^2 and 200 W/m^2 [3]. Thus a photodiode with an area of 10 μm $\times$ 10 μm and $\alpha = 0.5$ generates approximately 22 pA of photocurrent at office fluorescent lighting levels. The open-circuit forward-bias voltage developed in a photodiode is seen from Figure 11.2 (c) to be the forward-bias voltage needed

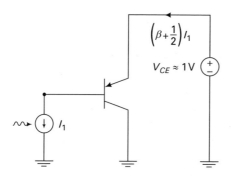

Figure 11.3. A phototransistor. The base terminal is left floating. Incident light generates a photocurrent I_l at the base that is then multiplied by the transistor's current gain β.

to generate a forward current in the diode that balances its reverse short-circuit current. Thus, if the diode has a characteristic pre-exponential constant of I_S, the open-circuit voltage, v_{OC}, developed by the photodiode is given by

$$v_{OC} = \frac{kT}{q} \ln\left(\frac{i_{PHOTO}}{I_S}\right) \tag{11.7}$$

A phototransistor can be built with a bipolar transistor that has a floating base as shown in an example circuit in Figure 11.3. Each pn junction of the transistor generates a floating photocurrent that attempts to forward bias the collector-to-base junction or the emitter-to-base junction respectively. Since each floating current source can be decomposed into two grounded current sources, the phototransistor may be well approximated by the circuit of Figure 11.3 with a light-dependent base current that is amplified by the β of the bipolar transistor. A phototransistor typically has more input capacitance at its base than does a photodiode generating the same base current. When it is used in a transimpedance-amplifier topology to generate photocurrent, the resulting bandwidth is determined by the slow dynamics at its base rather than by any speedup that may be achieved at its emitter (pnp photocurrent) or at its collector (npn photocurrent). The amplification of the photocurrent by β in the transistor does not improve the signal-to-noise ratio of the photodiode since the dominant source of noise in a bipolar transistor is usually base-current shot-noise, which is amplified by the same factor of β as the signal photocurrent. For the same light-collection area, photodiodes typically occupy a smaller area than phototransistors as well. For all of these reasons, photodiodes are usually preferable to phototransistors in most applications.

Figure 11.4 shows the host of pn junctions available in a normal CMOS process that can be used to create photodiodes or phototransistors. Figure 11.5 shows the host of pn junctions available in a BiCMOS process that can be used to create photodiodes or phototransistors. BiCMOS processes include an additional p-base layer, which increases the number of available photodevices. Not all of these junctions are equally effective in transducing light to electrons. Figure 11.6 (a)

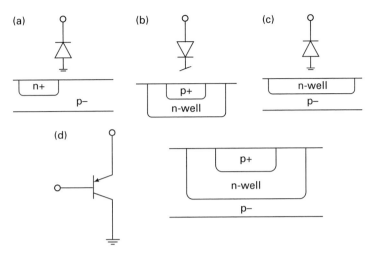

Figure 11.4a, b, c, d. Photodevices that can be fabricated in an ordinary CMOS process.

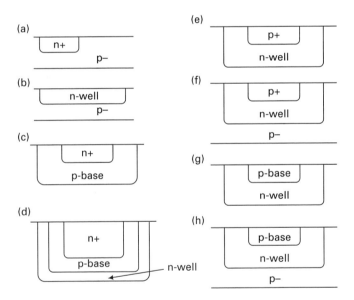

Figure 11.5a, b, c, d, e, f, g, h. Photodevices that can be fabricated in a BiCMOS process. There are more devices than in Figure 11.4 due to the presence of an additional p-base layer (normally used for making vertical npn transistors).

shows that the average optical absorption depth for photons of different wave-length is different, with longer-wavelength photons likely to penetrate deeper into silicon than shorter-wavelength photons. Light penetrating an absorbing medium typically decays exponentially in intensity as it travels along the medium since there is a constant probability of absorption of a photon per unit depth at any depth, similar to that described by a Poisson process. Figure 11.6 (b), adapted

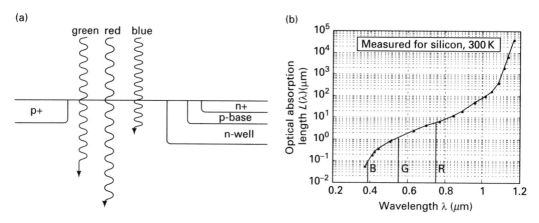

Figure 11.6a, b. The penetration depth (or absorption length) of light in any material (a) varies as a function of wavelength and (b) is plotted for silicon at 300 K (data from W. C. Dash and R. Newman. Intrinsic optical absorption in single-crystal germanium and silicon at 77 K and 300 K. *Physical Review*, 99 (1955), 1151–1155.).

from Dash and Newman, reveals that the optical absorption depth, the depth required to attenuate the intensity of light by an e-fold is different for red, blue and green light [4]. Thus, well-based pn junctions, which are typically 2–3 μm deep, are more likely to absorb red photons than active junctions, which are typically 0.1 μm to 0.2 μm deep in processes suited for efficient phototransduction. Well-substrate junctions are a popular choice for creating photodiodes in silicon because of their large depths and relatively low doping concentrations, which lowers their depletion capacitance. Not surprisingly, measurements reveal a uniformly good quantum efficiency for well-substrate junctions over a wide range of optical wavelengths compared with other junctions [5].

Minority carriers in a region can diffuse and go home to their native type at junctions that are remote from the location at which they were generated. Thus, photocurrents that are intentionally generated at one location in an imager can cause unwanted parasitic photocurrents at another location, thus effectively worsening an imager's resolution by blurring the signal. The diffusion lengths are determined by the minority carrier diffusion length, which is $\sqrt{D_n\tau_n}$ for electrons and $\sqrt{D_h\tau_h}$ for holes, where D_n and τ_n are the diffusion constant and recombination lifetime for electrons and holes, respectively. Depending on the process, these lengths can range up to 30 μm. An n-well guard ring that surrounds the photostructure and that is tied to the positive rail can help mitigate electron diffusion in the substrate by collecting stray electrons flowing to and from the photostructure.

Figure 11.7 (a) illustrates an active-substrate pn photodiode and its associated equivalent circuit. Since light forward-biases a junction, the output voltage of the n+ (active) region is forced below the grounded substrate voltage, an undesirable characteristic. The open-circuit voltage of this photodiode falls by 60 mV for every decade of increase in light intensity such that the overall characteristic is

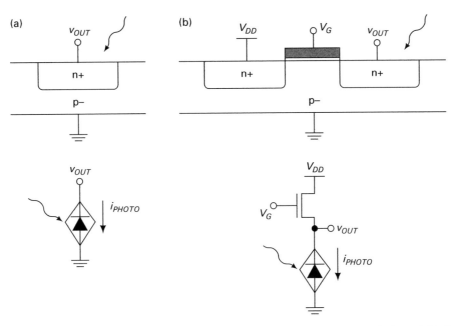

Figure 11.7a, b. Two simple logarithmic photoreceptors: (a) a single photodiode and (b) the source-substrate junction of an MOS transistor that acts as a source-follower.

logarithmic, as predicted by Equation (11.7). Figure 11.7 (b) illustrates the same circuit except that the pn photodiode is implicitly created at the source of the MOSFET. The overall equivalent circuit is then analogous to a source follower with the photocurrent forming the biasing source. The output voltage of the source follower still falls by 60 mV for every decade increase in light intensity such that the overall characteristic is logarithmic; however, if the voltage at the gate and power-supply voltage at the drain are appropriately high such that the MOSFET is in saturation and can equilibrate its current to the photocurrent, the output voltage is always between the two power supply rails and never falls below ground. Thus, we can show that

$$v_{OUT} = \kappa_S V_G - \frac{kT}{q} \ln \left(\frac{i_{PHOTO}}{I_{os}} \right) \tag{11.8}$$

Note that the small photocurrent ensures that the circuit always operates in the subthreshold regime even though the gate voltage v_G is high: The source voltage of the transistor, v_{OUT}, rises to ensure that the gate-to-source voltage of the transistor in Figure 11.7 (b) corresponds to subthreshold operation.

11.3 A transimpedance-amplifier-based photoreceptor

The logarithmic photoreceptor of Figure 11.7 (b) suffers from two problems. Firstly, the time constant at low light intensities is too slow to be useful for many

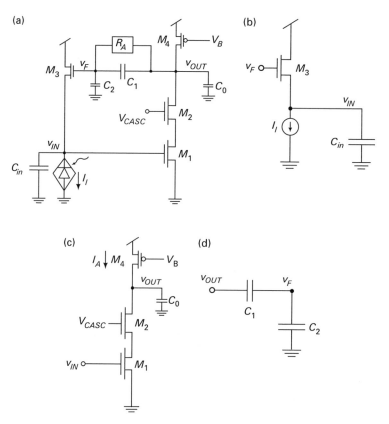

Figure 11.8a, b, c, d. The entire adaptive photoreceptor circuit is shown in (a). It can be split into three simpler subcircuits: (b) a source-follower, (c) a high-gain amplifier, and (d) a capacitive divider.

applications (e.g., 1.6 Hz with a 100 fF capacitive load at 1 mW/m²). Secondly, dc offsets that have little relationship to the light signal due to variations in the threshold voltage of the MOSFET, or equivalently I_{os} in Equation (11.8), affect the output significantly. We will now discuss a photoreceptor that attempts to address both of these problems. It uses a logarithmic transimpedance-amplifier topology to obtain faster responses. The feedback network in the transimpedance amplifier is configured such that it amplifies ac signal transients with higher gain than dc offsets and steady-state light levels. Its logarithmic nature ensures scale-invariant response, wide-dynamic-range operation over a wide range of light levels, and amplification of relative $\Delta I / I$ contrast-based signals rather than signals related to the absolute light level I. The photoreceptor is shown in Figure 11.8 and was first described in [5]. It mimics the operation of biological photoreceptors that have an approximately logarithmic dc response but that exhibit higher ac gain than dc gain at any light level. This photoreceptor has been used in several motion-sensing chips for low-power robotic applications, e.g., see [6], [7]. An advanced version of this photoreceptor has been used to construct an ultra-low-power pulse oximeter, a medical instrument for noninvasive measurement of the oxygen saturation level of hemoglobin in the blood via the use of infrared and red LEDs and photoreceptors [2]. The pulse oximeter is further described in Chapter 20.

Figure 11.8 (a) reveals the circuit of the photoreceptor. The transistor M_3, the light-dependent current I_l, and C_{in} form a source follower, which is shown in Figure 11.8 (b) as a subcircuit. The transistors M_1, M_2, and M_4 form a high-gain amplifier with a load capacitance C_0, which is shown in Figure 11.8 (c) as a subcircuit. The capacitors C_1 and C_2 form a capacitive divider, which is shown in Figure 11.8 (d) as a subcircuit. The element R_A in Figure 11.8 (a) represents a very large adaptation resistance whose conductance is insignificant compared with the admittance of capacitance C_1 at all but the lowest frequencies. We shall approximate it as having an infinite resistance unless otherwise mentioned. We shall discuss how R_A is implemented with two back-to-back diode-like resistances in parallel after discussing the rest of the circuit.

The overall topology of Figure 11.8 (a) is similar to that of the logarithmic transimpedance amplifier of Figure 11.1 (b) with I_l as an input, M_3 as the feedback transistor, and the high-gain amplifier comprised of M_1, M_2, M_4 and C_0. However, the reference voltage V_{REF} is implicitly established by the high-gain amplifier subcircuit. Furthermore, the capacitive-divider and R_A form a frequency-shaping attenuation network in the feedback path that causes the ac gain to be higher than the dc gain at all but the lowest frequencies. At equilibrium, the negative-feedback loop establishes operating-point parameters such that I_l flows through M_3 and the voltage v_{IN} is at an implicit reference value such that the current through M_1 and M_2 is equal to that of the bias current of M_4.

The transistor M_2 in Figure 11.8 (c) functions as a cascode transistor to prevent the C_{gd} capacitance of M_1 from loading the input node V_{in} with a Miller capacitance of value $C_{gd}(1 + A)$, where A is the gain of the amplifier. Such loading is detrimental to the bandwidth of the photoreceptor. It also increases the output impedance and the gain of the amplifier through the usual mechanism of a cascoding transistor (Equation (10.36) in Chapter 10). We can show that the small-signal equivalent circuit of the high-gain amplifier of Figure 11.8 (c) is that of Figure 11.9 with an output impedance r_o, gain A, and input-referred noise power per unit bandwidth $v_n^2(f)$ given by

$$r_0 = r_0^{M_4} // (r_0^{M_2}(1 + g_s r_0^{M_1}) + r_0^{M_1}) \approx r_0^{M_4}$$

$$g_A = \kappa g_s = \frac{\kappa I_A}{\phi_T} = g_m \text{ of } M_1$$

$$A = \frac{v_{out}}{v_{in}} = -g_A r_0 \qquad (11.9)$$

$$v_n^2(f) = \frac{4q I_A}{g_A^2}$$

We have assumed that all transistors operate in saturation. Only the $2q I_A$ shot-noise contributions of subthreshold transistors M_1 and M_4 affect the noise of the

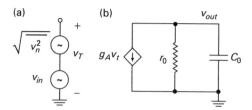

Figure 11.9a, b. The small-signal equivalent of the high-gain amplifier sub-circuit present within the adaptive photoreceptor.

amplifier since the noise generator of the cascode transistor, M_2, contributes almost no noise to the amplifier's output: the noise current at the source of a cascode transistor (here M_2) flows mostly through its own low source impedance $(1/g_s)$ compared with the high drain impedance of the current-source transistor that it cascodes $(r_0^{M_1})$. The current flowing through the source of the cascode transistor arrives at its drain and gets subtracted from that of the noise generator such that the current that is output to the external circuit (the difference between the noise-generator's current and that arriving at the drain of the cascode) is negligible. In effect, cascode transistors shunt out their own noise generators because the resistance to self-shunting is much smaller than the resistance of the external circuit attached to the cascode transistor. We can arrive at the same conclusion for the cascode transistor M_2 from return-ratio analysis and Equation 10.22 as well. The high return ratio of the g_s or g_m generator of transistor M_2 makes the circuit of Figure 11.8 (c) highly insensitive to variations in the g_s or g_m of M_2 via Equation (10.22) with $TF_0 = 0$. Such robustness translates to high insensitivity to noise currents in the cascode as well as increase in the output impedance of the M_1-M_2 configuration via Blackman's formula (as expressed in Equation (10.36) for the cascode).

The source-follower circuit of Figure 11.8 (b) can be described by

$$v_{in}(s) = \left(\frac{\kappa g_s}{g_s + sC_{in}}\right)v_f(s) \pm \sqrt{\frac{4qI_l}{g_s^2}\left(\frac{1}{1 + \tau_{in}s}\right)}$$

$$\text{where } g_s = \frac{I_l}{\phi_t}, \quad \tau_{in} = \frac{C_{in}}{g_s} \tag{11.10}$$

and the $2qI_l$ shot-noise contributions of the photodiode and M_3 have been added.

The capacitive-divider circuit of Figure 11.8 (d) is described by

$$v_f = \left(\frac{C_1}{C_1 + C_2}\right)v_{out} \tag{11.11}$$

11.4   Feedback analysis of photoreceptor

We may incorporate the relationships of Equations (11.9), (11.10), and (11.11) into the simple block diagram shown in Figure 11.10 (a) and into its successively

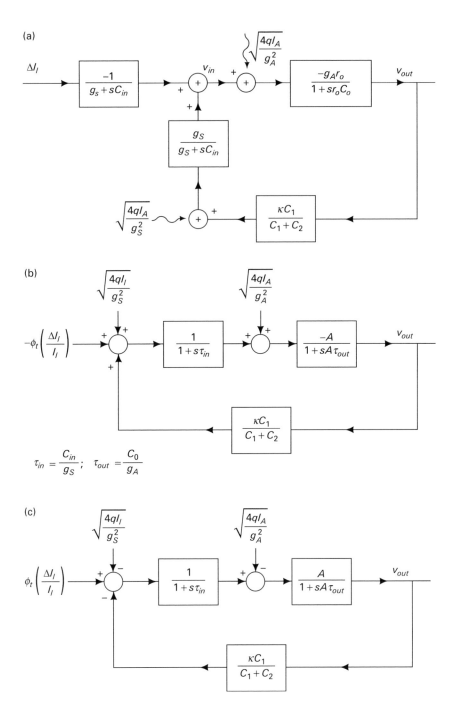

Figure 11.10a, b, c. Feedback block diagrams of the adaptive photoreceptor: (a) original form, (b) and (c) two simplified forms.

simpler forms shown in Figure 11.10 (b) and Figure 11.10 (c), respectively. We have defined

$$\tau_{out} = \frac{C_o}{g_A} \tag{11.12}$$

Note that the signs of the noise generators in Figure 11.10 (c) do not matter since we will eventually be squaring them. If the loop transmission of the feedback loop is sufficiently high, then the input-output transfer function is given by the reciprocal of the feedback path to be

$$v_{out} = \frac{\phi_t}{\kappa} \left(\frac{C_1 + C_2}{C_1} \right) \frac{\Delta I_l}{I_l} \tag{11.13}$$

Thus, the photoreceptor has a contrast amplification response as would be expected from a logarithmic photoreceptor. If we denote the closed-loop gain and dc loop gain to be

$$A_{cl} = \left(\frac{C_1 + C_2}{\kappa C_1} \right) \quad \text{and} \quad A_{lp} = \left(\frac{A}{A_{cl}} \right) \tag{11.14}$$

respectively, we can show from Black's formula that

$$H(s) \equiv \frac{v_{out}(s)}{\phi_t \left(\dfrac{\Delta I_l}{I_l} \right)} = \frac{\left(\dfrac{1}{1 + \tau_{in} s} \right) \left(\dfrac{A}{1 + A \tau_{out} s} \right)}{1 + \left(\dfrac{1}{1 + \tau_{in} s} \right) \left(\dfrac{A}{1 + A \tau_{out} s} \right) \dfrac{1}{A_{cl}}} \tag{11.15}$$

Algebraic simplification then yields the expression for $H(s)$ described in Equation (11.16) below. The poles are geometrically located in the s-plane as shown in Figure 11.11 (a) and have a closed-loop magnitude frequency response as shown in Figure 11.11 (b) and Figure 11.11 (c).

$$H(s) = \frac{A_{cl}}{\dfrac{A_{cl}}{A}(\tau_{in} s + 1)(A \tau_{out} s + 1) + 1}$$

$$H(s) = \frac{A_{cl}\left(\dfrac{A_{lp}}{A_{lp} + 1} \right)}{\left(\dfrac{A \tau_{out} \tau_{in}}{A_{lp} + 1} \right) s^2 + \left(\dfrac{\tau_{in} + A \tau_{out}}{A_{lp} + 1} \right) s + 1}$$

$$\tag{11.16}$$

$$H(s) = \frac{A_{cl}\left(\dfrac{A_{lp}}{A_{lp} + 1} \right)}{\tau^2 s^2 + \dfrac{\tau s}{Q} + 1}$$

$$\text{where} \quad \tau = \sqrt{\frac{A \tau_{out} \tau_{in}}{A_{lp} + 1}}, \quad Q = \frac{\sqrt{1 + A_{lp}} \sqrt{A \tau_{out} \tau_{in}}}{\tau_{in} + A \tau_{out}}$$

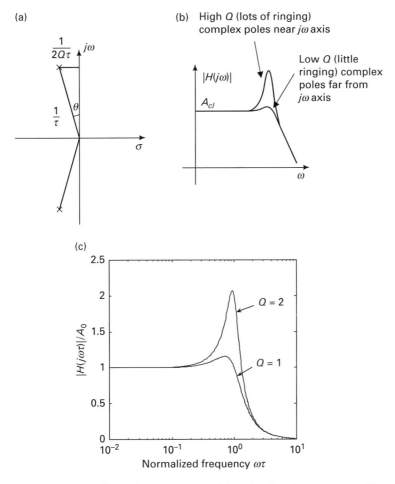

Figure 11.11a, b, c. Closed-loop responses of the adaptive photoreceptor: (a) s-plane plot, with high-Q poles near the $j\omega$ axis; (b) and (c) frequency responses. (b) Illustrates qualitative features of the frequency response for high and low values of Q while (c) plots the actual response for $Q = 1$ and $Q = 2$.

The value of Q from the s-plane plot of Figure 11.11 (a) is given by

$$\sin(\theta) = \frac{1}{2Q} \tag{11.17}$$

and is a measure of how near the $j\omega$ axis the poles of $H(s)$ are. High-Q closed-loop systems have lots of peakiness in their frequency response as illustrated in Figures 11.11 (b) and 11.11 (c) and as we have outlined in Chapter 9. They also have considerable ringing in their step response and are generally avoided in feedback-system design unless one is trying to explicitly build an oscillator. The value of $1/\tau = \omega_n$ in Figure 11.11 (a) is a measure of the distance of the poles from the origin and is reciprocally related to the 10% to 90% rise time of the closed-loop step response.

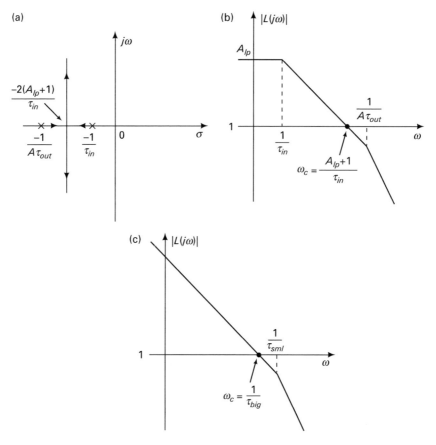

Figure 11.12a, b, c. (a) Root-locus plot, and (b) Bode plot of the adaptive photoreceptor. (c) shows the Bode plot of a two-pole system where the first pole occurs at zero frequency, thereby acting as a perfect integrator.

The root-locus plot of Figure 11.12 (a) reveals that as the root-locus gain parameter $A_{lp} = A/A_{cl}$ increases, the poles at $-1/(A\tau_{out})$ and $1/\tau_{in}$ approach each other and the Q of the system increases. When the system is critically damped, i.e., both poles are at the same location on the real axis, $Q = 0.5$. As A_{lp} increases, the two poles depart the real axis and begin to make increasingly smaller acute angles with the $j\omega$ axis, thus increasing the Q even further; Q increases to $1/\sqrt{2} = 0.707$ when the poles of the root-locus plot have equal real and imaginary parts, i.e., $\theta = 45°$ in Figure 11.12 (a) or in Figure 11.11 (a).

The Bode plot of the loop transmission shown in Figure 11.12 (b) reveals a crossover frequency near $(A_{lp} + 1)/\tau_{in}$ with the amplifier pole located at $1/(A\tau_{out})$, which is a factor of, say, $\eta > 1$ beyond the crossover frequency. For a loop transmission comprised of two widely separated poles, e.g., the low-frequency $1/\tau_{in}$ pole and the high-frequency $1/(A\tau_{out})$ pole, the resulting Q of the closed-loop system is, to an excellent approximation, $1/\sqrt{\eta}$, i.e., equal to the square-root of the ratio of the loop-transmission crossover frequency to the frequency of the pole beyond crossover, where the crossover frequency is determined by pretending that only the single-pole dynamics of the slow pole determines it. Thus, for the case shown in Figure 11.12 (b), if $\eta = 4$ we would predict that $Q = 1/\sqrt{4} = 0.5$; therefore we have critical damping. We shall call this approximation the closed-loop two-pole Q-approximation since it is typically an excellent approximation of the exact Equations (9.15) or (9.16) of Chapter 9 on feedback systems.

Thus, for our system, we have critical damping when

$$\frac{(1 + A_{lp})}{\tau_{in}} = \frac{1}{4(A\tau_{out})}$$

$$\tau_{in} = 4A\left(1 + \frac{A}{A_{cl}}\right)\tau_{out} \approx \frac{4A^2}{A_{cl}}\tau_{out}$$

(11.18)

If the closed-loop system is to be well behaved, the resulting Q must be within a factor of two of this critically damped value, e.g., a Q of 1.0 corresponding to 45° of phase margin results when

$$\tau_{in} \approx \frac{A^2}{A_{cl}}\tau_{out}$$

(11.19)

Thus, for good closed-loop dynamic performance (corresponding to phase margins of 45° or more), the bias current of the high-gain amplifier, which directly determines τ_{out} from Equations (11.9) and (11.12), needs to almost be a factor of A^2 larger than the input light current I_l, which directly determines τ_{in} from Equation (11.10). Such a wide separation of the bias-current value of the amplifier and the sensed input current is frequently seen in transimpedance-amplifier feedback loops that are dominated by two poles.

The closed-loop two-pole Q-approximation works even if the second pole is before crossover, i.e. if we have $\eta < 1$: we pretend that the first slow pole still determines crossover with a single-pole roll-off; the application of the rule then results in a Q greater than 1 since the fictitious crossover frequency occurs after the location of the second pole. As an example, if we have a slow pole at 10 Hz, a dc loop gain of 10^5, and a fast pole at 10^5 Hz, the fictitious crossover frequency is nearly at 10^6 Hz and the second pole occurs before the crossover frequency, i.e., at 10^5 Hz. Thus, the closed-loop Q of the system is $\sqrt{10^6/10^5} \approx 3.2$. A regular Bode plot of the loop transmission would reveal poor phase margin and a crossover frequency below 10^6 Hz.

Another useful rule allows us to determine $1/\tau = \omega_n$, the natural frequency of the closed-loop system. For a loop transmission with only two poles, e.g., the low-frequency $1/\tau_{in}$ pole and the high-frequency $1/(A\tau_{out})$ pole, the resulting $1/\tau = \omega_n$ natural frequency of the closed-loop system is the geometric mean of the loop-transmission crossover frequency and the frequency of the pole beyond crossover, where the crossover frequency is determined by pretending that only the single-pole dynamics of the slow pole determines it. This closed-loop two-pole τ-rule applies to all two-pole loop transmissions and is an exact application of

Equations (9.15) or (9.16) of Chapter 9. Thus, at critical damping in our photoreceptor, we would predict that the natural frequency is a factor of $\sqrt{4} = 2$ beyond the crossover frequency, i.e.,

$$\omega_n = \frac{1}{\tau} = \frac{2(1 + A_{lp})}{\tau_{in}} \qquad (11.20)$$

Similarly, with 45° of phase margin, corresponding to $Q = 1$, the natural frequency is at the crossover frequency, i.e., at $(1 + A_{lp})/\tau_{in}$. The net settling time to a step input, which is approximated by $1/(Q\omega_n)$ for moderate $Q < 1$, is then nearly $(1 + A_{lp})/\tau_{in}$ in both cases. In the $Q = 0.5$ case, ω_n is doubled but Q is halved compared with that in the $Q = 1$ case. Thus, we have a speedup of nearly $1 + A_{lp}$ in both cases.

The closed-loop two-pole τ-rule is an exact application of Equations (9.15) or (9.16). The closed-loop two-pole feedback Q-approximation is a special case of these equations when the two time constants are widely separated. In fact, Equation (9.15) shows that in the very worst case when the poles are identical and not widely separated, the maximal value of Q is given by

$$Q_{\max} = \frac{\sqrt{1 + A_{lp}}}{2} \qquad (11.21)$$

Figure 11.12 (c) shows the loop transmission corresponding to Equation (9.16) for which both the two-pole τ-rule and the two-pole Q-approximation are exact. The plot is shown for $A_{lp} = 1$.

11.5   Noise analysis of photoreceptor

Figure 11.10 (c) illustrates how the white noise per unit bandwidth is filtered by the feedback loop of the photoreceptor. The transfer function of the source-follower's noise $4qI_l/g_s^2$ to the output is the same as that of the input such that it is filtered by a second-order filter with a corner frequency at the closed-loop bandwidth of the system. The amplifier's noise per unit bandwidth has a closed-loop transfer function with an additional zero due to the presence of the $1/(1 + \tau_{in}s)$ block in its feedback path. This zero increases the noise contribution of the amplifier to the output past the $1/\tau_{in}$ angular frequency until the closed-loop natural frequency is reached, a factor of A_{lp} from the zero, after which it is filtered out by the two-pole roll-off of the loop. It is important to ensure that the amplifier's bandwidth is not too large such that the closed-loop system has a significant region of operation with just a one-pole roll-off; in this case, the amplifier's noise will contribute to the output past the closed-loop bandwidth of the loop until it is filtered out by the amplifier's own corner frequency. Of course, such excess noise could be filtered out by a post-processing filter operating on the outputs of the photoreceptor. Typically, unless the amplifier's bandwidth $1/(A\tau_{out})$ is very large and/or if its excess noise has not been filtered out, the contribution of its noise to the output is usually small. Since g_A^2 needs to be almost A^2/A_{cl} greater than g_s for good phase margin, the noise per unit bandwidth of the amplifier is almost A_{cl}/A^2 times smaller than that due to the source follower. The increase of the amplifier's noise by the zero increases its contribution by a factor of $A_{lp} = A/A_{cl}$ at most, such that the amplifier's noise per unit bandwidth over all frequencies within the closed-loop bandwidth is still only $1/A$ times that of the source-follower's noise per unit bandwidth at most. *Thus, in most practical situations, the noise of the photoreceptor is dominated by the source-follower noise and the amplifier's noise may be neglected.*

From the analysis of the source-follower's noise in Chapter 8, we can compute that the minimum detectable contrast for a simple source-follower photoreceptor with no transimpedance-amplifier speedup is given by

$$\frac{\Delta I_l^2}{g_s^2} = \frac{kT}{C_{in}}$$

$$\text{i.e. } \frac{\Delta I_l^2}{I_l^2} = \frac{kT}{C_{in}} \times \frac{1}{\phi_t^2} = \frac{q}{C_{in}\phi_t} \qquad (11.22)$$

With a speedup of $(1 + A_{lp})$, the closed-loop bandwidth is increased such that the total noise is increased. Thus, the minimum detectable contrast is increased and is given by Equation (11.23) below:

$$\frac{\Delta I_l^2}{I_l^2} = \frac{q}{C_{in}\phi_t}\left(1 + A_{lp}\right)$$

$$\text{i.e. } \left(\frac{\Delta I_l}{I_l}\right) = \sqrt{\frac{q}{C_{in}\phi_t}\left(1 + A_{lp}\right)} \qquad (11.23)$$

Figure 11.13 and the associated caption, which are adapted from [1], present measurements of the output noise of the photoreceptor of Figure 11.8 and the source follower of Figure 11.7. The photoreceptor simply amplifies the noise per unit bandwidth of the source-follower by about as much as it amplifies an intentional periodic LED light input (the tone inputs in Figure 11.13). Thus, it is hardly increasing the input-referred noise per unit bandwidth but it does increase the bandwidth of the system, and therefore the total input-referred noise and minimum detectable contrast. The bandwidth of the noise is the largest when the cascode transistor M_2 is used.

Equation (11.23) suggests that a good photoreceptor with a small minimum detectable contrast must have a large capacitance. Such a photoreceptor can be built by scaling up the area of the photodiode until the intrinsic bandwidth of the photodiode self limits (since I_l, g_s, and the depletion capacitance C_{dep} of the photodiode all scale with its area); A_{lp} is chosen to yield the desired speedup needed at the lowest light intensities and I_A is chosen to yield stability at the highest light intensities; then, from Equation (11.23), the size of the photodiode needed to yield the minimum detectable contrast can be computed. A 100 pF value of C_{in} and $A_{lp} = 35$ yields a minimum detectable contrast of nearly 0.16%.

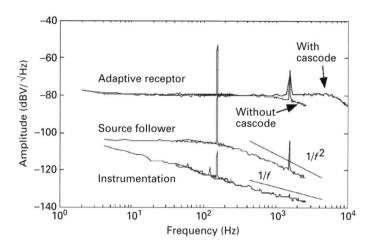

Figure 11.13. Experimentally measured output noise spectra of the adaptive photoreceptor and the simple source-follower photoreceptor. The input-referred noise of the two receptors is almost identical. The noise spectrum of the instrumentation used for making the measurement is also shown for convenience. Reprinted with kind permission from [1] (©1994 IEEE).

11.6 The adaptation resistor R_A

Figure 11.14 shows how the adaptation resistor R_A, thus far assumed to be infinite, is constructed with two back-to-back diode-like elements [1]. The device is made with a single well transistor such that an MOS transistor is operative when $V_{out} = V_0 > V_f$ in Figures 11.8 (a) and 11.14, respectively, and the parasitic bipolar transistors present in every MOS transistor are operative when $V_0 < V_f$. The lateral bipolar transistor revealed in Figure 11.14 typically plays a much smaller part than the vertical bipolar transistor due to the vertical bipolar's significantly better β. Thus the circuit model shown in Figure 11.15 (a) only analyzes the functioning of R_A with one vertical bipolar transistor. It also explicitly adds parasitic photocurrents that appear in any real photoreceptor due to unwanted light leakage onto the pn junctions of R_A. The large $I_{well}{}^P$ parasitic photocurrent due to the well-substrate junction is absorbed into the bias current of M_4 in Figure 11.8 and does not affect the circuit's operation much, one of the benefits of this configuration. Any stray electrons that diffuse into the n-well region are also absorbed. Figure 11.15 (b) reveals the exponential I–V curves (with slopes differing by a factor of κ) that result from the subthreshold MOS transistor and the bipolar transistor.

Simple analysis of the circuit of Figure 11.15 (a) yields:

$$
\begin{aligned}
i &= i_M - i_B + I_{par} \\
\Delta V &= v_{OUT} - v_F \\
i_M &= I_0 e^{\kappa \Delta v / \phi_t} \left(1 - e^{-\Delta V / \phi_t} \right) \\
i_B &= \frac{I_s}{\beta} \left(e^{-\Delta V / \phi_t} - 1 \right) \text{ at } v_{OUT} \text{ side} \\
&= I_s \left(e^{-\Delta V / \phi_t} - 1 \right) \text{ at } v_F \text{ side}
\end{aligned}
\tag{11.24}
$$

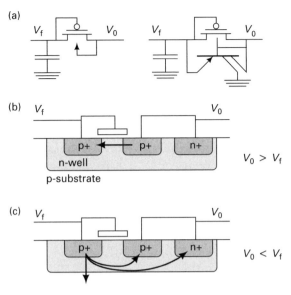

Figure 11.14a, b, c. The adaptive element used in the photoreceptor: (a) shown in two schematic forms, (b) conducting as an MOS transistor and (c) conducting as a bipolar transistor with two collectors. Reprinted with kind permission from [1] (©1994 IEEE).

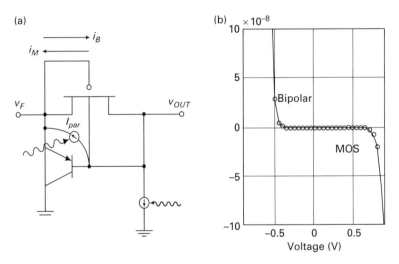

Figure 11.15a, b. The adaptive element: (a) transistor-level schematic including parasitic photocurrents and (b) experimentally measured $I - V$ curves, showing that the element behaves like two back-to-back diodes. Reprinted with kind permission from [1] (©1994 IEEE).

From Figures 11.15 (a) and 11.8 (a), we can conclude that the gate-node (the v_F node) conductance at equilibrium $G_A = 1/R_A$ determines the value of a zero located at $-1/R_A C_1$ beyond which the attenuation of the $R_A - C_1 - C_2$ frequency-shaping lead network begins and the location of a pole at $1/R_A(C_1\|C_2)$ at which it asymptotes. The value of G_A can be determined from small-signal

Taylor expansions of the exponential in Equation (11.24) for the two cases where I_{par} is assumed to be zero and where it is assumed to be finite.

$$G_A = \left(\frac{I_s}{\phi_t} + \frac{I_0}{\phi_t} \right), \quad \text{for small light levels}$$

$$G_A = \frac{I_{par}}{\phi_t}, \quad \text{for large light levels}$$

(11.25)

In the case where I_{par} is finite, we merely assume that the bipolar transistor is turned on. With moderate-sized values of $C_1 \approx 50\,\text{fF}$ and $C_2 \approx 1\,\text{pF}$ adaptation time constants of 2 s in the dark and 0.2 s at 5 W/m^2 have been achieved.

Since R_A behaves like a short at dc in Figure 11.8 (a), the attenuation of the capacitive-divider disappears at dc. Thus, the closed-loop dc gain of the photoreceptor may be described by removing the presence of the capacitive divider in Equation (11.13), to obtain

$$v_{out} = \frac{\phi_t}{\kappa} \frac{\Delta I_l}{I_l}$$

(11.26)

The adaptation resistor helps us implement a very slow time constant that is more immune to parasitic light than several other back-to-back diode configurations. It uses no junctions in the substrate, so it is immune to stray electrons present in the substrate and is shielded from them by the well. The parasitic photocurrent of the well-substrate junction is also effectively shielded due to its absorption by the amplifier bias current. It is a relatively compact high-resistance element and has found use in neural amplifiers and cardiac amplifiers where a very slow adaptation time constant is necessary [8], [9]. Low-power neural amplifiers are described in Chapter 19 and low-power cardiac amplifiers are described in Chapter 20.

11.7   Experimental measurements of the photoreceptor

Figure 11.16 shows that, as predicted by Equations (11.13) and (11.26), for a wide dynamic range of light levels, the contrast response of the photoreceptor to ac transients is higher by a factor of $(C_1 + C_2)/C_1$ than its dc contrast response. Figure 11.16 reveals that the photoreceptor successfully amplifies ac signals riding over a dc background over six orders of magnitude in intensity. At low light levels, the speedup inherent in the transimpedance topology permits it to sense a changing input current even though the source-follower photoreceptor, which has lower bandwidth, filters out the signal. The data was gathered by modulating the brightness of an LED shining on a photoreceptor fabricated on-chip. The dc background intensity was attenuated by interposing neutral density filters, designed to provide exact order-of-magnitude light attenuation, between the LED light and the chip.

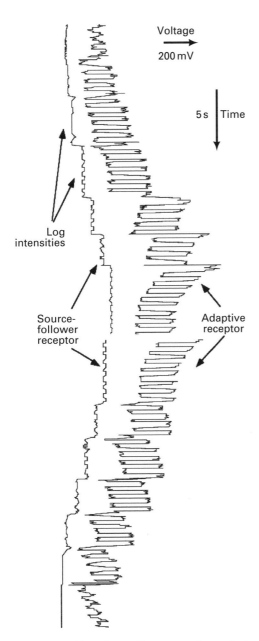

Figure 11.16. Experimentally measured transient responses of the source-follower and adaptive photoreceptors over six decades of background illumination level. Reprinted with kind permission from [1] (©1994 IEEE).

11.8   Adaptive biasing of I_A for energy efficiency

Adaptive biasing that scales I_A with I_l is significantly more energy efficient for the photoreceptor: the photoreceptor then does not need to be designed to handle the entire dynamic range of light levels at once with I_A being biased at a needlessly high value for the highest light setting, necessary for maintaining stability. In this application, biasing for worst case is a particularly expensive proposition due to the A^2/A_{cl} scaling in I_A with I_l. Thus, feed-forward biasing of I_A by sensing the average value of I_l on a dummy source-follower photoreceptor or by mirroring current flowing through M_3 to bias M_4 via a current amplifier and a filtering capacitance is beneficial. The adaptation circuit must change I_A sufficiently quickly compared with the time scale over which I_l changes. The overall light level in most real-world applications changes fairly slowly such that the latter requirement is not hard to fulfill. The value of A_{lp} can also be adapted with changes in light level to yield light-invariant closed-loop bandwidths, which are necessary for robust operation in pulse oximetry [2].

11.9   Zeros in the feedback path

In practice, the closed-loop performance of the photoreceptor exhibits better phase margin than may be expected from a two-pole analysis with a high value of A_{lp}. A left-half-plane zero in the feedback path due to the feed through C_{gs} capacitance in M_3 in Figure 11.8 (a) (the source-follower transistor) provides improved phase margin. This zero also limits the speedup that is possible by forcing one of the poles of the loop transmission to be always near it. Figure 11.17 (a) illustrates a block diagram of this feedback loop with ε given by C_{gs}/C_{in}. Figure 11.17 (b) reveals the root-locus plot for this configuration and Figure 11.17 (c) shows the Bode plot. Note that the zero affects the loop transmission but not the closed-loop response since it is in the feedback path. The maximum possible speedup of this configuration is then κ/ε independent of light level. Since C_{gs} is determined only by parasitics in sub-threshold operation, it is advantageous to operate in this regime as far as is practical. The source-follower noise in this configuration integrates over a bandwidth given by the zero-limited frequency, i.e. $\kappa/(\varepsilon\tau_{in})$. The amplifier noise integrates over this bandwidth as well, but in addition also integrates over a band between the zero-limited speedup frequency and the closed-loop crossover frequency.

Schemes to avoid the speedup-limiting zero implement a feedback loop that replaces M_3 with a pFET in Figure 11.8 (a). The source terminal of this pFET is driven by a low-impedance (source-follower-buffered) version of the voltage source v_F in Figure 11.8 (a). The pFET's gate is fixed at a constant value and its drain is tied to the input node (the input node marked with a v_{IN} label in Figure 11.8 (a)). The constant gate voltage prevents C_{gs} and or C_{gd} from causing feed through effects since it acts as a small-signal ground, thus converting both potential feed through (floating) capacitances to grounded capacitances [2].

The pulse oximeter described in [2] uses an advanced version of the photorecep-tor described in this chapter to improve its sensitivity to dim modulated light signals. These signals are observed over a normal cardiac cycle by passing LED light through one side of a subject's finger and detecting it with a photoreceptor on

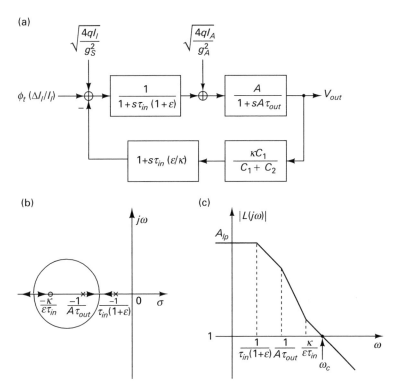

Figure 11.17a, b, c. Effects of including the feed through zero in our model of the adaptive photoreceptor circuit: (a) modified feedback block diagram, (b) root-locus plot, and (c) Bode plot.

the other side of the subject's finger (see Chapter 20). The use of a more sensitive photoreceptor enables the LED light intensity, and consequently LED power, to be reduced. Since LED power is the main component of power in a well-designed portable pulse-oximeter system without a display, the overall power of the pulse oximeter is then significantly reduced.

The feedback and noise analysis of transimpedance amplifiers that have been discussed in this chapter can be adapted to several current-sensing systems in a straightforward fashion to compute their bandwidth, dynamics, minimum detectable signal, and power.

References

[1] T. Delbruck and C. A. Mead, Adaptive photoreceptor with wide dynamic range. *Proceedings of the IEEE International Symposium on Circuits and Systems (ISCAS)*, London, UK, 339–342, 1994.

[2] M. Tavakoli, L. Turicchia and R. Sarpeshkar. An ultra-low-power pulse oximeter implemented with an energy efficient transimpedance amplifier. *IEEE Transactions on Biomedical Circuits and Systems*, in press (2009).

[3] David J. C. MacKay. *Sustainable Energy – Without the Hot Air* (Cambridge, UK: UIT Cambridge, Ltd., 2009).

[4] W. C. Dash and R. Newman. Intrinsic optical absorption in single-crystal germanium and silicon at 77 K and 300 K. *Physical Review*, **99** (1955), 1151–1155.

[5] T. Delbruck and C. A. Mead. Analog VLSI phototransduction by continuous-time, adaptive, logarithmic photoreceptor circuits. *CalTech CNS Memo*, **30** (1994).

[6] R. Sarpeshkar, J. È. Kramer, G. Indiveri and C. Koch. Analog VLSI architectures for motion processing: From fundamental limits to system applications. *Proceedings of the IEEE*, **84** (1996), 969–987.

[7] R. R. Harrison and C. Koch. A silicon implementation of the fly's optomotor control system. *Neural Computation*, **12** (2000), 2291–2304.

[8] R. R. Harrison and C. Charles. A low-power low-noise CMOS amplifier for neural recording applications. *IEEE Journal of Solid-State Circuits*, **38** (2003), 958–965.

[9] W. Wattanapanitch, M. Fee and R. Sarpeshkar. An energy-efficient micropower neural recording amplifier. *IEEE Transactions on Biomedical Circuits and Systems*, **1** (2007), 136–147.

12 Low-power transconductance amplifiers and scaling laws for power in analog circuits

The obvious is that which is never seen until someone expresses it simply.

Kahlil Gibran

In this chapter, we will discuss low-power transconductance amplifiers and circuits built with them such as first-order filters. Transconductance amplifiers are one of the most widely used building-block analog circuits. They implement a controlled linear conductance, which along with a set of capacitors, suffice to create any linear system with a finite number of state variables. Nonlinear transconductance amplifiers and capacitors can create any similar nonlinear dynamical system if the nonlinearity can be easily implemented by the transconductance amplifier. For example, tanh, sigmoid, and sinh nonlinearities are easily implemented in the subthreshold domain and are useful in a wide variety of computations based on statistical-mechanical exponential primitives.

We will begin by focusing on how to use feedback-linearization techniques to construct a low-power transconductance amplifier capable of wide-linear-range operation in the subthreshold and above-threshold domains. These techniques will involve using the well of the transistor as an input, using a well-known technique called source degeneration, a novel technique that we term gate degeneration, and a technique called bump linearization. We shall analyze how extending the linear range of an amplifier affects its offset and noise. We shall find that extensions of linear range arrive with extensions of dynamic range in thermal-noise-limited filters constructed with transconductance amplifiers but not in similar filters constructed with $1/f$-noise-limited amplifiers. Our mathematical analysis of noise, offset, and dynamic range in a simple first-order filter will lead us to one of the most important insights in low-power circuit design: *The five issues of task complexity, technology, topology, speed (bandwidth), and precision of a circuit determine its power consumption*. The principles on the low-power hand of Figure 1.1 in Chapter 1 will manifest in their first concrete instantiation in this chapter.

In our circuit, or in any linear circuit, the precision or signal-to-noise ratio (SNR) is equivalently parameterized by its dynamic range. The five determinants of power hold true for all circuits including amplifiers, filters, analog-to-digital converters, RF circuits, digital circuits, sensor circuits, biological circuits, chemical circuits, and also our circuit. For the wide-linear-range transconductance amplifier (WLR) circuit discussed in this chapter, we shall find a concrete formula for its power consumption as a function of these five variables, which will shed insight into the scaling laws of low-power design for all analog circuits. A general discussion of power in analog circuits is presented in Chapter 22.

This chapter also includes a brief discussion of general principles for low-voltage analog design with transconductance amplifiers being used as an example of their application.

12.1   A simple ordinary transconductance amplifier (OTA)

Figure 12.1 (a) reveals the circuit of a simple ordinary transconductance amplifier operated in the subthreshold regime, which we discussed briefly in Chapter 8. The input-output tanh curve of Figure 12.1 (b) is derived from the following relationships between currents and voltages marked in Figure 12.1 (a):

$$i_+ = I_0 e^{\frac{\kappa_s v_+ - v_S}{\phi_t}} \ ; i_- = I_0 e^{\frac{\kappa_s v_- - v_S}{\phi_t}} \ ; i_+ + i_- = I_B; i_+ - i_- = i_{out}$$

$$\boxed{i_{OUT} = I_B \tanh\left(\frac{\kappa_s(v_+ - v_-)}{2\phi_t}\right)} \tag{12.1}$$

If we define $v_{IN} = v_+ - v_-$, then we can define the transconductance of the amplifier as

$$G_m = \frac{\partial i_{OUT}}{\partial v_{IN}} = \frac{\kappa_s I_B}{2\phi_t}\bigg|_{V_{IN}=0} = \frac{I_B}{V_L}$$

$$V_L = \frac{2\phi_t}{\kappa_s} \tag{12.2}$$

where the linear range V_L can be obtained by extrapolating the G_m slope at the origin until the saturation current of I_B is reached. Alternatively, the linear range of the transconductance amplifier can by computed by realizing that $\tanh(x) \approx x$ if $|x| < 1$. The G_m of the transconductance amplifier is the small signal g_m of M_3 or M_4: if we apply $v_{in}/2$ and $-v_{in}/2$ to their respective gate terminals such that $v_s = 0$, the output current $i_{out} = g_m(v_{in}/2) - (-g_m v_{in}/2) = g_m v_{in}$. With $\kappa_S = 0.7$ and $\phi_t = 26\,\text{mV}$, the linear range of the transconductance amplifier is $75\,\text{mV}$. The linear range of $75\,\text{mV}$ is too small for several applications where we would like the power-supply voltage of a volt or a few volts to limit the dynamic range of operation of our circuit rather than the limited linear range of a transconductance amplifier. Operating in the above-threshold regime with V_L being the $(V_{GS} - V_{TS})$ of M_3 or M_4 yields more linear range but at the price of being less power efficient and the price of losing common-mode dc operating range due to larger stacked V_{DSAT} drops across transistors. It is easy to achieve wide linear range while losing common-mode dc operating range: stacking several diode-configured transistors in series in the sources of differential-pair transistors can yield a wide linear range but is not practical due to the severe loss of dc operating range resulting from accumulating voltage drops across stacked diodes. Thus, there is a need for

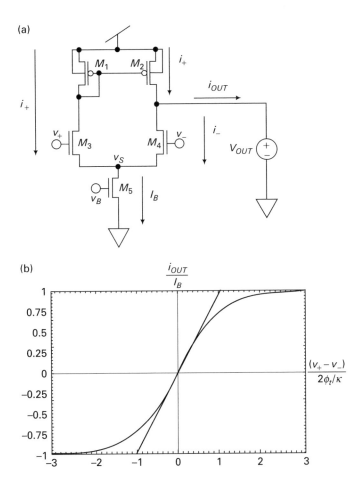

Figure 12.1a, b. (a) A simple transconductance amplifier with voltage source, and (b) and its $I-V$ tanh curve with 75 mV of linear range.

creating a low-power transconductance amplifier capable of having values of V_L that are 1 V or more in subthreshold operation without losing significant amounts of common-mode dc operating range. We shall now illustrate how linearization techniques can be used to meet this need. At first, we shall not focus on low-voltage amplifiers. Then, we shall discuss how linearization techniques can be ported to low-voltage amplifiers as well.

12.2   A low-power wide-linear-range transconductance amplifier: the big picture

To increase V_L for a given bias current I_B, we need to reduce the G_m/I_B ratio of the transconductance amplifier such that changes in its output current with changes in

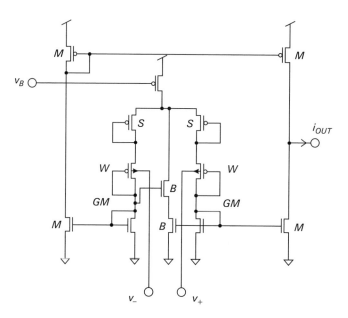

Figure 12.2. The wide-linear-range (WLR) transconductance amplifier. Reproduced with kind permission of Springer Science and Business Media from [2].

its input differential voltage are gentle and gradual. The gentler slope of the I–V curve then ensures that the saturation current I_B is reached at a larger V_L.

Figure 12.2 reveals a well-input wide-linear-range transconductance amplifier that exploits four techniques for reducing the G_m/I_B ratio, namely, the use of well inputs, source degeneration, gate degeneration, and bump linearization. These techniques are implemented in each of the two arms of the differential-pair transistors of Figure 12.2, i.e., in the W, S, GM, and B transistors respectively. The M transistors serve to mirror the currents in each differential arm to the output such that the output current is the difference in current between one differential arm and the other.

The first technique exploits the use of well inputs rather than gate inputs for reducing G_m: the W transistors in Figure 12.2 have their wells tied to the inputs of the amplifier. Since well inputs have a lower transconductance of g_{mb} for a given current than gate inputs, which have a transconductance of g_m, there is a gentler increase in current with differential voltage that reduces G_m. A second technique for reducing G_m exploits the well-known negative-feedback technique of source degeneration: any increases in current in a differential arm of the amplifier cause increases in voltage across the S transistors, which then reduce the source voltages of the pFET W transistors, thus attenuating the original current increase via negative feedback. The third technique for reducing G_m exploits a novel scheme that we term gate degeneration in analogy with source degeneration: any increases in current in a differential arm of the amplifier cause increases in voltage across the GM transistors, which then increase the gate voltages of the pFET W transistors, thus attenuating the original current increase via negative feedback.

The fourth technique for reducing G_m exploits the technique of bump linearization [1]: for small differential voltages, the transconductance of each differential arm is reduced because the B transistors steal current from them reducing the g_m of the transistors in the arms and thus the slope at the origin. At large differential voltages, the B transistors steal no current since the current flowing through these transistors in series is zero if the current in either differential arm is zero; consequently the saturation current of the differential pair, I_B, is unaffected by the B transistors. In between these extremes of differential voltage, as we move from a small differential voltage to a large one, the B transistors continually reduce their current-stealing properties as the current flowing through these transistors in series continues to reduce with the current in either differential arm reducing. If the B transistors are ratioed appropriately, the gradual return of stolen current actually serves as an expansive nonlinearity that counteracts the compressive saturation nonlinearity of differential pairs, thus serving to create a more linear characteristic than is possible with a tanh function alone. Indeed we will show that if the W/L of the bump transistors is twice that of the GM transistors, the third-order Taylor series term of a bump-linearized differential pair is ideally zero unlike that of a tanh function.

12.3   WLR small-signal and linear-range analysis

To start with, we shall analyze the operation of the WLR without the effects of bump linearization. Figures 12.3 (a), 12.3 (b), and 12.3 (c) reveal how the small-signal circuit of any pFET transistor can be reduced to a normalized block diagram where all small-signal currents are normalized by I_{DS}, all small-signal voltages are normalized by ϕ_t, and all small-signal conductances or transconductances are normalized by I_{DS}/ϕ_t. Thus, the usual un-normalized small-signal relationship of a saturated transistor leads to a normalized relationship with primed dimensionless variables:

$$
\begin{aligned}
i_{ds} &= g_m(v_s - v_g) + g_{mb}(v_s - v_w) \\
\frac{i_{ds}}{I_{DS}} &= \frac{g_m}{I_{DS}}(v_s - v_g) + \frac{g_{mb}}{I_{DS}}(v_s - v_w) \\
i'_{ds} &= \kappa_S \left(\frac{v_s}{\phi_t} - \frac{v_g}{\phi_t}\right) + (1 - \kappa_S)\left(\frac{v_s}{\phi_t} - \frac{v_w}{\phi_t}\right) \\
i'_{ds} &= \kappa_S \left(v'_s - v'_g\right) + (1 - \kappa_S)\left(v'_s - v'_w\right)
\end{aligned}
\tag{12.3}
$$

The block diagram of Figure 12.3 (c) expresses the normalized relationship between small-signal variables in Equation (12.3), but, for convenience, the primes on the variables have been dropped. Hereafter in the analysis we shall drop the primes throughout and work with variables that are understood to be normalized and dimensionless. Such variables are particularly convenient in balanced-differential-pair analysis, since the normalizing current I_{DS} is the same

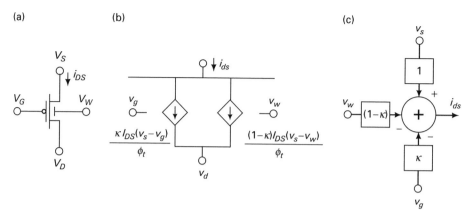

Figure 12.3a, b, c. Small-signal circuit of a transistor and its reduction into a block diagram (a), (b), and (c). Reproduced with kind permission of Springer Science and Business Media from [2].

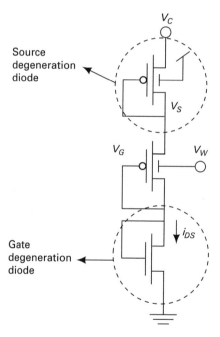

Figure 12.4. Source-degeneration and gate-degeneration diodes used in the WLR. Reproduced with kind permission of Springer Science and Business Media from [2].

in all transistors as is ϕ_t. It is because $\kappa_S > 0.5$ typically that we expect $g_{mb} < g_m$ such that we can exploit its gentler transconductance.

Figure 12.4 shows one differential arm with the source-degeneration and gate-degeneration diode-configured transistors circled. The common-mode voltage v_C has no differential effect on the small-signal currents in either arm; thus it may be assumed to be a small-signal ground for differential analysis. As in a regular differential pair, the effective transconductance of one arm of the differential pair with a grounded v_c is then the G_m of the overall amplifier. The gate-degeneration

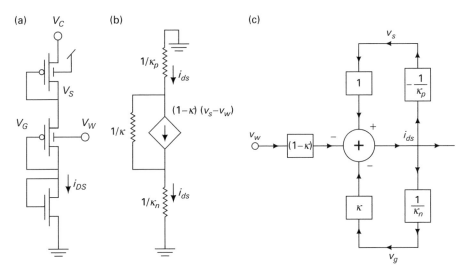

Figure 12.5a, b, c. Degeneration circuit used in the WLR (a), small-signal circuit (b), and a feedback representation (c). Reproduced with kind permission of Springer Science and Business Media from [2].

diode-configured transistor also serves a dual mirroring role in the WLR and is thus effectively free in terms of net transistor count.

Figure 12.5 (b) is a small-signal circuit model of the differential-arm circuit of Figure 12.5 (a). By using dimensionless variables for the two diode resistances and the block diagram of Figure 12.3 (c) for the W transistor, we can reduce Figure 12.5 (b) to its block-diagram equivalent shown in Figure 12.5 (c). Figure 12.5 (c) transparently reveals the negative-feedback source-degeneration and gate-degeneration loops. Since both negative-feedback loops are in parallel with a common feedforward path, the net feedback path transmission is the sum of each feedback path transmission such that by Black's formula we find that the normalized dimensionless transconductance of the amplifier $g = G_m/(I_B/2)$ is given by

$$|g| = \frac{1 - \kappa}{1 + \dfrac{1}{\kappa_p} + \dfrac{\kappa}{\kappa_n}} \tag{12.4}$$

We have used the magnitude of $|g|$ in Equation (12.4) because g is negative in each arm as increases in the well voltage reduce the current in that arm. Thus, without including the effects of bump linearization, we may conclude that

$$G_m = \frac{I_B}{2\phi_t} \left(\frac{1 - \kappa}{1 + \dfrac{1}{\kappa_p} + \dfrac{\kappa}{\kappa_n}} \right)$$

$$G_m = \frac{I_B}{V_L} \tag{12.5}$$

$$V_L = \frac{2\phi_t}{1 - \kappa} \left(1 + \frac{1}{\kappa_p} + \frac{\kappa}{\kappa_n} \right)$$

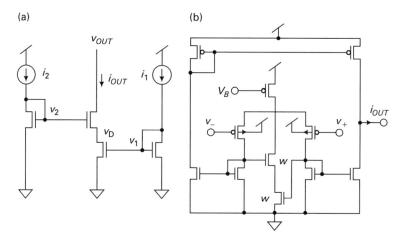

Figure 12.6a, b. Bump linearization circuit (a) and its use in an OTA with gate inputs (b).

To understand the effects of bump linearization on G_m we shall begin by understanding the bump subcircuit of Figure 12.6 (a) and its use in a pFET OTA in Figure 12.6 (b). From Figure 12.6 (a), we find that

$$i_{OUT} = e^{\kappa v_1/\phi_t} \left(e^{-0} - e^{-v_D/\phi_t} \right)$$
$$i_{OUT} = e^{\kappa v_2/\phi_t} \left(e^{-v_D/\phi_t} - e^{-v_{OUT}/\phi_t} \right)$$

(12.6)

By adding the two expressions of (12.6), we find that

$$i_{OUT} \left(\frac{1}{e^{\kappa v_2/\phi_t}} + \frac{1}{e^{\kappa v_1/\phi_t}} \right) = 1 - e^{-v_{OUT}/\phi_t}$$

$$\approx 1 \quad (\text{if } v_{OUT} \geq 4\phi_t)$$

(12.7)

Finally,

$$\frac{1}{i_{OUT}} = \frac{1}{i_1} + \frac{1}{i_2}$$
$$i_{OUT} = \frac{i_1 i_2}{i_1 + i_2}$$

(12.8)

By using telescopic series summation in Equations (12.6) and (12.7), the relationships of Equation (12.8) can be generalized to show that if we have N transistors in series and if $v_{OUT} > 4\phi_t$, then the output current of such a stack is the parallel of all input currents with

$$\frac{1}{i_{out}} = \frac{1}{i_1} + \frac{1}{i_2} + \frac{1}{i_3} + \cdots + \frac{1}{i_N}$$

(12.9)

The bump circuit actually exploits the exponentials normally present in the linear region of operation of the transistor such that only the final voltage v_{OUT} needs to be greater than $4\phi_t$ for it to function. Thus, it allows very-low-voltage operation even in a large series stack.

In Figure 12.6 (b), the bump transistors have been ratioed to have a W/L that is w times larger than those of their diode-connected input transistors. The derivation below then reveals that

$$i_{OUT} = i_+ - i_-; i_B = i_+ + i_- + w\frac{i_+ i_-}{i_+ + i_-}; i_+ = e^{\frac{\kappa v_+}{\phi_t}}; i_- = e^{\frac{\kappa v_-}{\phi_t}}$$

(12.10)

$$\boxed{i_{OUT} = \frac{\sinh x}{\beta + \cosh x} \text{ where } \beta = 1 + \frac{w}{2} \text{ and } x = \frac{\kappa(v_+ - v_-)}{\phi_t}}$$

If $w = 2$ such that $\beta = 2$ as well, then i_{OUT} is maximally linear in that its Taylor series expansion with x has no third-order term:

$$\frac{\sinh x}{2 + \cosh x} = \frac{x}{3} - \frac{x^5}{540} + \frac{x^7}{4536} - \frac{x^9}{77760} + \cdots$$

(12.11)

In comparison, a conventional differential pair without the bump transistors would have an output current described by

$$\tanh\frac{x}{2} = \frac{x}{2} - \frac{x^3}{24} + \frac{x^5}{240} - \frac{17x^7}{40320} + \cdots$$

(12.12)

By comparing the coefficients of the linear term in Equations (12.11) and (12.12), we find that a bump-linearized differential pair with $w = 2$ has lower transconductance by a factor of $3/2$ at the origin and is also significantly more linear than a tanh function. The lowered factor for transconductance is intuitive since at $w = 2$, the bump arm and each of the two differential arms each have a current of $I_B/3$ rather than $I_B/2$ in a differential pair.

If w is extremely large, the transconductance at the origin reduces considerably (by a factor of $4/(4 + w)$) but the saturation current is unchanged such that the I–V curve of the circuit of Figure 12.6 (b) exhibits a flat expansive dead zone at the origin followed by a compressive return to output saturation. If w is small, the transconductance is attenuated throughout without any changes from expansivity to compressivity. At $w = 2$, the expansive nature of the curve at the origin is exactly compensated for by the compressive nature of saturation such that maximally linear operation is achieved.

The net result of bump linearization at $w = 2$ is that the effective linear range is increased by $3/2$. This linear range increase is also seen in the WLR circuit of Figure 12.2, such that the overall linear range is given by modifying Equation (12.5) by a factor of $3/2$:

$$V_L = \frac{3}{2} \times \frac{2\phi_t}{1 - \kappa}\left(1 + \frac{1}{\kappa_p} + \frac{\kappa}{\kappa_n}\right) = \frac{3\phi_t}{1 - \kappa}\left(1 + \frac{1}{\kappa_p} + \frac{\kappa}{\kappa_n}\right)$$

(12.13)

Figure 12.7 (a) reveals experimental data for a WLR circuit for various different values of w. We see that, at $w = 0$, the I–V curve of the WLR is well fit by a tanh curve with $V_L = 1.16$ V. Even large w's only cause gentle distortion suggesting that achieving $w = 2$ exactly is not critical and consequently that the WLR circuit is well tolerant of mismatch in w. Figure 12.7 (b) reveals that the $3/2$ increase in

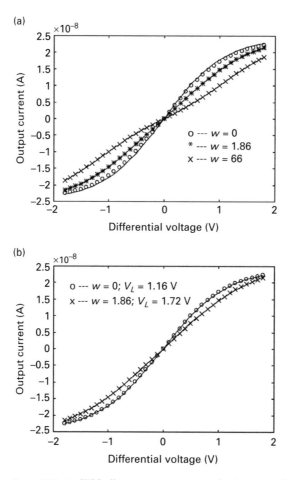

Figure 12.7a, b. Wide-linear-range transconductor: experimental $I-V$ data (a) and (b). Reproduced with kind permission of Springer Science and Business Media from [2].

linear range due to bump linearization is such that the $I-V$ curve is still well fit by a tanh except that V_L is increased to nearly 1.72 V.

It is worth noting that the greater-than-23-fold improvement in linear range from 75 mV to nearly 1.72 V does not arise from one effect but from a synergistic combination of several effects, each of which contributes to increasing V_L as shown in Equation (12.13). *In general, distributing improvements over multiple dimensions rather than in only one dimension is almost always advantageous in circuit design since extreme tradeoffs in any one dimension become unnecessary and do not compromise performance.*

12.4   WLR dc characteristics

Figure 12.8 (a) reveals the device topology of a pFET. From this topology, we see that every MOSFET actually has two parasitic vertical bipolar transistors at its

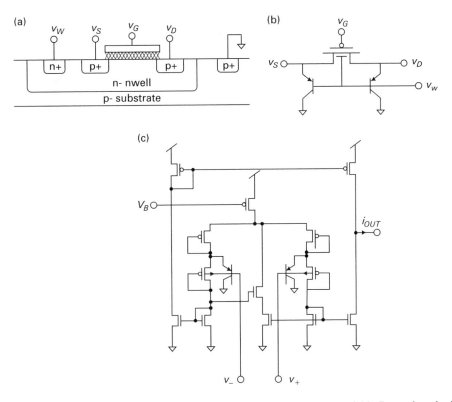

Figure 12.8a, b, c. Parasitic BJT's present inside the WLR (a), (b), and (c). Reproduced with kind permission of Springer Science and Business Media from [2].

source and drain as illustrated in Figure 12.8 (b). These bipolar transistors are normally reverse biased and turned off since the well is usually at the highest voltage, the power-supply rail. Since we use the well as an input, however, there is a possibility that these devices may be turned on. The possibility of turning on a bipolar transistor is significantly higher at the source since the source is at a higher voltage than the drain. Figure 12.8 (c) shows a bipolar transistor that could potentially turn on and shunt current away from the arms of the differential pair thus lowering its output current and transconductance. The turning on of parasitic bipolar transistors in CMOS chips is often viewed with alarm because of the potential for latchup. In the WLR circuit, however, even if the bipolar transistor turns on, the current through it is limited to be the tail current of the differential pair, a small value in low-power circuits. Thus, as long as a ground contact is not too distant from the W transistor's well, the few micro-amps of current that may be directed to the substrate by the turn on of the bipolar transistor is not of much cause for concern. Latchup is therefore not expected, and was not experimentally observed.

For the parasitic bipolar transistor in either differential arm in Figure 12.8 (c) to turn on, its emitter-to-base voltage must exceed V_{bip}. The emitter voltage of the

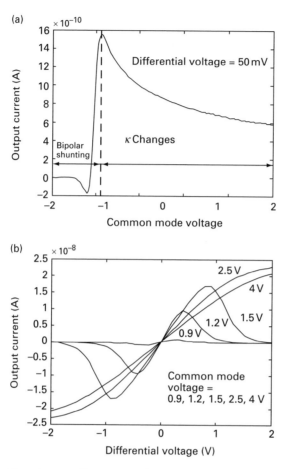

Figure 12.9a, b. Common-mode characteristics of the WLR (a) and (b). Reproduced with kind permission of Springer Science and Business Media from [2].

bipolar transistor is approximately at $2V_{DM}$, where V_{DM} represents the subthreshold bias voltage dropped across the diode-connected *GM* or the diode-connected *W* transistor (see Figure 12.2). Hence, from Figure 12.8 (c), we can predict that the current in an arm of the WLR is shunted by the bipolar if the well-input voltage of that arm is less than $2V_{DM} - V_{bip}$. Figure 12.9 (a) experimentally confirms that, when the dc common mode voltage is sufficiently low, the output current for a fixed differential voltage of 50 mV begins to fall. In the $2\,\mu$m process in which these data were collected, the threshold voltages and consequently subthreshold diode drops are fairly high such that the shunting appears to begin for voltages less than about 1 V. In more advanced processes with lower threshold voltages, higher doping, and larger bipolar turn-on voltages, bipolar shunting action occurs at lower voltages. The fall in current with rising common-mode voltage in Figure 12.9 (b) occurs because larger well-to-gate voltages increase the width of the depletion region, lower κ, increase V_L according to Equation (12.13), lower the G_m and thus the output current for a fixed differential voltage. Figure 12.9 (b) reveals that the G_m is systematically lower at higher common-mode voltages and results in more linear behavior. Figure 12.9 (b) also confirms the findings of Figure 12.9 (a): if the common-mode voltage is sufficiently high, the shunting bipolar transistor does not turn on for any value of differential voltage. As the common-mode voltage is lowered, the shunting bipolar transistor turns on in one arm or the other when the differential voltage is of sufficient magnitude such that the well-input voltage in that arm falls below $2V_{DM} - V_{bip}$. For positive differential voltages, one arm manifests bipolar shunting, and for negative differential voltages, another arm manifests shunting such that a symmetric I–V characteristic is observed. Quantitative predictions with regard to the change in κ with well-to-gate voltage and with regard to the turn-on voltage at which bipolar shunting begins are presented in more detail in [2] and were found to be consistent with experimental measurements.

If gate degeneration is used, as in Figures 12.9 (a) and 12.9 (b), the dc common-mode operating range of the WLR extends from the positive rail to within $2V_{DM} - V_{bip}$ of ground, where V_{DM} is a subthreshold diode drop and V_{bip} is a parasitic bipolar turn-on voltage. If the gate is grounded and gate degeneration is not used, the dc common-mode input operating range is nearly rail to rail, i.e., from the positive rail to within $(V_{DM} - V_{bip})$ of ground. Not using gate degeneration reduces the linear range by about 2/3, but with bump linearization the overall linear range is still nearly 1.2 V suggesting that such a strategy may be a good compromise in some applications. Thus, the linear range improvement in the WLR is obtained without compromising dc input operating range significantly.

If all techniques of linearization are used, the power supply needs to be at least $3V_{DM} + V_{DSAT} = 3V_{DM} + 4\phi_t$. If the gate is grounded and gate degeneration is not used, the power-supply voltage needs to be at least $2V_{DM} + 2V_{DSAT} = 2V_{DM} + 8\phi_t$. Thus, relatively low-power-supply voltage operation, which also implies that fewer techniques of linearization may be needed, is possible. In cases of very-low-power-supply voltage, source degeneration and gate degeneration may not be needed since it makes no sense to have V_L exceed V_{DD}, thus improving dc operating range. In low-voltage processes, threshold voltages are lower as well, such that the subthreshold diode drop V_{DM} is reduced. Figure 12.19 (a) shows a WLR topology suited for low-voltage implementations, which we shall discuss in a later section.

Figure 12.10 (a) shows that the saturation current I_B of the WLR can range over a few orders of magnitude as it ranges from deep subthreshold operation to above-threshold operation. The linear range V_L, which was measured with and without bump linearization and plotted in Figure 12.10 (b), is constant in the subthreshold regime and rises in above-threshold operation as the $(V_{GS} - V_{TS})$ overdrive voltage increases. Linear ranges in excess of 2.5 V in above-threshold operation were observed.

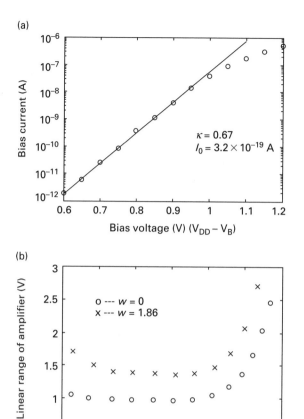

Figure 12.10a, b. Bias current (a) and linear range (b) characteristics of the WLR. Reproduced with kind permission of Springer Science and Business Media from [2].

The dc gain A_V of the WLR can be computed from the Early voltages of the output transistors of the WLR, defined as V_O, which set its output resistance, and by G_m. It is invariant of whether bump linearization is present (A_{VB}) or absent (A_V) as shown below.

$$V_O = V_O^P || V_O^N$$

$$A_V = \frac{I_B/V_L}{I_B/2V_O} = \frac{2V_O}{V_L}$$

(12.14)

$$A_{VB} = \frac{I_B/(3V_L/2)}{I_B/3V_O} = \frac{2V_O}{V_L}$$

Figure 12.11 (a) shows that the dc gain falls with rising common-mode voltage due to the fall in G_m with common-mode voltage shown in Figure 12.9 (b).

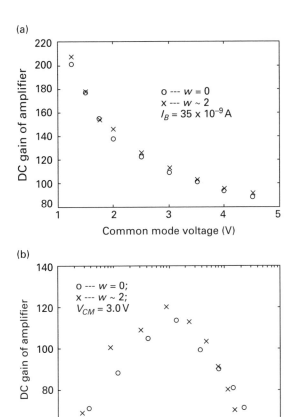

Figure 12.11a, b. The dc gain of the WLR (a) and (b). Reproduced with kind permission of Springer Science and Business Media from [2].

Figure 12.11 (b) shows that the dc gain peaks in moderate inversion: as $|V_{GS}|$ increases within subthreshold operation, the gate exerts stronger control of the field lines near the drain such that V_O and consequently the dc gain rise. As we move from subthreshold operation to above-threshold operation, V_L rises more quickly than V_O causing the dc gain to fall.

Figure 12.12 illustrates how the dimensionless block diagram of Figure 12.5 (c) is useful for analyzing offsets in the WLR. Since all currents are normalized to I_{DS}, we can represent mismatches in I_{DS} due to transistor geometry and threshold-voltage variations in a differential arm as small-signal deviations with a fractional and dimensionless mismatch ratio of $\Delta I_{DS}/I_{DS}$. Thus, in the dimensionless block diagram of Figure 12.12, they automatically appear as small-signal disturbances in the feedback loop of the circuit. The mismatch between various transistors in the WLR then creates disturbances at various points in the feedback loop such that the same value of disturbance can result in a different input-referred offset

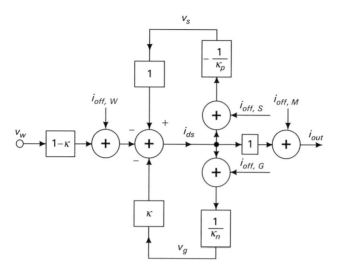

Figure 12.12. Offset and mismatch feedback loop block diagram of the WLR.

depending on where it occurs. Thus, from Figure 12.12, we find the input-referred offset from the block diagram to be

$$
\frac{V_{OFFS}}{\phi_t} = \frac{\left(1+\dfrac{1}{\kappa_p}+\dfrac{\kappa}{\kappa_n}\right)}{(1-\kappa)}\frac{\Delta I_M}{I_M} + \frac{1}{\kappa_p(1-\kappa)}\frac{\Delta I_S}{I_S} + \frac{\kappa}{\kappa_n(1-\kappa)}\frac{\Delta I_G}{I_G} + \frac{1}{(1-\kappa)}\frac{\Delta I_W}{I_W}
$$

$$
V_{OFFS} = \phi_t\left[\frac{\left(1+\dfrac{1}{\kappa_p}+\dfrac{\kappa}{\kappa_n}\right)}{(1-\kappa)}\frac{\Delta I_M}{I_M} + \frac{1}{\kappa_p(1-\kappa)}\frac{\Delta I_S}{I_S} + \frac{\kappa}{\kappa_n(1-\kappa)}\frac{\Delta I_G}{I_G} + \frac{1}{(1-\kappa)}\frac{\Delta I_W}{I_W}\right]
$$

(12.15)

Equation (12.15) reveals that a given percentage mismatch in the M mirror transistors of Figure 12.2 results in the highest input-referred offset since the full effect of the G_m reduction is referred back to the input. A given percentage mismatch in the W transistors results in the least input-referred offset. Source degeneration results in slightly more input-referred offset than gate degeneration. Thus, if we are to size the WLR for low offset, we must size the M transistors to have the largest area, then the S transistors, then the GM transistors, and then the W transistors. Thus, the well-input transistors, which may appear to be the most important transistors in the WLR, are actually the least important as far as offset is concerned. In fact, we could use the coefficient ratios in Equation (12.15), i.e., the various offset gains, to create sizing ratios for the respective transistor pairs in the WLR. It is worth noting that increases in V_L in any amplifier will always result in more input-referred offset in general. Thus, amplifiers with a large V_L must be built with larger transistors if the absolute offset performance is to remain the same as that of amplifiers with small V_L. If the sizing is maintained the same across

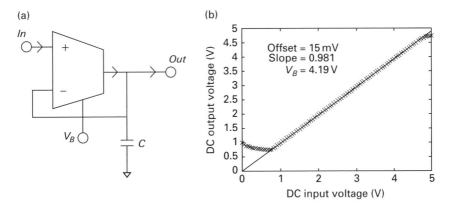

Figure 12.13a, b. (a) A $G_m - C$ integrator circuit and (b) its dc characteristics. Reproduced with kind permission of Springer Science and Business Media from [2].

low-V_L and high-V_L designs, then the relative offset performance, i.e., V_{OFFS}/V_L can be expected to be similar if the transistor count has not altered greatly between the two designs: intuitively, because V_{OFFS} and V_L are both proportional to $1/G_m$, V_{OFFS}/V_L is invariant with G_m or equivalently, with V_L.

12.5 Dynamic characteristics of the WLR

Figure 12.13 (a) shows a $G_m - C$ follower-integrator or first-order lowpass filter circuit constructed with the WLR. Figure 12.13 (b) reveals that, as expected, its dc characteristics are such that the output follows the input except at small values of the input where bipolar shunting is present. Figure 12.14 (a) shows that while the lowpass filter gain characteristic for an audio filter constructed with this $G_m - C$ filter is ideal for two orders of magnitude beyond its corner frequency, the phase characteristic in Figure 12.14 (b) exhibits the effects of parasitic poles and zeros just an order of magnitude past the filter's corner frequency. Such behavior is normal since phase transfer characteristics begin to exhibit noticeable effects an order of magnitude before a parasitic pole or zero frequency location while magnitude transfer characteristics begin to exhibit noticeable effects at the parasitic pole or zero frequency location. In this case, the parasitic dynamics can be attributed to an RHP zero due to the well-to-drain capacitance of the W transistor and an LHP pole due to the gate-to-bulk capacitance of the GM transistor.

12.6 Noise analysis

The noise analysis of the $G_m - C$ filter is similar to that of the $G_m - C$ filter described in Chapter 8. If we define α_B, α_S, α_W, α_G, and α_M to be the current gain

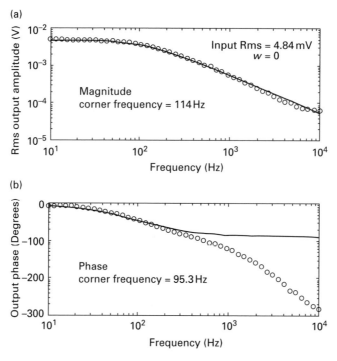

Figure 12.14a, b. A simple $G_m - C$ follower-integrator built with the WLR: (a) schematic, (b) gain response, and (c) phase response. Reproduced with kind permission of Springer Science and Business Media from [2].

between the input drain-to-source noise-current generator of a B, S, W, G, or M transistor of the small-signal circuit of Figure 12.2 and the short-circuit output current of the amplifier respectively, then we can show that

$$\alpha_B = 0$$

$$\alpha_M = 1$$

$$\alpha_S = \frac{\kappa_n}{\kappa_n + \kappa_n \kappa_p + \kappa \kappa_p}$$

$$\alpha_W = \frac{\kappa_n \kappa_p}{\kappa_n + \kappa_n \kappa_p + \kappa \kappa_p} \tag{12.16}$$

$$\alpha_G = \frac{\kappa_n + \kappa_n \kappa_p}{\kappa_n + \kappa_n \kappa_p + \kappa \kappa_p}$$

When deriving these results, it is useful to note that the output current for any given noise generator is a function of the common node voltage v_c, indicated explicitly in Figure 12.4, and the noise-current generator by superposition. However, in a balanced small-signal circuit where each half is symmetric, the transfer function from v_c to the output current is zero. Thus although v_c will itself be a function of the noise generator, we may assume that it is zero when performing the noise analysis. Another useful trick when computing current transfer functions is to replace floating current sources by two grounded current sources, one being a sink from which the floating current source originates, and the other being a source to which the floating current source flows. The transfer function α_B is zero because all common-mode currents that divide symmetrically through both arms of the amplifier yield no output current such that the B transistors and the bias-current transistor contribute no noise current.

Thus, the effective number of noise sources in the amplifier is not 13, due to there being 13 transistors in the amplifier but given by N where

$$N = 2\alpha_S^2 + 2\alpha_W^2 + 2\alpha_G^2 + 4\alpha_M^2 \tag{12.17}$$

For $\kappa = 0.85$, $\kappa_n = 0.7$, and $\kappa_p = 0.75$, values that are consistent with linear range and other experimental measurements, $N = 5.3$. The mirror transistors alone contribute a value of 4 to N while the differential-pair transistors, whose noise is largely filtered by the feedback topology of the circuit, only contribute a value of 1.3 to N. If we replace the offset generators in Figure 12.12 with thermal-noise generators, then this figure reveals how feedback attenuation of the noise in the differential pair occurs. Certain mirrors with feedback in them contribute significantly less noise, e.g., series-diode-stacked cascode mirrors and Wilson mirrors contribute approximately 1 device worth of noise even though they may be built with 4 transistors. Feedback attenuation of the noise in transistors is also a manifestation of return-ratio-based attenuation of g_m variations in transistors (see Chapter 10).

If we model flicker noise with

$$\overline{i_{flicker}^2} = \frac{K_f(I_{Bias})^2}{f}\Delta f \tag{12.18}$$

then we can define the total output current noise to be

$$\overline{i_o^2} = \int_{f_l}^{\infty} N\left(2q\left(\frac{I_B}{2}\right) + \frac{K_f(I_B/2)^2}{f}\right) df \tag{12.19}$$

We assume that there is low-frequency adaptation in the amplifier, so frequencies below f_l, typically around 1 Hz, are not passed through. This assumption is necessary if we are to prevent the $1/f$ noise from growing without bound at low frequencies. As we have pointed out in Chapter 8, the logarithmic scaling in $1/f$ noise with f_l implies that the value of f_l is not too critical. The corner frequency of our filter is given by

$$f_c = \frac{1}{2\pi} \cdot \frac{G_m}{C} = \frac{I_B}{2\pi C V_L} \tag{12.20}$$

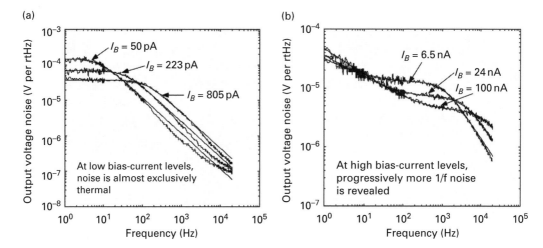

(a)

(b)

Figure 12.15a, b. Experimental WLR noise curves: (a) at low bias current and (b) at high bias current. Reproduced with kind permission of Springer Science and Business Media from [2].

To complete our noise calculations, two integrals will prove useful:

$$\int_0^\infty \frac{dx}{1+x^2} = \frac{\pi}{2}$$

$$\int_{f_l}^\infty \frac{dx}{x(1+x^2)} = \frac{1}{2}\ln(1+\frac{1}{f_l^2})$$

(12.21)

Thus the total output voltage noise of our filter is given by

$$\overline{v_i^2} = \int_{f_l}^\infty \frac{N\left(qI_B + \frac{K_f I_B^2}{4f}\right)}{(I_B/V_L)^2}\left(\frac{1}{1+(f/f_c)^2}\right)df$$

$$= \left(\frac{NqV_L^2}{I_B}\right)\frac{\pi f_c}{2} + \frac{NK_f V_L^2}{8f}\ln\left(1+(f_c/f_l)^2\right)$$

(12.22)

$$= \frac{NqV_L}{4C} + \frac{NK_f V_L^2}{8}\ln\left(1+\left(\frac{I_B}{2\pi f_l C V_L}\right)^2\right)$$

Figure 12.15 (a) illustrates that at low bias current values, the noise is mostly thermal and independent of I_B since the input-referred noise per unit bandwidth and the bandwidth scale like $1/I_B$ and I_B respectively to keep the total noise, i.e., the area under all curves in Figure 12.15 (a), constant. Figure 12.15 (b) reveals that at higher bias current values, progressively more $1/f$ noise is revealed as predicted by Equation 7.39 in Chapter 7. Nevertheless, we found

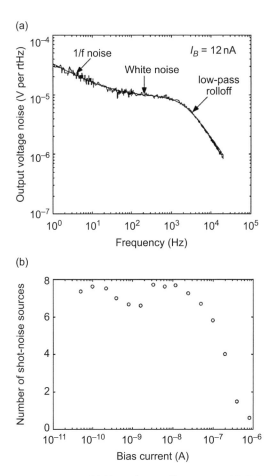

Figure 12.16a, b. (a) Experimentally measured WLR noise fit and (b) equivalent number of shot-noise sources. Reproduced with kind permission of Springer Science and Business Media from [2].

that even for large bias currents, the area under the $1/f$ noise portions on a linear scale was such that thermal noise was dominant. Figure 12.16 (a) reveals that the noise power spectra are well fit by the theory of Equation (12.22). Figure 12.16 (b) reveals that the equivalent number of shot-noise sources in the amplifier was measured to be $N = 7.0$ rather than $N = 5.3$ as predicted by theory in subthreshold. The equivalent number of shot-noise sources reduces above threshold in Figure 12.16 (b) because the drift current begins to significantly exceed the diffusion current. The diffusion current causes the noise such that it appears that the number of shot-noise sources have been reduced, i.e., we observe $2qNI_{diff}$ for our power spectral density, not $2qNI_{drft}$, so it appears that we are observing $2qN_{eq}I_{drft}$ with $N_{eq} = NI_{diff}/(I_{diff} + I_{drft})$. Chapter 7 discusses this issue in more depth (see Equation 7.43 and the discussion preceding it and after it).

12.7   Distortion analysis

For the WLR to remain linear in a first-order G_m–C filter, we must require its differential input voltage to be less than V_L, i.e.,

$$\left| V_{in}(s) \left(1 - \frac{1}{\tau s + 1} \right) \right| < V_L$$

$$V_{in}(s) < \left| \frac{\tau s + 1}{\tau s} \right| V_L \tag{12.23}$$

$$|V_{in}(j\omega)| < V_L \sqrt{1 + \frac{1}{\omega^2 \tau^2}}$$

Thus, at very low frequencies, the input amplitude to the filter can be significantly above its linear range without much distortion since its v_+ and v_- inputs will be near each other due to negative feedback as in an operational amplifier. As we near the corner frequency of the filter and pass it, the input amplitude required to prevent distortion asymptotes at the linear range of the transconductor.

Figure 12.17 shows the distortion characteristics of our filter, obtained by measuring the fundamental-frequency amplitude, second-harmonic amplitude, and third-harmonic amplitude versus frequency for an input sinusoid with an amplitude that is nearly $V_L = 1\text{V}$. The measurements were obtained with a spectrum analyzer. Bump linearization was not present in this WLR. We observe that as the input frequency rises and nears the corner frequency of the filter, the relative ratios of the harmonic amplitudes with respect to the fundamental rise. These ratios are then nearly constant past the corner frequency for an order of magnitude in frequency until the parasitic poles and zeros begin to exert their effects. The distortion is dominated by the second harmonic due to nonlinearities in the depletion capacitance which alters κ and thus the V_L and G_m of the amplifier.

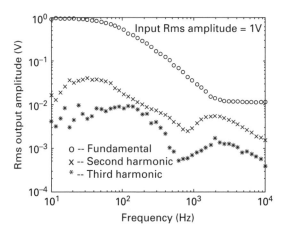

Figure 12.17. Experimentally measured WLR $G_m - C$ filter distortion characteristics. Reproduced with kind permission of Springer Science and Business Media from [2].

A fully differential implementation of the WLR significantly attenuates second-harmonic distortion. Thus, third-harmonic distortion is dominant in fully differential implementations.

From Equation (12.23), it is natural to define the maximum undistorted input signal to the $G_m - C$ filter to be a signal with an amplitude that generates a differential voltage across the WLR of V_L at its corner frequency, i.e., the input amplitude is $\sqrt{2}V_L$. At the corner frequency, the output signal is then attenuated to an amplitude of V_L and thus has an r.m.s. output power of $V_L^2/2$. Beyond this corner frequency, we are not really too interested in the output signal power, since presumably the action of the filter is to remove unwanted signals past the corner frequency. Thus, we can expect the maximum signal power available at the output to be $V_L^2/2$ if we are to operate in a regime without too much distortion.

12.8 Signal-to-noise ratio and power analysis

From our analysis of distortion and noise in the filter, we can expect the maximum signal-to-noise ratio of output power to noise power in the filter to be given by

$$
\begin{aligned}
SNR &= \frac{V_L^2/2}{v_{noise}^2} \\[2mm]
&= \frac{V_L^2/2}{\dfrac{NqV_L}{4C} + \dfrac{NK_f V_L^2}{8}\ln\left(1 + \left(\dfrac{I_B}{2\pi f_l C V_L}\right)^2\right)} \\[2mm]
&= \frac{1}{\dfrac{Nq}{2CV_L} + \dfrac{NK_f}{4}\ln\left(1 + \left(\dfrac{I_B}{2\pi f_l C V_L}\right)^2\right)}
\end{aligned}
\tag{12.24}
$$

In this linear system, the dynamic range of the filter, i.e., the ratio of the power in the maximal undistorted input to the minimum detectable signal, computed at the corner frequency, where the *SNR* of Equation (12.24) has been computed as well, are identical. In a nonlinear system with gain control, the dynamic range can greatly exceed the *SNR* because the system adapts its gain to handle a wide range of input levels. Thus, at any given gain setting, the *SNR* and instantaneous dynamic range are identical, but over its whole range of gain settings, the dynamic range greatly exceeds the *SNR*. For example, an oscilloscope may have a maximal signal-to-noise ratio corresponding to 4 orders of magnitude at any given gain setting (e.g. a 1 mV noise floor on a 100 mV full scale), but may be able to handle 8 orders of magnitude in dynamic range (1 mV to 10 V inputs) over its entire range of gain settings. In the biological cochlea, the input dynamic range corresponds to 12 orders of magnitude in intensity but the output signal-to-noise ratio as measured by maximal and spontaneous noise-induced firing rates for any given auditory nerve fiber is only 2 orders of magnitude in intensity [3]. The cochlea adapts its gain with sophisticated automatic-gain-control circuitry such that we do not have to manually turn a knob to sense soft signals versus loud signals as we do in an oscilloscope.

We find experimentally that the white-noise term dominates over the $1/f$ noise term in determining the output *SNR* of our filter because the use of large mostly pFET transistors reduces the value of K_f, because white noise is often dominant in ultra-low-power subthreshold circuits, and because frequencies below 1 Hz are filtered out by offset-adaptation circuits. Thus, if we only keep the white-noise term in Equation (12.24), we find that

$$SNR = \frac{2CV_L}{Nq} \qquad (12.25)$$

Thus, *if white noise dominates our system, more linear range implies more dynamic range.* If the bias current of the amplifier is I_B, the bandwidth of the filter f_c is given by

$$f_c = \frac{I_B}{2\pi CV_L} \qquad (12.26)$$

We can then compute the power needed to attain a given bandwidth and *SNR* to be

$$P = 2V_{DD}I_B$$
$$P = 2V_{DD}(2\pi f_c V_L) \qquad (12.27)$$
$$\boxed{P = (V_{DD}\pi Nq)(f_c)(SNR)}$$

Thus, we see that it costs power to attain precision (SNR) *and bandwidth, a lesson true in all circuits.* The cost of getting a better signal-to-noise ratio with a wider linear range amplifier V_L is not free. The wider linear range improves signal-to-noise ratio because the maximal signal power scales like V_L^2 while the noise power scales like V_L such that there is a net improvement in *SNR* that scales like V_L. But the reason that the noise power only scales like V_L even though the input-referred noise per unit bandwidth scales like V_L^2 is that the bandwidth f_c which scales like I_B/V_L is reduced thus leading to more averaging. Thus, if I_B is kept constant between two designs, one of which has a large V_L and the other of which has a small V_L, the large-V_L design will have large *SNR* and a small bandwidth while the small-V_L design will have small *SNR* and large bandwidth. Thus, to maintain bandwidth at the value possessed by a small-V_L design, we need to increase I_B proportionately with V_L in a large-V_L design. The bandwidth increase achieved by increasing I_B does not increase the noise and maintains the *SNR* at that given by Equation (12.25). Though there is less averaging in the latter case, the input-referred noise per unit bandwidth falls such that the total output noise is only determined by the capacitance and topology and independent of I_B as we've seen on multiple occasions.

Figure 12.18 (a) illustrates that, in a thermal-noise-limited design, if V_L is increased, the noise per unit bandwidth increases proportionately with V_L^2 and

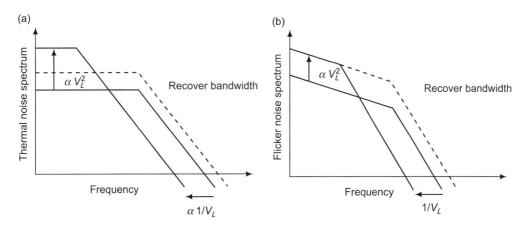

Figure 12.18a, b. Bandwidth recovery by burning more power: (a) the case for thermal noise and (b) $1/f$ noise.

the bandwidth scales with $1/V_L$ such that the total output noise scales with V_L. When we increase I_B to restore the bandwidth to its value before V_L is increased the noise power does not change because it is independent of I_B. The net improvement in *SNR* arises because the output signal power scales like V_L^2 while the noise scales like V_L. When we are all done, our bandwidth is as large as it was at a low value of V_L, but we have a better *SNR*, and we have achieved an overall improvement in performance at the cost of consuming more power. Thus, there is no free lunch. Attaining higher speed (bandwidth) and higher precision (*SNR*) always means more power.

Figure 12.18 (b) repeats the graphical analysis of Figure 12.15 (a) for the case where we are $1/f$-noise dominated. The total noise in this case is given by

$$\overline{v_i^2} \approx \frac{NK_fV_L^2}{8}\ln\left(1 + \left(\frac{I_B}{2\pi f_l CV_L}\right)^2\right) \qquad (12.28)$$

and increases like V_L^2 if we neglect the logarithmic change in noise due to a bandwidth reduction. If we increase I_B to restore our bandwidth to its low-V_L value, the *SNR* scales like V_L^2/V_L^2 and is unchanged. *Thus, if we are $1/f$-noise-limited, increased linear range implies no improvement in SNR.* We simply spend more power to maintain bandwidth at a large V_L but this power is wasted: it does not yield an *SNR* improvement.

12.9 Scaling laws for power in analog circuits

Similar analyses of noise and signal-to-noise ratio in other analog circuits presented in a more general context in Chapter 22 yield an equation very similar to that of Equation (12.27) for our $G_m - C$ filter:

$$P_{GmC} = (V_{DD}\pi Nq)(f_c)(SNR)$$
$$P_{ana} = (V_{DD}qC_{tchngy})(N_{tskcmplx})(N_{tpgy})(f_c)(SNR)$$

(12.29)

In all analog circuits the power P_{ana} depends on technology constants like V_{DD} and q, the mobility and f_T, on the bandwidth f_c (or speed), and on the SNR (or output precision or bit fidelity). The value of N in the $G_m - C$ filter is a measure of the complexity of the task of filtering and the number of devices needed to implement it in our topology. In general, this value of N will be different for circuits that are performing tasks of varying complexity, e.g., A-to-D conversion, energy extraction, current-to-voltage conversion, amplification, demodulation etc. It will also be different depending on whether such tasks are implemented with intelligent low-noise topologies with good signal-versus-noise amplification and a good mapping of the task to the transistors of the circuit or whether such tasks have been implemented with a topology that is noisy, has poor signal-versus-noise amplification, and has not been well mapped to the transistors of the circuit. The terms $N_{tskcmplx}$ and N_{tpgy} decompose N into the product of two terms that attempt to abstract these terms as preconstants in the bandwidth-SNR or speed-precision scaling law of Equation (12.29). Similarly the term C_{tchngy} abstracts other technology preconstants that may be relevant in other tasks and topologies.

In a thermal-noise-limited A-to-D converter, the SNR term of Equation (12.29) would be parametrized as 2^{2N}, where N is the output bit precision, or as 2^N in a non-thermal-noise-limited version. In an amplifier, f_c would be parameterized by GBW, the gain-bandwidth product. The SNR may be an implicit specification, for example because the input-referred noise and maximum input to the analog system are quoted, or because the input dynamic range is quoted and the system is linear so the SNR is the dynamic range. As long as the f_T of the topology is significantly in excess of the carrier frequency in RF circuits, it is the baseband bandwidth that parameterizes the f_c term of Equation (12.29).

The take-home message is that there are five determinants of power as shown by the low-power hand in Figure 1.1 in Chapter 1: the task complexity, the topology, the technology, the bandwidth or speed, and the output signal-to-noise ratio or output precision. The speed-precision scaling law of Equation (12.29) captures these determinants of power for analog systems. For digital systems, the scaling law is similar except that the SNR term is replaced by a term that is a polynomial function of $\log_2(1 + SNR)$ as we discuss in Chapter 22 on mixed-signal design.

12.10   Low-voltage transconductance amplifiers and low-voltage analog design

The transconductance amplifier of Figure 12.2 needs a minimum V_{DD} of three diode drops plus a saturation voltage of $4kT/q$ to operate with good power efficiency. In deep submicron processes, V_{DD} is lower, e.g., it can often be as low as 0.5 V, such that the design of Figure 12.2 is not feasible. Fortunately, in such

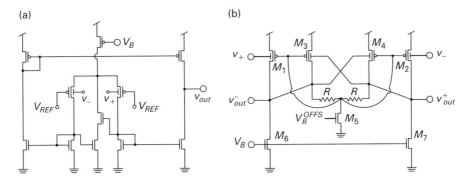

Figure 12.19a, b. A low-voltage subthreshold transconductance amplifier is shown in (a) and a low-voltage above-threshold transconductance amplifier is shown in (b). Reproduced with kind permission from [4] (©2005 IEEE).

processes, a linear range that exceeds V_{DD} is not needed: a linear range of $V_{DD}/2$ provides rail-to-rail operation with a common mode voltage of $V_{DD}/2$. Such a linear range can be obtained by using a subset of the linearization techniques described in this chapter rather than by using all of them. Figure 12.19 (a) reveals a subthreshold design suited for low-voltage operation and Figure 12.19 (b) reveals an above-threshold design suited for low-voltage operation [4]. Before we describe either design, we shall briefly digress to discuss seven general principles for low-voltage analog design. We shall then show how these principles manifest themselves in the particular circuits of Figure 12.19 (a) and 12.19 (b).

1. Since each transistor in a series stack causes a loss in saturated operating range of at least V_{DSAT} (V_{GS} if the transistor is diode connected), *series stacking of transistors should be minimized* in low-voltage designs.
2. *Weak-inversion operation* minimizes V_{DSAT}; therefore, unless high-speed requirements do not prohibit its use, it is preferable in low-voltage operation.
3. It is advantageous to not set the well-to-source voltage $v_{WS} = 0$ but to *use v_{WS} as a degree of freedom in low-voltage design*: the well can serve to modulate the dc biasing current while the gate serves as the primary ac input, or vice versa. The use of two control inputs rather than one is always advantageous as we have seen in the example of gate degeneration but it is particularly so in low-voltage design. If only one control input is used, say v_{GS}, the dc voltage V_{GS} must equal $V_{DD}/2$ to maximize voltage headroom; deviations in V_{GS} from $V_{DD}/2$, which are needed to alter the dc biasing current, will then compromise voltage headroom on the v_{gs} ac input at either the top or the bottom rail; however, if V_{WS} is also available as a control input, V_{GS} can be fixed at $V_{DD}/2$ maximizing input ac voltage headroom while V_{WS} can be varied to alter the dc biasing current.
4. If V_{DD} is small and less than the junction turn-on voltage, V_{WS} *can be biased* to be $-V_{DD}$ without danger of turning on the parasitic bipolar transistor in Figure 12.8. Thus, V_{T0} can be reduced to improve overdrive gate voltage in strong inversion, and rail-to-rail operation on the well voltage may be possible.

5. *Voltage drops across passive devices can serve to perform level-shifting functions* that alter dc operating points for active devices and enable them to function correctly. For example, differential-amplifier topologies can exploit passive devices like resistors to level shift the v_- and v_+ inputs to operational amplifiers without altering input or output common-mode operating levels or overall closed-loop gains. Chapter 20 provides an example of a classic differential instrumentation amplifier in an electrocardiogram-sensing context.

6. Due to Early-voltage limitations, dc gain is usually compromised in low-voltage processes, although the f_T of transistors is usually quite high. Therefore, *stable positive-feedback configurations that enhance dc gain at the expense of some loss in bandwidth are advantageous.*

7. As the use of cascode transistors compromises dc operating range in low-voltage processes, *cascaded-gain topologies* with relatively low dc gain per stage are often necessary. In certain contexts, such as general-purpose low-voltage low-power operational amplifiers, cascaded topologies require sophisticated frequency-compensation techniques such as Nested-Miller compensation to ensure unity-gain stability. Such techniques trade dc gain for some loss in overall closed-loop bandwidth [5]. Such a tradeoff is frequently worthwhile in low-voltage processes where bandwidth is more abundant than dc gain. For example a three-stage Nested-Miller operational amplifier only reduces the 3 dB-bandwidth by a factor of two over that of a two-gain-stage amplifier while improving gain by the gain of a complete amplification stage.

Figure 12.19 (a) illustrates a topology that removes source degeneration and gate degeneration from the topology of Figure 12.2 to make it more suitable for low-voltage operation. Note that gate-degeneration fed back is no longer present in Figure 12.19 (a) because changes in differential-arm current are not fed back to the gate, which is held constant at V_{REF}. In this topology, the linear range $V_L = (3/2)(2\phi_t/(1-\kappa))$. If $\kappa > 0.7$ as it often is, V_L is at least 250 mV, which implies that rail-to-rail operation on a 0.5 V supply with a common-mode input voltage of 0.25 V is possible. In Figure 12.19 (a), the minimum V_{DD} required such that all transistors operate in saturation is $2V_{DSAT} + V_{T0} = 8\phi_t + V_{T0}$ in weak inversion. Thus, if $V_{T0} = 0.3$ V, operation on a 0.5 V supply is possible. The value of V_{REF} should be near V_{DSAT}.

Figure 12.19 (b) is a 0.5 V topology built for strong-inversion operation and described in [4], which contains several exemplary circuits for low-voltage design. The design is fully differential with no more than two devices in series in any of its differential paths such that even differential pairs are absent; nevertheless, good common-mode rejection is attained via a common-mode feedback loop. Well inputs v_+ and v_- on M_1 and M_2 along with the current sources M_6 and M_7 implement a fully differential common-source amplification stage with outputs v_{out}^- and v_{out}^+ respectively. The two balanced resistors marked R extract the value of the common-mode output voltage $(v_{out}^- + v_{out}^+)/2$ and feed it back to the gates of M_1 and M_2. Thus, a negative-feedback common-mode loop equalizes currents in M_1 and M_6 and M_2 and M_7

respectively when no differential input is present. The bias voltage V_B^{OFFS} and M_5 implement a weak current source that causes a small IR drop across the R resistors such that the gate bias voltages of M_1 and M_2 are at a lower value than their drain bias voltages, enabling higher speed operation. The transistors M_3 and M_4 enhance the gain of the overall amplifier via cross-coupled positive-feedback action. Further details of these circuits, including replica-biasing techniques that ensure robust biasing of currents, voltages, and of the positive-feedback gain, are described in [4].

12.11   Robust operation of subthreshold circuits

Although we have pretended to use voltage biases to generate bias currents in this chapter and in Chapter 11, such biasing is highly non-robust to temperature variations and to power-supply noise. *Such biasing must never be used in subthreshold operation, where there is exponential sensitivity to temperature and voltage.* Robust biasing techniques for subthreshold operation that are insensitive to temperature variations and power-supply noise, and that digitally compensate for analog transistor mismatch, are described in Chapter 19 in the context of a large chip-level bionic-ear processor for the deaf. Figure 19.12 and an associated description in Chapter 19 illustrate how corner frequencies of filters can be made temperature invariant and highly immune to power-supply noise, and how voltage and current biases are generated in chips intended for practical applications. *Throughout this book, although we will frequently use voltage biases for brevity, it should be understood that underlying bias circuits actually generate bias currents and/or voltages.*

References

[1] T. Delbruck, Bump circuits for computing similarity and dissimilarity of analog voltages. *Proceedings of the International Joint Conference on Neural Networks*, Seattle, WA, 475–479, 1991.

[2] R. Sarpeshkar, R. F. Lyon and C. A. Mead. A low-power wide-linear-range transconductance amplifier. *Analog Integrated Circuits and Signal Processing*, **13** (1997), 123–151.

[3] R. Sarpeshkar, R. F. Lyon and C. A. Mead. A low-power wide-dynamic-range analog VLSI cochlea. *Analog Integrated Circuits and Signal Processing*, **16** (1998), 245–274.

[4] S. Chatterjee, Y. Tsividis and P. Kinget. 0.5-V analog circuit techniques and their application in OTA and filter design. *IEEE Journal of Solid-State Circuits*, **40** (2005), 2373–2387.

[5] R. G. H. Eschauzier and J. H. Huijsing. *Frequency Compensation Techniques for Low-Power Operational Amplifiers* (Dordrecht: Springer, 1995).

13 Low-power filters and resonators

The laws of nature are but the mathematical thoughts of God.

Euclid

In this chapter, we shall discuss techniques for the design of active transconductor-capacitor ($G_m - C$) filters, which are power efficient compared with several other topologies. We shall particularly focus on second-order systems and resonators, since they are a good vehicle for illustrating general issues in filter design and are ubiquitous in RF design. We shall present an intuitive $-s^2$ plane geometry for understanding second-order and first-order systems that allow them to be analyzed quickly without significant algebra and then generalize these methods to higher-order systems. Since cascades of first-order and/or second-order systems or summed first-order and/or second-order systems can architect all linear systems, such methods are useful in filter design, feedback system design, and all linear systems.

Filters have been built with various topologies. Five kinds include passive-RLC filters, op-amp-RC, $G_m - C$, MOSFET-C, and switched-capacitor filters. Passive-RLC filters are the most power efficient since the use of only passive components results in no power dissipation apart from that required to supply signal currents; these filters use only bidirectional components. Inductors above 100 nH, capacitors above 100 pF, and resistors above 10 MΩ are rarely practical in integrated-circuit design due to their large size. On-chip inductors and resistors are subject to significant parasitics that degrade their operation. For example, resistors often have significant distributed parasitic capacitance, and inductors frequently have significant resistance and parasitic capacitance. Passive components cannot be electronically tuned. Thus, active topologies that replace the inductor (op-amp RC) or that replace the inductor and resistor (switched-capacitor, MOSFET-C, or $G_m - C$) with active components are useful in making practical filters. Even at some RF frequencies, where inductor sizes are practical, active circuits can enhance the performance of passive filters, e.g, enhance their quality factor Q. Most active-filter topologies create complex filters by using integrators as building-block circuits.

Op-amp-RC filters are highly linear and use an op-amp-RC-based topology to build closed-loop-integrator building blocks, which can then be used to architect any filter. They have a limited frequency range since they require the operational amplifier's gain-bandwidth product to be well in excess of the bandwidth of the integrator if an ideal filter response is desired over a wide range of frequencies.

The power costs of achieving a high gain-bandwidth product in the operational amplifier make such topologies relatively power hungry. Typically, the op-amp's gain-bandwidth product is not adapted to that of the integrator that it implements such that worst-case power is consumed independent of the desired frequency range of the filter. MOSFET-C filters have less linearity than op-amp-RC filters since they exploit the linear above-threshold regime of the MOSFET to implement tunable linear resistances. A large linear range requires a high overdrive $\kappa_S(V_{GS} - V_{TS})$ and high power consumption. Switched-capacitor filters replace the resistance with a switched capacitor of value $R = 1/(f_S C_S)$ where f_S is the switching frequency and C_S is the switching capacitance. These filters also use closed-loop-integrator building blocks. The switching frequency has to be significantly higher than the integrator bandwidth to smooth switching noise. Since the corner frequencies of switched-capacitor filters are defined by well-controlled values of f_S and capacitance ratios, such filters have precise frequency responses. They are relatively power hungry since they are subject to op-amp-RC-filter-based tradeoffs, and in addition, anti-aliasing and smoothing constraints.

$G_m - C$ filters use controlled transconductance amplifiers and capacitors to architect filters via open-loop integrator building blocks. Since transconductors can be built with relatively few devices and the power burned per open-loop integrator is directly related to its bandwidth, there is good power efficiency. The limited linear range of operation of transconductors and the lack of closed-loop-distortion-reducing integrator-building blocks implies that these topologies have to be carefully architected to avoid distortion. Since we are interested in ultra-low-power design, we shall focus on $G_m - C$ architectures. In today's high-speed processes, they are capable of creating practical integrated-circuit filters with corner frequencies ranging from below 1 Hz to over 1 GHz if subthreshold operation, above-threshold operation, and capacitance scaling are used.

13.1 G_m–C filter synthesis

One easy way to synthesize $G_m - C$ filters is via state-space synthesis. We write the desired transfer function in terms of a set of integrator state variables, draw a block diagram (if needed), and replace each integrator in the block diagram of value $1/\tau_i s$ with open-loop $G_m - C$ integrator building blocks such that $1/\tau_i = G_i/C_i$. Thus, for example, if we want to synthesize a first-order filter, we perform the following manipulations

$$\frac{V_{out}(s)}{V_{in}(s)} = \frac{1}{\tau s + 1}$$
$$V_{out}(\tau s + 1) = V_{in}$$
$$\tau s V_{out} = V_{in} - V_{out} \tag{13.1}$$
$$V_{out} = \frac{1}{\tau s}(V_{in} - V_{out})$$

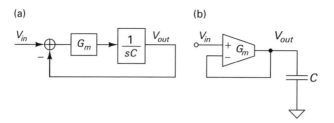

Figure 13.1a, b. A first-order $G_m - C$ low-pass filter: (a) block diagram and (b) circuit implementation.

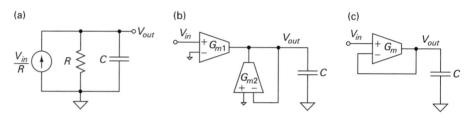

Figure 13.2a, b, c. Element-replacement example for a first-order $G_m - C$ filter: (a) passive prototype, (b) element-replaced form, and (c) after simplification.

and obtain the block diagram shown in Figure 13.1 (a). The circuit of Figure 13.1 (b) implements the block diagram with $\tau = C/G_m$.

Another procedure to synthesize $G_m - C$ filters is via element-replacement synthesis. We obtain an RLC filter topology from a filter design book, for example, [1]. We replace resistors with transconductors, typically WLRs, i.e., wide-linear-range transconductors, capacitors with capacitors, and inductors with gyrated capacitors. A technique for creating gyrated capacitors with a gyrator is described in the next section. Since this technique yields one-sided grounded inductors and needs to be repeated twice to create floating inductors, we strive to avoid working with RLC topologies with floating inductors, converting them to their duals with floating capacitors whenever possible. We then reduce or elimi-nate transconductors based on symmetries and simplifications in the circuit to get the final circuit. We shall first begin with a first-order system example of synthesis by element replacement that does not require gyrators. Then we shall discuss gyrators, second-order systems, and synthesis of second-order $G_m - C$ filters.

Figure 13.2 (a) shows a simple prototype RC filter. With the element-replacement technique, the current source is replaced with a transconductor, the resistor is replaced with a transconductor whose output is tied to its negative input such that it behaves like a resistor of value of $1/G_m$, and the capacitor is replaced with a capacitor to yield the circuit of Figure 13.2 (b). Since $G_{m1} = G_{m2} = 1/R$ from Figures 13.2 (a) and 13.2 (b), we can reduce both transconductors to one transcon-ductor of value $G_m = 1/R$ as shown in Figure 13.2 (c). In general, two transconductors with a common output terminal, a common value of transconductance, and each with a grounded input that is different from that of the other, can always be reduced to one differential transconductor as in the above example. Thus, the transfer function of the prototype filter of Figure 13.2 (a) is converted to the transfer function of the element-replacement filter of Figure 13.2 (c) according to the expressions

$$\frac{V_{out}(s)}{V_{in}(s)} = \frac{1}{RCs + 1}$$
$$\frac{V_{out}(s)}{V_{in}(s)} = \frac{1}{\dfrac{C}{G_m}s + 1}$$

(13.2)

13.2   Gyrators

Figure 13.3 (a) illustrates that, when two transconductors of value G_{m1} and G_{m2} are configured back to back in a negative-feedback 'gyrator' topology, the current I_{in} measured at the input node when a voltage V_{in} is applied to it is given by

$$I_{in}(s) = G_{m2}(G_{m1}Z_L)V_{in}(s)$$

(13.3)

The current flowing out of the voltage source V_{in} flows into the transconductor G_{m2} and thus corresponds to that seen in a normal positive impedance. The input conductance and impedance are therefore given by

$$G_{in} = \frac{I_{in}(s)}{V_{in}(s)} = G_{m2}(G_{m1}Z_L)$$
$$Z_{in} = \frac{1}{G_{in}} = \frac{1}{G_{m1}G_{m2}Z_L} = \left(\frac{1}{G_{m1}G_{m2}}\right)\frac{1}{Z_L}$$

(13.4)

Thus, from the point of view of the input, the gyrator transforms the impedance Z_L into an inverted impedance. Thus, if $Z_L(s) = 1/(Cs)$

$$Z_{in}(s) = \frac{sC}{G_{m1}G_{m2}} = s\left(\frac{C}{G_{m1}G_{m2}}\right) = sL$$

(13.5)

and we have converted a capacitance to an inductance. Similarly, if $Z_L(s) = 1/(C_L s + G_L)$

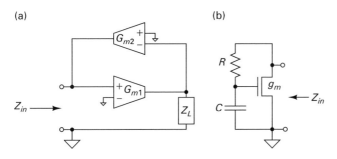

(a)   (b)

Figure 13.3a, b. Two gyrator circuits: (a) using two transconductors and (b) using a single transistor.

$$Z_{in}(s) = \frac{(C_L s + G_L)}{G_{m1} G_{m2}} = s\left(\frac{C_L}{G_{m1} G_{m2}}\right) + \left(\frac{G_L}{G_{m1} G_{m2}}\right) = sL + R \qquad (13.6)$$

and we have converted a parallel R and C into a series L and R.

Figure 13.3 (b) illustrates a single-transistor gyrator-like circuit. The impedance can be computed and approximated by a gyrator-like expression below

$$Z_{in} = \frac{sRC + 1}{sC + g_m} \approx \frac{sC}{\left(\frac{g_m}{R}\right)} \qquad (13.7)$$

with the inductor approximation holding true if

$$\frac{1}{RC} \ll \omega \ll \frac{g_m}{C} \qquad (13.8)$$

The conductance $G = 1/R$ in Figure 13.3 (b) is analogous to G_{m1} in Figure 13.3 (a), $1/(C_L s)$ is analogous to Z_L, and g_m is analogous to G_{m2}. Equation (13.8) is a statement of the condition that we are well past the corner frequency of the RC circuit of Figure 13.3 (b) such that we almost have an integrator in a feedback loop, and that the current at the drain of the transistor is dominated by the transistor current rather than by the current flowing through the RC circuit.

13.3   Introduction to second-order systems

Second-order systems are ubiquitous in science and engineering. For example, many resonators are well modeled by second-order systems including LCR circuits in electrical engineering and spring-mass resonators in mechanics. The most common negative-feedback systems can often be modeled by two dominant poles in a feedback loop and lead to second-order dynamics.

Figure 13.4 (a) illustrates a parallel LCR circuit. If we define

$$\tau = \sqrt{LC} = \frac{1}{\omega_n}, \quad Q = \frac{R}{\sqrt{\frac{L}{C}}} = \frac{1}{2\zeta} \qquad (13.9)$$

then the transfer functions from the input current to the current in the inductor, resistor, and capacitor are resonant lowpass, resonant bandpass, and resonant highpass transfer functions respectively and given by

(a)   (b)

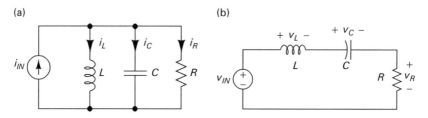

Figure 13.4a, b. Parallel (a) and series (b) passive LCR circuits.

$$\frac{I_L(s)}{I_{in}(s)} = \frac{1}{\tau^2 s^2 + \dfrac{\tau}{Q} s + 1}$$

$$\frac{I_R(s)}{I_{in}(s)} = \frac{\dfrac{\tau}{Q} s}{\tau^2 s^2 + \dfrac{\tau}{Q} s + 1}$$ (13.10)

$$\frac{I_C(s)}{I_{in}(s)} = \frac{\tau^2 s^2}{\tau^2 s^2 + \dfrac{\tau}{Q} s + 1}$$

$$I_{in}(s) = I_L(s) + I_R(s) + I_C(s)$$

Similarly, for the dual series LCR circuit of Figure 13.4 (b), if we define

$$\tau = \sqrt{LC} = \frac{1}{\omega_n}, \quad Q = \frac{\sqrt{\dfrac{L}{C}}}{R} = \frac{1}{2\zeta}$$ (13.11)

then the transfer functions from the input voltage to the voltage across the capacitor, resistor and inductor are resonant lowpass, resonant bandpass, and resonant highpass and given by

$$\frac{V_C(s)}{V_{in}(s)} = \frac{1}{\tau^2 s^2 + \dfrac{\tau}{Q} s + 1}$$

$$\frac{V_R(s)}{V_{in}(s)} = \frac{\dfrac{\tau}{Q} s}{\tau^2 s^2 + \dfrac{\tau}{Q} s + 1}$$ (13.12)

$$\frac{V_L(s)}{V_{in}(s)} = \frac{\tau^2 s^2}{\tau^2 s^2 + \dfrac{\tau}{Q} s + 1}$$

$$V_{in}(s) = V_C(s) + V_R(s) + V_L(s)$$

It is worth noting that Q is the magnitude of the susceptance-to-conductance ratio at $\omega = \omega_n$ for a parallel LCR circuit, as seen from Equation (13.9), and that Q is the magnitude of the reactance-to-resistance ratio at $\omega = \omega_n$ for a series LCR circuit, as seen from Equation (13.11). In a parallel LCR circuit, large resistors yield high Q such that, in the limit as Q goes to ∞, they act like an open circuit and consequently do not affect the currents in the inductor or capacitor in parallel with them. In a series LCR circuit, small resistors yield high Q such that, in the limit as Q goes to ∞, they act like a short circuit and consequently do not affect the voltages across the inductor or capacitor in series with them. A large Q in a parallel/series LCR circuit implies that the admittance/impedance changes sharply with frequency since a small difference between two large susceptance/ reactance values, whether inductive or capacitive, is small compared with the conductance/resistance of the circuit for only a narrow range of frequencies.

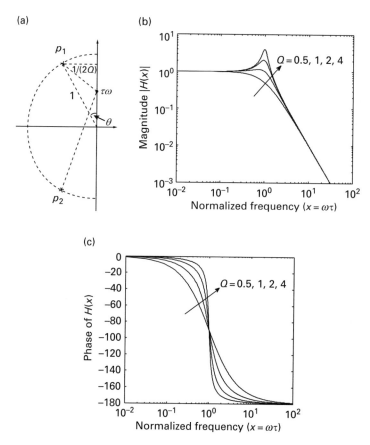

Figure 13.5a, b, c. (a) Under-damped poles on the s-plane, (b) their magnitude response, and (c) phase response.

By computing the roots of the lowpass resonant transfer function below,

$$H(s) = \frac{1}{1 + \dfrac{\tau s}{Q} + \tau^2 s^2}$$

$$H(s) = \frac{1}{(\tau s - p_1) \cdot (\tau s - p_2)}$$

$$p_{1,2} = -\frac{1}{2Q} \pm j \sqrt{1 - \left(\frac{1}{2Q}\right)^2}$$

$$|H(j\omega)| = \frac{1}{|j\tau\omega - p_1| \cdot |j\tau\omega - p_2|}$$

(13.13)

we see that the two poles of the transfer function are located as shown in Figure 13.5 (a). The geometric location of the poles is on a unit circle because of the normalized nature of the transfer function, i.e., $H(0) = 1$. We see that Q is a way to parameterize the angle θ made by the pole vectors with respect to the $j\omega$ axis with $\sin(\theta) = 1/2Q = \zeta$.

The resonant magnitude transfer functions have sharp-and-peaky gains near ω_n with the sharpness and peakiness increasing with increasing Q as shown in Figure 13.5 (b) for the lowpass resonant case. The resonant phase transfer functions have sharp transitions around ω_n with the slope at ω_n increasing with increases in Q as shown in Figure 13.5 (c) for the lowpass resonant case. The bandpass resonant transfer function always has a peak gain of 1.0 at ω_n while the lowpass and highpass resonant transfer functions always have a gain of Q at ω_n. The phase is $-90°$ at ω_n for the lowpass resonant transfer function, is $0°$ at ω_n for the bandpass resonant transfer function, and is $+90°$ at ω_n for the highpass resonant transfer function. The Q corresponds to a settling time with approximately Q cycles of ringing in the resonator's impulse response before the energy at the ringing frequency, $\omega_n\left(\sqrt{1-1/(2Q)^2}\right)$ has decayed to $e^{-2\pi} \approx 0$ of its initial value.

13.4   Synthesis of a second-order G_m–C filter

We shall illustrate how a second-order $G_m - C$ filter may be synthesized by state-space synthesis, and then by element-replacement synthesis. Then we shall analyze the properties of the filter. Simple manipulations of the transfer function of a resonant lowpass filter parameterized by the transfer function

$$\frac{V_{out}(s)}{V_{in}(s)} = \frac{1}{\tau_1\tau_2 s^2 + \tau_1 s + 1}$$

$$V_{out}\left(\tau_1\tau_2 s^2 + \tau_1 s + 1\right) = V_{in}$$

$$\tau_1 s(\tau_2 s + 1)V_{out} = V_{in} - V_{out}$$

$$(\tau_2 s + 1)V_{out} = \frac{1}{\tau_1 s}(V_{in} - V_{out}) \tag{13.14}$$

$$\tau_2 s V_{out} = \frac{1}{\tau_1 s}(V_{in} - V_{out}) - V_{out}$$

$$V_{out} = \frac{1}{\tau_2 s}\left(\frac{1}{\tau_1 s}(V_{in} - V_{out}) - V_{out}\right)$$

yield the block diagram shown in Figure 13.6 (a), which is converted to the block diagram of Figure 13.6 (b), which is converted to the circuit of Figure 13.6 (c) with

$$\tau_1 = \frac{C_1}{G_1}$$

$$\tau_2 = \frac{C_2}{G_2} \tag{13.15}$$

Block diagram reduction shown in Figure 13.7 and application of Black's formula to the reduced block diagram give us confidence that we are

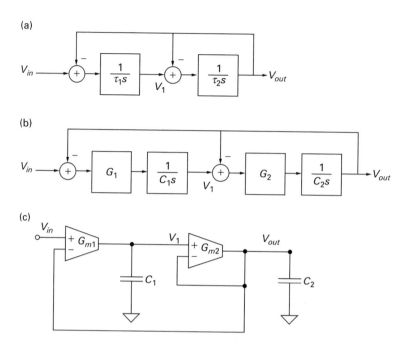

Figure 13.6a, b, c. A second-order $G_m - C$ lowpass filter: (a) abstract block diagram, (b) block diagram including transconductances, and (c) circuit implementation.

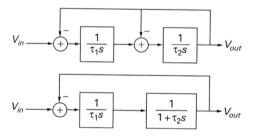

Figure 13.7. Block diagram reduction for a second-order lowpass filter.

indeed implementing the transfer function that we set out to implement in Equation (13.14):

$$\frac{V_{out}(s)}{V_{in}(s)} = \frac{\dfrac{1}{\tau_1 s} \cdot \dfrac{1}{1 + \tau_2 s}}{1 + \dfrac{1}{\tau_1 s} \cdot \dfrac{1}{1 + \tau_2 s}} = \frac{1}{1 + \tau_1 s + \tau_1 \tau_2 s^2} \tag{13.16}$$

Could we arrive at the same answer via element-replacement synthesis? Figure 13.8 (a) shows a transfer function described by

$$\frac{V_{out}(s)}{V_{in}(s)} = \frac{1}{LCs^2 + RCs + 1} = \frac{1}{\tau^2 s^2 + \dfrac{\tau}{Q} s + 1} \tag{13.17}$$

$$\tau = \sqrt{LC}, \quad Q = \sqrt{\frac{L}{C}} \cdot \frac{1}{R}$$

(a)   (b)

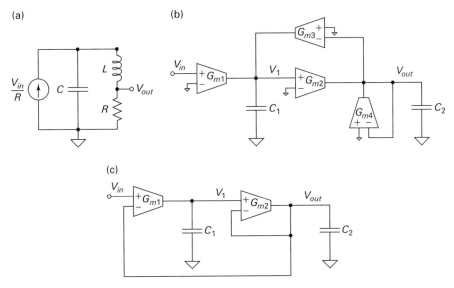

(c)

Figure 13.8a, b, c. A second-order lowpass filter using the element-replacement technique:
(a) the passive prototype, (b) after element replacement, and (c) after simplification.

Using element-replacement techniques, the current source and capacitor of Figure
13.8 (a) get replaced by the G_{m1} transconductor and C_1 in Figure 13.8 (b), respectively.
Since L and R are in series in Figure 13.8 (a), we gyrate a resistance and capacitance in
parallel to create them; i.e., in Figure 13.8 (b), $1/G_{m4}$ and C_2 in parallel are gyrated by
G_{m2} and G_{m3} to create $L = C_2/(G_{m2}G_{m3})$ and $R = G_{m4}/(G_{m2}G_{m3})$. If we now set
$G_{m1} = G_{m3}$ such that the two transconductors sourcing currents into the V_1 node
can be absorbed into one, and $G_{m2} = G_{m4}$ such that the two transconductors sourcing
currents into the V_{out} node can be absorbed into one, we obtain the circuit of Figure
13.8 (c), identical to that obtained in Figure 13.6 (c) by state-space synthesis. Thus,
both methods yield the same circuit and we may write

$$H(s) = \frac{V_{out}(s)}{V_{in}(s)} = \frac{1}{1 + \tau_1 s + \tau_1 \tau_2 s^2} = \frac{1}{1 + \dfrac{\tau s}{Q} + \tau^2 s^2}$$

$$\tau_1 = \frac{C_1}{G_{m1}}; \tau_2 = \frac{C_2}{G_{m2}}; \tau_1 = \frac{\tau}{Q}; \tau_2 = \tau Q \tag{13.18}$$

$$\tau^2 = \tau_1 \tau_2; Q = \sqrt{\frac{\tau_2}{\tau_1}}$$

13.5   Analysis of a second-order G_m–C filter

Figure 13.9 reveals a block diagram of the circuit of Figure 13.8 (c) incorporating
input-referred noise sources due to each transconductor. The transfer functions of
these noise sources to the output are given by

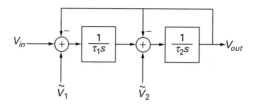

Figure 13.9. Block diagram with noise sources for a second-order filter section.

$$H_{1o}(s) = \frac{V_{out}(s)}{\tilde{V}_1(s)} = H(s)$$

$$H_{2o}(s) = \frac{V_{out}(s)}{\tilde{V}_2(s)} = \tau_1 s H(s) = \frac{\tau s}{Q} H(s)$$

(13.19)

If we assume that each transconductor has a linear range of V_L, N devices worth of shot noise, an input-referred noise per unit bandwidth denoted by $d\langle v_{n\ell}^2 \rangle$ where $l = 1,2$ depending on whether we are considering the G_{m1} or G_{m2} transconductor, and the output voltage noise per unit bandwidth is denoted by $d\langle v_{no}^2 \rangle$, then we can conclude that

$$d\langle v_{n\ell}^2 \rangle = \frac{NqV_L}{G_\ell} df = \frac{NV_L}{4\phi_t} \frac{4kT}{G_\ell} df, \ell = 1,2$$

$$d\langle v_{no}^2 \rangle = |H_{1o}(f)|^2 \, d\langle v_{n1}^2 \rangle + |H_{2o}(f)|^2 \, d\langle v_{n2}^2 \rangle$$

$$d\langle v_{no}^2 \rangle = |H(f)|^2 \frac{NqV_L}{G_1} df + \left| \frac{\tau}{Q}(j2\pi f)H(f) \right|^2 \frac{NqV_L}{G_2} df$$

(13.20)

$$\frac{1}{G_1} = \frac{\tau_1}{C_1} = \frac{\tau}{QC_1}, \frac{1}{G_2} = \frac{\tau_2}{C_2} = \frac{\tau Q}{C_2}$$

$$d\langle v_{no}^2 \rangle = \frac{NqV_L}{Q} \left(\frac{1}{C_1} + \frac{(2\pi\tau f)^2}{C_2} \right) |H(f)|^2 \tau df$$

Thus,

$$d\langle v_{no}^2 \rangle = \frac{NqV_L}{2\pi Q} \left(\frac{1}{C_1} + \frac{x^2}{C_2} \right) |H(x)|^2 dx \quad (x = \tau\omega)$$

$$\left| H(x)^2 \right| = \frac{1}{(1-x^2)^2 + \left(\frac{x}{Q}\right)^2}$$

$$\int_0^\infty |H(x)|^2 dx = \frac{\pi}{2}Q, \int_0^\infty x^2|H(x)|^2 dx = \frac{\pi}{2}Q$$

(13.21)

minimum when $C_1 = C_2 = C$: $\langle v_{no}^2 \rangle = \frac{NqV_L}{2C}$

$N = 5.3, V_L = 1\text{V}, C = 650 \text{ fF} : \sqrt{\langle v_{no}^2 \rangle} = 0.8 \text{ mV}$

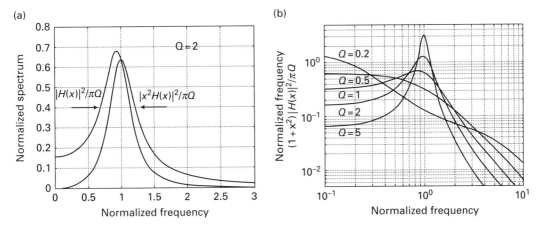

Figure 13.10a, b. (a) Noise transfer functions for the second-order lowpass filter and (b) for various values of Q.

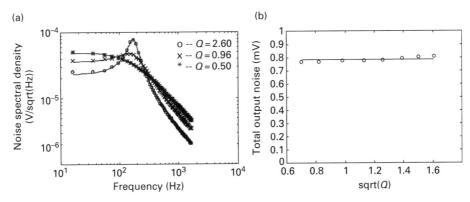

Figure 13.11a, b. Experimentally measured output noise of the second-order lowpass filter: (a) PSD for various values of Q and (b) total noise, showing its independence with respect to Q. Reproduced with kind permission of Springer Science and Business Media from [2].

Figure 13.10 (a) shows the normalized filter transfer functions due to each noise source and Figure 13.10 (b) reveals the total noise power density from both sources. The area under all the curves of Figure 13.10 (b) is the same. Thus, from Equation (13.21), we conclude that the total output voltage noise of the G_m-C filter is independent of Q and its bias currents and equal to $NqV_L/2C$. Figure 13.11 (a) shows experimental noise spectral density measurements from G_m-C filters for various Q [2]. The total area under each of the curves is the same and well predicted by Equation (13.21) to be 0.8 mV, almost exactly what is measured by experiment and plotted in Figure 13.11 (b).

Now that we have computed the total output noise of the filter, we need to compute the transfer functions from the input $V_{in}(s)$ to each of the differential input voltages of the G_{m1} and G_{m2} transconductors of Figure 13.8 (c). These transfer functions will allow us to estimate the maximal undistorted input before either transconductor begins to experience a differential input voltage of V_L such that distortion manifests.

$$H_{i1}(s) = \frac{V_{1,\text{diff}}(s)}{V_{in}(s)}, \ H_{i2}(s) = \frac{V_{2,\text{diff}}(s)}{V_{in}(s)}$$

$$\max|H_{i1}(s)| = \max|1 - H(s)| = \max\left|\left(\frac{\tau s}{Q} + \tau^2 s^2\right)H(s)\right| \approx Q \qquad (13.22)$$

$$\max|H_{i2}(s)| = \max|(1 + \tau_2 s) H(s) - H(s)| = \max|Q\tau s H(s)| = Q^2$$

In Equation (13.22), we have evaluated the maxima where $\omega\tau \approx 1$, i.e., at $\omega = \omega_n$, since $H(s)$ is peaked near this point (peaking in the frequency response arises only for $Q > 0.707$) and rolls off abruptly beyond this point. Thus, for $Q > 1$, it is the differential voltage across the G_{m2} transconductor that determines the maximal undistorted input; for $Q < 1$, it is the differential voltage across the G_{m1} transconductor that determines the maximal undistorted input. Since high-Q operation is typically more desirable and interesting than low-Q operation, in most cases, the maximal undistorted input to the filter is V_L/Q^2 at $\omega = \omega_n$ and determined by the G_{m2} transconductor. Thus, the maximal output signal amplitude at $\omega = \omega_n$ is given by

$$|V_{\max}(j\omega_n)| = \frac{V_L}{Q^2}|H(j\omega_n)| = \frac{V_L}{Q^2}Q = \frac{V_L}{Q} \qquad (13.23)$$

Thus, the maximal output SNR of the filter is given by the ratio of the output rms signal power to the noise power and given by

$$SNR_{\max} = \frac{\left(\frac{V_L}{Q}\right)^2 / 2}{\dfrac{NqV_L}{2C}} = \left(\frac{CV_L}{Nq}\right)\frac{1}{Q^2} \qquad (13.24)$$

Compared with a first-order filter, therefore, this second-order $G_m - C$ filter has an $SNR_{\max}$ that is $2/Q^2$ lower. Hence, a high-Q filter built with this topology suffers from an increasingly smaller input dynamic range and output signal-to-noise ratio as Q increases. This tradeoff with Q is seen in all active filters but the $1/Q^2$ factor is particularly expensive in this topology. Is it possible to build a $G_m - C$ filter with a different topology than can do better?

13.6   Synthesis and analysis of an alternative G_m–C filter

When synthesizing the filter of Figure 13.8 (c), we implemented a transfer function parameterized by τ_1 and τ_2 and used the minimal number of transconductors for implementing the filter that we could. The result was that extreme ratios of τ_1 and τ_2 were needed to implement a high-Q filter as $Q = \sqrt{\tau_2/\tau_1}$. These extreme ratios led to extreme imbalances in the differential voltage across various transconductors such that the maximally undistorted input was determined by the worst-case transconductor G_{m2}, i.e., the transconductor with the largest differential voltage across it, at a relatively small input amplitude, thus compromising the dynamic range of the filter and its maximal output *SNR*.

If we proceed with the alternative state-space synthesis parameterized by τ and τ/Q below, we find that

$$\frac{V_{out}}{V_{in}} = \frac{1}{\tau^2 s^2 + \dfrac{\tau s}{Q} + 1}$$

$$V_{out}\left(\tau^2 s^2 + \frac{\tau s}{Q} + 1\right) = V_{in}$$

$$\tau s\left(\tau s + \frac{1}{Q}\right)V_{out} = V_{in} - V_{out} \qquad (13.25)$$

$$\left(\tau s + \frac{1}{Q}\right)V_{out} = \frac{1}{\tau s}(V_{in} - V_{out})$$

$$\tau s V_{out} = \frac{1}{\tau s}(V_{in} - V_{out}) - \frac{1}{Q}V_{out}$$

$$V_{out} = \frac{1}{\tau s}\left(\frac{1}{\tau s}(V_{in} - V_{out}) - \frac{1}{Q}V_{out}\right)$$

Therefore we can create the block diagram of Figure 13.12 (a) and map it to the circuit of Figure 13.12 (b). Note that the first adder block of Figure 13.12 (a) is implemented by taking the difference of voltages in a transconductor in

(a)

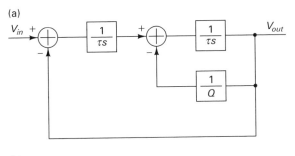

(b)

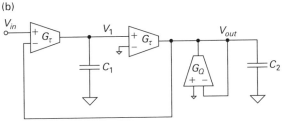

Figure 13.12a, b. An alternative second-order lowpass filter: (a) block diagram and (b) circuit representation.

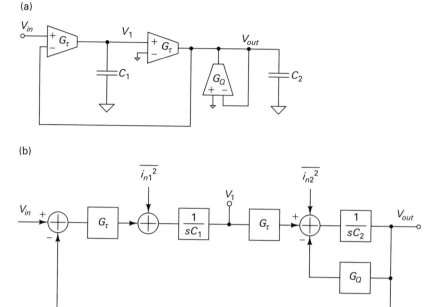

Figure 13.13a, b. (a) The alternative second-order lowpass filter topology and (b) its block diagram representation with noise sources included.

Figure 13.12 (b) while the second adder block of Figure 13.12 (a) is implemented by taking the difference of two currents by Kirchhoff's current law in Figure 13.12 (b). The transfer function is then computed to be

$$\frac{V_{out}(s)}{V_{in}(s)} = \frac{1}{\tau^2 s^2 + \dfrac{\tau s}{Q} + 1} = \frac{1}{\dfrac{C_1 C_2}{G_\tau^2} s^2 + \dfrac{G_Q C_1}{G_\tau^2} s + 1}$$

(13.26)

$$\tau = \frac{\sqrt{C_1 C_2}}{G_\tau}; Q = \sqrt{\frac{C_2}{C_1}} \cdot \frac{G_\tau}{G_Q}$$

Synthesis by element replacement also leads to the circuit of Figure 13.12 (b) if we set $G_{m1} = G_{m2} = G_{m3} = G_\tau$ and $G_{m4} = G_Q$ in Figure 13.8 (b).

The noise analysis of the alternate filter of Figure 13.12 (b) is similar to that performed on the filter of Figure 13.8 (c). The circuit of Figure 13.13 (a) is first transformed to the block diagram of Figure 13.13 (b) to incorporate noise sources. The analysis then proceeds via Black's formula, summing and squaring of magnitude transfer functions, and integration over all frequencies:

$$\frac{V_{out}(s)}{i_{n1}(s)} = \frac{1}{G_\tau} \cdot \frac{1}{\tau^2 s^2 + \dfrac{\tau s}{Q} + 1}; \frac{V_{out}(s)}{i_{n2}(s)} = \frac{1}{G_Q} \cdot \frac{\dfrac{\tau s}{Q}}{\tau^2 s^2 + \dfrac{\tau s}{Q} + 1};$$

$$\overline{i_{n1}^2} = N q G_\tau V_L; \overline{i_{n2}^2} = N q (G_\tau + G_Q) V_L$$

$$\overline{v_{no}^2} = \int_0^\infty \left(\left(\frac{V_{out}}{i_{n1}}\right)^2 \overline{i_{n1}^2} + \left(\frac{V_{out}}{i_{n2}}\right)^2 \overline{i_{n2}^2} \right) df$$

$$\overline{v_{no}^2} = \frac{1}{2\pi\tau} \left(NqG_\tau V_L \frac{\pi}{2} Q \cdot \frac{1}{G_\tau^2} + Nq(G_\tau + G_Q)V_L \frac{\pi}{2} Q \cdot \frac{1}{G_Q^2} \cdot \frac{1}{Q^2} \right)$$

$$= \frac{NqV_L G_\tau}{4\sqrt{C_1 C_2}} \left(\frac{1}{G_\tau} \cdot Q + \frac{1}{Q} \cdot \frac{1}{G_Q} \left(1 + \frac{G_\tau}{G_Q}\right) \right)$$

$$= \frac{NqV_L}{4\sqrt{C_1 C_2}} \left(Q + \frac{1}{Q} \cdot \frac{G_\tau}{G_Q} \left(1 + \frac{G_\tau}{G_Q}\right) \right) \qquad (13.27)$$

$$\tau = \frac{\sqrt{C_1 C_2}}{G_\tau}; \quad Q = \sqrt{\frac{C_2}{C_1} \cdot \frac{G_\tau}{G_Q}};$$

$$C_1 = C_2 = C$$

$$\boxed{\overline{v_{no}^2} = \frac{NqV_L}{4C}(2Q + 1) \approx \left(\frac{NqV_L}{2C}\right) Q}$$

Thus, the total output voltage noise power of this alternate filter increases with Q.

To analyze the maximal undistorted input to the alternate filter, we need to compute the transfer function from the input to each differential voltage of the transconductors of Figure 13.12 (b). If we label the differential voltages of each of these transconductors as $V_{1,diff}(s)$, $V_{2,diff}(s)$, and $V_{3,diff}(s)$ from left to right in Figure 13.12 (b), we would like to compute

$$H_1(s) = \frac{V_{1,\text{diff}}(s)}{V_{in}(s)}, \; H_2(s) = \frac{V_{2,\text{diff}}(s)}{V_{in}(s)}, \; H_3(s) = \frac{V_{3,\text{diff}}(s)}{V_{in}(s)} \qquad (13.28)$$

$$\max|H_1(s)| = \max|1 - H(s)| = \max\left|\left(\frac{\tau s}{Q} + \tau^2 s^2\right)H(s)\right| \approx Q$$

$$\max|H_2(s)| = \max\left|\frac{V_1(s)}{V_{in}(s)}\right| = \max\left|\left(\tau s + \frac{1}{Q}\right)H(s)\right| \approx Q \qquad (13.29)$$

$$\max|H_3(s)| = \max|-H(s)| = \max|H(s)| \approx Q$$

Thus, we note that, in this alternate filter the maximal undistorted input is simultaneously determined by all transconductors rather than by one worst-case transconductor. This input must have an amplitude of V_L/Q such that it is amplified by the transfer functions of Equation (13.29) to just create a differential voltage of V_L across any of the transconductor inputs. The maximum output *SNR* or input dynamic range is then given by

$$SNR_{\max} = \frac{V_L^2/2}{\frac{NqV_L}{2C}Q} = \left(\frac{CV_L}{Nq}\right)\frac{1}{Q} \qquad (13.30)$$

Table 13.1 Comparison between the two-transconductor (Figure 13.8 (c)) and three-transconductor (Figure 13.12 (b)) second-order filter topologies

Topology	V_{max}	Total output noise power	Total bias current (I_{TOT})	SNR_{max}	Power
Figure 13.8 (c)	$\dfrac{V_L}{Q}$	$\dfrac{NqV_L}{2C}$	$I_B\left(Q+\dfrac{1}{Q}\right)$	$\left(\dfrac{CV_L}{Nq}\right)\dfrac{1}{Q^2}$	$2\pi(qV_{DD})N(Q^3+Q)f_n(SNR_{max})$
Figure 13.12 (b)	V_L	$\approx\left(\dfrac{NqV_L}{2C}\right)Q$	$I_B\left(2+\dfrac{1}{Q}\right)$	$\left(\dfrac{CV_L}{Nq}\right)\dfrac{1}{Q}$	$2\pi(qV_{DD})N(2Q+1)f_n(SNR_{max})$

which is Q times higher than the filter of Figure 13.8 (c) whose maximal *SNR* or input dynamic range is given by Equation (13.24). If the bias current needed in a transistor to create $G_m/C = \omega_n$ is I_B, i.e.,

$$I_B = 2\pi f_n C V_L \qquad (13.31)$$

then Table 13.1 reveals that the power consumption of the filter of the two-transconductor filter of Figure 13.8 (c) versus the alternate three-transconductor filter of Figure 13.12 (b) is given by

$$\boxed{\begin{aligned} P_{2Gm} &= 2\pi(qV_{DD})N(Q^3+Q)f_n(SNR_{max}) \\ P_{2Gm} &= 2\pi(qV_{DD})N(2Q+1)f_n(SNR_{max}) \end{aligned}} \qquad (13.32)$$

Thus, the alternate three-transconductor filter has a significantly better speed-precision scaling law for power than the two-transconductor filter. Equation (13.32) is a dramatic example of how topology plays a key role in determining power. Here, the other four determinants of power, namely, the task complexity (high-Q active filter), bandwidth (f_n), precision (SNR_{max}), and technology (qV_{DD}) are identical in both topologies, but one topology does significantly better because of a better mapping of the task to the circuit implementing it.

To create bandpass resonant transfer functions, in the two-transconductor topology of Figure 13.8 (c), the ground terminal of C_1 is used as the input terminal and the input terminal is grounded; to create highpass resonant transfer functions, the ground terminal of C_2 is used as the input terminal and the input terminal is grounded. To create bandpass resonant transfer functions in the three-transconductor topology of Figure 13.12 (b), the ground terminal of C_1 is used as the input terminal and the input terminal is grounded; to create highpass resonant transfer functions, the ground terminal of C_2 is used as the input terminal and input terminal is grounded.

In deep submicron processes, the dc gain of $G_m - C$ filters can be fairly low due a small output resistance R_o and a large V_L, which yields a small transconductance G_m. The small value of R_o not only degrades the dc gain of the filter but its Q as well. Thus, for example, the two-transconductor topology of Figure 13.8 (c) yields

$$Q_{eff} \approx \frac{Q}{1+\left(\dfrac{1+Q^2}{G_mR_o}\right)} \qquad A_{dc} \approx \frac{1}{1+\left(\dfrac{1+Q^2}{G_mR_o}\right)} \qquad (13.33)$$

The three-transconductor topology of Figure 13.12 (b) scales better in this regard as well and yields

$$Q_{eff} \approx \frac{Q}{1 + \left(\dfrac{1 + 2Q}{G_m R_o}\right)} \; ; \; A_{dc} \approx \frac{1}{1 + \left(\dfrac{1 + 2Q}{G_m R_o}\right)} \qquad (13.34)$$

The solution in both cases is to use cascoding to improve R_o.

13.7 Higher-order G_m–C filter design

Higher-order $G_m - C$ filter design may proceed via two approaches: The first approach is to start with a high-order filter prototype such as a ladder filter, look up tabulated values for elements of the filter, and then use element-replacement techniques to synthesize a $G_m - C$ filter. The second approach is to look up tabulated pole and zero locations, decompose the poles and zeros into complex conjugate pairs, use element-replacement or state-space synthesis techniques to make first-order or second-order filters from these decompositions, and then to cascade the first-order or second-order filters. To obtain a good scaling law for power in terms of the band-width, precision, or Q of the filter, it is important to choose a $G_m - C$ topology with balanced differential-voltage transfer functions as we have illustrated for second-order filter design.

13.8 A $-s^2$-plane geometry for analyzing the frequency response of linear systems

Figures 13.14 (a), 13.14 (b), and 13.14 (c) illustrate how a simple $-90°$ rotation followed by a squaring of the frequency, i.e., a $(-js)^2 = -s^2$ mapping of the s-plane, is

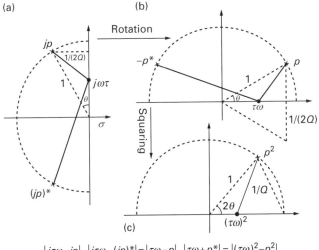

$$\left| j\tau\omega - jp \right| \cdot \left| j\tau\omega - (jp)* \right| = \left| \tau\omega - p \right| \cdot \left| \tau\omega + p* \right| = \left| (\tau\omega)^2 - p^2 \right|$$

Figure 13.14a, b, c. Two-pole to one-pole geometry transformation.

advantageous for computing magnitude transfer functions $|H(j\omega)|$. Figure 13.14 (a) is an exemplary s-plane depiction of poles in a second-order resonant lowpass transfer function $|H(j\omega)|$ except that we have used a normalized frequency axis $j\omega\tau$ such that the poles lie on a circle of unit radius. Figure 13.14 (b) is a $-90°$ rotated version of the s-plane of Figure 13.14 (a). Figure 13.14 (c) performs a s^2 mapping, i.e., a squaring of all the points in Figure 13.14 (b). The net result of these two sequential mappings is an overall $(-js)^2 = -s^2$ mapping. The overall $-s^2$ mapping transforms points on the s-plane in Figure 13.14 (a) to points in the squared-plane plot in Figure 13.14 (c). Why is the squared-plane geometry of Figure 13.14 (c) useful for computing transfer functions?

The squared-plane geometry is useful because it exploits certain symmetries in magnitude transfer functions that automatically lead to their easy geometric evaluation. The magnitude of a transfer function $|H(j\omega)|$ with two complex conjugate poles jp and $(jp)^*$ in Figure 13.14 (a) is evaluated as the reciprocal of **one distance** from the p^2 point to the $(\omega\tau)^2$ point in Figure 13.14 (c). Since only one distance is involved in the evaluation of the magnitude of $|H(j\omega)|$, several properties such as the location of the frequency of maximum gain of $|H(j\omega)|$, its value, and other properties become intuitive and obvious as we will show. In contrast, in the usual s-plane geometry of Figure 13.14 (a), we need the reciprocal of the product of **two distances** from each pole of the complex-conjugate pair to the $j\omega$ frequency point for evaluating $|H(j\omega)|$. Since one of these distances can increase as ω varies, while another can decrease as ω varies, it is not obvious where the location of the frequency of maximum gain in $|H(j\omega)|$ is, what this gain is. Several other properties require tedious algebra to compute in the s-plane.

The intuition and the proof behind the squared-plane geometry of Figure 13.14 (c) is shown at the bottom of Figure 13.14. The essence of the geometry and the proof rely on the fact that any set of four quantities p, p^*, $-p$, and $-p*$ in the s-plane all exhibit a nice four-fold mirror symmetry as shown in Figure 13.15: mirroring about the x-axis (conjugating) and then the y-axis is equivalent to a $180°$ rotation; similarly, mirroring about the y-axis and then the x-axis is also a $180°$ rotation. Therefore, the two complex poles and their mirror images about the y-axis all lie on a circle and

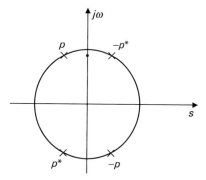

Figure 13.15. Symmetry of complex-conjugate poles on the s-plane.

there is a nice fourfold symmetry. Hence, from the symmetry of Figure 13.15, $|j\omega\tau - (jp)^*| = |j\omega\tau - (-jp)|$ in Figure 13.14. Normalization, i.e., the use of $\omega\tau$ rather than ω, will allow us to generalize to arbitrary transfer functions, as we will show. From Figure 13.14,

$$
\begin{aligned}
|H(j\omega)|^{-1} &= |(j\omega\tau - jp)|.|(j\omega\tau - (jp)^*)| \\
&= |(j\omega\tau - jp)|.|(j\omega\tau - (-jp))| \\
&= |(j\omega\tau - jp)(j\omega\tau + jp)| \\
&= |(\omega\tau - p)(\omega\tau + p)| \\
&= |\omega^2\tau^2 - p^2| \\
&= |p^2 - \omega^2\tau^2|
\end{aligned}
\tag{13.35}
$$

$$
\boxed{|H(j\omega)| = \frac{1}{|p^2 - \omega^2\tau^2|}}
$$

Note that p^2 is the location of the point jp after the $-s^2$ mapping has been applied to it and $\omega^2\tau^2$ is the location of the point $j\omega\tau$ after the $-s^2$ mapping has been applied to it. Thus, we are computing distances between correspondingly mapped points in the squared-plane geometry[1].

From Figure 13.14 (c), we immediately see that the peak gain of $|H(j\omega)|$ occurs when $\omega^2\tau^2 = \cos(2\theta)$ where $\sin(\theta) = 1/2Q$ because the $\omega^2\tau^2$ point is then right underneath the p^2 point and the distance between the two points is minimized. Similarly, the peak gain of $|H(j\omega)|$ at this location is just the reciprocal of the distance between the two points, i.e., $1/(\sin(2\theta))$. These results require a fair bit of algebra to obtain in the normal s-plane geometry. From a $+\theta$ rotation of the triangle denoted by dashed lines in Figures 13.14 (b) and (c), it is easy to derive that, when $\omega^2\tau^2 = 1$, i.e., at $\omega = \omega_n$ where the circle intersects the $\omega^2\tau^2$ axis, the gain is Q. Another easily obtainable result is that when $\omega^2\tau^2 = 2\cos(2\theta)$, i.e., we are a factor of $\sqrt{2}$ past the peak-gain frequency and outside the unit-radius circle, the gain is again unity. Thus, the unity-gain frequency is always half an octave past the peak-gain frequency independent of Q in a resonant low pass second-order system. A further result is that in order for peaking to occur, the point p^2 should have $2\theta < 90°$, i.e., $\theta < 45°$ in the original s-plane plot, or $1/(2Q) < 1/\sqrt{2}$, i.e., $Q > 1/\sqrt{2}$. The geometry also allows computation of the magnitude when $45° < \theta < 90°$. In this case, the p^2 point lies in the left half of the $-s^2$-plane on the unit circle and the magnitude is computed by taking the reciprocal of the distance from a point on the $\omega^2\tau^2$ line to the p^2 point as usual. At $Q = 0.5$, the p^2 point lies at -1 in Figure 13.14 (c).

---

[1] The simple proof listed here was shown to the author by Dr. George Efthivoulidis. The author had originally conceived of and proved results in the $-s^2$-plane geometry in a more complex fashion using trigonometry [3]. R. Sarpeshkar. *Efficient precise computation with noisy components: Extrapolating from an electronic cochlea to the brain.* PhD Thesis, Computation and Neural Systems, California Institute of Technology (1997).

What happens to real-axis poles and zeros when we perform a $-s^2$ mapping? A pole or zero located at $\pm\,\sigma$ is mapped to the point $-\sigma^2$. Thus, in the case of a simple zero,

$$
\begin{aligned}
|H(j\omega)| &= |(j\omega - \sigma)| \\
&= \sqrt{|(j\omega - \sigma)|.|(j\omega - (-\sigma))|} \\
&= \sqrt{|(j\omega - \sigma)(j\omega + \sigma)|} \\
&= \sqrt{|-\omega^2 - \sigma^2|} \\
&= \sqrt{|\omega^2 - (-\sigma^2)|} \\
&= \sqrt{|(j\omega)' - \sigma'|},
\end{aligned}
\tag{13.36}
$$

where $(j\omega)' = \omega^2$ is where the point $j\omega$ has been mapped to by the $-s^2$ mapping, and where $\sigma' = -\sigma^2$ is where the point σ has been mapped to by the $-s^2$ mapping. Thus, the only difference between Equation (13.36) for real-axis poles and zeros versus Equation (13.35) for complex-conjugate pairs is that we take the square-root of the distance from the squared pole or zero location in the $-s^2$-plane to the ω^2 (or $\omega^2\tau^2$ point if we normalize) point rather than the distance itself. In fact, the complex-conjugate case results from the product of two identical distances from two poles in the $-s^2$-plane, which are mirror images of each other about the ω^2 axis; the square root of the product of two identical distances then results in just the distance. Thus, the two cases are really one case. *Effectively we have mapped double complex-conjugate poles and single real-axis poles into single complex poles and square-root real-axis poles in the $-s^2$-plane.*

The σ' point is always on the negative x-axis in the $-s^2$-plane such that when we compute its distance to the ω^2 point, we end up computing

$$
d = \sqrt{\sigma^2 + \omega^2}
\tag{13.37}
$$

i.e., we've really just remapped Pythagoras' theorem for right-angled-triangle distances in the s-plane to the square root of a straight-line distance in the $-s^2$-plane.

If there are many poles and zeros and we want to compute a magnitude response in the $-s^2$-plane, then we always work with the normalized distances in the $-s^2$-plane. Normalized distances are useful for computing the magnitude in the $-s^2$-plane just as they are in the s-plane and as exemplified in our discussion of the normalized root-locus magnitude rule (see Equation (2.28) in Chapter 2). In the s-plane, distances for each pole or zero are normalized by their distance to the origin, $1/\tau$, or if the pole or zero is at the origin, by the integrator or differentiator constant, $1/\tau$. In either case, the normalization constant is $1/\tau$, a value associated with the singularity, wherever it is located. In the $-s^2$-plane, squared distances for each pole or zero are normalized by the squared distance to the origin, i.e., $1/\tau^2$, or if the pole or zero is at the origin, by the squared integrator or differentiator constant, $1/\tau^2$. In the example of Figure 13.15, since all the poles involved were normalized by the same constant, it was simply more convenient to work in normalized units throughout. In more complex situations, we work in the $-s^2$-plane and use different normalizing constants for each pole and zero.

Sometimes, by working in inverse-frequency $1/s$ units, analysis can be vastly simplified. For example, the highpass resonant transfer function $V_L(s)/V_{in}(s)$ and lowpass resonant transfer function $V_C(s)/V_{in}(s)$ of Equation (13.12) are both seen to be identical if we merely interchange τs for $1/(\tau s)$. Thus, all of the geometric analysis for lowpass resonant transfer functions in Figure 13.15 may be immediately applied to highpass resonant transfer functions: the peak gain for a highpass resonant transfer function occurs at $1/(\omega^2 \tau^2) = \cos(2\theta)$, the value of this peak gain is $1/\sin(2\theta)$, etc. All of the intuition and results that we obtained for the resonant lowpass transfer function from Figure 13.15 directly carry over to the resonant highpass transfer function case. The bandpass resonant transfer function can be rewritten as

$$H(s) = \frac{1/Q}{\tau s + \dfrac{1}{\tau s} + 1/Q} \tag{13.38}$$

from which we can see that it is maximized when the τs and $1/(\tau s)$ terms are equal. Thus, it always has its peak gain of 1 at $1/\tau = \omega_n$ independent of Q. In fact, bandpass resonant transfer functions are best viewed by summing $j\omega\tau$ and $1/(j\omega\tau)$ as vectors from the origin on the $j\omega$ axis with $1/Q$ on the σ axis to create a net vector whose reciprocated magnitude and inverted phase is $H(j\omega)$ rather than via a pole-zero plot. Thus, the conventional reactance-and-resistance phasor viewpoint is more useful in understanding bandpass resonant transfer functions than a pole-zero plot.

Figure 13.16 shows that it is also possible to use the $-s^2$-plane to obtain phase. In this figure, the phase of the resonant lowpass transfer function is $-(\pi - \psi)$. To obtain ψ, we first draw an arc of radius d with the center being the $\omega^2 \tau^2$ point of interest and find its intersection point with the vertical line at $\omega^2 \tau^2 = \omega_n^2$; the angle ψ is then the angle shown in Figure 13.16. We see that the phase must be $-90°$ at ω_n. The phase information is not as easily gleaned as the magnitude information

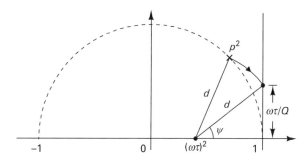

Figure 13.16. Estimating the phase of the transfer function from the one-pole geometry.

(a)

(b)

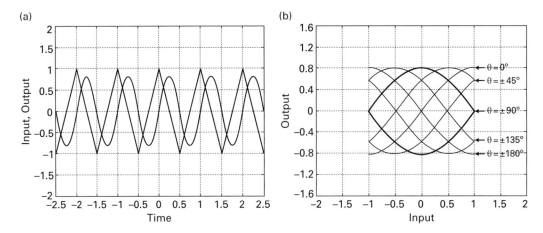

Figure 13.17a, b. Measuring ω_n and Q in a resonant lowpass or highpass transfer function.

that is more transparently present in the $-s^2$-plane, but is still more easily obtained than summing two angles in the s-plane.

The fact that the gain of a lowpass or highpass resonant transfer function is Q at the frequency where the phase is $-90°$ or $+90°$ is frequently useful in experiments when one wants to quickly compute Q and ω_n. If one plots the input and the output on a XY plot on an oscilloscope respectively while a function generator's sinusoidal input is being swept in frequency, a vertically oriented ellipse is obtained at $\omega = \omega_n$ and the ratio of the major-axis length of the ellipse to the minor-axis length of the ellipse (intercepts on the Y-axis and X-axis of the ellipse on the scope) is the Q of the system. Ellipses are rounded smooth functions, however, so the exact value of $\omega = \omega_n$ at which the ellipse just has a vertical orientation can be inexactly estimated by the eye. If the system has relatively low distortion, a useful approximation is to use a triangle-wave input with sharp corners to obtain the value of ω_n as shown in Figures 13.17 (a) and (b), and then switch to a sinusoidal input to obtain the Q. The resonant bandpass case yields a straight line with slope 1 at $\omega = \omega_n$ on a XY-plot but the Q can only be measured from the sharpness of the frequency response in this case.

Sometimes, in the literature, resonant transfer functions are parameterized by the forms that are not normalized, i.e., $H(0) \neq 1$. For example,

$$H(s) = \frac{1}{s^2 + 2\varsigma\omega_n s + \omega_n^2}$$

$$H(s) = \frac{1}{(s+a)^2 + b^2}$$

(13.39)

$$H(s) = \frac{1}{(s - p_1) \cdot (s - p_2)}$$

The reader is strongly discouraged from working with these forms since they are not physical and not intuitive. They are more suited to mathematical use on a

Table 13.2 Parameters of a resonant lowpass transfer function in various representations

	a and b	ω_n and Q	ω_n and Θ	ω_n and θ
Underamped region	$a>0, b>0$	$0.5 < Q < \infty$	$0 < \Theta < \pi/2$	$-\pi/2 < \theta < \pi/2$
Gain > 1 region	$0 < a < b$	$1/\sqrt{2} < Q < \infty$	$0 < \Theta < \pi/4$	$0 < \theta < \pi/2$
Damping	$a/\sqrt{a^2+b^2}$	$1/(2Q)$	$\sin\Theta$	$\sin(\theta/2)$
Peak frequency	$\sqrt{b^2-a^2}$	$\omega_n\sqrt{1-1/(2Q^2)}$	$\omega_n\sqrt{\cos 2\Theta}$	$\omega_n\cos\theta$
Peak gain	$(a^2+b^2)/2ab$	$Q/\left(\sqrt{1-1/(4Q^2)}\right)$	$1/(\sin 2\Theta)$	$1/(\sin\theta)$
$-90°$ frequency	$\sqrt{a^2+b^2}$	ω_n	ω_n	ω_n
$-90°$ frequency gain	$\sqrt{1+b^2/a^2/2}$	Q	$1/(2\sin\Theta)$	$\dfrac{1}{2\sin(\theta/2)}$
Unity gain frequency	$\sqrt{2(b^2-a^2)}$	$\omega_n\sqrt{2-1/(Q^2)}$	$\omega_n\sqrt{2\cos 2\Theta}$	$\omega_n\sqrt{2\cos\theta}$
3 dB frequencies	$\dfrac{b^2-a^2\pm 2ab}{\sqrt{a^2+b^2}}$	$\omega_n\left(1-\dfrac{1}{2Q^2}\pm\dfrac{1}{Q}\sqrt{1-\dfrac{1}{4Q^2}}\right)$	$\omega_n(\cos 2\Theta \pm \sin 2\Theta)$	$\omega_n(\cos\theta \pm \sin\theta)$

computer than to any practical engineering or physical system. Nevertheless, Table 13.2 parameterizes the properties of resonant lowpass transfer functions in various representations. The ω_n and θ representation is the $-s^2$-plane representation while the ω_n and Θ representation is the usual s-plane representation with $\sin(\Theta) = 1/(2Q)$. Thus, $\theta = 2\Theta$. All of these properties were first derived in the $-s^2$-plane representation and then mapped back to the other representations.

References

[1] M. E. van Valkenburg. *Analog Filter Design* (New York: Oxford University Press, Inc., 1982).
[2] R. Sarpeshkar, R. F. Lyon and C. A. Mead. A low-power wide-dynamic-range analog VLSI cochlea. *Analog Integrated Circuits and Signal Processing*, **16** (1998), 245–274.
[3] R. Sarpeshkar. *Efficient precise computation with noisy components: Extrapolating from an electronic cochlea to the brain.* PhD thesis, Computation and Neural Systems, California Institute of Technology (1997).

14 Low-power current-mode circuits

We aren't making the best products just because some customer suggested them to us, or even assured us of big orders, but because we have a passion to bring some art, in which we have a large personal investment, to the pinnacle of perfection.

Barrie Gilbert

In this chapter, we shall discuss circuits that use current inputs and current outputs to create static and dynamic linear and nonlinear systems. Current-mode circuits can operate at low power-supply voltages and over a wide range of currents with exponential subthreshold or bipolar transistors. For reasons that will become clear at the end of the chapter, current-mode signal processing implemented with exponential devices is also often termed log-domain signal processing.

We shall begin by discussing static translinear circuits, which were invented by Barrie Gilbert in the bipolar-circuit domain and that are easily generalized to the subthreshold-MOS domain [1]. Then, we shall discuss circuits for constructing linear current-mode dynamical systems analogous to $G_m - C$ filters in the voltage domain, which were invented and developed by Seevinck [2], Tsividis [3], Frey [4], Toumazou [5], Andreou [6], Vittoz [7], Minch [8] and others. We shall present gain-control techniques for achieving a nearly constant signal-to-noise ratio over a wide dynamic range of inputs, developed by Tsividis for use in linear current-mode input-output systems [9]. Such adaptive techniques separate signal-to-noise ratio (SNR) and dynamic-range variables enabling low-power operation over a wide dynamic range of inputs.

Then, we shall discuss a collective nonlinear current-mode circuit, the winner-take-all circuit, inspired by neuronal networks in biology, which were first invented by John Lazzaro and Carver Mead [10]. Such circuits identify the maximum amongst several current inputs in a nonlinear and highly parallel fashion. Finally, we shall discuss a distributed-feedback current-mode circuit inspired by the feedback networks present amongst neurons in the brain [11]. The latter circuit is an interesting example of how positive and negative feedback can interact to create simultaneous digital and analog behavior in a network: the behavior is a hybrid between that of a multi-input digital latch and a vector analog amplifier.

In Chapter 24, we shall show that polynomially nonlinear current-mode dynamical systems can enable extremely fast simulations of computationally intensive biochemical reaction networks including stochastics [12], [13]. Thus, these current-mode dynamical systems are useful for understanding, designing, and exploring the operation of molecular networks within cells, an important goal in the growing modern fields of systems biology and synthetic biology. Current mode circuits are also extensively used in low-power biomedical applications, examples of which are outlined in Chapter 19.

14.1   Voltage versus current

Voltage signal variables can be used an infinite number of times by simply hooking wires to the nodes at which they reside; overuse of any voltage signal variable inevitably leads to undesirable loading in the circuit such that buffering may become necessary. Transistor current signal variables, in contrast, can typically only be used twice in a circuit, once at the source end of the transistor and once at the drain end of the transistor; further copies of current signal variables require the use of current mirrors. Threshold-voltage and geometry mismatches in transistors lead to additive errors in voltage-mode circuits. Such offsets manifest as multiplicative gain errors in current-mode circuits.

Since current variables represent the dynamic range of the signal in current-mode circuits rather than voltage, current-mode circuits are amenable to low-voltage operation even in wide-dynamic-range applications. Addition in voltage-mode circuits can be implemented via Kirchhoff's voltage law (KVL). However, because no floating voltage sources are ever needed, the use of Kirchhoff's current law (KCL) to implement addition is more widely used in both current-mode and voltage-mode circuits, e.g., in summing junctions in operational-amplifier adder circuits. Multiplication in voltage-mode circuits requires stacked differential pairs in a Gilbert multiplier circuit, each of which has a limited linear range of operation. Four-quadrant precision multiplication (either multiplicative input can be positive or negative) is relatively easy to achieve. Multiplication and other nonlinear operations can often be implemented over a wide dynamic range of operation in current-mode translinear circuits. Four-quadrant current-mode multiplication requires current-splitting-and-recombining circuits and/or offset-current biasing circuits and is harder to achieve with precision. Linear current-to-frequency conversion is significantly easier than linear voltage-to-frequency conversion. Most voltage-mode circuits are Class A circuits and have constant power dissipation independent of signal strength while current-mode circuits necessarily always have signal-dependent power dissipation. Certain sensor outputs naturally generate currents, e.g., photocurrents in an imager, while others naturally generate voltage signals, e.g., cardiac or neural action potentials.

The pros and cons of voltage-mode versus current-mode operation are such that it is best to map the signal processing into whichever mode of operation is most advantageous and natural and to combine both modes of processing in hybrid voltage-and-current-mode circuits. Transimpedance or integrating-capacitive amplifiers and transconductance amplifiers are useful in converting from current to voltage and from voltage to current, respectively. They also perform signal amplification, and are often implicitly present in both voltage-mode circuits and current-mode circuits. For example, $G_m - C$ filters convert current to voltage linearly via capacitive integration and voltage to current linearly via a transconductor. Current-mode filters implement nonlinear compressive logarithmic current-to-voltage conversion in diode-like transistor circuits, process this voltage, and then reconvert the processed voltage into an output current via a nonlinear expansive exponential voltage-to-current conversion such that the overall input-output current-mode system is linear. Thus, a pure current-mode or voltage-mode circuit does not exist since the elementary devices in the circuit, i.e., transistors, are inherently transconductance devices that convert voltage to current and consequently require both voltage and current variables to operate. It is the abstractions and input-output signal variables we choose to focus on that give rise to current-mode or voltage-mode circuits.

14.2   Static translinear circuits

The existence and usefulness of translinear circuits resides in two key results, one mathematical and one physical. The mathematical result arises from the laws of addition and subtraction of logarithms:

$$\begin{aligned} \log(ab) &= \log(a) + \log(b) \\ \log\!\left(\frac{a}{b}\right) &= \log(a) - \log(b) \end{aligned} \tag{14.1}$$

The physical result is that a nearly ideal exponential function can be cheaply implemented in a single transistor operating in saturation with

$$I = \lambda I_s e^{\eta V/\phi_t} \tag{14.2}$$

where the variables are illustrated in Figure 14.1 (a) for a generic exponential transistor, which could be a subthreshold MOS transistor or a bipolar transistor. For real bipolar transistors $\eta = 1$, λ is proportional to the emitter area, and I_S is the pre-exponential constant of the bipolar transistor. Translinear circuits were

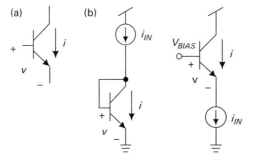

Figure 14.1a, b. (a) An elementary bipolar transistor circuit and (b) two simple circuits that have local negative feedback loops (a transistor diode and a source-follower).

first invented in bipolar-transistor technology. *For subthreshold MOS transistors* $\eta = \kappa_S$ *with the well/bulk tied to the source to ensure that there is no body effect,* λ *is proportional to the W/L of the transistor, and $I_S = I_{OS}$.* Note that n denotes $1/\kappa_S$. In processes without wells for pFETs and nFETs, only pFETs can be used to create such a generic bipolar transistor. In very-low-voltage processes, it is possible to tie the gate and well together to yield $\eta = \kappa_S + (1 - \kappa_S) = 1$ without forward biasing the source-to-well junction.

The presence of the body effect results in a κ-dependent exponential constant for the gate and a non-κ-dependent exponential constant for the source in a subthreshold MOS transistor such that the properties of logarithms may still be exploited to create translinear circuits. However, they often result in undesirable power laws. For example, a translinear circuit designed to compute $I_{out} = I_B{}^{1/2}I_{in}{}^{1/2}$ may yield $I_{out} = I_B{}^{1/(1+\kappa)}I_{in}{}^{\kappa/(1+\kappa)}$ instead. Thus, unless otherwise mentioned, we shall always tie the well to the source in this chapter when we are discussing translinear circuits in subthreshold MOS technology.

We are accustomed to exploiting the properties of the exponential in voltage-mode circuits where V is the input signal and I is the output signal. The alternate approach, which is used in current-mode circuits, is to use I as an input signal and magically set V such that I flows through the device. Then, from Equation (14.2) we have

$$V = \frac{\phi_t}{\eta} \ln\left[\frac{I}{\lambda I_s}\right] \tag{14.3}$$

In Figure 14.1 (b) local negative-feedback loops set V such that the transistors carry the current i_{IN} and create a voltage that is logarithmically related to i_{IN}. These simple circuits are an example of negative feedback creating an inverse logarithmic function from an exponential. If we somehow create one or more of these feedback-forced voltages and connect them in a loop or loops such that KVL holds for each loop, we can impose a set of algebraic constraints on the logarithms of the transistor currents, and thus on the products or ratios of these currents. For example, in the simple current-mirror circuit of Figure 14.2, KVL in a clockwise direction around the loop yields

$$
\begin{aligned}
& V_1 - V_2 = 0 \\
& V_1 = \frac{\phi_t}{\eta} \ln\left[\frac{I_{IN}}{\lambda_1 I_s}\right] \\
& V_2 = \frac{\phi_t}{\eta} \ln\left[\frac{I_{OUT}}{\lambda_2 I_s}\right] \\
& \frac{\phi_t}{\eta} \ln\left[\frac{I_{IN}}{\lambda_1 I_s}\right] = \frac{\phi_t}{\eta} \ln\left[\frac{I_{OUT}}{\lambda_2 I_s}\right] \\
& \left[\frac{I_{IN}}{\lambda_1 I_s}\right] = \left[\frac{I_{OUT}}{\lambda_2 I_s}\right] \\
& \frac{I_{IN}}{\lambda_1} = \frac{I_{OUT}}{\lambda_2}
\end{aligned}
\tag{14.4}
$$

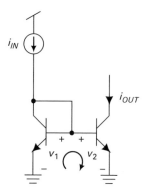

Figure 14.2. A simple current mirror circuit implemented with bipolar transistors.

where we have assumed that I_S, η, and the temperature are identical for each transistor in the mirror. Thus the current density of the two transistors of Figure 14.2 is the same and the currents in these transistors scale with their emitter areas or W/Ls.

The same technique may be applied to more general loops of transistors. As we travel around the loop in the direction of the arrow, we regard voltage rises as 'clockwise voltages' and voltage drops as counter-clockwise voltages. Then, KVL around the entire loop yields

$$\sum_{n \in CW} V_n = \sum_{n \in CCW} V_n$$

$$\sum_{n \in CW} \frac{\phi_t}{\eta} \ln\left[\frac{I_n}{\lambda_n I_s}\right] = \sum_{n \in CCW} \frac{\phi_t}{\eta} \ln\left[\frac{I_n}{\lambda_n I_s}\right]$$

$$\sum_{n \in CW} \ln\left[\frac{I_n}{\lambda_n I_s}\right] = \sum_{n \in CCW} \ln\left[\frac{I_n}{\lambda_n I_s}\right]$$

$$\ln\left[\prod_{n \in CW} \frac{I_n}{\lambda_n I_s}\right] = \ln\left[\prod_{n \in CCW} \frac{I_n}{\lambda_n I_s}\right] \qquad (14.5)$$

$$\prod_{n \in CW} \frac{I_n}{\lambda_n} = I_S{}^{CW-CCW} \prod_{n \in CCW} \frac{I_n}{\lambda_n}$$

$$\prod_{n \in CW} \frac{I_n}{\lambda_n} = \prod_{n \in CCW} \frac{I_n}{\lambda_n}$$

where the latter result is derived assuming that the number of clockwise and counter-clockwise elements in the loop are equal, and that the temperature, I_S, and η terms for each transistor are identical.

The translinear principle states that if the number of clockwise and counter-clockwise elements in a KVL translinear loop are equal, then the product of the current densities through the clockwise elements is equal to the product of the current densities through the counter-clockwise elements. In subthreshold MOS operation, the V_{SB} voltage

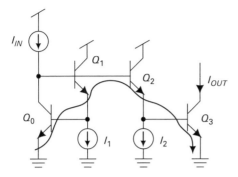

Figure 14.3. A static translinear circuit implemented with bipolar transistors, showing the translinear loop.

that determines $\eta = \kappa_S$ and I_S needs to be equal for all transistors in the translinear loop such that all η and I_S terms in Equation (14.5) cancel.

This condition is automatically fulfilled in the current-mirror circuit of Figure 14.2 but is not fulfilled in general subthreshold circuits. A simple way of fulfilling this condition in subthreshold translinear circuits is to use only pFET transistors and to set $V_{SB} = 0$ for all transistors in the loop by tying their wells to their respective sources. A disadvantage of tying the well to the source is that the well-to-substrate capacitance then manifests as a parasitic node capacitance at the source.

Figure 14.3 illustrates a simple one-quadrant translinear circuit with the translinear loop marked with a wavy arrow. Applying the translinear principle to the loop, we find that Q_0 and Q_1 have voltage rises in the direction of the loop and Q_2 and Q_3 have voltage drops in the direction of the loop; if all transistors have the same λ factors, we must then have

$$I_{IN} \cdot I_1 = I_2 \cdot I_{OUT}$$
$$I_{OUT} = I_{IN} \frac{I_1}{I_2} \tag{14.6}$$

Thus I_{OUT} is a gained version of I_{IN} with the gain determined by I_1/I_2. Figure 14.4 (a) illustrates a subthreshold MOS implementation of the translinear multiplier circuit of Figure 14.3. Figure 14.4 (b) reveals that I_{OUT} is indeed a gained version of I_{IN} with the gain determined by I_1/I_2 and equal to 2, 4, or 6 in this simulation example.

14.3   Dynamic translinear lowpass filters

Figure 14.5 (a) illustrates the same circuit as in Figure 14.3 except that there is now an explicit capacitor C at the base of the output transistor Q_3. From Figure 14.5 (b), we find that the current through this capacitor I_c is given by

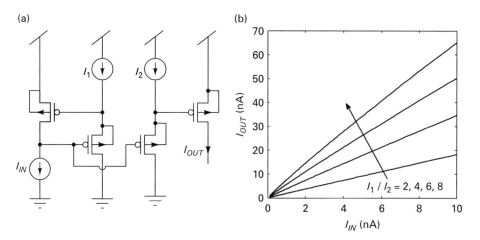

Figure 14.4a, b. (a) A static translinear circuit, implemented with subthreshold PMOS transistors and (b) simulation results.

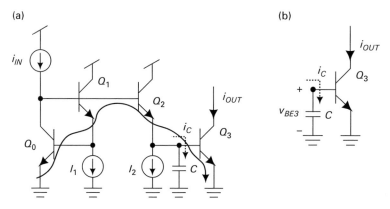

Figure 14.5a, b. (a) A dynamic translinear lowpass filter, circuit #1, showing the translinear loop, and (b) an isolated bipolar transistor with a capacitor at its base terminal.

$$i_C = C \frac{dv_{BE3}}{dt}$$
$$= C \frac{d}{dt} \frac{\phi_t}{\eta} \ln\left(\frac{i_{OUT}}{\lambda_3 I_S}\right)$$
$$= \frac{C\phi_t}{\eta} \frac{1}{i_{OUT}} \frac{di_{OUT}}{dt} \quad (14.7)$$

$$\boxed{i_c = \frac{C\phi_t}{\eta} \frac{1}{i_{OUT}} \frac{di_{OUT}}{dt}}$$

The current through a capacitor between the base and emitter of a transistor carrying current i is given by a similar expression to that in the boxed expression of Equation (14.7) with i_{OUT} replaced by i. The expression for the capacitive current is often referred to as yielding a dynamic translinear principle as the time derivative of the current i_{OUT} is the product of the two currents i_C and i_{OUT}.

The only difference between Figure 14.5 (a) and Figure 14.3 as far as the translinear loop is concerned is that a current of $i_C + I_2$ flows through Q_2 instead of just I_2. The translinear principle still applies. Therefore,

$$
\begin{aligned}
i_{IN}I_1 &= (I_2 + i_C)i_{OUT} \\
i_{IN}I_1 &= I_2 i_{OUT} + \frac{C\phi_t}{\eta} \frac{1}{i_{OUT}} \frac{di_{OUT}}{dt} i_{OUT} \\
i_{IN}I_1 &= I_2 i_{OUT} + \frac{C\phi_t}{\eta} \frac{di_{OUT}}{dt} \\
i_{IN}\frac{I_1}{I_2} &= i_{OUT} + \frac{C\phi_t}{I_2\eta} \frac{di_{OUT}}{dt}
\end{aligned}
\tag{14.8}
$$

Equation (14.8) is a large-signal input–output first-order linear differential equation in currents! If we take the Laplace transform of its signal variables, we find that

$$
\begin{aligned}
I_{in}(s)\frac{I_1}{I_2} &= (1 + \tau s)I_{out}(s) \\
\tau &= \frac{C\phi_t}{I_2\eta} \\
\frac{I_{out}(s)}{I_{in}(s)} &= \frac{I_1/I_2}{1 + \tau s}
\end{aligned}
\tag{14.9}
$$

We have created a lowpass filter with a dc current gain given by I_1/I_2. If we set $\tau = C = 0$ in Equation (14.9), we find a pleasing consistency: Equation (14.9), which describes the dynamic circuit of Figure 14.5 (a), yields the same answer as Equation (14.6), which describes the static circuit of Figure 14.3.

Figure 14.6 (a) shows a pFET subthreshold implementation of the translinear circuit of Figure 14.5 (a). For this circuit, the following parameters were extracted from a simulation of it: $C = 6$ pF; $I_1 = 2$ nA; $I_2 = 1$ nA; $\kappa \approx 0.7$; $\phi_t = 25.9$ mV. Thus, we would predict a time constant of 222 μs and a dc gain of 2 from Equation (14.9). Figure 14.6 (b) shows a simulated step response, which yields a time constant of 259 μs and a dc gain of 1.89. The discrepancy between theory and simulation may be attributed to parasitics and inaccuracy in the estimation of parameters, particularly κ: translinear dynamical circuits are subject to the presence of parasitic capacitances at all of their nodes, especially C_{gb} in subthreshold, not just intentional capacitances. Hence, practical translinear dynamical systems are actually of higher order and contain unwanted poles and zeros.

To understand how general translinear dynamical systems may be architected, it is worthwhile to first study other methods for creating lowpass filters. Figure 14.7 reveals another translinear lowpass filter circuit. From the marked translinear loop, we may conclude that

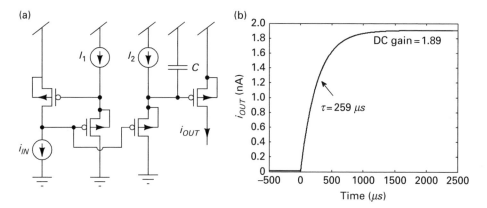

Figure 14.6a, b. (a) Dynamic translinear lowpass filter, circuit #1, implemented with subthreshold MOS transistors, and (b) its step response.

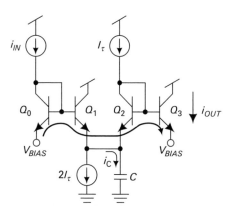

Figure 14.7. Dynamic translinear lowpass filter, circuit #2.

$$i_{IN} \cdot I_\tau = (I_\tau + i_C) \cdot i_{OUT}$$

$$i_C = C\frac{dv_{E1}}{dt} = C\frac{d(v_{E1} + V_{BE2})}{dt} = C\frac{dv_{B3}}{dt} = C\frac{dv_{BE3}}{dt} = \frac{C\phi_t}{\eta}\frac{1}{i_{OUT}}\frac{di_{OUT}}{dt}$$

$$i_{IN} \cdot I_\tau = \left(I_\tau + \frac{C\phi_t}{\eta}\frac{1}{i_{OUT}}\frac{di_{OUT}}{dt}\right)i_{OUT} \qquad (14.10)$$

$$i_{IN} = i_{OUT} + \frac{C\phi_t}{\eta I_\tau}\frac{di_{OUT}}{dt}$$

Equation (14.10) is a first-order differential equation identical to that of Equation (14.8) if $I_1 = I_2 = I_\tau$ and also implements a lowpass filter with a dc gain of 1. The crucial step in the derivation of Equation (14.10) arises from the fact

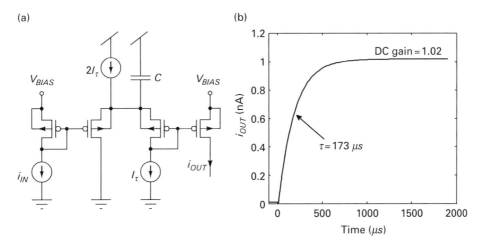

Figure 14.8a, b. (a) Dynamic translinear lowpass filter, circuit #2, implemented with subthreshold PMOS transistors and (b) its step response.

that V_{BE2} is a constant since the current I_τ is a constant such that the change in the capacitor's voltage is the change in v_{BE3}. The dc gain of this circuit may be increased beyond 1 by reducing the $2I_\tau$ current and/or increasing the I_τ current in Figure 14.7 to change the ratio between the dc current of Q_1 and Q_2. This change also lowers the time constant. The maximum practical dc gain attainable is limited by the precision of current matching in this circuit since the $2I_\tau$ current must always be at least as large as the I_τ current for the circuit of Figure 14.7 to operate properly. The circuit of Figure 14.5 (a) has independent control over the dc gain and time constant and is less sensitive to the precision of current matching.

Figure 14.8 (a) shows a pFET subthreshold implementation of the translinear circuit of Figure 14.7. For this circuit, the following parameters were extracted from a simulation of it: $C = 2$ pF; $I_\tau = 0.5$ nA; $\kappa \approx 0.7$; $\phi_t = 25.9$ mV. Thus, we would predict a time constant of 148 μs from Equation (14.10). Figure 14.8 (b) shows a simulated step response, which yields a time constant of 173 μs and a dc gain of 1.02. The use of a smaller explicit capacitance has made this circuit more sensitive to higher-order parasitic dynamics than the circuit of Figure 14.6 (a).

Figure 14.9 shows yet another dynamic translinear lowpass circuit. The generic bipolar transistors are transistors that form components of translinear loops and would be morphed to source-tied-to-well pFET transistors in a subthreshold MOS implementation. The other transistors are important as components of feedback loops that are integral to the functioning of the overall circuit and may be built with regular MOS transistors as shown. The circuit has two negative-feedback loops that are important for its function. A first minor negative-feedback loop regulates the current through one of the differential-pair transistors such that it is always I_B; thus, this differential-pair transistor, I_B, and the MOS transistor that sets the differential-pair tail current effectively behave like a source-follower with input v_{OUT}, output at the common-source node of the differential pair, v_C, and a

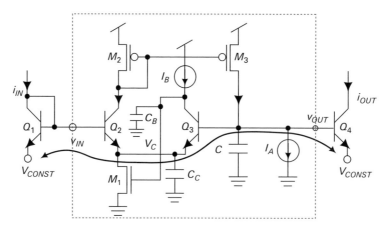

Figure 14.9. Dynamic translinear lowpass filter, circuit #3.

bias current of I_B; thus, the common-source node of the differential pair, v_C, quickly follows the voltage v_{OUT} with an offset voltage of $(kT/q)\ln(I_B/I_s)$. A second major negative-feedback loop changes v_{OUT} via capacitive charging in a slow fashion such that v_{OUT} and consequently v_C is driven to a value such that the current through Q_2 equilibrates at I_A. It is the slow charging of capacitor C that causes the circuit to function like a current-mode lowpass filter when the input current i_{IN} determines v_{IN} and the voltage v_{OUT} is used to create an output current i_{OUT}. The overall circuit from v_{IN} to v_{OUT} may be viewed as a dynamic floating voltage source with an equilibrium dc offset voltage of $v_{OUT} - v_{IN} = (kT/q)\ln(I_B/I_A)$ and a small-signal time constant of $C(kT/q)/I_A$. The additive dc offset between v_{IN} and v_{OUT} causes a multiplicative dc gain between i_{IN} and i_{OUT} of I_B/I_A and the dynamics of the floating voltage source manifest as a first-order lowpass filter in the i_{IN}-i_{OUT} current-mode input-output system.

The marked translinear loop in Figure 14.9 yields

$$i_{IN} \cdot I_B = (i_C + I_A) \cdot i_{OUT}$$
$$i_C = \frac{C\phi_t}{\eta} \frac{1}{i_{OUT}} \frac{di_{OUT}}{dt} \tag{14.11}$$

such that we may once again conclude that

$$i_{IN}I_B = I_A i_{OUT} + \frac{C\phi_t}{\eta} \frac{di_{OUT}}{dt}$$
$$\frac{I_{out}(s)}{I_{in}(s)} = \frac{I_B/I_A}{1 + s\left(\dfrac{C\phi_t}{I_A\eta}\right)} \tag{14.12}$$

It is interesting to note the similarities between the lowpass filter circuits of Figure 14.5 (a), Figure 14.7, and Figure 14.9: In all cases, a log-encoded input voltage created by the input current drives a source-follower-with-a-capacitor configuration whose output voltage is then exponentiated to create the output current. In Figure 14.5 (a), the source-follower-with-a-capacitor is formed by Q_2, I_2, and C. In Figure 14.7, the source-follower-with-a-capacitor is formed by Q_1, $2I_\tau - I_\tau$ and C. In Figure 14.9, the source-follower-with-a-capacitor is implicit because v_{OUT} and v_C differ by a nearly instantaneous dc offset such that I_A and C at node v_{OUT} may effectively be viewed as being present at v_C. From this point of view, the transistor Q_2 in Figure 14.7, which implements a dc offset, is analogous to transistor Q_3 in Figure 14.9. The nature of the translinear circuit is such that the nonlinear dynamics of a source-follower-and-a-capacitor are exactly linearized by log-compressing its input and exponentiating its output to create an input-output linear system in all three cases.

The circuit of Figure 14.9 may be viewed as the best current-mode lowpass filter. The circuit of Figure 14.5 (a) has a local feedback loop formed by Q_0, Q_1, i_{IN}, and I_1 in the log-encoding of its input, which leads to poor phase margin at large input currents. The circuit of Figure 14.7 has coupled dc gain and time-constant parameters and requires good current matching to achieve good time-constant control, especially at high dc gains. The circuit of Figure 14.9 has separable dc gain and time-constant parameters; its local feedback loop has a phase margin that is independent of its input current and only determined by I_B and I_A. It can be shown by simple application of the two-pole-feedback-Q rule (see Equation (9.16) in Chapter 9 with $A_{lp}=1$ and Figure 11.12 (c) in Chapter 11) that, as long as the parasitic node capacitance at the output of the I_B current source, C_B, is larger by $4I_B/(I_A + I_B)$ than the node capacitance at v_C, C_C, the local feedback loop of Figure 14.9 is always over-damped. For good filter dynamics we need $I_A/C \ll I_B/C_B$ such that parasitic time constants do not degrade the performance of the filter. In all three circuits, at sufficiently low input currents and/or high frequencies, parasitic time constants due to capacitances present at log-encoded input-voltage nodes degrade ideal filter dynamics. In all three circuits, above-threshold MOS operation or high-current bipolar operation causes distortion at large input-current levels.

14.4 Dynamic translinear integrators and high-order filters

Figure 14.10 illustrates a differential translinear integrator circuit that is an important building block for creating filters of arbitrary order. It is similar to the lowpass filter circuit of Figure 14.9 except for the presence of two differential inputs $i_{IN}{}^+$ and $i_{IN}{}^-$ and the absence of I_A. The translinear loops involving $i_{IN}{}^+$ and $i_{IN}{}^-$ in Figure 14.10 followed by simple algebra yield

$$i_{IN+} \cdot I_G = i_{Q1} \cdot i_{OUT}$$
$$i_{IN-} \cdot I_G = i_{Q2} \cdot i_{OUT}$$
$$(i_{IN+} - i_{IN-}) \cdot I_G = (i_{Q1} - i_{Q2}) \cdot i_{OUT}$$

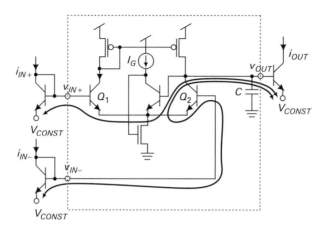

Figure 14.10. Dynamic translinear integrator circuit with two translinear loops.

$$i_C = i_{Q1} - i_{Q2}$$
$$(i_{IN+} - i_{IN-}) \cdot I_G = i_C \cdot i_{OUT}$$
$$i_C = \frac{C\phi_t}{\eta} \frac{1}{i_{OUT}} \frac{di_{OUT}}{dt} \qquad (14.13)$$
$$\boxed{(i_{IN+} - i_{In-}) \cdot \frac{\eta I_G}{\phi_t} = C\frac{di_{OUT}}{dt}}$$

The equation for the current-mode integrator derived in Equation (14.13) is exactly analogous to that derived for a voltage-mode $G_m - C$ integrator, i.e.

$$(v_{IN+} - v_{IN-}) \cdot G_m = C \cdot \frac{dv_{OUT}}{dt} \qquad (14.14)$$

if we replace currents with voltages and identify

$$G_m = \frac{\eta I_G}{\phi_t} \qquad (14.15)$$

With these voltage-to-current mappings, current-mode filter design can simply mimic $G_m - C$ filter design by using the circuit-replacement strategy shown in Figure 14.11 (a). Figure 14.11 (b) provides an example for how any filter in a $G_m - C$ topology can be rapidly transformed to a current-domain equivalent. We logarithmically encode the input current via a diode-based i_{IN}-to-v_{IN} mapping, use the current-mode circuit equivalent shown in Figure 14.11 (a) to replace the $G_m - C$ integrators in the boxed region of Figure 14.11 (b) with their current-mode equivalents, and finally exponentiate the output voltage v_{OUT} to create i_{OUT}. The boxed region of the current-mode filter is called its log-domain core since we are operating with logarithmically encoded voltages in this portion of the circuit. Current-mode filters are often referred to as log-domain filters.

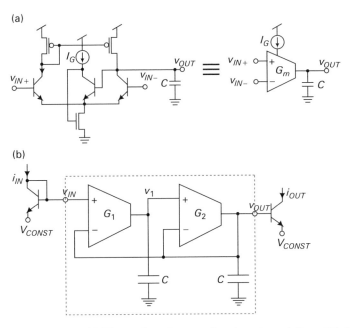

Figure 14.11a, b. (a) The analogy between log-domain and $G_m - C$ integrators, and (b) showing an example of a log-domain filter's 'core'.

14.5   Biasing of current-mode filters

Current-mode filters require that currents always be positive yet ac current inputs must be bidirectional. Figures 14.12 (a) and 14.12 (b) illustrate Class-A versus Class-AB biasing for current-mode filters respectively, which take two different approaches to solving this problem. In Class-A biasing shown in Figure 14.12 (a), we bias the input i_{IN} with an added dc current I_{MAX} such that

$$i_{IN} = I_{MAX} + i_{ac} \tag{14.16}$$

is always positive. The current I_{MAX} is chosen based on the expected largest amplitude of i_{ac} such that i_{IN} is always positive. Class-A biasing is similar to dc biasing of voltages with an ac component such that the total voltage always remains within the rails. In Class-AB biasing, the ac current is split into its positive and negative halves with a current splitter, processed separately, and then recombined via subtraction to create the output.

Class-A biasing requires the use of a quiescent current I_{MAX} that wastes power dissipation. The quiescent value of I_{MAX} establishes the conductance of the log-compressing input element of the filter at $g_{m,MAX}$ such that signal-dependent input voltage swings, which scale according to $v_{in,ac} \approx i_{ac}/g_{m,MAX}$, are small and distortion from the log-domain core is minimized. Class-AB biasing is more power efficient than Class-A biasing since it needs no quiescent I_{MAX}; however, the signal splitter often limits the performance of the overall filter; mismatch between the

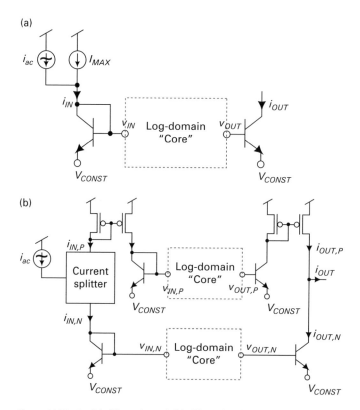

Figure 14.12a, b. (a) Class-A and (b) Class-AB biasing for log-domain circuits.

two log-domain halves causes harmonic distortion; and, the log-compressed voltage swing at the input element of the filter scales more strongly with the input current amplitude than in a Class-A scheme such that distortion from the log-domain core is increased.

Adaptive-Class-A biasing attempts to combine the benefits of the Class-A and Class-AB worlds without being strongly subject to their problems. The value of I_{MAX} in Class-A biasing is adaptively set to be proportional to that of envelope (peak amplitude) of i_{ac}, $I_{AC,ENV}$, e.g., at $1.2I_{AC,ENV}$. The factor of 1.2 provides a safety margin, thereby ensuring robust operation of the circuit. The signal $I_{AC,ENV}$ is expected to vary slowly and is extracted by a separate large-time-constant circuit. Thus, the quiescent average power dissipation is minimized since the value of $I_{AC,ENV}$ adapts to the ac signal level rather than being set at a worst-case large value to handle the largest expected ac signal. The precision requirements of envelope extraction are significantly more relaxed than the precision requirements of a current-splitting circuit since the exact value of the envelope is not critical. The log-compressed voltage swing at the input node of the filter is signal-independent since the g_m of the log-compressing input element of the filter scales with $I_{AC,ENV}$ such that $v_{in,ac} \approx i_{ac}/g_m$ is always fixed and independent of the

(a)

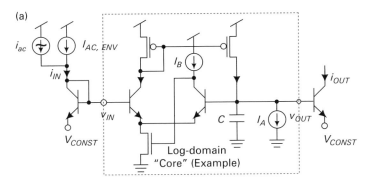

(b)

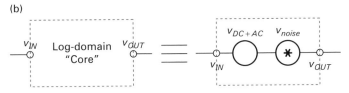

Figure 14.13a, b. Adaptive Class A log-domain biasing: (a) showing an example of a log-domain core and (b) showing the analogy with a floating voltage source.

amplitude of i_{ac}; thus, distortion from the log-domain core is minimized. Or more formally,

$$v_{in} = \frac{i_{ac}}{g_m} \propto \frac{\phi_t}{\eta} \left(\frac{i_{ac}}{I_{AC,ENV}} \right) \approx \text{invariant with signal amplitude} \qquad (14.17)$$

In essence, adaptive-Class-A biasing implements a form of slow gain control by proportionately scaling the g_m of the input and output elements of the current-mode circuit with i_{ac} and consequently $I_{AC,ENV}$. If signal amplitude increases, there is proportionately more attenuation at the input current-to-voltage conversion $(1/g_m)$ but also proportionately more gain at the output voltage-to-current conversion (g_m) such that overall input-output linearity is preserved $((1/g_m) \cdot g_m = 1)$. The log-domain core then experiences little change in v_{in} as Equation (14.17) shows. The log-domain core does experience dc shifts in V_{IN} as i_{ac} and consequently $I_{AC,ENV}$ change but it is insensitive to these shifts due to the usual common-mode rejection inherent in a differential pair, e.g., as in Figure 14.9. The presence of gain control in adaptive-class-A biasing enables several desirable properties including the ability to maintain a constant output signal-to-noise ratio (SNR) over a wide dynamic range of input currents as we discuss below. Figure 14.13 (a) shows examples of a log-domain core with adaptive-class-A biasing and Figure 14.13 (b) shows that this core can be well modeled as a dynamic floating voltage source with noise. The noise is due to noise in the devices in the log-domain core.

One potential problem with adaptive-Class-A biasing is that fast dynamic variations in the envelope, $I_{AC,ENV}$, might cause signals to unwantedly appear in or unwantedly disappear from the passband of the filter. Typically, these

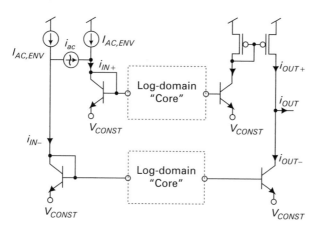

Figure 14.14. A generic pseudo-differential adaptive Class A log-domain circuit.

variations still ought to be slow enough to be outside the passband of the filter but sudden transients could potentially manifest themselves. If these signals do appear, then the pseudo-differential scheme shown in Figure 14.14 can help alleviate them. We first create two halves of the input signal with

$$
\begin{aligned}
i_{IN+} &= I_{AC,ENV} + i_{ac} \\
i_{IN-} &= I_{AC,ENV} - i_{ac}
\end{aligned}
\tag{14.18}
$$

These unwanted signals can then be processed and cancelled by subtraction according to

$$
\begin{aligned}
i_{OUT+} &= h * i_{IN+} = h * I_{AC,ENV} + h * i_{ac} \\
i_{OUT-} &= h * i_{IN-} = h * I_{AC,ENV} - h * i_{ac} \\
i_{OUT} &= i_{OUT+} - i_{OUT-} = h * 2i_{ac}
\end{aligned}
\tag{14.19}
$$

Here $h(t)$ is the impulse response of the filter, and $*$ denotes convolution. Automatic-gain-control (AGC) circuits with fast attack and slow release times for small-to-large and large-to-small transients respectively also help ensure that transients in $I_{AC,ENV}$ minimize passband signal distortion and/or signal loss. For examples of such circuits, see [14].

Adaptive-Class-A biasing is analogous to adapting the power-supply voltage in voltage-mode circuits according to the amplitude needs of the input signal. Such adaptation is impractical in most voltage-mode circuits but is quite practical in current-mode circuits.

14.6   Noise, SNR, and dynamic range of log-domain filters

Figure 14.15 (a) shows a log-domain filter circuit (in this case a lowpass filter) and Figure 14.15 (b) shows a simple model of its noise. Its input and output devices

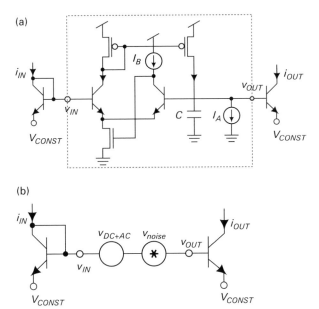

Figure 14.15a, b. (a) A log-domain lowpass filter circuit and (b) a simple model of its noise.

contribute noise. The noise due to its log-domain core may be modeled as the noise due to that of a dynamic floating voltage source. Thus, the total current noise spectral density of this circuit is given by

$$i_n^2(f) = 4qI_{OUT} + g_{m,out}^2 v_{n,core}^2$$

$$i_n^2(f) = 4qI_{OUT} + \eta^2 I_{OUT}^2 \frac{v_{n,core}^2}{\phi_t^2} \qquad (14.20)$$

where $v_{n,core}^2$ is the total noise spectral density of the core, which may be computed by voltage-mode input-referred-noise techniques as described in Chapters 7 and 8. We have assumed that each of the input and output devices exhibit thermal noise in Equation (14.20). Thus, we note that there is a term proportional to I_{OUT}^2 and a term proportional to I_{OUT}. The I_{OUT}^2 term dominates for all but the smallest output currents. This noise spectral density may be integrated over the bandwidth of the filter with the usual techniques for integration of noise described in the previous chapters to obtain the total output current noise. In the example of Figure 14.15 (a), the bandwidth is given by $\eta I_A/(C\phi_t)$. In Class-A circuits, $I_{OUT} = I_{MAX}$ and does not change with signal amplitude such that the total noise is invariant with signal level. Thus, the output *SNR* of a Class-A log-domain circuit improves with input signal level just like a voltage-domain Class-A circuit does.

In adaptive-Class-A operation, I_{OUT} is set by $I_{ENV,AC}$ and thus proportional to the amplitude of i_{ac}. Thus, the *SNR* improves with output signal level only if the I_{OUT} term is dominant in Equation (14.20) because the noise power grows linearly while the signal power grows quadratically. Once the I_{OUT}^2 term begins to

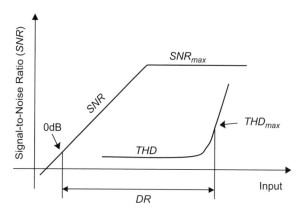

Figure 14.16. SNR and distortion of adaptive bias Class-A log-domain circuits versus input frequency.

dominate in Equation (14.20), which is almost always the case except for the very smallest output currents, the signal power and noise power grow at the same quadratic rate such that the output *SNR* becomes constant and invariant with output signal level. This maximal $SNR_{\max}$ is computed from noise analysis of the circuit of the log domain core of Figure 14.15 (a) to be

$$SNR_{\max} = \frac{C\phi_t}{Nq} \qquad (14.21)$$

where N is the equivalent number of shot-noise sources of value $2qI_A$ in the log-domain core. Figure 14.16 depicts the *SNR* saturation in adaptive-Class-A filters and also shows that the dynamic range of operation, *DR*, can be much greater than $SNR_{\max}$ because of the gain control present in an adaptive-Class-A filter. The upper limit of operation of the dynamic range is given by the total harmonic distortion present in the filter, usually limited by the input or output devices entering above-threshold MOS or high-current bipolar operation.

The operation of the filter is degraded at low input currents by the parasitic load capacitance C_{par} of the input device. The minimum value of bias current that must be present in parallel with the $I_{AC,ENV}$-dependent adaptive bias current to prevent this degradation is given by

$$I_{BIAS,\min} = 2\pi f_{\max} C_{par} \phi_t \qquad (14.22)$$

where $f_{\max}$ is the maximum frequency at which filter operation needs to be ideal. In adaptive-Class-A operation, the log-domain core does not typically limit the dynamic range of operation although it is the primary determinant of $SNR_{\max}$.

14.7 Log-domain vs. G_m–C filters

In our study of linear $G_m - C$ filters with a bandwidth of f_c, we found that their power consumption was given by a scaling law of the kind expressed by the equation below

$$P_{GmC} = (V_{DD}\pi Nq)(f_c)(SNR_{\max})$$
$$= (V_{DD}\pi Nq)(f_c)(DR) \tag{14.23}$$

where the $SNR_{\max}$ is equal to the dynamic range DR since the system is linear. In adaptive-Class-A biasing, the presence of gain control allows $SNR_{\max}$ to be much less than the DR as Figure 14.16 shows. The scaling law for power is based on the $SNR_{\max}$ as it is for $G_m - C$ filters in Equation (14.23). The $SNR_{\max}$ as expressed in Equation (14.21) can be expressed in terms of bias current, V_{DD}, and bandwidth to yield an expression similar to that in Equation (14.23). Since DR is typically much greater than $SNR_{\max}$ (e.g. 110 dB vs. 40–60 dB), adaptive-Class-A filters can implement very low-power wide-dynamic range systems. However, they do require large capacitances to attain a high $SNR_{\max}$ according to Equation (14.21). In contrast, $G_m - C$ filters can also implement linear-range V_L scaling to attain a higher $SNR_{\max}$ ($SNR_{\max} = CV_L/(Nq)$) and thus use smaller capacitances for the same $SNR_{\max}$. The power-supply voltage needed for a current-mode filter can be significantly less than that required for a voltage-mode filter since a large limiting voltage is not required to attain high dynamic range.

In summary, adaptive-Class-A log-domain filters are capable of low-power, low-voltage, wide-dynamic-range operation with a constant output signal-to-noise ratio. However, they need significantly larger capacitors to attain the same output signal-to-noise ratio as $G_m - C$ filters with a wide linear range.

14.8   Winner-take-all circuits

A two-channel winner-take-all circuit is shown in Figure 14.17 (a). Input currents $i_{IN,1}$ and $i_{IN,2}$ charge the gate nodes of differential-pair transistors M_1 and M_2 and cause them to compete for the bias current I_{BIAS}. The common-source-node voltage of the differential pair v_C gets established at a value that is determined by the larger of the gate voltages of M_1 or M_2. The voltage v_C then feeds back to create inhibition currents in M_0 and M_3 that subtract from the input currents $i_{IN,1}$ and $i_{IN,2}$ respectively and thus slow the charging of the input nodes. The winning channel with the larger of $i_{IN,1}$ and $i_{IN,2}$ has more say in setting v_C and establishes v_C at a value that allows its own input current to equilibrate with the feedback inhibition current. However, the value of v_C will then be larger than that needed to equilibrate the losing channel's input current; consequently, the high gain present at the input node causes the gate voltage of the losing channel's differential-input transistor to fall greatly, thus further weakening the losing channel's ability to compete for bias current. Thus, the winning channel takes almost all of the bias current leading to the name, winner-take-all. The net effect of the circuit for small-signal differences between $i_{IN,1}$ and $i_{IN,2}$ is to amplify these differences greatly in the output currents $i_{OUT,1}$ and $i_{OUT,2}$ respectively. The high gain present in this amplification process ensures that the winner-take-all output current of the winning channel saturates quickly at I_{BIAS} even when there is only a tiny difference

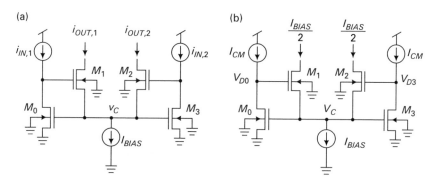

Figure 14.17a, b. (a) A two-channel winner-take-all circuit and (b) currents in the circuit at its symmetrical trip point.

between the values of $i_{IN,1}$ and $i_{IN,2}$. The distribution of currents in the winner-take-all circuit at its symmetrical balanced trip point is shown in Figure 14.17 (b).

The winner-take-all circuit is generalized to a multiple-input winner-take-all circuit by simply adding an input-inhibit transistor pair and input current for every channel and having several of these channels all compete for a common bias current at the common-source node. Figure 14.21 (a) shows an example of a six-channel winner-take-all circuit. The value of I_{BIAS} is scaled with the total number of channels in the winner-take-all circuit to maintain good phase margin in each channel's input-inhibit local feedback loop. If the load capacitance at v_C contributed by each channel is equal to the input-node capacitance of the channel, the common-source-node bias current available for each channel in a completely balanced symmetric configuration needs to be at least 4 times as large as the input current in that channel for critical damping in that channel's local feedback loop. The latter statement is a simple application of the two-pole-feedback-Q rule (see Equation (9.16) in Chapter 9 with $A_{lp} = 1$ and Figure 11.12 (c) in Chapter 11).

To understand the operation of an N-input winner-take-all circuit quantitatively, it is useful to compute the small-signal impedance of each of its component channels at the common-source node. Figure 14.18 shows one channel of the two-channel winner-take-all circuit of Figure 14.17 (a) as an example. We can see that

$$g_s = g_{mb2} + g_{m2}$$

$$r_{ch} = \frac{v_x}{i_x} = \frac{1}{g_s(1 + \kappa g_{m3} r_{03})} = \frac{1}{g_s A_m} \qquad (14.24)$$

The most interesting point of operation of the winner-take-all circuit lies at its balanced symmetrical trip point. At this point, suppose that all of its $N - 1$ input currents are equal except for that of one winning channel that has an infinitesimally larger current. We can understand the operation of the winner-take-all circuit by analyzing its small-signal operation at this trip point. To do so, we form a Norton or Thevenin equivalent of the winning channel as seen at the common-source node where all channels interact; we similarly represent each losing channel by its impedance at the common-source node. The winning channel is the only

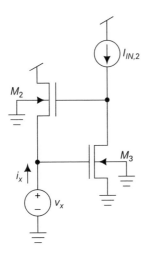

Figure 14.18. Input impedance of a single channel of the winner-take-all circuit.

channel with a small-signal input such that it is the only channel that is represented with a non-zero Norton/Thevenin source *and* an impedance; all other losing channels are represented with just an impedance since their Thevenin/Norton equivalent sources are zero.

Figure 14.19 (a) shows the small-signal schematic of the winner-take-all circuit at its trip point with a Norton equivalent for the winning channel and impedances for all other losing channels. Each of the impedances is marked with their *conductance* values of $A_m g_s$ in Figure 14.19 (a) for transparent understanding of this current-divider circuit. The Norton short-circuit current i_{sc} of the winning channel is seen from Figure 14.18 with a grounded v_x and a small-signal i_{in} to be

$$i_{sc} = i_{in} r_{0,M3} g_{m,M2}$$

$$i_{sc} = i_{in} \frac{V_{erly}}{I_{IN}} \frac{I_{BIAS}}{N(\phi_t/\kappa_S)}, \qquad (14.25)$$

where V_{erly} is the Early voltage at the input node of the channel due to transistor M_3 in Figure 14.18. Figure 14.19 (b) shows the Thevenin version of the Norton circuit of Figure 14.19 (a) but is otherwise identical to it. The current through the winning branch i_{out1} is the negative of the sum of the losing currents i_{sc}/N through each branch in Figure 14.19 (a) since the small-signal sum of the winning current and all losing currents must be equal to zero given that the total bias current I_B is a constant. Equivalently, the winning channel's current i_{out1} is the current emanating from the winning channel to v_c in Figure 14.19 (b). The current emanating from the winning channel towards the common-source terminal v_c corresponds to the small-signal current flowing through M_2 in Figure 14.18 because M_2 is the only transistor in the winning channel that communicates with the common-source node.

Note that the winning current is *NOT* the current i_{sc}/N through the winning channel's Norton impedance in Figure 14.19 (a). This current is an abstract current that represents the effects of several devices within the Norton black-box

(a)

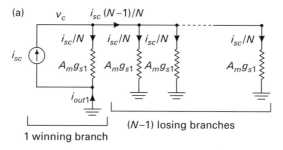

(b)

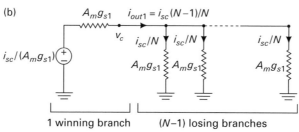

Figure 14.19a, b. (a) Norton and (b) Thevenin equivalent circuits for the winner-take-all circuit close to its trip point.

circuit, is meaningless from the point of view of the terminal, and cannot be equated as being due to that of some particular device in the black box. Similarly, the Thevenin voltage across the winning channel's impedance in Figure 14.19 (b) is meaningless and cannot be equated as being the voltage across some particular device in the Thevenin black box.

Thus, from Figure 14.19 (a) or Figure 14.19 (b) and Equation (14.25), we find that the winning channel's current is given by

$$
\begin{aligned}
i_{win} &= i_{sc}\left(\frac{N-1}{N}\right) \\
&= i_{in}\frac{V_{erly}}{I_{IN}}\frac{I_{BIAS}}{N(\phi_t/\kappa_S)}\frac{N-1}{N}
\end{aligned}
$$

$$
\frac{i_{win}}{I_{BIAS}} = \left(\frac{i_{in}}{I_{IN}}\right)\left(\frac{V_{erly}}{\left(\dfrac{N^2}{N-1}\right)(\phi_t/\kappa_S)}\right)
$$

(14.26)

Thus, we notice from Equation (14.26) that the winner-take-all circuit is amplifying the fractional input of the winning channel i_{in}/I_{IN} into a larger fractional output of the winning channel i_{win}/I_{BIAS} with a gain A_{wta}, where this gain is given by

$$
\frac{i_{win}}{I_{BIAS}} = A_{wta}\left(\frac{i_{in}}{I_{IN}}\right)
$$

$$
\boxed{A_{wta} = \frac{V_{erly}}{\left(\dfrac{N^2}{N-1}\right)\dfrac{\phi_t}{\kappa_S}}}
$$

(14.27)

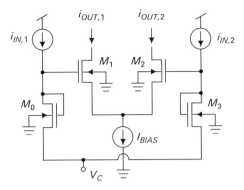

Figure 14.20. A current-mode differential pair.

A similar analysis may be performed for a multi-input current-mode differential pair, which is a generalization of the two-input current-mode differential pair shown in Figure 14.20. In this circuit, all input currents are converted to voltages on diodes and fed into a multi-input differential pair that then creates output currents as in a winner-take-all circuit but the feedback inhibition to each input node from the common-source node is absent. In this case, we find that

$$\left(\frac{i_{win}}{I_{BIAS}}\right) = \left(\frac{i_{in}}{I_{IN}}\right)\frac{1}{\left(\frac{N^2}{N-1}\right)} \quad (14.28)$$

By comparing Equations (14.26) and (14.28), we see that the winner-take-all circuit is providing an extra factor of $V_{erly}/(\phi_t/\kappa_S)$ in its amplification of fractional input changes due to the presence of feedback inhibition, which is absent in Figure 14.20 or its multi-input generalizations.

Note that we must scale I_{BIAS} with N to maintain good phase margin in the winner-take-all circuit. If we thus set

$$I_{BIAS} = NI_{BIAS}{}^{unit} \quad (14.29)$$

where $I_{BIAS}{}^{unit}$ is the unit bias current due to one channel that is kept constant as more channels are added to the winner-take-all circuit, then the absolute gain rather than relative gain of the winner-take-all circuit may be computed from Equation (14.26) to be

$$\frac{i_{win}}{i_{in}} = \left(\frac{I_{BIAS}{}^{unit}}{I_{IN}}\right)\left(\frac{V_{erly}}{\phi_t/\kappa_s}\right)\left(\frac{N-1}{N}\right) \quad (14.30)$$

If the same unit-bias-current scheme is applied to the circuit of Figure 14.20, its absolute current gain may be shown to be

$$\frac{i_{win}}{i_{in}} = \left(\frac{I_{BIAS}{}^{unit}}{I_{IN}}\right)\left(\frac{N-1}{N}\right) \quad (14.31)$$

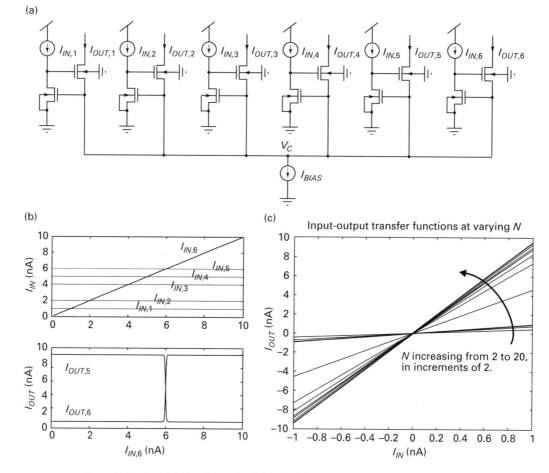

Figure 14.21a, b, c. (a) Six-channel subthreshold winner-take-all circuit, (b) simulation results, and (c) variation in gain with the number of inputs, N.

Figure 14.21 (a) shows a six-channel winner-take-all circuit. Figure 14.21 (b) shows that we keep five of the input currents, $I_{IN,1}$ through $I_{IN,5}$, constant while we sweep $I_{IN,6}$. Figure 14.21 (b) also shows that at the point where $I_{IN,6}$ exceeds $I_{IN,5}$, the output current $I_{OUT,5}$ from the prior winning channel goes to zero while the present winning channel takes all the current such that $I_{OUT,6}$ is maximized. Figure 14.21 (c) illustrates that the absolute current gain of the winner-take-all circuit increases with N as N goes from 2 to 20 with $V_{erly}/(\phi_t/\kappa_S) = 10$ and asymptotes at nearly 10 in accord with Equation (14.30) if $I_B{}^{unit} = I_{IN}$; the absolute current gain of the multiple-input current-mode differential pair of Figure 14.20 also increases with N as N goes from 2 to 20 but asymptotes at a gain of 1.0 in accord with Equation (14.31) if $I_B{}^{unit} = I_{IN}$.

It is instructive to compute the critical fractional input i_{in}/I_{IN} that is needed to just saturate the output current of the winning channel at I_{BIAS}. We may estimate this critical fraction by setting $i_{win}/I_{BIAS} = 1$ in the left hand side of Equation (14.27) and solving for the i_{in}/I_{IN} on the right hand side. We find that

$$\left.\frac{i_{in}}{I_{IN}}\right|_{crit} = \frac{\dfrac{N^2}{N-1}(\phi_t/\kappa_S)}{V_{erly}} \tag{14.32}$$

Thus, the critical fraction that just causes saturation of the winning channel, the equivalent of a linear range in the winner-take-all circuit, increases with the number of its inputs: it becomes harder for a single winner to take all the current in an N-input winner-take-all circuit as N is increased. To do so requires a fractional change in its input that increases like $O(N)$ for large N. In essence, winner-take-all circuits need larger changes in their inputs to completely over-whelm all other inputs as N increases because the effect of the winning channel on the common-source node reduces with N. The amplification of fractional changes is also decreased with N as revealed in Equation (14.27). This amplification decrease results in a concomitant increase in the critical fraction needed to saturate the winning channel; the situation is analogous to the linear range of a transconductor increasing as its transconductance gain is decreased such that saturation requires a larger differential input.

14.9   Large-signal operation of the winner-take-all circuit

The small-signal analysis of the winner-take-all circuit neglects nonlinear effects in its operation. The exponential dependence of the output current of a channel on the difference between its input gate voltage and V_C means the conductances in Figure 14.19 (a) or Figure 14.19 (b) are really nonlinear. Losing channels' nonlinear conductances fall rather than staying constant with V_C as assumed in a small-signal linear model. Similarly, the short-circuit current and nonlinear conductance at the V_C node of the winning channel rise with larger inputs as increasingly more bias current flows through the winning channel. These non-linear effects lead to a net *reduction* in the large-signal current gain compared with that predicted by the small-signal analysis used to derive Equations (14.26), (14.27), (14.30), and (14.32). In the limit of a very-large-signal input, the gain goes to zero since the winning channel's output is then saturated at I_{BIAS} and increases in i_{IN} have no effect on i_{OUT}.

We can get some intuition for why the gain is gradually reduced from the small-signal maximal value to zero at large-signal saturation by integrating up the effects of many small-signal changes with the models of Figure 14.19 (a) and Figure 14.19 (b). At any given operating point at which the I_{IN} of the winning channel exceeds the I_{IN} of all the other losing channels with identical inputs, the small-signal values of the Norton conductance and i_{sc} of the winning channel increase because of the increased g_s of the winning channel seen at the V_C node; as we move from small-signal operation to large-signal operation, the value of g_s

increases by a factor of N as the dc value of the output bias current changes from I_B/N to I_B. In contrast, the g_s of the losing channels seen at the V_C node decreases towards zero as the dc value of the output bias current through these channels changes from I_B/N to zero and as their feedback inhibition transistors enter the linear region. Thus, in Figure 14.19 (a), the small-signal current through all the losing channels systematically decreases at each large-signal operating point, thus lowering the winning channel's current and gain at this operating point. The net large-signal change in output current for a given change in input current may be viewed as the integration of all of these tiny small-signal changes as the input of the winning channel is infinitesimally increased from its previous operating point to the next operating point and all changes are integrated to determine the next operating point. The large-signal current gain is then necessarily less than the small-signal current gain because each small-signal change available for integration becomes progressively smaller as the winning channel becomes a stronger winner. In essence, the winner starts to saturate because the losers have progressively less to give.

Another interesting insight about large-signal operation can be obtained from Figure 14.19 (a). The small-signal conductance at the V_C node is relatively constant at all operating points since the sum of the g_s values of each channel at the node stays relatively constant at all operating points. In small-signal symmetrical operation with equal input currents $g_s = A_m I_B/(N\phi_t)$ for each channel and there are N such g_s conductances, one due to each channel, such that the sum of the g_s conductances is given by $A_m(I_B/(N\phi_t))N = A_m I_B/\phi_t$; in saturated operation with one large winning current, the sum of all the g_s conductances is still $A_m I_B/\phi_t$ with only the winning channel's conductance being nonzero; in between these two extremes, the sum of the g_s conductances stays relatively constant as the loss in the conductances of the losing channels is compensated for by the gain in the conductance of the winning channel. Thus, the time constant at the V_C node remains relatively invariant with input-signal level. If the winner-take-all circuit has good phase margin in the winner's local feedback loop at the largest expected winning input current, the largest crossover frequency of the winner's feedback loop is less than the characteristic frequency of the fast common-source node. Thus, at all other weaker input levels, the phase margin will be even better since the characteristic frequency of the common-source node remains relatively invariant with input level but the crossover frequency of the local feedback loop of the winner and that of all losers is reduced. Thus, if the winner's local feedback loop at the largest input level exhibits good phase margin, the winner-take-all circuit exhibits good phase margin in all of its local feedback loops.

14.10   Distributed-feedback circuits

Figure 14.22 (a) illustrates a current-mode circuit that models a pattern of synaptic connectivity often seen amongst neurons in cortex, the amazing sheet of tissue that

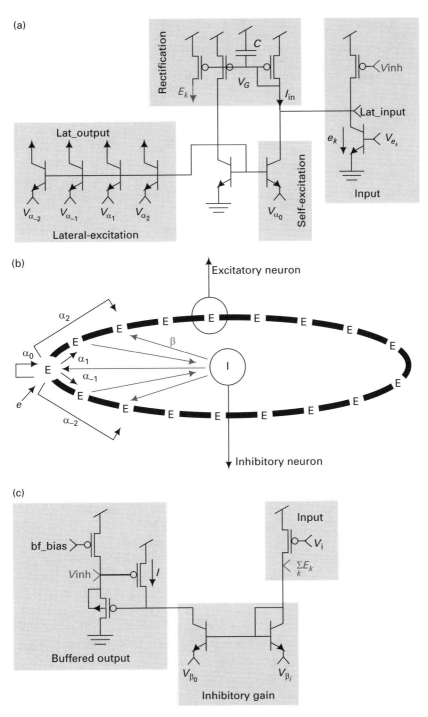

Figure 14.22a, b, c. A circuit model of excitation and inhibition in the cortex (featured on the cover of *Nature* in July 2000). Reprinted by permission from Macmillan Publishers Ltd: *Nature* [11] ©2000.

envelopes our brains. Figure 14.22 (b) illustrates that these neurons tend to excite themselves and their local neighbors leading to positive-feedback enhancement of their input currents. The neurons also all excite a central common neuron that then feeds back to inhibit each of them leading to global negative feedback. The central inhibitory neuron is reminiscent of the common-source node of a winner-take-all circuit that feeds inhibition currents back to each channel. The overall connectivity of this ring-of-neurons network is like that of a winner-take-all circuit with global inhibitory negative feedback and local excitatory positive feedback. In actual cortex, the space constant for inhibitory negative-feedback connections is larger than that for excitatory positive-feedback connections. In this model, the inhibitory space constant has been simplified to be global for all neurons for simplicity.

Figure 14.22 (a) shows that each neuron is modeled with a current-mode pFET MOS mirror whose rectifying diode input along with an integration capacitance forms a summing node for various current contributions; these contributions include a self-excitatory sink current from a current mirror, excitatory sink currents from neighboring lateral inputs, and an inhibiting source current from the global inhibitory neuron; the gains of the various excitatory contributions from oneself and neighbors that are one or two locations away in the ring are marked with values corresponding to α_0, α_1, α_{-1}, α_2, and α_{-2} in Figures 14.22 (a) and 14.22 (b) respectively. Figures 14.22 (b) and 14.22 (c) show that the inhibitory neuron sums the various excitatory currents, gains them via source voltages that determine β, the inhibitory gain, and creates a buffered voltage V_{inh} that is fed to all neurons to create their inhibition currents via an inhibitory transistor as in Figures 14.22 (b) and 14.22 (c). The voltage V_{inh} is the analogy of the common-source node of a classic winner-take-all circuit.

Figures 14.23 (a) and 14.23 (b) show that if two input currents of almost equal value are applied at nodes 9 and 13 in the 16-ring network, the winner-take-all effect with positive feedback functions to spread the input excitation to the nearest neighbors such that a selected group of neurons are active and the rest are shut off. The winner-take-all effect causes the center of mass of the excitation to be located at neuron 9 or 13. The digital selection of a group of active neurons is due to the nonlinear interaction between the strong inhibitory current and the rectification effect present in each neuron. Since the output of a rectifier is zero for negative inputs, only neurons with a net self, lateral, or input excitation that is strong enough to exceed the inhibition current will be active while others will be turned off. Those that are turned on are capable of amplifying their inputs in an analog fashion with positive-feedback-enhanced gain. Depending on the relative strength of the input currents at 9 or 13 the center of mass can switch hysteretically between the two locations as seen in Figure 14.23 (c), a classic sign of a positive-feedback-state-dependent threshold. Figure 14.23 (d) shows that the chip's outputs performed comparably to that of a computer model. This circuit made the cover of *Nature* in July 2000 for showing that a simple electronic current-mode circuit could mimic the feedback networks of the brain and exhibit many similar characteristics that had been experimentally observed and explained with such networks [11].

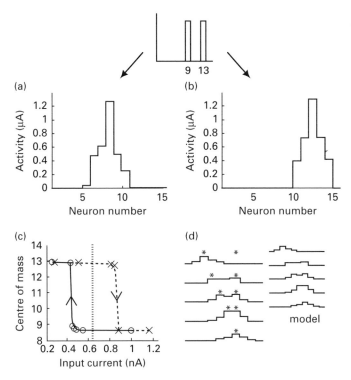

Figure 14.23a, b, c, d. Pattern of excitation in the electronic cortex circuit with the 9th (a) or 13th output (b) dominant in response to symmetric inputs at the 9th or 13th location. In (c), the hysteretic behavior between these two patterns of excitation are shown. In (d), we see that the chip and model behave similarly. Reprinted by permission from Macmillan Publishers Ltd: *Nature* [11] ©2000.

The simultaneous digital selection and analog amplification property of this circuit makes it a hybrid between a many-input digital latch and a classic analog amplifier. The circuit exploits the properties of distributed feedback in a sophisticated fashion to attend to only dominant inputs and amplify them while suppressing other weaker inputs and noise. Any real application of the circuit for noise suppression or attention will necessarily require additional feedback-and-calibration loops to suppress transistor mismatch; it will also require biasing circuits that are power-supply-noise immune and temperature insensitive. Such circuits are described in Chapter 19.

References

[1] B. Gilbert. Translinear circuits: An historical overview. *Analog Integrated Circuits and Signal Processing*, **9** (1996), 95–118.

[2] E. Seevinck. Companding current-mode integrator: A new circuit principle for continuous-time monolithic filters. *Electronics Letters*, **26** (1990), 2046–2047.

[3] Y. P. Tsividis, V. Gopinathan and L. Toth. Companding in signal processing. *Electronics Letters*, **26** (1990), 1331–1332.

[4] D. R. Frey. State-space synthesis and analysis of log-domain filters. *IEEE Transactions on Circuits and Systems II: Analog and Digital Signal Processing*, **45** (1998), 1205–1211.

[5] C. Toumazou, F. J. Lidgey and D. Haigh. *Analogue IC design: The Current-Mode Approach*. (Herts, United Kingdom: Institution of Engineering and Technology, 1990).

[6] A. G. Andreou, K. A. Boahen, P. O. Pouliquen, A. Pavasovic, R. E. Jenkins and K. Strohbehn. Current-mode subthreshold MOS circuits for analog VLSI neural systems. *IEEE Transactions on Neural Networks*, **2** (1991), 205–213.

[7] E. Seevinck, E. A. Vittoz, M. du Plessi, T. H. Joubert and W. Beetge. CMOS translinear circuits for minimum supply voltage. *IEEE Transactions on Circuits and Systems II: Analog and Digital Signal Processing*, **47** (2000), 1560–1564.

[8] B. A. Minch, Synthesis of multiple-input translinear element log-domain filters. *Proceedings of the IEEE International Symposium on Circuits and Systems (ISCAS)*, Orlando, FL, 697–700, 1999.

[9] Y. Tsividis, N. Krishnapura, Y. Palaskas and L. Toth. Internally varying analog circuits minimize power dissipation. *IEEE Circuits and Devices Magazine*, **19** (2003), 63–72.

[10] J. Lazzaro, S. Ryckebusch, M. A. Mahowald and C. A. Mead. Winner-take-all networks of $O(n)$ complexity. *Advances in Neural Information Processing Systems*, **1** (1989), 703–711.

[11] R. H. R. Hahnloser, R. Sarpeshkar, M. A. Mahowald, R. J. Douglas and H. S. Seung. Digital selection and analogue amplification coexist in a cortex-inspired silicon circuit. *Nature*, **405** (2000), 947–951.

[12] S. Mandal and R. Sarpeshkar, Log-domain Circuit Models of Chemical Reactions. *Proceedings of the IEEE Symposium on Circuits and Systems (ISCAS)*, Taipei, Taiwan, 2009.

[13] S. Mandal and R. Sarpeshkar, Circuit models of stochastic genetic networks. *Proceedings of the IEEE Biological Circuits and Systems Conference*, Beijing, China, 2009.

[14] M. W. Baker and R. Sarpeshkar. Low-Power Single-Loop and Dual-Loop AGCs for Bionic Ears. *IEEE Journal of Solid-State Circuits*, **41** (2006), 1983–1996.

15 Ultra-low-power and neuron-inspired analog-to-digital conversion for biomedical systems

Although nature commences with reason and ends in experience, it is necessary for us to do the opposite; that is, to commence with experience and from this to proceed to investigate the reason.

Leonardo da Vinci

An analog-to-digital converter converts real-world continuous analog signals into symbolic discrete digital numbers. It is often abbreviated as an ADC, A-to-D, or A/D. ADCs are ubiquitous in all electronic systems. A digital-to-analog converter performs the inverse function and is correspondingly abbreviated as a DAC, D-to-A, or D/A. Figure 15.1 shows the input-output curve of an ADC [1]. The input and output are equal to each other within a quantization error of $\pm \Delta/2$, a consequence of the fact that we need to round up or round down real numbers to the nearest integer to represent them digitally. The digital numbers are usually represented with binary digits or bits. If, because of the input statistics, any error between $[-\Delta/2, +\Delta/2]$ is equally likely, then, from evaluation of the second moment of a flat probability distribution, the rms error of the quantized representation of a real number can be shown to be $\Delta^2/12$. If the ADC samples its input at a sampling frequency f_S, the power spectrum of the quantization noise is well approximated as being white from 0 to $f_S/2$ and therefore having a noise per unit bandwidth of $\Delta^2/(12(f_S/2))$. If the precision of the converter is N bits, digital numbers between 0 and $(2^N - 1)$ represent analog signals between 0 and a full-scale voltage V_{FS} with V_{FS} corresponding to 2^N. Thus $V_{FS}/2^N = \Delta$. Hence a sine wave with amplitude $V_{FS}/2$ that spans the full $[0, V_{FS}]$ input range of the converter has a signal-to-noise ratio (SNR) given by

$$SNR = \frac{\frac{(V_{FS}/2)^2}{2}}{\Delta^2/12} = \frac{3}{2}2^{2N} \tag{15.1}$$

The latter relationship is frequently used to evaluate N, the equivalent number of bits or ENOB of the converter.

Several ADCs usually have their power consumption well described by

$$P_{ADC} \propto f_S 2^N$$
$$P_{ADC} = E_q f_S . 2^N \tag{15.2}$$

where E_q is relatively constant with N and represents the energy per quantization level. Such power scaling arises because ADCs that are more precise need to have

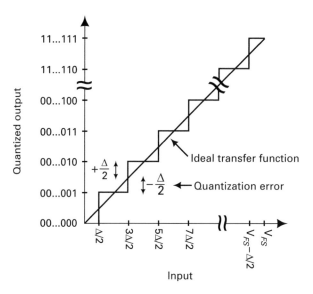

Figure 15.1. The basic input-output curve of an analog-to-digital converter. Reproduced with kind permission from [1].

larger transistors or capacitances to ensure that errors like offset, charge injection, finite loop-gain in a closed-loop amplifier, etc., do not limit their precision. The consequent increase in load then increases power consumption like 2^N for a fixed sampling frequency f_S. Alternatively, if voltage errors are fixed, the full-scale voltage can be increased to improve precision, which increases the power-supply voltage, which in turn increases power. In some architectures like flash ADCs, where 2^N reference levels and 2^N comparators are used to implement fast simultaneous comparisons in parallel, the power scales nearly like 2^N simply because there are 2^N more stages. In the literature, it is common to refer to a figure of merit (FOM) defined by

$$FOM = \frac{P_{ADC}}{f_S 2^N} \tag{15.3}$$

to evaluate the energy efficiency of an ADC [2]. This FOM is E_q, the energy per quantization level, and must be small if the ADC is energy efficient. Sometimes $1/E_q$ is used as a figure of merit and must be large if the ADC is energy efficient. The FOM of Equation (15.3) is inappropriate for a thermal-noise-limited ADC, where a value of 2^{2N} instead of 2^N in Equation (15.3) would yield an E_q or FOM that is constant with N: from the relationships in Chapters 7 and 8, to reduce thermal noise by a factor of 2 while preserving bandwidth requires four times more power consumption. Nevertheless, the energy efficiency of both thermal-noise-limited and non-thermal-noise-limited ADCs is evaluated with an FOM given by Equation (15.3) in the literature. We shall adopt this convention as well simply because it is in very wide usage now.

The precision and bandwidth requirements on A/D converters for most biomedical applications are rather modest: neural action-potential signals often have a minimum value of $5\,\mu V$ rms or more, an ac signal dynamic range that corresponds to 9 bits at most, and significant energy in a 40 kHz bandwidth. For cochlear-implant applications, due to the variability in speech waveforms, log spectral energy can be digitized at 7 bits and 100 Hz after analog preprocessing on the speech waveform; alternatively, in hearing aids and in cochlear implants for the deaf, the use of analog gain control in or before the ADC allows an audio signal with 16-bit dynamic range to be processed by an 11-bit converter sampled at 44 kHz. Low-power imagers may require a wide dynamic range of operation (over 16 bits, say), but the signal-to-noise ratio at the output is often fundamentally limited by the shot noise of collected electrons to be about 12 bits, even in bright sunlight. The pixel signal information is contained well within 60 Hz. Neural amplifiers used in brain implants, circuitry used in cochlear implants, and low-power imagers are described in Chapter 19. While electrocardiogram (ECG) and electroencephalo-gram (EEG) systems may require small minimum detectable signals for detecting subtle waveform differences ($5\,\mu V$ and $0.3\,\mu V$, respectively), the necessary dynamic range in a linear system corresponds to 8 bits and the high-frequency bandwidth requirements range from 100 Hz to 500 Hz at most [3]. The converter utilized in the acquisition of cardiac signals in implantable pacemakers requires only 8 bits of precision at a bandwidth of 250 Hz [4]. Due to the relatively slow rate of the heart, pulse-oximetry signals for measuring oxygen can be digitized at 7 bits and 100 Hz if analog preprocessing is used to preprocess the signal [5]. Circuits for pulse oximetry are described in Chapter 11 and in Chapter 20. The sensing of mechan-ical and chemical signals with MEMS or NEMS (micro or nano electro-mechan-ical systems) often requires ultra-low-noise analog circuitry (see Chapters 8 and 20) and may require a high dynamic range of operation, but the instantaneous SNR and bandwidth are usually less than 8–11 bits and 100 Hz.

In ultra-low-power systems, thermal-noise limitations alone can limit the precision of an ADC that can be built for a given bandwidth with a given power. For example if a thermal-noise-limited ADC operates at $E_q = 1\text{pJ}$ per quantiza-tion level efficiency at $N = 8$ bits and $f_S = 50\,\text{kHz}$, thus consuming 12.8 μW from Equation(15.2), an ADC operating at the same bandwidth but with 12 bits of precision will consume at least $(2^4)^2$ times more power, i.e., 3.27 mW. Thus, an implanted system that has a total power budget of 10 μW must exploit analog preprocessing to reduce the bandwidth and/or precision of the incoming sensor information before it is digitized.

Often high-precision converters are used in biomedical applications simply because non-informative common-mode ac signals can lead to a wider dynamic range of variation in the input. However, in any well-designed low-power system, such signals are most efficiently rejected via ac coupling or closed-loop feedback topologies that attenuate ac information as we describe in a microphone preampli-fier in Chapter 19, in a neural amplifier in Chapter 19, or in an ECG amplifier in Chapter 20. Through low-power analog preprocessing and intelligent gain control, almost all low-power biomedical systems can easily be designed to operate over more than 16 bits of dynamic range even though the output SNR corresponding to the inherent information in the signal is 8–11 bits at best. Frequently, dynamic range is confused to be identical with SNR and there is irrational fear that analog preprocessing will lead to a loss in flexibility and robustness, which is by no means true if the system is well designed for the application. Therefore, needlessly precise converters are used, which inevitably leads to wasteful power consumption: the power costs of the ADC scale exponentially with the increase in precision and then further power is wasted in the DSP crunching high-precision numbers to finally extract meaningful information in the input, which is almost always encoded at much lower SNR in a higher-level variable. An example of ultra-low-power programmable-and-robust analog preprocessing before digitization is described in Chapter 19 for a cochlear-implant or bionic-ear processor.

In ultra-low-power implanted systems that need to operate on a small implanted battery for decades with a limited number of wireless recharges such that resurgery is obviated, power efficiency is paramount. Power efficiency is important for minimizing tissue heating, size, weight, and cost in implanted systems as well. Thus, even systems capable of being infinitely wirelessly recharged on an ultra-capacitor, for example, have to operate with low power to ensure that the size of the ultra-capacitor is not large. In non-invasive medical applications such as cardiac monitoring or pulse oximetry, there is an increasing trend to operate via harvested RF, vibratory, or body-heat energy such that energy efficiency again becomes paramount. Energy harvesting is described in Chapters 17 and 26. Therefore, highly energy-efficient ADCs with modest speed and modest precision are important in several biomedical applications, and we shall now discuss how to architect them.

We will begin by reviewing some common ADC topologies that illustrate the basics of A-to-D conversion and that have good energy efficiency. These include algorithmic, pipelined, successive approximation, and sigma-delta topologies. Then, we shall focus on a highly energy-efficient bio-inspired ADC that is particularly suited for biomedical applications and that illustrates two important trends seen in low-power ADC conversion today. First, its processing is done with time as a signal variable rather than voltage. This property is advantageous for scaling to deep submicron processes where the dynamic range available with time is abundant but where the dynamic range available with voltage constantly shrinks with each generation. Second, it functions only with comparators avoiding relatively power-hungry operational amplifiers used in other topologies. Operational amplifiers are less energy efficient than comparators for two reasons. First, comparators are usually made with a low-gain amplifier stage followed by a positive-feedback latching stage such that their power for a given gain-bandwidth product is relatively low; in contrast, operational amplifiers are architected with relatively high-gain stages, which are less energy efficient. Second, in operational amplifiers, bandwidth for a given power is further reduced to ensure stability in the feedback loops that the amplifier may be used in. Furthermore, amplifiers scale less well with decreasing technology dimensions than comparators: low power-supply voltages and worsening Early voltages lead to dynamic-range and gain reduction that more severely impact high-gain-stage operational-amplifier topologies than comparator-based topologies that only use low-gain stages and positive feedback.

We will show how the bio-inspired ADC can easily be modified to build a computational ADC, i.e., an ADC that computes any analog polynomial function of its input before digitization, how it can be used for direct digitization of time-based cardiac or neural signals, and how it can be interleaved for higher-speed operation. Besides the bio-inspired ADC, we also discuss three other recent ADCs that provide further illustrations of the trend of using comparators and no closed-loop amplifiers for maximizing energy efficiency and the trend of using time to perform digitization. One of these ADCs relies on adiabatic or gradual charging of capacitances and the use of a low-static-power-dissipation comparator to achieve excellent energy efficiency [6]. The use of energy-efficient comparators can improve the energy efficiency of several ADCs including the bio-inspired ADC. Then, we discuss another trend in low-power ADCs that is useful if the ADC's precision is not limited by thermal noise: one uses inaccurate but energy-efficient analog circuits whose predictable gain and offset errors are corrected by digital techniques. Finally, we conclude by discussing similarities between neuronal processing and ADC conversion, of which our bio-inspired ADC represents one special case. We suggest that ADC conversion and mixed-signal processing in general may take further inspiration from neuronal processing in the future.

15.1   Review of ADC topologies

An algorithmic or cyclic ADC converter is architected to go through a cycle of N successive conversions to quantize its analog input to an N-bit output digital number. On each cycle, one quantization bit is obtained, starting from the most significant bit (MSB) of the digital number on the first cycle to the least significant bit (LSB) of the digital number on the last Nth cycle. On each cycle of the conversion, the output bit determines whether a conditional subtraction will be performed on the current analog input (or not). The analog value of the input after subtraction is gained by a factor of 2 and serves as a residue input for the next cycle of the conversion. The recursive application of conditional subtraction and gain-of-2 operations to each analog residue on a particular cycle generates a new analog residue that is processed identically on the next cycle. The process terminates after N cycles.

The details of the algorithm are essentially similar to those of a binary-search algorithm. To begin, we check to see if the analog input is within the top half or bottom half of the full dynamic range of the input. If it is in the top half, the output MSB bit is set to 1; if it is in the bottom half, the output MSB bit is set to 0. If the MSB is 1, we first subtract the bottom half of the full-scale dynamic range from the input. Then, we gain up the value remaining after this subtraction by a

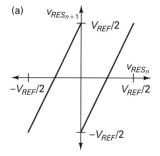

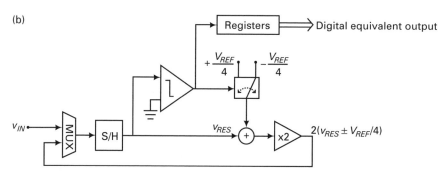

Figure 15.2a, b. The functional specification of an algorithmic ADC is shown in (a) together with an architectural implementation in (b). Reproduced with kind permission from [1].

factor of 2. Thus, we map the new residual input to the full dynamic range for the next cycle of the conversion. If the MSB is 0, we do not subtract the bottom half, but we still gain by a factor of 2 to remap the residue in one cycle of conversion to the full dynamic range in the next cycle. Thus, via this binary-search algorithm, we can quantize an input to finer and finer precision as we zoom in by successive factors of 2 on smaller and smaller portions of the input. Mathematically, if we define a function sgn( ) that yields $+1$ for a nonnegative number and -1 for a negative number, and if D_{n+1} denotes the digital bit and v_{n+1} denotes the analog residue on the $(n+1)^{th}$ iteration cycle respectively, then the algorithm can be formally described as follows:

$$v_{IN} \in [0, 1]$$
$$n \in \{0, 1, 2, 3, \ldots, N-1\}$$
$$v_0 = v_{IN}$$
$$D_{n+1} = \frac{1 + \mathrm{sgn}\left(v_n - \frac{1}{2}\right)}{2} \qquad (15.4)$$
$$v_{n+1} = 2\left(v_n - \frac{D_{n+1}}{2}\right)$$

Figure 15.2 (a) illustrates a residue-mapping scheme that is identical to that described by Equation (15.4) except that the residues lie in $(-V_{REF}/2, +V_{REF}/2)$

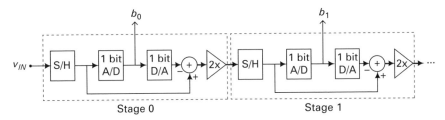

Figure 15.3. A pipelined algorithmic ADC. Reproduced with kind permission from [1].

and we have 1 and -1 bits rather than 1 and 0 bits. The residue v_{RESn} on the nth cycle of conversion, which lies within the top half $(0, V_{REF}/2)$ or within the bottom half $(-V_{REF}/2, 0)$, is recursively remapped to the residue v_{RESn+1} to span the full range $(-V_{REF}/2, V_{REF}/2)$ on the next cycle of the conversion. Figure 15.2 (b) illustrates how this algorithm may be implemented with a single comparator, amplifier, and sample-and-hold (S/H) to yield 1 quantization bit on each cycle. The analog multiplexer (MUX) is configured to output the input on the first cycle and analog residues on all cycles thereafter.

Figure 15.3 shows the architecture of a pipelined ADC. This ADC functions exactly like an algorithmic ADC except that the sequential operations are done in a pipelined assembly line fashion with N stages in cascade. Each stage in the pipeline processes inputs from its preceding stage in an identical fashion with the first stage in the pipeline generating the MSB and the last stage in the pipeline generating the LSB. The latency of the pipelined converter is identical to that of an algorithmic converter but the bandwidth of conversion is improved to the reciprocal of the time needed to process a single bit rather than the time needed to process N bits. The power consumption is correspondingly higher by a factor of N in the pipelined converter compared with the algorithmic converter.

Figure 15.4 (a) shows the architecture of a successive approximation ADC. The architecture needs one S/H, one comparator, one successive-approximation register (SAR), and one N-bit D/A converter. The SAR ADC digitizes the analog signal by adding successively fine partial approximations synthesized from a digital representation of the signal. These partial approximations, when added, sequentially bring the overall approximation nearer to the input with each iteration. The process is analogous to trying to represent a function with Taylor-series approximations to it that successively add higher-order terms to improve the nature of the approximation with each iteration. Thus the SAR ADC is an example of an analysis-by-synthesis architecture. In Figure 15.4 (a), we successively set bits of the register and consequently D/A bits to 1 in each iteration, working our way from the MSB to the LSB. In each iteration, we check if the best quantized approximation to the input so far is above it or below it; if it is below it, we leave the bit that we have set at 1; if not, we set it to 0 and repeat the procedure with the next bit until N cycles have terminated. The SAR ADC is a highly energy-efficient design since it needs no analog

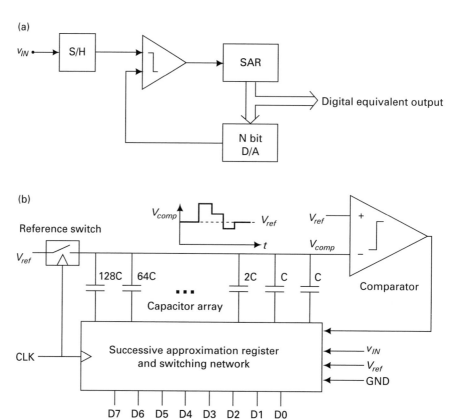

Figure 15.4a, b. The basic architecture of a successive approximation ADC (SAR). Reproduced with kind permission from [1] and [7] (©2003 IEEE).

amplifier. All of its components become more energy efficient as technology dimensions decrease.

Figure 15.4 (b) shows an example of a SAR ADC with $f_S = 100$ kHz, $N = \text{ENOB} = 7$, $P_{ADC} = 3.1\ \mu\text{W}$, and $E_q = 250$ fJ, a good number for energy efficiency at the time the ADC was built. It is described in detail in [7]. The energy efficiency arises primarily through the use of a small unit capacitor of $C = 12$ fF in the capacitive DAC along with the use of a 1 V power supply for V_{DD}, which also serves as V_{ref}. During an initial sample-and-hold phase, the value of V_{comp} is set to V_{ref} while the input V_{in} is sampled onto the bottom plates of all the DAC capacitors in the SAR. After this phase all bottom plates of all capacitors are tied to GND. On subsequent successive-approximation iterations, the bottom plates of the $128C$, $64C$, $32C$, $16C$...to *one* of the $1C$ capacitors are each sequentially tied to V_{ref} as we work our way from the MSB to the LSB in accord with the successive-approximation algorithm. During any such iteration, a capacitor plate is maintained at V_{ref} if the comparator output stays positive, which signifies that the bit corresponding to this particular measurement is a 1; the plate is returned to GND

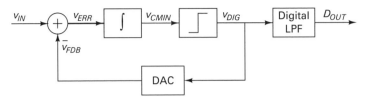

Figure 15.5. Architecture of a basic $\Sigma\Delta$ ADC.

if the comparator output goes negative, which signifies that the bit corresponding to this particular measurement is a 0. Advanced SAR topologies in 90 nm digital CMOS have achieved $E_q = 54$ fJ [8]. We shall discuss another extremely energy-efficient SAR ADC in more detail in a future section of this chapter.

Figure 15.5 shows the topology of a $\Sigma\Delta$ ADC. The integrator and comparator in the feed-forward path function like an operational amplifier with infinite DC gain. The 1-bit DAC in the feedback path provides crude 1 or 0 feedback from the feedback-loop output to the adder input. The comparator and DAC are clocked at a sample rate f_S that is extremely high compared with the bandwidth of v_{IN}, i.e., f_{IN}, and compared with the bandwidth of the digital lowpass filter (LPF), which is nominally near f_{IN} as well. To start, let us assume that v_{IN} is a constant and that $V_0 < v_{IN} < V_1$ where V_0 is the voltage level of a 0 and V_1 is the voltage level of a 1. If the comparator outputs a large number of 1s, there is a consistent discharging input to the integrator since $v_{IN} - V_1 = v_{ERR}$ is negative as $v_{IN} < V_1$. Thus, the input to the comparator falls. When it has fallen below its trip point, 0s start being produced by the comparator. If the comparator outputs a large number of 0s, there is a consistent charging input to the integrator since $v_{IN} - V_0 = v_{ERR}$ is positive as $v_{IN} > V_0$. Thus, the input to the comparator rises. When it has risen above its trip point, 1s again start being produced by the comparator. Thus the nature of the negative feedback serves to correct the number of 1s and number of 0s produced by the comparator until an equilibrium has been reached.

At equilibrium, the input to the comparator oscillates around its trip point in a hard-to-predict fashion. At this equilibrium, the number of 1s produced by the comparator, N_1, and the number of 0s produced, N_0, must be such that there is no net charging or discharging input to the integrator, i.e.,

$$\langle N_1(v_{IN} - V_1) + N_0(v_{IN} - V_0) \rangle = 0$$
$$\langle v_{IN} \rangle = \left\langle \frac{N_1 V_1 + N_0 V_0}{N_0 + N_1} \right\rangle \tag{15.5}$$

Thus, the duty cycle of the number of 1s produced, $N_1/(N_0 + N_1)$ provides information about the value of v_{IN}. The digital LPF in Figure 15.5 extracts this duty-cycle information easily by simply counting the number of 1s over a time window that is approximately $1/f_{IN}$ and thus obtains an accurate digital estimate of v_{IN} within the bandwidth f_{IN}. Intuitively, we might expect that the estimate will get more accurate if we oversample more, i.e., f_S/f_{IN} is large such that we have many 1s and 0s to average over. Intuitively, we might also expect that the time constant of the integrator, τ, should be near $1/(f_S/2)$. If so, the sampled-error inputs to the integrator are not just filtered out but cause significant error-correction inputs to the comparator. These error-correction inputs effectively serve to attenuate the comparator's quantization noise when averaged over the many samples within $1/f_{IN}$.

If we temporarily pretend that the comparator is just a noisy gain-of-1 block, the feedback topology of Figure 15.5 is analogous to that of a continuous integrator in a negative-feedback loop with unity-gain feedback. The noisy comparator may be modeled by adding a large quantization-noise disturbance input that adds to the comparator's regular output to create the final output that is fed back. Remarkably, this linear-systems intuition proves useful in understanding the highly nonlinear dynamical system of Figure 15.5. Since the disturbance input sees an integrator of $(1/(\tau s))$ in its feedback path, the quantization noise is filtered by the feedback loop with a squared-magnitude transfer function of $|\tau s/(\tau s + 1)|^2$. If $\tau \approx 1/(2\pi(f_S/2))$ and the quantization noise is represented as a white-noise generator with power per unit bandwidth of $\Delta^2/(12(f_S/2))$, then the integration of this noise over a bandwidth of f_{IN} by the digital LPF yields an estimate for the total output quantization noise of

$$v_q^2 \approx \left(\frac{\Delta^2}{12}\right)\left(\frac{2f_{IN}}{f_S}\right)^3 \tag{15.6}$$

In deriving Equation (15.6), we have approximated the quantization noise-transfer function $|\tau s/(\tau s + 1)|^2 \approx |\tau s|^2$ since $f_{IN} < f_S$, and integrated the quantization noise per unit bandwidth over the bandwidth of the input f_{IN}. Thus, from Equation (15.6) as the ratio f_{IN}/f_S falls and we have more oversampling, the inherently large $\Delta^2/12$ quantization noise is greatly reduced. Although the bandwidth of the integrator does need to be relatively high, the integrator's noise is digitally filtered as well, but via a lowpass transfer function rather than via a highpass transfer function. Such filtering implies that the integrator can be designed to be power efficient because it does not need to be simultaneously fast and low-noise. The subtractor block is typically implicitly implemented in a circuit by switching between two values on successive phases or via Kirchhoff's current law. If a 1-bit DAC is used, its linearity is irrelevant, which is one of the reasons for the popularity of a 1-bit version of the topology.

Since the $\Sigma\Delta$ topology uses only comparators and other components whose energy efficiency scales well with decreasing technology dimensions, its performance has improved over time. Oversampling bandwidth gets more abundant in advanced technologies, and switching power dissipation for a given value of f_S is lowered in smaller-dimension technologies as well. The topology is also consistent with the trends of using time-based (oversampling) and comparator-based strategies for energy-efficient conversion.

One issue with the topology is that the combination of clocking delays and the $-90°$ phase of the integrator can degrade phase margin such that the integrator's crossover frequency $1/\tau$ must be significantly below f_S to ensure well-behaved settling dynamics. This trade off causes the noise suppression through feedback to decrease such that we must trade off precision in order to gain better phase margin. Second-order $\Sigma\Delta$s use a double-integrator-plus-zero transfer function to replace the integrator transfer function in the loop. The double integration yields even stronger attenuation of the quantization noise within the f_{IN} band because the quantization noise power is now attenuated like $|\tau s/(\tau s + 1)|^4$ rather than like $|\tau s/(\tau s + 1)|^2$. The zero serves to ensure stability of the loop near the loop's crossover frequency, which must still be significantly below f_S. Third-order $\Sigma\Delta$s can have 3 integrators and 2 zeros but, due to slew-rate and saturating nonlinearities in the loop, the danger of instability at high input values can increase. Larger inputs effectively lower loop gain due to saturating nonlinearities and thus lower the crossover frequency as in many conditionally stable loops (see Chapter 9). If the effective crossover frequency for such inputs is lowered to a point where positive phase from the zeros has not yet cancelled the negative phase from the poles, instability can arise. Therefore, slew-rate and power need to be increased to maintain stability in such higher-order $\Sigma\Delta$ loops. A recent paper has dealt with this issue by reducing the effect of nonlinearities with a 4-bit quantizer and 4-bit DAC (rather than 1-bit versions) such that slew-rate power limitations are not severe; the ADC can also operate with a higher value of τ since it does not need to attenuate quantization noise as much, thereby further reducing power [9]. Concomitantly, digital power consumption is also lowered in this scheme since modest over-sampling ratios are effective in attenuating quantization noise. This $\Sigma\Delta$ converter achieves $E_q = 50$ fJ at audio bandwidths at 15-bit precision, and is among the most energy-efficient reported at this precision [9].

15.2   A neuron-inspired ADC for biomedical applications

Neurons, the computational cells of the brain, are known to be efficient at pattern recognition. Pattern recognition consists of classifying continuous analog vector inputs into a set of discrete digital output classes. Analog-to-digital conversion is a special scalar case of pattern recognition since we classify a scalar analog input signal into a set of 2^N output classes. A typical neuron in the human brain operates on a $\sim$6000-input vector with $\sim$0.5 nW of power such that the $\sim$22 billion neurons of the brain consume only about $\sim$15 W of power [10]. Neuronal pattern recognition is performed on continuous analog input currents using integrate-and-fire operations. The neuron spatially and temporally integrates its input currents onto a membrane capacitor within it. When the charging currents cause the capacitor's voltage to cross the neuron's threshold voltage at some point in time, the neuron outputs or fires an impulse-like voltage on the capacitor via positive-feedback action as in a comparator. After this spike has been fired, the neuron resets the capacitor voltage back to a resting value. Thus, the neuron behaves like a current-integrating comparator with a pulsatile output and a built-in reset [11]. In scalar pattern recognition, i.e., A-to-D conversion, we focus only on the temporal-integration aspect and ignore the spatial-integration aspect which is important in vector pattern recognition. The similarity of neuronal computation to several aspects of A-to-D conversion has stimulated a lot of theoretical work, e.g., [12] and some circuit attempts at building ADCs inspired by neurons [13].

Unfortunately, most neuronally inspired ADCs have had orders-of-magnitude worse performance when compared to traditional engineering architectures. However, we shall describe one neuronally inspired ADC that exhibits good energy efficiency: its energy efficiency with modest circuit optimization was the state of the art ($E_q = 120$ fJ) when it was first published in a $0.18\,\mu m$ technology even though its architecture and algorithm were new and had not been optimized over several decades like other architectures [14]. The architecture of the neuronally inspired ADC is well suited for scaling to more aggressive technologies and for implementing polynomial analog computations on its input before digitization. Its use of time-based and comparator-based signal processing is echoed in other highly energy-efficient architectures that we describe later. Due to the similarities in the building blocks of neurons and energy-efficient ADCs, neuronally inspired ADCs can benefit from advances in these building blocks in the future. In turn, they can inspire revolutionary energy-efficient ADC and other mixed-signal architectures. We shall begin by describing the algorithm and architecture behind the neuron-inspired ADC.

15.2.1   Algorithm and architecture

Figure 15.6 (a) shows how the algorithm works in two distinct stages. During the first stage, we employ a technique for conversion that is similar to that employed in other time-based ADCs known as "dual-slope converters": the input current I_{in} charges a capacitor for one clock cycle with an input-current slope; then, a reference current I_{ref} charges another matched capacitor until its voltage equals that of the first capacitor with a reference-current slope; the number of clock cycles that elapse during this charging period digitizes I_{in}/I_{ref}. Unlike a dual-slope converter, however, where we simply keep counting the number of clock cycles that it takes for the two voltages to match, we limit I_{in} to $4I_{ref}$ to ensure that the number of elapsed clock cycles in this first stage of conversion is never greater than 4. This limitation ensures that the conversion time of the ADC does not scale as $O(2^N)$ clock cycles where N is the number of bits, which is highly undesirable, and which causes dual-slope converters to be extremely slow. Therefore, only the first two MSBs of the converter are obtained via this counting technique. Any time-based residue that remains after the counting stage that is a fraction of a clock cycle is digitized in the second stage of the algorithm. In the second stage, we use a technique that digitizes time-based analog residues recursively to obtain $O(N)$ scaling using successive subranging techniques or zooming as in an algorithmic converter. However, time is the signal variable rather than voltage. At the end of the first stage of conversion, a residual time, denoted $0.5 + \varepsilon$ clock cycles in

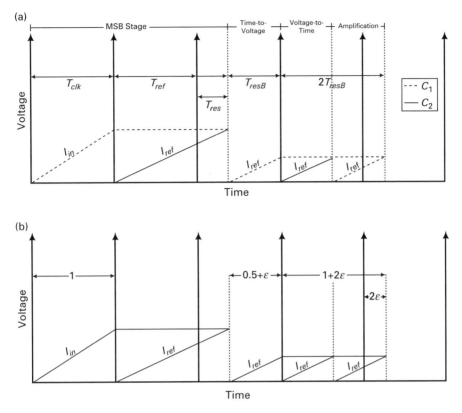

Figure 15.6a, b. Algorithm of the neuron-inspired ADC. Reproduced with kind permission from [15] (©2005 IEEE).

Figure 15.6 (b) encodes the remaining bits of I_{in}/I_{ref}. In this example, ε is positive such that the overall $0.5 + \varepsilon$ residue is greater than a half clock cycle.

In the second stage, we first convert the residual time, $0.5 + \varepsilon$ into a residual voltage during what we call a *Time-to-Voltage* conversion. Then we charge another matched capacitor until it reaches this residual voltage and repeat the process again such that the residual time before the clock edge, $0.5 + \varepsilon$, is doubled to $2(0.5 + \varepsilon) = (1 + 2\varepsilon)$ after the clock edge. We label the latter process a *Voltage-to-Time* conversion. In essence, we amplify the residue by 2 by doing things twice in time, i.e., we double the residue by integrating a capacitor to a threshold voltage twice. The 1 in $1 + 2\varepsilon$ provides quantization information revealing that the previous residue was greater than a half clock cycle, and the 2ε is automatically encoded as a residue referenced to this clock edge for the next stage of the conversion. Thus, subtraction of intermediate quantized bits is automatic because they occur as an integer number of clock edges, and amplified residues are always encoded with respect to the last seen edge. Hence the quantization and conditional-subtraction operations of Equation (15.4) are automatic due to the inherent modulo 2 nature of a clock and do not need to be explicitly performed! Figure 15.6 (b) shows that we follow a Time-to-Voltage phase, by a Voltage-to-Time phase twice, with the first Voltage-to-Time phase simply inverting timing durations encoded before the clock edge to timing durations encoded after the clock edge, and the second Voltage-to-Time phase providing the necessary factor-of-2 amplification. Since $1 + 1 = 2$, if the clock has very little phase noise, we obtain an open-loop gain of 2 reliably without the need for a closed-loop topology.

The amplification, quantization, and subtraction of the residues are repeated to obtain successive bits: the $2\varepsilon_n$ residue is converted to a $1 - 2\varepsilon_n$ residue by converting the time from the end of amplification to the next clock edge into a voltage. We can then recursively repeat the $\varepsilon_{n+1} \leftarrow (1 - 2\varepsilon_n)$ process to get successive conversion bits. The overall scheme ensures correct treatment of all residues whether quantization edges do or do not occur during amplification. Note that our subtraction-and-amplification routine produces alternating signs in the residues but we simply invert bits obtained on every other iteration to get the correct quantized answer: the 1st, 3rd, 5th, 7th, ... iterations after the MSB phase of the conversion invert bits while the even iterations do not; a 1 implies that a clock edge is seen during the amplification phase and a 0 means that it is not.

A problem common to many time-based algorithms is the possibility of an infinitesimally small temporal residue. We can guarantee a minimum residue by integrating for an extra clock period during the Time-to-Voltage conversion, a scheme we term the $1 + \varepsilon$ algorithm. The amplification process then not only amplifies the residue but generates two extra clock cycles as well. We simply ignore these two clock cycles when we digitally encode our output quantization bits. The $1 + \varepsilon$ scheme does however sacrifice efficiency in the number of clock cycles needed for conversion for greater robustness. Thus, it is typically only implemented for the first three bits of an 8-bit converter or five bits of a 12-bit converter. The cascaded amplification of the signal from early-residue stages makes later-residue stages quite insensitive to small errors making the $1 + \varepsilon$ algorithm unnecessary. This feature is true in all pipelined and algorithmic topologies including voltage-mode versions and we shall analyze it in more quantitative detail in a later section.

Figure 15.7 reveals a block diagram for the architecture of the overall ADC. The ADC consists of two LSB-matched capacitors, a reference current, an input current, an analog switching network to direct the currents to the capacitors or not, a comparator with a finite pulse width output, i.e., our spiking neuron, a state-machine, a clock divider, and N registers. No current matching is needed. The finite pulse width output of the comparator is useful in providing a minimum time to reset state variables. Transitions in the asynchronous state machine are triggered by the terminating negative edge of the neuronal pulse or by clock edges depending on whether we are in an MSB phase, a Time-to-Voltage phase, a Voltage-to-Time phase, or Amplification phase. The digital design of the control architecture is fairly straightforward from the description of the algorithm in Figure 15.6 (a) or Figure 15.6 (b) and is thus not presented here.

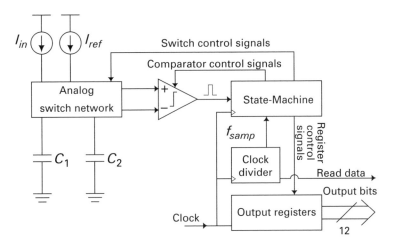

Figure 15.7. Architecture of the neuron-inspired ADC. Reproduced with kind permission from [15] (©2005 IEEE).

15.2.2   Error cancellation

Due to some inherent error-cancellation properties in the algorithm, comparator offset, comparator $1/f$ noise, comparator delay, and switching charge injection errors are also naturally minimized. Figure 15.8 (a) reveals the principle behind error cancellation in the converter. Any error, whether due to comparator delay (which includes that of the finite pulse width spike that it fires), offset, or $1/f$ noise or switching charge injection, causes a positive error during a Time-to-Voltage conversion. Due to the nature of the algorithm, this error is cancelled because it leads to a negative error when we convert back to time during a Voltage-to-Time conversion. For example, in Figure 15.8 (b) a delay in the comparator T_{cd} reduces the time to the next clock edge, thus decreasing the voltage used for comparison in the next phase, and reducing the time taken after the clock edge to reach this voltage; however, the comparator delay is now re-added to the smaller timing residue after the clock edge to bring it back to its correct value. Thus, after a single Time-to-Voltage and Voltage-to-Time phase have occurred, T_{resB} in Figure 15.8 (b) is exact. After the second Voltage-to-Time phase has occurred (the Amplification phase of Figure 15.6 (b)), we have an exact amplification of T_{resB} given by $2T_{resB}$, along with an extra T_{cd}. This T_{cd} error is cancelled out on the subsequent iteration of the conversion, i.e., we introduce errors on every iteration that then get taken out on the next iteration. On the very last iteration, the final error is not taken out but, in this iteration, any effect of error is small due to the large cascaded signal amplification from previous iterations.

Any $1/f$ noise is cancelled via this mechanism since it behaves like a slowly varying offset. If it changes slowly over the course of a clock cycle, it will be nearly completely removed. Equivalently, $1/f$ noise beyond $1/T_{clk}$ is not attenuated but $1/f$ noise below $1/T_{clk}$ is. Since charge injection is voltage dependent, it is not

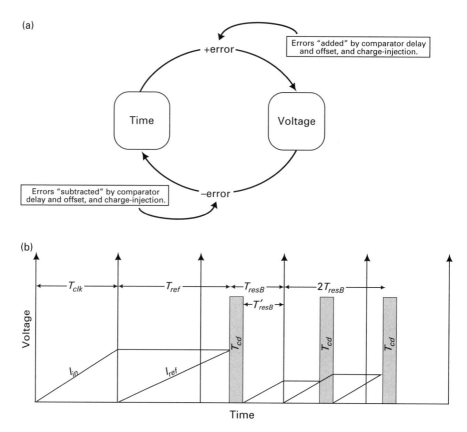

Figure 15.8a, b. Error-correction in the neuron-inspired ADC. Reproduced with kind permission from [15] (©2005 IEEE).

perfectly cancelled since the injection error introduced during one phase may not be the same as that on another phase if the comparator voltages are different on the two phases. Fortunately, with the sizes of capacitors used in typical ultra-low-power thermal-noise-limited converters, such second-order errors do not matter unless the precision of the converter exceeds 13–14 bits. Small variations in comparator delay due to common-mode variations across phases introduce second-order errors as well.

15.2.3   Circuit implementation and noise analysis

We shall now discuss the circuit implementation of the ADC and analyze its noise performance. The circuit implementation is fairly straightforward and not optimized so we shall not dwell on it too much. However, the noise analysis of the ADC will show how we can extend the voltage-noise-analysis techniques of Chapter 8 to signals encoded with time as a variable. In such systems, noise manifests as temporal jitter, rather than as amplitude fluctuations. Such temporal

(a)

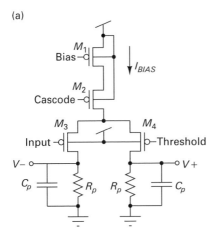

(b)

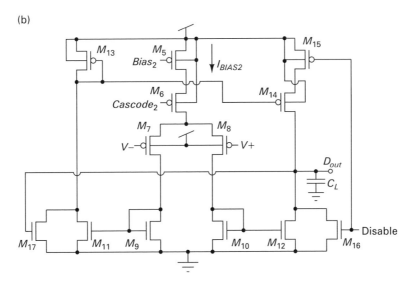

Figure 15.9. Comparator circuits in the neuron-inspired ADC. Reproduced with kind permission from [15] (©2005 IEEE).

noise analysis may prove useful for analyzing the temporal jitter of other low-power time-based ADCs in the future, so we shall flesh out the noise analysis in somewhat more detail than usual. The noise analysis is also useful in understanding how power consumption in different portions of an ADC impacts its overall precision.

The comparator is made up of a preamplifier stage followed by a positive-feedback gain/latch stage that yields a digital output. Figure 15.9 (a) shows the resistively loaded differential preamplifier and Figure 15.9 (b) shows the gain/latch stage. The transistors M_{17}, M_{13}, and M_{14} serve to create positive feedback in the gain/latch stage. The feedback can be disabled by turning the disable input high. The preamplifier serves to increase the signal level to a value that is sufficiently large such that the positive feedback is turned on primarily by signal rather than by noise.

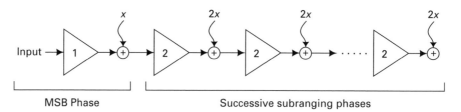

MSB Phase   Successive subranging phases

Figure 15.10. Noise contributions in the neuron-inspired ADC. Reproduced with kind permission from [15] (©2005 IEEE).

There are two primary sources of noise: the noise from the comparator and the noise from the integrating currents. The noise from the comparator has two contributions, one from the preamplifier and one from the gain/latch stage. We will begin by analyzing noise contributions due to the comparator and then due to the integrating currents. Since time is our primary variable, we analyze the noise and SNR of our converter in the time domain. If the converter's precision is thermal noise limited, the maximal SNR is given by

$$SNR = \left(\frac{4T_{clk}^2}{T_{NZ}^2}\right) \tag{15.7}$$

where T_{NZ}^2 is the total temporal jitter due to noise referred back to the output of the MSB stage. In Equation (15.7), the maximal SNR is the rms energy in a square wave input with a mean ac value of $2T_{clk}$ and excursions around this value of amplitude $\pm 2T_{clk}$.

Figure 15.10 shows the effect of comparator noise at various iterations in our conversion. During the MSB phase, the input is not gained; we have only one comparison in this phase such that the noise source is marked with an x to indicate that one unit's worth of comparator noise is present. During the successive subranging phases where we zoom in by factors of 2 on each iteration, we have two comparisons per phase; thus, the noise sources are marked with a $2x$ to indicate that two unit's worth of comparator noise is present on these phases. Hence the total input-referred contribution of comparator noise over all iterations is given by the sum of the mean-squared contributions from each iteration:

$$x^2 + 2\left(\frac{x}{2^1}\right)^2 + 2\left(\frac{x}{2^2}\right)^2 + 2\left(\frac{x}{2^3}\right)^2 + \ldots + 2\left(\frac{x}{2^N}\right)^2 = \boxed{\frac{5}{3}x^2} \tag{15.8}$$

The preamplifier's input-referred voltage noise from Figure 15.9 (a) is given by

$$V_{NZ,pa}^2 \approx (2)\left(\frac{5}{3}\right)\frac{4kT\left(\gamma g_{m3,4} + 1/R_p\right)}{g_{m3,4}^2}\left(\frac{1}{2\pi}\right)\left(\frac{\pi}{2}\right)\left(\frac{1}{R_pC_p}\right) \tag{15.9}$$

$$\approx \frac{10kT\left(\gamma g_{m3,4} + 1/R_p\right)}{3g_{m3,4}^2 R_pC_p}$$

To convert the preamplifier's input-referred noise to an equivalent time-domain jitter, we divide $V_{NZ,pa}{}^2$ by the square of the input charging slope I_{ref}/C_i, which is the gain of the time-to-voltage conversion process:

$$T_{NZ,pa}{}^2 = \frac{10kT\left(\gamma g_{m3,4} + 1/R_p\right)}{3g_{m3,4}{}^2 R_p C_p} \left(\frac{C_i}{I_{ref}}\right)^2$$

$$g_{m3,4} = \frac{I_{BIAS}}{V_{L3,4}}$$

$$Q_{pa}{}^2 = \frac{10kT\gamma C_i}{3} \tag{15.10}$$

$$I_{pach} = \frac{C_i V_{L3,4}}{R_p C_p}$$

$$T_{NZ,pa}{}^2 \approx \frac{I_{pach}}{I_{BIAS}} \frac{Q_{pa}{}^2}{I_{ref}{}^2}$$

So, burning power in the preamplifier to have low voltage noise and having a high input charging slope reduces the temporal noise due to the preamplifier. The parameter $V_{L3,4}$ represents the linear range of the input differential pair transistors of the preamplifier. In deriving the approximation of Equation (15.10), we have assumed that the noise due to M_3 and M_4 transistors overwhelms the noise due to the R_p resistors, which is an excellent approximation in any resistively loaded preamplifier with a gain greater than 5.

From Figure 15.9 (b), the gain/latch's output voltage noise and output temporal jitter are given by

$$V_{NZ,gl}{}^2 \approx \left(\frac{5}{3}\right) \frac{4kT\gamma(g_{m5} + g_{m9} + g_{m11} + g_{m13} + g_{m14})}{g_{o14}{}^2} \times \frac{g_{o14}}{C_L} \times \frac{1}{2\pi} \times \frac{\pi}{2}$$

$$T_{NZ,gl}{}^2 = \left(\frac{5}{3}\right) \frac{kT\gamma(g_{m5} + g_{m9} + g_{m11} + g_{m13} + g_{m14})}{C_L g_{o14}} \left(\frac{C_L}{I_{BIAS2}}\right)^2$$

$$Q_{gl}{}^2 = \left(\frac{5}{3}\right) \frac{kTC_L\gamma(g_{m5} + g_{m9} + g_{m11} + g_{m13} + g_{m14})}{g_{o14}} \tag{15.11}$$

$$T_{NZ,gl}{}^2 = \frac{Q_{gl}{}^2}{I_{BIAS2}{}^2}$$

So, burning power in the gain/latch stage to get a high output slew rate reduces the temporal noise.

In deriving Equation (15.11), we have assumed that only one side of the amplifier is turned on and that the noise contributions of M_6 and M_7 in Figure 15.9 (b) are negligible since they function as cascode devices that shunt their own noise. Thus, burning power in the gain/latch stage to get a high output slew rate reduces its temporal jitter. In deriving Equations (15.9) and (15.11), we have assumed that the preamplifier's input-referred voltage noise adds jitter when switching is initiated by a threshold crossing. After the crossing, the preamplifier gain is sufficiently high that the gain/latch stage's input differential pair is strongly turned on in one saturated half. The jitter in the gain/latch stage arises, because, even after its differential pair has been strongly turned on, there is jitter in the time at which the output charging current overwhelms the output noise.

The noise in the integrating currents derived from cascoded current sources is dominated by that of a single transistor and given by

$$\Delta V_{NZ,i}{}^2 = I_{NZ,i}{}^2 \left(\frac{T_i}{C_i}\right)^2 = \frac{8kTg_{mi}\Delta f}{3}\left(\frac{T_i}{C_i}\right)^2 \tag{15.12}$$

For a square window of integration of time T, the bandwidth Δf is well approximated by $1/(2T)$ due to the properties of its associated sinc^2 Fourier filtering function. Since each integration process is independent, the noise due to integration with the input current and then the reference current in the MSB phase is given by

$$\Delta V_{NZ,in}{}^2 + \Delta V_{NZ,ref}{}^2 = \frac{8kT}{3C_i{}^2}\left(g_{min}\frac{T_{clk}}{2} + g_{mref}\frac{T_{res}^{MSB}}{2}\right) \tag{15.13}$$

Since $T_{res}^{MSB} = T_{clk}I_{in}/I_{ref}$, we can substitute for it in Equation (15.13) to get the total voltage noise due to integrating currents in the MSB phase:

$$\Delta V_{NZ,MSB}{}^2 = \frac{4kT}{3C_i{}^2}T_{clk}\left(g_{min} + g_{mref}\frac{I_{in}}{I_{ref}}\right) \tag{15.14}$$

During each successive subranging phase in Figure 15.6 (b), we integrate three times. If we assume conservatively that the average residue T_{RESB} is $1.5T_{clk}$ because of the use of the $(1 + \varepsilon)$ algorithm, the voltage noise due to one successive subranging iteration is given by

$$3\Delta V_{NZ,res}{}^2 = 3\frac{4kTg_{mref}T_{res}}{3C_i{}^2} = 3\frac{4kTg_{mref}}{3C_i{}^2}\frac{3}{2}T_{clk} \tag{15.15}$$

If we input-refer the noise from all successive subranging phases to the output of the MSB phase, using reasoning similar to that employed for the comparator noise in Figure 15.10, we get the total noise for all these phases to be given by

$$\Delta V_{NZ,restot}{}^2 = 3\Delta V_{NZ,res}{}^2\left[\left(\frac{1}{2^1}\right)^2 + \left(\frac{1}{2^2}\right)^2 + \left(\frac{1}{2^3}\right)^2 + \ldots + \left(\frac{1}{2^N}\right)^2\right] = \Delta V_{NZ,res}{}^2 \tag{15.16}$$

By adding Equation (15.14) and (15.16), the total voltage noise due to integration on the MSB phase and on the successive subranging phases is then given by

$$V_{NZ,int}{}^2 = \frac{4kT}{3C_i{}^2}T_{clk}\left(g_{min} + g_{mref}\frac{I_{in}}{I_{ref}} + g_{mref}\frac{3}{2}\right) \tag{15.17}$$

The temporal noise due to integration is the voltage noise divided by the square of the integration rate $I_{ref}{}^2/C_i{}^2$.

$$T_{NZ,int}^2 = \frac{4kT}{3I_{ref}^2} T_{clk} \left(g_{min} + g_{mref} \frac{I_{in}}{I_{ref}} + g_{mref} \frac{3}{2} \right) \qquad (15.18)$$

An upper bound on this noise arises when the input is at full scale, i.e. $I_{in} = 4I_{ref}$. At this value, $g_{min} = 2g_{mref}$, if we assume above-threshold operation for the integrating currents. Thus, an upper bound on the integration noise is given by

$$T_{NZ,int}^2 = \frac{10kT g_{mref}}{I_{ref}^2} T_{clk}$$

$$g_{mref} = \frac{I_{ref}}{V_{L,ref}}$$

$$Q_{max} = I_{ref} T_{clk}$$

$$Q_{max} = \frac{C_i V_{max}}{4}$$

$$Q_{ref}^2 = 10q \left(\frac{\left(\frac{kT}{q} \right)}{V_{L,ref}} \right) Q_{max} \qquad (15.19)$$

$$\alpha_{ref} = \left(\frac{\left(\frac{kT}{q} \right)}{V_{L,ref}} \right)$$

$$Q_{ref}^2 = 10q \alpha_{ref} Q_{max}$$

$$T_{NZ,int}^2 = \frac{Q_{ref}^2}{I_{ref}^2}$$

In essence, we burn power to average over more electrons to reduce noise. The value of V_{max} is the maximum voltage that the maximal input $I_{in} = 4I_{ref}$ can charge C_i to within T_{clk} and $V_{L,ref}$ is the linear range of the transistors that supply I_{ref}.

From Equations (15.10), (15.11), and (15.19), we find the total temporal noise to be given by

$$T_{NZ}^2 = T_{NZ,pa}^2 + T_{NZ,gl}^2 + T_{NZ,int}^2$$

$$T_{NZ}^2 = \frac{I_{pach}}{I_{BIAS}} \frac{Q_{pa}^2}{I_{ref}^2} + \frac{Q_{gl}^2}{I_{BIAS2}^2} + \frac{Q_{ref}^2}{I_{ref}^2} \qquad (15.20)$$

Not surprisingly, we note that from Equation (15.20) that our total temporal noise is reduced if I_{BIAS}, I_{ref}, and I_{BIAS2} increase. The maximum SNR in dB from Equation (15.7) is then given by

$$SNR_{dB} = 10 \log_{10} \left(\frac{4 T_{clk}^2}{\frac{I_{pach} Q_{pa}^2}{I_{BIAS} I_{ref}^2} + \frac{Q_{gl}^2}{I_{BIAS2}^2} + \frac{Q_{ref}^2}{I_{ref}^2}} \right) \qquad (15.21)$$

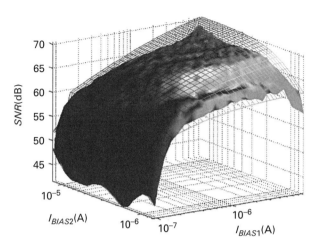

Figure 15.11. The SNR of the neuron-inspired ADC versus bias-current parameters.

If we substitute for T_{clk} and Q_{ref} from Equation (15.19) into Equation (15.21), we get

$$SNR_{dB} = 10\log_{10}\left(\frac{4\dfrac{Q_{max}^2}{I_{ref}^2}}{\dfrac{I_{pach}}{I_{BIAS}}\dfrac{Q_{pa}^2}{I_{ref}^2} + \dfrac{Q_{gl}^2}{I_{BIAS2}^2} + \dfrac{10\alpha_{ref}(qQ_{max})}{I_{ref}^2}}\right) \qquad (15.22)$$

For the parameter values given by $R_p = 180$ kΩ, $C_p = 40$ fF, $C_L = 1$ pF, $I_{in} = 40\ \mu$A, $I_{ref} = 10\ \mu$A, $T_{clk} = 0.5\ \mu$S, $C_i = 22$ pF and from Equations (15.10), (15.11), and (15.21), we can compute SNR_{dB} as we vary I_{BIAS} and I_{BIAS2}. Figure 15.11 shows that experimental measurements on a $0.35\ \mu$m converter operated with a 3 V power supply yield a good fit to theory: the grid represents theory while the experimental measurements are shown in a darker font. The SNR_{dB} was computed from Fourier transform measurements of a 4000-sample 1 kHz sine-wave full-scale input sampled at 31.25 kHz, which from a dB equivalent of Equation (15.1) yield the equivalent number of bits, the ENOB, of the converter:

$$\text{ENOB} = \frac{SNR_{dB} - 1.76}{6.02} \qquad (15.23)$$

This particular converter had 11 bits of thermal-noise-limited precision.

A converter's differential nonlinearity, or DNL, is the input error between any two successive digital codes from an ideal 1 LSB value. The DNL can and does vary from code to code. For the neuron-inspired ADC, it was less than 0.5 LSB for all codes with the LSB being defined at 12-bit precision with respect to full scale. A converter's integral nonlinearity, or INL, is its deviation from an ideal overall straight-line input-output function that is defined by joining the points corresponding to its lowest code and its highest code. An alternative measure of INL uses the deviation from a best-fit straight line that minimizes overall least-squared error across all codes. Like the DNL, the INL can and does vary from code to code. With the least-squares definition, for all codes of the neuron-inspired ADC, the measured INL was less than 1 LSB. Thus, this converter's nonlinearity would limit its precision to 12 bits but its thermal noise limits its precision to 11 bits. The INL of this converter is limited by the fact that the comparator delay is a weak function of its input voltage such that comparator delay is not perfectly cancelled during the MSB phase of the conversion.

15.2.4 Energy-efficient implementation

The 11-bit ADC described thus far consumes $75\,\mu W$ of total analog and digital power on a 3 V power supply. It was the first ADC that used time to perform analog-to-digital conversion rather than voltage or current but that achieved $O(N)$ scaling in conversion time [15]. In contrast, dual-slope converters, which also use time to perform analog-to-digital conversion, have $O(2^N)$ scaling. The ADC's E_q is 1.12 pJ, competitive with many ADCs that typically operate in the 6 pJ–10 pJ range. However, this ADC was not optimized for energy efficiency: it used the $1 + \varepsilon$ strategy over all iterations of the conversion rather than for just the first few iterations; it used an unnecessarily large power supply of 3 V for both analog and digital operation; its distribution of currents between I_{BIAS}, I_{BIAS2}, and I_{REF} was not optimized. A lower 8-bit-precise version of the neuron-inspired ADC in $0.18\,\mu m$ technology achieves $0.96\,\mu W$, 45 kHz conversion with $E_q = 120$ fJ [14]. The improved energy efficiency is attained by using a 1.2 V analog supply and 0.8 V digital supply, using the $1 + \varepsilon$ algorithm over only the first three iterations such that 8-bit conversion is achieved in 22 clock cycles, and implementing the comparator with just the gain-latch stage, a practical scheme at lower precision. The area of the latter ADC is only $130\,\mu m \times 160\,\mu m$, making it both area efficient and power efficient. Since two-thirds of the power of the ADC is digital, implementation in a 90 nm or deep submicron process, with further power-supply reduction and implementation of current-distribution optimizations, promise energy efficiencies less than 20 fJ at audio bandwidths and more than $3\times$ area reductions. The preamplifier-and-latch comparator design of Figures 15.9 (a) and 15.9 (b) are also relatively primitive proof-of-concept designs and can be greatly improved. We shall discuss other designs for comparators, which if combined with this algorithm with appropriate modifications of its control signals, could potentially lead to order-of-magnitude improvements. It is heartening to note that a completely new neuron-inspired topology without decades of optimization and study is already highly energy efficient, boding well for other neuron-inspired ADCs of the future. In fact, as we point out later, $\Sigma\Delta$ converters and other time-based converters themselves bear many similarities to the operations of adaptive neurons.

The speed of the neuron-inspired ADC can be improved by interleaving with little change in energy efficiency. Suppose the ADC takes M clock cycles for conversion. Since the ADC has a built-in sample-and-hold phase, the input current can be redirected to M converters that operate in parallel, with the i_{th} converter receiving the input current and staggering its onset of conversion to be on the ith clock cycle: each of the M converters begins its integration-of-the-input phase on a clock cycle that is sequentially staggered amongst the converters. Thus, the Mth converter will just be sampling its input on the Mth clock cycle when the first converter is in its terminating cycle of conversion.

15.3   Computational ADCs and time-to-digital ADCs

An interesting feature of the neuron-inspired ADC is that polynomial functions of the analog input current can be computed in a sequential Taylor-series-like fashion. We exploit successive *Charge = Current × Time* relationships to perform multiplicative operations. Figure 15.12 illustrates the process: ADC1 quantizes the first timing variable, T_{x1}, in the same manner as the neuron-inspired ADC. Therefore, we know that

$$T_{x1} = T_{clk} \frac{I_{in}}{I_{ref}} \tag{15.24}$$

Now ADC2 begins digitizing I_{in} after ADC1 has finished a sequence comprised of one Time-to-Voltage and one Voltage-to-Time operation. The effective integration time for ADC2's sample-and-hold phase is then T_{x1} as shown in Figure 15.12. The time interval T_{x1} that ADC2 operates on is free of comparator-delay and charge-injection errors; furthermore, T_{x1} is referenced to a following clock edge instead of a preceding clock edge for ADC2. Thus, the quantized output of ADC2 will reflect

$$T_{x2} = T_{x1} \frac{I_{in}}{I_{ref}} \tag{15.25}$$

From Equations (15.24) and (15.25), we get

$$T_{x2} = T_{clk} \left(\frac{I_{in}}{I_{ref}} \right)^2 \tag{15.26}$$

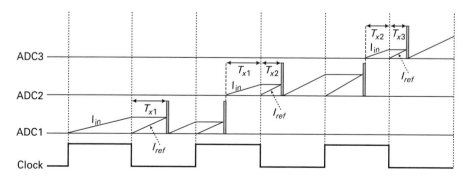

Figure 15.12. Computational ADCs. Reproduced with kind permission from [14] (ⓒ2006 IEEE).

The recursive application of the same procedure on successive ADCs as shown in Figure 15.12 then yields

$$T_{xm} = T_{clk} \left(\frac{I_{in}}{I_{ref}} \right)^m \tag{15.27}$$

The weighting of each term of the polynomial can be adjusted by changing the reference current for that converter. For instance, if the reference current of ADCm is $I_{ref}/3$ and all other ADCs till then have used a reference current of I_{ref}, Equation (15.27) is altered to

$$T_{clk} = 3T_{clk} \left(\frac{I_{in}}{I_{ref}} \right)^m \tag{15.28}$$

Another obvious generalization is to use different inputs for the different ADCs. Thus, if we use $I_{in1}, I_{in2}, \ldots, I_{inm}$ for the first m ADCs, Equation (15.27) is altered to

$$T_{xm} = T_{clk} \prod_{i=1}^{i=m} \left(\frac{I_{in}^i}{I_{ref}} \right) \tag{15.29}$$

Thus, we may compute *any* polynomial function of a single analog input or of multiple analog inputs and simultaneously have digital representations of these quantities to the level of precision that we desire. In effect, we have a new way of implementing translinear circuits in current mode by using *Charge = Current × Time* to perform multiplies rather than adding logarithms as we discussed in Chapter 14!

Computational ADCs could have several applications. For example, in pulse oximetry, which we describe in Chapter 20, oxygen saturation, which is a second-order polynomial function of a measured variable in the oximeter, can be directly computed. Networks of computational ADCs can implement large-scale hybrid analog-digital computers that have both polynomial and logical basis functions and both continuous and discrete representations of signals as neurons do: output bits from the ADCs can provide signal restoration; they can also condition the conversion of one ADC via logical combinations of signals from other ADCs or from the same ADC. Adaptive-precision computation on analog inputs, adaptive ADCs with feedback from bits to analog signals, and gated ADCs that only quantize when needed may then be imagined. Such architectures have the potential to be highly energy efficient and to blur the boundaries between analog and digital computation as nature appears to have done [16], [17]. We shall return to a discussion of such issues in a more general context in Chapter 22 and to bio-inspired computation in Chapters 23 and 24.

The neuron-inspired ADC is useful in biomedical applications that frequently need to digitize event-based signals. For example, instantaneous heart rates and instantaneous neural inter-spike intervals are event based, carry significant information, and are cheaper to digitize since a high dynamic range is not needed at the input, just the time of occurrence of a pulsatile event. The time-to-digital nature of the neuron-inspired ADC allows us to directly digitize the intervals between two such events by a simple modification to the algorithm: the integration phase of the conversion is defined by two successive events and we digitize the previous inter-event interval while sampling the next.

15.4   A time-based $\Sigma\Delta$ ADC

Figure 15.13 illustrates an architecture for an open-loop $\Sigma\Delta$ ADC that is based on time-domain computation [18]. The voltage v_{IN} serves as the bulk input of the pFET in the CMOS inverter of a ring oscillator. Modulation of v_{IN} alters the frequency of the oscillator, albeit in a weakly nonlinear fashion. Since phase is the integral of frequency, the voltage-to-phase transfer function automatically serves as the integrator in the feedforward portion of the $\Sigma\Delta$ in Figure 15.5. The phase of the output of the ring oscillator, which changes linearly over a period, however, is automatically digitized in an implicit or explicit comparator. The comparator is implicit, if square-wave-like outputs result from the ring oscillator itself. It is explicit if a logic stage that follows the ring oscillator restores the ring-oscillator output to logic values. The XOR stage with clocked registers in Figure 15.13 serves as such an explicit comparator but it implements another important function as well: it serves as a differentiator in the digital domain such that only sharp changing edges in its square-wave inputs result in 1s at its output after a clock cycle of delay while the constant portions of the square wave yield no outputs. Higher voltages lead to higher-frequency outputs from the ring-oscillator voltage controlled oscillator (VCO). Higher-frequency outputs lead to more edges and consequently more 1s reported by the XOR over a fixed number of clock cycles. Thus, the number of 1s reported by the XOR output over a fixed number of clock cycles is proportional to v_{IN} if the transfer function from v_{IN} to frequency is linear. Furthermore, the quantization noise from the square-wave digitization of the phase is highpass filtered by the XOR in the digital domain such that the noise transfer function gets differentiated, as in a closed-loop $\Sigma\Delta$ while the signal transfer function, which gets integrated and then differentiated, is unchanged. The topology is energy efficient because the ring oscillator can be run on a 200 mV power supply in subthreshold operation and almost all the operations are digital, and therefore scale relatively well with technology.

The challenge with this topology lies in its power-supply noise rejection, which is poor (the inverter's frequency is dependent on the power supply) and on

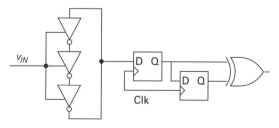

Figure 15.13. A time-based energy-efficient ADC. Reproduced with kind permission from [18] (©2006 IEEE).

nonlinearities in the transconductance of the bulk for large-signal inputs. Resistive source degeneration at the supply has been shown to improve linearity and is essential for this topology to be practical, especially at higher precision [18]. Unlike a closed-loop $\Sigma\Delta$ ADC, it is quite sensitive to errors in its components such that we have traded efficiency for robustness: we are not subject to the degradations in efficiency imposed by stability considerations in a closed-loop topology but we do not benefit from its inherent robustness either. Nevertheless, the topology of Figure 15.13 has achieved an E_q of 57 fJ in 90 nm and is worth keeping an eye on for the future.

15.5   Pipelined ADCs with comparators

Pipelined architectures such as those in Figure 15.3 require gain-of-2 amplifiers, which have been implemented in the past by closed-loop topologies employing operational amplifiers and gain-setting capacitors. Such topologies have relatively high power consumption since the gain-bandwidth product of the operational amplifiers has to be sufficiently large to ensure closed-loop gain accuracy within the desired bandwidth of the signal. In deep submicron technologies, device gain and power-supply voltages decrease making power-efficient operational-amplifier design difficult. A topology has been proposed in [19] that replaces operational amplifiers in switched-capacitor pipelined ADCs by comparators, once again echoing a common theme in this chapter.

Figure 15.14 illustrates the key idea. When we transition from a sampling phase to an amplification phase in a pipelined stage, the operational-amplifier-based

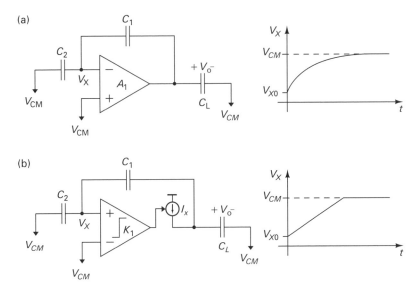

Figure 15.14a, b. Replacing amplifiers (a) with comparators (b) in pipelined ADCs. Reproduced with kind permission from [19] (©2006 IEEE).

topology and the comparator-driven current-source topology both cause the virtual-ground node V_X to traverse towards an equilibrium value of V_{CM}. The operational amplifier tries to force the virtual-ground condition while the comparator-driven current source merely detects the virtual-ground condition and shuts off. The shutting down of the current source defines the sampling instant such that the correct amplified value at the output is identical in both cases. Thus, comparator-based topologies can mimic operational-amplifier topologies but at lower power: it is more energy efficient to detect the virtual-ground condition with a comparator than to force a virtual-ground condition with an operational amplifier.

15.6   Adiabatic charging and energy-efficient comparators in ADCs

Recently, a SAR ADC has achieved an $E_q = 4.4\,\text{fJ}$ at an ENOB of 8 bits and a sample rate of 1 MHz in 65 nm technology, which is impressive [6]. Two of the key innovations behind this ADC are the use of adiabatic or slow charging in its capacitive DAC array to minimize capacitive switching energies, and the use of an energy-efficient comparator with very little static power dissipation. We shall discuss the use of adiabatic charging first and the comparator next.

Traditionally, the DAC voltage used for comparisons in an SAR ADC, e.g., the V_{comp} voltage in Figure 15.4, is altered by abruptly switching the bottom plates of the capacitors that feed it from one digital voltage to another. As we discuss extensively in Chapter 21 on low-power digital design, such switching results in dynamic digital power consumption due to dissipation of capacitive node energies in active resistive switches that are connected to these nodes. In Chapter 21, we also discuss how slow adiabatic alterations of digital node voltages such that very little voltage is dropped across these switches at any instant minimizes $I^2 R_{on}$ switch power dissipation and allows for reduced energy dissipation. If we switch the nodes very gradually most of the energy in the capacitive nodes is not dissipated and can be recovered and stored in an energy storage element and then recycled back to the capacitive node on another switching cycle. For example in Figure 21.8 in Chapter 21, the energy is stored on an inductor and recycled back to the switching nodes over an oscillatory clock cycle. A related version of the adiabatic scheme switches nodes from one power-supply voltage to a slightly higher power-supply voltage in small gradual steps with the different power-supply voltages implemented by a bank of large capacitors. The capacitor bank trades inductive storage of node energies for capacitive storage in the power-supply bank [20]. The latter reference shows how capacitive pad switching energies can be greatly reduced with adiabatic switching schemes. The capacitive power-supply bank has also been used for power-efficient current stimulation in retinal implants as we discuss in Chapter 19 [21]. The ADC described in [6] applies such ideas of adiabatic charging to lower switching power dissipation of capacitive node energies in the DAC of a SAR ADC as shown in Figures 15.15 (a) and 15.15 (b).

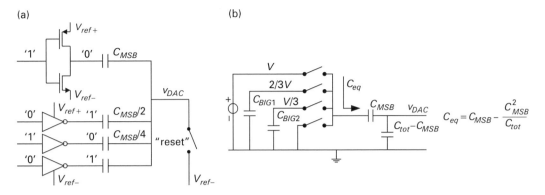

Figure 15.15a, b. Adiabatic charging in an energy-efficient SAR ADC. Reproduced with kind permission from [6] (©2008 IEEE).

Figure 15.15 (a) shows a traditional switching scheme in the DAC of an SAR ADC with digital inverter inputs switching from 1 to 0 to such that v_{DAC}, analogous to V_{comp} in Figure 15.4, is altered. The reset switch initializes v_{DAC} to a value that is within $\sqrt{kT/C_{tot}}$ of V_{ref}, where C_{tot} is the total capacitance at the v_{DAC} node from all the DAC capacitances, i.e., $C_{tot} = 2C_{MSB} + C_{par}$. Figure 15.15 (b) shows an adiabatic switching scheme where, rather than the traditional switching scheme of Figure 15.15 (a), we use the capacitive power-supply bank of Figure 15.15 (b) and appropriate control logic to switch DAC capacitances to the v_{DAC} node in gradual incremental steps, in this case with two additional power-supply levels besides 0 and V. In the example shown C_{MSB} is being switched and the equivalent switching capacitance that the capacitive power-supply bank sees is $C_{MSB}||(C_{tot} - C_{MSB})$ The switching energy of this equivalent capacitance is reduced by approximately a factor of 3 due to the use of three switching levels above ground. In practice, the number of levels that maximize energy savings has an optimum due to gate switching capacitance and control overhead. An interesting feature of Figure 15.15 (b) that is present in all such switching schemes is that the equilibrium values of $V/3$ and $(2/3)V$ automatically get established on the capacitances over a few cycles of DAC operation if the control logic times the switching correctly. The exact values of these equilibrium voltages are not too critical since the overall precision of the ADC is set by the precision of V, not by that of the intermediate levels. As long as

$$\left(\frac{\Delta C_{LSB}}{2C_{tot}}\right)V > 6\sqrt{\frac{kT}{2C_{tot}}} \tag{15.30}$$

the changes at the v_{DAC} node, even with capacitive DAC mismatch, ΔC_{LSB}, will be reliably greater than the random thermal noise at the v_{DAC} node established during its reset. The use of metal-plate capacitors, which matched relatively well in this process, allowed $V = 1\,\mathrm{V}$ and $C_{tot} = 600\,\mathrm{fF}$, to meet this specification even for 10-bit precision although the experimental demonstration of the ENOB

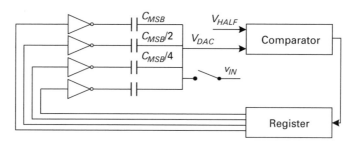

Figure 15.16. Switching architecture of the SAR ADC. Reproduced with kind permission from [6] (©2008 IEEE).

corresponded to 8 bits. Adiabatic capacitive charging is only implemented for the first 3 MSB bits since the savings in energy get $2\times$ smaller for each less significant bit, not making the added complexity worthwhile.

Figure 15.16 shows that the control logic for switching DAC capacitances is similar to that in Figure 15.4, since they both implement a successive-approximation scheme. In most standard SAR implementations, as in Figure 15.4, the bottom plates of the DAC capacitances are initialized to v_{IN} in a sampling phase, and then the DAC bits are sequentially changed from the MSB to the LSB in a conversion phase. In [6], since v_{DAC} is initialized at v_{IN}, the bottom plate of C_{MSB} can be initialized at 1 such that comparisons in the SAR ADC can begin right after initialization, saving switching energy in the bottom plate and also saving time. At the end of the conversion, the register in Figure 15.16 contains bits that have been set such that the initialized value of v_{IN} on v_{DAC} has been altered to be just near V_{HALF}.

Figure 15.17 shows the energy-efficient comparator composed of a subthreshold preamplifier with no static power dissipation in its first stage and a cross-coupled latch in its second stage. The nodes FN and FP are precharged to V_{DD} and the nodes SN and SP are discharged to ground when CLK is low. When CLK is high, the comparator is activated with MN_{tail} operating in its triode regime and the differential input transistors operating in their subthreshold regime. These transistors discharge the common-mode value of FN and FP with the larger of v_{IN+} or v_{IN-} determining which node has a lower voltage. FN and FP cause further preamplification in the latch, and when SN or SP have reached a sufficiently high value, they cause the latch to regenerate. The combination of two-stage preamplification, subthreshold operation, and minimal static-power dissipation lead to a very energy-efficient comparator. Such comparators could impact the energy efficiency of several topologies discussed in this chapter including the bio-inspired ADC.

The overall topology of the SAR ADC in [6] is also implemented in a fully differentially fashion such that common-mode noise on the reference voltages during comparisons have a negligible impact on the overall precision of the ADC. However, during sampling, these voltages do have to be quiet. The ADC achieved a DNL of 0.5 LSB and an INL of 2.2 LSBs due to the presence of a nonlinear parasitic capacitance at the v_{DAC} node.

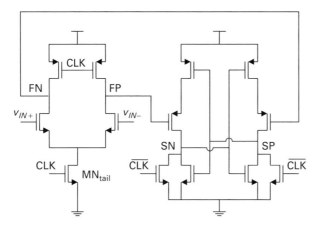

Figure 15.17. A highly energy-efficient comparator circuit. Reproduced with kind permission from [6] (©2008 IEEE).

15.7   Digital correction of analog errors

If an ADC is not thermal-noise limited, nonlinearities, gain, and offset errors in its components may limit its precision. Unlike thermal noise, these errors are predictable. Therefore, it is possible to correct for these errors on the digital output of the ADC if one can learn what these errors are [22]. To do so, we first form a parametric digital model of how various errors in its analog components cause error. For example, gain errors in a pipelined ADC from its amplifying stages have predictable effects on its digital output with errors in early amplification stages mattering exponentially more than errors in later amplification stages. Then, we experimentally determine the error of the ADC by sweeping its input and comparing its digital output with that from a slow highly precise ADC. We then learn the parameters of the model of the ADC that best describes the observed errors in a least-squares sense using standard learning techniques such as gradient descent on a multi-parameter space. Finally, we store the learned parameters digitally and use them and the error model to perform digital error correction on the outputs of the ADC while it is running. The learning can typically be done in a continuous online fashion with the calibrating ADC constantly running at a relatively low bandwidth in the background and constantly providing information to update the parameters of the error model. The ADC described in [23] uses such a strategy with a fast pipelined ADC being constantly calibrated in the background by a slow algorithmic ADC.

The overall scheme is energy efficient because one can use open-loop highly inaccurate, low-gain analog components that perform at high speed but with low precision in an energy-efficient fashion. If the known errors are well modeled and do not change often, the calibrating ADC can be slow and highly precise and perform calibrations relatively infrequently. Thus, the main ADC and the calibrating ADC may both be operated in an energy-efficient fashion because neither is simultaneously fast *and* precise. The digital error-correction power does now add to the overall power but the costs of such error correction are usually modest.

15.8   Neurons and ADCs

Intriguingly, highly energy-efficient ADCs appear to be converging to the digitization strategies used by $0.5\,\text{nW}$ neuronal cells in the brain that also use comparators and time-based strategies for performing vector pattern recognition. Vector pattern recognition is a generalization of A-to-D conversion from scalar inputs to vector inputs. The neuron-inspired algorithmic ADC that we described in this chapter was inspired by the operation of integrate-and-fire pulsatile or spiking neurons in the brain and naturally led to energy-efficient conversion, computational ADCs and time-to-digital architectures.

The highly energy-efficient ADC of Figure 15.13 also operates much like an integrate-and-fire neuron even though the authors do not explicitly claim to have arrived at their architecture through neuronal inspiration. The neuron is often well modeled as a linear current-controlled oscillator (CCO) since the frequency of pulsatile spike firing of the neuron is proportional to its input current. Thus, if we count the number of pulses fired by a neuron in a fixed time period or fixed number of clock cycles, we have a quantized representation of the current input. The topology of Figure 15.13 replaces the neuronal CCO with a VCO, and the neuronal spike generator with an edge-sensitive XOR. Both topologies quantize their input by simply counting the number of edges in a frequency input.

Closed-loop $\Sigma\Delta$ architectures have strong analogies to adaptive neurons. In certain neurons, the firing of spikes leads to the accumulation of calcium in the cell, which, in turn, then turns on a calcium-dependent potassium conductance that inhibits or subtracts current from the charging input current of the neuron [24]. The subtraction is analogous to the subtractive input of a $\Sigma\Delta$ and the feedback current is analogous to the feedback variable of a $\Sigma\Delta$ architecture. However, in most neurons, the time constant of the feedback is slow such that the presence of a slow integration in the feedback path leads to a differentiating $\Sigma\Delta$ A/D that only outputs events when the input changes.

In Chapter 22, we shall see that successive approximation ADCs are a special example of a more general analysis-by-synthesis architecture, where the synthesized DAC model of the input provides iterative 1 or 0 event feedback that helps us analyze the input. Such analysis-by-synthesis architectures can be efficiently generalized to create hybrid state machines with spiking neurons as we discuss in Chapters 22 and 23.

The numerous analogies between neurons and ADC architectures suggest that we may find further inspiration from neurons on how to build interesting computational and energy-efficient ADCs and in signal-to-symbol conversion in general. We shall return to such bio-inspired and other mixed-signal architectures in Chapters 22, 23, and 24.

References

[1] H. Y. Yang. *A time-based energy-efficient analog-to-digital converter*. Ph.D. Thesis, Electrical Engineering and Computer Science, Massachusetts Institute of Technology (2006).

[2] R. H. Walden. Analog-to-digital converter survey and analysis. *IEEE Journal on Selected Areas in Communications*, **17** (1999), 539–550.

[3] E. V. Aksenov, Y. M. Ljashenko, A. V. Plotnikov, D. A. Prilutskiy, S. V. Selishchev and E. V. Vetvetskiy, Biomedical data acquisition systems based on sigma-delta analogue-to-digital converters. *Proceedings of the 23rd Annual International IEEE Engineering in Medicine and Biology Conference (EMBS)*, Istanbul, Turkey, 3336–3337, 2001.

[4] A. Gerosa, A. Maniero and A. Neviani. A fully integrated two-channel A/D interface for the acquisition of cardiac signals in implantable pacemakers. *IEEE Journal of Solid-State Circuits*, **39** (2004), 1083–1093.

[5] M. Tavakoli, L. Turicchia and R. Sarpeshkar. An ultra-low-power pulse oximeter implemented with an energy efficient transimpedance amplifier. *IEEE Transactions on Biomedical Circuits and Systems* (2009).

[6] M. van Elzakker, E. van Tuijl, P. Geraedts, D. Schinkel, E. Klumperink and B. Nauta, A 1.9 μW 4.4 fJ/Conversion-step 10b 1MS/s Charge-Redistribution ADC. *Digest of Technical Papers IEEE International Solid-State Circuits Conference (ISSCC)*, San Francisco, CA, 244–610, 2008.

[7] M. D. Scott, B. E. Boser and K. S. J. Pister. An ultra-low-energy ADC for smart dust. *IEEE Journal of Solid-State Circuits*, **38** (2003), 1123–1129.

[8] V. Giannini, P. Nuzzo, V. Chironi, A. Baschirotto, G. Van der Plas and J. Craninckx, An 820 μW 9b 40 MS/s Noise-Tolerant Dynamic-SAR ADC in 90 nm Digital CMOS. *Proceedings of the IEEE International Solid-State Circuits Conference (ISSCC)*, San Francisco, CA, 238–239, 2008.

[9] S. Pavan, N. Krishnapura, R. Pandarinathan and P. Sankar. A Power-Optimized Continuous-Time Delta Sigma ADC for Audio Applications. *IEEE Journal of Solid-State Circuits*, **43** (2008), 351–360.

[10] L. C. Aiello. Brains and guts in human evolution: The expensive tissue hypothesis. *Brazilian Journal of Genetics*, **20** (1997).

[11] R. Sarpeshkar, L. Watts and C. A. Mead. Refractory neuron circuits. *Computation and Neural Systems Memo CNS TR-92-08* (1992).

[12] D. Tank and J. Hopfield. Simple 'neural' optimization networks: An A/D converter, signal decision circuit, and a linear programming circuit. *IEEE Transactions on Circuits and Systems*, **33** (1986), 533–541.

[13] H. Hamanaka, H. Torikai and T. Saito, Spike position map with quantized state and its application to algorithmic A/D converter. *Proceedings of the International Symposium on Circuits and Systems (ISCAS)*, Vancouver, BC, 673–676, 2004.

[14] H. Y. Yang and R. Sarpeshkar. A Bio-Inspired Ultra-Energy-Efficient Analog-to-Digital Converter for Biomedical Applications. *IEEE Transactions on Circuits and Systems I: Regular Papers*, **53** (2006), 2349–2356.

[15] H. Y. Yang and R. Sarpeshkar. A time-based energy-efficient analog-to-digital converter. *IEEE Journal of Solid-State Circuits*, **40** (2005), 1590–1601.

[16] R. Sarpeshkar. Analog versus digital: extrapolating from electronics to neurobiology. *Neural Computation*, **10** (1998), 1601–1638.

[17] R. Sarpeshkar and M. O'Halloran. Scalable hybrid computation with spikes. *Neural Computation*, **14** (2002), 2003–2038.

[18] U. Wismar, D. Wisland and P. Andreani, A 0.2 V 0.44 µW 20 kHz Analog to Digital $\Sigma\Delta$ Modulator with 57 fJ/conversion FoM. *Proceedings of the 32nd European Solid-State Circuits Conference (ESSCIRC)*, 187–190, 2006.

[19] J. K. Fiorenza, T. Sepke, P. Holloway, C. G. Sodini and L. Hae-Seung. Comparator-Based Switched-Capacitor Circuits for Scaled CMOS Technologies. *IEEE Journal of Solid-State Circuits*, **41** (2006), 2658–2668.

[20] L. Svensson, W. C. Athas and R. S. C. Wen, A sub-CV^2 pad driver with 10 ns transition time. *Proceedings of the International Symposium on Low Power Electronics and Design*, Monterey, California, 105–108, 1996.

[21] S. K. Kelly and J. Wyatt, A power-efficient voltage-based neural tissue stimulator with energy recovery. *Proceedings of the IEEE International Solid-State Circuits Conference (ISSCC)*, San Francisco, CA, 228–524, 2004.

[22] B. Murmann and B. E. Boser. A 12-bit 75-MS/s pipelined ADC using open-loop residue amplification. *IEEE Journal of Solid-State Circuits*, **38** (2003), 2040–2050.

[23] X. Wang, P. J. Hurst and S. H. Lewis. A 12-bit 20-Msample/s pipelined analog-to-digital converter with nested digital background calibration. *IEEE Journal of Solid-State Circuits*, **39** (2004), 1799–1808.

[24] Christof Koch. *Biophysics of Computation: Information Processing in Single Neurons*, (New York: Oxford University Press, 1999).

Section III

Low-power RF and energy-harvesting circuits for biomedical systems

16 Wireless inductive power links for medical implants

I do not think there is any thrill that can go through the human heart like that felt by the inventor as he sees some creation of the brain unfolding to success ... Such emotions make a man forget food, sleep, friends, love, everything.

Nikola Tesla

Implanted medical devices are rapidly becoming ubiquitous. They are used in a wide variety of medical conditions such as pacemakers for cardiac arrhythmia, cochlear implants for deafness, deep-brain stimulators for Parkinson's disease, spinal-cord stimulators for the control of pain, and preliminary retinal implants for blindness. They are being actively researched in brain-machine interfaces for paralysis, epilepsy, stroke, and blindness. In the future, there will undoubtedly be electronically controlled drug-releasing implants for a wide variety of hormonal, autoimmune, and carcinogenic disorders. All such devices need to be small and operate with low power to make chronic and portable medical implants possible. They are most often powered by inductive radio-frequency (RF) links to avoid the need for implanted batteries, which can potentially lose all their charge or necessitate re-surgery if they need to be replaced. Even when such devices have implanted batteries or ultra-capacitors, an increasing trend in upcoming fully implanted systems, wireless recharging of the battery or ultra-capacitor through RF links is periodically necessary.

Figure 16.1 shows the basic structure of an inductive power link system for an example implant. An RF power amplifier drives a primary RF coil which sends power inductively across the skin of the patient to a secondary RF coil. The RF signal on the secondary coil is rectified and used to create a power supply that powers internal signal-processing circuits, electrodes and electrode-control circuits, signal-sensing circuits, or telemetry circuits depending on the application. The power consumption of the implanted circuitry is eventually borne by external batteries that power the primary RF coil; if an RF link is energy efficient, most of the energy in the primary RF coil will be transported across the skin and dissipated in circuits in the secondary. It is also important for an RF link to be designed such that the power-supply voltage created in the secondary is relatively invariant to varying link distances between the primary and secondary, due to patient skin-flap-thickness variability, device placement, and device variability.

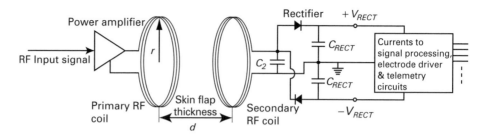

Figure 16.1. An example of a low-power bionic implant system. Reproduced with kind permission from [10] (©2007 IEEE).

RF power links for biomedical systems need to achieve good energy efficiency such that needless amounts of external power are not used to power an internal system. Small losses that are important in low-power systems may be insignificant in higher-power systems. For example, in milliwatt-level implanted systems such as in pacemakers or energy-efficient cochlear-implant processors [1], switching losses in the power amplifier and rectifier can be a significant portion of the overall power and hurt efficiency. In this chapter, we discuss how to design RF power links for maximum link energy efficiency and also to obtain acceptable robustness to inter-coil separation.

It is generally advantageous to separate the power and data transfer functions of a wireless link as some authors have done in the past [2], [3], [4]. Power signals carry no information, and power transfer efficiency is maximized for narrowband (high-Q) links that operate at low frequencies to minimize losses in body tissue. On the other hand, data signals carry information and therefore require larger link bandwidths, which are more easily obtained at higher operating frequencies. Separating the two functions therefore allows them to be independently optimized, improving overall performance.

In this chapter, we shall focus only on inductive power links and describe the design of inductive data links in a separate chapter, Chapter 18. We first discuss the theory of linear coupled resonators, a feedback analysis for understanding the power link, and derive expressions for its efficiency. Then, we discuss the design of a bionic implant power system, with attention to efficiency at low power levels. Finally, we present experimental results from a working system and discuss higher-order effects.

16.1   Theory of linear inductive links

A pair of magnetically coupled resonators is shown in Figure 16.2 and represents a model of our RF link with the primary external resonator on the left and the secondary implanted resonator on the right. The mutual inductance between the primary and secondary is represented by M. The resistances R_1 and R_2 are implicit resistances due to coil losses in the inductances L_1 and L_2 while C_1 and C_2 are

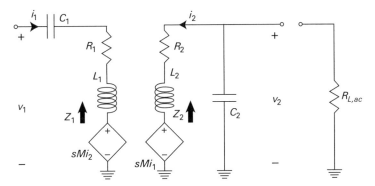

Figure 16.2. Inductively coupled series-parallel resonators. Coupling between the inductors is modeled with dependent voltage sources sMi_1 and sMi_2. Reproduced with kind permission from [10] (©2007 IEEE).

explicit capacitances used to create a resonance in the primary and secondary, respectively. Using a resonant secondary circuit amplifies the induced voltage and is helpful in overcoming the turn-on voltage of rectifier diodes. A series-resonant primary network requires lower voltage swings at its power-amplifier input because the phase of the inductor and capacitor voltage cancel at resonance.

The rectifier circuit that is in parallel with C_2 has been replaced with an equivalent linear resistance $R_{L,ac}$ that represents its effect at the RF frequency. If the ripple on the output of the rectifier circuit is small, achieved due to a large load capacitance C_{RECT} at its output as shown in Figure 16.1, then, since the rectifier output dc voltage is approximately the peak ac RF voltage at C_2, the ac rms energy at RF must be equal to the dc energy dissipated at the resistor by energy conservation. Thus,

$$R_{L,ac} = \frac{R_L}{2} \tag{16.1}$$

where R_L is the effective load of all the internal implanted circuits powered by the rectifier supply created at C_{RECT}. In Figure 16.2, $R_{L,ac}$ is shown as not being connected to C_2 because we shall first discuss the coupled resonator without its loading effect on C_2 and then connect it to C_2 to analyze the complete circuit.

The geometric coupling factor k between two coils of wire is a measure of the common flux linkage between the coils. If two coils of wire are placed near each other, the common flux between the two coils cannot exceed the total flux produced by either of the coils. Therefore, for a uniform dielectric environment the coupling factor $|k| < 1$. The mutual inductance M relates the change in current in one coil to the induced voltage in the other coil and is related to the coupling factor k and self inductances L_1 and L_2 by

$$M = k\sqrt{L_1 L_2} \tag{16.2}$$

We model the mutual inductance with the controlled sources Mi_1 and Mi_2 in Figure 16.2.

16.1.1   Dependence of *k* on geometric parameters

Suppose we have two on-axis circular coils of radius a_1 and a_2 where a_1, the radius of the primary, is typically much larger than a_2, the radius of the secondary. Such asymmetry provides some robustness to coil misalignment, and there is typically more room on the external primary side as well. Suppose the two coils have N_1 and N_2 turns respectively and are separated by an on-axis distance of r, which is typically the thickness of the skin. Then, the magnetic field $B(r)$ at the center of the secondary coil due to a current I in the primary coil is given by

$$B(r) = \mu N_1 I \frac{a_1}{2(r^2 + a_1^2)} \left(\frac{a_1}{\sqrt{(r^2 + a_1^2)}} \right) \tag{16.3}$$

$$= \mu N_1 I \frac{a_1^2}{2(r^2 + a_1^2)^{3/2}}$$

where the field is computed as the product of an inverse-square effect and a cosine on-axis projection factor (the bracketed term in Equation (16.3)). The flux $\Phi(r)$ in the secondary coil may then be approximated as

$$\Phi(r) = B(r) \times \pi a_2^2 = \mu N_1 I \left(\frac{\pi (a_1 a_2)^2}{2(r^2 + a_1^2)^{3/2}} \right) \tag{16.4}$$

such that the mutual inductance is given by

$$M = N_2 \frac{\Phi(r)}{I} = \mu N_1 N_2 \left(\frac{\pi (a_1 a_2)^2}{2(r^2 + a_1^2)^{3/2}} \right) \tag{16.5}$$

Since the self inductances L_1 and L_2 of the two coils may be shown to be approximately given by

$$L_1 = \mu a_1 N_1^2 \ln \left(\frac{a_1}{d_1} \right)$$

$$L_2 = \mu a_2 N_2^2 \ln \left(\frac{a_2}{d_1} \right) \tag{16.6}$$

where d_1 and d_2 are the coil wire diameters, the coupling factor k is then given by

$$k = \frac{M}{\sqrt{L_1 L_2}} = \frac{\pi}{2\sqrt{\ln \left(\frac{a_1}{d_1} \right) \ln \left(\frac{a_2}{d_2} \right)}} \left(\frac{a_1 a_2}{r^2 + a_1^2} \right)^{\frac{3}{2}} \tag{16.7}$$

Note that k is independent of N_1 and N_2 and only dependent on coil geometry and separation parameters. More accurate calculators for self inductance and coil self-resonance frequency can be found at [5] and [6], respectively.

(a)

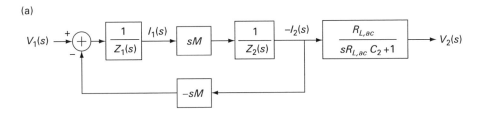

(b)

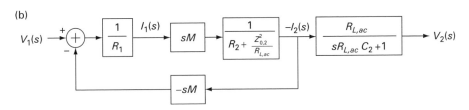

Figure 16.3a, b. Feedback diagrams for coupled resonators in Figure 16.2. (a) Shows the loop diagram under all frequencies; (b) shows an approximate loop diagram under resonant conditions $\omega_n = \frac{1}{\sqrt{L_1 C_1}} = \frac{1}{\sqrt{L_2 C_2}}$. Reproduced with kind permission from [10] (©2007 IEEE).

16.1.2   Feedback block diagram analysis

The block diagram in Figure 16.3 (a) models the circuit of Figure 16.2. The loop transmission that is given by

$$L(s) = \frac{M^2 s^2}{Z_1(s)Z_2(s)}$$

$$L(j\omega) = \frac{-\omega^2 k^2 L_1 L_2}{Z_1(j\omega)Z_2(j\omega)}$$

(16.8)

is of a form that appears to indicate positive feedback at dc: the loop transmission $L(j\omega)$ near $\omega = 0$ is $k^2 \omega^4 L_1 C_1 L_2 C_2$ implying a positive loop transmission at and near dc. While this fact may seem puzzling at first, the loop transmission is easily interpreted as the product of two Lenz's law expressions, each of which implements negative feedback from one coil to another, but whose product is positive. The magnitude of the loop transmission at $\omega = 0$ is 0, and at $\omega = \infty$ is k^2, which is less than 1. The phase and magnitude of the feedback changes with frequency in between these limits according to Equation (16.8).

The feedback effects can be viewed as creating an effective impedance in the primary circuit due to a reflection from the secondary if we evaluate the reciprocal of $(I_1(s)/V_1(s))$ from Figure 16.3 (a). Using the feedback diagram, we find that

$$Z_{in}(s) = 1 / \left(\frac{I_1(s)}{V_1(s)} \right)$$

$$= Z_1(s)(1 - L(s))$$

$$= Z_1(s) - L(s)Z_1(s)$$

$$= Z_1(s) - \frac{M^2 s^2}{Z_2(s)} \tag{16.9}$$

Thus, we may view the secondary impedance $Z_2(s)$ as being reflected into an impedance $Z_{refl}(s)$ in the primary that is in series with $Z_1(s)$ and given by

$$\boxed{Z_{refl}(s) = -\frac{M^2 s^2}{Z_2(s)}} \tag{16.10}$$

The *reflected impedance* is seen to be reciprocated or gyrated by the mutual coupling from the secondary into the primary. Note that the impedance transformations described by Equations (16.9) and (16.10) are merely a manifestation of a series-series feedback topology with current sensing at the output and voltage feedback to the input. That is,

$$Z_{in}(s) = Z_1(s)[1 - L(s)]$$

$$= Z_1(s) - Z_1(s)L(s) \tag{16.11}$$

$$= Z_1(s) + Z_{refl}(s)$$

As the coupling increases the loop transmission, the input impedance is dominated, first by the primary impedance $Z_1(s)$ and then by the impedance seen in the primary due to reflected secondary loading $Z_{refl}(s)$.

Figure 16.3 (b) shows a simplification of the block diagram of Figure 16.3 (a) under resonant conditions where $\omega_n \approx 1/\sqrt{(L_1 C_1)} = 1/\sqrt{(L_2 C_2)}$ and where $Z_{0,2}^2/R_{L,ac} = (L_2/C_2)/R_{L,ac}$ is the parallel-to-series impedance transformation of $R_{L,ac}$ due to the resonator formed by L_2 and C_2: we have simply computed the real part of $(1/(C_2 s))\|R_{L,ac}$ and evaluated it at the resonance frequency of the secondary $s_n = j\omega_n$. In RF circuits, it is frequently useful to convert a series combination of a resistance and reactance to a parallel combination of a resistance and reactance with the same impedance and vice versa as in Figure 16.3 (b). Therefore, we shall digress briefly to present the general formulas for such transformations, which will be of frequent use to us.

16.1.3 Series-to-parallel and parallel-to-series impedance transformations

The derivation below shows how series-to-parallel transformations or vice versa can be accomplished. The subscript S corresponds to series while the subscript P corresponds to parallel.

$$R_S + jX_S = R_P \| jX_P$$

$$= R_P \| \frac{jR_P}{Q}$$

$$= R_P \frac{1\dfrac{j}{Q}}{1+\dfrac{j}{Q}}$$

$$R_S + jX_S = R_P \frac{1}{Q^2+1} + jX_P \frac{Q^2}{Q^2+1}$$

$$\boxed{R_S = \frac{R_P}{Q^2+1}} \tag{16.12}$$

$$\boxed{R_P = R_S\left(Q^2+1\right)}$$

$$\boxed{X_S = X_P \frac{Q^2}{Q^2+1}}$$

$$\boxed{X_P = X_S \frac{Q^2+1}{Q^2}}$$

Thus, from the boxed terms in Equation (16.12), small series resistances transform to large parallel resistances in high-Q circuits and vice versa. The reactance is nearly unchanged in a series-to-parallel transformation in high-Q circuits. In the derivation of Equation (16.12), Q is merely a parameter that represents the ratio of $R_P/|X_P| = |X_S|/R_S$ at some particular frequency. Thus, Equation (16.12) is generally valid at all frequencies if a frequency-dependent $Q(\omega)$ that represents this ratio is used.

Often in resonant systems, one approximates $Q(\omega) \approx Q(\omega_{res})$ for all ω near ω_{res} with $Q(\omega_{res})$ being what is normally called the quality factor of the resonator. Then, the series-to-parallel transformation yields useful intuition on the behavior of the impedance at not just the resonance frequency where the transformation is exact but at frequencies near ω_{res} as well, i.e., it's a good approximation within the 3 dB bandwidth of the resonator.

16.1.4 Feedback root-locus analysis

A helpful simplification to the loop transmission in the feedback loop in Figure 16.3 (a) is to write the resonators as second-order systems with quality factor Q and natural frequency ω_n

$$Z = \frac{s^2 + \dfrac{sR'}{L} + \dfrac{1}{LC}}{\dfrac{s}{L}} = \frac{s^2 + s\dfrac{\omega_n}{Q} + \omega_n^2}{\dfrac{s}{L}} \tag{16.13}$$

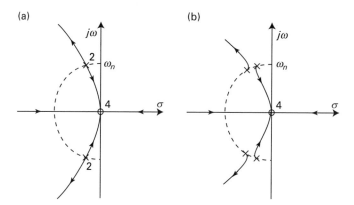

Figure 16.4a, b. Root locus diagrams are shown for two resonator cases. In (a), the resonators are identical. A loaded resonator case, (b), will have a higher damping factor due to energy lost to the rectifier. In this case, the poles move together before splitting from the constant ω_n circle. Reproduced with kind permission from [10] (©2007 IEEE).

where R' is the effective resistance due to all losses in the primary or secondary. Finding the simplified loop transmission using the bandpass definitions in Equation (16.13), we get

$$
L(s) = \frac{M^2 s^2}{Z_1(s)Z_2(s)} = \frac{s^2 k^2 L_1 L_2}{Z_1(s)Z_2(s)}
$$

$$
= k^2 \left[\frac{\dfrac{s^2}{\omega_{n,1}^2}}{\left(\dfrac{s^2}{\omega_{n,1}^2} + \dfrac{s}{\omega_{n,1}Q_1} + 1 \right)} \right] \left[\frac{\dfrac{s^2}{\omega_{n,2}^2}}{\left(\dfrac{s^2}{\omega_{n,2}^2} + \dfrac{s}{\omega_{n,2}Q_2} + 1 \right)} \right] \qquad (16.14)
$$

Here, the natural frequencies of the primary and secondary resonators are $\omega_{n,1}$ and $\omega_{n,2}$ respectively and the quality factors Q_1 and Q_2 are the quality factors of the primary and secondary resonant circuits.

Under feedback, the open-loop poles associated with each of the resonators move on the complex plane as the geometric coupling factor varies with inter-coil separation or misalignment and changes the root-locus gain. In a retinal implant, the eye moves during normal patient activity, changing k. Each of the resonators in Figure 16.2 is characterized by a natural frequency, ω_n, and damping factor, $\zeta = 1/(2Q) = R/(2\omega L)$. Ideally, the quality factor of each resonator is set by only the effective coil resistances, R_1 and R_2 and $R_{L,ac}$; however, in a real system, the resistance of the power amplifier, resistance in the capacitors and objects in the environment can all contribute to reducing the effective quality factor of the coils. Objects in the environment can unwantedly couple via induction to the coils such that their reflected impedance can affect the overall impedance of the system.

Figure 16.4 shows how the four open-loop complex poles move under positive feedback as the root-locus gain parameter k^2 is changed. The techniques used to create this plot are the standard techniques of root locus from feedback theory from Chapter 2. The open-loop poles in Figure 16.4 (a) are identical and model the case where no loading is applied to the secondary circuit. Now, if we add the resistor to model rectifier loading, the damping of the secondary resonators will be different from that of the primary. The open-loop secondary poles appear more damped on their common circle on the s-plane, as shown in Figure 16.4 (b). In the loaded case of Figure 16.4 (b), as k is increased, the poles move along the circle towards one another and then split along the trajectory loci as in the unloaded case of Figure 16.4 (a). The four asymptotes of the root-locus trajectories enter the origin along the positive and negative x-axis and positive and negative y-axis asymptotes at infinite root-locus gain. It is interesting to note that the root-locus trajectories predict that the lower frequency closed-loop pole pair that is radially closer to the origin and moving towards it will have a higher quality factor due to their extreme angle of entry, while the closed-loop pole pair that is radially further from the origin and moving away from it will have a lower quality factor due to their less extreme angle of entry. Our experimental measurements, presented in Figure 16.7 (a) in a later portion of this chapter, confirm this theoretical prediction.

16.1.5 Coupled-resonator voltage-transfer function

From the block diagram in Figure 16.3, we can write the transfer function between the primary input voltage $V_1(s)$ and the primary output voltage $V_2(s)$ as

$$\frac{V_2(s)}{V_1(s)} = \left(\frac{L(s)}{1-L(s)}\right)\left(\frac{1}{sM}\right)\left(\frac{R_{L,ac}}{sC_2R_L+1}\right)$$

$$= \left(\frac{k^2L'(s)}{1-k^2L'(s)}\right)\left(\frac{1}{sk\sqrt{L_1L_2}}\right)\left(\frac{R_{L,ac}}{sC_2R_{L,ac}+1}\right) \qquad (16.15)$$

$$= \left(\frac{kL'(s)}{1-k^2L'(s)}\right)\left(\frac{1}{s\sqrt{L_1L_2}}\right)\left(\frac{R_{L,ac}}{sC_2R_{L,ac}+1}\right)$$

where we have defined $L(s)=k^2L'(s)$ to explicitly indicate the dependence of the loop transmission on the parameter k. By differentiating the latter equation with regard to k, and equating it to zero, we can show that the voltage transfer function has an extremum when $k^2L'(s)=-1$, i.e., when the loop transmission $L(s)$ is -1. Physically, this extremum can be shown to be a maximum. When k is small, the voltage transfer function is small because there is little coupling of the current in the primary to the dependent voltage source in the secondary; when k is large, the input impedance seen in the primary $Z_{in}(s)=Z_1(s)(1-L(s))$ increases as $L(s)=k^2L'(s)$ increases with k such that the current in the primary reduces, decreasing the voltage transfer function. The optimal or critical coupling occurs when $k^2L'(s)=L(s)=-1$, at which point the voltage transfer function is maximized. At this value of $k=k_c$, the link is said to be critically coupled. Note that at this value $Z_{in}(s)=2Z_1(s)$ or equivalently, the reflected impedance of the secondary in the primary, $-Z_1 L(s)$, is equal to $Z_1(s)$. For $k > k_c$, the link is referred to as being overcoupled, and for $k < k_c$, the link is referred to as being undercoupled.

At resonance, the loop transmission can be further simplified by including the effect of the load resistance, $R_{L,ac}$, transformed by the secondary circuit

$$L(j\omega_n) = \frac{M^2(j\omega_n)^2}{Z_1(j\omega_n)(Z_2(j\omega_n)}$$

$$= -k^2 \left(\frac{\omega_n L_1}{R_1}\right)\left(\frac{\omega_n L_2}{R_2 + \dfrac{Z_{0,2}^2}{R_{L,ac}}}\right)$$

$$\tag{16.16}$$

$$R_2' = R_2 + \frac{Z_{0,2}^2}{R_{L,ac}}$$

$$Q_2' = \left(\frac{\omega_n L_2}{R_2'}\right)$$

$$L(j\omega_n) = -k^2 Q_1 Q_2'$$

where Q_2' is the loaded Q of the secondary and defined as shown. Thus, at critical coupling

$$k = k_c = \frac{1}{\sqrt{Q_1 Q_2'}} \tag{16.17}$$

The Nyquist plot of $L'(s) = L(s)/k^2$ is shown in Figure 16.5 such that $1/k^2$ may be viewed as a gain parameter of the plot. Since k^2 is always less than 1, the $1/k^2$ point always lies outside the contour that intersects the 1 point; therefore, in this positive-feedback version of the classic Nyquist plot, stability is guaranteed. Note that at the 180° phase point, each resonator contributes approximately 90° of phase. There is only one frequency at which the sum of the 90° phases from each resonator transfer function in Equation (16.14) is 180°, typically near $(\omega_{n,1} + \omega_{n,2})/2 \approx \omega_n$ if the resonators are fairly well matched in natural frequency. At this frequency, the voltage transfer function is non-monotonic in k with a peak magnitude at k_c as described by Equation (16.17). At other frequencies, there is non-monotonicity in the magnitude transfer function with k as well, but a closed-form expression for k_c as in Equation (16.17) is not as easy to compute.

Using values of experimental components listed in Table 16.1, and the experimental setup shown in Figure 16.6 we can obtain measurements from an RF link that illustrate the theoretical effects that we have been discussing. We shall describe the circuitry needed to drive the coils and obtain these measurements later.

Figure 16.7 (a) reveals the frequency transfer functions of coupled resonators as the distance r between the primary and secondary is varied, or equivalently k in Equation (16.7) is varied. We note that after the critical coupling distance at which the voltage transfer function is maximized, two frequency peaks are exhibited

Table 16.1 Efficiency test setup parameters

Parameter	Notes
Output power level $P_{R_{L,ac}}$	4 mW
Operating frequency f	$\simeq 4.5$ MHz
Coil separation distance d	1 mm – 30 mm
Class-E power nFET	Fairchild NDS351
	$R_{DS,on} \simeq 0.6\,\Omega$ for $V_{GS,on} = 3$ V
L_1	10 turns of 32 gauge $r = 15$ mm
	3.2 μH with $Q_1 \simeq 70$
C_1	150 pF
C_S	60 pF
L_2	15 turns of 32 gauge $r = 15$ mm
	4 μH with $Q_2 \simeq 70$
C_2	220 pF
L_{RFC}	Coilcraft 1812PS-223KL
	$L_{RFC} = 22\,\mu$H

(a)   (b)

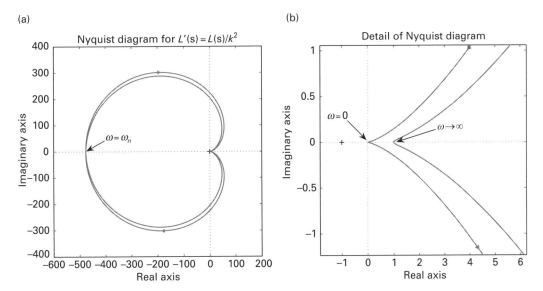

Figure 16.5a, b. An overall Nyquist diagram of $L(s)/k^2$ is shown in (a) and detail near the origin is shown in (b). As the frequency approaches ω_n the Nyquist plot approaches the 180° point with magnitude Q_1Q_2'. For $L(s) = -1$, and this phase condition we must have $k = k_c$. Reproduced with kind permission from [10] (©2007 IEEE).

with the lower frequency peak having higher Q and the higher frequency peak having lower Q in accord with the root-locus predictions of Figure 16.4. At ω_n, marked by the peak-frequency location of the bold curve in Figure 16.7 (a), we see that the amplitude of the voltage transfer function initially rises with k and then falls after the peak frequency splits. The nature of the frequency response near ω_n

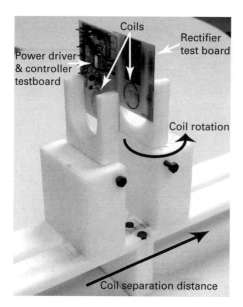

Figure 16.6. The test setup used for inductive power transfer system tests is shown. The coil separation can be adjusted and measured with millimeter accuracy. Coil rotation can also be explored for both the primary and secondary coils. The electronic boards are glued to Delrin scaffolding with Teflon screws used to bind the various portions of the scaffold. Reproduced with kind permission from [10] (©2007 IEEE).

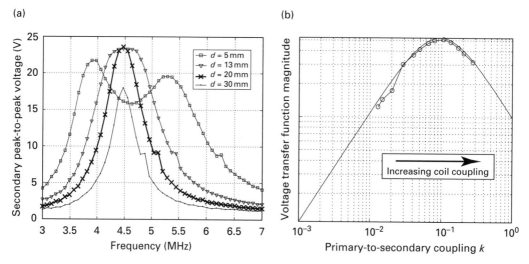

Figure 16.7a, b. The (a) plot shows how the closed-loop poles move as the coil distance is changed. At some coil separation, the poles begin to move apart and a low-frequency high-Q and high-frequency low-Q peak are observed. The (b) plot shows the voltage peak at resonance as we change the coil coupling factor. The peak occurs at $k_c \approx 0.1$. Reproduced with kind permission from [10] (©2007 IEEE).

then changes from a maximum to a minimum. Figure 16.7 (b) reveals the dependence of the voltage transfer function versus k at ω_n. It shows that the voltage peak at resonance is non-monotonic and peaks at the critical coupling $k_c \approx 0.1$. The critical coupling point, being a maximum, yields robustness in the voltage gain to variations in inter-coil distance or misalignment, which affect k.

16.1.6 The use of reflected impedance to analyze coupled resonators

The concept of reflected impedance as embodied by Equation (16.10) is often handy in understanding coupled resonators. For example, as we show below, it predicts why weak coupling broadens the peak at ω_n in Figure 16.7 (a). In the strongly coupled regime, it predicts the locations of the split peaks in Figure 16.7 (a) and explains why the quality factor of the low-frequency split peak is higher than that of the high-frequency split peak.

From Equations (16.10) and (16.11), the input impedance of the primary is given by

$$Z_{in}(s) = Z_1(s) - \frac{M^2 s^2}{Z_2(s)}$$

$$= Z_1(s) - \frac{M^2 s^2}{L_2 s + \frac{1}{C_2 s} + R_2'}$$

$$= Z_1(s) - \frac{M^2 C_2 s^3}{L_2 C_2 s^2 + R_2' C_2 s + 1} \tag{16.18}$$

$$= Z_1(s) - k^2 L_1 s \left(\frac{s^2/\omega_{n,2}^2}{s^2/\omega_{n,2}^2 + s/(Q_2'\omega_{n,2}) + 1} \right)$$

$$= Z_1(s) - k^2 (L_1 s) H_{hghps}(s)$$

$$\boxed{Z_{in}(s) = L_1 s (1 - k^2 H_{hghps}(s)) + \frac{1}{C_1 s} + R_1}$$

Thus, the net effect of the coupling is to scale the inductance in the primary via the frequency-dependent $\left(1 - k^2 H_{hghps}(s)\right)$ factor defined in Equation (16.18).

When the coupling is weak, there is little shift in the peak ω_{pk} such that $\omega_{pk} = \omega_n$ for the coupled resonator and $H_{hghps}(j\omega_n) = jQ_2'$. Substituting this value into Equation (16.18) and noting that $\omega_n L_1 = Q_1 R_1$, $s = j\omega_n$, we find that

$$Z_{in}^{weak}(s) = j\omega_n L_1 + R_1(1 + k^2 Q_1 Q_2') + \frac{1}{C_1 s} \tag{16.19}$$

Since the damping of the primary is affected in Equation (16.19) but neither the inductance nor capacitance are changed, the primary effect of the coupling is to broaden the peak or decrease its quality factor by $1 + k^2 Q_1 Q_2$ compared with the uncoupled case where $k = 0$. The peak location at ω_n is relatively unchanged. Indeed, we see in Figure 16.7 (a) that for weak coupling the primary effect of the coupling is to broaden the resonant peak but leave its location unchanged. The voltage gain in the secondary also increases in Figure 16.7 (a) because, for a given primary current, increasing k increases the current in the secondary and causes more voltage drop across C_2.

When the coupling is strong, there are significant peak shifts such that we may assume that $H_{hghps}(s)$ is well to the right (in the high-frequency peak-shift case) or well to the left (in the low-frequency peak-shift case) of the uncoupled peak location ω_n. Equivalently, we are gyrating the dominant inductive impedance (high-frequency shift case) or the dominant capacitive impedance in the secondary into a reflected impedance in the primary. We shall analyze the high-frequency peak shift first since it is the easier of the two. At frequencies significantly beyond ω_n, $H_{hghpss}(s) \approx 1$ such that $Z_{in}(s)$ is given from Equation (16.18) to be

$$Z_{in}(s) = L_1 s(1 - k^2) + \frac{1}{C_1 s} + R_1 \tag{16.20}$$

Since the net inductance is reduced by a factor of $1 - k^2$ in Equation (16.20), the resonance of the coupled resonator shifts to a higher frequency ω_{rght} and the quality factor is reduced to a value Q_{rght} such that

$$\omega_{rght} = \frac{1}{\sqrt{L_1(1 - k^2)C_1}} = \frac{\omega_n}{\sqrt{1 - k^2}}$$

$$Q_{rght} = \frac{\sqrt{L_1(1 - k^2)/C_1}}{R_1} = Q_1\sqrt{1 - k^2} \tag{16.21}$$

From Equation (16.18), at frequencies significantly below ω_n, $H_{hghpss}(j\omega) \approx (j\omega)^2/\omega_n^2$, such that, at these frequencies, from Equation (16.18), we find

$$Z_{in}(j\omega) = j\omega L_1\left(1 + k^2\frac{\omega^2}{\omega_n^2}\right) + \frac{1}{j\omega C_1} + R_1 \tag{16.22}$$

Equation (16.22) reveals that, for frequencies significantly below ω_n, the secondary impedance is gyrated into a reflected inductance in the primary whose inductance value increases quadratically with frequency. This inductance resonates with C_1 at the resonant peak ω_{lft} when

$$L_1\left(1 + k^2\left(\frac{\omega_{lft}}{\omega_n}\right)^2\right)C_1 = \frac{1}{\omega_{lft}^2}$$

$$\left(1 + k^2\left(\frac{\omega_{lft}}{\omega_n}\right)^2\right) = \frac{\omega_n^2}{\omega_{lft}^2} \tag{16.23}$$

The solution to Equation (16.23) is obtained via simple quadratic-equation methods to be

$$\omega_{lft} = \omega_n \left[\sqrt{\frac{\sqrt{1+4k^2}-1}{2k^2}} \right] \tag{16.24}$$

At $k = 0$, from a simple Taylor-series expansion, Equation (16.24) predicts that $\omega_{lft} = \omega_n$ as expected. The Q_{lft} corresponding to this peak is then given by

$$Q_{lft} = \frac{\sqrt{\dfrac{L_1 \left(1+k^2 \dfrac{\omega_{lft}^2}{\omega_n^2}\right)}{C_1}}}{R_1} \tag{16.25}$$

$$= Q_1 \left[\sqrt{\frac{\sqrt{1+4k^2}+1}{2}} \right]$$

by substituting from Equation (16.24). At $k = 0$, Equation (16.25) predicts that $Q_{lft} = Q_1$ as expected. At $k = 1$, Equations (16.24) and (16.25) predict that the minimum value of ω_{lft} and the maximum value of Q_{lft} are given by

$$\omega_{lft}^{min} = \omega_n \left[\sqrt{\frac{\sqrt{5}-1}{2}} \right] = 0.79\omega_n$$

$$\tag{16.26}$$

$$Q_{lft}^{max} = Q_1 \left[\sqrt{\frac{\sqrt{5}+1}{2}} \right] = 1.27Q_1$$

From the measurements of Figure 16.7 (a), we can estimate that $\omega_n = 4.5\,\text{MHz}$ and $\omega_{rght} = 5.3\,\text{MHz}$. By substituting these values of ω_{rght} and ω_n in Equation (16.21), we find that $k = 0.53$. If we substitute this value of k in Equation (16.24), we find that $\omega_{lft} = 4.0\,\text{MHz}$, which is very near the actual measured value in Figure 16.7 (a). Thus, the theoretical predictions of the peak locations appear to be in good accord with experimental measurements.

At high coupling, the maximal resonant current in the primary at the low-frequency peak of Figure 16.7 (a) leads to a voltage in the secondary that is largely all dropped across the dominant capacitive impedance of the secondary, whose voltage forms the output voltage. Thus, the output voltage maximum is well approximated by the primary-current maximum predicted by Equation (16.24). On the other hand, at high coupling, the maximal resonant current in the primary at the high-frequency peak leads to a voltage in the secondary that is largely all dropped across the dominant inductive impedance of the secondary rather than across the output voltage of the capacitive impedance. Thus, the predictions regarding the high-frequency peak in Equation (16.21) represent a higher degree

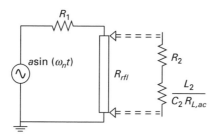

Figure 16.8. Equivalent coupled circuit at resonance for determining power-transfer efficiency using the concept of reflection impedance. The reflected impedance R_{rfl} is shown.

of approximation as they ignore frequency-dependent impedance-divider effects in the secondary. In fact, the high-frequency peak in the strongly coupled regime is seen to be more attenuated in Figure 16.7 (a). Hence the validity of Equation (16.21) for ks near 1.0 is not as accurate as that for smaller ks. Fortunately, most experimental situations do not involve such a high k. Even so, the data of Figure 16.7 (a) appear to be in accord with theory even for k as high as 0.53.

16.1.7   Limits to energy efficiency

At $\omega = \omega_n$, both the primary and secondary impedances are purely resistive due to the conjugate cancellation of the inductive and capacitive reactances at resonance, if we assume that $\omega_{n,1} \approx \omega_{n,2} \approx \omega_n$. Figure 16.8 shows how we compute the energy efficiency of the circuit at $\omega = \omega_n$, i.e., the fraction of the power or energy supplied by the source in the primary that is actually dissipated in the load. The reflected resistance and R_1 form a resistive divider in the primary such that only a fraction of the primary power is dissipated in the reflected resistance or equivalently dissipated in the secondary rather than in R_1; the secondary impedance R_2 and the impedance $Z_{0,2}^2/R_{L,ac} = L_2/(C_2 R_{L,ac})$ form a resistive divider in the secondary such that only a fraction of the power dissipated in the secondary is dissipated in the load rather than in R_2; the overall efficiency is then the product of the energy efficiency of the resistive-divider circuit in the primary and the energy efficiency of the resistive-divider circuit in the secondary.

The $Z_{0,2}^2/R_{L,ac} = L_2/(C_2 R_{L,ac})$ resistance in the secondary is obtained via a parallel-to-series transformation of the resistance $R_{L,ac}$ assuming that

$$Q_L = \omega_n R_{L,ac} C_2 \tag{16.27}$$

is the quality factor of the secondary capacitive load at resonance and that Q_L^2 is much greater than 1. The quality factor of a reactive element at a given frequency that has a resistive element in parallel with it is the ratio of its resistive to reactive impedance magnitude at that frequency. This transformation can be derived from Equation (16.12) to be given by

$$R_{srs} = \frac{R_{prll}}{Q_L^2 + 1} \tag{16.28}$$

In the secondary circuit of Figure 16.2, the parallel-to-series transformation of $R_{L,ac}$ simplifies analysis of the overall quality factor of the series resonant circuit.

At $\omega = \omega_{n,2}$, the power transfer efficiency of the primary circuit is maximized because the reflected resistive impedance in the primary is maximal at this frequency for two reasons: first, the reflected impedance from the secondary has a maximal magnitude at this frequency due to its minimal value in the secondary being transformed to a maximally gyrated magnitude in the primary as per Equation (16.10); second, the reflected impedance is solely resistive at this frequency, hence maximizing the resistive portion of the overall reflected impedance, and thus the power dissipated in the secondary.

Operating the primary circuit at the same series resonant frequency $\omega_{n,1}$ as the secondary resonant frequency $\omega_{n,2}$ is advantageous since it maximizes power transfer although it has no effect on the energy efficiency: the current in the primary circuit for a given R_1 and given amplitude of voltage input is maximized if the total impedance in the primary is minimized due to conjugate matching of its reactances such that there is maximal power transfer. The energy efficiency of the primary circuit, however, is determined by a resistive-divider ratio formed by R_1 and the reflected resistive impedance of the secondary and is independent of whether there is conjugate matching in the primary. Thus, the primary benefit of operating the primary circuit at the same resonant frequency as the secondary is that it reduces the input primary source voltage needed for a given desired level of power dissipation in the secondary. Equivalently, having the same resonant frequency in the primary and the secondary increases the voltage gain from the primary to the secondary and minimizes the input voltage at the primary needed to create a given secondary voltage.

The voltage-transfer functions between the primary and the secondary have maxima at the split peaks ω_{rght} and ω_{lft} and a minima at ω_n in the overcoupled case where $k > k_c$ as seen in Figure 16.7 (a). Nevertheless, the energy efficiency for power transfer at these frequencies is still inferior to that at ω_n. The ω_{lft} and ω_{rght} cases result in a lower reflected resistive impedance in the primary degrading efficiency and also result in the dominance of stored power rather than dissipated power in the secondary minimizing power transfer in the secondary. Hence, we shall only analyze the energy efficiency of our link at ω_n since operating the link at frequencies other than ω_n results in sub-optimal energy efficiency in the primary and minimizes power transfer in the secondary.

From Figure 16.8, we can compute the overall energy efficiency of the RF link at $\omega = \omega_n$ to be given by

$$\eta = \eta_{prm}\eta_{scnd}$$

$$= \left(\frac{R_{rfl}}{R_{rfl} + R_1}\right)\left(\frac{L_2/(R_{L,ac}C_2)}{L_2/(R_{L,ac}C_2) + R_2}\right) \tag{16.29}$$

where we have taken the product of the resistive-divider ratios of the primary circuit and the secondary circuit respectively. The reflected impedance in the primary is given by

$$R_{rfl} = \frac{\omega^2 M^2}{R_2 + (L_2/(R_{L,ac}C_2))}$$

$$= k^2(\omega L_1)\frac{(\omega L_2)}{R_2 + (L_2/(R_{L,ac}C_2))} \tag{16.30}$$

$$= k^2(Q_1 R_1)Q_2'$$

where Q_2' is the loaded quality factor of the secondary with the damping effects of the load and R_2. Thus, the efficiency of the primary circuit of Figure 16.8 is given by

$$\eta_{prm} = \frac{k^2 Q_1 Q_2'}{1 + k^2 Q_1 Q_2'} \tag{16.31}$$

The efficiency of the secondary circuit of Figure 16.8 is given by

$$\eta_{scnd} = \frac{(L_2/(R_{L,ac}C_2))}{L_2/(R_{L,ac}C_2) + R_2}$$

$$\eta_{scnd} = \frac{L_2/R_2}{L_2/R_2 + R_{L,ac}C_2}$$

$$\eta_{scnd} = \frac{(\omega_n L_2)/R_2}{(\omega_n L_2)/R_2 + \omega_n R_{L,ac}C_2} \tag{16.32}$$

$$\eta_{scnd} = \frac{Q_2}{Q_2 + Q_L}$$

Note that the loaded quality factor of the secondary is given by

$$Q_2' = \frac{Q_2 Q_L}{Q_2 + Q_L} \tag{16.33}$$

From Equations (16.31) and (16.33), we see that the efficiency of the primary circuit increases as Q_L increases. Physically, a large $R_{L,ac}$ leads to a large Q_L, which leads to a lower overall total impedance in the secondary, which is reflected into a larger value of R_{refl} in the primary, thus improving the efficiency of the primary resistive divider. From Equation (16.32), we see that the efficiency of the secondary circuit decreases as Q_L increases. Physically, a large $R_{L,ac}$ leads to a large Q_L, which leads to a smaller value of the transformed parallel-to-series resistance $L_2/(C_2 R_{L,ac})$, thus lowering the efficiency of the secondary resistive divider. Hence, there is a value of Q_L or equivalently of $R_{L,ac}$ that optimizes the product of the primary and secondary efficiency and thus the overall efficiency.

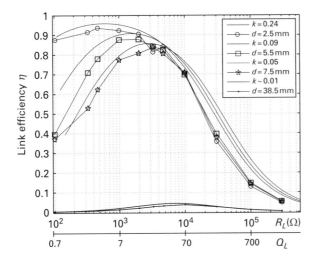

Figure 16.9. Efficiency of power transfer between coupled resonators for varying loads. This particular data were produced by ignoring switching losses in a discrete Class-E amplifier implementation. Reproduced with kind permission from [10] (©2007 IEEE).

The optimal Q_L that optimizes the overall efficiency

$$
\eta = \eta_{prm}\eta_{scnd}
$$
$$
= \left(\frac{k^2 Q_1 \dfrac{Q_2 Q_L}{Q_2 + Q_L}}{1 + k^2 Q_1 \dfrac{Q_2 Q_L}{Q_2 + Q_L}} \right) \left(\frac{Q_2}{Q_2 + Q_L} \right) \tag{16.34}
$$

is found by differentiating Equation (16.34) with respect to Q_L. We find that the optimum

$$
Q_{L,opt} = \frac{1}{k} \sqrt{\frac{Q_2}{Q_1}} \tag{16.35}
$$

and that the maximally achievable efficiency η_{MAX} attainable at this optimum is given by substituting $Q_{L,opt}$ for Q_L in Equation (16.34). We find that

$$
\eta_{MAX} = \frac{k^2 Q_1 Q_2}{(kQ_1 + 1)(kQ_2 + 1)} \tag{16.36}
$$

Figure 16.9 shows experimental results of the overall power transfer efficiency. These results were obtained from a coupled-resonator system with no power amplifier or rectifiers in Figure 16.1 such that we could focus on only the essential resonator circuit of Figure 16.2 for now. The data are shown for four separation distances of the RF link. We see that the peak efficiency shifts to lower load resistance as the coils are moved closer together in accord with the theoretical predictions of Equation (16.35). Intuitively, as the coils are moved closer together, k increases, increasing the reflected impedance for a given total impedance in the secondary; thus, we can decrease $R_{L,ac}$ to improve the secondary efficiency; the loss in primary efficiency created by the increase in total impedance in the secondary is compensated for by the higher coupling that maintains a significant reflected impedance and thus good efficiency in the primary. The maximal achievable efficiency also improves as k, Q_1, and Q_2 increase since the primary efficiency improves as $k^2 Q_1 Q_2'$ increases and the secondary efficiency improves as Q_2 becomes significantly greater than the optimal Q_L.

Even though the voltage transfer function degrades as $k > k_c$ as seen in Figure 16.7 (a), Figure 16.9 reveals that the energy efficiency continues to improve with k such that it is advantageous to exceed critical coupling and have k as near 1.0 as is allowable by geometry. At critical coupling, the energy efficiency of the primary η_{prm} is at 50% since the reflected impedance equals R_1 at this point. The overall energy efficiency at critical coupling can be significantly lower than 50% if $L_2/(C_2 R_{L,ac})$ does not significantly exceed R_2 or equivalently if Q_2 does not significantly exceed Q_L degrading η_{scnd}.

At $k = k_c$, the voltage-transfer function is at a maximum such that there is robustness in the secondary voltage to variations in inter-coil coupling due to the flatness exhibited near a maximum. Nevertheless, as we demonstrate later, even when $k > k_c$, if the link has been designed for optimal energy efficiency at a mean expected inter-coil distance, the robustness of the rectified secondary voltage to variations in k about this value can be quite good, thus ensuring a design that is both robust and efficient. Hence, it is advantageous to operate with a link that is overcoupled rather than critically coupled if coil Qs and geometric parameters allow one to do so. If there is room available in the external primary, a_1 should be larger or comparable to r such that the variation in k with r is reduced in Equation (16.7) to improve robustness.

16.1.8   Voltage gain at ω_n

At ω_n, the power dissipated in the reflected resistance of the primary must be identical to the power dissipated in the effective resistance of the secondary. Since the output voltage across the capacitor C_2 is Q_2' times the output voltage across the effective resistance in the secondary, we must then have that

$$\frac{v_{prmry}^2 (k^2 Q_1 Q_2') R_1}{R_1^2 (1 + k^2 Q_1 Q_2')^2} = \frac{(v_{out}/Q_2')^2}{(\omega_n L_2/Q_2')}$$

$$R_1 = \frac{\omega_n L_1}{Q_1} \tag{16.37}$$

$$\boxed{v_{out} = v_{prmry} \left(\sqrt{\frac{L_2}{L_1}} \right) \left(\frac{k Q_1 Q_2'}{1 + k^2 Q_1 Q_2'} \right)}$$

The boxed expression has a peak gain when $k = k_c$ as in (16.17) and this peak gain is given by

$$\left(\frac{v_{out}}{v_{prmry}}\right)\Big|\Big|_{MAX} = \frac{\sqrt{Q_1 Q_2'}}{2}\sqrt{\frac{L_2}{L_1}} \tag{16.38}$$

and is independent of k.

16.2 Experimental system design

Equation (16.36) predicts that the best unloaded quality factor for both primary and secondary coils maximizes the asymptotic efficiency of the link. Thus, it is important to have as high a Q_1 and Q_2 as is practically achievable. In ultra-low-power systems, loading in the secondary due to R_L and power-amplifier switching losses in the primary inevitably reduce Q_1 to $Q_{1,eff}$ and Q_2 to Q_2' respectively such that these losses are often the dominant determinant of the final quality factors of the primary and the secondary, and thus the link efficiency.

Coils generally have higher Q at higher operating frequency since electromagnetic skin-effect resistive losses scale like $\sqrt{\omega}$ while their reactance scales like ω. A large operating frequency does however lead to more sensitivity to parasitics, increased tissue absorption, and increased switching losses in the power amplifier due to finite rise-time and finite fall-time effects becoming a more significant fraction of the period of operation. The presence of the high dielectric tissue of the body surrounding the implant can also lower the self-resonant frequency of the coil due to inter-turn coupling. With these tradeoffs in mind, a good choice of operating frequency for RF power links may be near $\omega_n = 2\pi \times 6.78\,\text{MHz}$, an unlicensed ISM band.

Multi-stranded Litz wire mitigates skin-loss resistive effects by accumulating the increased skin-effect circumferential area available in each strand of wire rather than relying on the skin-effect circumferential area available in just one strand. In the band chosen, inductors from $1\,\mu\text{H}$ to $15\,\mu\text{H}$ and quality factors of 80–150 are easily available. More details on the influence of the number of turns, coil separation, and coil misalignment on k can be found in [7].

The design of an overall link for a given resonant frequency ω_n, given mean geometry parameters r, a_1, a_2, given Q_1 and Q_2, and given load R_L, is accomplished by a 'backwards procedure'. This procedure determines circuit parameters starting at the secondary output load and ending at the primary input source.

1. We first determine a C_2 that achieves an optimal Q_L for the expected value of Q_1 and Q_2 and with the calculated (or measured) k from Equations (16.35), (16.7), (16.27), and (16.1). The Schottky diode is designed to have as low an input capacitance as is practical. Any parasitics due to this diode or due to inter-turn coil coupling in L_2 contribute to C_2 such that the intentional value of C_2 is correspondingly reduced.

Table 16.2 Secondary resonator design

Parameter	Notes
$R_{L,ac}$	$2.5\,\text{k}\Omega$
C_2	93 pF Mica Q $\sim$ 400
Rectifier diode, d	$2 \times$ HBAT45C Schottky diode $C_{par} \simeq 13\,\text{pF}$
$C_{2,total}$	119 pF
L_2	8 turns of 22-strand Litz wire $r = 15\,\text{mm}$ $4.7\,\mu\text{H}$ with $Q \sim 140$
Unloaded Q_2	90
Q_L	12.6

2. The value of L_2 is then determined by requiring it to resonate with C_2 at ω_n.
3. The value of L_1 is then determined by voltage-gain requirements between the primary and the secondary using Equation (16.37). The voltage gain must be sufficiently high such that the secondary ac voltage is much greater than the Schottky diode turn-on voltage.
4. The value of C_1 is then determined by requiring it to resonate with L_1 at ω_n.
5. The power amplifier is then designed to have as low an output resistance as is practically possible and as described below for a Class-E example. In many ultra-low-power designs, the on resistance of the switches in the Class-E amplifier will determine the effective Q_1 of the primary, $Q_{1,eff}$, thus reducing efficiency.

Some iterations of this basic design methodology lead to convergence of an actual choice of component values.

16.2.1   Example design for the resonators

In many transcutaneous bionic implants, the skin-flap thickness can vary from 1 mm to 10 mm. For moderate robustness of k to variations in r, coils with $a_1 = a_2 = 14$ mm represent a good compromise between robustness and size. This size and range of separations yields $0.04 < k < 0.17$ for commercially available 22-strand Litz wire in experimental situations. Since a 6 mm skin flap is most common, the nominal k at which we would like to optimize the link design is near 0.084. Thus, assuming that the unloaded Q_1 and Q_2 of the coils is equal, $Q_{L,opt} = 1/k \approx 12$ from Equation (16.35). If we need to drive an effective load of $5\,\text{k}\Omega$, $R_{L,ac} = R_L/2 = 2.5$ kΩ. Thus, $C_2 \approx 12/(\omega_n R_{L,ac}) \approx 100\,\text{pF}$. To resonate at 6.78 MHz, L_2 must then be near $1/(\omega_n^2 C_2) \approx 5\,\mu\text{H}$. Table 16.2 reveals more exact values for this example in a real experimental system. We then pick $L_1 \approx 6.5\,\mu\text{H}$ to reduce the voltage gain due to relatively high quality factor of our resonators. With this choice of L_1, C_1 may then be chosen to resonate with L_1 at 6.78 MHz near 100 pF. To complete the design and determine $Q_{1,eff}$ in the primary we need to design a power amplifier. We shall discuss the design of a Class-E power amplifier since these amplifiers are capable of high efficiency at our operating frequency.

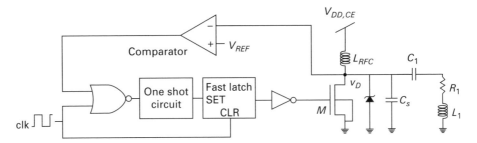

Figure 16.10. A switching power driver (Class-E) is shown with the primary circuit and a one-shot control loop. The Class-E supply voltage, $V_{DD,CE}$, controls the power level, while the clock, comparator, NOR gate, one-shot circuit, and latch control the gate drive timing. Reproduced with kind permission from [10] (©2007 IEEE).

16.2.2   Basics of Class-E power-amplifier design

The right half of Figure 16.10, which comprises a big switching transistor M, L_{RFC}, C_S, C_1, R_1, and L_1, forms a classic Class-E amplifier topology. The circuitry shown in the left half of Figure 16.10 is control circuitry for the Class-E amplifier. A Schottky diode that prevents undershoots below ground at the node marked v_D is also shown. To begin, we shall ignore the control circuitry and assume that the gate of the transistor M is periodically turned on and off by a square wave.

When M is turned on, v_D is quickly pulled to ground and L_{RFC}, a large inductor, is fluxed up by the large voltage across it to reflux its stored dc current to a value that is slightly greater than its average value I_{RFC}. Since L_{RFC} is fairly large, and consequently has high ac impedance compared with the circuitry in parallel with it, the current through L_{RFC} may be approximated as an almost purely dc current I_{RFC}. There are tiny variations about this dc value created as v_D goes above and below $V_{DD,CE}$ due to the activity of the switching transistor M. When M is turned off, the current I_{RFC} charges C_S and v_D begins to rise. The rising v_D excites activity in the resonant primary $L_1 C_1 R_1$ circuit, which filters the harmonic activity at v_D to create a nearly sinusoidal current i_{ac} in the primary. The difference between the nearly dc current I_{RFC} and the nearly sinusoidal current i_{ac} serves to charge capacitor C_S to values that cause it to first rise above $V_{DD,CE}$ and then return more gently and more sinusoidally towards ground. If the parameters of the circuit components are appropriately chosen, the return of v_D towards ground can be such that it has zero derivative and zero value (v_D is at ground) at exactly the right moment when the switch M turns on again to reflux the inductor. The cycle of fluxing, charging of C_S, and discharging of C_S then begins again over another cycle. Figure 16.11 shows waveforms for the Class-E amplifier of Figure 16.10 including the voltage in the secondary and a rectifier output voltage over a few cycles. When the circuit has equilibrated over many such cycles, the dc power flowing from $V_{DD,CE}$ is nearly all converted to ac power in R_1 with little loss of power due to switching dissipation in M.

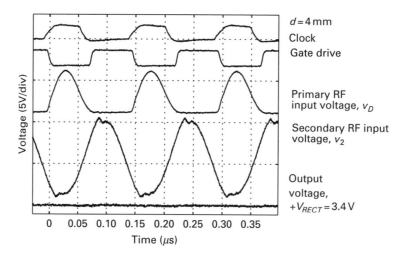

Figure 16.11. System waveforms, including the clock, gate-driver signal, Class-E drain voltage v_d, secondary voltage v_2, and rectified voltage V_{RECT} are shown for a 4 mm coil separation condition. Reproduced with kind permission from [10] (©2007 IEEE).

Dissipative losses in M are minimized during the fraction of the period when it is on because the drain-to-source voltage across M is nearly zero although its current i_M is maximal during this phase. Dissipative losses in M are minimized during the fraction of the period when it is off because the current through M is nearly zero although the voltage across it is maximal during this phase. This analysis assumes that the transistor M turns on and off infinitely fast in response to a sharp-edged square wave. However, even if there is a finite transition time between on and off states of the transistor, the Class-E topology is inherently robust to dissipative losses in M. The zero-value and zero-derivative return of v_D to ground around the point when M is turned on ensures that $i_M.v_D$ losses in the transistor are minimized during the off-to-on transition of M since v_D is still relatively low and flat around its minimum if M takes some time to turn on. The turn-on time of M may even have a slight positive or negative delay with respect to the minimum of v_D. The gentle rise in v_D due to the presence of C_S ensures that the $i_M.v_D$ dissipative loss is minimized near the on-to-off transition as well since, even if the transistor takes some time to shut off, v_D will still be near ground as its rise is relatively slow compared with the turn-off time of the transistor.

The exact nonlinear dynamics of Class-E amplifiers are fairly complex but empirical design criteria have been established to help with choosing component values [8], [9]. It has been found that if the primary circuit has a quality factor of Q_1 and operates at a frequency of ω_n, such that the $L_1C_1R_1$ circuit of the primary can be characterized by an impedance of $R_1 + jX_1$, and the C_S capacitance has a susceptance magnitude of $B = |j\omega_n C_S|$, then

$$Q_1 = \frac{\omega_n L_1}{R_1}; B = |j\omega_n C_S|$$

$$X_1 = \left(\omega_n L_1 - \frac{1}{\omega_n C_1}\right) = \frac{1.11 Q_1}{Q_1 - 0.67} R_1 \tag{16.39}$$

$$B = \frac{0.1836}{R_1}\left(1 + \frac{0.81 Q_1}{Q_1^2 + 4}\right),$$

for maximum energy efficiency. Other parameters at this optimum include

$$V_D^{peak} = 3.56 V_{DD,CE}; I_M^{peak} = 2.86 I_{RFC};$$

$$I_{RFC} = \frac{V_{DD,CE}}{1.734 R}$$

$$P_{ac} = \left(\frac{2}{1 + \frac{\pi^2}{4}}\right) \frac{V_{DD,CE}^2}{R} = 0.58 \frac{V_{DD,CE}^2}{R} \tag{16.40}$$

$$V_{R_1}^{peak} = \frac{2}{\sqrt{1 + \frac{\pi^2}{4}}} V_{DD,CE} = 1.074 V_{DD,CE}$$

$$\eta_{CLSE} \approx 1 - \frac{(\omega_n t_{swtch})^2}{12}$$

Most variables are self explanatory in Equation (16.40). The overall Class-E amplifier efficiency is denoted by η_{CLSE} and depends on the fractional switching time $\omega_n t_{swtch}$.

16.2.3 Practical Class-E amplifier design

In practice, there are three non-idealities in practical Class-E power amplifiers that lead to a finite energy efficiency in ultra-low-power systems.

1. Variations in the load of the secondary are manifested as changes in the effective value of R_1 in Figure 16.10, i.e., R_1 is really $R_1 + R_{refl}$. Changes in the load can thus cause the return of v_D to zero to not occur exactly when M turns on. The $f_n C_S v_D^2$ energy is then dissipated through discharging currents in M and leads to efficiency degradation in the power amplifier.
2. The finite on resistance of the switch M, $R_{ds,on}$, reduces Q_1 in the primary to $Q_{1,eff}$ and thus reduces the energy efficiency of the coupled resonator.
3. The gate capacitance of M is periodically charged and discharged in each cycle by gate-driver circuitry and leads to switching energy losses.

Although complex control schemes for addressing each of these problems may be imagined, in ultra-low-power systems, such schemes often consume high

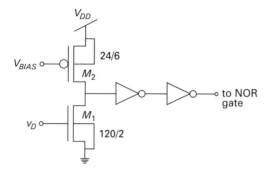

Figure 16.12. A simple comparator for the Class-E controller. The comparator triggers the NOR gate when v_d approaches zero volts. The device sizes were ratioed to obtain a low voltage threshold. The threshold is intentionally not exactly at ground to allow for some error compensation due to delays in our control scheme. Reproduced with kind permission from [10] (©2007 IEEE).

overhead power. The left half of Figure 16.10 illustrates a simple scheme that allows for some adaptation in the switching time and partly addresses the first issue. The second and third issues may be addressed by optimally sizing M to have a large enough W/L such that its on resistance is fairly low but not sizing it to be too large such that its gate capacitance adds significantly to switching energy losses.

The controller shown on the left half of Figure 16.10 achieves fixed-frequency operation set by a clock but adapts to changes in the driven network that speed up the return of v_D to ground: a comparator senses when v_D is near ground and generates a one-shot pulse that activates the fast latch to turn M on. If the comparator signal is not received because v_D does not get sufficiently near ground by the time a high clock signal arrives, the clock signal turns M on. Thus, the circuitry ensures that either the clock going high or the comparator going high turns M on, whichever comes first. The transistor M is always turned off by the clock signal going low.

Figure 16.12 shows that the comparator is a ratioed inverter, intentionally sized to have a threshold near ground but not exactly at ground such that delays in the control scheme automatically have some level of error compensation. Figure 16.13 illustrates the turn-on behavior of M in the circuit as it is manifested in the reset of v_D to ground. For $d = 1$ mm, there is more reflected resistance in the primary such that the ringing frequency of the resonant primary circuit slows due to the increased damping. The circuit adapts and resets v_D to ground before the clock signal goes high such that $f_n C_S v_D^2$ switching power is reduced. For $d = 4$ mm, the reset of v_D is more smooth since the switching time is nearly optimal. For $d = 10$ mm, the reduced reflected resistance leads to decreased damping and a faster ringing frequency. The Schottky diode and M together help return v_D to ground.

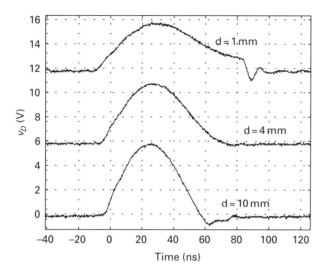

Figure 16.13. Three cases are shown for the Class-E drain voltage v_D. Note that the switch timing has adapted slightly to reduce the amount of $fC_s v_D^2$ wasted power under the 1 mm and 10 mm conditions. Furthermore, the action of the protection diode can be seen in the $d = 10$ mm case where the drain voltage has dropped below ground. Reproduced with kind permission from [10] (©2007 IEEE).

The primary coil L_1 can have an intrinsic quality factor of over 100. However, the effective quality factor if on-resistance losses in the power amplifier are included, $Q_{1,eff}$, is given by

$$Q_{1,eff} = \frac{\omega_n L_1}{R_1 + R_{DS,on}} \tag{16.41}$$

Such losses can reduce the quality factor significantly. If we assume that $Q_{1,eff}$ in the primary resonator is determined entirely by the on resistance of M, $R_{DS,on}$, then from the primary resistive-divider circuit of Figure 16.8 and Equation (16.31), we can show that the *loss* due to $R_{DS,on}$ is given by

$$P_{RDS,on} = P_{IN} \frac{1}{1 + k^2 Q_L \dfrac{\omega_n L_1}{R_{DS,on}}} \tag{16.42}$$

where we have assumed that $Q_2' \approx Q_L$ and that the coil-limited Q_1 contributes little to determining $Q_{1,eff}$ as $R_{DS,on}$ is significantly bigger than the coil resistance R_1 of the primary resonator. The parameter P_{IN} is the input power delivered by the amplifier. The principal effect of $R_{DS,on}$ in this design is captured by its effect on the primary resonator, i.e., by Equation (16.42) since the dc power loss, $I_{RFC}^2 R_{DS,on}$ is negligible in our design.

Equation (16.42) yields the resonator power loss as a function of $R_{DS,on}$ or equivalently with $1/(W/L)$ of the transistor M. If we sum this power loss with the increase in gate switching power, which rises with the WL value of M, an optimal value for W and L that minimizes the total power loss is found to be nearly $1200\,\mu m \times 0.5\,\mu m$ in the $0.5\,\mu m$ process in which this circuit was fabricated.

The Class-E driver must have a sufficiently low value of $V_{DD,CE}$ to not cause junction breakdown of the 11 V-limited devices of the technology since the maximum voltage at v_D is nearly $3.6V_{DD,CE}$ when the Class-E amplifier is operational. At low secondary power dissipation levels in the 1 mW–10 mW range, the effective magnetic field strengths are low enough for regulatory issues to not be a big concern as we shall quantitatively show in Chapter 18.

16.3　Experimental measurements

A complete RF power link including a Class-E driver, a coupled resonator, and a Schottky-diode-based rectifier was built with the component values shown in Table 16.3 with a 1.8 V supply voltage for the controller circuits and $V_{DD,CE}$ varying between 0.6 V–2.5 V depending on the desired RF output power level of the link. The comparator in Figure 16.10 was biased at $28\,\mu A$ to minimize delay, and consumed $56\,\mu W$ of power if wasted biasing power was also included. The gate driver and remaining circuits consumed $48\,\mu W$ from the supply. The $5\,\Omega$ value of $R_{DS,on}$ reduces $Q_{1,eff}$ to 35.

Table 16.4 summarizes system parameters and the overall efficiency of the power link when all sources of power are included at different output power levels.

Figure 16.14 reveals that the system has less than 16% variation in rectifier output voltage when the link distance is varied from 1 mm to 10 mm. Figure 16.15 reveals measurements of the overall efficiency of the link. The bold trace indicates the theoretical performance possible if the load is adapted for each coupling

Table 16.3 Final system specifications

Parameter	Notes
Output power level, $P_{R_{L,dc}}$	$1\,mW - 10\,mW + V_{RECT}, -V_{RECT}$ with $R_{L,dc} = 10\,k\Omega$ each
Operating frequency, f	6.785 MHz
Coil separation distance, d	1 mm – 10 mm
Class-E nFET	100 fingers $\times$ 12 μm $\times$ 0.5 μm $C_{gate} \sim 2.1$ pF
Class-E supply voltage, $V_{DD,CE}$	0.6 V – 2.5 V
L_1	22-strand Litz wire r = 15 mm
	10 turns 6.5 μH with $Q_1 \sim 94$
C_1	1000 pF Mica
C_S	30 pF
L_2	22-strand Litz wire r = 15 mm
	8 turns 4.7 μH with $Q_2 \sim 90$
C_2	93 pF
Rectifier diode, D	$2 \times$ HBAT54C Schottky diode $C_{par} \simeq 13$ pF
L_{RFC}	Coilcraft 1812PS-223KL $L_{RFC} = 22\,\mu H$

Table 16.4 Final system performance

Component	Performance @ $V_{DD} = 1.8$ V
Gate driver	$46\,\mu$W
Gate switching controller	$56\,\mu$W
Power switch, M	$R_{DS,on} \simeq 5\,\Omega$
$Q_{1,eff}$	35
Q_2	90
η_{total} @ 1 mW	66% @ $d = 1$ mm
	62% @ $d = 6$ mm
	51% @ $d = 10$ mm
η_{total} @ 10 mW	74% @ $d = 1$ mm
	66% @ $d = 6$ mm
	54% @ $d = 10$ mm

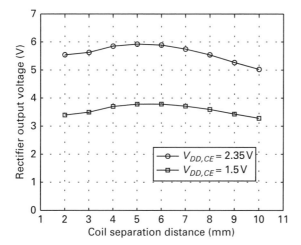

Figure 16.14. Variation in the rectified output voltage is shown for two power driver voltage levels. The rectified voltage variation is less than 16% for all cases. Reproduced with kind permission from [10] (©2007 IEEE).

factor, i.e., Equations (16.35) and (16.36). The dashed trace shows the theoretical performance for a fixed $Q_{L,opt}$, the experimental situation, and for which Equation (16.34) is predictive. The agreement of experimental measurements of efficiency with theory is better for a 10 mW output secondary power level, where the fixed power used by the controller and gate driver are not a significant fraction of the overall power. However, the agreement with theory even at the 1 mW output power level is fairly good. Figure 16.16 reveals measurements of various power components in the system. We see that, at the lowest power operating level, rectifier and gate-switching losses contribute the most because these fixed losses do not scale with power level like efficiencies in linear portions of the system, e.g., the primary-secondary coupling. Rectifier losses are well

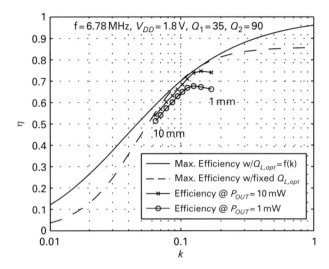

Figure 16.15. A comparison between the asymptotically efficient coupled resonator power transfer system ($Q_1 = 35$, $Q_2 = 90$) is shown with our system. The top curve indicates the maximum possible efficiency for an adapting load condition (Equation 16.36). The dashed theory curve indicates efficiency for our fixed loading condition, $Q_{L,opt}$, (Equation 16.34). Our system performance is shown for 10 mW and 1 mW operation. Reproduced with kind permission from [10] (©2007 IEEE).

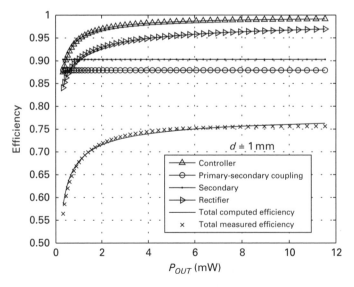

Figure 16.16. A plot of the efficiency is shown for a sweep in the power level by changing the Class-E supply voltage. The efficiency of each mechanism is shown for comparison. At low power levels, controller power and losses in the rectifier contribute most to inefficiency of the system. Reproduced with kind permission from [10] (©2007 IEEE).

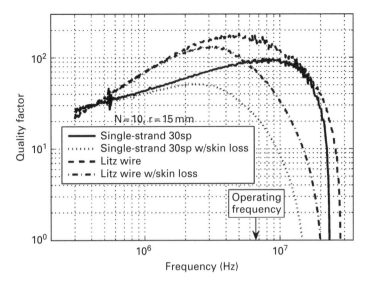

Figure 16.17. Effects of skin loss on coil Q and self-resonance frequency for single-stranded and Litz wire coils.

approximated as $I_{load} \cdot V_{turnon}$ where I_{load} is the dc load current of the rectifier and V_{turnon} is the turn-on voltage of the Schottky diode, near 0.25 V in many practical designs.

The efficiency of the system at 10 mW output power levels is 74% and 54% at 1 mm and 10 mm separation distances respectively, in good accord with theoretical predictions. At 1 mW output power levels, the efficiencies are 67% and 51%, also in good accord with theoretical predictions. A redesigned system with an $R_{DS,on} = 2 \; \Omega$ could achieve efficiencies near 85% in accord with Equation (16.36). Measurements with a bag of saltwater or Delrin between the primary and secondary coil, to simulate the properties of skin tissue, show a slight shift in resonance frequency, which degrades efficiency by 2–3%. The robustness of the design to such effects is because an explicitly large C_2 in the secondary ensures that stray capacitances and self-resonance effects in the coils do not affect performance significantly. Note that the resonant frequency of the primary circuit is determined by $(C_1 C_S)/(C_1 + C_S)$ rather than by C_1 alone when the Class-E driver circuit is used. Further details are provided in [10].

Figure 16.17 illustrates how the high dielectric constant of skin (40–80) can shift the self-resonant frequencies of coils, both for coils made from Litz wire and those made without. The Litz-wire coils have higher self-resonant frequencies and higher quality factors than those made from ordinary wire and show smaller shifts in resonant frequency as well.

Figure 16.18 reveals the typical conductivity and dielectric permittivity characteristics for heterogenous biological tissue versus carrier frequency [11], [12]. We see that operating at carrier frequencies beyond a few MHz is detrimental to power efficiency due to increase in conductivity σ and consequently increased

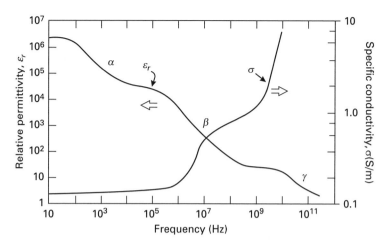

Figure 16.18. Relative permittivity and specific conductivity of skin as a function of frequency. Reproduced with kind permission from [11].

absorption at these frequencies. The three sigmoid-like changes in permittivity ε_r with frequency, marked α, β, and γ, are believed to be due to the motion of extracellular and cytoplasmic ions at membrane and tissue interfaces, the motion of dipoles fixed in cell membranes, and the motions of water dipoles respectively [11], [12]. Models for the electric properties of biological tissue may be found on the web at [13].

The overall characteristics of skin can be approximated by introducing a series-connected R-and-C bridging impedance element that couples the primary and secondary loops. The capacitive coupling due to this bridging element can sometimes help efficiency by introducing an additional coupling path although it often causes unpredictable changes in resonant frequency.

References

[1] R. Sarpeshkar, C. D. Salthouse, J. J. Sit, M. W. Baker, S. M. Zhak, T. K. T. Lu, L. Turicchia and S. Balster. An ultra-low-power programmable analog bionic ear processor. *IEEE Transactions on Biomedical Engineering*, **52** (2005), 711–727.

[2] L. Theogarajan, J. Wyatt, J. Rizzo, B. Drohan, M. Markova, S. Kelly, G. Swider, M. Raj, D. Shire and M. Gingerich, Minimally invasive retinal prosthesis. *Proceedings of the IEEE International Solid-State Circuits Conference (ISSCC)*, San Francisco, CA, 99–108, 2006.

[3] M. Ghovanloo and S. Atluri. A Wide-Band Power-Efficient Inductive Wireless Link for Implantable Microelectronic Devices Using Multiple Carriers. *IEEE Transactions on Circuits and Systems I: Regular Papers*, **54** (2007), 2211–2221.

[4] R. Sarpeshkar, W. Wattanapanitch, S. K. Arfin, B. I. Rapoport, S. Mandal, M. W. Baker, M. S. Fee, S. Musallam and R. A. Andersen. Low-Power Circuits for Brain-Machine Interfaces. *IEEE Transactions on Biomedical Circuits and Systems*, **2** (2008), 173–183.

[5] Available from: http://www.technick.net/public/code/cp_dpage.php?
aiocp_dp = util_inductance_calculator or

[6] R. J. Edwards. Available from: http://www.smeter.net/feeding/transmission-line-
choke-coils.php#Circuit.

[7] F. E. Terman. *Radio Engineers Handbook* (New York: McGraw-Hill, 1943).

[8] N. O. Sokal and A. D. Sokal. Class E – A new class of high-efficiency tuned
single-ended switching power amplifiers. *IEEE Journal of Solid-State Circuits*, **10** (1975),
168–176.

[9] F. Raab. Idealized operation of the class E tuned power amplifier. *IEEE Transactions
on Circuits and Systems*, **24** (1977), 725–735.

[10] M. W. Baker and R. Sarpeshkar. Feedback analysis and design of RF power links
for low-power bionic systems. *IEEE Transactions on Biomedical Circuits and Systems*,
1 (2007), 28–38.

[11] D. Miklavcic, N. Pavselj and F. X. Hart. Electric properties of tissues. In *Wiley
Encyclopedia of Biomedical Engineering*, ed. M Akay (New York: John Wiley & Sons;
2006).

[12] S. Gabriel, R. W. Lau and C. Gabriel. The dielectric properties of biological tissues:
III. Parametric models for the dielectric spectrum of tissues. *Physics in Medicine and
Biology*, **41** (1996), 2271–2293.

[13] Italian National Research Council and Institute for Applied Physics. Available from:
http://niremf.ifac.cnr.it/tissprop/.

17 Energy-harvesting RF antenna power links

I do not think that the wireless waves I have discovered will have any practical application.

Heinrich Rudolph Hertz

Ultra-low-power systems have the potential to operate in a battery-free fashion by harvesting energy from their environment. Such energy may take the form of solar energy in a solar-powered system, chemical energy from carbohydrates in an enzyme-based system, mechanical energy from vibrations in the system's platform, thermal energy in systems that exploit temperature differences between themselves and their environment, or radio-frequency electromagnetic energy in the environment. In Chapter 26, we shall discuss several forms of energy harvesting. In this chapter, we will focus on energy harvesting with radio-frequency antennas or *rectennas* as they are sometimes called.

Radio-frequency electromagnetic energy is increasingly becoming ubiquitous due to the growing presence of cellular phones, local area networks, and other wireless devices. Systems that operate by harvesting electromagnetic energy need an antenna for sensing electromagnetic waves and a rectifier for converting the sensed ac energy to a dc power supply. The created power supply can then be used to power an ultra-low-power system such as a radio-frequency identification (RF-ID) tag in a grocery store or a medical monitoring tag on the body of a person (see Chapter 20 on medical monitoring). In this chapter, we shall discuss important principles and building blocks for creating such RF-energy-harvesting systems including antennas and rectifier circuits. We shall discuss an example of a complete functioning experimental system to illustrate system-level tradeoffs. We begin by reviewing the fundamentals of antenna operation.

Antennas are complicated distributed circuits that serve to transmit or receive electromagnetic energy. When functioning as receivers, they sense distributed electromagnetic wave input signals in free space to create a local electrical signal between a pair of lumped output terminals. When functioning as transmitters, they convert a local electrical signal between a pair of lumped input terminals to an electromagnetic wave that is radiated in a distributed fashion into free space.

A quantitative and detailed understanding of antennas could and does form the subject of several books that analyze antennas using Maxwell's equations of electromagnetism. For example, see [1]. Rather than repeat such analysis here, we shall focus on providing an intuitive understanding that is useful for our purposes and that is practical for low-power circuit design.

17.1 Intuitive understanding of Maxwell's equations

To begin, we note that a spatially discrete approximation to Maxwell's equations can be simulated in an 'analog computer' made up of an infinite LCR network of infinitesimally small and invisible inductors, capacitors, and resistors in free space that are connected to each other in the mesh of Figure 17.1. This ingenious analog-circuit simulation of Maxwell's equations was formulated by Gabriel Kron as described in [2]. The formulation cleverly configures loop currents and node voltages in the mesh such that distributed parameters such as the electric field $\boldsymbol{E}$, magnetic field $\boldsymbol{H}$, or current density $\boldsymbol{J}$ are proportional to the voltage across the capacitors, the current in the inductors, or the current in the resistors, respectively, as shown; the constants of proportionality are given by ε, μ, and σ, which represent the capacitive, magnetic, and conductive properties of the medium. Kirchhoff's voltage law (KVL), Kirchhoff's current law (KCL), and

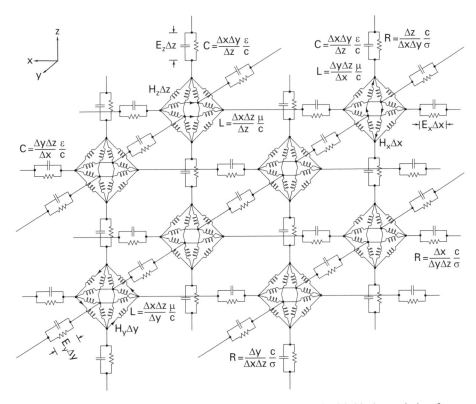

Figure 17.1. Circuit model of Maxwell's equations. Reproduced with kind permission from [2] (©1944 IEEE).

the constitutive current-voltage relations for capacitors, inductors, and resistors in Figure 17.1 lead to Maxwell's equations in the limit where the discrete becomes the continuous, i.e., the spatial increments, Δx, Δy, and Δz go to zero. Curl operators for fields are analogous to KVL, divergence operators for fields are analogous to KCL, and gradient operators on scalars are analogous to differences. Charge conservation, which is a consequence of Maxwell's equations, is also automatically satisfied. Figure 17.1 suggests that several properties of electromagnetic media may be understood by analogy to the properties of distributed LCR circuits. Therefore, we shall briefly review the properties of these circuits. In particular we will begin with a short introduction to the behavior of a very important distributed LC circuit, the non-lossy one-dimensional transmission line. These lines will later aid in our understanding of the properties of antennas.

17.2   The non-lossy, one-dimensional transmission line

An infinite mesh of one-dimensional series floating inductors between grounded shunt capacitors forms a discrete approximation to a continuous one-dimensional transmission line. Usually L denotes the lateral inductance per unit length of the line and C denotes the shunt capacitance per unit length of the line. Co-axial BNC cables that are ubiquitous in electronics are examples of such transmission lines. Such a transmission line is capable of propagating a wave without attenuation and with a speed of $c = 1/\sqrt{(LC)}$ at any frequency from left to right or from right to left [3]. Waves of frequency f and wavelength λ propagating from left to right or right to left are represented by expressions of the form

$$V_{L \to R} = V_f e^{j(\omega t - kx)}$$
$$V_{R \to L} = V_b e^{j(\omega t + kx)}$$

(17.1)

where the angular frequency ω and wave number k are given by

$$\omega = 2\pi f$$
$$\lambda = \frac{c}{f}$$
$$k = \frac{2\pi}{\lambda}$$
$$\omega = ck$$

(17.2)

The parameters V_f and V_b are constants and the x-axis is directed from left to right. Since energy put in at any location of an infinite transmission line simply flows away as a wave to infinity, and never returns, it appears as if the energy is 'dissipated'. The impedance looking in at any location of an infinite transmission line is hence found to be real and dissipative and related to a characteristic impedance $Z_0 = \left(\sqrt{(L/C)}\right)$. The ratio of the voltage $V(x, t)$ and current $I(x, t)$ at any location x for a wave propagating in the forward direction (left to right) in the transmission line is found to be given by

$$V(x,t) = Z_0 I(x,t) \tag{17.3}$$

and for a wave propagating in the backward direction (right to left) is found to be given by

$$V(x,t) = -Z_0 I(x,t) \tag{17.4}$$

If the one-dimensional transmission line is terminated with an impedance Z_L at its right end at $x=0$, then a forward-going wave characterized by $(V_f e^{j(\omega t - kx)}, I_f e^{j(\omega t - kx)})$ is reflected at the termination to generate a new backward-going wave whenever Z_L is not equal to Z_0. The reflection ensures that a weighted sum of the incident forward-going wave and the reflected backward-going wave satisfies the impedance condition at the termination boundary and also satisfies the properties of the transmission line. If we denote the reflection coefficient at the termination boundary by Γ and characterize the backward-going wave by $(V_b e^{j(\omega t + kx)}, I_b e^{j(\omega t + kx)})$ we find that

$$V_b = \Gamma V_f$$

$$I_b = -\Gamma I_f$$

$$V_f = Z_0 I_f \tag{17.5}$$

$$V_b = -Z_0 I_b$$

$$\Gamma = \frac{Z_L - Z_0}{Z_L + Z_0}$$

The set of relationships in Equation (17.5) is the only solution consistent with the properties of the transmission line and the boundary condition imposed by the terminating load impedance Z_L. A shorted termination corresponds to $Z_L = 0$ and an open termination corresponds to $Z_L = \infty$. In both of the latter cases, the magnitude of $\Gamma = 1$. Note that the reflection coefficient for current is negative with respect to the reflection coefficient for voltage in Equation (17.5) as current reverses direction in the backward-going wave.

For a transmission line terminated at $x=0$ with the impedance Z_L, the voltage, current, and impedance at any location x can then be described by equations of the form

$$V(x,t) = V_f e^{j\omega t} e^{-jkx} + \Gamma V_f e^{j\omega t} e^{+jkx}$$

$$I(x,t) = \frac{V_f}{Z_0} e^{j\omega t} e^{-jkx} - \frac{\Gamma V_f}{Z_0} e^{j\omega t} e^{+jkx} \tag{17.6}$$

$$Z(x,t) = \frac{V(x,t)}{I(x,t)} = Z_0 \left(\frac{1 + \Gamma e^{+2jkx}}{1 - \Gamma e^{+2jkx}} \right)$$

If we substitute Equation (17.5) for Γ in Equation (17.6), and perform some algebra, we find that at $x=-L$, i.e., a length of L to the left of the termination at $x=0$,

$$Z(-L) = Z_0 \left(\frac{1 + \Gamma e^{-2jkL}}{1 - \Gamma e^{-2jkL}} \right)$$

$$Z(-L) = Z_0 \left(\frac{Z_L \cos(kL) + j Z_0 \sin(kL)}{Z_0 \cos(kL) + j Z_L \sin(kL)} \right) \tag{17.7}$$

Equation (17.7) shows that the input impedance of a transmission line of finite length L has resistive and reactive parts. It is periodic in L with an overall periodicity described by $kL = n\pi$, where n is an integer, i.e., we have a periodicity of $\lambda/2$. The periodicity is described by $n\pi$ rather than $2n\pi$ because both the voltage and the current invert with a phase change of π such that their ratio, which describes the impedance, is invariant with phase changes of π.

If we evaluate Equation (17.7) at $kL = \pi/2$, we conclude that a transmission line of length $L = \lambda/4$ transforms a termination impedance of Z_L at its output into a 'gyrated impedance' at its input given by

$$Z_{in}^{\lambda/4} = \frac{Z_0^2}{Z_L} \tag{17.8}$$

Thus, an open at one end of the line is transformed to a short as seen at the other end of the line and vice versa.

An intuitive explanation of Equation (17.8) can be provided for the case of $Z_L = \infty$ (an open). A reflected wave for voltage from the end of the line returns to the input and adds with a net travel phase of $-90° + -90° = -180°$ to the incident wave at the input of the line that created it; a reflected wave for current, however, suffers an additional $-180°$ phase upon reflection, returns to the input, and adds with a net travel phase of $-360°$ to the incident wave at the input of the line that created it. Since the voltage waves then cancel perfectly while the current waves add and are finite, the infinite impedance at the end of the line appears to be transformed to that of a short at the input of the line, i.e., the voltage is 0 while the current is finite. A similar explanation for $Z_L = 0$ reveals that a short at one end of the line transforms to an open at the other end of the line.

Equation (17.7) also reveals that

$$Z(-L)|_{Z_L=\infty} = \frac{Z_0}{j \tan(kL)}$$

$$Z(-L)|_{Z_L=0} = j Z_0 \tan(kL) \tag{17.9}$$

Thus, a short length of an open-terminated transmission line ($L \ll \lambda$) has an impedance at its input given by

$$Z(-L)|_{Z_L=\infty} = \frac{Z_0}{jkL} \tag{17.10}$$

and appears to be capacitive. In contrast, a short length of a short-terminated transmission line ($L \ll \lambda$) has an impedance at its input given by

$$Z(-L)|_{Z_L=0} = j Z_0 kL \tag{17.11}$$

and appears to be inductive.

17.3   The impedance of free space

In Figure 17.1, if the medium is free space, $R=0$ everywhere except at very high fields. Thus, just as an infinite one-dimensional transmission line is capable of propagating waves in two directions, the three-dimensional LC mesh of Figure 17.1 is capable of propagating waves in all directions with a speed of propagation given by $c = 1/\sqrt{(\mu\varepsilon)}$, the speed of light, and has a characteristic impedance at any point given by $\eta = \sqrt{(\mu/\varepsilon)}$. The parameter η is referred to as the impedance of free space, which evaluates to $120\pi\,\Omega$. By analogy to the one-dimensional transmission-line case, we also find that $E(x,y,z) = \eta H(x,y,z)$ for a plane wave propagating in any direction.

The impedance of free space η is a constant that scales the impedance of several electromagnetic structures. For example, a co-axial BNC cable with a cylindrical inner conductor of radius r_1, a cylindrical concentric outer conductor of radius r_2, and an insulating material between the conductors characterized by ε and μ has a capacitance C per unit length, inductance L per unit length, and characteristic impedance Z_0 given by

$$C = \frac{2\pi\varepsilon}{\ln\left(\frac{r_2}{r_1}\right)}$$

$$L = \frac{\mu}{2\pi}\ln\left(\frac{r_2}{r_1}\right)$$

$$Z_0 = \sqrt{\frac{L}{C}} = \frac{1}{2\pi}\sqrt{\frac{\mu}{\varepsilon}}\ln\left(\frac{r_2}{r_1}\right)$$

$$= \frac{\eta}{2\pi}\ln\left(\frac{r_2}{r_1}\right)$$

(17.12)

17.4   Thevenin-equivalent circuit models of antennas

Accelerating charges or changing currents in a transmitting antenna radiate electromagnetic waves into free space represented by propagating E and H 'far fields' in free space. These radiated far fields, after a time delay determined by the speed of light, cause charges and currents in a receiving antenna to change such that we can say that an energy exchange has occurred between the transmitting antenna and the receiving antenna. The loss of energy in the transmitting antenna causes it to have 'damping' or an effective 'radiation resistance'. The gain in energy in the receiving antenna causes it to develop a source voltage or source current across its terminals that is proportional to the local field strength.

In addition to radiating or receiving electromagnetic energy, antennas also store inductive and capacitive energy in the E and H fields that are present near them, i.e., in their 'near fields'. Such fields are not radiated but effectively function like field lines that begin and end on antenna structures (capacitive E fields) or form closed loops near the antenna (inductive H fields). Equivalently, we can say that photon wave packets that are radiated by a transmitting antenna are recaptured in the near-field region by the antenna but escape from the antenna if they enter the far-field region of the antenna. Photon wave packets that lead to electrical signals in a receiving antenna are captured by the receiving antenna when they enter its near-field region while photon wave packets in its far field are not captured. We have simplified some technical issues in the exact boundaries of what is near field vs. what is far field, and what is radiated vs. what is not, since the actual situation is somewhat more analog than we have described. Nevertheless, our simplification will still allow us to capture and understand all the significant concepts behind antenna operation.

The properties of antennas are extremely complex because the source currents and source charges on antenna structures lead to field patterns with complex geometries. Such field patterns in turn alter the original sources and currents in the antenna structures that caused them, leading to a distributed feedback loop between the sources and the fields. The feedback loop is often broken by assuming plausible distributed boundary conditions for the source charges and currents on the antenna structures and then checking that they are indeed consistent with the field patterns. Fortunately we can abstract away much of the complexity inherent in antennas by creating lumped Thevenin-equivalent circuit models at their terminals near some operating frequency. Such models represent near-field energy storage and far-field energy radiation by reactive or resistive impedance components in the Thevenin self-impedance of the antenna, represent signals across the terminals of the antenna by Thevenin-equivalent sources, and represent coupling between antennas by mutual-impedance equivalents.

A simple two-port model between a transmitting antenna and a receiving antenna is shown in Figure 17.2. The signal source V_g with source impedance Z_g drives the transmitting antenna with self-impedance Z_{11} and a mutual coupling impedance to the receiving antenna of Z_{12}. The receiving antenna drives a load Z_l. The impedance Z_{11} corresponds to the voltage IZ_{11} seen across the transmitting antenna terminals when a current I is applied across these terminals and the receiving antenna is 'open', i.e., the load Z_l is infinite. The impedance Z_{12} corresponds to the voltage IZ_{12} developed across the terminals of the receiving antenna when Z_l is infinite as well, i.e., the open-circuit Thevenin voltage of the receiving antenna when I is applied to the transmitting antenna. The impedance Z_{22} corresponds to the voltage IZ_{22} developed across the terminals of the receiving antenna when a current I is applied across its terminals and the transmitting antenna is open, i.e., Z_g is infinite. Since space is symmetric with respect to propagation and reception of waves from the transmitter to the receiver and vice versa, and Maxwell's equations are time symmetric, if we switch the roles of the receiving and transmitting antennas, we find that $Z_{12} = Z_{21}$, a powerful property known as reciprocity. Reciprocity allows us to only analyze the properties of transmitting antennas and then automatically know that they will be valid for the antenna when it is used as a receiving antenna as well. Thus, for example, if an antenna

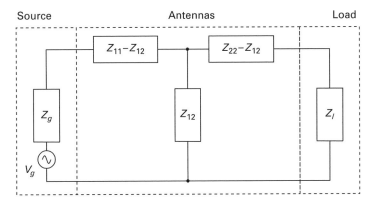

Figure 17.2. General Z_{11}, Z_{12}, and Z_{22} coupling parameters for near-field and far-field RF links.

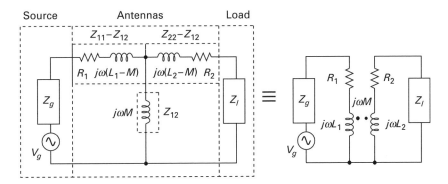

Figure 17.3. Near-field mutual-inductance RF link.

radiates more energy in a particular direction in space when it is a transmitter, it will also receive more energy from this particular direction in space when it is a receiver.

Figure 17.3 shows that the two-port model of a transmitting antenna and receiving antenna can even be used to analyze coupling between electromagnetic structures when there is no radiation. In fact, we have already analyzed such coupling in depth in Chapter 16 where we studied coupled-resonator mutual-inductance links. Such coupling corresponds to 'near-field' reactive coupling between electromagnetic structures and can be well modeled by the Thevenin-equivalent two-port circuit shown in Figure 17.3. Thus, in Figure 17.3, we have $Z_{12} = j\omega M$ where M is the mutual inductance, $Z_{11} = j\omega L_1 + R_1$, and $Z_{22} = j\omega L_2 + R_2$. The transmitting antenna corresponds to the primary coil in a coupled-resonator mutual-inductance link and the receiving antenna corresponds to the secondary coil in the same link.

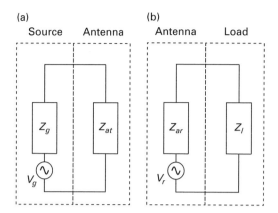

Figure 17.4a, b. Transmitting antenna (a) and receiving antenna (b).

The near-field reactive-coupling case is useful for wireless energy transfer when the transmitter and receiver are very close while the far-field radiative-coupling case is useful for wireless energy transfer when the transmitter and receiver are far apart. Both cases are useful for harvesting electromagnetic energy to power electronic systems that operate at low levels of power. First, we describe some relationships that are true in both the near-field and far-field cases because the general lumped circuit equivalent of Figure 17.2 is an abstract representation that applies to both cases.

By using the reflected-impedance concept described in Chapter 16 for mutual-inductance coupling (Equation 16.10) and noting that Z_{12} is analogous to $Ms = Mj\omega$, we can derive that the terminal impedance seen at the transmitting antenna by the source, Z_{at}, is given by

$$Z_{at} = R_{at} + jX_{at} = Z_{11} - \frac{Z_{12}^2}{Z_{22} + Z_l} \qquad (17.13)$$

The impedance Z_{at} is shown in Figure 17.4 (a). Similarly, the Thevenin impedance Z_{ar} seen by the load at the receiving antenna is given by

$$Z_{ar} = R_{ar} + jX_{ar} = Z_{22} - \frac{Z_{12}^2}{Z_{11} + Z_g} \qquad (17.14)$$

At the receiving antenna, the open-circuit Thevenin source voltage V_r seen by the load is given by

$$V_r = \left(\frac{V_g}{Z_{11} + Z_g}\right)Z_{12} \qquad (17.15)$$

The voltage V_r and Z_{ar} that represent the Thevenin equivalent circuit at the receiving antenna are shown in Figure 17.4 (b). We shall now apply these relationships to derive results for near-field antenna coupling identical to those derived in Chapter 16, and then discuss far-field coupling.

17.5   Near-field coupling

In the near-field case, shown in Figure 17.3,

$$Z_{11} = j\omega L_1 + R_1$$

$$Z_{12} = j\omega M$$

$$Z_{22} = j\omega L_2 + R_2 \tag{17.16}$$

$$k \equiv \frac{M}{\sqrt{L_1 L_2}}$$

where k is defined as the coupling constant. If we define the inductor quality factors as

$$Q_1 = \frac{\omega L_1}{R_1}$$

$$Q_2 = \frac{\omega L_2}{R_2} \tag{17.17}$$

and if a capacitive Z_g is used to resonate out the value of $j\omega L_1$, then

$$|V_r| = \left|\frac{Z_{12}}{Z_{11} + Z_g}\right| |V_g| = \frac{\omega M}{R_1} |V_g| = \frac{\omega k \sqrt{L_1 L_2}}{R_1} |V_g| = k Q_1 \sqrt{\frac{L_2}{L_1}} |V_g| \tag{17.18}$$

Since $L_2/L_1 = (n_2/n_1)^2$, the square of the turns ratio of the inductive link, we get

$$\left|\frac{V_r}{V_g}\right| = k Q_1 \sqrt{\frac{L_2}{L_1}} = k Q_1 \left(\frac{n_2}{n_1}\right) \tag{17.19}$$

The load Z_l typically resonates out the value of $j\omega L_2$ such that

$$Z_{at} = Z_{11} - \frac{Z_{12}^2}{Z_{22} + Z_l} = j\omega L_1 + R_1 + \frac{\omega^2 M^2}{R_2} = j\omega L_1 + R_1 \left(1 + k^2 Q_1 Q_2\right) \tag{17.20}$$

as we have derived earlier in Chapter 16.

17.6   Far-field coupling: the 'monopole' antenna

We shall now discuss the far-field radiative coupling case, the main focus of this chapter. In accord with the insightful discussion in [4], we begin by discussing how a hypothetical time-varying point charge, i.e., a monopole, leads to reactive near fields and radiative far fields. Such a monopole is an impossibility because, by charge conservation, the time variation of charge at a point must necessarily arise from the time variation of charges at neighboring points that supply charge to the monopole or sink charge from it. Thus, an elementary radiator must really be a dipole involving at least one other charge with a charge variation that is the negative of the monopole charge's variation. However, we can still obtain insight from the toy monopole case and illustrate conceptual issues with it. For example, the far-field electric field strength of this monopole and that of more complex real antennas both fall inversely with the distance r from the antenna. Similarly, the effective area for capturing photons in a monopole radiator is found to be the same as that of any hypothetical isotropic antenna.

The potential at a point a distance r from a monopole static charge source of value q is given by the well-known Coulomb relation

$$\phi = \frac{q}{4\pi\varepsilon r} \tag{17.21}$$

If this charge oscillates with angular frequency ω, i.e., $q = q_0 e^{j\omega t}$, then the potential at the point r, due to the finite light speed of propagation of electromagnetic waves of c is given by

$$\phi = \frac{q_0 e^{j\omega(t-r/c)}}{4\pi\varepsilon r}$$

$$= \left(\frac{q_0 e^{j\omega t}}{4\pi\varepsilon}\right) \frac{e^{-jkr}}{r} \tag{17.22}$$

where

$$k = \frac{\omega}{c} = 2\pi\frac{f}{c} = \frac{2\pi}{\lambda} \tag{17.23}$$

The parameter k is called the wave number and is related reciprocally to the wavelength of the wave. The electric field is then given by

$$E(r) = -\frac{\partial\phi}{\partial r} = \frac{q_0}{4\pi\varepsilon} \left(\frac{j}{kr} + \frac{1}{(kr)^2}\right) k^2 e^{+j\omega t} e^{-jkr} \tag{17.24}$$

The radiative far-field term corresponds to the $1/r$ term in Equation (17.24) while the reactive near-field term corresponds to the $1/r^2$ term. The differing phases of these terms, i.e., the j in one term but not in the other, corresponds to the near-field term begin reactive and non-dissipative and the far-field term being resistive and dissipative. The near-to-far-field transition is said to occur when the magnitude of these two terms are equal, i.e., when

$$kr_{crit} = 1$$

$$r_{crit} = \frac{\lambda}{2\pi} \tag{17.25}$$

We can think of r_{crit} as the monopole antenna's capture radius. The effective capture area for an isotropic antenna such as the monopole, which uniformly captures photon wave packets in all directions, is then given by

$$A_{eff} = \pi\left(\frac{\lambda}{2\pi}\right)^2$$

$$= \frac{\lambda^2}{4\pi} \tag{17.26}$$

Within this capture area determined by the edge of the near-field zone, there is a high probability that any photon wave packet near the antenna will be captured by it.

17.7 Far-field coupling: basics of dipole antennas

Figure 17.5 shows one embodiment of a dipole antenna, a vertical wire that is broken into two symmetric segments, one pointing left, and one pointing right with a small gap at the center. Short vertical sections of wire or a cable connect to or 'feed' the horizontal wires at the central gap at one end and form the input/ output terminals of the antenna at their other end. In Figure 17.5, the transmitting antenna is driven by a current source and said to be 'center fed'. Dipole antennas are among the most common antennas and are often discussed in textbooks as the first elementary example of a real antenna.

An oscillating current at the terminals of the antenna launches a forward wave of current in the left and right wires that moves from the center till it reaches the open ends of the wire segments. The left wire segment has a 'source' current wave propagating from the center while the right wire segment has a 'sink' current wave. Since both wires are horizontal, the current in both segments is leftward or rightward. The current wave is analogous to the forward wave in a transmission line that propagates at the speed of light. The transmission line is created by the distributed inductance of the wire segments and the distributed capacitance between the two segments. However, the two conducting portions of the trans-mission line are formed by vertical wire segments that carry current in the same horizontal direction in the antenna rather than by conductors that carry currents in opposing directions in transmission lines like co-axial cables. When the forward wave reaches the open end of the wire segments it is reflected and creates a backward current wave that ensures that the boundary condition of zero current at the ends of the wire segments is met. The forward and backward waves superpose to create a standing-wave pattern of current in the wire segments that is equal to the input at the center and that goes to zero at the ends. The oscillating currents in all portions of the wire segments have accelerating charges or changing currents and therefore radiate electromagnetic energy. The radiation serves to make the transmission line lossy even if the wire segments are made up of perfect conductors.

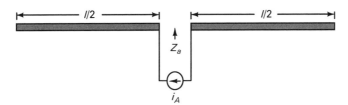

Figure 17.5. A dipole antenna.

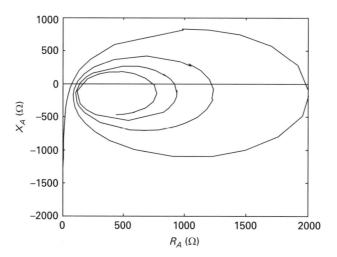

Figure 17.6. Input impedance of a dipole antenna as frequency is increased: a complex distributed transmission line with radiation resistance.

A particularly interesting case occurs when the length of each wire segment $l/2 = \lambda/4$ or equivalently $l/\lambda = 1/2$. This important case is often referred to as that of a $\lambda/2$ dipole. In this case, we have the equivalent of a $\lambda/4$ transmission line for the wire-segment pair, such that by Equation (17.8), the open end of the wire segments is perceived as a short between the input terminals of the antenna. If the transmission line were non-lossy, we would therefore expect to measure zero impedance at the antenna terminals. Since the transmission line is actually radiative and lossy, we expect to measure a purely resistive impedance, the radiation resistance. Since the input impedance of a transmission line is periodic in $\lambda/2$, purely resistive impedances would also be expected for wire segment lengths of $3\lambda/4, 5\lambda/4, 7\lambda/4, \ldots$

Figure 17.6 shows how the actual input impedance of a $\lambda/2$ dipole at its input terminals $Z_A = R_A + jX_A$ varies with l/λ or equivalently with angular frequency ω. For very small l/λ, $Z_A \approx jX_A$ with no resistive term, and X_A is a large negative number indicating a small capacitive input impedance. This finding is consistent with a short length of transmission line terminated with an open at its end having a capacitive input impedance as in Equation (17.10). Near $l/\lambda = 1/2$ (0.47λ to 0.48λ [5]) the impedance is almost purely resistive at the first (leftmost) crossing of the $X_A = 0$ line in Figure 17.6, and $Z_A = R_A \approx 73$ Ω. This finding corresponds to the $\lambda/4$ case. As l/λ varies, the impedance varies in a spiral fashion making successive upward-going zero crossings of the X_A line at nearly $\lambda/2$ intervals. Near the upward-going zero crossings, the impedance of the antenna behaves like that of a series LCR resonator, looking capacitive for frequencies below resonance (the zero-crossing frequency) and inductive for frequencies above resonance. Near the downward-going zero crossings, the impedance of the antenna behaves like that of a parallel LCR resonator, looking inductive for frequencies below resonance and capacitive for frequencies above resonance. The series and parallel resonances alternate as current waves or voltage waves in the transmission line are alternately nulled at the antenna terminals with changes in l/λ as one would expect for a transmission line.

The radiation impedance of a $\lambda/2$ dipole is nearly $73\,\Omega$. Thus, if we have a current of $I_0\sin(\omega t)$, the effective radiated power by the antenna over all directions is given by $(I_0^2/2)(73)$, where the factor of 2 arises from a peak-to-rms conversion. Two other practically useful radiation impedances correspond to those of a short dipole ($l/\lambda \ll 1/2$) and those of a 'small-loop dipole'. The small-loop dipole is the magnetic dipole equivalent of the short electric dipole that we have thus far discussed. A loop dipole is formed by simply using a loop of wire of circumferential length l and with N turns as an antenna. An analysis of the radiation impedance for these two cases shows that [5]

$$R_{rad}^{shrt-dpl} = 2\eta\frac{\pi}{3}\left(\frac{l}{\lambda}\right)^2$$

$$R_{rad}^{loop-dpl} = \eta N^2 \frac{\pi}{6}\left(\frac{l}{\lambda}\right)^4 \qquad (17.27)$$

$$\eta = \sqrt{\frac{\mu}{\varepsilon}} = 120\pi = 377\,\Omega$$

The radiated electric field E scales like l/λ in the short-dipole case since the radiated E field scales like ωl in general, increasing with more current radiators (larger l) and with higher values of charge acceleration (larger ω). Thus, the radiated power density, which scales like $(E^2/\eta)(1/r)$ at any point in space, when integrated over any spherical cross-section in space scales like $(l/\lambda)^2$ and leads to $R_{rad}^{shrt-dpl}$ in Equation (17.27). Just as the electric dipole has its first resonance when $l = \lambda/2$, magnetic dipoles have their first resonance when the total circumferential length of the loop is $\lambda/2$. The first resonance is a parallel resonance for a magnetic dipole rather than a series resonance.

Changing currents in a loop have opposing directions of flow (as the current loops around) such that only first-order differences in the current contributions to the field at a point establish the strength of the radiative field [6]. Hence, in the case of a magnetic dipole, the radiated field has an additional l/λ factor compared with that of an electric dipole. This additional factor, when squared to compute power, leads to an additional factor of $(l/\lambda)^2$ in the radiation resistance of a small-loop dipole compared with that of a short dipole in Equation (17.27). The radiated field strength of a magnetic dipole is proportional to N such that the radiation resistance is proportional to N^2.

17.8 Directional radiation and antenna gain

Real antennas do not radiate uniformly in all directions and by reciprocity are not equally sensitive to receiving photons from all directions. For example,

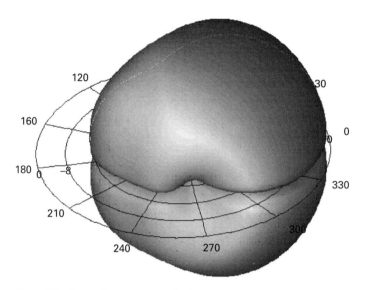

Figure 17.7. Example antenna radiation pattern.

Figure 17.7 shows the radiation pattern of a log-periodic dipole antenna simulated using the software package ADS. We see regions of maximum and minimum radiation in particular directions, which are represented by a gain factor with respect to a hypothetical isotropic antenna that radiates uniformly in all directions and is equally sensitive to receiving photons in all directions. Thus, compared with an isotropic radiator, which has an effective area of $\lambda^2/(4\pi)$ in all directions as we found in Equation (17.26), we can define a gain for, say, a receiving antenna, $G_r(\theta, \phi)$, that is dependent on the angle of reception with higher effective areas for capture of photons in certain directions than in others. Here, θ represents the angle with respect to the z-axis and ϕ represents the angle in the xy-plane as in standard spherical coordinates. That is,

$$A_{eff}(\theta, \phi) = \frac{G_r(\theta, \phi)\lambda^2}{4\pi} \tag{17.28}$$

If a transmitting antenna has a gain of G_t in some particular direction and radiates a power P_t into all of space, then the power received by an antenna with gain G_r in the same direction is given by

$$P_r = \left(\frac{P_t}{4\pi r^2}\right)(G_t(\theta, \phi))\left(\frac{\lambda^2}{4\pi}\right)(G_r(\theta, \phi))$$

$$\boxed{P_r = P_t G_t G_r \left(\frac{\lambda}{4\pi r}\right)^2} \tag{17.29}$$

The latter boxed equation is known as the Friis equation and is perhaps one of the most important equations in antenna design. It reveals that the only way to get large amounts of received power for a fixed power of transmission P_t and fixed distance of communication between transmitter and receiver of r is to increase the gains of either the transmitter or receiver in the direction that corresponds to the line between them or to decrease λ such that there is a larger effective area for capture. It is hard to get enormous gains in particular directions without highly complex or large antennas. Thus, for power-efficient long-range communication, the Friis equation tells us that we need to increase λ by operating at a lower carrier frequency. The tradeoff is that operating at lower carrier frequencies results in a lower maximal modulation bandwidth for data transmission. In practice, the situation is even worse than that predicted by the Friis equation since multi-path effects due to reflections and absorptions of radio waves often lead to a $1/r^n$ falloff with n ranging from 2 to 4 rather than just 2 as predicted by an inverse-square law for free space. We shall discuss other practical tradeoffs that become important in choosing a carrier frequency in the next chapter.

A short dipole has a $(3/2)\sin^2(\theta)$ gain such that maximal radiation occurs when θ is at $90°$ with respect to horizontal dipole axis in Figure 17.5 and no radiation occurs along the dipole axis. The far-field dipole radiation is also polarized with E fields oriented along the $\hat{\theta}$ direction and H fields oriented along the $\hat{\phi}$ direction. A $\lambda/2$ dipole is only marginally more directive than a short dipole with a maximum gain of 1.64 rather than 1.5 at $\theta = 90°$. Loop antennas have the roles of H and E reversed compared with that of an electric dipole. That is, H is oriented along the $\hat{\theta}$ direction and E is oriented along the $\hat{\phi}$ direction when the loop's rotational axis is along the horizontal. Both electric dipoles and loops have no radiation along the horizontal axis.

A well-designed antenna has the resistive component of its impedance dominated by its radiation resistance. Inevitably, however, because wires are not perfect conductors, some energy is wasted as heat dissipation in the resistance of the wire rather than being radiated. Thus, the antenna has a radiation efficiency η given by

$$\eta = \frac{R_{rad}}{R_{rad} + R_{cond}} \qquad (17.30)$$

where R_{rad} is the radiation resistance and R_{cond} represents the resistance due to conduction losses in the wire. In fact, in antenna literature, what we have called 'gain' is called directivity. The gain is strictly the product of the directivity and the radiation efficiency.

17.9 Derivation of far-field transfer impedance or Z_{12}

For the near-field coupling case, the derivation of Z_{11}, Z_{12}, and Z_{22} is straightforward and given by Equation (17.16). In the far-field case, since Z_{12} is typically very small for all but uninterestingly short communication distances, Equations (17.13) and (17.14) then predict that the transmitting and receiving antenna self-impedances are given by

$$Z_{at} = Z_{11}$$
$$Z_{ar} = Z_{22} \qquad (17.31)$$

to a very high degree of accuracy. Thus, unlike near-field coupling, feedback effects as manifested by reflection impedances between the transmitting and receiving antenna are generally small and unimportant in far-field coupling. The antenna's local impedance is mainly influenced by itself and will have real and imaginary parts as in Figure 17.6. But how do transmitting and receiving antenna's influence each other, i.e., what is Z_{12}?

If both the transmitting antenna and the receiving antenna are driven with or loaded with matched impedances respectively, i.e., with impedances whose values are conjugate to the impedances of the respective antennas, then there is maximal power transfer and

$$Z_g = Z_{at}^* \approx Z_{11}^*$$
$$Z_l = Z_{ar}^* \approx Z_{22}^*$$

(17.32)

in Figures 17.4 (a) and 17.4 (b). Thus,

$$P_t = \frac{V_g^2}{8\operatorname{Re}(Z_{11})}$$

$$P_r = \frac{V_r^2}{8\operatorname{Re}(Z_{22})}$$

(17.33)

The factor of 8 in Equation (17.33) arises partly because of a factor-of-2 resistive-divider effect in impedance-matched versions of the circuits of Figure 17.4 (a) and Figure 17.4 (b), which leads to a factor of 4 in power units, and partly because of an amplitude-to-rms conversion, which leads to another factor of 2. By the Friis transmission formula of Equation (17.29), we can write

$$P_r = G_t G_r \left(\frac{1}{2kr}\right)^2 P_t$$

(17.34)

Substituting Equation (17.33) into Equation (17.34) yields

$$\frac{V_r^2}{V_g^2} = G_t G_r \left(\frac{1}{2kr}\right)^2 \frac{\operatorname{Re}(Z_{22})}{\operatorname{Re}(Z_{11})}$$

(17.35)

The known relationship between V_g and V_r is given by

$$\frac{V_r}{V_g} = \left(\frac{Z_{12}}{Z_{11} + Z_g}\right) = \frac{Z_{12}}{2\operatorname{Re}(Z_{11})}$$

(17.36)

Solving Equations (17.35) and (17.36) then yields

$$\boxed{Z_{12} = \sqrt{G_t G_r \operatorname{Re}(Z_{11}) \operatorname{Re}(Z_{22})} \left(\frac{1}{kr}\right)}$$

(17.37)

The value of Z_{12} is purely real because power transfer occurs only via a resistive radiation impedance. Not surprisingly, large gains, large radiation impedances, and close proximity between the antennas increase coupling between them.

17.10   Impedance matching: the Bode-Fano criterion

To ensure that the electromagnetic energy collected by a receiving antenna is transferred to a load maximally, and to prevent this energy from being reflected, the impedance at the terminal of the antenna Z_{ar} should be conjugate to the impedance of the load Z_l. That is

$$Z_{ar} = Z_l^* \qquad (17.38)$$

This requirement is a consequence of Equation (17.5) and the well-known maximal power theorem. If the load has both reactive and resistive components, it is easy to have the antenna's terminal reactance at one frequency cancel the load's reactance at that frequency and have the antenna's radiation impedance at that frequency match that of the load's resistance to achieve such a conjugate match. However, since the load's and antenna's reactance and resistance vary with frequency, achieving a match at one frequency does not mean that such a match will be attained at other frequencies. Ideally, we would like such matching to occur over some bandwidth around a carrier frequency such that an antenna-based power link will function efficiently and robustly even if the carrier frequency shifts due to temperature, if the antenna's resonance shifts due to conductors in its environment, or because fabrication errors cause shifts in the resonant properties of the load and antenna. Also, if the antenna is doubling as a power link and as a data link, a common occurrence in RF-ID systems, reasonably good matching over the bandwidth of data transmission is essential for efficient operation. For example, RF-ID systems need at least 30 MHz of bandwidth to work robustly in the 902 MHz–928 MHz band in the United States.

To achieve relatively wideband matching, impedance-matching networks interposed between the antenna and the load can help transform the impedance of the load as seen by the antenna such that matching is achieved over a range of frequencies. In particular, for the important practical cases of parallel or series RC or RL loads, alternating lossless series and parallel LC impedances can provide an effective conjugate match to the load over a range of frequencies. Lossless transformers can help scale the residual resistive component of the load to that of the antenna. A fundamental theorem referred to as the Bode-Fano criterion states that for passive loads, such as series or parallel RC or RL loads, no matter how complex or high order the matching network is, there is a tradeoff between achieving good matching, i.e., achieving values of reflection coefficient $|\Gamma(\omega)|$ near zero, and achieving matching over a wide bandwidth [7]. If you want to be extremely well matched, i.e., you want $|\Gamma(\omega)|$ to be almost zero, then you can only attain such matching over a narrow bandwidth. If you relax the requirements on matching such that $|\Gamma(\omega)|$ is small but not very near zero, then you can attain such matching over a wider bandwidth. In fact, there is a net area tradeoff that limits the performance of all matching networks similar to that of a gain-bandwidth tradeoff: the integral of the logarithm of

$1/|\Gamma(\omega)|$ over all frequencies is bounded for any and all matching networks and given by

$$
\int_0^\infty \ln\left(\frac{1}{|\Gamma(\omega)|}\right) d\omega \le \frac{\pi\omega_0}{Q_{L0}}
$$
$$
= \frac{\pi}{RC}
$$
$$
= \frac{\pi R}{L} \tag{17.39}
$$

The first bound on the right-hand side of Equation (17.39) applies to all passive series or parallel RC or RL loads. The latter two bounds are specific instantiations of the more general formula to RC or RL loads. For passive loads, $0 < |\Gamma(\omega)| < 1$.

If the antenna outputs are fed to an integrated-circuit chip that rectifies it to create a power supply, the most common load that the chip presents to the antenna is an RC load. The capacitance arises from MOS gate capacitances, package, and bonding capacitances. The resistance arises from the RF current effectively drawn by the rectifier to power other circuits on the chip, from finite conductance effects in metal wires, and from polysilicon wires that form the gates of transistors or that are used for connectivity.

The Bode-Fano criterion of Equation (17.39) predicts that perfect matching, i.e., $|\Gamma(\omega)| = 0$, is only possible at discrete frequencies such that the sum of impulse areas at these frequencies adds up to a finite value. The best possible matching networks attempt to reach the Bode-Fano limit by making $|\Gamma| = \Gamma_{max}$, a maximally acceptable level of reflection over $\Delta\omega$, the bandwidth of interest around ω_0, and $|\Gamma| = 1$ everywhere else. In this ideal 'box-car' case, Equation (17.39) reduces to

$$
\frac{\Delta\omega}{\omega_0} Q_{L0} = \frac{\pi}{\ln\left(\frac{1}{\Gamma_{max}}\right)}
$$
$$
B_\infty Q_{L0} = \frac{\pi}{\ln\left(\frac{1}{\Gamma_{max}}\right)} \tag{17.40}
$$

The fractional bandwidth, B_∞, is so called because to realize the ideal Bode-Fano bound, we need a matching network made up of an infinite number of L's and C's. This requirement is analogous to the need for an infinite number of L's and C's needed to construct an ideal box filter or an infinite number of Fourier coefficients needed to construct an ideal square wave.

How close are simple first-order (one L and a C) and second-order matching networks (two L's and two C's) to the Bode-Fano bound? An analysis in [7] reveals that first-order and second-order matching networks have fractional bandwidths B_1 and B_2 given by

$$
B_1 Q_{L0} = \frac{2}{\frac{1}{\Gamma_{max}} - \Gamma_{max}}
$$
$$
B_2 Q_{L0} = \frac{2}{\frac{1}{\sqrt{\Gamma_{max}}} - \sqrt{\Gamma_{max}}} \tag{17.41}
$$

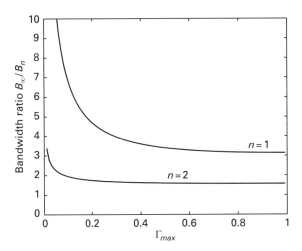

Figure 17.8. First-order and second-order matching network performance.

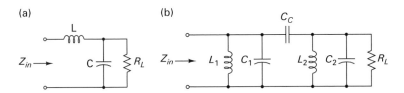

Figure 17.9a, b. First-order and second-order matching networks.

Figure 17.8 plots B_∞/B_n as a function of Γ_{max} for $n = 1$ and $n = 2$ corresponding to Equation (17.40) and Equation (17.41). It reveals that ideal Bode-Fano networks have a fractional bandwidth that is at least 4 times greater than those of first-order networks for $\Gamma_{max} < 0.25$. Ideal Bode-Fano networks have a fractional bandwidth that is $\pi/2$ to 2 greater than those of second-order networks for almost all Γ_{max} unless Γ_{max} is very small. Thus, second-order matching networks are a good compromise between moderate complexity and some loss of fractional bandwidth since they are already within a factor of 2 of what is theoretically achievable. Higher-order networks yield increasingly diminishing returns in fractional-bandwidth recovery at the cost of significantly higher complexity.

Figure 17.9 (a) shows a first-order matching network, a series LC network that transforms R_L from a large value in parallel with C to a small series value as seen at the antenna. The transformation is described by the following equations and is only valid for a small frequency range $\Delta\omega \approx \omega_0/Q$ around the carrier frequency ω_0 with Q being at least 4.

$$Z_{in} \approx \frac{R_L}{Q^2} = \frac{1}{R_L(\omega_0 C)^2}$$

$$Q = \frac{R_L}{\sqrt{\dfrac{L}{C}}}, \quad \omega_0 = \frac{1}{\sqrt{LC}} \tag{17.42}$$

The LC network provides resonant voltage amplification of approximately Q from the input to the output (voltage across R_L). Since no net power is dissipated in the reactive L and C elements, the power dissipated at the input must be dissipated in R_L. Thus, the RF current gain from input to output must be $1/Q$ by energy conservation. Hence, the resistive impedance seen at the input is just R_L transformed by the inverse ratio of the voltage gain to current gain of the LC network, i.e., $(1/Q)/Q = 1/Q^2$. Equation (16.12) in Chapter 16 provides a more general and more exact derivation of such parallel-to-series impedance transformations in resonant and nonresonant RF systems. In many practical cases, the L in this matching network typically arises from the antenna's reactive impedance itself and the C is set by the capacitance seen at the input of a chip that the antenna is connected to. Thus, the implicit matching network that is created transforms R_L to the radiation impedance of the antenna.

Figure 17.9 (b) shows an example of a second-order matching network consisting of two resonators coupled by a capacitance. Coupled resonators provide a broader bandwidth of match than first-order networks for the same gain, which is measured by $1/\Gamma_{max}$ in the context of impedance matching. Intuitively, the broadening arises because of their propensity to split resonances during coupling such that a flat band pass characteristic is created between the split resonances. Chapter 16 provided an example of a coupled-resonator architecture with such properties.

17.11   Making the antenna and the load part of the matching network

An antenna is a complex distributed circuit that is often well modeled by lumped-resonator analogs [8], [9]. For example, an electric dipole with length l can be modeled by a series LCR circuit with R being the radiation impedance given by Equation (17.27) for a short dipole or a value nearing $73\,\Omega$ for a $\lambda/2$ dipole, L being proportional to the inductance between the two wire segments, and C being proportional to the capacitance between the wire segments. For an electric dipole of total length l and wire thickness of w, a good approximation for L and C for $l < 0.3\lambda$ is given by

$$L \approx \frac{(1.5)\mu_0 l}{2\pi}\ln\left(\frac{l}{w}\right)$$

$$C \approx \frac{(1.3)\varepsilon_0 l}{\ln\left(\dfrac{l}{w}\right)}$$

(17.43)

Equation (17.43) reveals that the total capacitance and total inductance are proportional to the capacitance and inductance per unit length for a co-axial line (Equation (17.12)) times the length of the line. Equation (17.43) suggests that an electric dipole is indeed like a lossy co-axial transmission line whose conductors

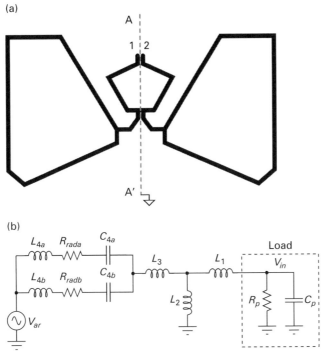

Figure 17.10a, b. Bugs Bunny antenna (a) and coupled-resonator analog (b). Figure (a) reproduced with kind permission from [10] (©2007 IEEE).

have been opened out from being near each other to being spread out and pointing away from each other. Note that the radiation resistance for a short dipole is proportional to the square of the frequency due to the inverse-squared dependence on λ in Equation (17.27). Thus, the resistance is not constant with frequency as in a traditional series LCR circuit.

A natural way to create a relatively broadband antenna is to let the distributed reactance of the antenna and the lumped reactance of the load form an implicit matching network between the antenna radiation resistance and the load resistance. The antenna, matching network, and load are then all part of one structure and appear less separable. While a dipole is well approximated by a simple series LCR circuit, more complex antennas that incorporate loops, dipoles, and distributed conductive structures within them can serve to create a more complex implicit matching network. We now provide an example of how a second-order matching network is implicitly created by an antenna and a load such that the radiation resistance of the antenna is matched to the resistive impedance of a load.

Figure 17.10 (a) shows the layout of a 'Bugs-Bunny-like' antenna composed of two 'ears', a small loop, and two curved chair-like wire segments that couple each of the ears to the small loop. The terminals of the antenna are marked as 1 and 2 in Figure 17.10 (a) and form the input to the small loop. The terminals of the antenna are connected to two pins of a surface-mounted chip on a printed circuit board that the antenna is also printed on. The chip houses the rectifier circuitry that converts the electromagnetic energy gathered by the antenna to create a dc power supply. The planar antenna is built on a 31 mil (1 mil $= 25\,\mu$m) thick FR-4 printed circuit board with antenna wire thicknesses being about 40 mil. The overall size of the antenna is 2.8 inches $\times$ 1.7 inches. The maximum gain of the antenna was measured to be 1.5 dBi at 900 MHz, near the intended carrier frequency of operation.

The overall radiation pattern of this antenna is similar to that of a $\lambda/2$ dipole antenna oriented perpendicular to AA$'$ with each ear forming a wire segment of the dipole. Nulls in the radiation pattern occur where the ears point as they would in such a dipole antenna. Maximum radiation occurs in the ground plane orthogonal to the ear-to-ear axis. Each ear has two halves that split from the chair-like wire segments at a fork and that then rejoin at a point that is at a distance of $\lambda/4$ from the fork. Each of these halves may be viewed as being a $\lambda/4$ dipole segment with the joining point being the open where the current in each half goes to zero. The two $\lambda/4$ half segments of the ears each serve to transform the infinite impedance at their joining points to a short between the fork points of the two ears.

From Equation (17.11), a small length of transmission line as formed by the two chair-like segments with a shorted termination impedance looks inductive at its input. Thus, the short-length chair-segment transmission line transforms the short between the fork points to an inductance between the connection points to the small loop. The small loop may be modeled as a distributed inductance with a small fraction of the inductance manifested between the connection points to the chair-like segments and the rest manifested as series inductances to terminals 1 and 2. The chip itself has a parallel R_L and C_L that form the load across terminals 1 and 2.

A circuit model of the antenna is shown in Figure 17.10 (b). A half circuit that represents one symmetric half of the antenna around the grounded AA$'$ axis is shown. An identical half circuit models the other mirrored half. The series $L_{4a}C_{4a}R_{4a}$ and $L_{4b}C_{4b}R_{4b}$ resonators model the electric dipole resonances due to the two forks in a ear with R_{4a} and R_{4b} representing the radiation resistance of each dipole. These resonators have slightly different parameters from each other due to the asymmetry of the two forks and their coupling leads to antenna bandwidth broadening due to coupled-resonator action. The coupling inductance due to the transmission line formed by the chair-segments with a shorted termination is represented by the L_3 element. The distributed inductance of the loop is represented by L_2 and L_1. The inductance L_2 is primarily due to a short length of wire that couples chair-segment ends in each antenna half to one another in and near its small loop. The inductance L_1 is primarily due to the small loop segment in each antenna half. The load impedance at the chip input is represented by R_p and C_p to emphasize that they are mainly created by chip parasitics. Note that all portions of the antenna contribute to the radiation resistance but the dominant contribution arises from the ears such that we have ignored the radiation resistance elsewhere.

In effect, we have two series resonators created by the antenna ear coupled via an inductive T-network to the RC impedance of the load. Thus, a high-order matching network matches the radiation resistance of the antenna to the resistance of the load. The presence of L_2 increases attenuation in the signal coupled by L_3, and reduces the effective coupling of the antenna to the chip. Thus the value of L_2 is critical in determining the impedance transformation ratio between $R_{rada} \parallel R_{radb}$ and the load R_p, and provides a design degree of freedom.

For the antenna shown in Figure 17.10, a reasonably good fit to the real and imaginary parts of Z_{11} and Γ, which was obtained from a 2.5 D electromagnetic simulation with ADS, occurred for $L_{4a} = 22\,\text{nH}$, $L_{4b} = 22.5\,\text{nH}$, $C_{4a} = 1\,\text{pF}$, $C_{4b} = 1\,\text{pF}$, $R_{rada} = 12\,\Omega$, $R_{radb} = 12\,\Omega$, $L_1 = 10\,\text{nH}$, $L_2 = 1.5\,\text{nH}$, $L_3 = 2\,\text{nH}$, $R_p = 750\,\Omega$, and $C_p = 2.7\,\text{pF}$. For these parameters, the antenna is resonant near 915 MHz.

17.12   Rectifier basics

After impedance matching, a fraction of all the available RF power P_{RF} at the antenna, $(1 - |\Gamma|^2)$, is delivered to the effective resistive load at the input of the rectifier, R_p, to create a voltage with RF amplitude V_{in}. That is

$$\frac{V_{in}^2}{2R_p} = \left(1 - |\Gamma|^2\right)P_{RF}$$

$$V_{in} = \sqrt{2\left(1 - |\Gamma|^2\right)P_{RF}R_p} \tag{17.44}$$

In the far field, the Friis transmission formula applies such that the received power depends on the gain of the receiving and transmitting antennas and on the power of the transmitting antenna.

$$P_{RF} = \frac{1}{2}G_rG_t\left(\frac{\lambda}{4\pi r}\right)^2 P_t \tag{17.45}$$

The factor of $(1/2)$ arises because P_{RF}, the available power, is only half of the total received power when the antenna impedance is matched to the load.

For RF-ID applications, the maximum transmission power P_t that is allowed in the 902–928 MHz band is 4 W or 36 dBm EIRP (equivalent isotropic radiated power). The gains G_t and G_r need to be fairly uniform in all directions such that the tag can detect the transmitted energy from any direction as long as it is sufficiently close to the transmitter.

Thus, if we assume that G_t and G_r are near 1, and the tag's available power is to be at least $10\,\mu\text{W}$, a 10 m read range is possible. A gain of 1.5 dB at the transmitting and receiving antennas can increase this range, and lower-power operation of the tag increases range even further. In an efficient rectifier a significant fraction of the available power is dissipated in the dc load at the output of the rectifier, which transforms to an effective dissipative RF load at the input of the rectifier. The rest of the available power is dissipated in the rectifier circuits and parasitic resistances.

It is hard to build a rectifier that operates at low RF power levels because rectification is a nonlinear process and at low signal levels most devices appear linear. Most rectifiers built with diodes and transistors with exponential non-linearities have an abrupt 'dead zone': if V_{in} is below a certain threshold, no output dc supply is created; if V_{in} exceeds the threshold, the dc output power increases with the RF input power. This dead-zone effect is a manifestation of the 'abrupt-turn-on' approximation often used to model the threshold voltage of diodes or of MOS transistors, one or both of which are present in most rectifiers. Equation (17.44) reveals that R_p needs to be large to ensure that V_{in} crosses the dead-zone threshold at low levels of received power P_{RF}. High values of R_p however, reduce the limiting $1/(R_pC_p)$ bandwidth of match for Γ given by the Bode-Fano criterion (Equation (17.39)). Thus, low values of C_p and high values of R_p are desirable for simultaneously attaining a low power-up threshold and high bandwidth of match. High values of R_p are practical in micropower systems where the reflected impedance of the output load is large at the input of the rectifier. Low values of C_p are practical in deep submicron processes where the input gate capacitances of transistors that comprise the rectifier are small.

Schottky diodes with low turn-on voltages near 0.2 V are popular for building rectifiers. However, many CMOS processes do not possess Schottky diodes. If the cost of an RF-ID tag is to be small, it is important that the tag function with circuits that are cheap and practical in any CMOS process. Therefore, we shall focus on an all-MOS rectifier that is efficient and that can be fabricated in any CMOS process.

Figure 17.11 shows a rectifier block diagram built out of N stages, where $N = 3$ in the figure. The rectifier receives an RF input V_{in} and creates an output dc voltage V_L on the load capacitor C_L. The cascading of N stages accumulates

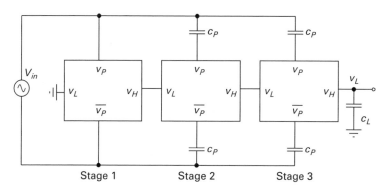

Figure 17.11. Rectifier block diagram. Reproduced with kind permission from [10] (©2007 IEEE).

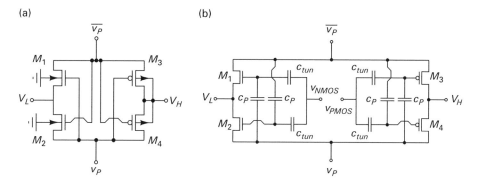

Figure 17.12a, b. CMOS H-bridge. The basic four transistor cell (synchronous rectifier) has a fully differential structure and is suitable for use with balanced antennas. The four transistor cell using floating gate transistors allows transistor threshold voltages to be adjusted via bidirectional tunneling through C_{tun}. Reproduced with kind permission from [10] (©2007 IEEE).

dc voltage across each stage of the rectifier in a charge-pump-like fashion such that the output dc voltage is increased. In the instantiation of the rectifier shown in Figure 17.11, the first stage is dc coupled to the RF voltage while the second and third stages of the rectifier are capacitively coupled. The output dc voltage V_L, input impedance Z_{in}, and output impedance Z_{out} of the rectifier may be approximated by

$$V_L \approx N(V_{in} - V_{drp})$$

$$Z_{in} \approx (NC_{stage})||(R_{in-stage}/N) \tag{17.46}$$

$$Z_{out} \approx NR_{out-stage}$$

The voltage V_{in} is the amplitude of the ac voltage. The parameter V_{drp} empirically models the fact that each stage of the rectifier is not ideal. Thus, it only converts the peak voltage of its input signal into a dc increment with some loss V_{drp} due to a dead-zone turn-on voltage, finite saturation voltages of transistors, and incomplete refreshing of capacitor voltages that have drooped due to load-current and reverse-current losses.

Figure 17.12 (a) shows the CMOS H-bridge circuit that is used to form a stage of the rectifier. The voltages v_P and $\overline{v_P}$ represent ac voltages of opposite phases. When v_P is high and $\overline{v_P}$ is low, M_1 turns on to drive V_L near the low value of $\overline{v_P}$ and M_4 turns on to drive V_H near the high value of v_P, while M_2 and M_3 are off. When v_P is low and $\overline{v_P}$ is high, M_3 turns on to drive V_H near the high value of $\overline{v_P}$ and M_2 turns on to drive V_L near the low value of v_P, while M_1 and M_4 are off. Thus, like in a full-wave rectifier, on any phase, V_H is always driven to the higher value of v_P or $\overline{v_P}$ while V_L is always driven to the lower value of v_P or $\overline{v_P}$. The actual operation is of course less digital than this intuitive description of the circuit's operation and v_P and $\overline{v_P}$ are actually sinusoidal voltages.

The dead-zone voltage for turn on is determined by the threshold voltages of the transistors. A small threshold voltage is beneficial in reducing the power-up threshold of the rectifier but results in more leakage current and higher undesirable reverse current in the rectifier when devices are meant to be off. Thus, there is an optimum threshold voltage for the transistors [10]. The use of a switch-like H-bridge configuration rather than a classic diode-based configuration for rectification is beneficial. When switching transistors conduct, when their gate-to-source voltage exceeds a threshold, they continue to conduct strongly to make their drain and source voltages equal in the switch until the drain-to-source voltage becomes smaller than a small saturation voltage, V_{DSAT}, at which point the conduction strength of the switch falls. In contrast, a diode-like configuration implemented with the same transistors has weak conduction as soon as the drain-to-source voltage drops below the gate-to-source threshold for turn on, which is at a relatively higher diode-drop voltage rather than at a small V_{DSAT}. Hence, diode-based configurations have slower charging times and suffer from larger dynamic and static voltage-drop losses than switch-based configurations as in Figure 17.12.

If v_P and $\overline{v_P}$ are capacitively coupled rather than direct coupled as in Figure 17.11 or in Figure 17.12 (b), and the dc value of V_L is established by a prior stage of the rectifier, then the value of V_H is given by

$$V_H \approx V_L + (V_{in} - V_{drp}) \tag{17.47}$$

where V_{in} is the amplitude of v_P.

Thus, a cascade of capacitively coupled H-bridge stages can function like a charge pump with the V_H of one stage becoming the V_L of the successive stage such that dc voltage is accumulated across several stages. The value of V_{drp} is higher than the drain-to-source saturation voltage of the transistor because load-current losses propagate back from the final rectifier stage and undesirable reverse-current losses cause incomplete charging of the V_L and V_H nodes in each rectifier stage. The body effect increases the threshold voltage of the NFET transistors of Figure 17.12 (a) such that each stage becomes less effective at rectification as the dc voltages in it rise. Thus, the average V_{drp} per stage rises with increasing V_{in}, increasing load current I_L, and with N. The dependence of V_{drp} on I_L is gentle since the $I-V$ curves of transistors are expansive (exponential or square law) such that the voltage drop needed to support a given current is compressive in the current (logarithm or square root).

Figure 17.12 (b) illustrates a floating-gate version of an H-bridge circuit where v_{NMOS} and v_{PMOS} are high-voltage signals that establish charge on the gate of the transistors via tunneling. Chapter 6 provides a brief discussion of tunneling. Such floating-gate circuits may be used to adjust the threshold voltages of the transistors and thus adjust their overdrive voltages and V_{DSAT} to an optimum value [10].

17.13   Rectifier analysis and optimization

To maximize rectifier efficiency, i.e., minimize the value of P_{RF} at which we can deliver the required dc power to the load $P_L = V_L I_L$, we apply Equations (17.44) and (17.46):

$$P_{RF} = \frac{V_{in}^2}{2\left(1 - |\Gamma|^2\right)R_p}$$

$$P_{RF} = \frac{\left(V_L/N + V_{drp}\right)^2 \left(NC_{stage} + C_{package}\right)}{2\left(1 - |\Gamma|^2\right)\left(\dfrac{R_{stage}}{N}\right)\left(NC_{stage} + C_{package}\right)} \tag{17.48}$$

$$P_{RF} \approx \frac{\omega_0 \left(V_L/N + V_{drp}\right)^2 \left(NC_{stage} + C_{package}\right)}{2\left(1 - |\Gamma|^2\right)Q_{L0}}$$

where $C_{package}$ is the package capacitance and Q_{L0}, the Q of the load, i.e., $1/(\omega_0 R_p C_p)$, is assumed to be relatively invariant with N: if N increases, and $C_{package}$ is small, the resistance $R_p = R_{stage}/N$ while the capacitance $C_p \approx N C_{stage}$ such that their product and thus Q_{L0} is relatively constant. The parameter Q_{L0} is only the unloaded Q, i.e. when the rectifier's output voltage looks into an open circuit. The presence of a load decreases Q_{L0} but in low-power circuits losses due to parasitic polysilicon gate resistances of transistors and polysilicon or metal resistances of capacitors are often more important. Such resistances are present in series with gate capacitances and pump capacitances and effectively lead to 'lossy capacitors' with a lowered quality factor of $Q_{cap} = \omega_0 R_{par} C_{cap}$. Changing the areal sizes of these capacitances increases capacitance and lowers resistance thus keeping Q_{cap} constant. However, these parasitic resistances manifest as a parallel equivalent in the input resistance of the rectifier R_p approximately like $R_{par} Q_{cap}^2$ and in the parallel capacitance C_p like C_{cap} (Equation 16.12 and Chapter 16 provide more discussion on series-to-parallel and parallel-to-series transformations in RF systems). Hence, as the areal sizes of capacitors increase, Q_{L0} in the rectifier is relatively constant as C_p increases and R_p decreases. The decrease in R_p, however, degrades the rectifier power-up threshold. Thus, it is important to keep capacitances small to avoid increasing the power-up threshold as Equation (17.48) suggests.

For a given load, i.e., a given desired dc output value of V_L and I_L, maximization of rectifier efficiency involves optimizing N, optimization of the W/L ratio and threshold voltages of the H-bridge transistors, and optimization of the H-bridge pump-capacitor sizes. Measured and simulated data that illustrate such tradeoffs are discussed in [10]. Here, we shall only summarize the key physical insights behind these tradeoffs, many of which are illustrated by Equation (17.48).

1. **Number of stages** – Larger values of V_L or higher load currents that increase V_{drp} require N to be larger such that dc voltage can be accumulated across several stages. However, if N is too large, the parasitic gate resistances in each stage that determine $R_p = R_{stage}/N$, lower R_p and thus increase the power-up threshold.

2. **Threshold voltages of transistors** – High transistor threshold voltages increase the dead-zone voltage, which increases V_{drp} in Equation (17.48) and increases the power-up threshold. However, very low transistor threshold voltages can hurt rectifier performance by increasing reverse-current losses when devices are meant to be off.

3. **Transistor W/L ratios** – A large transistor W/L ratio reduces the output impedance of the rectifier thus making its output voltage less sensitive to changes in its load current. However, very large W/L ratios attenuate signals from the pump capacitors via capacitive division, hurting performance; they also lead to a higher effective input capacitance degrading the power-up threshold. When transistor W/L ratios increase, the input capacitance C_p of the rectifier goes up while its effective input resistance falls due to lowered parasitic gate resistance losses that transform to a smaller value of R_p. The increase in C_p and decrease in R_p keeps Q_{L0} relatively constant in Equation (17.48).

4. **Pump capacitor sizing** – The pump capacitors in the H-bridge should be sized to be large enough to drive the input gate capacitances of the transistors without much capacitive attenuation. However, if they are needlessly large, polysilicon gate resistances will once again transform to a low value of R_p at the rectifier input and degrade its power-up threshold. In addition, bottom-plate parasitic capacitances then increase the value of C_p as well.

17.14   Output voltage ripple in rectifiers

The capacitance C_L at the final output of the rectifier must be large enough to ensure that its dc output voltage does not have too much ripple. The ripple ΔV_{pp} due to load-current (I_L) droop in between RF recharging cycles is given by

$$\Delta V_{pp} \approx \frac{I_L}{2 f_{RF} C_L} \qquad (17.49)$$

This ripple is often small as long as received RF power is always available. A more stringent constraint on the ripple arises if we desire to have low droop in the rectifier for a short period of time t_{low} when the tag may be returning data to a base station by modulating its impedance (impedance modulation is discussed in the next chapter). During this period of time, due to impedance mismatching, the rectifier may be receiving no RF power at all and may need to operate on stored capacitive energy in its load. In this case, the ripple is given by

$$\Delta V_{pp} \approx \frac{I_L t_{low}}{C_L} \qquad (17.50)$$

For $\Delta V_{pp} \leq 10\,\text{mV}$, $I_L = 2\,\mu\text{A}$, $f_{RF} \approx 900\,\text{MHz}$, $t_{low} = 1\,\mu\text{s}$, $C_L \geq 200\,\text{pF}$. Thus, the output capacitor C_L is often the most area-hungry component of a rectifier chip.

17.15   Latchup in CMOS rectifiers

Radio-frequency inputs can transiently forward-bias source and drain junctions in the rectifier's transistors and cause the chip on which it is present to 'latchup' via positive-feedback parasitic-bipolar action. Rectifiers with no backup battery that are solely powered by harvested RF energy are particularly susceptible to this problem: they do not have the luxury of a battery that keeps all substrate and well potentials at well-defined low and high values to prevent latchup.

Two techniques are helpful in combating latchup:

1. The use of well-separated NMOS and PMOS transistors with lots of substrate contacts and guard rings for isolation.
2. The use of a carrier frequency f_{RF} that is significantly greater than the f_T of the parasitic bipolar transistors that cause latchup. In this situation, the positive-feedback loop gain that is necessary for latchup is well below 1.0 such that latchup is prevented. At low levels of received RF power (10 μW or less), and relatively high carrier frequencies (900 MHz), this condition is usually fulfilled.

17.16   Rectifier modeling

Figure 17.13 illustrates the typical features of the input-output characteristic of a rectifier. The input ac voltage V_{in} is rectified to an output dc voltage V_L. At larger values of load current I_L, the load voltage V_L falls due to the finite output resistance of the rectifier, manifested by increasing voltage drops in each rectifier stage. The RF power-up threshold V_{T0}, i.e., the minimum input value of V_{in} needed to create a minimum supply voltage V_{L0}, rises with I_L. The dead zone V_D below which the rectifier appears to have little output voltage also rises with V_L.

Figure 17.14 (a) shows a model of the antenna, matching network, and rectifier. We may define a rectifier gain $G(V_{in})$ and an output resistance $R_{out}(V_{in})$ that represent small-signal slopes of the rectifier input-output curve around the operating point V_{in}, which varies with the available RF power. The input RF energy from the antenna is represented by a Norton equivalent source composed of I_{RF} and Z_0. The parameter Z_0 represents the radiation resistance of the antenna. The reactive impedances in the antenna are absorbed into the matching network that interfaces with the load composed of R_p and C_p in parallel. The value of R_p is determined by parasitic gate resistance losses and the value of C_p is determined by the input capacitance due to pump capacitances and input gate capacitances of transistors.

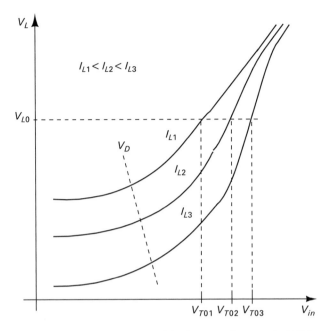

Figure 17.13. Rectifier load model where V_{in} denotes the RF input voltage and V_L denotes the dc output voltage. Reproduced with kind permission from [10] (©2007 IEEE).

The total load current I_L can be modeled as being composed of an analog portion I_a and a digital portion I_d such that

$$
\begin{aligned}
I_L &= I_a + I_d \\
&= I_{b0}\left(1 - \exp\left(\frac{V_L}{a}\right)\right)(1 + bV_L) + \beta C_T V_L^2
\end{aligned}
\tag{17.51}
$$

assuming subthreshold operation for low-power load circuits. The values of a and b are empirically determined but are near V_{DSAT} and the Early voltage for transistors. The value of β is set by the frequency of switching of digital circuits with C_T representing the total switching capacitance. Such switching could arise due to impedance-modulation circuits that are modulating the impedance of the tag in order to be visible to and to send data to a remote reading circuit.

The rectifier transforms the loss resistance of $R_L + R_{out}$ into an effective shunt resistance at its input R_{eff} given by

$$
R_L = \frac{V_L}{I_L}, \quad R_{eff} = \frac{R_{out} + R_L}{2[G(V_{in})]^2}
\tag{17.52}
$$

The $2G(V_{in})^2$ term arises from inverting the rectifier's ac input to dc output curve while conserving rms power. The additional shunt capacitance at the input of the rectifier lowers the effective quality of the load seen by the matching network to

$$
Q_L = \frac{Q_{L0}}{1 + \dfrac{R_p}{R_{eff}}}
\tag{17.53}
$$

(a)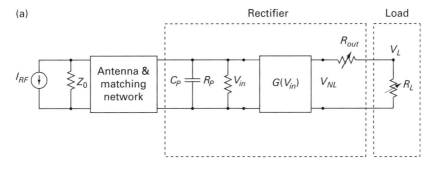

(b)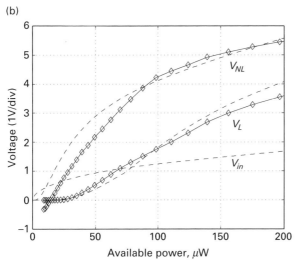

Figure 17.14a, b. Rectifier load figure (a) and data (b). Reproduced with kind permission from [10] (©2007 IEEE).

where Q_{L0} is the no-load quality factor and Q_L is the loaded quality factor. The value of Q_L decreases as the RF input amplitude rises and R_{eff} decreases. The loading due to R_{eff} thus acts as a natural gain control feedback loop in the system. The power conversion efficiency of the rectifier is given by

$$\eta = \frac{V_L I_L}{P_A} \qquad (17.54)$$

Since V_{in} affects R_{eff} and R_{eff} affects V_{in}, the values of V_{in}, V_{NL} (V_L when $I_L = 0$), V_L and R_{eff} are determined through a self-consistent forward-backward iteration procedure until consistent values are found for all parameters as a function of the available RF power P_A at the antenna. We assume a value for V_{in}, determine R_{eff} from known properties of the rectifier and load at V_{in}, use R_{eff} to determine Q_L, use the found Q_L to modify our initial estimate of V_{in} and continue iterating until there is convergence.

17.17   Experimental measurements

Figure 17.14 (b) displays experimental measurements of the no-load output voltage of the rectifier V_{NL} and loaded output voltage V_L ($R_L \approx 500 \, \mathrm{k\Omega}$) versus available antenna power P_A for a complete tag composed of a PCB antenna coupled to a rectifier on a surface-mounted chip. The data shown in Figure 17.14 (b) corresponds to a chip built in a $0.5 \, \mu\mathrm{m}$ technology. The expected square-root dependence of V_{in} on P_A is also shown. Assuming $C_p = 1.2 \, \mathrm{pF}$, $\omega_0 = 2\pi \times 900 \, \mathrm{MHz}$ and $B = 2\pi \times 1 \, \mathrm{MHz}$ (this bandwidth is low enough that the type of matching network used is irrelevant), the predictions of the model built from Equations (17.52), (17.53), and (17.54) are shown as dashed curves in Figure 17.14 (b). The fit of theory to experiment is reasonably good, even with the approximations and simplifications made in the theoretical modeling. The existence of a dead zone of around $25 \, \mu\mathrm{W}$ for the loaded V_L is seen in the flat response of V_L for low values of P_A. The no-load output voltage V_{NL} is higher than V_L as expected. For the data of Figure 17.14 (b), P_A was measured with a network analyzer to estimate actual path loss from the transmitting antenna rather than assuming an inverse-square Friis law.

Equation (17.48) suggests that processes with low parasitic gate resistance due to saliciding, low threshold voltages that allow for smaller dead zones and lower power supply operation, and low values of input capacitance will enable a lower power-up threshold. Thus, deep submicron and small-channel-length processes lead to lower power-up thresholds than in less aggressive processes such as $0.5 \, \mu\mathrm{m}$. Indeed, Figure 17.15 (a) shows that sub-10 $\mu\mathrm{W}$ operation is possible if a two-stage rectifier is built in $0.18 \, \mu\mathrm{m}$ technology. The presence of two peaks in the frequency-response curves are a manifestation of a second-order matching network in the antenna. The frequency-response curves for the rectifier correspond to states where a back-scatter impedance-modulator was active on the chip (MOD $= 0.5 \, \mathrm{V}$) decreasing the performance of the antenna due to intentional impedance mismatching or inactive (MOD $= 0$) such that the antenna had better performance due to existent impedance matching. A $4 \, \mu\mathrm{A}$ load was used for the measurements of Figure 17.15 (a). Thus, if we define the power-up threshold as the minimum RF power needed to create a 0.5 V supply, allowing $2 \, \mu\mathrm{W}$ power consumption, the power-up threshold of this tag is approximately $8.5 \, \mu\mathrm{W}$. If the load is reduced to $1 \, \mu\mathrm{W}$ ($2.5 \, \mu\mathrm{A}$ at $0.4 \, \mathrm{V}$ operation), the power-up threshold of the tag is measured to be $6 \, \mu\mathrm{W}$. We have used the (MOD $= 0$) values, since a sufficiently large C_L in Equation (17.50) ensures that during brief states of impedance mismatch, the tag is still powered up.

Both measured values of the power-up threshold agree well with predictions from theory. The input impedance of the rectifier chip may be measured by directly wire bonding it to a printed circuit board without the antenna. Such measurements yield R_p and C_p from the real and imaginary components of the

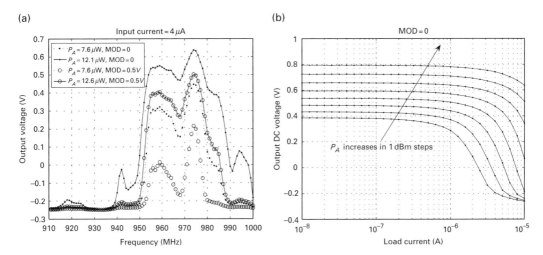

Figure 17.15a, b. Frequency response (a) and load resistance (b) in 0.18 μm. Reproduced with kind permission from [10] (©2007 IEEE).

impedance respectively. At $\Gamma = 0$, simulations with the rectifier model of Figure 17.14 (a) predict $V_{in} = 0.36$ V to just power up a 2 μW load and $V_{in} = 0.29$ V to just power up a 1 μW load. Thus, from the first line of Equation (17.48), the power-up threshold P_{th} is given by

$$P_{th} = \frac{V_{in}^2}{2R_p} \qquad (17.55)$$

With R_p measured to be nearly 8.2 kΩ in both cases, and with the values of V_{in} obtained from simulation, we predict power-up thresholds of 8 μW and 5.2 μW respectively and measure 8.5 μW and 6 μW, respectively.

Figure 17.15 (b) shows measured values of V_L for different values of load current I_L for several different parametric values of received power P_A. For this measurement, the RF frequency was fixed at 970 MHz and MOD was set at 0. We note that as I_L varies, V_L drops due to the finite output resistance of the rectifier. Since the output resistance of the rectifier drops as P_A increases and the switches in the rectifier turn on more strongly, the curves corresponding to larger values of P_A are relatively flat with I_L in Figure 17.15 (b). Decreasing P_A makes the no-load voltage smaller and R_{out} larger which shifts the load curves downward and to the left. The lowest value of P_A in Figure 17.15 (b) corresponds to 3.4 μW and the highest value corresponds to 17.2 μW.

For a 2 μW load, the overall power efficiency of this tag at threshold is about 16.7% and at 23.5% for a 1 μW load. The theoretical read range for a 4 W EIRP transmitter was found to be 21.6 m and 25.7 m for the 2 μW and 1 μW cases.

17.18   Summary

Since we have covered a lot of ground in this chapter, it is worth summarizing its main points. We have discussed how to harvest energy from an RF antenna and rectify it to create a power supply suitable for battery-free ultra-low-power electronics. Such systems are important for medical monitoring in body sensor networks and in telemedicine (see Chapter 20). We have shown experimental measurements from a $\sim$900 MHz RF-ID tag that powers up at 6 μW of received RF power in a 0.18 μm traditional CMOS process with no special devices. These measurements agree well with theories of antenna operation and Bode-Fano broadband impedance-matching networks, and with a general model of rectifier operation that was specifically applied to our CMOS H-bridge rectifier. We will return to themes of energy harvesting in Chapter 26.

References

[1] J. D. Kraus and R. J. Marhefka. *Antennas for All Applications* (New York: McGraw-Hill, 2002).

[2] G. Kron. Equivalent circuit of the field equations of Maxwell-I. *Proceedings of the IRE*, **32** (1944), 289–299.

[3] Thomas H. Lee. *The Design of CMOS Radio-Frequency Integrated Circuits*. 2nd ed. (Cambridge, UK; New York: Cambridge University Press, 2004).

[4] W. Gosling. *Radio Antennas and Propagation* (Oxford, UK: Newnes, 1998).

[5] Constantine A. Balanis. *Antenna Theory: Analysis and Design*. 3rd ed. (Hoboken, NJ: John Wiley & Sons, Inc., 2005).

[6] Carver Mead. *Collective Electrodynamics: Quantum Foundations of Electromagnetism* (Cambridge, MA: MIT Press, 2000).

[7] R. M. Fano. Theoretical limitations on the broadband matching of arbitrary impedances. *Technical Report (Research Laboratory of Electronics, Massachusetts Institute of Technology)*, **41** (1948).

[8] L. J. Chu. Physical limitations of omnidirectional antennas. *Technical Report (Research Laboratory of Electronics, Massachusetts Institute of Technology)*, **64** (1948).

[9] T. G. Tang, Q. M. Tieng and M. W. Gunn. Equivalent circuit of a dipole antenna using frequency-independent lumped elements. *IEEE Transactions on Antennas and Propagation*, **41** (1993), 100–103.

[10] S. Mandal and R. Sarpeshkar. Low-power CMOS rectifier design for RFID applications. *IEEE Transactions on Circuits and Systems I: Regular Papers*, **54** (2007), 1177–1188.

18 Low-power RF telemetry in biomedical implants

I could never accept findings based almost exclusively on mathematics. It ain't ignorance that causes all the trouble in this world. It's the things people know that ain't so.

Edwin Armstrong

Biomedical implants such as cochlear implants, retinal implants, brain implants, cardiac pacemakers, implanted defibrillators, and electronic pills require information to be wirelessly communicated from outside the body to inside the body and vice versa. Wired links from an external electronic unit to an implanted unit inside the body are prone to infection in the long term and are thus unlikely to meet approval by regulatory agencies such as the Food and Drug Administration (FDA). Thus, wireless communication is essential in such implants. In Chapter 16, we discussed how to wirelessly transmit power to such implants via an RF inductive near-field link. Near-field communication is important when the distance of communication, often 1 mm to 10 mm across the skin of the patient, is considerably less than the RF carrier wavelength. In contrast, in far-field communication systems used in most traditional radios, the communication distance is significantly in excess of the carrier wavelength. The relationship between near-field and far-field communication is discussed in quantitative depth in Chapter 17 on antennas and RF energy harvesting.

In this chapter, we shall first focus on how to communicate data via ultra-low-power near-field RF telemetry in such implants. Near-field links can form the final stage of a hybrid link that is composed of a far-field portion to communicate over relatively large distances with a low-power Bluetooth, Ultra Wide Band (UWB), Medical Implant Communication Service (MICS), or Zigbee system followed by a near-field link that efficiently communicates through the body. Near-field links are also the only links, as in current cochlear implants, retinal implants, and some brain implants. Far-field antenna links are well known in RF design, e.g., see any standard text on RF design [1], [2], and their modification for medical applications is relatively straightforward. Therefore, we shall focus more on the critical near-field link in this chapter, which is extremely useful in biomedical implants, and whose design is less well known. We shall exploit the coupled-resonator near-field system described in depth in Chapter 16 to design our data link. However, our focus will be on communicating data rather than power with this system. A review of Chapter 16 is helpful in understanding this chapter in more depth, and we encourage the reader to review this chapter. We conclude with a brief review of antenna-based RF links that communicate with implants through a relatively thick section of biological tissue, e.g., for applications such as electronic pills for gastrointestinal monitoring, or for implants that monitor, say, tumors in the future. Such links work like traditional antenna links but the transmission loss through the body must be taken into account.

It is generally advantageous to separate the power and data transfer functions of a wireless link. Power signals carry no information, and, as discussed in Chapter 16, power transfer efficiency is maximized for narrowband (high-Q) links. Power links that operate at low carrier frequencies to minimize losses in body tissue are also more efficient. Data signals carry information, and therefore data transfer efficiency is maximized for broadband (low-Q) links that are more easily attained at higher operating frequencies. Separating power and data transfer functions, therefore, allows for independent optimization of these functions. In implantable applications where a battery has been implanted and the power link is only operative during periodic bouts of wireless recharging, the functions of the data link and the power link are naturally separable. We shall only focus on the data link in this chapter, assuming that a power link as described in Chapter 16 has already been designed. The techniques described in this chapter can be combined with those described here to design data and power links with one pair of coils, although having separate power and data coils and separate operating frequencies for power and data links should always be explored if space constraints in the implant allow it.

A carrier frequency that is a few tens of MHz represents a good tradeoff between maximizing the bandwidth of the information that can be conveyed via a data link and power-efficient communication that does not result in high loss due to absorption of the RF carrier energy by the conductive body (see Figure 16.18 in Chapter 16). At such carrier frequencies, near-field links are significantly more power-efficient than far-field links, which explains their ubiquitous use in implants.

In implantable biomedical systems, it is beneficial to ensure that as much power is dissipated outside the body in the external portion of the system rather than in the implanted portion of the system that communicates with the external portion via the wireless link. Heat dissipation within the implanted portion can cause potential tissue damage, fever, if the implant is within the brain, and also stresses the cooling mechanisms of the body as the implanted portion is not exposed to the atmosphere. Large power dissipation within the implanted unit leads to the need for large implanted batteries or large implanted coils to provide the needed power and drastically increases the cost and reduces the viability of the implant. Therefore, if it is possible to trade power in the internal unit for power in the external unit, one must do so.

In this chapter, we discuss a wireless link that exploits impedance modulation, a technique that is particularly effective in near-field telemetry, to communicate information from inside the body to outside the body. Changes in implanted-unit impedance accomplished by shorting or opening the implanted resonator cause reflected impedance changes in the external-unit resonator that may be sensed to extract data. Since shorting and opening in the implanted unit may be accomplished with almost zero-power dissipation, power in the implanted unit is minimized as the costs of *both* transmitter and receiver power are borne by the external unit rather than by the implanted unit. Impedance modulation is used in RF-ID far-field systems as well to minimize power consumption at the RF-ID tag such that the costs of *both* transmission and reception are borne by the RF-ID reader and power consumption is minimized at the tag. RF-ID systems have been discussed in Chapter 17 where impedance modulation was referred to as back-scatter modulation, another common term for it. In fact, the MOD $= 0$ and MOD $= 0.5$ V plots in Figure 17.15(a) in Chapter 17 correspond to impedance modulation of the RF antenna. Load modulation is another common term for impedance modulation.

The idea of impedance modulation is similar to that used in sonar, radar, or light imaging of an object. There, one senses the modulated intensity of a reflected incident wave to extract information about the object except that the modulation due to the object occurs passively in space rather than actively in time, as in our system. The power for transmission and detection is borne by the emitting system while the object scattering the incident wave has minimal power consumption. The external unit is analogous to the emitting system and the implanted unit is analogous to the object being 'imaged'.

To send data *to* an implanted unit a receiver must be present in the implanted unit to detect the transmitted data. One may minimize power in the implanted unit by using higher power in the external-unit transmitter and coding the information from the external transmitter in such a fashion that the internal-unit receiver may be built with low power consumption. In particular, pulse-width modulation of transmitted data may be cheaply sensed by a low-power envelope detector in the implanted unit without the need for relatively power-hungry circuits such as phase-locked loops or clock-recovery circuits.

The low-power transceiver described in this chapter uses impedance modulation for the forward-telemetry *uplink*, which extracts data from the implant, and pulse-width amplitude modulation for the back-telemetry *downlink*, which downloads data into the implant. The particular transceiver that we describe here is primarily intended for a brain-machine implant intended to cure paralysis or to wirelessly extract data from the brain for experimental- neuroscience applications. It can be easily modified for use in other biomedical implants by altering design parameters of the link such as the uplink and downlink data bandwidths.

We shall begin by describing a slightly different version of the coupled-resonator system used for power transfer in Chapter 16. We use parallel resonators in both the external primary resonator and the implanted secondary resonator rather than series resonators due to the simplicity of this topology for low-power data transfer. We first discuss the basics of impedance modulation for this topology. Then, we describe circuits of the external-unit impedance-modulation transceiver and circuits of the internal-unit pulse-width-demodulation receiver. We present experimental results for a complete transceiver system that achieves an energy efficiency of 1 nJ/bit for data rates that are a few Mbps. An important effect regarding the asymmetry of rising and falling edges that is inherent to impedance modulation is predicted by theory. Such theory leads to predictions of the bit-error rate in such links, which are confirmed by experiment [3].

We then analyze the fundamental limits of energy efficiency of an impedance-modulation RF communication link. Such limits determine the minimal energy needed to transmit a bit of information. For a given distance of communication, they can be used to evaluate a figure of merit for RF communication links. A low value of transmitter power increases the power cost needed to detect faint signals at the receiver while reducing transmitter power. In contrast, a high value of transmitter power reduces the power cost needed to detect strong signals at the receiver while increasing transmitter power. Not surprisingly, transceiver power is minimized when the power used for transmission and reception is balanced at an optimum value such that one does not spend too much power in either transmission or reception. We shall compute this optimum value. We show how this optimum depends on the strength of the coupling in the link, i.e., whether the resonators are near each other or far apart. Links composed of resonators that are relatively near each other have well balanced transmitter and receiver power dissipation at their optimum and a relatively low energy per bit of information communicated. Links composed of resonators that are relatively far apart require most of the power to be spent in the transmitter at their optimum and a relatively high energy per bit of information communicated at this optimum.

We also outline the pros and cons of using incoherent (no fine carrier phase information is present) envelope-detector-based receivers versus coherent (fine carrier phase information is present) mixer-based receivers. Systems with either kind of receiver minimize total power when transmission and reception power costs are balanced. Coherent receivers are relatively more efficient in strongly coupled near-field links that are capable of a relatively high bandwidth of information flow while incoherent receivers are relatively more efficient in weakly coupled near-field links that are capable of a low bandwidth of information flow.

We use our RF link as an example that illustrates how to evaluate whether a designed system obeys Federal Communications Commission (FCC) regulations. In very-low-power RF links, there is considerably more freedom in choosing an optimal carrier frequency if one can operate at very low power levels that do not violate such regulations.

We then describe seven considerations that determine the choice of RF carrier frequency in a wireless telemetry system. We review RF antenna-based links for implants that operate at relatively high RF carrier frequencies. Such links utilize one far-field RF link to establish communication between an implant within the body and an RF receiver a short distance away from it. They do so primarily because they can exploit small antenna sizes for the implant, are convenient, and have the potential to operate at higher bandwidths. Their primary disadvantage is that transmission of RF energy through the body at these high frequencies is subject to significant transmission loss such that the data bandwidths and energy efficiency of such links have thus far been modest. We conclude by estimating the skin depth for RF wave propagation through the body at various carrier frequencies.

18.1 Impedance modulation in coupled parallel resonators

The coupled parallel resonator topology is shown in Figure 18.1. The inductance L_1 and capacitance C_1 form a parallel resonator in the external primary while L_2 and C_2 form a parallel resonator in the implanted secondary. The resistances R_1 and R_2 are parasitic series coil resistances that determine the quality factor $Q_1 = (\omega_{res}L_1)/R_1$ and $Q_2 = (\omega_{res}L_2)/R_2$ of the primary and secondary resonators respectively at the resonant RF operating frequency ω_{res}. The transistor with a dc bias of V_{BIAS} and a small-signal RF input of v_{in} has a small-signal transconductance of g_m and serves along with the RF load to create a common-source amplifier that amplifies v_{in} to a larger voltage $v_1 = i_{in}R_{eff}$ where R_{eff} is the effective resistance seen at the primary. The mutual inductance $M = k\sqrt{(L_1L_2)}$ serves to provide bidirectional coupling between the primary and secondary such that a voltage v_2 is created in the secondary.

Impedance modulation is accomplished by shorting or opening C_2 in the secondary with '0' or '1' data bits respectively such that the impedance in the

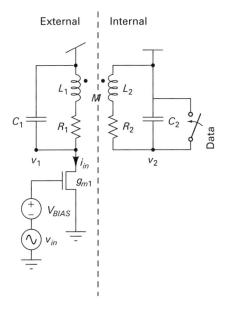

Figure 18.1. A coupled parallel resonator topology. Reproduced with kind permission from [3] (©2008 IEEE).

secondary varies from an inductive value to a finite resonant resistive value. This variation in impedance in the secondary is reflected as a variation in a reflected impedance in the primary that is in series with L_1 and R_1 as we discussed in Chapter 16. From the discussion in Chapter 16, the reflected impedance from the secondary to the primary is given by

$$Z_{prim}^{refl} = -\frac{M^2 s^2}{Z_{scnd}(s)} = \frac{k^2 \omega_{res}^2 L_1 L_2}{Z_{scnd}(s)} \qquad (18.1)$$

Thus, when the secondary is shorted, such that $Z_{scnd}(j\omega_{res}) \approx j\omega_{res} L_2$ the reflected impedance in series with L_1 and R_1 is given by $-j\omega_{res} k^2 L_1$. When the secondary is open, $Z_{scnd}(j\omega_{res}) \approx R_2$, such that the reflected impedance in series with L_1 and R_1 is given by $k^2 Q_1 Q_2 R_1$. Thus, when the secondary is shorted, the net effect in the primary is that

$$L_1 \rightarrow L_1(1 - k^2) \qquad (18.2)$$

When the secondary is opened, the net effect in the primary is that

$$R_1 \rightarrow R_1\left(1 + k^2 Q_1 Q_2\right) \qquad (18.3)$$

Since $k^2 Q_1 Q_2$ is much larger than k^2, the relative impedance modulation of the primary as described by Equation (18.3) when the secondary is open is much larger than the relative impedance modulation of the primary as described by Equation (18.2) when the secondary is shorted. Since k^2 is typically small ($0.02 < k < 0.2$), we can, to good approximation, neglect impedance modulation in the shorted case and assume that the dominant impedance-modulation effect occurs only in the open case. This approximation is a good one except at extremely high Q's. Henceforth, we shall assume that impedance modulation is only present during the open case and described by Equation (18.3) unless otherwise mentioned. The *modulation index m* is defined from Equation (18.3) to be

$$m = k^2 Q_1 Q_2$$
$$R_1 \rightarrow R_1(1 + m) \qquad (18.4)$$

With this definition, and our approximation, we can conveniently set $m = 0$ in the shorted case such that there is no impedance modulation and $m = k^2 Q_1 Q_2$ in the open case. Then, the second row of Equation (18.4) is valid in both cases, and both cases can be analyzed with the use of a single parameter m. We can then set the effective R_1 in the primary to just be $R_1(1 + m)$.

In Figure 18.1, it is useful to convert the series impedance formed by L_1 and $R_1(1 + m)$ to an equivalent parallel combination such that the impedance of the primary transforms to that of a classic parallel $R_{prll} L_{prll} C_1$ resonator and the effective resistance in the primary, R_{eff}, is then easily determined. With the series-to-parallel transformation formula in Chapter 16, Equation (16.12), we can represent $R_1(1 + m)$ and L_1 in the primary resonator by L_{eff} and R_{eff} in parallel, which are given by

$$\boxed{L_{eff} \approx L_1}$$

$$R_{eff} = R_1(1+m)\left[\left(\frac{\omega_{res}L_1}{R_1(1+m)}\right)^2 + 1\right]$$

$$R_{eff} = R_1(1+m)\left[\left(\frac{Q_1}{(1+m)}\right)^2 + 1\right] \qquad (18.5)$$

$$\boxed{R_{eff} \approx R_1 Q_1^2 \frac{1}{(1+m)}}$$

We have assumed that $(Q_1/(1+m))^2$ is much greater than 1 in deriving Equation (18.5), an approximation that is usually valid in practical situations. We notice that shorting the secondary leaves R_{eff} unchanged at $R_1 Q^2 (m = 0)$ while opening the secondary reduces the impedance at the primary by a factor of $1 + m = 1 + k^2 Q_1 Q_2$. Thus, the effective modulation depth, i.e., $(\Delta v_1/v_1)$ with Δv_1 evaluated between the condition of a 0 and the condition of a 1 in the secondary, is given by

$$m_{eff} = \frac{v_1\left(1 - \frac{1}{(1+m)}\right)}{v_1} \qquad (18.6)$$

$$m_{eff} = \frac{m}{1+m}$$

If we want to send data *to* the implant, the switch in the secondary is maintained open while the voltage in the primary is modulated in accord with the data. The voltage v_2 in the secondary then changes as the voltage v_1 in the primary varies. We can derive that the transfer function between v_2 and v_1 is given by

$$m_d = \frac{v_2}{v_1}$$

$$m_d \approx \frac{Ms}{L_1 s} Q_2 \qquad (18.7)$$

$$m_d \approx kQ_2\sqrt{\frac{L_2}{L_1}}$$

If $Q_1 = Q_2$, kQ_2 is $\sqrt{m}$ in Equation (18.7). Thus, the downlink transfer function is proportional to $\sqrt{m}$ while the uplink transfer function, given by Equation (18.6), is proportional to m_{eff} or nearly m if m is much less than 1.0. Since m is proportional to k^2 and k^2 falls with distance steeply as given in Chapter 16, the linear sensitivity of the uplink to distance is much stronger than the square-root sensitivity of the downlink to distance.

18.2 Impedance-modulation transceiver

18.2.1 RF oscillator

To transmit data via the uplink or downlink we must first create an RF carrier frequency to modulate. An oscillator on the external side of an implanted system

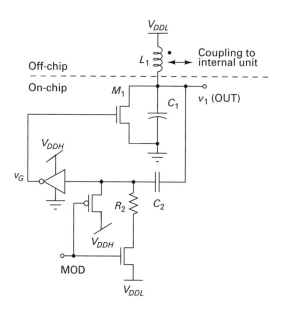

Figure 18.2. The external RF oscillator. Reproduced with kind permission from [3] (©2008 IEEE).

can create such a carrier. The primary L_1C_1 resonator can be incorporated as part of the oscillator circuit such that a separate oscillator circuit and a circuit that drives the primary resonator at this oscillation frequency are not needed. In this fashion, matching issues between the driving circuit's operating frequency and the resonant frequency do not arise. Power is also saved since two separate circuits are not needed.

Figure 18.2 shows how such an oscillator circuit can be built. In this oscillator, the use of positive feedback via M_1 and the inverter overcome the loss in energy due to damping in the resonator. The differentiator-like configuration implemented with R_2 and C_2 provides a predictive positive phase shift that nearly cancels the negative phase shift created by the inverter such that the positive feedback is almost exactly in phase with v_1 [4]. If the inverter delay t_{inv} is to be exactly cancelled by the C_2-R_2 differentiator at the oscillation frequency $1/\sqrt{(L_1C_1)}$, then from simple phase relationships of linear circuits it can be shown that

$$R_2C_2 = \frac{\sqrt{L_1C_1}}{\tan\left(\dfrac{t_{inv}}{\sqrt{L_1C_1}}\right)} \qquad (18.8)$$

Downlink data modulation occurs by turning the oscillator on and off. In Figure 18.2, such modulation occurs by turning MOD on or off via switches such that the oscillation is activated or deactivated by activating or deactivating the positive feedback respectively. The value of V_{DDL} is maintained near $V_{DDH}/2$ such that the inverter operates in its high-gain region. The voltages V_{DDL} and V_{DDH} are easily obtained via stacking of two batteries. By trading various pros and cons as we describe later, the oscillation frequency is chosen to be near 25 MHz. The amplitude of the oscillation v_1 increases with the g_m or equivalently with the W/L of M_1 but is limited by V_{DDH}. The power consumption of the oscillator increases with v_1 as well.

This oscillator circuit of Figure 18.2 is only one example of a tuned oscillator that is useful for this application. Other well-known configurations like the low-phase-noise Colpitts oscillator may also be easily modified to incorporate the primary resonator into the oscillator circuit [1]. The Colpitts oscillator allows the amplitude of the oscillation to be determined by a bias current. Systems with a current reference and switches may then be used to create a simple digital-to-analog (DAC) converter that allows for a programmable bias current and thus programmable amplitude in the oscillator. In contrast, the oscillator of Figure 18.2 requires programmable supply voltages for amplitude control, which is usually less convenient. The focus of this chapter is on issues that involve the design of a complete low-power RF telemetry system for medical implants. Thus, this oscillator or any oscillator will still be useful in understanding the various system tradeoffs that we shall outline.

18.2.2 Impedance-modulation receiver

Figure 18.3 shows an impedance-modulation receiver with the oscillator of Figure 18.2 represented by the OSC block. When TX is low, the RF oscillator shown in Figure 18.2 or 18.3 runs continuously and we receive data from the

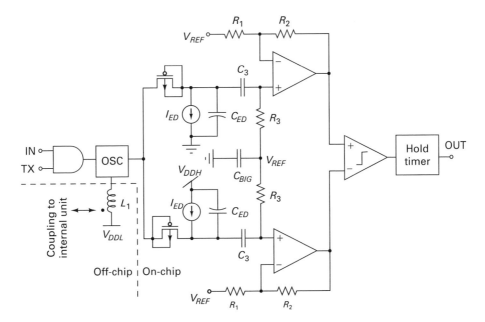

Figure 18.3. An impedance-modulation receiver with the oscillator of Figure 18.2 represented by the OSC block. Reproduced with kind permission from [3] (©2008 IEEE).

implant. The shorting and opening of the secondary resonator in the implant modulates the value of v_1 in Figure 18.2 via impedance modulation. The receiver circuitry that is attached to OSC in Figure 18.3 demodulates and amplifies the amplitude variations in v_1 such that data from the implant is detected.

When TX is high, input data modulate the MOD input of OSC shown in Figure 18.2 and transmit data to the implant, which is demodulated by an implanted receiver that is described later. When TX is high, the circuitry to receive data from the implant to the right of OSC and beyond in Figure 18.3 is shut off to save power.

Since this link can either receive data from the implant (when TX is low) or transmit data to the implant (when TX is high) at any given time but not both, it is referred to as a half-duplex link. It is capable of bidirectional (duplex) communication to and from the implant but not simultaneously as in a full-duplex link. Half-duplex links are efficient for applications where data transfer mostly occurs in one direction, which is true in several implants. For example, external transceivers in paralysis prosthetics mostly receive neural data from the implant and only need to transmit programming and control parameters to the implant occasionally. External cochlear-implant transceivers, on the other hand, typically mostly transmit data to the implant via 'forward telemetry' and periodically receive diagnostic or 'back telemetry' information from the implant.

The circuitry shown in Figure 18.3 contains a pair of differential diode-capacitor envelope detectors that extract the upper and lower envelopes of the impedance-modulated waveform v_1. The extracted envelopes are then ac coupled to a differential amplifier, which amplifies its differential envelope input and feeds its output to a comparator for digitization. A hold timer ensures that noisy digital signals from the comparator that switch too frequently are rejected. The overall differential topology improves the signal-to-noise power by a factor of 2 (3 dB) compared to a single-ended scheme that only tracks one half of the modulated RF signal.

The envelope detectors are implemented by the diode-connected transistors, I_{ED} and C_{ED} in each differential half with the time constant of the envelope detectors given by

$$\tau_{ED} = \frac{C_{ED}}{g_{m,diode}} = \frac{C_{ED}\phi_t}{\kappa I_{ED}} \qquad (18.9)$$

with $C_{ED} = 2.6\,\text{pF}$, $I_{ED} = 2.6\,\mu\text{A}$, and the subthreshold exponential constant $\kappa \approx 0.7$. The time constant τ_{ED} then gets set to a value that is large enough such that most of the 25 MHz RF carrier is filtered out but small enough such that the modulated envelope signal is not significantly attenuated as long as its frequency content is not substantially above 4 MHz.

The high-pass filters formed by C_3 and R_3 in each differential half set a highpass corner frequency of $1/(C_3 R_3)$ with $C_3 = 10\,\text{pF}$ and $R_3 = 1\,\text{M}\Omega$. On a 25 MHz carrier, these settings imply that the envelope cannot remain at a constant '1' or '0' for more than approximately 1000 cycles or the modulated envelope will not be successfully ac coupled to the differential amplifier formed by the operational amplifiers, R_1, and R_2 in each half. The use of ac coupling is beneficial in removing dc offsets from the envelope detectors.

The differential amplifier yields a gain of $G_A = 1 + R_2/R_1 = 6$ with $R_1 = 20\,\mathrm{k\Omega}$ and $R_2 = 100\,\mathrm{k\Omega}$. The standard two-stage operational amplifiers each consume nearly $45\,\mu\mathrm{A}$. An asynchronous comparator that consists of a wide-output-swing operational transconductance amplifier (OTA) digitizes the amplified output and consumes nearly $37.5\,\mu\mathrm{A}$ [5].

The hold timer eliminates pulses that are shorter than a certain duration and provides resilience to noise-induced data transitions. The hold timer is implemented with a current-starved inverter loaded by an output capacitor that is in turn connected to a Schmitt trigger [3]. Pulses of short duration are not successful in charging or discharging the capacitor by amounts that cause the Schmitt trigger to transition between its output states. Thus, such pulses are rejected. The hold time is set to 100 ns for low-to-high and high-to-low transitions.

18.2.3   Phase-locked loop

The demodulated waveform from the hold timer is input to a phase-locked loop (PLL) to extract clock and data information from the waveform. The use of a PLL is beneficial whenever data is encoded with a NRZ (non-return-to-zero) scheme. The use of NRZ encoding maximizes the data rate for a given link bandwidth since at most one transition is needed to encode a data bit on any clock cycle. Examples of NRZ encoding include simple level-based 'high' or 'low' representations of a '1' or '0' bit, respectively. In contrast, RZ (return-to-zero) encoding requires two transitions to encode a data bit on every clock cycle. Examples of RZ encoding include modulation of the width of a pulse on every clock cycle with short pulses representing a '0' and long pulses representing a '1'. We use NRZ envelope-amplitude-modulated encoding in our uplink to maximize information bandwidth. We use RZ pulse-width-modulated encoding in our downlink to minimize power in the implanted receiver.

Since data transitions may not necessarily occur on every clock cycle in NRZ encoding, the data may contain 'missing clock edges', which must be restored by the PLL when it performs clock and data recovery (CDR), necessitating some complexity in the decoding. Since data transitions occur on every clock cycle in RZ encoding, clock and data recovery are simple in RZ-data decoding, require no PLL or complex decoding circuit, and therefore may be performed with low power consumption. In our transceiver, we use NRZ encoding for transmitting high-bandwidth impedance-modulation data from the implant since the higher-power costs of demodulating NRZ data with a PLL are borne by the external unit. We use RZ encoding for transmitting lower-bandwidth pulse-width modulation data to the implant such that the costs of demodulating RZ data are relatively cheap in the implanted unit, which is more power constrained.

PLLs are described in several textbooks, e.g., [1]. The loop that we use is shown in Figure 18.4 and integrates a PLL with a frequency-locked loop (FLL).

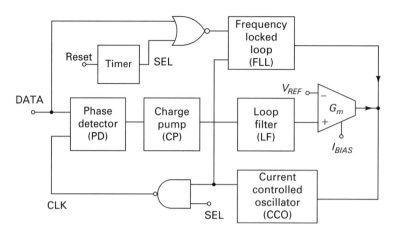

Figure 18.4. The block diagram of the phase-locked loop (PLL) in our system. Reproduced with kind permission from [3] (©2008 IEEE).

At first, we shall describe the PLL and ignore the FLL circuitry, which comprises the timer, NOR gate, and FLL block in Figure 18.4. In PLL mode, the output of the digital current controlled oscillator (CCO) is fed back through the NAND gate with SEL active to the phase detector (PD) to create a classic PLL feedback loop. In FLL mode, SEL is inactive and the PLL feedback is turned off while the FLL feedback is turned on.

The PLL in Figure 18.4 is configured such that the CCO frequency and consequently its rate of phase change is reduced if the oscillator's phase leads the phase of the data. The CCO frequency and consequently its rate of phase change is increased if the oscillator's phase lags the phase of the data. When this negative-feedback loop has locked, the CLK signal tracks the phase of digital signals in the DATA input such that positive edges in the DATA are aligned with positive edges in the CLK signal. The phase lead or lag is measured and extracted by the phase detector (PD) which outputs pulses that cause the charge pump (CP) to increase or decrease its current input to a loop filter (LF). The current inputs of the charge pump are filtered by the loop filter to create a voltage. The voltage is reconverted to a current by the G_m block, and this current is used as a control input to change the frequency of the CCO. The phase detector is designed such that, when there are missing edges in the DATA input, no corrective feedback action is output by it. Thus, the CCO holds its oscillator frequency for a time inversely proportional to the closed-loop crossover frequency of the PLL and regenerates these 'missing edges' to perform 'clock recovery' in its CLK output. The closed-loop crossover frequency of the PLL is primarily determined by the cascaded gain of the PD, CP, LF, CCO and the time constants of the loop filter. The PD is designed such that it also provides 'retimed data' outputs, which are synchronized to the positive edges of CLK, but not shown in Figure 18.4.

Frequency-locked loops use frequency-difference information in addition to phase information to speed up tracking as well. PLLs that do not use frequency

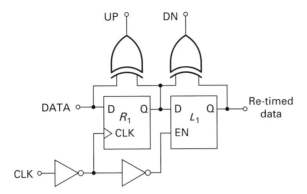

Figure 18.5. The Hogge-type phase detector. Reproduced with kind permission from [3] (©2008 IEEE).

information often converge slowly when the frequency of the CCO and that of the implicit clock in the DATA are significantly different. Purely phase-based PLLs can be confounded by phase-wrapping errors that flip the sign of feedback from negative to positive over some portions of a tracking cycle. That is the reason for the FLL block in Figure 18.4.

Figure 18.5 shows the Hogge phase detector (PD) used in the PLL to sense phase differences between the incoming data and the CCO's output, the CLK signal of Figure 18.4. In Figure 18.5, the PD outputs an UP pulse whenever a data edge in DATA occurs: the XOR goes high only if its two inputs are different and they are always equal except when DATA changes and until the first R_1 register samples and conveys this change to its output. The width of the UP pulse is therefore equal to the difference in arrival time between the data edge in DATA and the succeeding negative edge in CLK, the CCO output. Similarly, the L_1 register in Figure 18.5 outputs a DN pulse whose width is equal to the difference in arrival time between the negative edge and next positive edge of the CLK, i.e., it is always equal to half a clock cycle. The UP and DN pulses command the charge pump to speed up or slow down the CCO such that, when the PLL attains lock, the UP and DN pulses are both equal in width to half a clock cycle, and the data transitions are aligned with positive edges of CLK. The retimed data in Figure 18.5 emerges after a full clock-cycle delay and is aligned with the positive edge of CLK. Missing edges in DATA result in no UP or DN pulses since both XORs in Figure 18.5 output '0' in this case, such that the charge pump provides no corrective action to the loop filter. The loop filter simply holds the prior integrated oscillator command inputs on a capacitor, which therefore cause CLK to maintain its previously established value. Since edges in DATA are aligned with positive edges of CLK, when the loop has locked, the frequency of CLK is twice the maximal rate of the inter-edge arrival times in DATA. Thus, we have a maximum of one DATA edge per clock cycle.

The cascoded charge pump and a third-order loop filter are shown in Figure 18.6. The output voltage of the loop filter is converted to current by a

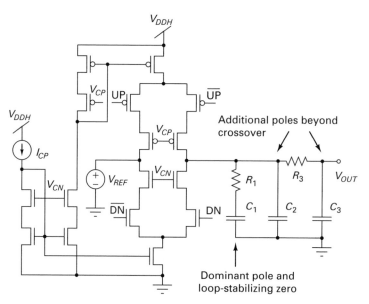

Figure 18.6. A cascoded charge pump and a third-order loop filter. Reproduced with kind permission from [3] (©2008 IEEE).

wide-linear-range (WLR) transconductor with a 1.5 V linear range as shown in Figure 18.4 [6]. The output current of the WLR, I_{CCO}, is fed to a current-starved ring oscillator, the CCO, that can implement a nearly linear current-to-frequency converter with $f_{CCO} = I_{CCO}/Q_{CCO}$. The loop transmission $L(s)$ of the overall loop of Figure 18.4 with the Hogge phase detector of Figure 18.5 and the charge pump and loop filter of Figure 18.6 scales approximately like

$$L(s) \approx \frac{1}{2\pi}\left(\frac{I_{CP}}{Q_{CCO}s}\right)\left(\frac{G_m}{C_1 s}\right)\left(\frac{R_1 C_1 s + 1}{\left(\frac{R_1 C_1}{1+b}s + 1\right)(R_3(C_2 \| C_3)s + 1)}\right); b = \frac{C_1}{C_2 + C_3} \quad (18.10)$$

One of the integrator poles in Equation (18.10) results from frequency getting integrated to phase and the other arises from the integration of the current from the charge pump on the charge pump capacitor C_1. The double integrator at the origin ensures zero steady-state error for phase and frequency tracking. The loop zero, whose location is set by $R_1 C_1$, ensures good phase margin near crossover, in spite of the two integrators at the origin. Two of the loop filter's poles, determined by R_1, R_3, C_1, C_2, and C_3 are placed, with appropriate component values, beyond the crossover frequency to filter high-frequency ripple in the CCO input. Such filtering reduces jitter in the output clock. The reference voltage V_{REF} is set to V_{DDL}. The PLLs crossover frequency is set at 20 kHz, which is nearly 1% of the nominal 2 MHz clock. The loop bandwidth is low enough to ensure good phase margin with the sampling clock delay and to ensure that even long runs of

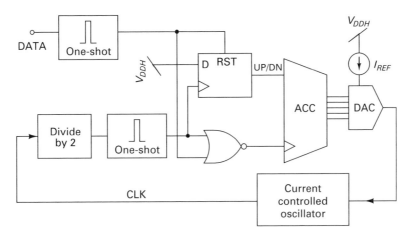

Figure 18.7. The block diagram of the frequency-locked loop (FLL) used in our system. Reproduced with kind permission from [3] (©2008 IEEE).

consecutive 0s or consecutive 1s, which lead to data edges not being present in every clock cycle, do not cause the loop to lose lock.

The rest of the circuitry in Figure 18.4 implements a frequency-locked loop (FLL) that speeds up the lock time of the main PLL during initial startup conditions when its CLK and DATA inputs may have very different frequencies. A timer activates the FLL for a fixed number of clock cycles after a system reset. During this period, the main PLL is disabled: the FLL, loop filter, transconductor, and CCO form a feedback loop and the main PLL loop involving the phase detector and charge pump is inactive. The implanted unit transmits a synchronization sequence consisting of an alternating number of '0's and '1's. The FLL, shown in Figure 18.7, then counts clock and data edges and sets the initial bias current of the CCO via a DAC such that the number of data edges is nearly half the number of clock edges. This initial setting of the CCO bias current to nearly the right value by the FLL is done via a relatively coarse digital adjustment in the DAC that is then refined to more precise locking by the analog PLL when final lock is achieved.

The FLL is shown in Figure 18.7. The digital accumulator's output is used to determine bits in a current DAC that sets the initial CCO bias current. The accumulator is incremented by either rising or falling edges in DATA while the divide-by-2 circuit ensures that only rising clock edges effectively decrement it. The accumulator and DAC have 5 bits each such that the initial frequency error that must be handled by the PLL is less than $1/2^5 \approx 3\%$ of the data frequency output by the implant.

18.3 Pulse-width modulation receiver

Data transmitted to the implant is encoded using a 25/75% pulse-width modulation (PWM) scheme with a counter-based chip modulator. Return-to-zero schemes like

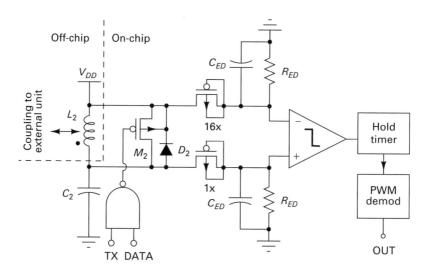

Figure 18.8. The implanted pulse-width modulation receiver. Reproduced with kind permission from [3] (©2008 IEEE).

PWM and PPM (pulse position modulation) are spectrally inefficient but allow for easy low-power decoding in the implanted unit. On-off keying (OOK) is used to turn on or turn off the oscillator in the external unit by modulating the MOD input in Figure 18.2 to create the amplitude-modulated RF pulses in v_1. This modulation is conveyed via the transfer function of Equation (18.7) to create changes in its resonator voltage v_2.

Figure 18.8 shows the implanted pulse-width modulation receiver. The switch M_2 is used for shorting and opening the L_2–C_2 resonator when data is transmitted from the implant but is off when data is being received by the implant. In this section, we will assume that it is always off. A differential envelope detector formed by the diode-connected transistors, R_{ED}, and C_{ED} on each differential half are used to detect modulations in the voltage on C_2. Due to the geometric scaling of the two transistors in the envelope detectors, the comparator is triggered whenever the envelope of the received voltage on C_2 exceeds $(\phi_t/\kappa)\ln(16) \approx 96\,\text{mV}$, which is considerably higher than the typical 10–20 mV offset of the comparator. The parasitic diode D_2 clamps this voltage to be less than approximately 0.6 V. With $R_{ED} = 500\,\text{k}\Omega$ and $C_{ED} = 5.2\,\text{pF}$, modulation envelopes with a bandwidth of nearly 100 kHz can be detected while the high-frequency RF carrier is filtered. The comparator is a wide-output-swing OTA and is biased at a current of nearly 0.5 μA. A hold timer with a hold time of nearly 1μs is used to remove spurious comparator transitions due to noise. The output of the hold timer yields a pulse-width modulated waveform V_{PWM} that is then demodulated by a pulse-width demodulation circuit.

Figure 18.9 shows the pulse-width demodulation circuit and associated waveforms. The charging current I_A charges C_A when V_{PWM} is high while charging current I_B

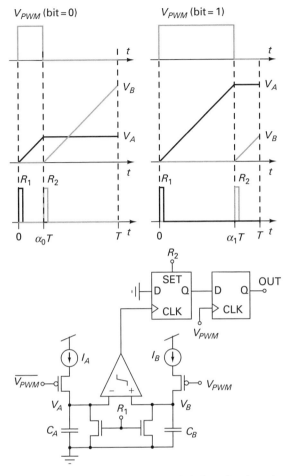

Figure 18.9. The pulse-width demodulation circuit and associated waveforms. Reproduced with kind permission from [3] (©2008 IEEE).

charges C_B when V_{PWM} is low. Both C_A and C_B are reset by a digitally generated one-shot R_1 pulse at the positive edge of V_{PWM}. The one-shot pulse R_2 that is generated at the negative edge of V_{PWM} causes the output of the demodulator to reset to a default that is then changed if V_B exceeds V_A during the charging period of C_B. For 25–75 % PWM, setting $I_A/C_A = I_B/C_B$ ensures an identical magnitude in the differential voltage between V_A and V_B in the case of a '0' or '1'.

18.4   Dynamic effects in impedance modulation

The top row of Figure 18.10 reveals how the envelope of the implanted resonator's voltage v_2 changes when we transition from a shorted configuration to an open configuration and then return to a shorted configuration. We notice that the transition from a short to an open is characterized by an RC-like approach to a

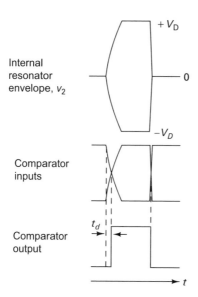

Figure 18.10. The envelope of the implanted resonator's voltage v_2 changes when we transition from a shorted configuration to an open configuration and then return to a shorted configuration. The comparator output transitions as shown in the last row occur when the two differential waveforms are equal and at V_{REF}. Reproduced with kind permission from [3] (©2008 IEEE).

final equilibrium voltage. The RC time constant is given by $2Q_2/\omega_{res}$. The equilibrium is never reached if the voltage v_2 is large enough to asymptote at the diode-clamp voltage V_D. Although diode clamping in Figure 18.8 only occurs when v_2 is larger than V_{DD} by V_D, the bandpass nature of the L_2C_2 circuit transforms asymmetric clamping to symmetric clamping in v_2: the inductor L_2 and capacitor C_2 in Figure 18.8 together form a high-Q filter centered around the resonant frequency when the switch M_2 is open, thus ensuring that the positive and negative envelopes of the voltage across the internal resonator are both equal to V_D as shown in Figure 18.10. Figure 18.10 also shows that the transition from an open configuration to a shorted configuration exhibits no RC-like dynamics and is nearly instantaneous since M_2 rapidly discharges C_2 due to the its low on resistance. In essence, transitions to the high-Q open or '1' configuration of the implanted resonator are slow and RC-like while transitions to the very-low-Q shorted or '0' configuration of the implanted resonator are rapid.

The asymmetric transitions in v_2's envelope in the secondary are reflected as similarly asymmetric transitions in the envelope of v_1 in the primary. This asymmetry leads to inputs to the comparator of Figure 18.3 that appear as shown in the middle row of Figure 18.10. The dc value of the envelope waveforms coupled from each differential half in Figure 18.3 to the input of the differential amplifiers must be at V_{REF} and the ac variations of these differential waveforms around V_{REF} at any instant have equal magnitude but opposite signs. Thus, the comparator output transitions, as shown in the last row of Figure 18.10, occur when the two differential waveforms are equal and at V_{REF}. The asymmetry in the rising and falling edges of the inputs to the comparator leads to the positive edge of the comparator output being delayed with respect to the data edge in the secondary. However, the negative edge of the comparator output is undelayed as shown in Figure 18.10. Note that the dynamics in the primary impose further RC-like dynamics on the secondary envelope with RC being $2Q_1/\omega_{res}$. However, the latter dynamics are symmetric for rising and falling edges and lead to symmetric delays for both positive and negative edges, so we have simply ignored them.

The net effect is that 0-to-1 transitions in the secondary data lead to a delayed response in the comparator output in the primary while 1-to-0 transitions lead to an undelayed response. The length of the first '1' bit in a continuous sequence of 1-bits is consequently shortened. Isolated 1-bits have a shorter pulse width compared with isolated 0-bits. The diode clamping serves to reduce the 0-to-1 delay because RC waveforms start to look like square waves.

From Figure 18.5, we see that the Hogge phase detector samples the input data during the low phase of the recovered clock from the CCO and stores it in the D-register R_1. The stored value appears at the output of the D-latch L_1 half a bit period later when the clock goes high. Therefore, when the PLL is in lock, data is sampled during the low-phase of the clock and data edges become aligned with high-going clock edges. If, due to erroneous delays or noise on some cycle, a data edge becomes delayed with regard to the clock edge by more than half a bit period, it will only get sampled on the next low phase of the clock such that the sampled and reported bit in the current clock cycle becomes erroneous.

If there is no duty cycle distortion in the data, the time window t_w between data transitions and negative clock edges is normally $T/2$, where T is the period of the clock. The PLL is configured with negative feedback such that such that the UP and DN pulses in Figure 18.5 attempt to be equal and there is no average correction signal to the CCO, i.e., the PLL actively tries to make the time between a data edge and the negative clock edge equal to the time between the negative clock edge and the positive clock edge, i.e., to make both times equal to $T/2$. In our system, however, 0-to-1 data edges suffer a delay t_{dr} while 1-to-0 data edges do not. This asymmetric situation is resolved by causing the feedback correction signal to the CCO to be different on each data edge but to average to zero: Figure 18.11 shows that, when the PLL is in lock, the data becomes delayed with regard to the clock such that the distance from the 0-to-1 data edge to the following clock edge is $T/2 - t_{dr}/2$ while the distance from the 1-to-0 data edge to the following clock edge is $T/2 + t_{dr}/2$. On average, the data edge to negative clock edge is then still $T/2$ such that the VCO clock frequency locks to twice that of the maximal data-transition frequency. However, the noise margin for correctly sampling a data transition is reduced from $t_w = T/2$ to

$$t_w = \frac{T}{2} - \frac{t_{dr}}{2} \qquad (18.11)$$

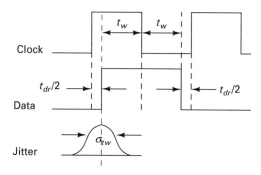

Figure 18.11. When the PLL is in lock, the data becomes delayed with regard to the clock such that the distance from the 0-to-1 data edge to the following clock edge is $T/2 - t_{dr}/2$ while the distance from the 1-to-0 data edge to the following clock edge is $T/2 + t_{dr}/2$. Reproduced with kind permission from [3] (©2008 IEEE).

because data is more likely to be sampled on the wrong clock cycle. Hence, as Figure 18.11 shows, if there is some standard deviation σ_{tw} due to jitter in the clock or data caused by noise, the probability of a bit error, P_e, given by

$$P_e = \frac{1}{2}\operatorname{erfc}\left(\frac{t_w}{\sqrt{2}\sigma_{tw}}\right) \tag{18.12}$$

increases because t_w is reduced. Equation (18.12) is correct if we assume Gaussian temporal jitter, which has been shown to be a valid assumption [3]. While this reduction in noise margin is not important for modest data rates, it can be important if we want to push the limits of data bandwidth of an impedance-modulation system to that set by fundamental theory. A simple solution is to have a 'soft switch turn on' during 1-to-0 transitions, i.e., M_2 is switched on in a gradual way in Figure 18.8 rather than abruptly such that 0-to-1 and 1-to-0 transitions have symmetric delays. An inverter that is current starved in its pull-down transistor can switch M_2 on gradually. The value of the pull-down current can be set via a feed-forward or feedback technique to ensure equal delay on rising or falling edges.

A detailed analysis of the bit error rate in impedance-modulation systems including a quantitative prediction of t_{dr} has been performed in [3]. Measurements of the bit error rate have been found to be quite consistent with Equations (18.11) and (18.12).

18.5   Experimental results for a complete transceiver

Separate external and internal transceiver chips, each 1.5 mm × 1.5 mm in size, were fabricated in a 0.5 μm process implementing the architectures of Figures 18.3 and 18.8 respectively with $V_{DDH} = 2.8$ V and $V_{DDL} = 1.4$ V. Printed circuit boards with 3.5 cm × 3.5 cm square two-turn coils with an inductance of 500 nH and a simulated quality factor of 30 at 25 MHz were fabricated. Packaged chips were

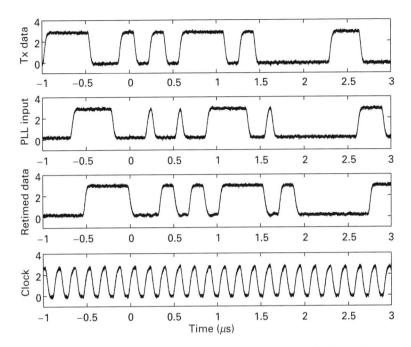

Figure 18.12. Transmitted and received data and recovered clock waveforms measured for the uplink at 5.8 Mb/s with coils 2 cm apart. Reproduced with kind permission from [3] (©2008 IEEE).

surface mounted on the printed circuit boards and aligned parallel to each other at various separations for testing. Implanted coils are typically less than 2 cm on a side and operate at link distances between 0.3 cm and 1 cm (the typical skin flap thickness). Therefore, to reduce the coupling constant to more typical values the link was tested over larger distances between 1.5 cm and 5 cm. A square coil of these dimensions is well fit by formulas for circular coils with radius 2 cm and the same square area as the 3.5 cm coils. The finite output impedance of the M_1 transistor in Figure 18.2 reduces Q_1 to 10 while Q_2 reduces from its simulated value of 30 to an actual experimental value of 25.

Figure 18.12 shows the transmitted and received data and recovered clock waveforms measured for the uplink at 5.8 Mb/s with coils 2 cm apart. The PLL synchronizes data edges to the rising edges of the clock. The isolated '1' bits at the output of the comparator are significantly narrower than the '0' bits due to the asymmetry of the 0-to-1 and 1-to-0 transitions. At this separation, the PLL locks between 1 Mb/s and 5.8 Mb/s. The upper end of the lock range is set by the loop filter, which saturates at V_{DDH} in Figure 18.6. The lower end of the lock range is set by the loop-filter output voltage driving the well-input voltage of the G_m WLR transconductor in Figure 18.4 to a sufficiently low value such that its shunting bipolar input transistors are activated (see Chapter 12). The rms clock jitter was measured to be 7.2 nS, t_{dr} varied between 65 ns and 140 ns with link distance

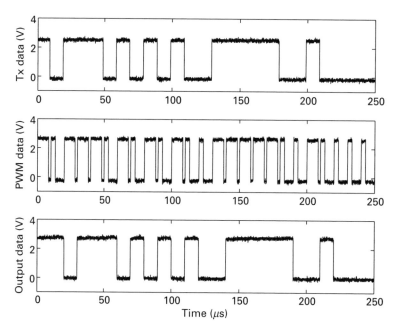

Figure 18.13. Transmitted and received data waveforms (before and after pulse-width demodulation) measured for the downlink at 200 kb/s. Reproduced with kind permission from [3] (©2008 IEEE).

varying from 2 cm to 4 cm, while σ_{tw} varied from 17 ns to 27 ns as the link distance varied from 2 cm to 4 cm. A good fit to Equation (18.12) was obtained for the 2 cm case with the bit error rate (BER) being below 10^{-6} for data rates less than 2.8 Mb/s and unmeasurable below this value, and at 10^{-3} for data rates of 4.8 Mb/s. Cycle slipping in the PLL increases as the link distance increases and the signal-to-noise ratio falls such that for link distances larger than 2 cm, the BER is larger than that predicted by Equation (18.12). The value of m_{eff} varies from 0.225 to 0.07 as the link distance varies from 2 cm to 4 cm. Equivalently, k varies from 0.034 to 0.017 as the link distance changes from 2 cm to 4 cm.

Figure 18.13 shows the transmitted and received data waveforms (before and after pulse-width demodulation) measured for the downlink at 200 kb/s with the coils 2 cm apart. Recovered data transitions are aligned with the rising edges of the PWM signal. The downlink is functional for data rates between 15 kb/s and 300 kb/s. The lower limit is set by the voltages V_A and V_B in Figure 18.9 hitting V_{DD}. The upper limit is determined by the bit period becoming smaller than the minimum allowable pulse width of 1 μs in the hold timer. No bit errors were observed for a 2 cm link distance and a 200 kb/s data rate.

The downlink and uplink function with almost no change in performance if a 0.9 % saline solution, that mimics the properties of body tissue, is placed between the coils: magnetic properties are not greatly altered by body tissue near 25 MHz.

18.6   Energy efficiency of the uplink and downlink

The total power consumption of the impedance-modulation link is 2.5 mW on the external side and 0.1 mW on the internal side for the uplink. The total power consumption of the PWM link is 1.5 mW on the external side and 0.14 mW on the internal side for the downlink. Thus, the energy efficiency of the impedance-modulation uplink is 2.6 mW/ (2.8 Mb/s) for a bit error rate less than 10^{-6}, which is nearly 0.96 nJ/bit. The energy efficiency of the uplink is quite good indeed, and can be improved even further if the value of V_{DDL} and V_{DDH} are adapted with k. In the next two sections, we analyze the fundamental limits of energy efficiency that may be possible in impedance-modulation systems of the future.

The energy efficiency of the downlink is similar to that of the uplink if we only use internal power consumption in computing the efficiency: 0.14 mW/(0.2 Mb/s) = 0.7 nJ/bit. The use of total power consumption (external and internal) yields an energy efficiency for the downlink of 1.64 mW/(0.2 Mb/s) = 8.2 nJ/bit. The external power consumption in the downlink case is relatively high because of overhead and because we have purposely made it large such that the implanted receiver can function with very low power consumption. The imbalanced power consumption leads to a total energy efficiency for the transceiver that is further from its optimum and, therefore, not as good as that of the impedance-modulation link. While overall energy efficiency is not optimized in the downlink, implanted power is minimized, which is significantly more important.

Since the costs of both receiver and transmitter power are borne by the external unit in an impedance-modulation system, implanted power is already minimized in such systems. Therefore, we can focus on minimizing external power consumption in impedance-modulation links, which is important in reducing the cost, size, weight, and battery life of the external unit.

18.7   Scaling laws for power consumption in impedance-modulation links

We shall now discuss issues in the balance between receiver power and transmitter power in impedance-modulation links that lead to optimal energy efficiency. At this optimum, we shall derive scaling laws for power consumption as the signal-to-noise ratio, SNR, and the bandwidth, B, vary in the impedance-modulation link. Then, in the following section, we shall use these results to compute the optimal energy per bit in an impedance-modulation system.

Figure 18.14 shows a simple and minimal impedance-modulation communication system. A 'power amplifier' composed of a single transistor with transconductance g_{mp} drives a parallel coupled resonator. Modulations in the amplitude of v_1 are detected by a Gilbert differential mixer composed of a multiplier and a lowpass filter. The multiplier is composed of a differential pair with transistors having a transconductance g_{m2} that implement one of its inputs. Switching transistors driven by a local oscillator (LO) square wave implement another of

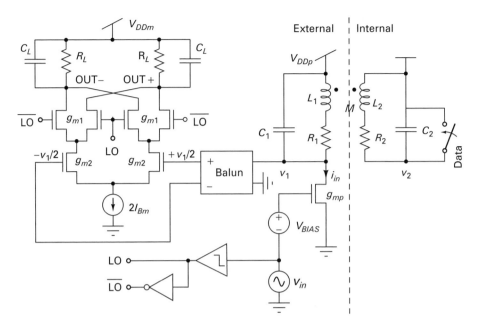

Figure 18.14. A simple and minimal impedance-modulation communication system.

its inputs. The square wave is generated from the voltage v_{in} that drives the power amplifier via simple thresholding as shown. A balun, implemented for example by a center-tapped transformer, performs single-ended to differential conversion that converts v_1 into differential inputs suitable for the g_{m2} transistors. The lowpass filter is implemented via the passive $R_L - C_L$ network. The power amplifier and mixer supply voltages V_{DDp} and V_{DDm} are different and kept as low as possible to maximize energy efficiency. At the low power levels that we focus on, V_{DDp} is typically much less than V_{DDm}.

The configuration of Figure 18.14 is simpler than that composed of the oscillator-driven resonator of Figure 18.2 and the receiver circuit of Figure 18.3. The simpler configuration allows us to focus on issues that determine the fundamental limits of energy efficiency without getting bogged down in circuit details that differ from topology to topology and that obscure the insights that generalize across topologies. The power versus signal-to-noise (SNR) tradeoffs of the simple system of Figure 18.14 for a fixed demodulation signal bandwidth of operation B are similar to those of more complex transceiver configurations except for scaling constants. We shall first summarize these key insights before we explain how they are derived in more depth.

In a well-designed power amplifier with even modest voltage gain, its output noise is dominated by that due to its active transistor with transconductance g_{mp} in Figure 18.14. In this scenario, increasing the power of the power amplifier increases its output impedance-modulation signal power proportional to v_1^2 and its output noise power proportional to v_1, i.e., proportional to P_{PA}^2 and P_{PA}, respectively, where P_{PA} is the power consumption of the amplifier. In a well-designed mixer with even modest voltage gain, its output noise is dominated by the g_{m2} differential-pair transistors in Figure 18.14. In this scenario, increasing power in the mixer decreases its input-referred noise proportional to $1/P_{mxr}$. Thus, for a given total SNR at the output of the power amplifier, which determines the net bit error rate (BER) of the communication system, P_{PA} and P_{mxr} must be balanced such that the total power consumption $P_{TOT} = P_{PA} + P_{mxr}$ needed to achieve this SNR is minimized. That is, for a given value of SNR equal to $P_{PA}^2/(C_{PA}P_{PA} + C_{mxr}/P_{mxr})$, where C_{PA} and C_{mxr} are topology-dependent constants, P_{TOT} must be minimized. Remarkably, independent of several details of the topology, this optimization problem leads to the following conclusions. If the mixer noise dominates as it often does in large-k strongly coupled impedance-modulation scenarios, where the signal v_1 in Figure 18.14 is relatively low, then the optimal balance occurs when $P_{PA} = 2P_{mxr}$; if power-amplifier noise dominates as it often does in small-k weakly coupled scenarios, where the signal v_1 is necessarily relatively high, then optimization requires that the power-amplifier power be increased until the desired SNR is achieved while mixer power is maintained at a constant value. Not surprisingly, strongly coupled scenarios lead to a lower power consumption for a given bandwidth B and $\text{SNR} = S_N$ than weakly coupled scenarios and therefore a lower energy per bit as well. Furthermore, the increase in power with increasing S_N and B follows a gentle one-third power law in both S_N and B in strongly coupled scenarios. In contrast, the increase in power with increasing S_N and B is proportional to both S_N and B in weakly coupled scenarios.

The derivation of the results in the following sections is conceptually simple but involves a fair bit of algebra and attention to mathematical detail. Readers who are more interested in the key ideas rather than these mathematical details should pay attention to certain results that we shall highlight as boxed formulas.

18.7.1   Power and noise in the power amplifier

In Figure 18.14, suppose that I_{PA} denotes the dc current in the power amplifier such that its average power consumption is $P_{PA} = I_{PA}V_{DDp}$. In this amplifier, if the amplitude of the ac input $v_{in} = V_{Lp}$, where V_{Lp} is the linear range of operation of the transistor with transconductance equal to g_{mp}, then its power consumption and distortion are minimized: values of v_{in} larger than V_{Lp} lead to distortion in the carrier and degrade spectral efficiency; values of v_{in} smaller than V_{Lp} imply that V_{Lp}, I_{PA} and thus P_{PA} can be reduced to save power while maintaining $g_{mp} = I_{PA}/V_{Lp}$. Hence, we shall assume that $v_{in} = V_{Lp}$. When the resistive impedance seen at the v_1 node is at its maximum value of R_1Q^2 during a '0' bit, suppose that the amplitude of the carrier is denoted by v_1. Then,

$$g_{mp}(R_1Q_1^2)v_{in} = v_1$$
$$g_{mp}(R_1Q_1^2)V_{Lp} = v_1$$
$$I_{PA}(R_1Q_1^2) = v_1 \qquad (18.13)$$
$$I_{PA} = \frac{v_1}{R_1Q_1^2}$$

From Equation (18.5), because the resistive impedance at the v_1 node falls to $(R_1Q^2)/(1+m)$ through the mechanism of impedance modulation, during a '1' bit, the amplitude of the voltage at the v_1 node will fall from its maximum v_1 to $v_1/(1+m)$. For maximum Class-A power efficiency, v_1 must equal V_{DDp} such that 50% of the dc power dissipation of V_{DDp} from Equation (18.13), $(v_1/(R_1Q^2))$, is dissipated as ac power in $R_1Q_1{}^2$ and given by $v_1^2/2(R_1Q_1^2)$; the voltage at the v_1 node barely touches ground when this maximal efficiency is achieved. In practice, since the voltage at the v_1 node falls to $V_{DDp}/(1+m)$ during a '1' bit, and because of finite saturation voltages in transistors, the efficiency is lower during a '1' bit than during a '0' bit. In general, if P_1 is the probability of a '1' bit, the efficiency of the power amplifier can be shown to be given by

$$\eta = \frac{1}{2}\left[\left((1-P_1)+\frac{P_1}{1+m}\right)\left(\frac{v_{in}}{V_{Lp}}\right)\left(\frac{v_1}{V_{DDp}}\right)\right] \tag{18.14}$$

Equation (18.14) quantitatively shows that the efficiency is maximized at 50% if $P_1=0$, $m=0$, $v_{in}=V_{Lp}$, and $v_1=V_{DDp}$, i.e., if all bits are '0's. In general, we may assume some average efficiency for the amplifier based on signal statistics. Regardless, if $v_{in}=V_{Lp}$ the average power consumption of the amplifier is given by

$$P_{PA} = V_{DDp}\left(\frac{v_1}{R_1Q_1^2}\right) \tag{18.15}$$

and only a fraction η of this power consumption will be dissipated as ac power in the carrier.

The noise at the output node of v_1 is dominated by that from transistor g_{mp} in most situations: a gain of $g_{mp}R_1Q_1^2/(1+m)$ of even 5 means that the transistor's noise is 5 times more significant than the thermal noise due to the resistance $R_1Q_1^2/(1+m)$. Assuming that m is small such that the bandwidth is relatively unchanged during a '1' and a '0', the output noise of the power amplifier over the demodulation signal bandwidth B can be approximated by

$$\begin{aligned}
v_{nPA}^2 &\approx 4kT\gamma g_{mp}\left(\frac{R_1Q_1^2}{1+m}\right)^2 B \\
&= 4kT\gamma\frac{I_{PA}}{V_{Lp}}\left(\frac{R_1Q_1^2}{1+m}\right)^2 B \\
&= 4kT\gamma\frac{P_{PA}}{V_{DDp}V_{Lp}}\left(\frac{R_1Q_1^2}{1+m}\right)^2 B
\end{aligned} \tag{18.16}$$

where γ is 2/3 in above-threshold operation, $1/(2\kappa)$ in subthreshold operation, and in general can incorporate excess noise factors as well.

18.7.2   Power and noise in the mixer

The power of the mixer in Figure 18.14 is given by

$$P_{mxr} = 2V_{DDm}I_{Bm} \tag{18.17}$$

Since the transistors with LO inputs operate as switches in Figure 18.14, they may be assumed to contribute little noise in the mixer. If the mixer gain is even moderate, the noise contribution of the R_L resistors is negligible compared with those of the g_{m2} transistors. The mixer's input-referred noise at the v_1 node due to the two g_{m2} transistors is then given by

$$
\begin{aligned}
v_{nmxr}^2 &= 8kT\frac{\gamma}{g_{m2}}B \\
&= 8kT\gamma\frac{V_{Lm}}{I_{Bm}}B \\
&= 16kT\gamma\frac{V_{Lm}V_{DDm}}{P_{mxr}}B
\end{aligned}
\tag{18.18}
$$

where V_{Lm} is the linear range of the g_{m2} transistors and I_{Bm} is the dc current flowing through each half of the mixer.

18.7.3   Optimization of total power consumption for a given SNR

The amplitude at the v_1 node in Figure 18.14 changes from v_1 to $v_1/(1+m)$ when we modulate the secondary from a '0' to a '1' and vice versa. Thus, the strength of the modulation signal is given by $v_1 - v_1/(1+m)$. The total SNR, S_N, at the v_1 node for a modulation signal is then given by

$$
\begin{aligned}
S_N &= \frac{\left(v_1 - \dfrac{v_1}{1+m}\right)^2}{2\left(v_{nPA}^2 + v_{nmxr}^2\right)} \\
S_N &= \frac{v_1^2\left(\dfrac{m}{1+m}\right)^2}{2\left(v_{nPA}^2 + v_{nmxr}^2\right)}
\end{aligned}
\tag{18.19}
$$

Substituting for v_1, v_{nPA}, and v_{nmxr} from Equations (18.15), (18.16), and (18.18), respectively, and after some straightforward but tedious algebra, we find that

$$
\boxed{
S_N = \frac{\dfrac{P_{PA}^2}{P_{th}}}{P_{PA} + \dfrac{P_{char}^2}{P_{mxr}}}
}
\tag{18.20}
$$

where we define the P_{th} and P_{char} power quantities as

$$
\boxed{
\begin{aligned}
P_{th} &= \left(\frac{8kT\gamma B}{m^2}\right)\left(\frac{V_{DDp}}{V_{Lp}}\right) \\
P_{char} &= 2(1+m)\frac{\sqrt{V_{DDp}V_{DDm}}\sqrt{(V_{Lm}V_{Lp})}}{R_1Q_1^2} \\
m &= k^2Q_1Q_2
\end{aligned}
}
\tag{18.21}
$$

For small values of m, i.e., for weak coupling or equivalently for small k, the P_{th} term is large while the P_{char} term is small. For large values of m, strong coupling, or large k, the P_{th} term is small while the P_{char} term is large. For a fixed SNR $= S_N$, Equation (18.20) is a quadratic equation in P_{PA} and can be solved to yield P_{PA} as a function of P_{mxr}. We get

$$P_{PA} = P_{th}\frac{S_N}{2}\left(1 + \sqrt{1 + \frac{4P_{char}^2}{S_N P_{mxr} P_{th}}}\right) \tag{18.22}$$

At large values of P_{mxr}, S_N, or P_{th}, when $P_{char}^2 \ll S_N P_{mxr} P_{th}$, Equation (18.22) predicts that

$$P_{PA} = P_{th}S_N \tag{18.23}$$

independent of P_{mxr}. In these situations, the SNR $= S_N$ requirement is met by simply increasing the power in the power amplifier with the power amplifier determining both the signal strength and the noise strength. This situation typically occurs when we have a weakly coupled link (P_{th} is large because m is small) such that the power-amplifier signal has to be very strong to yield any discernible modulation signal, at which point the power-amplifier noise increases and swamps the mixer's input-referred noise, making the mixer's power insignificant. Substituting for P_{th} from Equation (18.21) in Equation (18.23) we get

$$P_{TOT} \approx P_{PA} = \left(\left(\frac{8kT\gamma}{k^4 Q_1^2 Q_2^2}\right)\left(\frac{V_{DDp}}{V_{Lp}}\right)\right)(S_N)(B) \tag{18.24}$$

For small values of P_{mxr}, S_N, or P_{th}, Equation (18.22) predicts that

$$P_{PA} = P_{char}\sqrt{\frac{S_N P_{th}}{P_{mxr}}} \tag{18.25}$$

In these situations, the SNR $= S_N$ requirement is met by a balance between mixer power and power amplifier, with the power amplifier primarily determining the signal strength and the mixer power primarily determining the noise strength. This situation typically occurs when we have a strongly coupled link such that the power-amplifier signal does not need to be very strong to yield a discernible modulation signal, and therefore the power amplifier's noise is well below the mixer's input-referred noise. In this scenario, there is an optimum balance of power between the power amplifier and mixer. By taking logarithms on both sides of Equation (18.25) and differentiating, we get

$$\frac{dP_{PA}}{P_{PA}} = -\frac{1}{2}\frac{dP_{mxr}}{P_{mxr}} \tag{18.26}$$

To minimize the total power consumption $P_{TOT} = P_{PA} + P_{mxr}$, we must, at an optimum, also have

$$dP_{PA} + dP_{mxr} = 0 \tag{18.27}$$

The simultaneous solution of Equations (18.26) and (18.27) then leads to a minimum total power consumption when the power amplifier and mixer power are balanced such that

$$\boxed{P_{PA} = 2P_{mxr}} \tag{18.28}$$

At this optimum, we can substitute Equation (18.28) into Equation (18.25) to solve for the optimum value of P_{PA} and then take $3/2$ of this value to get the total power consumption. We find that

$$P_{TOT} = \frac{3}{2} P_{char}^{2/3} (2S_N P_{th})^{1/3} \tag{18.29}$$

Substituting from Equation (18.21) for P_{th} and P_{char} we get

$$\boxed{P_{TOT} = \left(3 \left(\frac{(1+k^2Q_1Q_2)^{2/3}}{k^{4/3}Q_1^2Q_2^{2/3}} \right) \left(\frac{V_{DDp}^2 V_{LM} V_{DDm}}{R_1^2} \right)^{1/3} (8kT\gamma)^{1/3} \right) \left(S_N^{1/3} \right) \left(B^{1/3} \right)} \tag{18.30}$$

Equation (18.30) indicates that, for strong coupling, the power scaling with increasing S_N and increasing bandwidth B is a gentle one-third power law, in fact the gentlest that we have seen amongst the numerous *power* $= f(S_N, B)$ relations in this book. In contrast, for weak coupling where Equation (18.24) is valid, and power-amplifier power is dominant, a typical linear dependence of power on both the SNR and bandwidth is obtained. Since k depends on distance like r^{-3} (Equation (16.7) in Chapter 16), the dependence of the minimum power on distance is r^4 in Equation (18.30) compared with r^{12} in Equation (18.24), indicating that impedance modulation is viable only for short distances and that the balanced-power regime is where it is more efficient. If m is small, the dependence on Q_1 is an inverse square law in both cases but the dependence on Q_2 in Equation (18.30) is a gentler $-2/3$ power law compared with a -2 power law in Equation (18.24). Equation (18.30) predicts lower power with reduced V_{DD}'s, which is not surprising. The parameter V_{Lp} is absent in Equation (18.30) because power-amplifier noise is unimportant in this regime. However, V_{Lm} is present in Equation (18.30) since increasing mixer noise implies higher amplifier power to maintain SNR. In Equation (18.24), increasing V_{DDp} not surprisingly increases power. Surprisingly, decreasing V_{Lp} increases the total power in this regime. The explanation is that, for a fixed v_1, decreasing V_{Lp} increases the noise seen at v_1 due to increased amplification in the power amplifier degrading SNR. Thus, v_1 needs to increase further to maintain SNR, which turns up amplifier power.

Equations (18.20) and (18.21), which formed the foundation for the analysis in this section, are only valid for values of Q_1 that are not too large since they assume that v_1 keeps increasing with Q_1 according to Equation (18.15). They predict that performance in an impedance-modulation link will continue to improve without bound if Q_1 is raised since the voltage in the '0' bit state keeps increasing like Q_1^2 even though the voltage in the '1' state saturates to a small value near ground consequently improving the net SNR with an increase in Q_1. The increase in amplitude in the '0' bit state is predicted to occur in spite of filtering effects that only attenuate the signal like $1/Q_1$. In practice, at very high values of Q_1 where the link becomes strongly overcoupled, our assumptions in deriving these equations are no longer valid: we move off the high-Q_1 resonant frequency in the '0' bit state as predicted by Equation (18.2) such that the voltage on v_1 no longer increases in the '0' bit state if Q_1 increases but merely saturates to a Q_1-independent value. Thus, the equations of this section are only an approximation valid for Q_1's that are not too large. In fact, both Q_1 or Q_2 must lie within the limit set by $\max(Q_1, Q_2) > \omega_{res}/(2B)$ to avoid filtering effects from coming into play. Our experimental link was always operated under conditions below critical coupling ($m < 1$) and with Q_1 and Q_2 both sufficiently small such that filtering of the modulation input was insignificant. Thus, our theoretical analysis is valid in this regime.

18.8   The energy per bit in impedance-modulation links

In digital communication theory where bits are being sequentially transmitted over a channel, it is convenient to write the SNR S_N as

$$S_N = \frac{E_b}{N_0} \times \frac{R}{B} \tag{18.31}$$

where E_b is the integrated power of the signal over a clock cycle, i.e., the energy per bit, the clock period is $1/R$, N_0 is the noise power spectral density, and B is the bandwidth of the system. The quantity R represents the number of bits being transmitted per second: 1 bit transmitted over a time of $1/R$ yields a rate of R bits being transmitted per second. The quantity $E_b \times R$ represents the signal power: energy per bit times the number of bits per second. The quantity $N_0 B$ is the noise power. Hence, the right-hand side of Equation (18.31) is indeed the ratio of signal power to noise power, i.e., the SNR S_N on the left-hand side. The sinc-like spectral energy of rectangular data pulses of width $1/R$ with energy distributed over an equivalent rectangular bandwidth of $R/2$ is significantly filtered if $R/2$ exceeds B. Thus, the bit rate R must be significantly less than $2B$ to avoid attenuation of the signal, or equivalently to avoid inter-symbol interference (ISI), i.e., interference between the bits due to overlapping filtered waveforms from adjacent cycles. The parameter E_b/N_0 is called the *SNR per bit* while the parameter R/B is called the *link spectral efficiency*.

If $R/B \ll 1$, i.e., inter-symbol interference is not an issue, then the SNR per bit, E_b/N_0, becomes independent of R/B. The probability of a bit error, P_e, is then determined by the probability that Gaussian noise with noise amplitude $\sqrt{(N_0 R)}$ is sufficiently large such that a '0' rises from its normal value of $-\sqrt{(E_b R)}$, becomes positive and is wrongly interpreted as a '1', or a '1' falls from its normal value of $+\sqrt{(E_b R)}$, becomes negative and is wrongly interpreted as a '0'. Of course, we are assuming that the noise is additive and Gaussian, which is often but not always the case. Our analysis is performed on the demodulated signal at v_1 assuming that this signal is well approximated by a digital square-wave-like signal. If the noise is additive and Gaussian, the probability of error can then be shown to be

$$P_e = \text{erfc}\left(\sqrt{\frac{E_b}{2N_0}}\right) \tag{18.32}$$

Thus, if we are weakly coupled and Equation (18.24) applies, we can invert and square Equation (18.32) to obtain E_b/N_0, substitute E_b/N_0 to compute the SNR in Equation (18.31), substitute this SNR into Equation (18.24) to obtain P_{PA}, and finally divide P_{PA} by R to compute the minimum energy per bit, $E_{bit}^{weak} = P_{PA}/R$:

$$E_{bit}^{weak} = \left(\frac{8kT\gamma}{k^4 Q_1^2 Q_2^2}\right)\left(\frac{V_{DDp}}{V_{Lp}}\right)\left(\text{erfc}^{-1}(P_e)\right)^2 \tag{18.33}$$

$$\boxed{E_{bit}^{weak} = E_{0w}\left(\text{erfc}^{-1}(P_e)\right)^2}$$

In deriving Equation (18.33), there is one subtle factor of 2: the rms energy in a square wave of amplitude v_1 is v_1^2 while that in a sine wave of amplitude v_1 is $v_1^2/2$. Thus, an equation derived for sine waves is converted into one valid for square waves if we simply substitute $\text{SNR}_{square}/2$ for the SNR in a sine-wave equation. Since Equation (18.24) was derived for sine waves, while Equation (18.33) applies to square waves, E_{bit}^{weak} is consequently a factor of 2 lower than we would have gotten by blindly substituting the SNR in Equation (18.31) into Equation (18.24).

If we are strongly coupled and Equation (18.30) applies, the exact same procedure applied to Equation (18.30) instead of Equation (18.24) yields

$$E_{bit}^{strong} = \left(3\left(\frac{(1+k^2 Q_1 Q_2)^{2/3}}{k^{4/3} Q_1^2 Q_2^{2/3}}\right)\left(\frac{V_{DDp}^2 V_{Lm} V_{DDm}}{R_1^2}\right)^{1/3}(8kT\gamma)^{1/3}\right)\left((\text{erfc}^{-1}(P_e))^{2/3}\right)\left(\frac{1}{R^{2/3}}\right) \tag{18.34}$$

$$\boxed{E_{bit}^{strong} = E_{0s}\left((\text{erfc}^{-1}(P_e))^{2/3}\right)\left(\frac{1}{R^{2/3}}\right)}$$

The minimum energy per bit in the weakly coupled case of Equation (18.33) is a strong square-law function of the erfc error-probability function and independent of R. The minimum energy per bit in the strongly coupled case of Equation (18.34) is a gentle two-third law function of the erfc error-probability function and decreases as the bit rate R increases. Equations (18.33) and (18.34) are simply the single-bit digital versions of the analog Equations (18.24) and (18.30) respectively with $2\left(\text{erfc}^{-1}(P_e)\right)^2$ analogous to the SNR S_N, R analogous to B, and $E_{bit}R$ analogous to signal power. The energy per bit decreases with R in the strongly coupled case because the power only scales like $B^{1/3}$ in Equation (18.30) and therefore only like $R^{1/3}/R$ in Equation (18.34).

The analysis thus far has neglected *inter-symbol interference* (*ISI*) that occurs when R nears and exceeds the modulation bandwidth B. Figure 18.15 shows that its main effect is to attenuate the energy per bit due to RC-like filtering, which lowers E_b from its value E_{b0} when R/B is much less than 1 to a value given by

$$E_b = E_{b0}f\left(\frac{R}{B}\right) \tag{18.35}$$

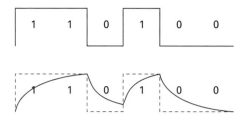

Figure 18.15. Effects of filtering by the link for square (transmitted) pulses.

when R/B nears and exceeds 1. The function $f(R/B)$ can be easily analytically computed in closed form but it is tedious to compute. It is found by integrating the signal energy in an RC-like output wave over a time period of $1/R$ with the RC time constant $\tau = 1/(2\pi B)$. In essence, a sharp square wave of period $1/R$ that alternates between $+\sqrt{(E_{b0}R)}$ and $-\sqrt{(E_{b0}R)}$ in amplitude is applied as input to a lowpass filter with bandwidth B and we compute the signal energy of the output square wave in a period. The square wave has period $1/R$ as well. It is flat when $R/B \ll 1$ and attenuates like a lowpass filter as R/B nears and exceeds 1. We can approximate $f(R/B)$ by the power-attenuation function of a lowpass filter:

$$E_b = E_{b0} f\left(\frac{R}{B}\right)$$

$$E_b \approx \frac{E_{b0}}{1 + (R/B)^2} \tag{18.36}$$

$$f\left(\frac{R}{B}\right) \approx \frac{1}{1 + \left(\frac{R}{B}\right)^2}$$

Since the signal power $(E_{b0}R)$ is attenuated to (E_bR) while the noise power $(N_{b0}B)$ is unchanged, the SNR drops. To compensate and maintain the same error rate, we have to effectively increase our unfiltered SNR, such that after filtering it is identical to the SNR we would have had in our earlier analysis. Equations (18.20), (18.24), and (18.30) are effectively still correct if we replace SNR by $S_N/(f(R/B))$. Similarly, Equations (18.33) and (18.34) are effectively still correct if we replace $\left(\mathrm{erfc}^{-1}(P_e)\right)^2$ by $\left(\mathrm{erfc}^{-1}(P_e)\right)^2/f(R/B)$. Thus, we can compute the energy per bit including the effects of filtering and ISI by writing

$$E_{bit}^{weak} = \left(\frac{8kT\gamma}{k^4 Q_1^2 Q_2^2}\right)\left(\frac{V_{DDp}}{V_{Lp}}\right)\frac{\left(\mathrm{erfc}^{-1}(P_e)\right)^2}{f(R/B)}$$

$$\boxed{E_{bit}^{weak} = E_{0w}\left(\frac{\left(\mathrm{erfc}^{-1}(P_e)\right)^2}{f(R/B)}\right)} \tag{18.37}$$

Similarly, we may write

$$E_{bit}^{strong} = \left(3\left(\frac{(1+k^2 Q_1 Q_2)^{2/3}}{k^{4/3} Q_1^2 Q_2^{2/3}}\right)\left(\frac{V_{DDp}^2 V_{Lm} V_{DDm}}{R_1^2}\right)^{1/3}(8kT\gamma)^{1/3}\right)\left(\frac{(\mathrm{erfc}^{-1}(P_e))^{2/3}}{f(R/B)^{1/3}}\right)\left(\frac{1}{R^{2/3}}\right)$$

(18.38)

$$\boxed{E_{bit}^{strong} = E_{0s}\left(\frac{(\mathrm{erfc}^{-1}(P_e))^{2/3}}{f(R/B)^{1/3}}\right)\left(\frac{1}{R^{2/3}}\right)}$$

Although we have approximated the weakly coupled and strongly coupled regimes through approximations of the more exact Equations (18.20) and (18.22), it is possible to formulate our analysis more exactly without approximations for the purposes of numerical computation: *Minimize $P_{TOT} = P_{PA} + P_{mxr}$ subject to the constraint that*

$$\frac{\frac{P_{PA}^2}{P_{th}}}{P_{PA} + \frac{P_{char}^2}{P_{mxr}}} = \left((\mathrm{erfc}^{-1}(P_e))^2\right)\left(\frac{R/B}{f(R/B)}\right)$$

(18.39)

At the minimum, compute P_{TOT}/R to find the minimum energy per bit. Figure 18.16 (a) plots the minimum energy per bit derived from such numerical computation as a function of k and for various values of data rate R. The curves saturate at high values of k since the modulation depth $m_{eff} = m/(1+m)$ cannot exceed 1. The values used for numerical computation were $Q_1 = Q_2 = 25$, $f_{res} = 25\,\mathrm{MHz}$, $B = f_{res}/(2Q_1)$, $V_{DDp} = 2V_{Lm} = 0.4\,\mathrm{V}$, $V_{Lp} = 35\,\mathrm{mV}$, $V_{DDm} = 2.0\,\mathrm{V}$, $\gamma = 1$, $P_e = 10^{-4}$, $L_1 = 0.5\,\mu\mathrm{H}$. The choice of $V_{DDp} = 2V_{Lm}$ ensures that saturation

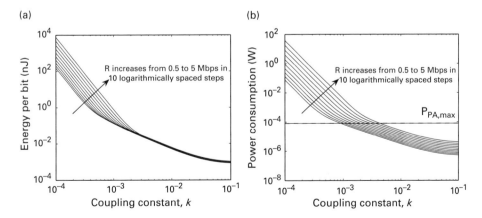

Figure 18.16a, b. (a) Plots of the minimum energy per bit derived as a function of the coupling constant k and various values of data rate R. (b) Total link power, P_{TOT}, as a function of k for various R.

is reached in the mixer and the power amplifier simultaneously. The maximum possible amplifier power consumption is given by such a saturation condition to be

$$V_{DDp} = 2V_{Lm}$$

$$v_1 \leq 2V_{Lm}$$

$$\tag{18.40}$$

$$P_{PA} \leq \frac{(2V_{Lm})(2V_{Lm})}{R_1 Q_1^2}$$

The value of P_{PA}^{max} allowed by this equation is marked in Figure 18.16 (b), which plots the optimal power consumption versus k for various values of R. The link is only usable for link distances that are less than a critical value such that $k > k_{min}$ where k_{min} is the value of k at which the P_{PA} curve intersects the $P_{PA,max}$ line. The weakly coupled regime corresponds to small values of k where the approximation of Equation (18.37) is valid. The dependence on k is a strong $1/k^4$ power law, which is reflected in Figures 18.16 (a) and Figures 18.16 (b). In Figure 18.16 (a), the energy per bit is relatively constant until R/B nears 1, and then it starts to increase in the weakly coupled regime due to the $f(R/B)$ term in Equation (18.37). The strongly coupled regime corresponds to large values of k where the approximation of Equation (18.38) is valid. The dependence on k is a gentler $1/k^{4/3}$ power law. The energy per bit improves as R increases because of the $1/R^{2/3}$ dependence and then saturates at a constant value due to the $(R/B)^{2/3}$ dependence of the $f(R/B)$ term in Equation (18.38) that arises from Equation (18.36). At very large values of R, however, the large increases in P_{PA} needed to maintain a low error probability will eventually cause the power-amplifier noise to swamp the mixer noise such that the scaling laws of the weakly coupled regime begin to apply rather than those of the strongly coupled regime. The energy per bit then begins to increase again but such large values of R are not shown in Figure 18.16 (a).

Figure 18.16 (a) suggests that there is plenty of room for improvement in the future. The limits of energy efficiency for coupling factors that are practical in medical implants are near 1 pJ/bit to 10 pJ/bit. Thus, even though the 1 nJ/bit relatively unoptimized system discussed in this chapter has good energy efficiency, future systems could be even more energy efficient. Such impedance-modulation systems could enable very high bandwidths at modest amounts of power to become practical in biomedical implants.

18.9   Incoherent versus coherent RF receivers

When k is low, the energy per bit is relatively high such that for a given power consumption, only low data rates are feasible. In this weakly coupled regime, v_1 is large such that receiver noise is insignificant. A large v_1 leads to signals that easily exceed the dead zone of rectifier-based envelope detectors. Thus a low-power passive envelope-detector-based receiver built with diodes and capacitors like those in Figures 18.3 and 18.8 is more power efficient than an active-mixer topology. Incoherent envelope detectors are not sensitive to fine carrier phase and therefore capable of decoding less information in the carrier signal, but this failing is not a big issue when data rates are expected to be low. Hence, an incoherent receiver is optimal under such conditions.

When k is high, the energy per bit is relatively low such that for a given power consumption, higher data rates are feasible. In this strongly coupled regime, v_1 is small and receiver noise is significant. The small v_1 leads to signals that may not necessarily robustly exceed the dead zone of rectifier-based envelope detectors. However, a small v_1 is compatible with energy efficiency in a mixer whose linear range V_{Lm} may only be 35 mV if it is to be energy efficient. A small v_1 is then beneficial since the linear range of the mixer can then be matched to the amplitude of v_1 without wasting power in the mixer or causing distortion. Furthermore, mixer-based detectors are sensitive to fine carrier phase and therefore capable of decoding more information in the carrier signal, a desirable property if data rates are expected to be high. Thus, a coherent mixer-based receiver is optimal under such conditions.

18.10   Radiated emissions and FCC regulations

None of the unlicensed Industrial-Scientific-and-Medical (ISM) bands below 433 MHz have enough bandwidth to support uplink data rates larger than a few hundred kb/s. For example, the widest band centered at 27.12 MHz is only 670 kHz wide. Hence biomedical links with relatively high data rates, such as the one described in this chapter, must fit outside an ISM band but operate below permissible radiation limits for such bands. Part 15 of the FCC rules (Title 47 of the code of Federal Regulations) requires that radiated emissions for unlicensed bands between 1.705 MHz and 30.0 MHz must not exceed 30 μV/m at a distance 30 m from the device with the exception of certain restricted frequency bands.

To check if the FCC specification is met, we first need to compute the radiation resistance of the coil in our external unit. From Chapter 17, we recall that the radiation resistance of a coil with N turns and circumferential length l is given by

$$R_{rad}^{loop\_dpl} = \eta N^2 \frac{\pi}{6} \left(\frac{l}{\lambda}\right)^4 \tag{18.41}$$

where $\eta = 120\pi$ is the radiation resistance of free space, and λ is the wavelength of the radiated electromagnetic wave. At 25 MHz, $\lambda = \left(3 \times 10^8\right)/\left(25 \times 10^6\right) = 12$ m. If we conservatively approximate the radiation resistance of our square 3.5 cm two-turn coil with that of an equivalent 2 cm two-turn radius loop with the same area, $l = 2\pi \times 2$ cm. The radiation resistance from Equation (18.41) is then found to be 9.5 $\mu\Omega$. The radiated power is given by $P_{rad} = I^2 R_{rad}$. With $I = 1$ mA, a typical value for our design, we get $P_{rad} = 9.5$ pW. The maximum radiated power density (W/m^2) at a distance R from the coil is given by

$$P_{dens} = \frac{D_0 P_{rad}}{4\pi R^2} = \frac{E_{rad}^2}{Z_0} \qquad (18.42)$$

where $D_0 = 1.5$ is the maximum gain produced by a small loop antenna, E_{rad} is the intensity of the radiated electric field, and $Z_0 = 120\pi$ is the impedance of free space. Thus, from Equation (18.42), we find for $R = 30$ m that $E_{rad} = 0.69$ μV/m, which is well below the FCC specification. Higher-frequency operation increases R_{rad} as does the use of more turns or a bigger coil.

18.11   Seven considerations in choosing a carrier frequency

When choosing a carrier frequency for an RF link, seven considerations are often balanced against each other:

1. **Bandwidth** – Higher frequency provides more bandwidth at constant Q.
2. **Power** – Higher baseband bandwidth, usually obtained at a higher carrier frequency, consumes more power. In a given technology, the power consumption of certain components of RF blocks, e.g., active transistors that must have adequate power gain at the RF frequency, also increases with carrier frequency.
3. **Antenna size** – Higher carrier frequencies require smaller antennas since the effectiveness of an antenna in transducing electromagnetic energy is small if its total length is significantly less than half the carrier wavelength.
4. **Communication range** – For a given power consumption, higher carrier frequency leads to a smaller communication range in accord with the Friis formula that we discussed in Chapter 17.
5. **Device Q** – Inductors generally have better Q's with increasing carrier frequency, due to steeper increase of inductive reactance with frequency than that of skin-effect or other resistive losses. Capacitors generally have better Q's with decreasing carrier frequency due to the steep increase of capacitive reactance with decreasing frequency and a decrease of skin-effect and other resistive losses with decreasing carrier frequency.
6. **The channel** – The transmission characteristics of the channel can result in higher or lower loss. For example, human tissue is extremely lossy near 433 MHz and significantly less so near 25 MHz. Diffraction, scattering, and directionality effects are wavelength dependent with higher carrier frequencies being more prone to reflection, scattering, or absorption by obstacles of a fixed size in their path than lower carrier frequencies that diffract more easily around obstacles.
7. **Federal Communications Commission (FCC) regulations** – FCC regulations on RF spectral bandwidths and intensities at various carrier frequencies are complex and need to be followed in any design.

18.12   RF antenna links for implants

Applications such as electronic pills are used for diagnosis of gastrointestinal (GI) problems. In such applications, a patient swallows a pill, and for a few hours imaging (via implanted LEDs and imagers within the pill), temperature, and pH information about the patient's GI tract can be wirelessly monitored. Such monitoring is useful in diagnoses of Crohn's disease and Celiac disease. In applications such as these, the size budget for the receiving coil/antenna in the pill is very small, and the distance between the transmitter and receiver r is relatively large. Thus, k, evaluated from Equation (16.7) in Chapter 16, is quite small, and inductive RF links become relatively inefficient. Systems that operate with far-field links at higher carrier frequencies and smaller antennas have the potential to be advantageous. Therefore, even though transmission of RF energy through the body beyond a few tens of MHz is quite lossy because of the highly conductive nature of the body at these frequencies, antenna-based systems have been and are being explored at such frequencies.

There are currently four frequency bands that have been or are being investigated. The **MICS, 402–405 MHz band**, has been the most successful thus far with low-power 0.8 Mbps commercial systems and chips for electronic pills already having been developed [7]. The latter chips also work in the **433 MHz ISM band**. Antenna, matching, fading, and body losses for 2 m communication ranges with the swallowed pill can lead to losses of 40 dB. Thus, ~99% of the power is dissipated within the body. For frequencies in the 300 MHz to 6 GHz range, the US specification is that the specific absorption rate (SAR) of this power in the body cannot exceed 1.6 W/kg in spatial peaks, e.g., in the head, say, and the whole-body average cannot exceed 0.08 W/kg. A system reported in [8] has achieved 2 Mbps at a carrier frequency of 144 MHz. Because of their promise of small antenna sizes, high bandwidth, good power efficiency, and low-cost transceivers, **3.1 GHz–10.6 GHz UWB** impulse systems have also been researched for such applications. The system reported in [9] has a 25 dB loss for every 2 cm of meat-tissue thickness in the 3 GHz–5 GHz band. UWB systems for cochlear-implant applications, which require shorter link distances, have been tested by bringing the knuckled fists of each hand together [10]. A rectenna system for transmitting RF power at 915 MHz in the **902–928 MHz band** (see Chapter 17) has achieved an RF link gain of −33 dB across 1.5 cm of bovine tissue [11]. A chip for reporting neural data through 2 cm of skin and skull in the brain and ~2 cm in air has achieved 0.33 Mbps in the 433 MHz band [12]. An integrated inductive powering system at 2.64 MHz on this chip has thus far achieved 1% power-link efficiency.

18.13   The skin depth of biological tissue

A useful parameter that characterizes RF plane-wave propagation with frequency f through a medium with conductivity σ, dielectric permittivity ε, and magnetic

(a)   (b)

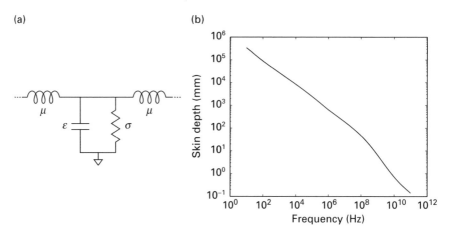

Figure 18.17a, b. A transmission-line model of biological tissue is shown in (a). The values of ε and σ vary with frequency as shown in Figure 16.18 for typical heterogeneous biological tissues. The variation in the skin depth for with frequency is shown in (b) for such tissues.

susceptibility μ is the skin depth $d(f)$, the distance over which there is exponential attenuation in the wave's amplitude. A medium of thickness $5d(f)$ will attenuate the wave amplitude by e^{-5} and the energy by e^{-10}, i.e., by about 40 dB. An intuitive way to understand skin depth is to imagine that the medium is composed of an infinitesimal set of inductances, capacitances, and conductances arranged in a transmission-line configuration as shown in Figure 18.17 (a) with the conductance, capacitance, and inductance per unit length in the line equal to σ, ε, and μ respectively. This transmission line models wave propagation in the medium just as Figure 17.1 models wave propagation in a 3D structure, as predicted by Maxwell's equations. For such a transmission line, the wave number k is given by [1], [2]

$$k = \sqrt{Z(f)Y(f)} \qquad (18.43)$$

where $Z(f) = j2\pi f \mu$ is the lateral impedance per unit length and $Y(f) = j2\pi f \varepsilon(f) + \sigma(f)$ is the shunt conductance per unit length in the line. We shall discuss such wave propagation in more depth in Chapter 23. Since such a wave propagates according to $\exp(kx)\exp(-j2\pi ft)$, the skin depth is given by

$$d(f) = \frac{1}{\text{Re}\{k(f)\}} \qquad (18.44)$$

We use the real part in Equation (18.44) because only this part of $k(f)$ causes attenuation in the wave as it propagates. The imaginary part of $k(f)$ causes phase shifts characteristic in wave propagation but has no effect on the wave amplitude. By using values of $\sigma(f)$ and $\varepsilon(f)$ given in [13] for biological tissue, which are similar to those shown in Figure 16.18, and which are available on the web at [14], we can get an estimate for the skin depth as a function of frequency. Figure 18.17 (b) shows that the skin depth decreases with frequency f such that higher-frequency RF waves are significantly more attenuated. The values of $\sigma(f)$ and $\varepsilon(f)$ vary depending on whether the biological tissue is skin, muscle, fat, or bone. The values shown in Figure 16.18, which are similar to the ones that we have used, represent a typical average for heterogeneous tissue. Figure 18.17 (b) illustrates that beyond approximately 100 MHz, the skin depth falls steeply such that there is significantly more attenuation. Indeed almost all reported systems with data bandwidths exceeding 2 Mbps have operated with carrier frequencies below 100 MHz.

In general, due to the steep attenuation of frequencies beyond a few tens of MHz in biological tissue, when transmitting data through such tissue, higher data bandwidths are not necessarily achieved at higher carrier frequencies. That is one of the reasons that the impedance-modulation system discussed in this chapter operates near a carrier frequency of 25 MHz and achieves good energy efficiency even at relatively high data bandwidths.

References

[1] Thomas H. Lee. *The Design of CMOS Radio-Frequency Integrated Circuits*. 2nd ed. (Cambridge, UK: New York: Cambridge University Press, 2004).

[2] Behzad Razavi. *RF Microelectronics* (Upper Saddle River, NJ: Prentice Hall, 1998).

[3] S. Mandal and R. Sarpeshkar. Power-Efficient Impedance-Modulation Wireless Data Links for Biomedical Implants. *IEEE Transactions on Biomedical Circuits and Systems*, **2** (2008), 301–315.

[4] A. C. H. MeVay and R. Sarpeshkar. Predictive comparators with adaptive control. *IEEE Transactions on Circuits and Systems II: Analog and Digital Signal Processing*, **50** (2003), 579–588.

[5] Carver Mead. *Analog VLSI and Neural Systems* (Reading, Mass.: Addison-Wesley, 1989).

[6] R. Sarpeshkar, R. F. Lyon and C. A. Mead. A low-power wide-linear-range transconductance amplifier. *Analog Integrated Circuits and Signal Processing*, **13** (1997), 123–151.

[7] P. D. Bradley. An ultra low power, high performance medical implant communication system (MICS) transceiver for implantable devices. *IEEE Biomedical Circuits and Systems Conference (BioCAS)*, London, UK, 158–161, 2006.

[8] J. Thoné, S. Radiom, D. Turgis, R. Carta, G. Gielen and R. Puers. Design of a 2 Mbps FSK near-field transmitter for wireless capsule endoscopy. *Sensors & Actuators: A. Physical*, (2008).

[9] M. R. Yuce, T. Dissanayake and H. C. Keong. Wireless telemetry for electronic pill technology. *Proceedings of the IEEE Conference on Sensors*, Christchurch, New Zealand, 2009.

[10] T. Buchegger, G. Oßberger, A. Reisenzahn, E. Hochmair, A. Stelzer and A. Springer. Ultra-wideband transceivers for cochlear implants. *EURASIP Journal on Applied Signal Processing*, **2005** (2005), 3069–3075.

[11] S. O'Driscoll, A. Poon and T. H. Meng. A mm-sized implantable power receiver with adaptive link compensation. *Digest of Technical Papers of the IEEE International Solid-State Circuits Conference (ISSCC)*, San Francisco, California, 294–295,**295a**, 2009.

[12] R. R. Harrison, P. T. Watkins, R. J. Kier, R. O. Lovejoy, D. J. Black, B. Greger and F. Solzbacher. A low-power integrated circuit for a wireless 100-electrode neural recording system. *IEEE Journal of Solid-State Circuits*, **42** (2007), 123–133.

[13] S. Gabriel, R. W. Lau and C. Gabriel. The dielectric properties of biological tissues: III. Parametric models for the dielectric spectrum of tissues. *Physics in Medicine and Biology*, **41** (1996), 2271–2293.

[14] Italian National Research Council and Institute for Applied Physics. Available from: http://niremf.ifac.cnr.it/tissprop/.

Section IV

Biomedical electronic systems

19 Ultra-low-power implantable medical electronics

When one door of happiness closes another opens; but we often look so long at the closed one that we do not see the one which has opened for us.

Helen Keller

Implantable electronics refers to electronics that may be partially or fully implanted inside the body. Several implanted electronic systems today have revolutionized patients' lives. For example, more than 130,000 profoundly deaf people in the world today have a cochlear implant in their inner ear or cochlea that allows them to hear almost normally [1]. Some cochlear-implant subjects have word-error recognition rates in clean speech that are better than those of normal hearing subjects, and they understand telephone speech easily. Cochlear implants electrically stimulate the auditory nerve, the nerve that conducts electrical impulses from the ear to the brain, with an ac current. Cochlear implant subjects are so profoundly deaf that even the feedback-limited $\sim 1000\times$ gain of a hearing aid is not large enough to help them hear. Therefore, an implant that directly stimulates their auditory nerve is necessary. Cochlear implants today are partially implanted systems: the electrodes and a wireless receiver are implanted inside the body while a microphone, processor, and wireless transmitter are placed outside the body.

Patients with Parkinson's disease have had their quality of life significantly improve because of a deep brain stimulator (DBS) that has been fully implanted inside their bodies [2]. This stimulator provides ac electrical current stimulation to a highly specific region in their brain, which is most commonly the subthalamic nucleus (STN). The stimulation is an effective treatment for their uncontrollable shaking and inability to initiate movement. Some Parkinson's patients recover so well that they are even capable of dancing. Pacemakers, which were introduced several decades ago, have helped patients with disorders in their heart's rhythms lead normal and safe lives. Implanted cardiac defibrillators have been similarly beneficial.

While these and other FDA-approved clinical treatments are already in the market place, many believe that we have only scratched the surface of what will be possible in the future. For example, treatments to help cure paralysis by recording from electrical signals in the motor regions of the brain, and decoding these electrical signals to stimulate a muscle or robot arm, have already been demonstrated in patients [3]. Retinal implants for the blind function by electrically stimulating cells in the eye. They have been researched for decades and are beginning to show promise [4], [5]. Brain implants that stimulate regions of the deep brain (the lateral geniculate nucleus (LGN)) or regions at the back of the brain (V1) are actively being explored as an alternative method to treat blindness [6], [7]. Electrical stimulation of the deep brain and other regions of the brain are being researched for treatments of epilepsy, stroke, chronic pain, treatment-resistant depression, obsessive compulsive disorder, chorea, and other disorders. Functional electrical stimulation (FES) of peripheral muscles has been used to help patients with paralyzed limbs. The space of applications grows every day. New research on using vestibular implants to treat balance disorders, brain implants to treat speech disorders, implants to treat facial paralysis, spinal-cord implants for controlling chronic back pain, implants for treating urinary incontinence, and vagal-nerve implants for the treatment of immune disorders appears every day.

The brain, heart, and muscle are natural regions for electrical treatments but the applications of implantable electronics are not restricted to these areas. For example, electrical actuation of implanted drug capsules via external control, e.g., when a heart attack is detected, or to regulate drug dosage in other regions of the body is an obvious application. All cells in the body are electrically charged, and diseased tissue often has significantly different electrical characteristics from healthy tissue. Electrochemical sensing and actuation with chemically functionalized electrodes will undoubtedly impact closed-loop medical treatments in the future [8]. For example, feedback control of implanted insulin pumps via glucose sensing for diabetes or, more generally, drug release based on electrochemical sensing of drug efficacy will make medical treatments more effective [9]. Biology and medicine are benefiting greatly from an increasingly quantitative understanding of biological system operation and from increasingly better measurement techniques that derive their origins from other disciplines like physics, chemistry, and engineering. In turn biology is inspiring new engineering designs. The world of electronics is no exception. The coming years will undoubtedly usher in a new era of explosive growth in the border between biology and electronics and between medicine and electronics.

Implanted systems need to be small since space is always at a premium within the body. Therefore, the size of RF coils needed for wireless powering of the implant or the size of batteries in the implanted unit must also be small. Independent of space, power dissipation must be small in an implanted system to ensure that there is no tissue heating or damage. The heating concern is especially stringent in the brain where small increases in temperature can cause a fever. For both of these reasons, ultra-low-power operation is a very stringent constraint in implantable systems. Power dissipation is also a strong determinant of cost: since a low-power system can be sized smaller (batteries and/or wireless recharging coils are smaller), the cost of expensive hermetic and biocompatible materials and packaging is reduced, and the complexity of the surgery needed for implantation reduces as well. With reduced power, the cost of the battery, which can be a significant portion of the cost and weight of the implant, reduces as well.

The intertwined reasons of size, heating, weight, and cost make ultra-low-power electronics extremely important in implantable systems. Therefore, we shall focus on such systems in this chapter, providing concrete examples of electronic systems that have actually been built and tested in humans or in animal subjects.

We shall begin by studying a cochlear-implant example in depth, which is useful in understanding the broader field since concerns of low-power telemetry, efficient wireless recharging, low-power sensing, low-power processing, and low-power stimulation are common to all implants. Chapter 16 discusses how to build efficient wireless recharging circuits for implants in depth. Chapter 18 focuses on how to build low-power telemetry circuits for implants in depth. Therefore, in this chapter, we shall focus on sensing circuits, processing circuits, and stimulation circuits.

We first describe an ultra-low-power cochlear-implant processor. Then we describe how to architect low-power and miniature electrode-stimulation circuits, which are useful in all neural and cardiac implants. Following these descriptions, we outline how brain implants for the blind could potentially be built in the future, by learning from the better-established domain of cochlear implants. Such implants could be built with either deep-brain stimulation hardware, or with surface-brain implants. In fact, a general brain-machine interface with common hardware could be useful for several clinical applications including the treatment of paralysis, stroke, and epilepsy. However, the sensors required for various applications vary. We shall describe how to create ultra-low-power microphone front ends for cochlear implants, ultra-low-power imagers for brain implants for the blind, and ultra-low-power neural amplifiers for brain-machine interfaces for the treatment of paralysis and stroke. This chapter will suggest that, with ongoing research and development, there may be significantly fewer deaf, blind, dumb or paralyzed people in the future.

We shall defer our discussion of cardiac applications to Chapter 20, which focuses on ultra-low-power noninvasive medical electronics with a special focus on the heart. Implantable cardiac stimulation or recording shares many functional characteristics of neural recording or stimulation, so it is easy to generalize from one to the other. However, the bandwidth needed for cardiac applications is usually significantly less than that needed for neural applications.

In this chapter, for reasons of space, we shall only model electrodes as abstract impedances. Chapter 25 on batteries provides an in-depth discussion of electrodes and electrochemistry which is useful for understanding how electrical stimulation actually affects the nervous system.

The extra-cellular salt-like solution in the body functions as an electrolyte that a stimulating or recording electrode is immersed in. The positive or negative terminal of a battery is also comprised of an electrode immersed in an electrolyte. Chemical reactions with a net motion of electronic charge *must* occur at battery electrodes to provide energy. Thus dc current is always present at battery electrodes when a battery is functional. In neural or implanted cardiac electrodes, net chemical reactions and a net exchange of electronic charge are always *avoided* through careful charge-balanced current stimulation or with passive high-impedance electrical recording.

19.1   Cochlear implants or bionic ears

Figure 19.1 illustrates the operation of a cochlear-implant or bionic-ear system with seven labeled parts: 1) A microphone transduces sound to electrical signals; 2) the signals are conveyed via a cable to a speech processor, which is often part of a behind-the-ear (BTE) unit (see Figure 1.3 in Chapter 1 for an example); 3) the signals are processed by the speech processor to extract the logarithm of the spectral energy in 8–22 filter bands, and these energy coefficients are compressed to reduce the electrical dynamic range of stimulation; 4) the processed outputs are conveyed back via the cable to a headpiece transmitter for wireless transmission through the skin, which minimizes the risk of infection as compared with a wired link from inside the body to outside the body; 5) an implanted wireless receiver rectifies the wireless transmission to create a power supply and demodulates the wireless transmission to extract signals for electrode stimulation; 6) charge-balanced ac current signals are conveyed to electrodes implanted in the inner ear or cochlea with spectral intensity mapped to current intensity and frequency mapped to place: loud high-frequency spectral energy causes intense stimulation of the basal or shallow electrodes and soft low-frequency spectral energy causes

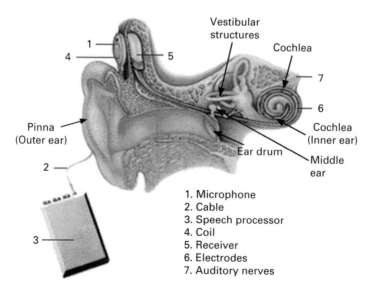

1. Microphone
2. Cable
3. Speech processor
4. Coil
5. Receiver
6. Electrodes
7. Auditory nerves

Figure 19.1. Configuration of a cochlear implant or bionic ear. (Courtesy of Advanced Bionics Corporation, Valencia, California.) Figure 1.3 shows a more compact configuration with a behind-the-ear (BTE) unit. The main parts of the ear, the outer ear, the middle ear, and the inner ear or cochlea are also marked. The vestibular structures are organs important for balance.

gentle stimulation of the apical or deep electrodes; 7) the electrical stimulation excites the auditory nerve and eventually reaches the brain, which then interprets the electrical stimulation as sound.

Cochlear implants in the future will be fully implanted inside the body, making deaf subjects indistinguishable in appearance from normal-hearing subjects, and allowing them even to swim while wearing a cochlear implant. Such implants cannot be constantly wirelessly powered and therefore need an implanted battery to function. The battery can be wirelessly recharged at periodic intervals. However, the best lithium-ion batteries today can only be recharged about 1000 times. If a fully implanted cochlear implant is to operate for 30 years in the body on a small 100 mA h battery with 1000 wireless recharges without the need for re-surgery to replace the battery, its total power consumption must be less than 1 mW. If 16 electrodes (a typical number for good performance in a cochlear implant and one that is representative of commercial designs) consume 750 μW of power for electrical stimulation alone, the power budget for wide-dynamic-range microphone transduction and processing must be less than 250 μW. This is a very stringent power budget indeed. A current highly energy-efficient audio ADC operating at 16 bits and 44 kHz will alone consume at least 50 fJ $\times 2^{16} \times$ 44 kHz = 145 μW [10]; with a low-power 106 μW microphone preamplifier that is described later, the microphone front end and ADC alone just make up the entire power budget. The sampling rate and precision of sampling in the ADC are needed for the quantization noise, roundoff noise, and temporal-aliasing noise to not significantly degrade the ~80 dB dynamic range of operation. The leakage power in highly energy-efficient commercially available DSPs alone at 25°C and 1.6 V operation is about 180 μW and increases to about 277 μW at 37°C in the body [11], [12], [13]. Such energy-efficient DSPs today operate at nearly 250 μW/ MIP when chip and board-level parasitics are considered in a practical cochlear implant (MIP = million instructions per second). Thus, even a ~20 MIPS algorithm in a cochlear-implant processor will consume nearly 5 mW today. Hence, a traditional ADC-then-DSP solution has a power consumption of 5 + 0.14 mW = 5.14 mW that is nearly 35 times the 145 μW power budget available for processing. We have neglected the power consumption of anti-alias filters in computing this estimate. The use of 100 μW/MIP for the DSP still requires a 15× power reduction.

One solution is to delay digitization with programmable analog preprocessing to a more optimal point such that significantly lower-bandwidth, lower-precision digitization may be performed on the speech spectral envelopes. Such envelopes have a bandwidth of 100 Hz or less due to the rate of speaker articulation and need less than 7-bit precision due to inherent variability in a speaker's voice and environmental variability. Deaf patients can resolve 5 bits of precision in a spectral envelope channel at best. The ultra-low-power programmable analog bionic-ear processor that we describe below meets the power budget by delaying digitization, while allowing 86 patient parameters to be programmed. It exploits 'leakage' power, i.e., it uses subthreshold analog processing, rather than treating such power as a nuisance. The processor that we describe implements all of the processing in the analog domain and digitizes its output, all within $145\,\mu W$, which is about $35\times$ lower in power than an ADC-then-DSP processing solution. Due to the use of low-power subthreshold analog techniques, it can meet the power budget of a fully implanted cochlear implant in a $1.5\,\mu m$ technology today that would not be possible with a conventional ADC-then-DSP solution even at the end of Moore's law: the microphone preamplifier and ADC alone comprise the total budget such that even the use of a $0\,\mu W/MIP$ DSP in the future does not help. Thus, it is advantageous to think outside the box of traditional ADC-then-DSP electronic architectures.

Some of the analog processor's power consumption is thermal-noise limited and does not scale with technology while some of its power consumption is parasitic-capacitance limited and does scale with technology. The same is true of all analog systems including ADCs. Hence, the power of analog preprocessing and ADC power scale significantly more weakly with technology than digital power consumption. Analog systems and ADCs of the future will need to exploit low-voltage-compatible current-mode processing as described in Chapter 14 or time-based architectures as described in Chapter 15 to stay efficient, if they are implemented in digital CMOS processes.

Delaying digitization with low-power analog preprocessing not only reduces the power costs of the ADC but it also reduces the power costs of post processing digital bits. A digital system that works at lower bandwidth can be designed to be significantly more power efficient than one that works at higher bandwidth as we discuss in Chapter 21. There is therefore an optimal point to digitize: too late, and the costs of maintaining SNR with analog preprocessing increase; too early and ADC and digital power increases. We discuss the optimal point for digitization in a more general context in Chapter 22.

The ultra-low-power programmable analog bionic-ear processor that we now describe was tested on a cochlear-implant subject and worked on its very first trial: The patient was able to understand speech by removing her traditional digital processor and replacing it with this processor. The analog processor contained several DAC, calibration, and configuration bits (373 bits in total) that were digitally programmed to match the parameter settings of her DSP, enabling 86 patient-specific parameters to be changed. *Through the use of feed-forward and feedback calibration and robust-biasing techniques, the processor was designed to be very insensitive to power-supply noise, temperature variations, $1/f$ noise, and transistor mismatch, which are essential features in practical ultra-low-power subthreshold analog systems.*

It is worth noting that the use of ultra capacitors, which have low energy density, and good power density, but that can be recharged at least 10,000 times does not alleviate low power concerns. It may appear that we can potentially just wirelessly recharge an implanted ultra capacitor to power our implant at the end of each day and not worry about power. However, the low energy density of ultra capacitors means that they usually have to be used in conjunction with a battery which is charging them in the background such that they can deliver a flash of high energy when needed. If they are used alone without the battery, they increase the size of the implant that is needed significantly, drastically turning up costs and reducing medical viability. Therefore, ultra-low-power electronics will always be important in implanted medical systems, not just for reasons of power, but also because of the impact of power on size, weight, and cost. We shall discuss such issues in more technical depth in Chapters 25 and 26, which focus on batteries, energy harvesting, and energy generation. We shall see that energy density increases when power density decreases. Thus, ultra-low-power operation is an important part of an efficient system solution in all implantable electronics including the bionic-ear processor, which we shall now describe.

19.2   An ultra-low-power programmable analog bionic ear processor

Figure 19.2 shows the overall architecture of the single-chip analog bionic ear processor. An FG3329 electret microphone's output is amplified by 20 dB with an on-chip digitally programmable audio front end (AFE), which is architected to have low power and high power-supply rejection in the noisy mixed-signal-and-RF environment of a typical cochlear implant. A digitally programmable automatic-gain-control circuit (AGC) compresses the 77 dB input-referred dynamic range of sound intensity ($\sim$30 dB SPL to $\sim$110 dB SPL) to a 57 dB internal dynamic range (IDR) at which the rest of the analog processing operates. The AGC reduction in dynamic range enables both lower-power operation for the rest of the analog processing as well as improved sensitivity in the spectral analysis

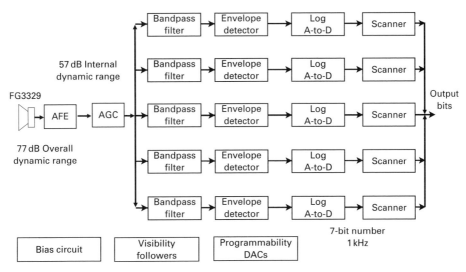

Figure 19.2. Block diagram of the ultra-low-power programmable analog bionic ear processor. Reproduced with kind permission from [17] (©2005 IEEE).

to changes in sound intensity. The instantaneous dynamic range of speech is less than 60 dB even if talker-to-talker variability and talker effort are included. The output of the AGC is fed to a bank of sixteen spectral-analysis channels that span the audio spectrum over a user-programmable frequency range that can be anywhere from 100 Hz to 10 kHz.

Each channel is composed of a programmable band pass filter, a programmable envelope detector to extract the spectral energy, and a programmable log ADC. Each log ADC outputs 7-bit numbers at sampling rates that can be programmed to be 500 Hz, 1000 Hz, 1500 Hz, or 2000 Hz or set by a user-supplied external current. Such high-rate sampling is rather excessive for sampling speech spectral envelopes but may provide redundant information and may excite auditory nerves in a more efficacious manner in some patients. The parallel bits from each of the 16 channels are scanned onto a serial bus and reported off chip for use in a sequential-and-cyclic channel-by-channel stimulation scheme known as continuous interleaved sampling (CIS). The CIS scheme is highly effective in alleviating electrode interactions in subjects and is the most popular stimulation paradigm in cochlear implants today. The biasing circuit generates power-supply-noise-immune voltage references for various cascode and voltage references needed on the chip. It also generates power-supply-noise-immune proportional-to-absolute-temperature (PTAT) reference currents that are distributed throughout the chip to bias the programmability DACs. These DACs program patient parameters throughout the chip. The PTAT biasing ensures that all frequency parameters on the chip are invariant to temperature because they are proportional to a small-signal g_m and g_m is proportional to $I_{bias}/(kT/q)$, where I_{bias} is a PTAT current. The visibility followers allow seven waveforms from any channel to be reported off chip for debugging of patient waveforms if needed, a feature that is available in DSP-based processing as well. We shall now describe the microphone AFE, the AGC, the bandpass filter, the envelope detector, the log ADC, the CIS scanner, the programming and visibility circuits, the robust-biasing circuits, and the experimental performance of the processor in more detail.

19.2.1   Low-power microphone AFE

Figure 19.3 shows the architecture of the microphone AFE. A schematic for the miniature FG3329 Knowles microphone, which is popular in hearing aids and cochlear implants, is drawn within the boxed square. A piezoelectret capacitor with a fixed charge responds to sound-pressure variations to create a time-varying voltage. The voltage is usually buffered via a source-follower created by the low-$1/f$-noise JFET and R_S and reported at the 'output terminal' of the microphone as v_{JF}; the drain of the JFET is usually tied to V_{DD}. However, tying the drain of the JFET to V_{DD} makes the output of the microphone very prone to power-supply noise due to the finite output conductance of the JFET and is undesirable in a cochlear implant. Therefore, in the design of Figure 19.3, the drain of the JFET is maintained at a quiet virtual reference, $V_{MIC,REF}$, via negative feedback

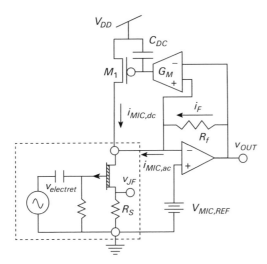

Figure 19.3. A low-power microphone preamplifier. Reproduced with kind permission from [17] (©2005 IEEE).

and the *current* from the microphone is sensed and amplified via a transimpedance-amplifier configuration. The feedback resistor of the transimpedance amplifier is R_f. Transimpedance amplifiers are described in depth in Chapter 11. With respect to the source-follower-buffered output voltage, the output voltage of the trans-impedance amplifier has a gain of $-R_f/R_S$. Since the thermal noise voltage of the resistor R_f scales as $4kTR_f$ while the output signal power increases like the square of the gain, and consequently as R_f^2, increasing the value of R_f is advantageous for increasing gain and increasing output SNR. However, the FG3329 has an in-built dc bias current of approximately $20\,\mu\text{A}$ such that increasing the value of R_f to get even a gain of 10 saturates the output of the amplifier on the 2.8 V power supply. The solution is to shunt the dc current of the microphone, which carries no audio information, away from the transimpedance amplifier such that it does not flow through R_f and is not amplified. However, we do want to amplify ac variations in current, which carry audio information. The transconductor G_M, C_{DC}, and M_1 form a negative-feedback loop that helps us accomplish these goals. Any increase in the current input to the transimpedance amplifier increases the voltage drop across R_f, which increases the source-to-gate voltage on M_1, which increases $i_{MIC,dc}$, which attenuates the original increase in current by shunting it away from the transimpedance amplifier. If the gain of the operational amplifier is sufficiently large, the loop transmission of this negative-feedback loop is very nearly $-G_M/(sC_{DC})\left(g_m^{M1}R_f\right)$. Thus, with an appropriate choice of G_M and C_{DC} we can arrange the closed-loop bandwidth of the feedback loop to be nearly 100 Hz such that it is ineffective at attenuating desired currents beyond 100 Hz but attenuates unwanted dc and low-frequency currents below 100 Hz. The negative-feedback loop therefore attenuates $1/f$ noise currents of M_1 and the JFET source-follower below 100 Hz as well.

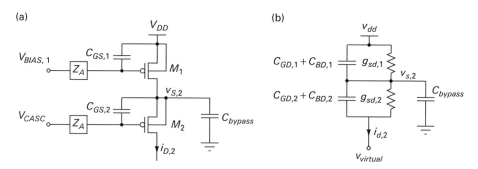

Figure 19.4a, b. A strategy for improving power-supply rejection is shown in (a) and its small-signal equivalent is shown in (b). Reproduced with kind permission from [14] (©2003 IEEE).

One issue with the circuit of Figure 19.3 is that the finite output conductance of M_1 couples in power-supply noise from V_{DD}, which was the reason for using a current-sensing topology in the first place, so we are back to the problem we started with! We can, of course, simply increase the channel length of M_1 greatly, a freedom we do not possess with the JFET, which is built into the microphone. But we can do even better. Figure 19.4 illustrates a series-transistor circuit that is useful for building a current source with high power-supply rejection. It is useful for improving the power-supply rejection of M_1, and for improving the power-supply rejection of the operational amplifier as well. Therefore, we will take a brief diversion to discuss how this circuit works.

In Figure 19.4 (a), the transistor M_1 functions like a current source and the transistor M_2 functions as a cascode transistor. The elements marked Z_A represent high-impedance elements such as a set of parallel diodes, such as the 'adaptive element' described in Chapter 11, or a subthreshold transconductor. The value of Z_A is sufficiently large such that it sets dc operating points on the gates of transistors M_1 and M_2 but may otherwise be assumed open at all but extremely low frequencies that are not of interest. For both transistors, we intentionally add large capacitances C_{GS} in parallel with the transistor's intrinsic C_{gs} such that $C_{GS} \gg C_{gs}$, C_{gd}. The small-signal equivalent circuit of the series transistors is shown in Figure 19.4 (b). The value of g_{sd} for each transistor is given by

$$g_{sd} = g_m \frac{C_{gd}}{C_{gd} + C_{GS}} + g_o \qquad (19.1)$$

Since $C_{GS} \gg C_{gd}$ any change in the source voltage of the transistor is coupled to its gate with a gain of almost 1 such that v_{gs} barely changes and the effects of the g_m generator in the transistor are nearly cancelled. Thus, g_{sd} is then dominated by g_o, the output conductance of the transistor due to channel-length modulation. Equation (19.1) quantitatively predicts how small g_{sd} can be made to be. Mere increases in the gate length of a transistor to decrease g_o without simultaneously increasing C_{GS} cause the g_m generator's effects to eventually dominate g_{sd} in

Equation (19.1), which is undesirable for reducing power-supply coupling. Thus, the use of a large C_{GS} is essential. The capacitances in parallel with g_{sd} are well approximated as $C_{gd} + C_{bd}$ if $C_{GS} \gg C_{gd}$ as is the case. If the power supply has a small-signal variation v_{dd}, g_{sd}, $C_{gd} + C_{bd}$, and C_{bypass} form a lowpass network that filters out this variation, such that $v_{s,2}$ barely moves. The small motion in $v_{s,2}$ causes only a small current $i_{d,2}$ since g_{sd2} and $C_{GD,2}$ are intentionally designed to be quite small. Hence this current source is relatively insensitive to variations in the power-supply voltage at all but extremely low frequencies.

The current source topology of Figure 19.4 (a) is used for biasing *all* current sources in the operational amplifier to improve its power-supply rejection with Z_A being the adaptive element described in Chapter 11. Similarly, in Figure 19.3, the transistor M_1 is replaced by the series-transistor configuration of Figure 19.4 (a) to improve power-supply rejection. However, the gate of the non-cascode top transistor of this configuration is driven by the high-impedance output of the G_M transconductor as in Figure 19.3 rather than by a dc bias: the feedback loop in Figure 19.3 must still determine this bias. The gate of the cascode transistor of the configuration is coupled to a dc bias via an adaptive element just as in Figure 19.4 (a).

The amplifier of Figure 19.3 with the high-power-supply-rejection-biasing strategies of Figure 19.4 achieves 20 dB of gain, 80 dB dynamic range (30 dB SPL – 110 dB SPL), 100 μW of power consumption *including* the in-built 60 μW power consumption of the microphone, and an input-referred noise floor of 5 μV rms in the 100 Hz–10 kHz bandwidth. The noise of the AFE is dominated by that of the microphone, falling by more than an order of magnitude when the microphone is off. An auxiliary signal, e.g., from a CD output, can be coupled into the virtual-reference input of the transimpedance amplifier via a simple series RC circuit, which slightly increases the input-referred noise to 8 μV rms. The power-supply rejection ratio (PSRR), i.e., $(v_{out}/v_{if})/(v_{out}/v_{dd})$, was measured to be over 90 dB at 300 Hz. It falls with frequency as the loop gain of the operational amplifier falls, and is 50 dB at 10 kHz, where the operational-amplifier gain and consequently the gain of the AFE begins to roll off. The power-supply transfer function continues to attenuate from -45 dB at 10 kHz to asymptote at nearly -70 dB at 26 MHz, a feature useful in attenuating power-supply noise at RF frequencies. At very low frequencies (100 Hz and below), because the feedback loop decreases the finite output impedance of the transconductor G_M and prevents variations in V_{DD} from coupling through to the gate of the bias-setting transistor M_1 in Figure 19.3, there is a rise in the power-supply transfer function. Experimental tests with large RF coils that encircle the microphone show that the operation of the AFE is extremely robust to power-supply noise. Further details of the AFE are described in [14]. The output of the AFE is fed to an AGC circuit that compresses the 80 dB output dynamic range of the AFE (50 μV rms to 500 mV rms) to a nearly 60 dB internal dynamic range at which the spectral-analysis channels operate (500 μV rms to 500 mV rms).

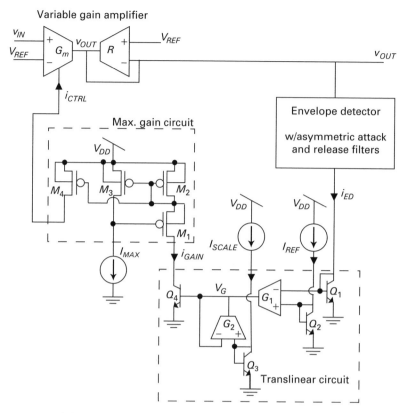

Figure 19.5. A low-power automatic gain control (AGC) circuit. Reproduced with kind permission from [18] (©2005 IEEE).

19.2.2 Low-power AGC circuit

Figure 19.5 shows the schematic of the AGC circuit used in the bionic-ear processor. A variable gain amplifier (VGA) is architected with a transconductor-resistor $G_m - R$ topology with both the transconductor and resistor implemented with wide-linear-range transconductors as shown. Wide-linear-range transconductors are described in Chapter 12. The gain of the VGA and consequently the AGC is modulated by changing the bias i_{CTRL} of the G_m transconductor via negative feedback such that i_{CTRL} decreases if v_{OUT} increases and vice versa. A compression characteristic is achieved because i_{CTRL} is large at the smallest input or output levels, providing a gain of nearly 10 at these levels, and small at the largest input or output levels, providing a gain of nearly 1 at these levels. The envelope detector senses the envelope of the output voltage v_{OUT} and converts it to a current i_{ED} using circuits that are described later. Increasing changes in envelope are sensed with a faster 'attack' time constant than decreasing changes in envelope, which are sensed with a slower 'release' time constant. The asymmetry mimics the gain-control response seen in the real auditory system, which responds more quickly to soft-to-loud transients than loud-to-soft ones. The attack and release time constants are digitally programmable. The translinear circuit shown in Figure 19.5 transforms i_{ED} to an output i_{GAIN} given by

$$ i_{GAIN} = I_{SCALE} \left(\frac{I_{REF}}{i_{ED}} \right)^{\frac{I_1}{I_2}} \tag{19.2} $$

where I_1 and I_2 are bias currents of the G_1 and G_2 transconductors respectively. Since I_1 and I_2 are digitally programmable, the power-law compression characteristic of the AGC that results from Equation (19.2) also is digitally programmable and is described by

$$ v_{OUT} \propto v_{IN}^{1/(1+I_1/I_2)} \tag{19.3} $$

The 'maximum gain' circuit is an adapted Wilson-mirror circuit with two inputs i_{GAIN} and I_{MAX} and an output i_{CTRL} that implements a soft minimum according to

$$ i_{CTRL} \approx \min(I_{MAX}, i_{GAIN}) \tag{19.4} $$

The circuit uses negative feedback to automatically reduce the current through both M_1 and M_3 to be the smaller of I_{MAX} or i_{GAIN}: if i_{GAIN} exceeds I_{MAX}, Q_4 is driven into its saturation regime to reduce i_{GAIN}; if I_{MAX} exceeds i_{GAIN}, the current-source supplying I_{MAX} is driven below its crowbar voltage to reduce I_{MAX} and M_1 operates as a switch. Thus, the AGC turns up the gain at smaller amplitudes until it reaches a maximum value for small signals beyond which the gain is no longer increased. Consequently, the input-output curve of the AGC functions with a compression characteristic based on Equation (19.2) for large inputs. For inputs smaller than a certain 'knee', it reverts to a linear characteristic. The knee of the AGC is digitally programmable since I_{MAX} is digitally programmable. One of the advantages of this AGC is that the knee is 'soft' and not hard as in most software implementations, a feature appreciated by subjects who do not like their AGCs jittering between two values across a hard threshold. A hysteretic threshold can avoid jitter in these implementations, but it still results in an abrupt transition between the two regimes unlike this AGC. A knee is always necessary in an AGC to ensure that its gain is determined by the input and not by noise when the input is very small. The human auditory system is linear at small inputs as well. The AGC of Figure 19.5 consumes $30\,\mu W$ and is carefully designed such that the noise of the envelope detector and its other circuits do not degrade the output SNR and the input dynamic range. Further details of this AGC and interesting properties of its feedback loop dynamics are described in [15]. For example, one interesting property of this AGC is that its feedback loop gain is independent of intensity and only depends on the compression ratio which is advantageous for stability. The reference [15] also describes how more complex software AGCs with dual time constants and hold times may be architected through the use of simple analog nonlinear dynamical systems.

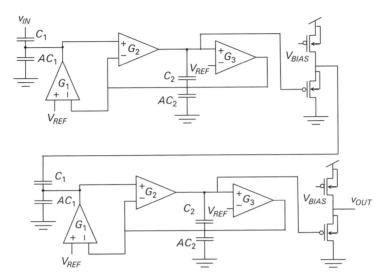

Figure 19.6. A capacitive-attenuation programmable bandpass filter. Reproduced with kind permission from [17] (©2005 IEEE).

19.2.3   Micropower bandpass filter

Figure 19.6 shows the circuitry of the micropower bandpass filter. The filter is a straightforward implementation of two identical stages of a bandpass $G_m - C$ filter coupled to each other by a source follower to avoid inter-stage loading effects. However, rather than attain wide linear range through the use of the well input and feedback-degeneration techniques, as described in Chapter 12, these transconductors attain wide linear range via simple capacitive attenuation. As we discussed in Chapter 12, in thermal-noise-dominated systems, widening the linear range increases the input-referred noise per unit bandwidth quadratically but reduces the bandwidth linearly such that there is a net linear increase in the noise power. Since the maximum signal noise power increases quadratically with a widening in linear range, there is a net linear increase in the dynamic range or maximal SNR of the filter. However, the price that is paid for attaining this increase is that we need to increase power consumption to maintain bandwidth. These same tradeoffs are true in the system of Figure 19.6 as well. Nodes that are tied to dc inputs in the filter do not require capacitive attenuation, which saves area. Nodes that share inputs may share one capacitive attenuator, which also saves area. Nodes that have no outputs tied to them 'float' and require very-low-frequency circuits that establish their dc value. The G_3 transconductors in Figure 19.6 are biased with very low currents and accomplish this function in Figure 19.6. The bias currents of the bandpass filter are digitally programmable such that the center frequency and quality factor of the filters can be altered over a large range. At $Q = 4$, which is a good setting for speech processing that trades off spectral and temporal resolution, the filters consume 147 nW at a center frequency of 150 Hz and 22 μW at a center frequency of 5 kHz with a dynamic range of 65.5 dB. If filters beyond a Q of 6 are desired, the superior topologies described in Chapter 13 should be used since their power scaling with Q for a given SNR and center frequency is linear with Q, rather than cubic like the simple filter of Figure 19.6. Nevertheless, the simple filter of Figure 19.6 works quite well in the bionic-ear processor and has a power consumption that is low enough. Further details of the filter are described in [16], [17], [18]. Future bionic-ear processors could reduce power even more aggressively through the use of current-mode techniques and the use of topologies with good power scaling with Q as we have described in Chapters 13 and 14.

19.2.4   Micropower envelope detector

Figure 19.7 (a) shows the envelope detector. The input v_{IN} from the output of the bandpass filter drives the negative terminal of a wide linear range transconductor (WLR) labeled G_m. The transconductor's output current i_{IN} is fed to a current rectifier that causes the positive and negative halves of i_{IN} to flow through M_p and M_n, respectively, that flips the sign of each half current via the mirrors connected to M_p and M_n, and that finally recombines these mirrored current halves to create i_{OUT}. Thus, i_{OUT} is simply $-i_{IN}$. The current i_{OUT} charges the capacitance C, whose output voltage then feeds back to the positive terminal of the WLR. Due to the sign transformation between i_{IN} and i_{OUT}, the feedback to the positive terminal is actually negative feedback. The overall configuration just implements a $G_m - C$ lowpass filter with transfer function $1/(\tau s + 1)$ and $\tau = G_m/C$ in what, at first, appears to be a strange fashion. The *current* that charges the capacitor in a

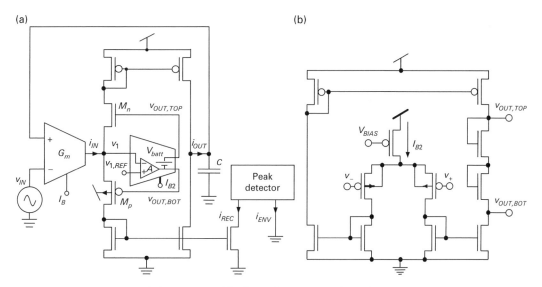

Figure 19.7a, b. A low-power wide-dynamic-range envelope detector. Reproduced with kind permission from [19] (©2003 IEEE).

lowpass $G_m - C$ filter has a highpass-filter transfer function $(\tau s/(\tau s + 1))G_m$ such that i_{IN} is only a faithful unattenuated replica of v_{IN} if the frequency of v_{IN} exceeds $1/(2\pi\tau)$. If we set $1/(2\pi\tau)$ to around 100 Hz, any dc and low-frequency information in v_{IN} or due to transconductor offsets are strongly attenuated, while the actual audio information from the filter beyond 100 Hz is maintained in i_{IN} with $i_{IN} = G_m v_{IN}$. In essence, the $G_m - C$ filter allows us to do V-to-I conversion with automatic auto-zeroing such that only ac information in v_{IN} is ported into i_{IN}. In addition to the auto-zeroing and V-to-I conversion, the current rectifier that is embedded in the $G_m - C$ filter allows us to rectify v_{IN} and eventually perform peak detection, such that the energy of a spectral-analysis channel can be detected.

The current rectifier in Figure 19.7 (a) consists of three parts: 1) the transistor M_n that conducts current if i_{IN} is a sink current and the transistor M_p which conducts current if i_{IN} is a source current; 2) the two current mirrors; 3) the feedback amplifier, which drives the gates of M_n and M_p to $v_{OUT,TOP}$ and $v_{OUT,BOT}$ respectively and maintains v_1 at a virtual reference voltage $V_{1,REF}$ through transimpedance-amplifier negative-feedback action. The feedback amplifier drawn in Figure 19.7 (a) as an operational amplifier with gain A and a 'battery' is shown in transistor-level detail in Figure 19.7 (b). The 'battery', which sets a constant value for $v_{OUT,TOP} - v_{OUT,BOT}$, is implemented by having the bias current of the operational transconductance amplifier flow through two diodes. The battery and amplifier serve to reduce the dead-zone of the rectifier, i.e., the minimum threshold value of i_{IN}, I_{in}^{th}, that just leads to rectification. The battery pre-biases the gate-to-source voltage across M_n and the source-to-gate voltage across M_p such that the feedback-amplifier output swing needed to turn on the abrupt-exponential subthreshold transistors is reduced. The turn-on voltage varies only logarithmically with input current. If we approximate it as a constant, V_D, the battery voltage as V_{batt}, and the impedance present at the V_1 node as $Z_{in}(s)$, then the threshold current needed to just cause rectification is given by

$$I_{in}^{th}(s)Z_{in}(s)(1 - (-A)) = V_D - V_{batt}$$

$$I_{in}^{th}(s)\frac{1 + A}{C_1 s + (1 + A)C_{gs}s} = V_D - V_{batt}$$

$$I_{in}^{th}(s) \approx sC_{gs}(V_D - V_{batt}) = sC_{gs}V_D^{eff} \qquad (19.5)$$

$$\boxed{\begin{aligned} I_{in}^{th}(f) &= 2\pi fC_{gs}V_D^{eff} \\ V_{in}^{th}(f) &= \frac{2\pi fC_{gs}V_D^{eff}}{G_m} \end{aligned}}$$

Here, C_1 is the parasitic capacitance to ground at the V_1 node and C_{gs} is the capacitance across the input and output terminals of the feedback amplifier, composed mainly of the C_{gs} capacitances of M_n and M_p. If A is sufficiently large, C_1 is made irrelevant by the amplifier since the Miller capacitance $(1 + A)C_{gs}$ dominates the overall capacitance as we have assumed in deriving

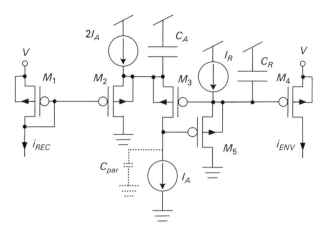

Figure 19.8. A peak-detector circuit. Reproduced with kind permission from [19] (©2003 IEEE).

Equation (19.5). Since C_{gs} is minimal in subthreshold operation, it is a good region of operation to obtain a low value of $I_{in}^{th}(f)$. Minimum size devices must be used for M_n and M_p to lower C_{gs}. The threshold current of the rectifier increases linearly with frequency and leads to a input voltage threshold of $V_{in}^{th}(f) = I_{in}^{th}(f)/G_m$. The use of V_{batt} to reduce V_D to V_D^{eff} is an example of Class-AB biasing that we discussed in Chapter 14. We shall now describe the black-box peak detector of Figure 19.7 (a) in more detail.

The current-mode peak-detector circuit is shown in Figure 19.8. If we pretend for a moment that the source follower comprised of M_5, and I_R may be replaced by a short and that $C_R = 0$, then M_1, M_2, M_3, and M_4 form a classic current-mode lowpass filter, which we have described in Chapter 14 (see Figure 14.8 (a)). The current mode lowpass filter merely averages the rectification input current i_{REC} fed to it from the mirror in Figure 19.7 (a) to generate an output current i_{ENV} with an attack time constant given by

$$\tau_A = \frac{C\phi_t}{\kappa I_A} \tag{19.6}$$

The 'attack' time constant corresponds to the fact that this analysis is valid in the case where sudden increases in input current cause M_5 to be strongly turned on such that the source-follower dynamics are quick and may be approximated as being instantaneous, i.e., we have a short in the overall lowpass-filter dynamics. When we have sudden decreases in input current, however, M_5 is turned off and there is only slow recovery at the C_R capacitor due to slow charging of C_R by I_R. The linear charging at C_R is converted to an exponentially decaying current at i_{ENV} such that the release dynamics may be approximated by a release time constant given by

$$\tau_R = \frac{C_R\phi_t}{\kappa I_R} \tag{19.7}$$

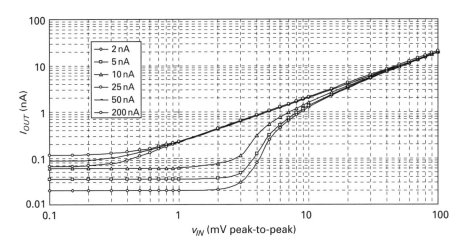

Figure 19.9. Stochastic resonance in an envelope detector. Reproduced with kind permission from [19] (©2003 IEEE).

where τ_R is significantly less than τ_A. Thus, the current-mode peak detector extracts the average energy in the rectification input current i_{REC} and reports it at i_{ENV} with faster responses to increasing values of i_{REC} than to decreasing values of i_{REC}.

Equation (19.5) predicts that if $V_{batt} = V_D$, and V_D^{eff} is zero, then the threshold current of the rectifier will be zero. However, Equation (19.5) has ignored the thermal noise of the transconductor, which also plays a role in setting a threshold current for the rectifier. In fact, Figure 19.9 reveals what happens if we keep increasing V_{batt} by increasing the value of I_{B2} in Figure 19.7 (a) from 2 nA to 200 nA and measure the output current i_{ENV} of the envelope detector as a function of its input v_{IN}. Initially, the increase in V_{batt} does reduce the threshold V_{in}^{th} as evidenced by the dead-zone-limited sharp decreases in the output current of the rectifier occurring at smaller and smaller values of v_{IN} in Figure 19.9. However, as we lower V_{batt} we notice that the output offset current of the envelope detector, i.e., the current when the input v_{IN} is small and below threshold, begins to increase. This mysterious increase in the offset current of the rectifier proves to have a simple explanation: thermal-noise currents with a Gaussian amplitude distribution from the transconductor get rectified and converted to a dc output current. It is more likely that Gaussian noise with a certain variance will exceed a lower threshold and get rectified than that it will exceed a higher threshold and get rectified. Thus, lowering of the dead-zone-limited threshold barrier makes it easier for a smaller input to get rectified but it also makes it easier for the noise to get rectified such that the offset current increases. In fact, as Figure 19.9 shows, there is an optimal threshold: at some ideal value of V_{batt}, or equivalently I_{B2}, the dead-zone-limited threshold leads to an output current from the rectifier that is just equal to the offset current due to the thermal noise at V_{in}^{th}. At that point, the dead-zone-limited minimum detectable signal and the thermal-noise-limited minimum detectable signal become identical. Decreasing V_{batt} beyond this optimum leads to more thermal-noise-caused offset current degrading the minimum detectable signal. Increasing V_{batt} beyond this point leads to lower thermal-noise-caused offset current but a larger dead-zone-limited threshold degrading the minimum detectable signal. Thus, we have a *stochastic resonance*: there is an optimal amount of noise that leads to the smallest minimum detectable signal, a well-known effect that is observed in other physical systems with noise and a threshold barrier. In our envelope detector, stochastic resonance manifests itself as an optimal threshold since the noise is fixed and we vary the threshold, while in other physical systems, the threshold is fixed and the noise varies. In both cases, it indicates that we don't want to make it too easy for the signal to get through, or the noise will get through as well; we don't want to make it too hard for the signal to get through, since neither signal nor noise will get through. We shall now analyze the stochastic-resonance effect quantitatively.

At low frequencies, Equation (19.5) reveals that the dead-zone threshold is small; therefore, the probability that the noise will exceed the dead-zone threshold and contribute to the rectifier current is low. At high frequencies, Equation (19.5) reveals that the dead-zone threshold is high; therefore, the probability that the noise will exceed the dead-zone threshold and contribute to the rectifier current is high. Thus, the capacitance C_{gs} and the frequency-dependent dead-zone threshold create an effective equivalent rectangular bandwidth f_0 for the noise: for frequencies less than f_0, all the noise power contributes, while for frequencies larger than f_0, the noise is perfectly filtered out and does not contribute. Suppose that the transconductor has N effective devices worth of white noise and operates in subthreshold. Then its current noise power is given

$$\sigma^2 = N\left(2q\frac{I_B}{2}\right)f_0 = NqI_Bf_0 \tag{19.8}$$

The rectified output noise current I_{noise} that we observe is the average of the positive part of the noise current of a Gaussian distribution with a standard deviation of σ. Thus, it should be given by

$$I_{noise} = \int_{0}^{+\infty} I \cdot \frac{1}{\sqrt{2\pi} \cdot \sigma}e^{-\frac{I^2}{2\sigma^2}} \cdot dI = \frac{\sigma}{\sqrt{2\pi}} = \sqrt{\frac{NqI_Bf_0}{2\pi}} \tag{19.9}$$

The current output by the rectifier will fall once the frequency-dependent dead-zone threshold $2\pi f C_{gs}V_D^{eff}$ becomes comparable to σ. If we assume that f_0 can be approximated by the point at which this frequency-dependent threshold becomes equal to σ, we get

$$
\begin{aligned}
2\pi f_0 C_{gs} V_D^{eff} &= \sigma \\
2\pi f_0 C_{gs} V_D^{eff} &= \sqrt{NqI_Bf_0} \\
\sqrt{f_0} &= \frac{\sqrt{NqI_B}}{2\pi C_{gs}V_D^{eff}}
\end{aligned} \tag{19.10}
$$

Substituting this value of f_0 into Equation (19.8), we can obtain σ and then get I_{noise} from Equation (19.9). We find that

$$I_{noise} = \frac{NqI_B}{2\pi\sqrt{2\pi}(C_{gs}V_D^{eff})} \qquad (19.11)$$

Indeed, the offset current of the rectifier was experimentally found to depend linearly on I_B, the bias current of the G_m transconductor, and was quantitatively consistent with Equation (19.11) on several chips [19]. Thus, the observed offset current is not a parasitic leakage effect and is indeed due to thermal-noise rectification. By noting that the maximum current output by the rectifier is related to I_B while the minimum current is given by Equations (19.5) and (19.11), we can compute two dynamic ranges for the rectifier, one determined by the dead-zone threshold, D_R^{dead}, and one determined by the thermal noise, D_R^{noise}:

$$D_R^{dead} = \frac{I_B}{2\pi f_{max} C_{gs} V_D^{eff}}$$

$$D_R^{noise} = \frac{I_B/\pi}{\left(\dfrac{NqI_B}{2\pi\sqrt{2\pi}(C_{gs}V_D^{eff})}\right)} = \frac{2\sqrt{2\pi}(C_{gs}V_D^{eff})}{Nq} \qquad (19.12)$$

The factor of π in the second line of the numerator of Equation (19.12) arises from the dc average of a half-wave rectified sinusoid. The frequency $f_{max} = 10$ kHz in our application. When the two dynamic ranges are equal, we have the optimal maximal dynamic range. Thus, at the optimum, we can equate the two dynamic ranges in Equation (19.12) to get

$$(C_{gs}V_D^{eff})_{OPTIMUM} = \sqrt{\frac{NqI_B}{4\pi\sqrt{2\pi}f_{max}}} \qquad (19.13)$$

We can substitute this optimum back into Equation (19.12) to get the optimal dynamic range

$$D_R^{opt} = \sqrt[4]{\frac{2}{\pi}}\sqrt{\frac{I_B}{Nqf_{max}}} \qquad (19.14)$$

Under optimal conditions, the technology-dependent parameter $C_{gs}V_D^{eff}$ is selected to match technology-independent parameters in Equation (19.13) determined by thermal-noise considerations. Thus, the optimal dynamic range is technology independent! Equation (19.14) was experimentally verified in a 75 dB (1.7 V pp to 0.3 mV pp), 2.8 μW envelope detector with $I_B = 1$ μA and $V_{DD} = 2.8$V, which is described in [19]. The detector was dc-offset free and operated from 100 Hz to 10 kHz. The bionic-ear processor only requires an envelope detector with 60 dB dynamic range that is dead-zone limited (D_R^{dead} in Equation (19.12)). The power consumption of such a dead-zone-limited envelope detector is lower (1 μW in 1.5 μm) than that of a thermal-noise-limited detector and reduces with

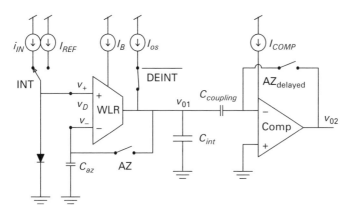

Figure 19.10. A micropower logarithmic ADC with offset and temperature compensation [20] (©2004 IEEE).

technology as C_{gs} gets smaller (0.3 μW in 0.5 μm technology) in accord with the first line of Equation (19.12).

19.2.5   Micropower log ADC

Output current from the envelope detector is digitized by the logarithmic ADC. This ADC uses a dual – slope algorithm which exploits a successive integration–de-integration strategy to quantize its input. During the initial integration phase, we charge a capacitor from a starting value for a fixed period of time with a current proportional to a voltage input that we want to quantize. Then, on the following de-integration phase, the charged value is discharged back to its starting point with a fixed discharging current. The time taken to return to the starting point from the beginning of the de-integration phase is evaluated by counting clock cycles in a counter. The digital value of the counter at the point of return then yields a quantized representation of the input: a larger input in the integration phase will lead to a proportionally larger value of maximum charge, which will lead to a proportionally longer time to return to the starting point in the de-integration phase. Three advantages of the dual-slope algorithm are its automatic cancellation of capacitive nonlinearity due to the fact that it operates with a charge-based strategy, its ability to share the counter amongst many converters running in parallel, and its simplicity.

Figure 19.10 illustrates the logarithmic dual-slope ADC. The logarithmic properties are achieved by converting the input current to a logarithmic voltage on a diode. The WLR and C_{int} capacitor function as the open-loop $G_m - C$ integrator and de-integrator. The capacitor $C_{coupling}$ ac couples the results of the integration or the de-integration to a comparator. The ADC operates by performing auto-zeroing, integration, and de-integration operations on three successive phases. During the auto-zeroing phase when *AZ* and $AZ_{delayed}$ are closed, a reference current I_{REF} is steered to the input diode, the WLR transconductor auto-zeros to a negative differential offset to equilibrate with the output current I_{os}, and the comparator auto-zeros its offset as well. The negative offset in the WLR allows us to exploit a larger $[-V_L, V_L]$ linear range of operation on the following integration phase rather than just a $[0, V_L]$ linear range, thus increasing the dynamic range of operation by a factor of two for the same power. It also allows simple implementation of de-integration as we discuss later. The auto-zeroing phase ends when AZ and $AZ_{delayed}$ have been opened, with $AZ_{delayed}$ being opened slightly later than AZ to ensure that the charge injection from AZ's opening does not affect the comparator. The capacitors C_{az} and $C_{coupling}$ store the auto-zeroed offset values.

When the integration phase begins, the input diode is connected to the input current i_{IN}, which is larger than I_{REF}, such that C_{int} is charged by a current given by $G_m(kT/q)\ln(i_{IN}/I_{REF})$ for a fixed period of time to increase v_{0I}. The parameter G_m is the transconductance of the WLR. Since G_m has an inverse PTAT dependence (it is proportional to qI_B/kT as discussed in Chapter 12), any temperature dependence due to the diode is cancelled by an identical inverse-temperature dependence in the WLR subthreshold transistors. Since we require the input to vary by about 3 orders of magnitude (60 dB), the input voltage on the diode changes by about $\phi_t\ln(10^3) \approx 180$ mV. From Chapter 12, we can derive that the output noise power of the WLR $G_m - C$ transconductor is $NqV_L/(4(C_{az} + C_{int}))$ where N is the effective number of shot-noise sources. Since this noise is sampled onto C_{az} at the end of the auto-zeroing phase, a linear range well in excess of the needed 180 mV increases the thermal noise amplitude sampled at the input and degrades performance. Therefore, the linear range of the WLR, V_L, is architected to be near 240 mV through the use of a gate input, bump linearization and source degeneration. The use of I_{os} nearly equal to I_B yields a safety factor of nearly 2. Since the sampled value can be $\pm 3.0 \, \sigma_{rms}$ where σ_{rms} is the standard deviation of the Gaussian thermal noise distribution, it can significantly limit the precision of the converter. With $C_{int} = 30$ pF, $C_{az} = 10.0$ pF, and $C_{coupling} = 10.0$ pF, the total noise is 30 μV rms and is lower than the 70 μV rms noise contribution of the diode and implicit capacitor at the positive input of the WLR. Such a noise floor with a $\pm 3.0 \, \sigma$ spread yields a precision of 180 mV/420 μV $\approx 8 - 9$ bits at a 300 Hz sampling rate per spectral channel. Note that the comparator barely contributes to the noise because the integration gain $(G_m T_{int})/C_{int}$ amplifies the noise of the first stage of the ADC to overwhelm that of the second. Since most cochlear-implant patients can barely discriminate 5 bits in a given spectral channel, and there are advantages to high-rate sampling, it is significantly better to lower precision and increase sampling rate while preserving constant power. Rescaling capacitances such that $C_{int} = 7.5$ pF, $C_{az} = 2.5$ pF, and $C_{coupling} = 2.5$ pF yields a 7-bit log ADC sampled at 1 kHz with 3.3 μW total power consumption, of which 2.5 μW is highly scalable digital power. The comparator's power consumption is determined by I_{COMP}, which is usually significantly lower than the WLR's, which is set by I_B.

De-integration is accomplished by simply switching I_{os} off and setting the diode current to I_{REF}. The inherent large negative offset in the WLR then leads to

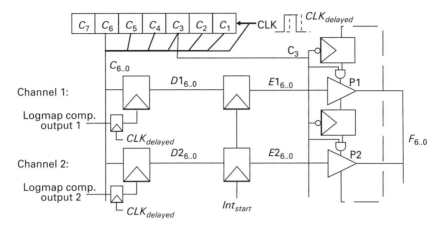

Figure 19.11. An architecture for continuous interleaved sampling (CIS). Reproduced with kind permission from [17] (©2005 IEEE).

a discharging current. The number of clock cycles T_{deint} that quantizes the input is given by

$$T_{\text{deint}} = \frac{I_{\text{int}} T_{\text{int}}}{I_{\text{deint}}} \qquad (19.15)$$

Since both I_{int} and I_{deint} are proportional to I_B, any temperature dependence in I_B is automatically cancelled. This log ADC achieved a 2.5 bit improvement in precision at the same power over that of prior dual-slope topologies. Its use of local feedback loops rather than global feedback loops during auto-zeroing helps allocate power more efficiently based on noise considerations rather than on stability considerations. Further details are described in [20]. The output of the comparator, v_{O2}, is fed to the CIS scanner.

19.2.6 CIS scanner

Figure 19.11 shows the CIS scanner. An 8-bit counter is implemented with only 7 register bits by using the clock itself as the C_0 LSB counter bit. The C_7 bit is the MSB counter bit. The de-integration phase occurs when the counter has values $\{0, 1, \ldots, 127\}$, auto-zeroing occurs when the counter has values $\{128, 129, \ldots, 191\}$, and integration for the next conversion occurs when the counter has values $\{192, 193, \ldots, 255\}$. Only two channels are shown in Figure 19.11 but there are actually 16 such channels. The comparators from each log ADC asynchronously trigger during some portion of the de-integration cycle, which initiates latching of the C counter bits into the D registers. The latching is done by sampling the comparator transitions with a slightly delayed clock, which ensures that the counter bits have settled by the time they are latched into the D registers. To ensure that the digitized bits are valid for a whole CIS sampling period, the

D registers are latched into the E registers at the start of integration (192 or when C_6 and C_7 go high). A 16-stage shift-register, clocked by the C_3 bit, causes a 'high' token to sequentially and cyclically pass between the stages, one at a time, such that, on a given clock cycle of C_3, only one channel's E registers are output onto the serial F bus. The C_3 bit runs 16 times faster than the C_7 sampling clock: thus, if the main clock runs at 128 kHz, and C_7 runs at 1 kHz, C_3 will run at 16 kHz. Each shift register enables a tri-state driver to multiplex the E bits onto the F bus when that shift register has the token. To prevent tri-states from fighting each other during shift-register transitions, a tri-state is deactivated as soon as C_3 goes low (via the *AND* gate); it is activated when C_3 goes high a half clock cycle later, when we are confident that all shift registers have settled and that only one shift register has the token. The output data of the F bus is guaranteed to be valid after the falling edge of C_3; hence C_3 is outputted as a "CIS clock" for latching the data in subsequent electrode-stimulation circuits. In addition, the de-integration pulses from each channel are also reported off chip such that external A/D conversion may be performed on these pulses. Alternatively simple processing can directly transform these pulses into electrode-stimulation pulses.

19.2.7   Programming and visibility

A programmability-and-visibility clock sequentially shifts an active token in a shift-register-based scanner on the chip. When this token is activated, a particular channel can be programmed and its visible outputs can be examined for debugging. The active token can also activate no channel, the default mode when the processor is not being programmed and in use. If a programming line is enabled, any of the 22 programmable bits of each channel can be altered and stored in on-chip latches. In each channel, there are 7 bits to alter the center frequency, 7 bits to alter the quality factor of the filter, 3 bits to alter the envelope detector attack time, 2 bits to alter the envelope detector release time, and 3 bits to alter I_{REF} for fine transistor-mismatch offset compensation of the channel. In addition to these channel-specific bits, the log ADC's sampling may be globally set to be 500 Hz, 1000 Hz, 1500 Hz, 2000 Hz or to any analog value determined by a user-supplied external current with 3 global bits. There are 16 global bits that allow programmability of attack, release, knee, and compression ratio in the global AGC that follows the microphone. The visibility outputs include output and intermediate-node waveforms from the bandpass filter, rectifier and peak-detector outputs from the envelope detector, the diode input, C_{int} output, and comparator output from the log ADC. Internal current waveforms are transduced to a voltage with an instrumentation transconductor with a bias current that is electronically adjustable. Two bits allow us to scale up or scale down the envelope detector outputs by a factor of 2 or 0.5 respectively. Thus, 373 total bits on the chip allow for a great deal of programmability and visibility in the analog bionic-ear processor including the setting of 86 subject-specific parameters.

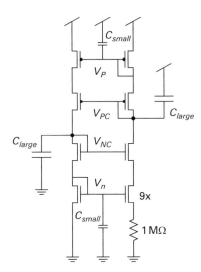

Figure 19.12. A power-supply-noise-immune and temperature-invariant constant-G_m reference for subthreshold operation. The reference also generates voltage biases that are useful for setting cascode bias voltages on a chip. Reproduced with kind permission from [17] (©2005 IEEE).

19.2.8 Robust current-and-voltage biasing

Temperature-independent biasing is *crucial* for robust operation in subthreshold circuits where currents have an exponential sensitivity to temperature. This is true even within the body where the temperature can fluctuate by 1.5 degrees centigrade in a healthy subject between day and night, and by up to 5 degrees centigrade in a subject with a fever. Figure 19.12 illustrates the PTAT biasing circuit, which leads to constant G_m-biasing in subthreshold, and therefore leads to temperature-independent biasing of all bandwidth and time-constant parameters throughout the chip. Any residual linear PTAT variation in the absolute value of a bias current (rather than exponential) has little effect on the operation of other parameters on the chip because it is automatically cancelled out, e.g., in the log ADC, or creates small variations in the noise and power that do not affect operation significantly, e.g., in the envelope detector. The relative constancy of body temperature makes such higher-order effects inconsequential.

The nominal reference current generated to bias DACs on the chip at equilibrium is

$$I_{DACbias} = \frac{25 \text{ mV} \times \ln(9)}{10^6 \ \Omega} = 57 \text{ nA} \tag{19.16}$$

In practice, the current is rarely exactly that predicted by Equation (19.16), due to parameter variations and mismatch. For example, we measured 45 nA. Such variability in the absolute value of the reference does not matter as long as there are a sufficient number of DAC bits to alter currents on the chip to what is needed. The reference current is scaled up and down with mirroring in a current-distribution network, which distributes currents to the DACs that bias transconductors, amplifiers, comparators, and current-mode circuits throughout the chip. Care must be taken to not have too many stages of mirroring between the reference and the actual point of usage to avoid random-gain-error accumulation. Care must also be taken to have big bias currents generated only at the point of usage via a large local mirror to avoid power overhead in the biasing circuits. For example, the extra power consumed by the entire biasing network on the bionic-ear processor is only 4.2 μW. Bias-current determining gate voltages must *never* be distributed on a subthreshold chip since they can lead to mismatch variations by factors of 2 or 3 across different corners of a chip. Getting all portions of a chip to work with mismatched worst-case values can then lead to power consumptions that are non-optimal by factors of 2 and 3.

Power-supply immunity with respect to ground and V_{DD} is achieved by the $C_{large} = 100$ pF and $C_{small} = 10$ pF bypass capacitors in Figure 19.12. They function in a manner similar to that described for the microphone AFE in Figure 19.4, i.e, they help keep gate-to-source voltages constant independent of power-supply variations in the reference circuit. The use of cascode mirrors in Figure 19.12 helps limit Early-effect variations. The circuit also generates four voltage references that are used for biasing cascode voltages on the chip. For the process that the chip was built in, with a 2.8 V supply, the V_N, V_{NC}, V_{PC}, and V_P voltages were at 0.45 V, 1.05 V, 0.95 V, and 2.0 V, respectively.

This chip operated extremely robustly primarily because of this bias circuit. An intentional capacitively coupled 49 MHz carrier modulated at 1 kHz onto the power supply of the chip to create a 430 mV pp waveform on the supply with a 100% modulation depth had barely any effect on the output bits of the chip. In addition, tests of RF interference performed with a large RF coil surrounding the test board revealed that even very high levels of wirelessly coupled RF (1 V rms on a 10-turn coil of dimensions 30 cm $\times$ 30 cm encircling the chip, and inducing a carrier amplitude of 400 mV pp on the power supply) did not interfere with the operation of the chip.

19.2.9   Experimental performance

Figure 19.13 reveals the architecture of the 9.23 mm $\times$ 9.58 mm bionic ear processor chip. The envelope detectors have zero-crossing detection incorporated into them as well to allow phase information *and* envelope information to be extracted in the spectrum. In Figure 19.7, transitions in the common-mode value of $v_{OUT,TOP}$ or $v_{OUT,BOT}$ as we switch from conduction in M_n to conduction in M_p signal zero-crossings in the fine center-frequency signals of the bandpass filters. By detecting the common-mode value of $v_{OUT,TOP}$ and $v_{OUT,BOT}$ with a simple inverter-like circuit addition to the envelope detector [18], we can detect the time of occurrence of such zero-crossings cheaply. Such phase information is important for listening to music. Figure 19.14 (a) shows the bits produced by the 16 channels of the chip in response to varying frequency inputs. Figure 19.14 (b) shows the bits

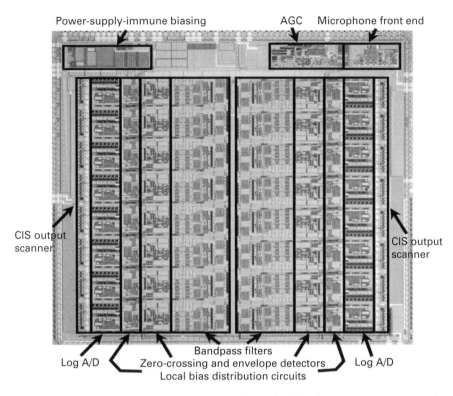

Power-supply-immune biasing

AGC   Microphone front end

CIS output scanner

CIS output scanner

Bandpass filters

Log A/D   Zero-crossing and envelope detectors   Log A/D

Local bias distribution circuits

Figure 19.13. The ultra-low-power programmable analog bionic ear processor. Reproduced with kind permission from [18] (©2005 IEEE).

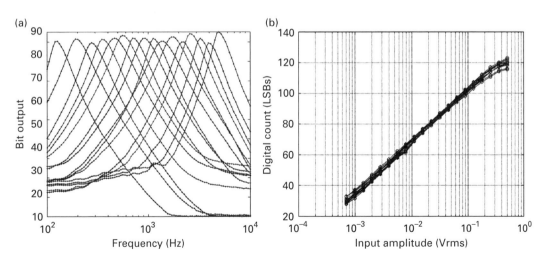

Figure 19.14a, b. Output bits from the bionic ear processor chip for different input frequencies in (a) and versus input amplitude in (b). (a) Reproduced with kind permission from [18] (©2005 IEEE). (b) Reproduced with kind permission from [17] (©2005 IEEE).

produced in all 16 channels as the amplitude of the input at the center frequency of the channel is varied. The overall power consumption of the chip is 251 μW, with the microphone AFE consuming 106.4 μW, the AGC consuming 28 μW, the 16 bandpass filters consuming 37.8 μW with the filter quality factors being set at 4 (a value that represents a good tradeoff between temporal and spectral resolution for speech recognition and that is generally well liked by subjects), the 16 envelope detectors consuming 18.2 μW, the 16 log ADCs consuming 56 μW, and the robust biasing consuming 4.2 μW. The chip functioned very robustly and a deaf patient successfully understood speech on her first try once the chip's parameters were programmed to match those of her DSP. It is interesting that several deaf patients often prefer analog hearing aids over digital hearing aids because they say that they are more natural and sound less machine like. The bionic ear processor has also found application in portable speech-recognition applications. The chip can easily be modified for use in electroencephalogram (EEG) and phonocardiogram (PCG) applications, where spectrum analysis of $\sim 100\times$ lower-bandwidth signals implies that sub microwatt analog processing should be possible.

The use of current-mode analog processing (described in Chapter 14) could lead to significant further reductions in power even in low-voltage processes where voltage-mode techniques are not suitable. Some of the power consumption of the various blocks scales with technology and some does not, as we have described. It is worth pointing out that the power consumption of the processor is low for three reasons: 1) Delayed digitization such that 16 slow-and-parallel ADCs operating at 7-bit precision, and 1 kHz sampling rate on high-level speech spectra (that require far less precision and bandwidth) consume far less power than a single fast 16-bit precise, 44.1 kHz ADC operating on low-level microphone data (that require far more precision and bandwidth); 2) the use of an AGC to reduce the internal dynamic range from ~ 80 dB to ~ 60 dB such that the spectral analysis is power efficient; 3) the use of several robust, efficient, and programmable low-power analog preprocessing circuits and topologies. Every device and every signal in the analog processor are not robust as in a digital processor. But the overall output is robust. The wise allocation of power, the use of robust biasing, feed-forward calibration, feedback calibration, and inherently noise-robust topologies minimize the effects of thermal noise, power-supply noise, $1/f$ noise, transistor mismatch, and temperature variations. We shall return to issues of robustness and flexibility vs. efficiency in our discussion of low-power mixed-signal design in Chapter 22.

19.3   Low-power electrode stimulation

Electrode stimulation is widely useful in all neural, muscular, and cardiac implants. Such stimulation is implemented with a biphasic current source driving the electrode, with the negative current and positive current supplied to the electrode on two successive phases. The phases may have a brief zero-current gap in between them. The FDA clinical requirement is that the charge supplied on the negative phase $I_- T_-$ be exactly equal to that supplied on the positive phase $I_+ T_+$ such that there is no net dc charge supplied to the body. Net charge imbalance of even small amounts can cause toxic effects in subjects in the long term. From the discussion in Chapter 25, we note that an electrode can often be well approximated by a resistance R_S and a capacitance C_{dl} in series. The resistance represents the electrode spreading resistance due to finite extracellular ionic conductivity and the capacitance primarily represents the double-layer Helmholtz capacitance. The larger the surface area of the electrode, the lower is the value of R_S and the higher is the value of C_{dl}. In addition, there is a resistance R_f in parallel with C_{dl} due to the effect of diffusive mass-transport and Faradaic chemical reactions that involve charge transfer (see Chapter 25 for an in-depth quantitative discussion). Typically R_f is large enough that it may be ignored. If it cannot, there may be serious issues with the toxicity of stimulation if the current through R_f is not perfectly charge balanced. We shall generally assume that R_f is large unless otherwise mentioned.

The threshold electrical-stimulation current I_{th} needed to elicit a neural response is often well described by

$$I_{th} = I_{rhbs} \left(1 + \frac{\tau_{chraxie}}{t_{pw}} \right) \tag{19.17}$$

where I_{rhbs} is the *rheobase*, the minimal threshold current when the pulse width of a stimulation pulse, t_{pw}, is very large, and $\tau_{chraxie}$ is the *chronaxie*, the value of pulse width at which the threshold current drops by half. Thus, larger pulse widths lead to lower values of current and lower power dissipation. Most values of $\tau_{chraxie}$ in clinical applications range from $100\,\mu s$ to $10\,ms$, while values of I_{rhbs} are rarely less than $10\,\mu A$. Equation (19.17) suggests that resistive effects in neurons are dominating stimulation at large pulse widths while capacitive effects in neurons are dominating stimulation at small pulse widths. The values of t_{pw} in clinical applications range from $10\,\mu s$ to $10\,ms$.

To lower the power needed for electrical stimulation requires approaches at the electrode, waveform, circuit, and algorithmic levels. At the electrode level, the primary focus is often on creating high-surface-area electrodes with low impedance. At the waveform level, one can attempt to introduce gaps between the negative and positive phases to help refractory neurons recover, or have larger positive phases. We shall focus primarily on the circuit and algorithmic levels since the other levels, while extremely important, are often outside the control of the circuit designer.

Work described in [21] has taken an adiabatic approach to lowering the wasted power of electrode stimulation. The magnitude of the power-supply voltages, V_{DD}, that most stimulating current sources are tied to is rather large and designed for worst-case stimulation of the highest-impedance electrode with the highest current. An energy of $\int (V_{DD} - v_{ELECTRODE}(t)) i_{ELECTRODE} dt$ is wasted across such current sources during each stimulating pulse. The solution is to have V_{DD} always only slightly larger (or smaller) than $v_{ELECTRODE}$ such that little power is wasted across the current source when charging (or discharging) the electrode. One way to accomplish this feat is to have a bank of capacitors with quantized values of V_{DD} stored on them. To charge an electrode adiabatically, one connects a capacitor with a V_{DD} slightly larger than the current voltage on the electrode to it; to discharge an electrode adiabatically, one connects a capacitor with a V_{DD} slightly smaller than the current voltage on the electrode to it. If this scheme is followed sequentially with increasingly larger V_{DD}s for charging and decreasingly smaller V_{DD}s for discharging on each electrode, one can simulate a nearly constant current-source drive in a quantized fashion without wasting much power across the current source.

The bank of capacitors in [21] is itself adiabatically charged from a secondary implanted coil in a similar fashion: the secondary coil only replenishes the charge lost on a capacitor in the bank when the secondary's ac voltage is just above (or just below) the voltage value of the capacitor in the bank. To ensure that the comparators needed to perform such synchronous rectification have little delay but do not burn too much power, predictive comparators that are described in [22] are used. A predictive comparator uses knowledge about the derivative of its input waveform to predict where its input will be in the future; this prediction allows it to set a threshold for triggering that compensates for its own delay and causes it to trigger at exactly the right value of its input with no overshoot in spite of its delay; in effect, the prediction acts like a negative delay that cancels the positive delay of the comparator and thus prevents overshoot. Consequently, the comparator can have a fairly large delay, which reduces its power significantly, and it can yet operate in a seemingly fast and accurate fashion. The predictive comparator in [22] actually has adaptive control to minimize overshoot or undershoot but the scheme in [21] only uses a fixed predictive value for simplicity. Nevertheless, it achieves good energy efficiency. Such exploitation of the knowledge of the waveform to perform comparisons is possible because the secondary waveform is relatively fixed and does not vary. Overall, more than a factor of two savings in electrode power per electrode was achieved.

The concept of adiabatic charging is shown in Figure 15.15 (b) in Chapter 15 in the context of analog-to-digital conversion and is also discussed in Chapter 21 on ultra-low-power digital design. The concept of predictive negative delays is used in Figure 18.2 in Chapter 18. There, we exploit negative $-C_2R_2$ delays to cancel the positive delay of an inverter, t_{inv}, in a positive-feedback loop, thus enabling oscillations to occur.

Algorithmic strategies can also be used to save power. We provide an example below. High-rate stimulation that can encode fine phase information in the bandpass filter spectral-channel waveforms of a cochlear implant is essential for perceiving music and is also important in the perception of speech in noise. However, to provide such high-rate stimulation, the value of t_{pw} must decrease, and the threshold needed for current stimulation increases as per Equation (19.17). Envelope information with randomized phase, the current CIS processing paradigm, has limited the ability of the deaf to perceive music and to hear in noise. However, unless stimulation power constraints are overcome, we simply cannot provide such information to them. A change in the algorithmic strategy used for stimulation can potentially come to the rescue. An asynchronous interleaved sampling (AIS) algorithm has been proposed to encode fine phase information and envelope information but with a scheme that lowers stimulation power [23]. Half-wave rectified waveforms from the bandpass filters of each channel are used as charging currents to neurons in their respective channels. The neurons from all channels compete in a 'race-to-first-spike'. The highest intensity channel with the highest charging current wins the race and is awarded the winner-take-all prize of stimulating the electrode corresponding to its channel. The stimulation is a charge-balanced pulse with a current proportional to the envelope of the spectral energy of the channel. The pulse occurs at the time of the neuronal spiking such that it is correlated with the fine phase of the half-wave rectified bandpass waveform. All neurons are then reset and the race proceeds again except that a winning channel is handicapped from winning again with an inhibition current right after it has won, limiting its maximal rate of firing; it is also handicapped each time after it wins to give other channels a chance to win as well. If a very-high-intensity channel wins multiple times in a row, its handicap grows with each win such that eventually, weaker-intensity channels also get a chance to stimulate their electrodes. Thus, on average high-intensity channels are sampled more often than low-intensity channels, and each time a channel is sampled, phase information is preserved. A stimulus reconstruction shows that there is remarkably high correlation between such an AIS output and input [23]. Such reconstructions have, in the past, been shown to be predictive of actual subject performance. Yet, the *average* stimulation power across all channels is typically six times lower than would be expected if all channels are sampled at high rate, not just the higher-intensity ones. Like CIS sampling, the AIS scheme prevents electrode interactions because only one electrode is active at a time. A programmable analog bionic ear processor capable of implementing such an asynchronous scheme only consumes $357\,\mu W$ in spite of the increased computational complexity [24]. The computational needs of such a strategy are high in a synchronous digital paradigm due to the need for high-rate clocking in all channels. This example illustrates that the combination of algorithmic strategies (worst case is NOT average case) and circuit strategies (efficient analog asynchronous implementation) can be a potentially powerful way to lower stimulation power. Furthermore, in this particular case, the bio-inspired sampling is stochastic and synchronized to the input, much like the sampling of the actual nervous system, rather than being synchronized to an artificial digital clock, which is not beneficial for subjects. Sampling with an AIS-like strategy may have applications in the natural and power-efficient stimulation of other implants as well, e.g., in brain implants for the blind: higher-intensity or higher-contrast pixels that convey more information may be sampled more frequently as they are in the nervous system but the average stimulation rate may be still maintained at a low value to conserve power.

19.4   Highly miniature electrode-stimulation circuits

Imbalanced current sources on negative and positive phases lead to a net charge imbalance in stimulation. To prevent net dc current stimulation of the electrode, it is customary to use 'dc blocking capacitors' that ac couple the outputs of the current sources to the electrode. To prevent large drops in voltage across this series coupling capacitor from significantly increasing the voltage compliance of the current sources needed for stimulation, the magnitude of the impedance of the dc blocking capacitor must be significantly lower than the magnitude of the electrode impedance that it is coupled to. A cochlear-implant electrode with $R_S = 10$ kΩ and $C_{dl} = 10$ nF that is stimulated with a current of 1 mA for 30 μs can have as much as 30 nC of charge per phase for a large stimulation current. These values drop 10 V across R_S and 3 V across C_{dl}. To ensure that the voltage drop across the coupling capacitance C_{BC} is not more than a tenth of that across C_{dl}, C_{BC} must be at least 100 nF. Since one such C_{BC} is needed for every electrode, the size and consequent cost of the implant are greatly impacted by these rather large capacitances. Thus, there is great motivation to do away with these blocking capacitances to reduce the size of an implant. The situation is particularly stringent in retinal implants for the blind since there is a very high premium on space in the eye. However, if we do away with these capacitances, any charge imbalance must be such that any net dc current coupled to the electrode, which results in potentially toxic chemical reaction products in the electrolyte, must be significantly less than 100 nA to prevent tissue damage [25]. The use of a dc blocking capacitance with a 5 GΩ parasitic resistance across it easily meets this specification (less than 1 nA of dc current flows through this resistance). However, it is not so easy to achieve 10 nA or less of net dc error without this coupling capacitance: if the full-scale current is 1 mA, the matching between the positive and negative current sources must be better than 1 part in 10^5 to achieve a 10 nA error.

One might imagine that the positive and negative phases of current could just be calibrated to the needed $\sim$17-bits of precision, and then there would be no need to have any blocking capacitors. However, high-precision calibration consumes power and area due to the need to minimize charge injection, offset, noise, and yet settle quickly. Another possibility is to simply short the electrode between stimulation pulses, removing any residual charge that has accumulated due to current imbalance (the shorting current effectively re-establishes current balance). However, shorting only reduces error in accord with an exponential RC decay between stimulation pulses. If the stimulation rate is relatively high (e.g. 1 kHz), there can still be significant dc current error in a shorting scheme: the RC decay does not have sufficient time to reach a nearly zero value between stimulation pulses. A practical method for achieving such stringent balance without significantly increasing the power consumed per electrode is to first balance the positive and negative phases of current stimulation to good precision via feed-forward or feedback calibration techniques; the electrode is then shorted after nearly charge-balanced stimulation to remove any residual charge due to inevitable calibration errors. The combination of calibration and shorting is much more efficient than either technique alone: the error is reduced through a multiplicative combination of the two techniques rather than trying to push either technique to its extreme limit. Indeed, this combination of techniques has been used to achieve 6 nA of dc current error with 1 mA of full-scale stimulation without the use of any blocking capacitors [26].

Figures 19.15 (a) and 19.15 (b) illustrate a stimulation-circuit topology that uses a combination of calibration and shorting. After the sampling phase in Figure 19.15 (a) is complete, the sink stimulation current i_{IN} is accurately sampled and held such that M_1's current is a source current nearly equal to the i_{IN} sink current. During the hold phase in Figure 19.15 (b), the sink and source currents successively stimulate the electrode respectively, followed by shorting. During calibration, the technique of dynamic-current mirroring is used to achieve good matching (0.4%) between the positive and negative phases of current stimulation with only 47 μW of calibration power per channel. The residual 120 pC error after calibration for 1 mA full-scale stimulation is attenuated by a factor of 20 by shorting with an electrode: the electrode has an RC time constant of 330 μs and is shorted during inter-pulse intervals between asynchronous stimulation pulses that are at least 1ms apart (three time constants apart such that $e^{-3} \approx 1/20$). The resultant 6 pC error over 1 ms implies a 6 nA average dc-current error.

In Figure 19.15 (a), M_2 and I_{BS} form a source follower that diode connects M_1 for sampling and representing i_{IN} on v_P in a current-mirror topology. The source follower also creates a dc shift that helps the active cascode formed by M_4, M_3, and I_{BCP} to operate with all its transistors in saturation and thus create a high-impedance current source. The value of v_P is stored on capacitor C while the transconductor G_{MH} auto zeros its offset; G_{MH} and C implement a classic sample-and-hold topology. When we transition to the hold phase in Figure 19.15 (b), SAMP1 is first opened, then SAMP, and finally HOLD is closed. This sequence of opening ensures that the charge injection of SAMP does not lead to an error in v_P because it is attenuated via negative-feedback action that maintains v_H at nearly the value it had right before SAMP1 opened. The charge injection due to the opening of SAMP1 occurs at the input rather than the output of the feedback loop and does lead to an error. The critical SAMP1 switch is actually an ultra-low-leakage switch that has been shown to be capable of less than 5 electrons per second of leakage [27], [28]. This switch, implemented with a 4-transistor circuit that is not shown, minimizes four sources of leakage in the SAMP1 switch that degrade precision during the hold phase: it minimizes subthreshold leakage by making gate-to-source voltages negative, minimizes diode-junction leakage by zeroing pn junction voltages, minimizes lateral bipolar leakage by making drain-to-source voltages equal, and minimizes well-to-substrate leakage by using a differential topology [27], [28]. The ultra-low-leakage switch has been applied to creating very-long-hold-time analog memories with 8-bit precision over 5 hours and 12-bit precision for 45 minutes. Further details of the use of the sample-and-hold circuit

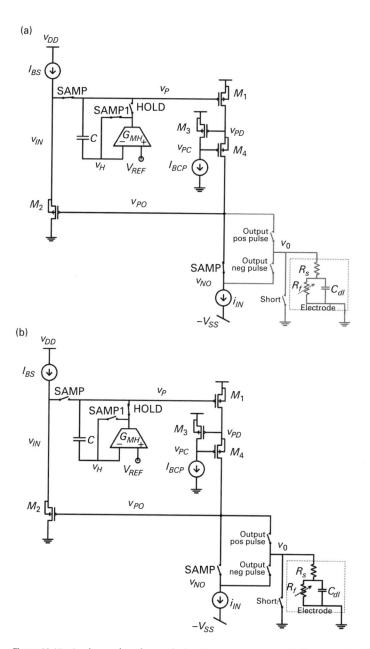

Figure 19.15a, b. A precise charge-balancing current-stimulation circuit that does not require blocking capacitors. The same circuit is shown in (a) and (b) but in a sampling configuration in (a) and in a hold configuration in (b). Reproduced with kind permission from [26] (©2007 IEEE).

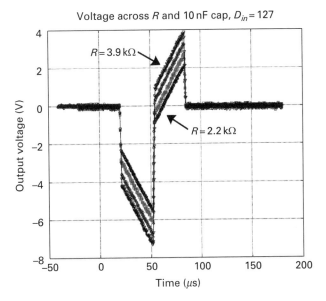

Figure 19.16. Data taken for the circuit of Figure 19.15. Reproduced with kind permission from [26] (©2007 IEEE).

in the context of Figures 19.15 (a) and 19.15 (b) are described in [26]. Figure 19.16 shows the voltage on a series RC circuit after charge-balanced current stimulation from the circuit of Figure 19.15 has been applied. The current stimulation is composed of a negative-current rectangular pulse followed by a positive-current rectangular pulse. The abrupt drops and rises in the series RC are due to the R (which is varied to simulate different electrode resistances) and the linear charging and discharging is due to the C (maintained at 10 nF). In all cases, we see that the voltage on the RC is nearly perfectly charge balanced at full-scale stimulation (DAC bits are set to 127 in an 8-bit DAC for maximal full-scale stimulation of i_{IN}). Further circuit details including a stability and noise analysis are provided in [26]. Chapter 25 also has a deeper discussion of the actual electrochemistry at an electrode-electrolyte interface, which determines the electrode's impedance.

19.5 Brain-machine interfaces for the blind

Cochlear implants represent a highly successful and relatively mature technology. Their success and the success of deep-brain stimulation for helping treat Parkinson's disease is leading to active research and development in other applications that can benefit from very similar technology. For example, retinal implants for the blind share a very similar paradigm to cochlear implants except that images rather than sounds need to be processed, and the optic nerve rather than the auditory nerve needs to be stimulated. The technologies of wireless recharging, wireless telemetry, electrodes, implantable batteries, hermetic sealing,

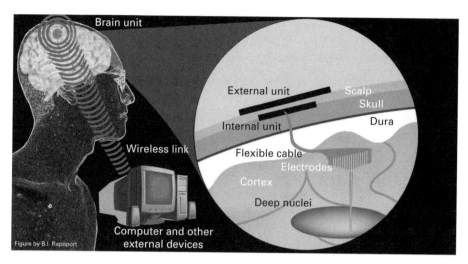

Figure 19.17. A flexible configuration for a brain-machine interface (BMI) or brain implant.

bio-compatible packaging, and electrical stimulation are very similar in both applications. However, several specifications get more stringent due to the severe space available in the eye, because of the delicacy of the retina, because the eye moves, and because the ionic environment inside the eye is challenging for long-term implantation. The rapid success of deep-brain stimulation hardware for the treatment of Parkinson's and other motor disorders has naturally led researchers to ask whether the brain, which is a much bigger and less delicate organ than the eye, may be a more suitable site for treating blindness via electrical stimulation than the eye. Experiments performed in the lateral geniculate nucleus (LGN), which is in the deep-brain region known as the thalamus, and the primary visual cortex (V1), which is at the back of the head, suggest that such visual prosthetics may indeed be promising [6], [7]. Such prosthetics would be applicable to all common causes of blindness including retinititis pigmentosa, macular degeneration, and glaucoma, and could help improve the lives of over 37 million people in the world.

Figure 19.17 shows what such a general brain-machine interface that could be configured to treat blindness (and other disorders) may look like in the future.

Electrodes that are implanted in the brain may stimulate one of its regions, in this case, say, the LGN. The electrodes connect to an implanted unit between the skin and the scalp via a flexible cable that can take up the slack caused by brain motion. The internal unit can receive power and data wirelessly from an external unit as in cochlear implants and as discussed in Chapters 16 and 18 respectively. The external unit connects in a wired (or wireless) fashion with an imager and a processor, which can be worn on a set of 'glasses'. The external unit is capable of wireless (or wired) communication to a computer via a standard Zigbee or Bluetooth interface for programming, monitoring, and debugging functions. The external unit and internal unit are mechanically aligned via electronic

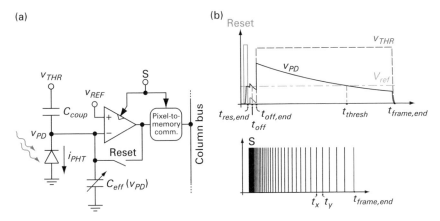

Figure 19.18a, b. Operation of a low-power wide-dynamic-range imager. The basic architecture is shown in (a) and waveforms that are important in its operation are shown in (b). Reproduced with kind permission from [30].

magnets. The overall configuration is one of several but presents several advantages including stable coil coupling, a relatively benign bio-environment between skin and scalp for the implanted unit, minimization of tissue heating within the skull, the ability to be relatively stable during brain and head motion, and good RF link efficiency.

In order to develop a visual prosthesis we need to address four critical issues: 1) the overall system has to operate with low power to be small, portable, and to minimize tissue heating; 2) the imager used in such a system needs to have a wide dynamic range and needs to operate with very low power to be able to sense real-life images in an efficient fashion; 3) effective coding strategies need to be developed to deliver image information to the subject using relatively few electrodes; and 4) stimulation strategies need to be developed to minimize electrode interactions commonly found during stimulation of neural tissues. The first issue can be addressed through low-power techniques for wireless recharging (Chapter 16), wireless telemetry (Chapter 18), neural recording (to be covered in this chapter), stimulation (covered in this chapter), and processing (covered in this chapter). The description in [29] also provides a review. We shall now discuss the second issue (the low-power imager). Then we shall suggest how the third and fourth issues may potentially be solved by taking inspiration from the solution of similar problems in cochlear implants.

19.5.1 Low-power wide-dynamic-range imager

Figure 19.18 (a) shows the pixel block diagram of the imager. Figure 19.18 (b) shows the waveforms of the global v_{THR}, Reset, and S signals that, along with the local photodiode current, i_{PHT}, determine the comparator input voltage v_{PD}. The sampling signal for the comparator, S, is shown at the bottom most part of Figure 19.18 (b). If we temporarily ignore the apparent complexity of Figures 19.18 (a) and 19.18 (b), the core idea behind the operation of the imager is simple. At each pixel, we count the number of clock cycles that it takes photocurrent i_{PHT} to discharge a fixed amount of charge on C_{coup} and thus digitize i_{PHT}; the fixed charge on C_{coup} is established near the time of reset at the start of the imaging frame. The time of discharge is evaluated by the value of a global clock-cycle counter if and when the pixel comparator triggers and is stored in an associated memory array. Thus, the overall operation of the imager is similar to that of the de-integration phase of a dual-slope ADC. The embellishments around this core idea that lead to the complexity of the various waveforms have to do with the fact that the imager has four novel additions around this basic idea that enable it to achieve low-power wide-dynamic-range performance: 1) the use of a dual-threshold mechanism to automatically eliminate thermal reset noise, fixed pattern noise (FPN) due to comparator offsets, and $1/f$ noise of the comparator; 2) the use of synchronous comparisons with S rather than asynchronous comparisons that save power by not having the pixel comparators on all the time; 3) the use of optimal variable-rate sampling with S to minimize the number of comparisons needed, which saves power, and yet ensures that the quantization noise is well below the pixel noise at all times; 4) the use of a capacitively coupled pixel topology that enables highly linear operation and minimizes the common-mode range of operation needed by the comparator, thus simplifying its design. In addition, at very low light levels, the fixed charge is lowered to ensure that discharge always occurs over a finite fixed frame interval, independent of light level. We now delve into the details of Figure 19.18 (a) and Figure 19.18 (b) to explain how these improvements are actually accomplished.

When the reset pulse is activated, auto-zeroing resets v_{PD} to a value near V_{REF}, which depends on the offset of the comparator and the thermal kT/C_{eff} reset noise. After this reset, v_{THR} rises briefly and then falls again in a right-angled-triangle fashion as shown in Figure 19.18 (b). The capacitive coupling of v_{THR} causes v_{PD} to rise and then fall as well. The voltage v_{PD} falls partly because v_{PD} falls but also because i_{PHT} discharges C_{coup} a little. The time at which v_{PD} reaches the trip point of the comparator again, t_{off}, is then a deterministic function of the falling slope of v_{THR}, i_{PHT}, C_{coup}, the thermal reset noise, and any slight changes in comparator offset due to $1/f$ noise. The time t_{off} represents the first-of-two threshold crossings of V_{REF} by v_{PD} which is why the algorithm is referred to as a dual-threshold algorithm. The digital value of t_{off}, as evaluated by the sampling clock S, is later used to eliminate the effects of thermal reset noise, which would otherwise cause random FPN from pixel to pixel. The use of auto-zeroing and t_{off} help alleviate FPN due to comparator offset as well. After t_{offend}, v_{THR} rises again to initiate a rectangular-pulse-like waveform as shown in Figure 19.18 (b). The value of v_{PD} rises and slowly discharges due to i_{PHT} until a second threshold crossing at t_{thresh} is reached. Note that when t_{thresh} is digitally evaluated, it will be quantized to be somewhere between t_x and t_y as shown in Figure 19.18 (b). From measurements of t_{off} and t_{thresh} and our knowledge of v_{THR} from a 12-bit DAC we can digitally evaluate the pixel photo current according to

$$i_{PHT} = C_{coup} \frac{v_{THR}(t_{thresh}) - v_{THR}(t_{off})}{t_{thresh} - t_{off}} \qquad (19.18)$$

Since all threshold crossings occur at V_{REF}, the charge on C_{eff} is unchanged, any nonlinearity in C_{eff} becomes irrelevant, linearity is determined by a controllable C_{coup}, and the needed common-mode range of operation of the comparator is reduced. Capacitive coupling allows low-voltage operation of the comparator to become possible in future deep submicron processes and not limit imager dynamic range. At very low values of i_{PHT}, t_{thresh} will exceed $t_{frame,end}$ if v_{THR} is maintained at a high level. To avoid this scenario, a little before $t_{frame,end}$ is reached, v_{THR} is abruptly lowered to terminate the rectangular pulse.

The sampling signal S is extremely frequent during the initial offset-compensation phase (between $t_{res,end}$ and $t_{off,end}$) and during the ending pulse-termination phase (near $t_{frame,end}$). In these regions, time is short, and therefore high temporal resolution is needed to maintain precision. In addition, near the end of the brief offset-compensation phase, the small numbers of electrons collected leads to a small absolute value of charge shot noise, and the sampling is made even more frequent to ensure that the quantization noise is below the intrinsic shot noise of the pixel. Similarly, the sampling is made even more frequent near the end of the pulse-termination phase. In the region where v_{THR} is almost constant, there is relatively constant SNR since we always collect the same numbers of electrons. The sampling in this region changes from fine to coarse as shown in Figure 19.18 (b); such sampling saves power but still ensures that the quantization noise is well below that of the pixel's intrinsic noise.

A prototype 150×256 pixel imager employing this algorithm experimentally achieved 95.5 dB dynamic range, 37 dB peak SNR, and a highly linear transfer characteristic while consuming 1.79 nJ per pixel per frame, making it one of the most energy-efficient wide-dynamic-range imagers reported. The individual pixels experimentally achieved 98.8 dB dynamic range and 44 dB peak SNR. The array performance lags slightly behind that of the individual pixels due to the additional noise power contributed to the array data by residual pixel-to-pixel mismatch effects, attributed primarily to gain and dark-current FPN. The SNR is nearly shot-noise limited, i.e., limited by the number of electrons that are generated by light. The prototype imager implements pixels and their associated 18-bit timing memories in two on-chip arrays linked by a 200 MHz time-domain-multiplexed communication bus, enabling a pixel pitch of 12.5 μm with 42.7% fill factor in a 0.18 μm, 1.8 V CMOS process. To save switching energy, the pixel comparators signal to the memory that a threshold crossing has occurred only during a brief time following the threshold crossing. Figure 19.19 (a) shows the image gathered from an intentionally low-dynamic-range (60 dB) version of the algorithm with no improvements and from the high-dynamic-range version with improvements. It is easy to see that the outdoor scene and indoor scene are visible in Figure 19.19 (b) but not in Figure 19.19 (a). Further details of the imager are described in [30].

(a)   (b)

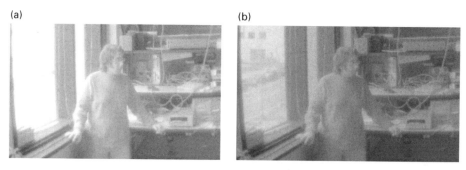

Figure 19.19a, b. The image obtained with a low-dynamic-range imager in (a) saturates for bright external light while reproducing indoor image information. The image obtained with the wide-dynamic-range imager in (b) reproduces both indoor and outdoor information. Reproduced with kind permission from [30].

19.5.2   Coding and compression strategies for visual prosthetics

It is remarkable that our ear has 3,500 spectral channels of information but that deaf subjects with a cochlear implant can almost hear perfectly with 8–16 channels. How is this possible? It occurs because the 16 spectral channels form a *complete* basis for describing sound with the bandpass impulse-response kernels being the basis vectors. The information about the sound is contained in these basis coefficients, such that if the brain explicitly or implicitly learns the kernels, the sound can be explicitly or implicitly reconstructed. Given time, the remarkable brain does learn to reconstruct the sound because the information to reconstruct it is there. The learning is not instantaneous and often takes 3 months or more, yet, as mentioned previously, some cochlear-implant subjects have even better word recognition scores than normal-hearing subjects. In fact, experience with cochlear implants has revealed two consistent findings. 1) If the information about sound is represented by an algorithm or encoding, the brain will eventually learn it. In fact, reconstructions of the sound by an encoding-decoding pair that are presented to a normal-hearing person are highly predictive of how a deaf subject will eventually perform with an encoding algorithm [31]. Thus, cochlear-implant simulations are now common. 2) An encoding that is nearer the natural biological encoding is usually more successful. The basis kernels that have been used successfully in cochlear implants correspond to constant-Q filters, which are similar to the basis kernel functions of the ear. A good general discussion of encoding and decoding of information by neural systems may be found in [32].

How do we encode a million pixels worth of information in a few tens or few hundreds of electrodes? The situation is very analogous to that in cochlear implants such that we may be able to leverage off the insights gained there. We do not encode information with pixel-based basis information, which is highly expensive in the needed number of electrodes and rather impractical. Such a scheme is analogous to encoding sound by simply sampling it in time and reporting these sampled values: we would then need a lot of samples since impulses form an

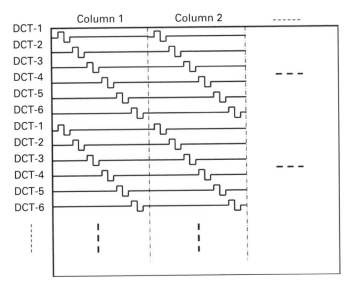

Figure 19.20. A scheme for stimulating electrodes for a visual prosthesis that minimizes electrode interactions and that encodes visual information in a compressed fashion in a relatively small multi-electrode array. Reproduced with kind permission from [30].

inefficient basis! Instead, we encode the information about an image in a set of kernel basis coefficients that more efficiently and compactly represent the image. In fact, we already know from neurobiological studies that our own brains use center-surround and Gabor-kernel basis functions to represent and analyze images: such kernels form the 'receptive fields' of visually sensitive neurons in the brain. Thus, by analyzing an image with spatial filters that correspond to these kernels, and reporting these basis coefficients, we may be able to provide the brain with the information that it needs to represent an image efficiently, and that is similar to its own representation. The brain may then learn from the information that we provide it if we give it time, as has been proven in cochlear implant subjects, or in the normal learning and development of a child parsing its sensory world. We must also make sure that the information is provided in a fashion that prevents electrode interactions as in cochlear implants.

Figure 19.20 reveals an example of how a simple compression-and-encoding strategy may work in even a 60-electrode visual prosthesis. The discrete-cosine-transform (DCT) basis coefficients and kernels are used in this example for simplicity since such DCT coefficients are routinely used in .jpg encodings of an image. However, the key idea is not altered if an encoding with a more complex and biologically plausible basis like that derived from a Gabor filter is used instead. A down-sampled image with 80×80 pixels is divided into 10 rows and 10 columns of 8×8 image sub-blocks to create a 10×10 block image [30]. Each 8×8 sub-block in the block image is represented by 6 DCT coefficients that create a compressed set of basis coefficients for representing it; these basis coefficients are quantized to 8 bits. Information regarding the first column of this block image is encoded into a set of 60 electrodes as shown in Figure 19.20. Inspired by CIS, electrodes 1, 7, 13, 20,...,54, which are sufficiently far apart from each other such that electrode interactions are small, are simultaneously stimulated with charge-balanced information corresponding to the first DCT coefficient. Then, electrodes 2, 9, 16,...,55 are simultaneously stimulated with information corresponding to the second DCT coefficient. After all 6 DCT coefficients for the first column have been presented in this fashion, we turn our attention to the second column and repeat the same procedure. Now, we use information from the second column to stimulate the 60 electrodes rather than information from the first column. After all 10 columns of the block image have been represented via stimulation onto the electrodes, we pause briefly between frames, and then repeat the procedure for the next frame. If we devote 10 ms to each column, i.e., we have 100 pulses per second on any electrode with 600 DCT coefficients per second on 6 simultaneous electrodes, and create a frame pause of 20 ms, the overall frame period is 10×10 ms $+$ 20 ms $= 120$ ms and corresponds to a frame rate of 8 Hz, which is quite adequate for vision. Jumping spiders and humans can reconstruct images by column scanning as is evident by the fact that a person walking past a door that is slightly ajar can form a complete image of what is on the other side of the door [33]. Humans constantly scan images via directed eye movements known as saccades. Non-columnar scanning patterns can also of course be implemented. Figure 19.21 shows that image reconstruction with this simple scheme for letters and faces is not perfect but reasonable. More electrodes (600 electrodes are now possible [34]), more DCT coefficients, and better basis functions could certainly improve the image representation. Time and experiments on real subjects will eventually tell if such an approach will actually succeed on blind patients. The success of similar strategies in cochlear implants does give us hope. Low-power image-processing circuits inspired by biology are discussed in Chapter 23.

19.6 Brain-machine interfaces for paralysis, speech, and other disorders

Experiments using brain-machine interfaces (BMIs) have shown that it is possible to predict intended limb movements by decoding simultaneous recordings from 10–100 neurons. See [3] for a report of the first human trials of such devices, and the references therein for recent reviews of this field. These findings have suggested a potential approach for treating paralyzed patients by recording and decoding neural data from the motor regions of their brain and using the decoded results to stimulate a prosthetic arm or muscle, or to control a computer. Such approaches could generalize to treating patients with speech disorders by recording and decoding neural data from the speech regions of their brain and using the decoded results to control their vocal tract or an artificial vocal tract, e.g., like the one described in [35]. For all such recording prosthetics, low-power neural decoding is very important. Low-power analog decoding architectures that are potentially $50\times$ more power efficient than digital decoders have been described in [36]. Since the

Figure 19.21. Original image and its reconstruction on the right with a 60-electrode version of the visual-prosthesis algorithm. Reproduced with kind permission from [30].

needed computation is inherently slow and parallel and compresses information from many electrodes to 2–3 output motor parameters, the mapping to analog filtering-and-learning circuits is particularly good, making such circuits highly power efficient. Analog decoding can potentially reduce the bandwidth needed for wireless telemetry of raw neural data by over a factor of 100,000 in such neural prosthetics, drastically reducing telemetry and digital power in such a prosthetic [36]. Once again, the benefits of delaying digitization through analog preprocessing become evident. Analog decoders in the implanted unit can be synergistically

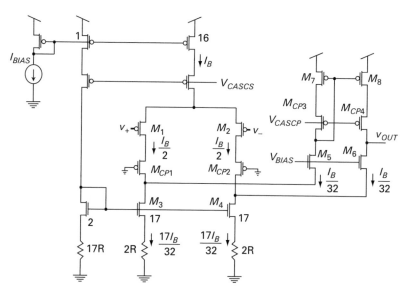

Figure 19.22. A low-noise micropower differential transconductance amplifier designed for neural-recording applications. Reproduced with kind permission from [38] (©2007 IEEE).

combined with digital processing in the external unit such that flexible configuration of the implanted analog system through external digital analysis becomes possible [29]; the flexibility of a higher-power digital system, is then only periodically necessary, saving power. However, the full flexibility of a high-power external digital system is always present if needed. As another example of the use of analog preprocessing with delayed digitization, a novel low-power spectral-analysis IC for deep-brain stimulation with local-field-potential-based recordings has been recently reported [37].

Ultra-low-power neural amplifiers are described in [38]. They are built with low-power OTAs, capacitive gain-setting elements, and adaptive elements (Chapter 11) to set floating-node voltages in a closed-loop inverting-amplifier topology. The neural amplifier described in [38] has 40 dB of gain, 3.05 μV rms input-referred noise in a 45 Hz to 5.3 kHz bandwidth, with a power consumption of 7.56 μW in a 2.8 V process. It is currently one of the most energy efficient and lowest power differential neural amplifiers. The key circuit of this neural amplifier is shown in Figure 19.22. Due to the distribution of currents in this folded-cascode topology, the negligible contribution of noise of the 2R resistors because $g_m^{M1,2} R_{1,2}$ is $\gg 1$, and the self shunting of noise in cascode devices, this amplifier's noise is almost entirely due to the input devices M_1 and M_2, the lowest it can be in a differential amplifier. The transistors M_3 and M_4 ensure that the impedance looking into their drains is high such that most of the differential pair's output current is shunted to the folded output cascode.

Figure 19.23 shows that, in typical multi-electrode neural systems, the input-referred noise of recording electrodes has a probability distribution with some electrodes having low noise floors and others having high noise floors. The power needed to build an amplifier with a fixed bandwidth and input-referred noise

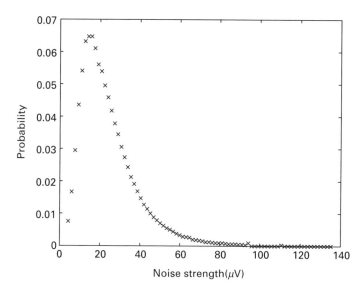

Figure 19.23. Probability distribution of noise measured in a neural-recording multi-electrode array. Reproduced with kind permission from [29] (©2008 IEEE).

v_n^2 scales like $1/v_n^2$. Therefore, significant power savings can be attained if we adaptively bias amplifiers in an array rather than bias them all at one high-power lowest-noise setting. One high-power low-noise neural amplifier can periodically measure and set the power consumption of each amplifier in an array such that its power consumption is just high enough to achieve a noise floor slightly lower than its input (say 3× lower). For the probability distribution in Figure 19.23, adaptive power biasing leads to a 12× power savings in the 64-electrode array. The details are described in [29]. This example illustrates dramatically that in large complex systems *adaptive power biasing* can lead to significant power savings over *worst-case* design since the mean power for a large system is significantly below the *worst-case* power determined by one of its outlier parts. An overview of low-power techniques for BMIs is presented in [29]. A description of a highly miniature wireless neural stimulator tested on the brain of a zebra-finch song bird for neuroscience experiments is described in [39]: the BMI could successfully stop the bird from singing in mid-song by stimulating certain of its neurons in a remote-controlled, wireless fashion. The wireless operation also enabled tether-free stimulation such that the birds were happier and more likely to sing. The latter BMI could be adapted to design a highly miniature next-generation deep-brain stimulator in the future.

19.7   Summary

We discussed low-power circuits, algorithms, and strategies for sensing (micro-phone amplifiers, imagers, neural amplifiers, adaptive power biasing), processing (bionic ear processor for the deaf, compression and encoding for the blind, analog decoders for paralysis), and electrical stimulation (adiabatic circuits). We discussed algorithms for low-power stimulation (AIS) and highly miniature blocking-capacitor-free circuits. In Chapter 16, we discussed energy-efficient wireless recharging circuits for implants, and in Chapter 18, we discussed power-efficient wireless telemetry circuits. In real implants, such electronic design must be synergistically combined with electrode, mechanical, biocompatible, and hermetic design [40] to lead to a system that is robust, efficient, and practical in real patients.

References

[1] P. C. Loizou. Mimicking the human ear. *IEEE Signal Processing Magazine*, **15** (1998), 101–130.

[2] M. L. Kringelbach, N. Jenkinson, S. L. F. Owen and T. Z. Aziz. Translational principles of deep brain stimulation. *Nature Reviews Neuroscience*, **8** (2007), 623–635.

[3] L. R. Hochberg, M. D. Serruya, G. M. Friehs, J. A. Mukand, M. Saleh, A. H. Caplan, A. Branner, D. Chen, R. D. Penn and J. P. Donoghue. Neuronal ensemble control of prosthetic devices by a human with tetraplegia. *Nature*, **442** (2006), 164–171.

[4] J. Wyatt and J. Rizzo. Ocular implants for the blind. *IEEE Spectrum*, **33** (1996), 47–53.

[5] J. D. Weiland and M. S. Humayun. Visual prosthesis. *Proceedings of the IEEE*, **96** (2008), 1076–1084.

[6] J. S. Pezaris and R. C. Reid. Demonstration of artificial visual percepts generated through thalamic microstimulation. *Proceedings of the National Academy of Sciences*, **104** (2007), 7670–7675.

[7] E. M. Schmidt, M. J. Bak, F. T. Hambrecht, C. V. Kufta, D. K. O'rourke and P. Vallabhanath. Feasibility of a visual prosthesis for the blind based on intracortical micro stimulation of the visual cortex. *Brain*, **119** (1996), 507.

[8] M. Roham, J. M. Halpern, H. B. Martin, H. J. Chiel and P. Mohseni. Wireless amperometric neurochemical monitoring using an integrated telemetry circuit. *IEEE Transactions on Biomedical Engineering*, **55** (2008), 2628–2634.

[9] M. Ho, P. Georgiou, S. Singhal, N. Oliver and C. Toumazou, A bio-inspired closed-loop insulin delivery based on the silicon pancreatic beta-cell. *Proceedings of the IEEE International Symposium on Circuits and Systems (ISCAS), Seattle, Wash.*, 1052–1055, 2008.

[10] S. Pavan, N. Krishnapura, R. Pandarinathan and P. Sankar. A Power-Optimized Continuous-Time Delta Sigma ADC for Audio Applications. *IEEE Journal of Solid-State Circuits*, **43** (2008), 351–360.

[11] Texas Instruments Incorporated TMS320C55x Technical Overview, *Literature Number SPRU393*. Technical Report, February 2000.

[12] Texas Instruments Incorporated TMS320VC5510 Power Consumption Summary, *Literature Number SPRA972*. Application report, November 2003.

[13] Texas Instruments Incorporated BGA *Mechanical Data, Document MPBG021C*. Technical Report, May 2002.

[14] M. W. Baker and R. Sarpeshkar. A low-power high-PSRR current-mode microphone preamplifier. *IEEE Journal of Solid-State Circuits*, **38** (2003), 1671–1678.

[15] M. W. Baker and R. Sarpeshkar. Low-Power Single-Loop and Dual-Loop AGCs for Bionic Ears. *IEEE Journal of Solid-State Circuits*, **41** (2006), 1983–1996.

[16] C. D. Salthouse and R. Sarpeshkar. A practical micropower programmable bandpass filter for use in bionic ears. *IEEE Journal of Solid-State Circuits*, **38** (2003), 63–70.

[17] R. Sarpeshkar, C. D. Salthouse, J. J. Sit, M. W. Baker, S. M. Zhak, T. K. T. Lu, L. Turicchia and S. Balster. An ultra-low-power programmable analog bionic ear processor. *IEEE Transactions on Biomedical Engineering*, **52** (2005), 711–727.

[18] R. Sarpeshkar, M. W. Baker, C. D. Salthouse, J. J. Sit, L. Turicchia and S. M. Zhak, An analog bionic ear processor with zero-crossing detection. *Proceedings of the IEEE International Solid State Circuits Conference (ISSCC)*, San Francisco, CA, 78–79, 2005.

[19] S. M. Zhak, M. W. Baker and R. Sarpeshkar. A low-power wide dynamic range envelope detector. *IEEE Journal of Solid-State Circuits*, **38** (2003), 1750–1753.

[20] J. J. Sit and R. Sarpeshkar. A micropower logarithmic A/D with offset and temperature compensation. *IEEE Journal of Solid-State Circuits*, **39** (2004), 308–319.

[21] S. K. Kelly and J. Wyatt. A power-efficient voltage-based neural tissue stimulator with energy recovery. *Proceedings of the IEEE International Solid-State Circuits Conference (ISSCC)*, San Francisco, CA, 228–524, 2004.

[22] A. C. H. MeVay and R. Sarpeshkar. Predictive comparators with adaptive control. *IEEE Transactions on Circuits and Systems II: Analog and Digital Signal Processing*, **50** (2003), 579–588.

[23] J. J. Sit, A. M. Simonson, A. J. Oxenham, M. A. Faltys, R. Sarpeshkar and C. MIT. A low-power asynchronous interleaved sampling algorithm for cochlear implants that encodes envelope and phase information. *IEEE Transactions on Biomedical Engineering*, **54** (2007), 138–149.

[24] J. J. Sit and R. Sarpeshkar. A Cochlear-Implant Processor for Encoding Music and Lowering Stimulation Power. *IEEE Pervasive Computing*, **1** (2008), 40–48.

[25] R. K. Shepherd, N. Linahan, J. Xu, G. M. Clark and S. Araki. Chronic electrical stimulation of the auditory nerve using non-charge-balanced stimuli. *Acta Oto-Laryngologica*, **119** (1999), 674–684.

[26] J. J. Sit and R. Sarpeshkar. A low-power, blocking-capacitor-free, charge-balanced electrode-stimulator chip with less than 6 nA DC error for 1mA full-scale stimulation. *IEEE Transactions on Biomedical Circuits and Systems*, **1** (2007), 172–183.

[27] M. O'Halloran and R. Sarpeshkar. A 10-nW 12-bit accurate analog storage cell with 10-aA leakage. *IEEE Journal of Solid-State Circuits*, **39** (2004), 1985–1996.

[28] M. O'Halloran and R. Sarpeshkar. An analog storage cell with 5 electron/sec leakage. *Proceedings of the IEEE International Symposium on Circuits and Systems (ISCAS)*, Kos, Greece, 557–560, 2006.

[29] R. Sarpeshkar, W. Wattanapanitch, S. K. Arfin, B. I. Rapoport, S. Mandal, M. W. Baker, M. S. Fee, S. Musallam and R. A. Andersen. Low-Power Circuits for Brain-Machine Interfaces. *IEEE Transactions on Biomedical Circuits and Systems*, **2** (2008), 173–183.

[30] L. Turicchia, M. O'Halloran, D. P. Kumar and R. Sarpeshkar, A low-power imager and compression algorithms for a brain-machine visual prosthesis for the blind. *Proceedings of the SPIE*, San Diego, CA, 7035101–7035113, 2008.

[31] L. M. Friesen, R. V. Shannon, D. Baskent and X. Wang. Speech recognition in noise as a function of the number of spectral channels: comparison of acoustic hearing and cochlear implants. *The Journal of the Acoustical Society of America*, **110** (2001), 1150.

[32] Chris Eliasmith and C. H. Anderson. *Neural Engineering: Computation, Representation, and Dynamics in Neurobiological Systems.* (Cambridge, Mass.: MIT Press, 2003).

[33] M. F. Land. Movements of the retinae of jumping spiders (Salticidae: Dendryphantinae) in response to visual stimuli. *Journal of Experimental Biology,* **51** (1969), 471–493.

[34] Second Sight Medical Products Inc. Second Sight Completes U.S. Phase I Enrollment and Commences European Clinical Trial for the Argus II Retinal Implant. [Press Release] Available from: http://www.2-sight.com/press-release2-15-final.html.

[35] K. H. Wee, L. Turicchia and R. Sarpeshkar. An Analog Integrated-Circuit Vocal Tract. *IEEE Transactions on Biomedical Circuits and Systems,* **2** (2008), 316–327.

[36] B. I. Rapoport, W. Wattanapanitch, J. L. Penagos, S. Musallam, R. Andersen and R. Sarpeshkar. A biomimetic adaptive algorithm and low-power architecture for implantable neural decoders. *Proceedings of the 31st Annual International Conference of the IEEE Engineering in Medicine and Biology Society (EMBC),* Minneapolis, MN, 2009.

[37] A. T. Avestruz, W. Santa, D. Carlson, R. Jensen, S. Stanslaski, A. Helfenstine and T. Denison. A $5\,\mu$W/channel Spectral Analysis IC for Chronic Bidirectional Brain-Machine Interfaces. *IEEE Journal of Solid-State Circuits,* **43** (2008), 3006–3024.

[38] W. Wattanapanitch, M. Fee and R. Sarpeshkar. An energy-efficient micropower neural recording amplifier. *IEEE Transactions on Biomedical Circuits and Systems,* **1** (2007), 136–147.

[39] S. K. Arfin, M. A. Long, M. S. Fee and R. Sarpeshkar. Wireless Neural Stimulation in Freely Behaving Small Animals. *Journal of Neurophysiology,* (2009), 598–605.

[40] K. D. Wise. Silicon microsystems for neuroscience and neural prostheses. *IEEE Engineering in Medicine and Biology Magazine,* **24** (2005), 22–29.

20 Ultra-low-power noninvasive medical electronics

It has long been an axiom of mine that the little things are infinitely the most important.

Sir Arthur Conan Doyle

Noninvasive medical electronics refers to electronics for medical instruments that do not invade or penetrate the body. The sensors in these instruments can and often do contact the body. Examples of such sensing include:

- Electrocardiogram (EKG or ECG) measurements of heart function.
- Photoplethysmographic (PPG) measurements of blood-oxygen saturation, or pulse oximetry.
- Phonocardiogram (PCG) measurements of heart sounds.
- Electroencephalogram (EEG) measurements of brain function.
- Magnetoencephalogram (MEG) measurements of brain function.
- Electromyogram (EMG) measurements of muscle function.
- Electrooculogram (EOG) measurements of eye motion.
- Electrical impedance tomography (EIT): measurements to infer composition of the body's tissues. Impedance cardiography (ICG) is a further specialization within the field.
- Temperature measurements.
- Blood-pressure (BP) measurements.
- Pulmonary auscultation (lung-sound) measurements.
- Biomolecular detection of small molecules, DNA, proteins, cells, viruses, or microorganisms for point-of-care or lab-on-a-chip applications, which often exploit BioMEMS (Bio Micro Electro Mechanical Systems) and microfluidic technologies, mostly in a noninvasive fashion thus far.

When such sensing is done chronically, for example as heart tags on patients with a high risk for myocardial infarction (MI), i.e., a heart attack, it is often called *wearable electronics*. The various sensors on the body may form a body sensor network (BSN) or body area network (BAN) that communicate with each other and/or the patient's cell phone, or with an RF-ID, Bluetooth, Zigbee, MICS, UWB, or other wireless receiver in the home, hospital, or battlefield [1]. Such sensors have applications in emergency, surgical, intensive-care, bedside, ambulatory, athlete, farm-animal, soldier, infant, home-care, or elderly monitoring. The need for chronic, wireless, and portable monitoring makes low-power operation important. Chronic monitoring is helpful for alarm situations, e.g., for automatically calling 911 after sudden cardiac arrest (SCA): defibrillation within the first six minutes after a heart attack is critical to saving life. Chronic monitoring is also important for providing long-term diagnostic information at a relatively cheap cost. It is widely recognized that medical costs could be greatly impacted through intelligent, networked, wireless medical-monitoring systems.

In this chapter, we will study some examples of ultra-low-power electronics for noninvasive medical monitoring applications, especially those involving cardiac function for body sensor networks or wearable systems. We shall then review developments in biomolecular sensing. For such applications, high-precision ultra-low-noise electronics for sensing, e.g., the 23-bit-precise MEMS capacitance sensor of Chapter 8, rather than ultra-low-power electronics is currently more important. Chapter 8 discusses how noise from the sensor affects the minimum detectable signal and sensitivity of an overall system including sensors and electronics. The reader may find it helpful to review the chapter since the noise-analysis and block-diagram methods of the chapter apply to all sensing systems. As we have discussed throughout the book, ultra-low-power analog design is all about minimizing noise within a tight power constraint. Thus, a good ultra-low-power analog designer is automatically a good ultra-low-noise high-precision designer, just without the power constraint. Many of the ultra-low-power circuits that we describe in this chapter such as an EKG amplifier rely on principles for good low-noise circuit design.

We shall begin by describing an analog electronic chip that models the heart and circulatory system with electrical circuit analogs: pressure corresponds to voltage and the volume velocity of blood flow corresponds to current. Such analogs can help electrical engineers rapidly and intuitively understand mechanical systems like the heart, as we will show. Understanding the heart is useful for understanding several noninvasive measurements including the EKG, PPG, PCG, and BP measurements. We then describe how the EKG arises in the body and how it relates to normal cardiac function. Our understanding of the heart and EKG will then tie together EKG, PCG, and BP measurements and reveal the relationships between them. Then, we shall describe how to build a micropower EKG amplifier and make it insensitive to 60 Hz noise. This noise typically overwhelms EKG signals that are extremely tiny compared with it. We shall then describe how pulse oximeters work (based on the PPG) and relate them to other measures of cardiac function like the EKG. We briefly describe how low-power pulse oximeters may be built and present PPG measurements from such an oximeter. Then, we shall study an example of an ultra-low-power battery-free medical tag that harvests RF energy to function. It exploits PCG-based cardiac sensing for ultra-low-power operation, is useful for measurements of heart-rate and blood-pressure variations, and can enable audio localization and audio alarms in case the patient needs attention. We shall then describe work on a low-power communication system that uses the conductive body itself as a 'wire' to transmit information between sensors implanted within the body or on it.

We shall conclude by reviewing work on sensors for biomolecular detection in the optical, mechanical, and electrical domains. In our brief review, rather than focus on circuit techniques, many of which are further instantiations of low-noise analog techniques that we have already described, and which can be found in the cited papers, we shall focus more on the key principles behind these schemes. We shall discuss 'label free' and 'labeled' detection of molecules, including examples for DNA detection.

20.1 Analog integrated-circuit switched-capacitor model of the heart

Figure 20.1 shows an artist's depiction of the heart, which is about the size of one's fist. The heart is a four-chambered organ with upper right and left atria and lower right and left ventricles. By convention, the right and left refer to the heart of the person that the reader is viewing, not the reader's right and left, so they appear flipped in Figure 20.1. The right atrium receives blood from the upper regions of the body via the superior vena cava, and from the lower regions of the body via the inferior vena cava. It pumps this blood to the right ventricle. The right ventricle in turn pumps this blood to the left and right lungs via the pulmonary artery, where the blood's hemoglobin molecules bind oxygen. The left atrium receives blood from the lungs via the pulmonary veins and pumps the blood to the left ventricle. The left ventricle in turn pumps this blood to the rest of the body via the aorta. The left ventricle is the biggest, most important, most contractile, and most muscular chamber. It pumps 75 ml to 90 ml of blood with every heart beat, wringing the blood out in a twisting helical action due to muscles that surround its walls. In typical, healthy individuals who are not super-athletic, the healthy heart beats about once every second at rest. The heart is itself sustained by blood from coronary arteries that branch off the aorta.

From the description above, it may appear that the heart is a serial four-phase pump with the right atrium, right ventricle, left atrium, and left ventricle being activated in turn. A healthy heart actually operates on two phases known as diastole and systole. During diastole, the right and left atria contract and pump in synchrony while the ventricles relax and fill with blood. During systole, the right and left ventricles contract and pump in synchrony while the atria relax and fill with blood. During diastole, blood from the atria is conveyed to the ventricles via the opening of valves between the atrial and ventricular chambers; these valves are otherwise closed. During systole, blood from the ventricular chambers is conveyed to body organs via the opening of valves in the output arteries of the heart; these valves are otherwise closed. The valves ensure that the flow of blood is always in the right direction much like diodes regulate the direction of current flow in an electrical circuit.

Figure 20.2 illustrates an analog electrical circuit model of the heart. Fluid pressure of the blood is mapped to voltage while volume velocity of blood is mapped to current. With this mapping, compliance, the inverse of stiffness, is

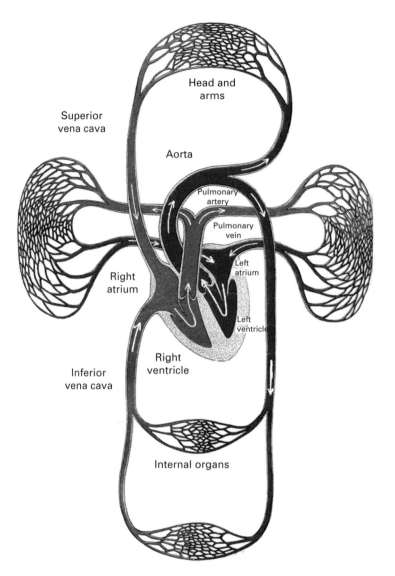

Figure 20.1. An artist's rendering of the heart. Reprinted with permission. C. J. Wiggers. The heart. *Scientific American*, 195. Copyright © 1957 by Scientific American, Inc. All rights reserved.

mapped to capacitance; viscous flow resistance is mapped to electrical resistance; and inertial mass is mapped to inductance. The pumping action of the heart chambers is modeled via a time-varying Thevenin equivalent voltage (pressure) source in series with a capacitance (compliance). Note that a Norton equivalent with a current (volume-velocity) source in parallel with a capacitance may also be used. The compliance of the ventricular chambers is larger during diastole and smaller during systole such that the capacitances of Figure 20.2 may be time varying as well. Four such Thevenin equivalents with different parameters are

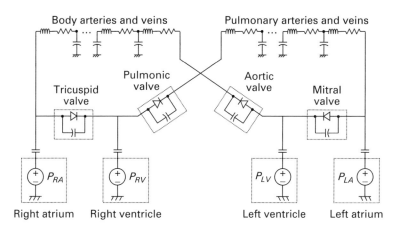

Figure 20.2. Analog electric circuit model of the heart. Reproduced with kind permission from [3] (©2008 IEEE).

shown in Figure 20.2 corresponding to each chamber of the heart. The valves of the heart are modeled as ideal diodes (zero turn-on voltage) with a small capacitance across them to represent the fact that the valves are not perfectly stiff. The tricuspid and mitral (bicuspid) diodes correspond to the valves between the right atrium and right ventricle and between the left atrium and left ventricle respectively. The pulmonic and aortic diodes correspond to the valves at the output pulmonary and output aortic artery of the right ventricle and the left ventricle respectively. The circulatory system of the arteries, veins, and capillaries can be modeled by a dendritic network of flow tubes of varying radius, i.e., a branched, distributed, spatially varying transmission-line RLC network; R corresponds to the local viscous resistance per unit length of the artery, capillary or vein and scales inversely with the fourth power of the radius of the local flow tube; L corresponds to the local inertial mass per unit length of the blood and scales inversely with the area of cross-section of the flow tube; and C corresponds to the local compliance per unit length of the flow-tube walls and scales proportionately with the radius of the flow tube; the value of C is slightly pressure dependent such that we have a slightly nonlinear capacitance.

Modeling the full complexity of the heart and circulatory system are rarely needed for understanding its function or even for matching experimental measurements of blood pressure and blood flow. For our purposes, therefore, several simplifications can be made that enable an analog electronic model of the heart to be easily implemented on a single chip together with a few off-chip discretes. Figure 20.3 shows the simplified model. If we map seconds to seconds, 1 ml/s of blood flow to 1 nA of electrical current, and 1 mmHg of pressure to 1 mV of electrical voltage, then the capacitances needed in Figure 20.3 to model mechanical compliances range from $0.5\,\mu$F to $5\,\mu$F, which can be implemented with small off-chip tantalum capacitors. The reset switches establish an initial condition on the capacitors. Time-varying pressure sources are mimicked by rectifying

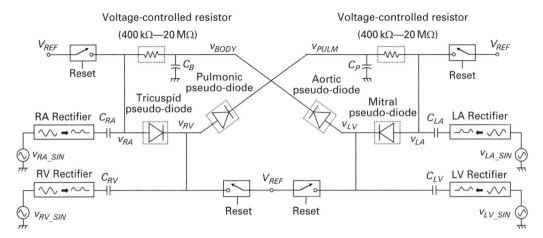

Figure 20.3. Simplified electrical-circuit model of the heart used in a simple integrated-circuit implementation. Reproduced with kind permission from [3] (©2008 IEEE).

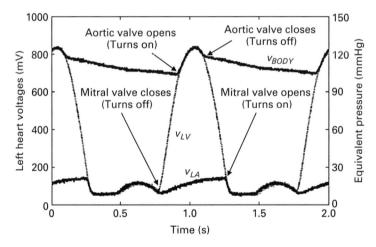

Figure 20.4. Waveforms from the heart chip. Reproduced with kind permission from [3] (©2008 IEEE).

sinusoids to half-wave-rectified voltage waveforms with diodes, resistors, and a buffer. The pseudo-diodes are implemented with comparators that gate a MOSFET switch to be closed or open depending on the voltage across its drain and source. Electronically tunable MOSFET resistors with guaranteed zero current at zero voltage can be implemented as described in [2] or with an approximate proof-of-concept back-to-back transconductor scheme as described in [3]. Figure 20.4 shows waveform measurements of the dominant left ventricle and left atrium of the heart obtained by implementing the simplified switched-capacitor model of Figure 20.3 on a 0.5 μm chip [3]. Waveforms from the chip were gained by a factor of about 5 to enable easy viewing on an oscilloscope. We notice that the left ventricular pressure, v_{LV}, rises and falls as the heart contracts and relaxes. This rise and fall is transformed by the aortic diode and body capacitance C_B of Figure 20.3 into a peak-detector-like output voltage, v_{BODY}. This voltage v_{BODY} mimics blood-pressure variations in the body with a typical high/low in the left arm of a healthy individual being near 120/80 mmHg even though the left ventricular pressure is varying between 120 mmHg and a very small value.

The timing of the mitral and aortic valves are also shown in Figure 20.4. The closing of these two valves generates the 'lub-dub' heart sounds, while valve openings are silent: the 'lub' or S_1 sound, as it is technically known, occurs when the mitral valve closes, and the 'dub' or S_2 sound, as it is technically known, occurs when the aortic valve closes. Systole is often defined to occur between mitral valve closing and aortic valve closing, i.e., between the S_1 and S_2 sounds; diastole occurs when systole does not occur. The chip waveform corresponding to v_{LA} in Figure 20.3 exhibits the *a* wave rise in atrial pressure due to its small contraction during diastole. It also exhibits the *v* wave rise in atrial pressure as C_{LA} is charged during systole, i.e., as the left atrium fills with blood increasing its pressure. In the real heart there is also an additional *c* wave peak in the atrial pressure between the *a* and *v* waves. The *c*-wave is due to capacitive back coupling of v_{LV} to v_{LA} across the mitral pseudo-diode in Figure 20.3. The capacitance of the mitral diode is represented in the complex model of Figure 20.2 but not implemented in the simpler chip model of Figure 20.3, so the chip shows no *c*-wave. Another effect that is seen in the real heart but not in Figure 20.4 is a small rising 'glitch' in v_{BODY} right after the aortic valve closes (a tiny and sharp increase in blood pressure right after aortic valve closing or right after S_2). This glitch arises because the discharging current on C_B when v_{LV} and v_{BODY} are both falling in Figure 20.4 abruptly terminates when the aortic valve closes; the presence of a small inductance in series with the resistance connected to C_B then yields an $L(dI/dt)$ voltage glitch when the discharging current goes to zero. The more complex model of Figure 20.2 with inductance can model this effect but the simpler model of Figure 20.3 does not. The addition of small capacitances across pseudo-diodes and the addition of a small gyrator-based inductance connected to C_B can make the model of Figure 20.3 reproduce these higher-order effects with only a modest increase in complexity.

20.2   The electrocardiogram

A biological pacemaker in the sino-atrial node (SA node), which is located near the upper right corner of the right atrium, serves as the heart's master clock. The pacemaker generates an electrical pulse that travels down a conductive pathway in the heart in a wave-like fashion. The traveling pulse excites the heart's muscular contraction, first in its atria where the wave begins, and then in its ventricles as the wave propagates. The electrical wave happens on each heart beat and causes the muscles in the wall of the heart to contract on each beat. On each electrical pulse during a heart beat, calcium ions (Ca^{2+}) rush into the muscle cells in the walls of the heart. The inrush of Ca^{2+} causes these cells to charge up positively or depolarize, and because of Ca^{2+}-dependent mechano-chemistry, to contract. When the pulses of the electrical wave terminate, Ca^{2+} rushes back out of the cells, the cardiac muscles relax and repolarize, and get ready to repeat this action again on the next heart beat. Artificial pacemakers and subthreshold pacemaker circuits for treating slow heart rates are described in [4].

Since the overall heart must be charge balanced and neutral, any accumulation of charge in one region of the heart must be balanced by a loss in a nearby region. The motions of charge in the heart thus lead to lots of tiny dipole current vectors due to charge moving away from one location and appearing at a nearby location such that overall charge is conserved. The average effect of all these dipole current vectors, which have different orientations and magnitudes in different regions of the heart, may be approximated by one large mean current dipole vector at the heart's origin, that we term the heart vector. That is, from an electrical point of view, the rest of the conductive body surrounding the heart sees the complex motions of charge within the heart as that due to an equivalent large floating current source located near the heart's origin. As the dominant charge motions in the heart shift to different locations within it over the course of a heart beat, the magnitude and orientation of this floating current source with respect to the body changes in a stereotypical time-varying fashion. Most charge motions occur from locations behind the advancing wave-front of depolarization/repolarization to locations ahead of the wave-front. The charge distribution in regions well ahead or well behind the front is relatively unchanged since cells are fully depolarized or fully repolarized here and static charge distributions do not generate currents. Therefore, most of the elementary dipole current vectors that contribute to the mean heart vector are on the advancing wavefront of polarization or repolarization. In the heart, the contribution to the mean heart vector is initially dominated by contributions from the right atrium, then from the septum, the wall between the left and right ventricles, and finally from the left ventricular walls.

If we approximate the body as having some average conductivity σ, the potential measured at a point on the body that is located at r, a vector from the origin of the heart to the point on the body, is then given by

$$\phi(r) \approx \frac{(\int J_{hrt\_dpl}dV).r}{4\pi\sigma|r^3|} = \frac{H.r}{4\pi\sigma|r^3|}, \tag{20.1}$$

where J_{hrt_dpl} is an infinitesimal current dipole vector and H is the average heart dipole vector obtained by integration. Equation (20.1) is the solution of Poisson's equation for distributed dipole current sources in a conducting medium. The units of H are current × length. Equation (20.1) is of course a very gross approximation and is approximately valid when r is on the torso and not too far from the heart. In the arms and legs, the potential will not vary strongly as we move along the limb since most of the heart's floating current will flow near the heart and mostly in the torso region of the body; thus, the potential near the beginning and end of

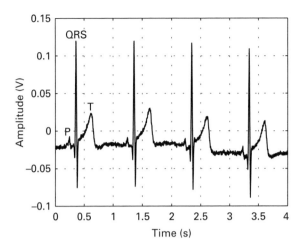

Figure 20.5. An example of an electrocardiogram trace (EKG) recorded from a micropower EKG amplifier. Reproduced with kind permission from [10] (©2009 IEEE).

the limb only differ by a small *IR* drop but *I* is small and *R* is small (the body is quite conductive). If we assume that this *IR* drop is 0, the electrocardiogram potential (EKG or ECG potential) between the left arm r_{LA} and the right arm r_{RA} can then be approximated by

$$v^{LR}_{EKG}(t) = \frac{\boldsymbol{H}(t)}{4\pi\sigma} \cdot \left[\frac{\boldsymbol{r}_{LA}}{r^3_{LA}} - \frac{\boldsymbol{r}_{RA}}{r^3_{RA}} \right]$$

$$v^{LR}_{EKG}(t) \approx \frac{r_{LR}}{4\pi\sigma r^3_{Hsh}} [\boldsymbol{H}(t).\hat{\boldsymbol{r}}_{LR}]$$

(20.2)

where $\hat{\boldsymbol{r}}_{LR}$ is the horizontal unit vector from left shoulder to right shoulder of the body, r_{Hsh} is the average distance of the heart to either shoulder, and r_{LR} is the distance between the shoulders. Equation (20.2) should not be taken too seriously except for three broad implications that do generalize:

1. The instantaneous heart vector $\boldsymbol{H}(t)$ changes in magnitude and direction during the beat such that its projection onto the electrode-vector direction ($\hat{\boldsymbol{r}}_{LR}$ in this example) varies. Therefore, the measured EKG potential depends on the geometric placement of electrodes on the body, i.e., between which two points on the body the potential is measured.
2. The potential increases with increasing inter-electrode distance (r_{LR} in this example).
3. The potential is larger if the average electrode distance is nearer the heart (r_{Hsh} in this example).

Figure 20.5 shows an example of a healthy EKG trace recorded by a micropower EKG amplifier that we will soon describe and Figure 20.6 shows a corresponding cartoon version of it. The ~1 mV peak EKG in Figure 20.5 has been amplified by about 100 to yield a larger voltage that is easily visible on

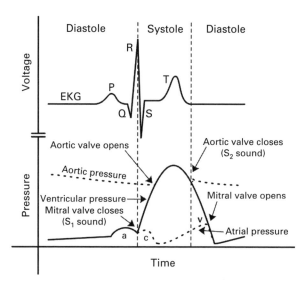

Figure 20.6. The timing relationships between electrical (EKG), blood pressure (BP), and heart-sound (S_1 and S_2) waveforms.

an oscilloscope. The recording was obtained by using skin electrodes called 'derma-trodes'. These electrodes are typically made with a gel electrolyte that permits a 'salt-bridge' connection between electronic conduction in artificial instruments and ionic conduction in the body's tissues. This particular trace corresponds to a difference in voltages between the left arm and right arm like in the example above. Referring to Figures 20.5 and 20.6, the P wave is the initial small bump in the EKG trace and corresponds to atrial depolarization. The P wave is very tiny in Figure 20.5 and more exaggerated in Figure 20.6. The large QRS complex refers to the undershoot-overshoot-undershoot (Q, R, and S, respectively) that follow the small P wave and correspond to ventricular depolarization; in general, the structure of the complex depends on the electrode configuration and the subject. The T wave is the broad bump after the QRS complex and corresponds to ventricular repolarization. Figure 20.6 shows the temporal relationships between the EKG, PPG (S_1 and S_2 sounds), and BP variations. The mechanical BP responses initiate near the peaks of the EKG waveforms such that the EKG is predictive of them: the ventricular pressure begins to rise near the QRS peak and the ventricular pressure begins to fall near the T peak; the atrial contraction begins near the P peak. The S_1 sound usually occurs soon after the QRS peak and the S_2 sound usually occurs after the T-wave.

In healthy subjects, when the Ca^{2+} ions depolarize the atrial walls, $\boldsymbol{H}(t)$ is small and points downward and to the left leading to the P wave component of the electrocardiogram (keep in mind, all coordinates are in body coordinates of the subject facing the reader). The vector $\boldsymbol{H}(t)$ then swings back to the right as Ca^{2+} ions begin to charge up the left-ventricular septal wall; it then grows large in magnitude and swings back to the left as Ca^{2+} ions begin to depolarize the outside wall of the left ventricle, leading to the QRS complex of the electrocardiogram. During the repolarization phase, $H(t)$ generally points towards the left as Ca^{2+} ions slowly leave the left ventricular heart cells, leading to the T wave of the electrocardiogram.

When $H(t)$ is maximal, it actually points towards the back of the body, because the outside wall of the left ventricle has a slightly posterior orientation. Thus precordial leads or electrodes that straddle the heart from front to back, and that are near it, often pick up the largest EKG signals. Six electrodes are generally used with the first, V_1, being slightly to the right of the sternum on the fourth rib, the second, V_2, being slightly to the left of the sternum on the fourth rib, and the others, V_3–V_6, being placed to move from front to back near the fourth and fifth ribs. Unipolar recordings correspond to referring the potential of a given electrode to the average of all the other electrodes and are conventional for precordial V-electrode recordings. Precordial leads provide valuable information about the behavior of $H(t)$ in the horizontal plane in normal and pathological conditions. Other lead placements include those on the left arm, right arm, and left leg, which provide information about the orientation of $H(t)$ with respect to the vertical; they are also useful in diagnosis. The example that we have discussed, i.e., left arm–right arm, is an example of a bipolar recording since it is a purely differential recording between two electrodes. It is termed the I recording in EKG literature. Similarly, (left leg–right arm) corresponds to II and (left leg–left arm) corresponds to III. The vertical leads are also used to create three unipolar recordings termed aV_R, aV_L, and aV_F which correspond to the positive electrode being the right arm, left arm, or left leg respectively, while the negative electrode is composed as an average of the other vertical leads. The full set of EKG recordings corresponding to I, II, III, aV_L, aV_R, aV_F, V_1, V_2, V_3, V_4, V_5, and V_6 constitute a standard 12-lead EKG even though there are only 9 electrodes. The right leg is sometimes used for passive or active grounding. We discuss active grounding in the following section on micropower EKG amplifiers.

EKGs are tremendously useful for relatively cheap diagnosis [5]. For example, elevations of the segment between the S and T and large-magnitude Q-waves signal that a heart attack has occurred. The use of information from various leads can help pinpoint what region of the heart has had an infarction. Atrial fibrillation is often signaled by chaotic and irregular P waves and irregular rates of the QRS complex. Often, simple heart-rate monitoring of R-to-R intervals may suffice in chronic applications.

For such applications, it has been shown that a simple two-lead placement near the precordial leads, but with a recording axis that is oriented downward and leftward, provides good SNR [6]: $H(t)$ is relatively well aligned with this axis when it is large, the inter-electrode distance is not too small, and the leads are near the heart. Thus all three factors in Equation (20.2) are maximized. To attenuate 60 Hz common-mode noise on the body, a third grounding lead is highly advantageous as we explain in the next section. Well-positioned two-lead placement along with a grounding lead may represent the best compromise between simplicity and utility for chronic monitoring applications.

20.3   A micropower electrocardiogram amplifier

One of the banes of EKG recording, and of recording any other bio-potentials (EEG, EMG etc.) from the surface of the body is the presence of a large common-mode component of 60 Hz noise that is approximately equal at all points on the body [7]. The body is a relatively high impedance node so it is susceptible to pickup and noise. For design purposes, the interference due to 60 Hz in typical indoor environments may be well approximated by a Norton equivalent of an ac current source, I_{60}, in parallel with a capacitance of C_{body} to ground. The value of this capacitance is dominated by that of the body's capacitance to ground (earth ground). With $I_{60} = 0.5\,\mu A$ and $C_{body} = 200\,pF$, we obtain a common-mode voltage on the body of nearly $0.5\,\mu A/(2\pi \times 60 \times 200\,pF) \approx 6.6\,V$ amplitude. Potentials of this magnitude are easily measured by holding one end of an un-insulated wire with one's finger and sticking the other end of the wire into an oscilloscope BNC input. If a finger in the other hand touches the outside of the exposed BNC connector, the 60 Hz signal on the oscilloscope is greatly reduced because both fingers have a similar common-mode signal such that the differential signal measured by the oscilloscope is attenuated. For comparison, the QRS-complex peak between two points on the body may only be 1 mV. The P-wave peak may only be $25\,\mu V$ as in Figure 20.5. Thus, the attenuation of 60 Hz noise with high common-mode-rejection-ratio (CMRR), low-noise amplifiers is essential.

Constructing a CMRR amplifier with a common-mode rejection ratio of $6\,V/6\,\mu V$, i.e., 120 dB, requires incredibly good matching, only possible with expensive trimming. Furthermore, the need to have a large common-mode input operating range requires a large power-supply voltage, which is deleterious to low-power operation. One solution to this problem is to connect the ground or a reference terminal in the amplifier to a third electrode on the body that forces the common-mode voltage on the body to a well-defined value. While such passive grounding is helpful, it is limited by the fact that the grounding electrode itself has a significant impedance Z_{gnd} due to the difficulty of forming a good low-impedance connection to the body with a reasonably sized low-cost electrode. Hence, the 60 Hz common-mode signal is reduced but still has a magnitude given by

$$V_{CM}^{body} = I_{60}\left(Z_{gnd} \,||\, \frac{1}{C_{body}s}\right) \tag{20.3}$$

Winter and Webster showed that *active grounding*, wherein the third grounding electrode is actively driven to be at the common-mode value of the main differential-input electrodes, yielded reliably superior performance for measuring bio-potential signals [8]. This advantage arises because the common-mode signal is now attenuated to a value given by

$$V_{CM}^{body} = I_{60}\left(\frac{Z_{gnd}}{1+A_{lp}} \middle\| \frac{1}{C_{body}s}\right)$$

$$\approx I_{60}\frac{Z_{gnd}}{1+A_{lp}}$$

(20.4)

where A_{lp} is the loop gain of the common-mode feedback loop at 60 Hz. The impedance of the grounding electrode is effectively reduced by feedback to a low value such that the interfering 60 Hz signal does not affect the common-mode voltage on the body much.

The benefit of such active grounding is that the power-supply voltage and CMRR requirements of the differential-input amplifier needed to obtain a given SNR can then be greatly reduced. The reduction in power-supply voltage obviously reduces power. Since large-size components need more power to be consumed to maintain bandwidth, the decrease in the size of components needed for good matching also lowers power consumption. The symmetry of a truly differential topology is also beneficial for attaining good matching and CMRR. For all of these reasons, we shall use the technique of active grounding in the micropower EKG amplifier that we now describe.

Figure 20.7 shows the schematic of the micropower EKG amplifier, which is based on a classic two-gain-stage instrumentation-amplifier topology with

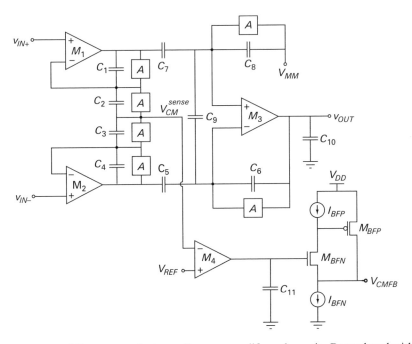

Figure 20.7. Micropower electrocardiogram amplifier schematic. Reproduced with kind permission from [10] (©2009 IEEE).

M_1 and M_2 forming the first gain stage, and M_3 forming the second gain stage. The differential inputs from the EKG electrodes are sensed at v_{IN+} and v_{IN-} and the amplified output is reported at v_{OUT}. The amplifier M_4, C_{11}, and the super-buffer circuit [9], which is formed by the elements M_{BFN}, M_{BFP}, I_{BFP}, and I_{BFN}, drive the active-ground electrode at the V_{CMFB} output via an integrator-based common-mode feedback loop. The four M_1, M_2, M_3, and M_4 amplifiers are standard 5-transistor pFET-differential-pair OTAs. The input transistors in these OTAs have a high W/L and operate in subthreshold, while the mirror transistors have a low W/L and operate above threshold such that input-referred noise of the OTAs is mainly due to input-differential-pair transistors. The use of pFETs also minimizes $1/f$ noise. The M_1 and M_2 amplifiers have larger devices and sizing since their noise is critical to determining the noise of the overall amplifier. Capacitor ratios rather than resistor ratios determine gains, and the high-impedance adaptive A elements (see Figure 11.14 in Chapter 11) ensure that there are no floating nodes. The M_1–M_2 first-stage amplifier and the super buffer consume the most current ($\sim 0.6\ \mu A$ each) and operate off a half-rail supply that is nominally at 1.5 V. The first-stage amplifier needs to be low-noise and the super buffer needs to ensure that its output impedance is significantly lower than that of the active grounding electrode. Therefore, the first-stage amplifier and the super-buffer circuit are the most power-hungry stages. The M_3 common-mode OTA and the second-stage amplifier M_4 consume the least current ($\sim 0.1\ \mu A$ each) as common-mode noise is inherently well rejected and the noise of the second-stage amplifier is attenuated by the gain of the first stage, and therefore uncritical. The OTAs M_3 and M_4 operate on a 3 V rail. In battery-powered systems, the center connection of two 1.5 V batteries in series can generate the 1.5 V supply while the series addition generates the 3 V supply. If the EKG amplifier is powered by RF as we show is possible in a later example (also see Chapter 17 for in-depth descriptions of RF-powered systems), inter-mediate outputs of the rectifier charge pump may be used to generate the 1.5 V supply while the final output of charge pump may be used to generate the 3 V supply.

The block diagram of Figure 20.8 shows all differential signal inputs, common-mode inputs, noise sources (crooked-line inputs) and signal outputs for the amplifier of Figure 20.7. This comprehensive block diagram is literally *all* that is needed to analyze every aspect of the amplifier of Figure 20.7 by simple application of Black's formula and feedback analysis (Chapters 2, 9, and 10). The electrode admittance G_{elec} is shown to be conductive for simplicity though all real electrodes have a susceptive component as well. The output conductance of the super buffer G_{BUF} is given by $g_m^{BFP}(g_s^{BFN}r_o^{BFN} + 1)$ where g_s and r_o are the small-signal source transconductance and small-signal drain-to-source resistance parameters of transistor M_{BFN} in Figure 20.7 respectively. The low output impedance of the super buffer ensures that even though its bias current is only $0.6\ \mu A$, which is necessary to ensure that it can quench interfering common-mode currents on the body, its output impedance is still significantly lower than the $\sim 50\ k\Omega$ output

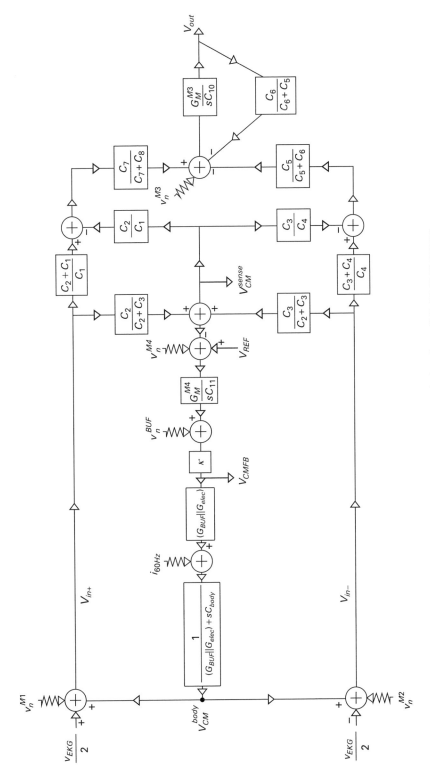

Figure 20.8. Block diagram of EKG amplifier. Reproduced with kind permission from [10] (©2009 IEEE).

impedance of the grounding electrode. The κ parameter in Figure 20.8 corresponds to the subthreshold exponential constant of transistor M_{BFN}.

From the block diagram of Figure 20.8, we can derive that the differential gain v_{out}/v_{EKG} is

$$A_{diff} = A_{1ststg}A_{2ndstg}$$

$$= \left(\frac{C_2 + C_1}{C_1}\right)\left(\frac{C_7}{C_6}\right) \tag{20.5}$$

$$= \left(\frac{C_3 + C_4}{C_4}\right)\left(\frac{C_5}{C_6}\right)$$

if there is perfect capacitor matching between the two differential halves of the signal processing in both stages of amplification. If capacitor matching limits the CMRR, then a slightly more involved analysis that can also be derived from the block diagram shows that [10]

$$CMRR = \frac{v_{out}/v_{EKG}}{v_{out}/v_{CM}^{body}}$$

$$= \frac{\dfrac{C_7}{C_7 + C_8}\dfrac{C_1 + C_2}{C_1} + \dfrac{C_5}{C_5 + C_6}\dfrac{C_3 + C_4}{C_4}}{2\left(\dfrac{C_7}{C_7 + C_8} - \dfrac{C_5}{C_5 + C_6}\right)} \tag{20.6}$$

From the denominator of Equation (20.6), we observe that it is the matching between capacitances in the second stage of the instrumentation amplifier that is the key determining factor in the CMRR of the overall topology. From Figures 20.7 and 20.8, the M_1 and M_2 amplifiers both drive their negative-input voltages to the common-mode input voltage such that independent of the matching of C_1/C_2 or C_3/C_4 or C_2/C_3 the gain to V_{CM}^{body} is 1 in both differential halves of the first stage; thus, it is only the second stage of the instrumentation amplifier that determines the CMRR. Finite-loop-gain effects in the M_1 and M_2 amplifiers, and finite common-mode effects in M_1 and M_2, if mismatched can also limit the CMRR of the topology. Similarly, if M_3 has finite CMRR, even if all other parameters are perfectly matched, the CMRR of M_3 will determine the CMRR of the overall topology. A small mismatch between C_2 and C_3 leads to V_{CM}^{sense} having a small differential v_{EKG} component in addition to the desired V_{CM}^{body}. This component has a small effect on the amplifier output since it is mostly attenuated by the good common-mode rejection of the topology.

The loop transmission of the common-mode feedback loop is given by

$$L_{CMFB}(s) = \kappa\left(\frac{G_M^{M4}}{sC_{11}}\right)\frac{G_{BUF}\,\|\,G_{elec}}{G_{BUF}\,\|\,G_{elec} + sC_{body}} \tag{20.7}$$

From Figure 20.8, at frequencies $s = j\omega$ including 60 Hz, where the loop gain is sufficiently high, the transfer function from i_{60Hz} to V_{CM}^{body} is well approximated by the reciprocal of the common-mode feedback path transmission, the sum of the common-mode loop transmissions in Figure 20.7, such that

$$
\begin{aligned}
v_{CM}^{body} &= \frac{i_{60Hz}}{G_{BUF} \| G_{elec}(j.2\pi.60)} \frac{C_{11}.2\pi.60}{\kappa G_M^{M4}} \\
&\approx \frac{i_{60Hz}}{G_{elec}(j.2\pi.60)} \frac{60}{\kappa f_{CMFB}}
\end{aligned}
\tag{20.8}
$$

where $G_{elec}(j.2\pi.60)$ is the electrode conductance at 60 Hz and f_{CMFB} is the crossover frequency of the common-mode feedback loop. From Equation (20.8), we see that it is advantageous to have a high f_{CMFB} to attenuate 60 Hz noise. However, $2\pi f_{CMFB}$ must be at least a factor of 4 less than $(G_{BUF}\|G_{elec})/C_{body}$ for the common-mode feedback loop to be overdamped and exhibit no ringing. A large value of G_M^{M4} is also harmful to low-power consumption. Thus, the choice of G_M^{M4} was informed by these considerations. The capacitor C_{11} can be fairly small since common-mode noise is inherently attenuated by the instrumentation-amplifier topology. More sophisticated compensation strategies in the common-mode feedback loop (such as those described in Chapter 9) can further improve performance.

From Figures 20.8 and 20.9, and our discussion of the W/L sizing in each OTA, the input-referred noise of the amplifier can easily be derived from that of its component M_j amplifiers according to

$$
\begin{aligned}
|i_n^{Mj}|^2 &= (4qI_b^{Mj} + \frac{16}{3}kT g_m^{Mj})\Delta f \\
|v_{ni}|^2 &= \frac{|i_n^{M1}|^2}{(G_m^{M1})^2} + \frac{|i_n^{M2}|^2}{(G_m^{M2})^2} + \frac{|i_n^{M3}|^2}{(G_m^{M3})^2} \frac{1}{(A_{1ststg}(1 + 1/A_{2ndstg}))^2}
\end{aligned}
\tag{20.9}
$$

It is dominated by the input-referred noise in M_1 and M_2, which is why they are configured to have the highest power consumption.

Measurements of the amplifier in a $0.5\,\mu m$ process revealed a total power consumption of $2.76\,\mu W$ at 45.3 dB gain, a 290 Hz bandwidth, $8.1\,\mu V$ rms input-referred noise, a 41.8 dB dynamic range, and a CMRR at 60 Hz of 90 dB. The voltage of the first-stage amplifiers and the buffer had to be increased to 1.8 V to accommodate a larger-than-expected common-mode range. Figure 20.5 reveals an *EKG* trace recorded with *FS-TB*1 hydrogel electrodes from Skintact. The connecting wires were twisted together as much as possible to reduce interference. In this implementation, feedback techniques improved the inherent CMRR from 57 dB to 90 dB. Further details are described in [10].

20.4   Low-power pulse oximetry

Pulse oximeters provide information about the level of oxygen saturation in the blood, i.e., what fraction of the hemoglobin molecules in the blood have oxygen bound to them. Oxygen saturation is widely used in anesthesiology as a feedback

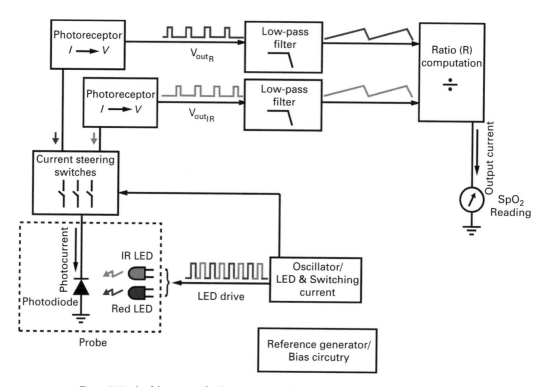

Figure 20.9. Architecture of a low-power pulse oximeter. Reproduced with kind permission from [11] (©2009 IEEE).

variable to regulate the level of an anesthetic. It is also an important vital sign. If oxygen saturation is not in the 80%–100% range, it indicates that the heart and/or the lung is not functioning as one would normally expect. Increasingly, pulse oximeters are ubiquitous in several clinical settings as *the* vital-sign monitor, for example, in intensive care, operating rooms, emergency care, birth and delivery, neonatal and pediatric care, sleep studies, and veterinary care. They are likely to be an important part of wireless medicine in the future. Their popularity stems from the fact that they are easy to use and provide valuable information: at one glance, they can provide vital-sign information about heart rate, oxygen saturation, and indirectly about blood-pressure variations. They are also noninvasive. It is remarkable that one can extract oxygen-saturation information noninvasively without having to prick a finger and do a blood test. The basic idea behind the operation of the pulse oximeter is that well-oxygenated blood is composed mostly of oxyhemoglobin, which is 'blood red' and does not absorb red light much, while de-oxygenated blood is composed mostly of deoxyhemoglobin, which is 'dark red' and absorbs red light more. Therefore, by shining red LED light through the index finger, where there is a good artery that is not buried deeply beneath the skin, and monitoring absorption of the light on the other side of the finger with a photoreceptor, we can obtain information about the oxygen content of the blood. There are, however, three challenges that have to be met. The first challenge is to distinguish absorption of light by the blood from absorption by other background tissues like the skin and bone. The second challenge is to ensure that any measurement is not confounded by absolute light level: variations in absolute light level received at the photoreceptor due to varying LED brightness, varying photoreceptor sensitivity, varying darkness or thickness of the skin of the subject must not confound the measurement. The third challenge is to ensure that the measurement does not depend on the absolute concentration of hemoglobin molecules or on arterial properties of the subject.

The first two challenges are easily met because, by a stroke of good fortune, the heart beats! As the blood-pressure rises and falls, the arteries dilate and contract respectively such that there is a modulation in the absorption of light by blood: more light is absorbed when the arteries are thick than when they are thin. By monitoring the relative modulated absorption over time rather than the absolute absorption, we can distinguish background tissue from blood and get rid of any absolute-light-level dependence in our measurement. The third challenge is solved by monitoring relative modulated absorption at two wavelengths, i.e., at 660 nm with a red LED, and at 940 nm with an infrared LED: the absorption properties of oxyhemoglobin and deoxyhemoglobin at these two wavelengths are known and fixed; absorption at both wavelengths depends on the absolute concentration of hemoglobin molecules and on arterial variation in an identical way; therefore, the ratio of relative modulated absorption at the two wavelengths cancels out the common dependency on absolute concentration of hemoglobin and on the arterial variation, and provides true information about oxygen saturation.

Beer's law relates the incident light and transmitted light through a medium and states that

$$i_{TRNS} = i_{INC} e^{-\varepsilon(\lambda)cd} \tag{20.10}$$

where i_{INC} is the intensity of incident light, c is the concentration of the substance, d is the optical path length traveled, $\varepsilon(\lambda)$ is the extinction coefficient or absorptivity of the substance at a given wavelength λ, and i_{TRNS} is the intensity of transmitted light. The transmittance T is defined to be i_{TRNS}/i_{INC} and the unscattered absorbance A is defined to be $-\ln(T)$. The total absorbance in a medium with n absorbing substances is given by

$$A_t = \varepsilon_1(\lambda)c_1 d_1 + \varepsilon_2(\lambda)c_2 d_2 + \ldots + \varepsilon_n(\lambda)c_n d_n = \sum_{i=1}^{n} \varepsilon_i(\lambda)c_i d_i \tag{20.11}$$

Thus, the total light absorbance of blood at a given wavelength λ, $A_{bld}(\lambda)$ is given by

$$A_{bld} = [\varepsilon_{HbO2}(\lambda)c_{HbO2} + \varepsilon_{Hb}(\lambda)c_{Hb}]d \tag{20.12}$$

where ε_{HbO2} is the extinction coefficient for oxyhemoglobin and ε_{Hb} is the extinction coefficient for deoxyhemoglobin. Thus, if we look at the ratio of transmitted light at its maximum, i_H, where the arteries have a thickness of d_{min}, to the transmitted light at its minimum, i_L, where the arteries have a thickness of d_{max}, we get

$$\frac{i_H}{i_L} = \frac{i_{INC}\exp[-\varepsilon_{DC}c_{DC}d_{DC}] \times \exp[-(\varepsilon_{Hb}c_{Hb} + \varepsilon_{HbO2}c_{HbO2})d_{min}]}{i_{INC}\exp[-\varepsilon_{DC}c_{DC}d_{DC}] \times \exp[-(\varepsilon_{Hb}c_{Hb} + \varepsilon_{HbO2}c_{HbO2})d_{max}]}$$

$$= \exp[+(\varepsilon_{Hb}c_{Hb} + \varepsilon_{HbO2}c_{HbO2})\Delta d] \qquad (20.13)$$

where Δd is the arterial variation in thickness, $d_{max} - d_{min}$ between the maximum value in systole and the minimum value in diastole. If we now take the log of the quantity in Equation (20.13), we get

$$\ln\left(\frac{i_H}{i_L}\right) = +(\varepsilon_{Hb}c_{Hb} + \varepsilon_{HbO2}c_{HbO2})\Delta d$$

$$\ln\left(\frac{I_{DC} + \dfrac{i_{ac}}{2}}{I_{DC} - \dfrac{i_{ac}}{2}}\right) = (\varepsilon_{Hb}c_{Hb} + \varepsilon_{HbO2}c_{HbO2})\Delta d \qquad (20.14)$$

$$A_{rlt-bld} \approx \frac{i_{ac}}{I_{DC}} = (\varepsilon_{Hb}c_{Hb} + \varepsilon_{HbO2}c_{HbO2})\Delta d$$

In deriving Equation (20.14), we have made the assumption that the modulation of the received photoreceptor output, i_{ac}, is much smaller than I_{DC}, the absolute level, which is a good assumption in practice: the modulation depth in pulse oximetry $A_{rlt-bld} \approx i_{ac}/I_{DC}$ is usually 0.5–2.0%. We term $A_{rlt-bld}$ the relative absorbance of the blood. The relative absorbance of the blood will differ at the red and infrared wavelengths. Thus, if we define the ratio of relative absorbances at the red and the infrared wavelengths as R, we get

$$R = \frac{\ln\left(i_{H,R}/i_{L,R}\right)}{\ln\left(i_{H,IR}/i_{L,IR}\right)} \approx \frac{\left(i_{ac}^{R}/I_{DC}^{R}\right)}{\left(i_{ac}^{IR}/I_{DC}^{IR}\right)}$$

$$= \frac{(\varepsilon_{Hb}(\lambda_R)c_{Hb} + \varepsilon_{HbO2}(\lambda_R)c_{HbO2})\Delta d}{(\varepsilon_{Hb}(\lambda_{IR})c_{Hb} + \varepsilon_{HbO2}(\lambda_{IR})c_{HbO2})\Delta d} \qquad (20.15)$$

$$= \frac{(\varepsilon_{Hb}(\lambda_R)c_{Hb} + \varepsilon_{HbO2}(\lambda_R)c_{HbO2})}{(\varepsilon_{Hb}(\lambda_{IR})c_{Hb} + \varepsilon_{HbO2}(\lambda_{IR})c_{HbO2})}$$

The oxygen saturation SpO_2 is defined as

$$SpO_2 = \frac{c_{HbO2}}{c_{Hb} + c_{HbO2}} \times 100\% \qquad (20.16)$$

and is the quantity that we seek in pulse oximetry, i.e., the percentage of the total hemoglobin that exists in oxygenated form. Substitution of Equation (20.16) in Equation (20.15), followed by some algebra yields

$$SpO_2 = \frac{\varepsilon_{Hb}(\lambda_R) - \varepsilon_{Hb}(\lambda_{IR})R}{\varepsilon_{Hb}(\lambda_R) - \varepsilon_{HbO2}(\lambda_R) + [\varepsilon_{HbO2}(\lambda_{IR}) - \varepsilon_{Hb}(\lambda_{IR})]R} \times 100\%$$

$$\simeq \frac{0.81 - 0.18R}{0.63 + 0.11R} \times 100\% \qquad (20.17)$$

where we have substituted known values for the extinction coefficients of oxyhemoglobin and deoxyhemoglobin at the red and infrared wavelengths. Thus a measurement of R can be directly mapped to a measurement of SpO_2.

Beer's law does not account for the reflection of light at the surface of the skin and the scattering of light in human tissue. Thus, SpO_2 as a function of R looks more linear rather than bilinear as predicted by Equation (20.17). It is often well approximated by

$$SpO_2 = 110 - 25R \qquad (20.18)$$

Commercial pulse oximeters derive a more exact relationship via calibration of blood samples derived from invasive measurements, usually fitting an empirical second-order polynomial for SpO_2 vs. R.

The relative absorbance $A_{rlt\text{-}bld}$ defined in Equation (20.14) is well approximated by i_{ac}/I_{DC} and can be plotted as a function of time as the photoreceptor current $i_{ac}(t)$ varies and I_{DC} is fixed. Equivalently, we can plot its exact expression, defined by the logarithm in Equation (20.14), which also varies as Δd varies with time. If we plot $A_{rlt\text{-}bld}(t)$ at the red and infrared wavelengths, we get two traces that together form what is known as the photoplethysmogram (PPG). The PPG provides information about $\Delta d(t)$, which provides indirect information about blood-pressure variations: changes in blood pressure at the finger tip acting on the compliance of the arterial walls lead to $\Delta d(t)$.

Figure 20.9 shows the architecture of a low-power single-chip pulse oximeter. Signals from a red and infrared LED that alternately emit light through the finger are used to create red and infrared photocurrents in a photodiode. To save power, both LEDs are duty cycled at a low duty cycle ($\sim$3%) with brief sampling pulses. Current information in the photodiode is converted to voltage information via transimpedance-amplifier-based photoreceptors similar to those described in Chapter 11. Current steering switches ensure that the photocurrent is directed to a red photoreceptor during the red LED phase and to the infrared photoreceptor during the infrared LED phase. Processing and lowpass filtering of the over-sampled information lead to the extraction of PPG information. The use of envelope detectors (described in Chapter 19) and current-mode processing (described in Chapter 14) lead to the ratio computation of R in Equation (20.15), which then directly yields SpO_2 information. To ensure low-power operation, several innovations were made to create a sensitive energy-efficient photo-receptor [11], which allows one to turn down the LED power: a sensitive receiver implies that transmitter power can be turned down as in our discussion of RF telemetry in Chapter 18.

The use of programmable and robust analog processing and biasing using principles similar to those described for the bionic ear processor in Chapter 19 enables a compact single-chip solution. Figure 20.10 shows the layout of this single-chip pulse oximeter. Figure 20.11 shows red and infrared PPG traces from the oximeter. The sharp falls in both traces corresponds to the sharp rise in blood pressure (v_{BODY} in Figure 20.4 or the blood pressure in Figure 20.6) during the

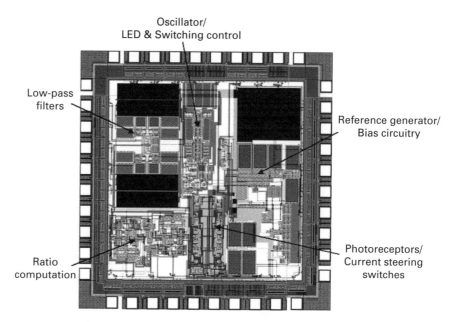

Figure 20.10. Pulse-oximeter chip. Reproduced with kind permission from [11] (©2009 IEEE).

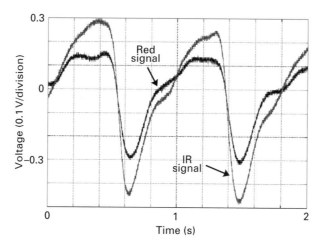

Figure 20.11. Pulse-oximeter waveforms. Reproduced with kind permission from [11] (©2009 IEEE).

systolic phase of the heart; in this phase, arterial thickness increases causing increased absorption and a fall in the received photoreceptor signal. The slow rise in both traces corresponds to the slow fall in blood pressure (v_{BODY} in Figure 20.4 or the blood pressure in Figure 20.6) during the diastolic phase of the heart; in this phase arterial thickness decreases causing decreased absorption and a rise in the received photoreceptor signal. The abrupt change in rising slope in both traces corresponds to the closing of the aortic valve (see Figures 20.4 or 20.6) and the end of systole. The onset of the falling slope in both traces corresponds to the onset of systole. Both traces exhibit similar variations in time because the arterial thickness variations, given by Δd, are the same for the red and infrared photoreceptors. This particular oximeter lowered power dissipation by almost an order of magnitude from those in several measured commercial implementations (the lowest-power implementations consumed 30 mW–50 mW if display power is not included in these measurements), primarily because of innovations in the energy-efficient photoreceptor and secondarily because of the efficiency of analog processing [11]. Further power reductions by optimizing the distribution between photo-receptor power and LED power, and optimizing photoreceptor design are possible. The outputs of the pulse oximeter have an accuracy of -1.2% and a standard deviation of 1.5% with regard to commercial implementations, which is adequate for the 2% accuracy needed in medical systems. The minimum detectable contrast is well below the 0.5–2% needed for pulse oximetry, even at LED light levels significantly below those used.

20.5   Battery-free tags for body sensor networks

While the EKG has been widely explored for chronic monitoring, its poorer cousin, the PCG, which measures heart sounds, has been relatively neglected. If only heart-rate information or proof-of-life information is needed for chronic monitoring, a PCG has three advantages over the EKG. First, it does not require any electrical contact with the body, which is often difficult to obtain with dry skin. Second, it can be implemented with a cheap low-power microphone: the heart is easy to sense acoustically as it is the loudest organ in the body. Third, it requires little maintenance, unlike the EKG. A low-power RF-ID tag that monitors heart sounds to provide such information is shown in Figure 20.12. The tag operates in a battery-free fashion using harvested RF energy gathered with an antenna as described in Chapter 17. The antenna shown in Figure 20.12 is very similar to the one described in Chapter 17. It has a back-scatter modulator that allows far-field impedance-modulation communication with the RF-ID receiver. The tag has two microphones, one facing towards the heart and body, which we term the heart microphone, and one facing the environment and away from the body, which we term the environmental microphone. The environmental microphone is usually switched off to save power. It is only turned on if the heart microphone does not detect heart sounds and a disconnection or patient emergency is suspected. If both microphones pick up similar environmental sounds, a disconnection alarm is generated since it is probable that the tag is no longer in proximity to the skin. A patient-emergency alarm is generated if the heart microphone does not pick up environmental sounds, but the environmental microphone does. Finally, if neither microphone picks up any sounds, the tag is probably malfunctioning such that a device-malfunction alarm is generated. The difference in

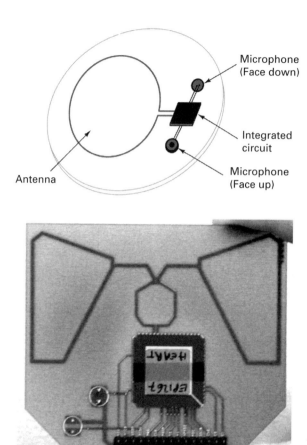

Figure 20.12. A low-power battery-free cardiac medical tag that functions through RF energy harvesting. A cartoon version with two microphones is shown at the top while the actual experimental tag is shown in the bottom figure. Reproduced with kind permission from [12] (©2009 IEEE).

arrival times between sounds from two speakers separated by 12 ft at a microphone can help localize where the microphone is located. Even when the loudspeakers output relatively low-frequency 230 Hz tone bursts, a localization accuracy of 0.6 m with a standard deviation of 0.4 m is achievable [12]. The loudspeakers can turn on only when alarm signals are received from a tag. The loudspeaker tones themselves then provide an audible alarm and can trigger other power-hungry sensors such as video cameras to turn on.

Commercial microphones contain built-in JFET preamplifiers as we have described in Chapter 19. Since heart sounds are relatively loud and low in bandwidth (typically 10–250 Hz), the microphone can be biased with currents far below those specified by the manufacturer to save power. The JFET then operates in its insensitive linear regime. For example, a Panasonic omni-directional condenser electret microphone (WM-63PR) is a small cheap thin device (thickness = 1.3 mm,

(a)

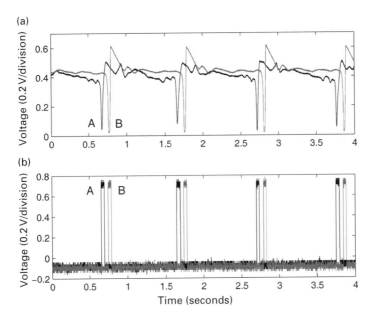

(b)

Figure 20.13a, b. Recorded PCG waveforms at the neck and wrist in (a) and corresponding event waveforms in (b). Reproduced with kind permission from [12] (©2009 IEEE).

diameter = 6 mm) and can be biased at 30 μA at 0.5 V. Microphone membranes cannot vibrate freely if they are directly attached to the skin. Therefore a small (1 mm thick) air chamber below the microphone is needed for the heart microphone. The vent-free chamber can reduce the amount of ambient noise and performs some low-frequency filtering.

Besides an RF rectifier and RF communication circuitry, the chip shown in the tag of Figure 20.12 contains four channels for sensing. Each channel has a programmable-gain preamplifier for sensing signals, a programmable-threshold event generator, and can supply programmable external bias currents (from 0.5 μA to 128 μA) to operate sensors from a 0.5 V supply if needed. The core power supply operates at 0.8 V. It also contains a programmable logical-combination-of-events generator that can generate any of the 2^{16} logical functions of 4 input events. Such logic is useful in the kind of dual-microphone sensor scenario that we have described. The static power consumption is 1 μW if no sensors are used, with sensor power being application dependent. Some details of this chip as well a summary of low-power circuits for medical monitoring are discussed in [12], [13].

Figure 20.13 (a) shows PCG information obtained from microphones placed at the neck and wrist. High-frequency information that is present in traditional PCG recordings is absent in these recordings because low-power microphone operation and the presence of an air chamber implement a lowpass filter. Nevertheless heart-rate information, which mainly resides in the loud 10–80 Hz frequency range is easily discernible. The PCG recordings generate A and B events at the neck and wrist in Figure 20.13 (b) that correspond to the A and B signals in Figure 20.13 (a) respectively. The delay between these events can be used to obtain information about blood pressure. Higher blood pressure leads to more arterial dilation, which reduces the mean transmission-line time delay between a systolic pulse at the neck and a systolic pulse at the wrist. It is these systolic pulses which lead to the PCG recordings of Figure 20.13 [14].

Another example of a low-power noninvasive system includes an EEG system intended to be powered by solar cells and body heat (thermoelectric generation) [15]. An amplifier with chopper stabilization uses clever variations on the lock-in techniques described in Chapter 8 to achieve low $1/f$ noise performance at little power for EEG applications [16].

20.6   Intra-body galvanic communication networks

The presence of multiple sensors on the body, e.g., microphones at the neck and wrist, with wires needed for communication between sensors or wires needed between the various sensors and a central 'base-station' tag on the body, is awkward. A low-power wireless solution is desirable. One innovative idea proposes to use the conductivity of the body itself to communicate, i.e, since the conductive body already 'wires' its various parts together via a distributed three-dimensional RC-transmission line, we may as well use it to communicate [17]. Thus sensors can be galvanically coupled via the body itself. These approaches use charge-balanced alternating current in the 10 kHz to 1 MHz carrier-frequency range to stimulate sites on the body, much like charge-balanced current stimulation is used to stimulate nerves in medical implants. Charge-balanced current stimulation has been described in depth in Chapter 19. The use of carrier frequencies above 10 kHz has been proposed to avoid interference with regular electrical communication between tissues in the body.

In [18], a peak ac current of 1 mA amplitude was applied at various locations on the thorax and limbs of several subjects with a transmitter and voltages were measured by a receiver at different portions of the body. The results revealed a 6 to 9 dB attenuation for every 5 cm increase in transmitter-receiver distance in the arm, 8 dB attenuation across joints, and 50 dB attenuation from front to back in the thorax. For the latter measurement, a 20 dB SNR at a channel bandwidth of 30 kHz suggests that a 130 kb/s data rate may represent an upper bound on the data rate possible with such communication. Prior work by the same group has demonstrated a 4.8 kb/s data rate on an 'elliptical phantom' that attempts to simulate the electrical characteristics of human tissue. Electrical measurements of human tissue may be found in [19]. Activity reduces the attenuation somewhat since the wetting of the skin reduces skin resistance. Increases in electrode transmitter size reduce attenuation as well. A reduction in muscle resistance, which is the most conductive tissue, short-circuits transmitter current at the transmitter and prevents its propagation. A reduction in fat resistance, in contrast, allows more of this current to stay in the fat layer rather than the muscle layer and improves propagation.

20.7   Biomolecular sensing

Biomolecular sensing has grown explosively in recent years due to rapid advances in miniaturization and fabrication in the fields of BioMEMS and microfluidics. We shall discuss some principles behind biomolecular sensing and some recent developments. The brief account below should not be seen as a review of the field but rather as a collection of examples that paint a picture that is relevant to the themes of this book. This picture is painted from the perspective of a bioelectronic engineer. It is too early to tell where this young and still maturing field will have its highest impact.

The word BioMEMS is used as a catch-all phrase to describe just about any system for biological applications that exploits MEMS or nanotechnology to function. Therefore, it is sometimes used to describe microfluidic systems as well. A good overview of BioMEMS for sensing and medical diagnostic applications may be found in [20]. The review in [21] discusses possibilities for therapeutics and drug delivery. The description in [22] provides examples of the potential of BioMEMS for tissue engineering. We shall focus almost exclusively on sensing.

Six common biological sensors include sensors for whole cells, viruses, micro-organisms like bacteria, DNA, proteins, and small molecules. Examples of small molecules include glucose, lactate (a common by-product of anaerobic metabolism), H^+ ions (pH sensing), urea, or neurotransmitters such as dopamine. The sensing is invariably accomplished in the following three steps, which are summarized below:

1. The binding of one unit of a specific target species to one unit of a complementary capture species changes the analog state of a mechanical, optical, or electrical variable. The target species can be labeled with a molecular tag or fluorescence marker to facilitate optical detection when the binding event occurs; or it can be labeled with a magnetic bead to facilitate electrical detection when the binding event occurs. Certain forms of sensing that we shall discuss do not require the use of such labels and are therefore termed *label free*.
2. The change in the mechanical, optical, or electrical state is passively transduced to yet another domain, e.g., a mechanical state change is passively transduced to an electrical state change, or a mechanical state change is passively transduced to an optical state change. In most sensors, 0 or 1 such passive conversions between domains is typical.
3. The final transduced signal is then actively amplified, usually in the electrical or optical domain.

The detection of a species with high specificity (only the target species triggers detection), high sensitivity (a few molecules or units of the target species are detectable), in a short time (a second), over a wide dynamic range in concentration (e.g., fM to 0.1M), and in a low-cost fashion is challenging. Label-free sensing is more convenient than labeled sensing but performance and cost are key determinants in measuring the pros and cons of one sensing scheme versus another.

As an illustration of the three steps, we discuss a typical method for DNA detection. The detection of DNA exploits capture probes of single-stranded DNA with a base-pair sequence that is complementary to that of the base-pair sequence of the target DNA that we want to detect. The capture probes are immobilized via thiolation bonds to a silicon dioxide or gold substrate. The binding of the complementary single-strand target DNA to the capture probe is known as hybridization. If the target DNA has been labeled with an appropriate optically active molecule, such hybridization can change the fluorescence signal at the particular spot where the complementary capture probe has been immobilized. Thus, in a DNA micro array [23], [24], with different capture probes at different locations, only certain locations may exhibit particular optical signals because only certain target DNAs (or mRNAs) are present in the solution being analyzed. The optical signals can be detected via photodetectors or sensitive photomultiplier tubes. The basics of phototransduction are outlined in Chapter 11. Magnetic-bead labels rather than fluorescence labels can cause DNA hybridization to affect the reflected inductance of an LC resonator (see Chapter 16), and thus move its resonant frequency by a few parts per million [25].

Other examples of antibody-antigen binding complexes, which are ubiquitous in the immune system, are being exploited in protein microarrays using techniques very similar to those employed for DNA detection [26]. DNA detection is significantly more advanced than protein detection partly because a polymerase chain reaction (PCR) can amplify even minute amounts of DNA via a cascaded-gain-of-2 amplification system [27]. Analogous general-purpose methods for protein amplification that work with such efficacy do not yet exist. Examples of protein-detection-and-amplification systems include magnetic-bead-labeled 'nanotags' that indirectly attach to bound cancer biomarkers and then lead to magneto-resistive sensing and amplification [28]. A wireless immunosensor that detects magnetic-bead-labeled tags that attach to bound antigens on a chip uses a Hall-effect detection scheme [29]. The latter chip has a wireless data and power link.

The rate of molecular detection (number of molecules detected per second) in a biomolecular sensing system can be mass-transport limited and governed by diffusion, or chemical-reaction limited, and governed by kinetic rate constants for binding [30]. We shall not describe these issues in further depth in this chapter. However, Chapter 24 discusses chemical reaction dynamics including the equations that determine molecular binding and unbinding. Chapter 25 discusses how mass transport and chemical-reaction dynamics can conspire to affect large-signal and small-signal dynamics at a battery electrode, both of which can be modeled by equivalent electrical circuits. A reader with knowledge of electrical engineering but with little knowledge of chemistry will find the material in Chapters 24 and 25 useful for understanding such effects in biomolecular sensing systems as well.

20.7.1   Advantages of small structures for sensing

What is the advantage of using small MEMS or microfluidic structures to perform such sensing? There are four primary advantages:

1. Smaller volumes are needed for detection and analysis, which saves costs and reduces required sample sizes of fluids or tissues gathered from patients.
2. Systems can be portable and miniature. Thus, they can be employed in contexts that bring medicine to the home, emergency room, or trauma location for higher-speed, lower-cost health care.
3. Increases in the concentrations of species caused by their confinement in smaller spaces speeds up the flux rates of binding reactions due to an increased probability of interaction between species (see Chapter 24). The time needed for species travel in smaller fluidic systems is reduced compared with that in larger fluidic systems such that analysis time is reduced. Just as deep submicron transistors operate at faster time scales than transistors with large channel lengths, BioMEMS and microfluidic systems that exploit fabrication advances at small scales can operate at faster time scales.
4. The reduction in size of the structure to match the size of the species being detected offers better sensitivity because the fractional minimum detectable signal that is required for sensing in the sensor is raised. The fractional minimum detectable signal f needed to sense a tiny change ΔS in a variable S with nominal value S_0

$$ f = \frac{\Delta S}{S_0} \tag{20.19} $$

is increased because S_0 is lowered. As we discussed in Chapters 7 and 8 on noise, it is significantly easier to detect a 1-part-per-million change in a variable (20-bit precision) than a 1-part-per-billion change in a variable (30-bit precision). Intuitively, it is easier to detect a 1 nm change between two plates of a capacitor if they are 1 μm apart than if they are 100 μm apart, to detect a fg increase in mass in a structure that weighs a ng rather than a μg, or to detect a localized change in an optical signal not confounded by interference from numerous sources in a large system. The increase in sensitivity is usually accompanied by a loss in the maximum detectable signal since smaller sensors saturate at a smaller ΔS than larger sensors. Therefore, the sensor must have gain control if it is to operate over a wide dynamic range. For examples of gain control circuits that enable operation over a wide dynamic range, see Chapter 19.

There is usually an optimal size to the sensing structure since a very small S_0 can make the design of the amplifying system difficult. The amplifier must itself be small to prevent degradation of the signal via loading of the sensor, but it is difficult to build a sensitive amplifying system in a small area or volume. For example, a low-noise high-g_m transistor requires a large W/L to lower its input-referred thermal noise and a large W/L to lower its $1/f$ noise. However, large transistors can add significant parasitic capacitance that degrades the minimum detectable signal of the sensor. Thus, in all sensing systems, for optimal sensitivity, there must be good impedance matching between the sensor's impedance and the impedance of the amplifying system.

20.7.2 Mechanical sensing

Mechanical MEMS sensors for sensitive molecular sensing have primarily used label-free cantilever approaches [31], [32], [33]. The cantilevers are functionalized with chemical capture probes. Binding of the target species to these probes can cause surface-energy changes in the cantilevers that cause them to bend. Binding of the target species can also simply increase the mass of the cantilever. Cantilevers function like 'diving boards' that stretch or vibrate, with their stiffness k and mass m determining a $\sqrt{k/m}$ angular resonant frequency ω_{res}. The stretching of the cantilever can be coupled to piezo-resistive changes in materials that support or form part of the cantilever. Such resistance changes can then be electrically amplified and sensed. Changes in mass can shift the cantilever's resonant frequency, ever so slightly. Such resonant-frequency shifts can be sensed by configuring the cantilever to operate as a MEMS electro mechanical oscillator with positive feedback provided by electrostatic actuation of the cantilever. Electrical measurement of the oscillator frequency by sensitive electronics then provides information about the mass change. Optical measurements of cantilever displacements via an interference technique or electrical measurements of capacitance changes to substrate caused by cantilever displacement provide alternative measurement techniques [33].

Note that, in these examples, we have used a chemical→mechanical→optical or chemical→mechanical→electrical sensing scheme. In the optical scheme, amplification is obtained by having changes in the light energy caused by motions of the peaks and troughs of an interference pattern greatly exceed corresponding changes in cantilever mechanical energy such that they are easily sensed in a photodetector. Optical techniques generally provide excellent sensitivity but tend to be relatively large and are hard to integrate and scale up into an array format. In contrast, electrical techniques tend to be relatively smaller and easier to integrate and scale up into an array format.

Piezoresistive techniques have the advantage that sensing is based on displacements that are naturally referenced to the cantilever's own nominal length and are thus not subject to artificial constraints imposed by the distance of the substrate from the cantilever. Such constraints can hurt sensitivity in capacitance sensing or require the need for high electrostatic actuation voltages in resonant-frequency sensing. However, piezoresistors do require special materials, and these can exhibit $1/f$ noise. Furthermore, whole-cell sensing does not appear to cause surface-energy changes in the cantilever, such that a mass-based approach must be used in this case [20].

In the MEMS vibration-sensor or accelerometer example of Chapter 8, we discussed how low values of minimum detectable accelerations required a high Q

to minimize damping noise, required a large mass m for small accelerations to create large forces, and required a small resonant frequency ω_{res} such that small accelerations could cause large displacements (see Equation (8.32)). In the case of a resonant mass sensor, for low values of minimum detectable molecular sensing, we still need high levels of Q to minimize damping noise but the mass must be *small* such that tiny changes in mass lead to a large fractional shift in the resonant frequency. Recent work has shown that vacuum-packaged MEMS resonators with the species-containing fluid *inside* the cantilever have a Q of 15,000 and 100 ng and therefore allow even relatively light pathogens to be weighed [33]. Indeed, if the mass of the resonator is light enough, one can even weigh a fraction of a gold atom in a 1 Hz bandwidth in a carbon-nano-tube-based nanomechanical resonator, as has recently been demonstrated [34]!

20.7.3   Electrical sensing

Many biomolecules are electrically charged in the near-neutral pH of biological fluids. For example, the phosphate backbone of DNA is negatively charged such that the overall negative charge of a DNA strand is proportional to its nucleotide count. Acidic amino-acid components of proteins, such as glutamate or aspartate, are negatively charged, while basic amino-acid components of proteins, such as lysine, arginine, or histidine, are positively charged. Most proteins have a net charge in body fluids. All cells in the body maintain a resting negative electric potential between their intracellular compartments and the extracellular world around them that is a few tens of mV. The most common ions in intracellular and extracellular fluids are potassium (K^+), sodium (Na^+), magnesium (Mg^{2+}), calcium (Ca^{2+}), chloride (Cl^-), and bicarbonate (HCO_3^-) ions. The latter two are negatively charged while the rest are positively charged. Not surprisingly, electrical and electrochemical methods to directly sense biological molecules are important. Even uncharged molecules such as glucose can be made to undergo redox chemical reactions through the use of catalytic enzymes such as glucose oxidase that are present at an electrode. The net result of such a redox reaction is that an electron current that is stoichiometrically related to the chemical reaction flux can be measured at the functionalized electrode. Chapter 25 provides a deeper discussion of batteries and electrochemistry for the interested reader including how redox chemical reactions, the energy source of batteries, generate anodic and cathodic currents at the anode and cathode terminals, respectively.

Electrical sensing of biomolecules can be current based or amperometric, voltage based or potentiometric, or impedance based or conductometric. Amperometric sensing uses transimpedance-amplifier configurations described in Chapter 11 to maintain a fixed potential between two electrodes in an electrolyte via negative feedback. The redox current flowing between these electrodes is sensed and converted to a voltage as in the transimpedance amplifiers of Chapter 11. One of the electrodes is functionalized with incorporated enzymes or with a hydrogel-conductive-polymer combination to create one half of the redox reaction, while the other half of the redox reaction occurs at the other electrode, which may be functionalized as well. Amperometric sensing has been used to measure glucose, lactate, urea, and dopamine concentrations [35]. With the incorporation of ferrocenyl derivatives in DNA, electrochemical sensing of DNA via this labeled technique can be performed as well [36]. Other electrochemical sensing techniques for DNA are reviewed in [37]. Potentiometric sensing uses ion-sensitive FETs (ISFETs) or chemical FETs (chem-FETs) that sense ionic charges of molecules that affect the gate of a transistor that has been appropriately functionalized [38]. Transistors with metal source and drain terminals and nanowire channels built with carbon nano tubes and with silicon are extensively being researched for such applications [39]. Conductometric sensing simply measures conductivity or impedance changes between two electrodes due to changes in the ionic composition of the electrolyte between them or due to species-induced changes in the electrode-electrolyte interface impedance. Impedance models for electrodes are discussed in Chapter 25. Conductometric techniques are usually the least specific of the three electrical-measurement techniques, but they are also the easiest to implement. Non-functionalized metal electrodes can be used to get a gross sense of how the electrolyte environment surrounding cells changes as cellular activity changes within it. However, specificity can be achieved via functionalization in a similar manner to amperometric or potentiometric techniques: for example, the binding of DNA oligonucleotides (relatively short DNA strands with a few nucleotides) functionalized with gold nanoparticles leads to conductivity changes associated with hybridization [40], which can then be measured.

The charged nature of DNA has lead to several efforts to create label-free sensing schemes for its detection. One example can be found in [41], which uses electrostatic immobilization of probe DNA on a poly-L-lysine layer and FET sensing. Another example can be found in [42], which exploits hybridization-sensitive changes in capacitance between the probe-DNA electrode and an adjacent electrode. Capacitance measurements are increasingly being explored for lab-on-a-chip microfluidic applications as well [43].

20.7.4   Microfluidics

Biological cells exist in fluids and are mostly made up of fluids. Therefore, techniques for high-throughput processing of biofluids via automation and miniaturization of fluid technologies at the microscale, i.e., microfluidics, promise to revolutionize biology [44]. The transparency, gas permeability, and biocompatibility of the plastic PolyDiMethlySiloxane (PDMS), along with soft lithographic techniques to process it [45], have led to complex microfluidic chips based on this technology. Advanced microfluidic chips can contain channels for fluid flow, air-pressure-activated fluid pumps to move the fluids, fluid-mixing compartments to mix fluids and enable reactions, electric-field-based flow actuators, pressure-activated valves, pH-activated hydrogel valves, cell-sorting compartments, compartments for cell lysing with detergents, functionalized antibody-based protein detection, on-chip thermal systems for PCR-based DNA amplification, integrated nanowire or cantilever electronic detection, and integrated fluorescence-based optical detection [46]. The dream of creating a lab-on-a-chip or a μTAS (Micro Total Analysis System) on a cheap diposable CD (to avoid cross-contamination between samples) will likely be realized in the near future. The four advantages of reducing size that we outlined previously also apply to microfluidic technologies. In addition, phenomena such as surface tension, electrophoresis and electrosmosis conspire to enable more efficient fluid flow than would be possible at larger scales [47], [48]. Fluid flow at such small scales is laminar and well controlled with diffusion playing a more important role than it would at larger scales. For further details on microfluidic technologies, we refer interested readers to reviews and books on the subject [44], [49], [50], and [48].

20.7.5   Molecular computation

In Chapter 24, we will return to biomolecules in the context of cellular hybrid analog-digital computation. Such computation is implemented by interacting DNA and protein molecules in biochemical reaction networks. Due to the similarity of Boltzmann-exponential dynamics in both chemistry and subthreshold electronics, we shall see that we can model and mimic highly computationally intensive biomolecular networks including noise and stochastics in a very detailed, ultrafast, and compact fashion. Thus, *cytomorphic electronics*, a novel form of electronics inspired by cell biology and introduced in this book, can be potentially useful in designing and analyzing complex biological systems in the future. Such systems are the focus of the newly emerging and rapidly growing fields of systems biology and synthetic biology, which attempt to take a systems-and-engineering approach to analyzing and designing complex biological systems. Such approaches are essential for understanding and treating complex diseases like diabetes and cancer, which are rarely caused by single-gene or single-protein malfunction, but by feedback network malfunction. In turn, cytomorphic electronics may lead to novel ultra-low-power hybrid analog-digital computational architectures that could be useful in nonbiological applications.

References

[1] Guang Zong Yang. *Body Sensor Networks* (London: Springer-Verlag, 2006).
[2] K. H. Wee and R. Sarpeshkar. An electronically tunable linear or nonlinear MOS resistor. *IEEE Transactions on Circuits and Systems I: Regular Papers*, **55** (2008), 2573–2583.
[3] J. Bohorquez, W. Sanchez, L. Turicchia and R. Sarpeshkar. An integrated-circuit switched-capacitor model and implementation of the heart. *Proceedings of the First International Symposium on Applied Sciences on Biomedical and Communication Technologies (ISABEL)*, Aalborg, Denmark, 1–5, 2008.

[4] L. S. Y. Wong, S. Hossain, A. Ta, J. Edvinsson, D. H. Rivas, H. Naas, C. R. M. Div and C. A. Sunnyvale. A very low-power CMOS mixed-signal IC for implantable pacemaker applications. *IEEE Journal of Solid-State Circuits*, **39** (2004), 2446–2456.

[5] G. S. Wagner. *Marriott's Practical Electrocardiography.* 10th ed. (Philadelphia: Lippincott, Williams & Wilkins, 2001).

[6] M. Puurtinen, J. Hyttinen and J. Malmivuo. Optimizing bipolar electrode location for wireless ECG measurement–analysis of ECG signal strength and deviation between individuals. *International Journal of Bioelectromagnetism*, **7** (2005), 236–239.

[7] R. A. C. Metting van Rijn, A. Peper and C. A. Grimbergen. High-quality recording of bioelectric events. *Medical and Biological Engineering and Computing*, **28** (1990), 389–397.

[8] B. B. Winter and J. G. Webster. Reduction of interference due to common mode voltage in biopotential amplifiers. *IEEE Transactions on Biomedical Engineering*, **30** (1983), 58–62.

[9] P. Gray, P. Hurst, S. Lewis and R. Meyer. *Analysis and Design of Analog Integrated Circuits.* 4th ed. (New York: Wiley, 2001).

[10] L. Fay, V. Misra and R. Sarpeshkar. A Micropower Electrocardiogram Recording Amplifier. *IEEE Transactions on Biomedical Circuits and Systems*, vol.3, No. 5, pp. 312–320, October 2009.

[11] M. Tavakoli, L. Turicchia and R. Sarpeshkar. An ultra-low-power pulse oximeter implemented with an energy efficient transimpedance amplifier. *IEEE Transactions on Biomedical Circuits and Systems*, **in press** (2009).

[12] S. Mandal, L. Turicchia and R. Sarpeshkar. A Low-Power Battery-Free Tag for Body Sensor Networks. *IEEE Pervasive Computing*, **in press** (2009).

[13] L. Turicchia, S. Mandal, M. Tavakoli, L. Fay, V. Misra, J. Bohorquez, W. Sanchez and R. Sarpeshkar. Ultra-low-power Electronics for Non-invasive Medical Monitoring. *Proceedings of the IEEE Custom Integrated Circuits Conference (CICC)*, Invited Paper, 6–5, San Jose, CA, 2009.

[14] J. C. Bramwell and A. V. Hill. The velocity of the pulse wave in man. *Proceedings of the Royal Society of London. Series B, Containing Papers of a Biological Character*, **93** (1922), 298–306.

[15] R. F. Yazicioglu, P. Merken, R. Puers and C. Van Hoof. A 200 μW Eight-Channel EEG Acquisition ASIC for Ambulatory EEG Systems. *IEEE Journal of Solid-State Circuits*, **43** (2008), 3025–3038.

[16] T. Denison, K. Consoer, A. Kelly, A. Hachenburg, W. Santa, M. N. Technol and C. Heights. A 2.2 μW 94nV/$\sqrt{\text{Hz}}$ Chopper-Stabilized Instrumentation Amplifier for EEG Detection in Chronic Implants. *Digest of Technical Papers of the IEEE International Solid-State Circuits Conference (ISSCC 2007)*, San Francisco, CA, 162–594, 2007.

[17] T. Handa, S. Shoji, S. Ike, S. Takeda and T. Sekiguchi. A very low-power consumption wireless ECG monitoring system using body as a signal transmission medium. *Proceedings of the International Conference Solid-State Sensors and Actuators*, Chicago, IL, 1003–1007, 1997.

[18] M. S. Wegmueller, A. Kuhn, J. Froehlich, M. Oberle, N. Felber, N. Kuster and W. Fichtner. An Attempt to Model the Human Body as a Communication Channel. *IEEE Transactions on Biomedical Engineering*, **54** (2007), 1851–1857.

[19] S. Gabriel, R. W. Lau and C. Gabriel. The dielectric properties of biological tissues: III. Parametric models for the dielectric spectrum of tissues. *Physics in Medicine and Biology*, **41** (1996), 2271–2293.

[20] R. Bashir. BioMEMS: State-of-the-art in detection, opportunities and prospects. *Advanced Drug Delivery Reviews*, **56** (2004), 1565–1586.

[21] A. C. R. Grayson, R. S. Shawgo, A. M. Johnson, N. T. Flynn, Y. Li, M. J. Cima and R. Langer. A bioMEMS review: MEMS technology for physiologically integrated devices. *Proceedings of the IEEE*, **92** (2004), 6–21.

[22] S. N. Bhatia and C. S. Chen. Tissue engineering at the micro-scale. *Biomedical Microdevices*, **2** (1999), 131–144.

[23] Affymetrix. Available from: http://www.affymetrix.com/index.affx.

[24] Harvey F. Lodish. *Molecular Cell Biology*. 6th ed. (New York: W.H. Freeman, 2008).

[25] H. Wang, Y. Chen, A. Hassibi, A. Scherer and A. Hajimiri, A frequency-shift CMOS magnetic biosensor array with single-bead sensitivity and no external magnet. *Digest of Technical Papers IEEE International Solid-State Circuits Conference (ISSCC)* San Francisco, CA, 438–439, 2009.

[26] B. B. Haab, M. J. Dunham and P. O. Brown. Protein microarrays for highly parallel detection and quantitation of specific proteins and antibodies in complex solutions. *Genome Biology*, **2** (2001), 1–13.

[27] Kary B. Mullis, François Ferré and Richard Gibbs. *The Polymerase Chain Reaction* (Boston: Birkhäuser, 1994).

[28] S. J. Osterfeld, H. Yu, R. S. Gaster, S. Caramuta, L. Xu, S. J. Han, D. A. Hall, R. J. Wilson, S. Sun and R. L. White. Multiplex protein assays based on real-time magnetic nanotag sensing. *Proceedings of the National Academy of Sciences*, **105** (2008), 20637.

[29] T. Ishikawa, T. S. Aytur and B. E. Boser. A wireless integrated immunosensor. *Complex Medical Engineering*, (2007), **555**.

[30] T. M. Squires, R. J. Messinger and S. R. Manalis. Making it stick: convection, reaction and diffusion in surface-based biosensors. *Nature Biotechnology*, **26** (2008), 417–426.

[31] A. Gupta, D. Akin and R. Bashir. Single virus particle mass detection using microresonators with nanoscale thickness. *Applied Physics Letters*, **84** (2004), 1976–1978.

[32] K. L. Ekinci, X. M. H. Huang and M. L. Roukes. Ultrasensitive nanoelectromechanical mass detection. *Applied Physics Letters*, **84** (2004), 4469–4471.

[33] T. P. Burg, M. Godin, S. M. Knudsen, W. Shen, G. Carlson, J. S. Foster, K. Babcock and S. R. Manalis. Weighing of biomolecules, single cells and single nanoparticles in fluid. *Nature*, **446** (2007), 1066–1069.

[34] K. Jensen, K. Kim and A. Zettl. An atomic-resolution nanomechanical mass sensor. *Nature Nanotechnology*, **3** (2008), 533–537.

[35] R. Hintsche, C. Kruse, A. Uhlig, M. Paeschke, T. Lisec, U. Schnakenberg and B. Wagner. Chemical microsensor systems for medical applications in catheters. *Sensors & Actuators B.: Chemical*, **27** (1995), 471–473.

[36] C. J. Yu, Y. Wan, H. Yowanto, J. Li, C. Tao, M. D. James, C. L. Tan, G. F. Blackburn and T. J. Meade. Electronic detection of single-base mismatches in DNA with ferrocene-modified probes. *Journal of the American Chemical Society*, **123** (2001), 11155–11161.

[37] T. G. Drummond, M. G. Hill and J. K. Barton. Electrochemical DNA sensors. *Nature Biotechnology*, **21** (2003), 1192–1199.

[38] M. Lehmann, W. Baumann, M. Brischwein, H. J. Gahle, I. Freund, R. Ehret, S. Drechsler, H. Palzer, M. Kleintges and U. Sieben. Simultaneous measurement of

cellular respiration and acidification with a single CMOS ISFET. *Biosensors and Bioelectronics*, **16** (2001), 195–203.

[39] E. Stern, J. F. Klemic, D. A. Routenberg, P. N. Wyrembak, D. B. Turner-Evans, A. D. Hamilton, D. A. LaVan, T. M. Fahmy and M. A. Reed. Label-free immunodetection with CMOS-compatible semiconducting nanowires. *Nature*, **445** (2007), 519–522.

[40] S. J. Park, T. A. Taton and C. A. Mirkin. Array-based electrical detection of DNA with nanoparticle probes. *Science*, **295** (2002), 1503–1506.

[41] J. Fritz, E. B. Cooper, S. Gaudet, P. K. Sorger and S. R. Manalis. Electronic detection of DNA by its intrinsic molecular charge. *Proceedings of the National Academy of Sciences*, **99** (2002), 14142–14146.

[42] C. Stagni, D. Esposti, C. Guiducci, C. Paulus, M. Schienle, M. Augustyniak, G. Zuccheri, B. Samor, L. Benini and B. Ricco. Fully electronic CMOS DNA detection array based on capacitance measurement with on-chip analog-to-digital conversion. *Proceedings of the IEEE International Solid-State Circuits Conference (ISSCC)*, San Francisco, CA, 69–78, 2006.

[43] E. Ghafar-Zadeh and M. Sawan. A hybrid microfluidic/CMOS capacitive sensor dedicated to lab-on-chip applications. *IEEE Transactions on Biomedical Circuits and Systems*, **1** (2007), 270–277.

[44] T. M. Squires and S. R. Quake. Microfluidics: fluid physics at the nanoliter scale. *Reviews of Modern Physics*, **77** (2005), 977–1026.

[45] Y. Xia and G. M. Whitesides. Soft lithography. *Annual Review of Materials Science*, **28** (1998), 153–184.

[46] M. A. Burns, B. N. Johnson, S. N. Brahmasandra, K. Handique, J. R. Webster, M. Krishnan, T. S. Sammarco, P. M. Man, D. Jones and D. Heldsinger. An integrated nanoliter DNA analysis device. *Science*, **282** (1998), 484.

[47] B. Zhao, J. S. Moore and D. J. Beebe. Surface-directed liquid flow inside microchannels. *Science*, **291** (2001), 1023–1026.

[48] Nam-Trung Nguyen and Steven T. Wereley. *Fundamentals and Applications of Microfluidics* (Boston, MA: Artech House, 2002).

[49] D. J. Beebe, G. A. Mensing and G. M. Walker. Physics and applications of microfluidics in biology. *Annual Review of Biomedical Engineering*, **4** (2002), 261–286.

[50] H. Andersson and A. Van den Berg. Microfluidic devices for cellomics: a review. *Sensors and Actuators: B. Chemical*, **92** (2003), 315–325.

Section V

Principles for ultra-low-power analog and digital design

21 Principles for ultra-low-power digital design

A small leak will sink a great ship.

Benjamin Franklin

In this chapter, we shall review important principles for ultra-low-power digital circuit and system design. We shall focus on operation with extremely low power-supply voltages and on subthreshold operation, although we shall provide some analysis of moderate-inversion and strong-inversion operation with the EKV model as well. As Chapter 6 on deep submicron effects in transistors discussed, because threshold voltages scale significantly less strongly than power-supply voltages, subthreshold operation is an increasingly dominant fraction of the voltage operating range. Subthreshold operation has become and will continue to get increasingly fast such that ultra-low-power operation in this regime does not sacrifice bandwidth in many applications. In biomedical and bioelectronic applications, subthreshold operation is ideal since bandwidth requirements are typically modest while energy efficiency is of paramount importance. An insightful paper by Meindl, that was way ahead of its time, pioneered subthreshold digital design [1]. An analysis by Burr and Peterson analyzed the optimal energy efficiency of ultra-low-power subthreshold circuits [2]. A more recent publication by Vittoz [3], the pioneer of subthreshold analog design, has analyzed issues in subthreshold digital design using his EKV model. Through such pioneering and other work, subthreshold digital design has been revived and is an active field of research in several academic and industrial institutions.

We shall begin by discussing the operation of a subthreshold CMOS inverter. Operation in the subthreshold regime is highly subject to transistor mismatch. We present equations that help quantify transistor sizing and a lower limit to the power-supply voltage needed for robust subthreshold operation. The CMOS inverter serves as a good vehicle for understanding the three kinds of power dissipation in digital CMOS circuits, namely, dynamic power, static power, and short-circuit power. Dynamic power is dominant during switching and is due to the dissipation caused by on current flowing in a PMOS transistor charging a node capacitance or by on current flowing in an NMOS transistor discharging a node capacitance. Static power or 'leakage power' is caused by off current or subthreshold leakage current or background current present in an NMOS or PMOS transistor that has been turned off. Static power dominates during static operation when there is no switching and node voltages are static. Short-circuit power is caused by charging and discharging pathways being simultaneously active in a circuit such that on currents flow in a short-circuit fashion from V_{DD} to ground rather than to or from node capacitances. Short-circuit power contributes during switching when the NMOS and PMOS devices have comparable currents. Short-circuit power can typically be neglected unless the rise or fall times of the input are significantly larger than the rise or fall times of the output.

At an optimal value of the power-supply voltage, we shall show that the balance between dynamic energy and static energy leads to the lowest total energy dissipated per cycle of operation. For many computations, this optimal point causes the power-supply voltage to be in the subthreshold regime where the on-current to off-current ratio is maximum. At this optimum power-supply voltage, the threshold voltage may be set by body biasing to fulfill frequency-of-operation requirements set by the bandwidth of the task. Alternatively, if thresholds are fixed, the optimal power-supply voltage and threshold voltage determine an optimal frequency of operation. Increased switching activity causes more dynamic power dissipation and shifts the optimal point to occur at lower power-supply voltages. Reduced duty cycles in the computation cause more static power dissipation and shift the optimal point to occur at higher power-supply voltages.

The dynamic power dissipation of nodes with high activity factors (frequent switching activity) such as clock nodes can be considerable. We shall discuss how gated clocks and adiabatic clocks can greatly reduce such power dissipation. Adiabatic clocks are based on a more general computational paradigm termed adiabatic computing. While the paradigm of adiabatic computing is completely general and applies to all logic nodes, not just to clocked nodes, its high overhead has thus far made it of practical importance only in circuits with high node capacitances.

We then briefly review the key ideas behind architectural and algorithmic techniques for power reduction including parallelism, pipelining, ordering, symmetry, and algorithmic-efficiency improvements. Since many of these techniques have been reviewed previously [4], and in texts on low-power digital design [5], our treatment will focus only on presenting the key insights. Recent examples of systems that embody the principles described in this chapter include a sensor processor in [6], a microcontroller in [7], and a JPEG co-processor in [8].

21.1   Subthreshold CMOS-inverter basics

Figure 21.1 (a) shows a classic CMOS inverter circuit and Figure 21.1 (b) shows its input-output characteristic. The inverter inverts its digital inputs from 1 (V_{DD}) to 0 (GND or ground) and vice versa. In our case, the power-supply voltage V_{DD} is sufficiently small such that both the NMOS and PMOS transistors operate with subthreshold currents for all inputs. That is, V_{DD} is less than either $|V_{TP}|$ or V_{TN}. If we desire the switching transition point of the inverter to be at a symmetrical $V_{DD}/2$, as shown in Figure 21.1 (b), then we must require that

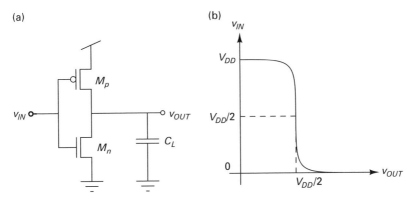

Figure 21.1a, b. (a) A basic CMOS inverter. (b) Input-output waveforms in a balanced CMOS inverter.

$$\mu_n C_{ox}\phi_t^2 \frac{W_n}{L_n}\frac{1-\kappa_n}{\kappa_n} e^{\kappa_n\left(\frac{V_{DD}}{2}-V_{Tn}\right)/\phi_t}\left(1-e^{-V_{DD}/(2\phi_t)}\right)$$
$$= \mu_p C_{ox}\phi_t^2 \frac{W_p}{L_p}\frac{1-\kappa_p}{\kappa_p} e^{\kappa_p\left(\frac{V_{DD}}{2}-|V_{Tp}|\right)/\phi_t}\left(1-e^{-V_{DD}/(2\phi_t)}\right) \tag{21.1}$$

Some algebraic simplification of Equation (21.1) then yields

$$\frac{W_p/L_p}{W_n/L_n} = \left(\frac{\mu_n}{\mu_p}\right)\left(\frac{\frac{\kappa_p}{1-\kappa_p}}{\frac{\kappa_n}{1-\kappa_n}}\right)\left(e^{(\kappa_p|V_{Tp}|-\kappa_n V_{Tn})/\phi_t}\right)\left(e^{(\kappa_n-\kappa_p)V_{DD}/(2\phi_t)}\right) \tag{21.2}$$

The off currents in the NMOS and PMOS transistors are given by

$$I_{OFF}^{NMOS} = \mu_n C_{ox}\phi_t^2 \frac{W_n}{L_n}\frac{1-\kappa_n}{\kappa_n} e^{-\kappa_n V_{Tn}/\phi_t}$$
$$I_{OFF}^{PMOS} = \mu_p C_{ox}\phi_t^2 \frac{W_p}{L_p}\frac{1-\kappa_p}{\kappa_p} e^{-\kappa_p |V_{Tp}|/\phi_t} \tag{21.3}$$

The relative sizing of the PMOS and NMOS transistors given in Equation (21.2) will also automatically equalize off currents in the NMOS and PMOS transistors if $\kappa_n = \kappa_p = \kappa = 1/n$.

If we temporarily assume that there are no DIBL or Early-voltage effects in transistors, the maximal gain, A_{inv}, of the inverter occurs at the switching threshold of $V_{DD}/2$ and can be computed from small-signal analysis to be

$$A_{inv} = \frac{g_m^{NMOS} + g_m^{PMOS}}{g_d^{NMOS} + g_d^{PMOS}}$$

$$A_{inv} = \frac{\left(\dfrac{\kappa_n I_{SAT}}{\phi_t} + \dfrac{\kappa_p I_{SAT}}{\phi_t}\right)\left(1-e^{-V_{DD}/(2\phi_t)}\right)}{\left(\dfrac{I_{SAT}}{\phi_t} + \dfrac{I_{SAT}}{\phi_t}\right)e^{-V_{DD}/(2\phi_t)}} \tag{21.4}$$

$$A_{inv} = \left(\frac{\kappa_n + \kappa_p}{2}\right)\frac{1-e^{-V_{DD}/(2\phi_t)}}{e^{-V_{DD}/(2\phi_t)}}$$

The minimum power-supply voltage at which digital circuits can function with signal restoration occurs when $A_{inv} > 1$. Some algebra on Equation (21.4) then yields that

$$V_{DD}^{min} = 2\phi_t \ln\left(\frac{\kappa_n + \kappa_p + 2}{\kappa_n + \kappa_p}\right) \tag{21.5}$$

If κ_n and κ_p are at least 0.6, V_{DD}^{min} is almost exactly $2\phi_t$. The $2\phi_t$ value (50 mV) represents an absolute lower limit on the power-supply voltage and provides no noise margin. Even a modest increase of V_{DD} to $4\phi_t$ leads to a relatively good noise margin in an inverter. By equating currents in the NMOS and PMOS transistors of the balanced inverter sized according to Equation (21.2), with its input at V_{DD}, with the PMOS transistor nearly saturated, with the NMOS transistor in its linear regime, and assuming that $\kappa_n \approx \kappa_p$, we can show that such an input voltage produces an output voltage of

$$\begin{aligned} v_{OUT} &= \phi_t \ln\left(\frac{e^{\kappa V_{DD}/\phi_t}}{e^{\kappa V_{DD}/\phi_t} - 1}\right) \\ &\approx \phi_t e^{-\kappa V_{DD}/\phi_t} \\ &\leq 0.5\,\text{mV if } V_{DD} = 4\phi_t; \kappa \geq 0.6 \end{aligned} \tag{21.6}$$

Thus, a subthreshold inverter with a power-supply voltage of even $4\phi_t$ can be used for general-purpose digital computation. While such a low power-supply voltage seems very power efficient, in practice transistor mismatch requires us to sacrifice efficiency to attain robustness: the relative sizing of the PMOS and NMOS transistors in Equation (21.2) is exponentially sensitive to the nominal values of the PMOS and NMOS threshold voltages. The voltage V_{DD} must be sufficiently large such that, if we set the input of the inverter at V_{DD}, even at the worst process corner, a weak NMOS transistor with a $+3\sigma_{vthN}$ high threshold-voltage magnitude can just overwhelm a strong PMOS transistor with a low $-3\sigma_{vthP}$ threshold-voltage magnitude to force the output of the inverter to just fall below $V_{DD}/2$. Here, σ_{vthN} and σ_{vthP} represent the standard deviation of the threshold-voltage of the NMOS and PMOS transistors respectively. Assuming that the nominal mean threshold voltages for the NMOS and PMOS transistors are such that Equations (21.1) and (21.2) apply, we can compute deviations from this balanced $V_{DD}/2$ midpoint, with a $+V_{DD}/2$ deviation for the NFET and a $-V_{DD}/2$ deviation for the PFET, to find that

$$e^{\kappa_n((V_{DD}/2) - 3\sigma_{vthN})/\phi_t} \geq e^{\kappa_p((-V_{DD}/2) + 3\sigma_{vthP})/\phi_t}$$

$$V_{DD} \geq 6\left(\frac{\kappa_n\sigma_{vthN} + \kappa_p\sigma_{vthP}}{\kappa_n + \kappa_p}\right) \tag{21.7}$$

If we require the NMOS current to be at least a factor of 10 larger than the PMOS current in Equation (21.7) such that the logic noise margin is not significantly degraded by the inverter even in this worst case, i.e., v_{OUT} is still quite close to GND, we get

$$V_{DD} \geq 6\left(\frac{\kappa_n \sigma_{vthN} + \kappa_p \sigma_{vthP}}{\kappa_n + \kappa_p}\right) + \phi_t \ln(10) \qquad (21.8)$$

From Chapter 6, in a 0.1 μm process, the A_{vth} parameter, which describes the threshold-voltage variation is $\sim$3.2 mV/μm such that a minimum-size $\sim$0.1 μm device has σ_{vthN} and σ_{vthP} at $\sim$32 mV. Equation (21.8) then predicts that $V_{DD} \geq 250$ mV. A symmetric analysis can be performed for the other case where we have a weak PMOS transistor and a strong NMOS transistor, which also yields Equation (21.8). Thus, robustness to transistor mismatch sets a minimum value of V_{DD} that is far in excess of that predicted by gain (50 mV) and noise-margin (100 mV) considerations alone. Of course, the use of non-minimum-size devices will lower σ_{vthN} and σ_{vthP} such that V_{DD} can be lowered at the cost of some increase in switching-capacitance-based and static power consumption (see discussion in next section). As we discuss in Chapter 22, in both the analog and digital domains, robustness and efficiency always trade off with each other.

For most practical power-supply voltages in excess of 100 mV, the gain of a subthreshold CMOS inverter is set by short-channel DIBL effects rather than by saturation effects (see Chapter 6 for a description of DIBL in short-channel transistors). If the DIBL coefficient is given by η, then, from a small-signal analysis similar to that in Equation (21.4), we can compute the inverter gain to be

$$A_{inv} = \frac{\kappa_n + \kappa_p}{\eta_n + \eta_p} \qquad (21.9)$$

since, typically, $\kappa > 0.6$ and $\eta < 0.1$, A_{inv} is at least 6.

If V_{DD} is sufficiently large (greater than 100 mV), and we apply a step input to the CMOS inverter, to an excellent approximation, we may assume that only the NMOS transistor is on for a positive step input and that only the PMOS transistor is on for a negative step input. Thus, if we assume that $\kappa_n = \kappa_p = \kappa$ for simplicity, the delay of the CMOS inverter to a step input, T_{inv}, can be computed by calculating the time taken by the on current of either the NMOS or PMOS transistor to discharge or charge its load capacitance C_L by a voltage differential of magnitude V_{DD} respectively:

$$I_{OFF} = \left(\mu_n C_{ox} \frac{W_n}{L_n} \frac{1-\kappa}{\kappa} \phi_t^2\right) e^{-\kappa V_{Tn}/\phi_t} = \left(\mu_p C_{ox} \frac{W_p}{L_p} \frac{1-\kappa}{\kappa} \phi_t^2\right) e^{-\kappa |V_{Tp}|/\phi_t}$$

$$I_{OFF} \approx I_{T0} e^{-\kappa V_{T0}/\phi_t}$$

$$T_{inv} = \frac{C_L V_{DD}}{I_{ON}}$$

$$I_{ON} = I_{OFF} e^{\kappa V_{DD}/\phi_t} \qquad (21.10)$$

$$T_{inv} = \frac{C_L V_{DD}}{I_{OFF}} e^{-\kappa V_{DD}/\phi_t}$$

$$\boxed{T_{inv} = \frac{C_L V_{DD}}{I_{T0}} e^{-\kappa(V_{DD} - V_{T0})/\phi_t}}$$

From differentiation of Equation (21.10) with respect to V_{T0}, we can derive that the variance in inverter delay, σ_{Tinv}, is given by

$$\frac{\sigma_{Tinv}}{T_{inv}} = \frac{\sigma_{VT0}}{\phi_t/\kappa} \tag{21.11}$$

From the discussion and numbers presented right after Equation (21.8), Equation (21.11) predicts a large variation in the inverter delay if minimum-size devices are used. Therefore, minimum-size devices must be avoided in subthreshold operation.

From Equations (21.1) and (21.2), and the laws of statistics, we can derive that the variance in the switching threshold of the inverter, σ_{invTH}, deviates from $V_{DD}/2$ as a root-mean-square function of the threshold-voltage mismatch:

$$\sigma_{invTH} = \sqrt{\sigma_{vthN}^2 + \sigma_{vthP}^2} \tag{21.12}$$

21.2   Sizing and topologies for robust subthreshold operation

Circuits that rely on ratioing, e.g., certain latch configurations where the latch state is written with a powerful write transistor but the latch state is maintained with a weak feedback transistor, perform unreliably in subthreshold operation. Such unreliability arises because currents are exponentially dependent on threshold-voltage variations such that small mismatches in threshold voltage between devices can cause failure in ensuring that the on current in one device is strong enough to overwhelm the on current in another. The consequent logic failures can reduce circuit yields. Hence, logic circuits that operate by explicitly turning off devices whose functions compete with each other must be the norm. Similarly, as Equation (21.2) shows, small differences between NMOS and PMOS threshold voltages of a few ϕ_t translate to large differences in the W/L ratios of the NMOS and PMOS transistors needed to maintain a balanced switching threshold of $V_{DD}/2$ in a CMOS inverter. Fortunately, Equation (21.12) shows that threshold-voltage mismatch results in a modest change in the input-referred threshold for switching in a CMOS inverter. Thus, if V_{DD} is sufficiently high and there are good noise margins, robust operation can be preserved. However, the time needed to finish a logic operation is not robust as per Equation (21.11), which implies that increasing device sizes to improve temporal matching is important. Alternatively, asynchronous architectures that are inherently robust to such timing mismatch may represent a promising research direction in the future [9].

To preserve robustness, efficiency must be sacrificed, as is true in analog systems as well: To yield better matching of thresholds across all process corners, devices must be sized larger than minimum size, which necessarily increases load capacitances, switching power dissipation, and leakage power. One benefit of larger-sized devices, however, is that they are better able to drive wiring parasitic capacitances such that modest speed improvements can result.

In subthreshold logic circuits, increased off currents caused by many parallel devices or decreased on currents caused by many devices stacked in series can also cause logic circuits to malfunction. The collective strength of several off currents can overwhelm one on current and lead to malfunction, e.g., in reading one cell from a memory on a common bitline where there are a large number of off devices. Thus topologies and architectures must be reconfigured to avoid excessive parallelism and excessive series stacking.

21.3   Types of power dissipation in digital circuits

There are three types of power dissipation in digital circuits, namely dynamic, static, and short-circuit power dissipation. We shall describe each kind in turn initially using the CMOS inverter as our exemplary digital circuit.

21.3.1   Dynamic power dissipation

In Figure 21.1 (a), suppose the input of the inverter is at a digital 1, i.e., at V_{DD}, and its output is at a 0, i.e., at ground. If the input to the inverter now changes abruptly from V_{DD} to 0 with a relatively fast risetime, C_L will be charged via M_p from 0 to V_{DD}. If the input signal is abrupt enough, M_n will be turned off at all times during this charging process. During the charging process, current only flows through M_p to charge C_L. The net charge provided by the supply is the current flowing through M_p integrated over the entire charging process. By charge conservation, the charge flowing from the supply is the charge stored on the capacitor. This integrated charge must therefore be $C_L V_{DD}$ and the net energy provided by the supply during charging is then $C_L V_{DD} V_{DD} = C_L V_{DD}^2$. Of this energy, $(1/2)C_L V_{DD}^2$ is stored on the capacitor C_L. Since all the current that charges C_L flows through M_p, by energy conservation, the remaining energy of $(1/2)C_L V_{DD}^2$ must have been dissipated in M_p as E_{Mp}. This remarkable result is true independent of the details of how the current in M_p, $i_{Mp}(t)$, changes with time during the charging process. It can be formally derived as follows:

$$i_{Mp}(t) = C_L \frac{dv_{OUT}(t)}{dt}$$

$$E_{Mp} = \int_0^\infty (V_{DD} - v_{OUT}(t))i_{Mp}(t)dt$$

$$E_{Mp} = \int_0^\infty (V_{DD} - v_{OUT}(t))C_L \frac{dv_{OUT}(t)}{dt}dt \tag{21.13}$$

$$E_{Mp} = \int_0^{V_{DD}} (V_{DD} - v_{OUT}(t))C_L dv_{OUT}$$

$$E_{Mp} = C_L V_{DD}^2 - \frac{1}{2}C_L V_{DD}^2 = \frac{1}{2}C_L V_{DD}^2$$

Now suppose that the input changes from 0 to V_{DD} in a similarly abrupt fashion such that C_L is only discharged through M_n from V_{DD} to 0 and M_p is completely turned off during the discharging process. Then, by almost identical reasoning to that in Equation (21.13), we can derive that the stored energy of $(1/2)C_L V_{DD}^2$ on C_L is dissipated in M_n. Thus, an energy of $C_L V_{DD}^2$ is consumed from the supply on a $0 \rightarrow 1$ output transition, and only stored capacitive energy is dissipated in a $1 \rightarrow 0$ transition. Hence we can say that only $0 \rightarrow 1$ transitions consume energy from the supply. If $0 \rightarrow 1$ output transitions occur with probability $\alpha_{0 \rightarrow 1}$ during each clock cycle f_{clk}, then the mean dynamic power dissipation, P_{dyn}, is given by the rate of mean energy consumption from the supply:

$$\boxed{P_{dyn} = \alpha_{0 \rightarrow 1} f_{clk} C_L V_{DD}^2} \tag{21.14}$$

If P_0 is the probability that the output is at a 0 and P_1 is the probability that the output is at a 1, then the probability $\alpha_{0 \rightarrow 1}$ is given by the probability that the current state is a 0 times the probability that the next state is a 1:

$$\begin{aligned}
\alpha_{0 \rightarrow 1} &= P_0 P_1 \\
\alpha_{0 \rightarrow 1} &= P_0(1 - P_0) \\
\alpha_{0 \rightarrow 1} &= (1 - P_1)P_1
\end{aligned} \tag{21.15}$$

If the input signal statistics to the inverter are completely random such that P_0 and P_1 are both 1/2, then $\alpha_{0 \rightarrow 1}$ is at its maximal value of 1/4.

In general, for other logic circuits, there are four factors that affect α: 1) the logic function, 2) the input signal statistics, 3) the logic style, and 4) the circuit topology. We shall briefly describe each of these in turn in the next four paragraphs [4].

If all inputs have a 1 probability of a 1/2, logic functions such as AND and NOR have $P_1 = 1/4$; similarly, logic functions such as NAND and OR have $P_1 = 3/4$. Hence, $\alpha_{0 \rightarrow 1}$ is 3/16 for such functions rather than 1/4 for inverter or XOR logic functions. Consequently switching power dissipation is logic-function dependent. The output column of the function's truth table must have a balanced number of 0's and 1's for $\alpha_{0 \rightarrow 1}$ to have a maximum value of 1/4.

Input signals to a logic function may be correlated and not exhibit random probabilities. For example, if two signals to an AND gate have a correlation coefficient of 1, the AND gate will exhibit output switching statistics like that of an inverter. If two signals to an AND gate have a correlation coefficient of -1, the AND gate will never switch. In general, for any single-output multi-input logic function described by $Y = f(X_1, X_2, X_3, X_4, \ldots)$, there is a truth table, and every X-input-vector entry, $\overline{X}$, in this truth table has a corresponding probability for its occurrence, which takes correlations amongst the components of X into account. The value of $\alpha_{0 \rightarrow 1}$ for such a logic function taking input statistics into account is given by

$$\alpha_{0 \rightarrow 1} = \left(\sum_{f(\overline{X}) = 0} P(\overline{X}) \right) \left(1 - \sum_{f(\overline{X}) = 0} P(\overline{X}) \right) \tag{21.16}$$

Certain logic styles such as dynamic logic *always* have a precharge cycle to 1 followed by a conditional discharge to 0 that is based on a logic evaluation. In this case, the value of $\alpha_{0\to1}$ is simply the probability that the state on the previous clock cycle was a 0, i.e., P_0. Thus a NAND gate, which is usually 1 for random equiprobable inputs, will rarely discharge and have $\alpha_{0\to1} = 0.25$, while a NOR gate, which is usually 0 for random equiprobable inputs, will frequently discharge and have $\alpha_{0\to1} = 0.75$. Another important difference between static and dynamic logic is that if the input does not change, static logic exhibits no dynamic power dissipation; dynamic logic can exhibit dynamic power dissipation even with unchanging inputs since its power dissipation is based on signal probability, not on signal-transition probability.

When the input vector to a logic circuit changes, even if its output is not supposed to change, as per its truth table, there can be transient or glitching power dissipation at its output and at its intermediate nodes. Glitching dissipation arises because imbalanced logic propagation delays in the circuit may cause certain intermediate nodes and the output to incorrectly evaluate based on the prior value of some inputs rather than on their current value; eventually, the arrival of the delayed information alters the incorrect evaluation to a correct one, but power is then wasted in a transient glitch. Glitching power dissipation is minimized by attempting to balance propagation delays through all paths by making circuits as symmetric as possible. However, such balancing can then lead to higher ordinary (non-glitching) power dissipation because the circuit is then not necessarily implemented in a probability-reducing chain. For example, an OR gate that computes $(A + B + C + D)$ can be implemented with three 2-input OR gates in a series $(((A + B) + C) + D)$ chain-like configuration, or as three 2-input OR gates in a balanced $((A + B) + (C + D))$ tree-like configuration. In the chain configuration, a $(0001) \to (1000)$ transition for $(ABCD)$ will cause the output to go to 0 soon after the last input D switches to 0; information from the output of the $(A + B)$ gate will then make its way from input to output in a delayed fashion such that the output will later return to 1; thus, a glitch from 1 to 0 and then back to 1 is seen at the output. In the balanced tree-like configuration, the same configuration will cause no glitches if the $(A + B)$ and $(C + D)$ gates have symmetric delays. However, the chain-like configuration has lower summed non-glitching $\alpha_{0\to1}$ activity over its three gate outputs $(91/256)$ than the tree-like configuration does $(111/256)$. Clearly, if certain glitch-causing transitions are rare in the data, the chain-like configuration is favorable, while if they dominate the switching statistics, the tree-like configuration is preferable. Dynamic logic has no glitching power dissipation since all evaluations result in a one-way conditional discharge.

21.3.2 Static power dissipation

Static power dissipation in a digital circuit is caused by current in devices that are supposed to be off when the circuit is static and not switching. Although reverse-biased diode-junction leak current does contribute to this power, the most important source of static power dissipation is the subthreshold current present in transistors that have a gate-to-source voltage of nearly zero-volt magnitude. Thus, in Figure 21.1 (a), when the input to the inverter is at 0 and the output is near V_{DD}, transistor M_n conducts a current of I_{OFF}^{NMOS} as computed in Equation (21.3), and transistor M_p will have a small V_{DS} drop across it to support this current, i.e., the output of the inverter will be slightly lower than V_{DD}. Similarly, if the input to the inverter is at V_{DD}, and the output is near 0, transistor M_p conducts a current of I_{OFF}^{PMOS} as computed in Equation (21.3), and transistor M_n will have a small V_{DS} drop across it to support this current, i.e., the output of the inverter will be slightly higher than ground. Since, in Equation (21.3), the off currents are exponentially sensitive to the threshold voltage, and the threshold-voltage magnitude is lowered by DIBL, the off current increases with an increase in V_{DD}. The increase in off current with V_{DD} can be significant. With a DIBL coefficient of 0.1 V/V, a 0.5 V increase in V_{DD} can increase the leakage current by a factor of 7. When the CMOS inverter in Figure 21.1 (a) is dissipating static power, the transistors M_n or M_p have $V_{DS} \approx V_{DD}$ such that the absolute thresholds of the leak transistors are at their lowest values, and the leakage currents are consequently at their highest values. The static power consumption can be summarized by the simple equation

$$\boxed{P_{stc} = I_{OFF} V_{DD}} \tag{21.17}$$

21.3.3   Short-circuit power dissipation

If the input to the CMOS inverter in Figure 21.1 (a) is a step from 0 to V_{DD}, transistor M_p is instantly inactivated and the dynamics at the output of the inverter are dominated by the on current of M_n discharging C_L. Similarly, if the input to the CMOS inverter is a step from V_{DD} to 0, transistor M_n is instantly inactivated and the dynamics at the output of the inverter are dominated by the on current of M_p charging C_L. In this ideal scenario, only one transistor is active during switching such that the dynamic power dissipation of Equation (21.14) applies. But what if the input to the inverter is not infinitely fast and has a finite rise time or fall time? Then, during the transition, there will be periods where M_p and M_n are simultaneously conducting current such that there is some short-circuit current from V_{DD} to ground, not just from one supply to C_L. The power dissipated by this short-circuit current is termed short-circuit power dissipation. Note that, strictly speaking, static power dissipation is also short-circuit power dissipation since leakage current flows from V_{DD} to ground through M_n and M_p simultaneously. However, the short-circuit current that we will focus on is significantly higher than this leakage current and is present only for brief periods during switching transitions. Short-circuit power dissipation is analytically challenging to compute since it requires knowledge of the nonlinear input-output waveforms of the digital circuit. Thus, even sophisticated treatments of ultra-low-power digital design just compute

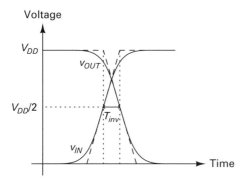

Figure 21.2. Waveforms in a CMOS inverter along with asymptotic approximations.

it numerically, or just ignore it by flatly stating that its impact is small [3]. We shall take an approach that approximates short-circuit power in an analytic fashion. Although our analysis is only approximate, it can provide insight that brute-force simulations and numerical approaches do not. We begin by analyzing short-circuit power dissipation in the context of a circuit composed only of CMOS inverters, a ring oscillator.

A ring oscillator is a daisy chain of an odd number of CMOS inverters (at least 3) hooked to each other with the final CMOS inverter's output hooked to the first CMOS inverter's input to form a ring. A ring oscillator is an excellent digital circuit to perform analyses on because the input and output load capacitances for each logic block are identical, which is representative of many complex digital circuits with nearly equal fan-in and fan-out. Figure 21.2 illustrates one edge of the square-wave-like input and output waveforms of a CMOS inverter in a subthreshold ring oscillator. The asymptotic behavior of the waveforms as well as their actual shapes are shown. These waveforms are representative of situations where $V_{DD} > 4\phi_t$ such that nearly rail-to-rail operation is obtained, and of situations where there are many stages in the ring such that each of the two square-wave transitions in the waveform are well separated. We find, remarkably, that even though the input is not a perfect step, the delay between input and output for a CMOS inverter is almost exactly T_{inv}: the value of T_{inv} is given by the boxed expression of Equation (21.10) for an ideal step input [3].

Figure 21.2 shows the input v_{IN} and output v_{OUT} waveforms in a CMOS inverter in a ring oscillator during a 0 to V_{DD} transition in v_{IN}. The asymptotes in Figure 21.2 provide four valuable clues for generating approximations to the voltage waveforms in a CMOS inverter, which will help us estimate its short-circuit power dissipation under typical conditions of use: 1) the asymptotic output starts to discharge only when the asymptotic input reaches $V_{DD}/2$; 2) when the asymptotic input has reached V_{DD}, the asymptotic output has linearly discharged to $V_{DD}/2$; 3) the time spent by the input or output below and above $V_{DD}/2$ is symmetric; 4) the input and output voltage waveforms have the same slopes and are mirror symmetric about an axis of time that passes through the point of their crossing.

These clues provide us with a simple physical picture that approximately describes inverter dynamics. The PMOS transistor M_p overwhelms M_n until the V_{GS} across M_n exceeds that across M_p, such that the output barely changes until the input reaches the $V_{DD}/2$ switching threshold. The steep nature of exponential functions allows the abrupt turn-on approximation at $V_{DD}/2$ to have validity. When the input is below $V_{DD}/2$, a small V_{DS} and large V_{GS} drop across M_p causes its current to balance the current of M_n, which, in contrast, has a small V_{GS} and large V_{DS} drop. Once the input crosses $V_{DD}/2$, M_n starts to discharge the output at an increasingly fast rate as M_p exponentially turns off and M_n exponentially turns on; this rate reaches CV_{DD}/I_{ON} when the input is at V_{DD}. Given the symmetry between the input and the output, and the fact that the output does not start to change until the input reaches $V_{DD}/2$, the output is at the halfway point, $V_{DD}/2$, when the input has reached V_{DD}. Past this halfway point at the output, M_n simply discharges the output at a constant rate of CV_{DD}/I_{ON}, with M_p completely turned off until M_n eventually comes out of saturation as its V_{DS} falls below V_{DSAT}. The asymptotic approximation is equivalent to assuming that V_{DSAT} is 0 for both M_p and M_n. If we assume that V_{DSAT} is 0, the time taken by M_n to discharge the output from $V_{DD}/2$ to 0 is $C(V_{DD}/2)/I_{ON}$. From waveform symmetry in Figure 21.2, the time taken to discharge the output from V_{DD} to $V_{DD}/2$ is then also $C(V_{DD}/2)/I_{ON}$. Thus, the total time taken to discharge the output is CV_{DD}/I_{ON}, i.e., T_{inv}. From the waveforms of Figure 21.2, the total time taken to discharge the output is the delay between the input and the output. Thus, the delay between ring-oscillator stages turns out to be remarkably close to T_{inv} even though we do not have a step input! The asymptotic approximation works well because underestimates in the delay time that arise by neglecting the influence of V_{DSAT} in M_n's operation as v_{OUT} nears 0 are compensated for by symmetric overestimates in the delay time that arise by neglecting the influence of V_{DSAT} in M_p's operation when v_{OUT} is near V_{DD}.

When the input changes from 0 to $V_{DD}/2$, there is an exponential increase in the current of M_n, which is nearly equal to that in M_p such that v_{OUT} only drops by a small amount. To first approximation, we can estimate the short-circuit charge during this period as being the integrated current in M_n from 0 to $T_{inv}/2$ and pretend that none of the current of M_n contributes much to discharging the capacitor. When the input changes from $V_{DD}/2$ to V_{DD} the current in M_p starts to get significantly weaker than that in M_n and most of the current in M_n goes to discharge the capacitor. However, a small fraction of the current in M_n must still supply the weakening current in M_p. By symmetry, the short-circuit charge during the period from $T_{inv}/2$ to T_{inv} is the integrated current in M_p during this period. In the balanced symmetrical inverter that we have been discussing, the short-circuit charge in the period from 0 to $T_{inv}/2$ is equal to the short-circuit charge in the period from $T_{inv}/2$ to T_{inv}. Thus, the total short-circuit charge, Q_{sc}, during an inverter transition is given by

$$v_{IN}(t) = V_{DD} \frac{t}{T_{inv}}$$

$$Q_{sc} \approx 2 \int_0^{T_{inv}/2} I_{OFF} e^{\kappa v_{IN}(t)/\phi_t} dt \tag{21.18}$$

$$Q_{sc} \approx 2 I_{OFF} T_{inv} \frac{\phi_t}{\kappa V_{DD}} \left(e^{\kappa V_{DD}/(2\phi_t)} - 1 \right)$$

Substituting the value of T_{inv} from Equation (21.10) in Equation (21.18) leads to

$$Q_{sc} \approx 2 C_L \frac{\phi_t}{\kappa} \left(e^{-\kappa V_{DD}/(2\phi_t)} - e^{-\kappa V_{DD}/\phi_t} \right) \tag{21.19}$$

Since short-circuit charge is provided by the power supply on both $0 \to 1$ and $1 \to 0$ transitions, the short-circuit power dissipation may be estimated to be

$$\boxed{P_{sc} \approx 2(\alpha_{0\to1} + \alpha_{1\to0}) f_{clk} C_L V_{DD} \frac{\phi_t}{\kappa} \left(e^{-\kappa V_{DD}/(2\phi_t)} - e^{-\kappa V_{DD}/\phi_t} \right)} \tag{21.20}$$

Note that Equation (21.20) is only meaningful for V_{DD} greater than about $4\phi_t$ (100 mV) where nearly rail-to-rail operation occurs. If other sources of leakage besides subthreshold leakage are significant such that non-rail-to-rail operation, low on-off ratios, and leakage power dissipation are significant below $4\phi_t$, V_{DD} must be at least $8\phi_t$ for Equation (21.20) to be a good approximation. Simulations in a $0.18\,\mu$m process indicate a good fit of Equation (21.20) to the data for $V_{T0} < V_{DD} < 8\phi_t$. For $\kappa \approx 0.6$, and $V_{DD} = 250$ mV, it predicts that short-circuit power dissipation is less than 5% of the dynamic power dissipation of Equation (21.14). For most practical values of V_{DD}, the second exponential term of Equation (21.20) is negligible compared with the first. As V_{DD} decreases, the short-circuit charge increases due to the increase of $I_{OFF}T_{inv}$ as predicted by Equation (21.19); as V_{DD} decreases, the short-circuit power P_{sc} in Equation (21.20), which is proportional to the product of the short-circuit charge and V_{DD}, exhibits an optimum near $3\phi_t/\kappa$. However, the approximations that lead to Equation (21.20) are only valid for $V_{DD} > 4\phi_t$ such that, in all cases where Equation (21.20) is meaningful, we are to the right of this optimum, and power dissipation falls with increasing V_{DD}.

Short-circuit power dissipation is generally important only if input transition times are significantly slower than output transition times in a digital circuit. In such cases, the charge and discharge paths quickly alter the output voltage to a value that equalizes their currents; thus, they create a short-circuit conduction path for a significant portion of the input transition. When this conduction path is established, the device in the conduction path that is supposed to be 'off' at the end of the transition can have significant current flowing through it during the transition because it has a simultaneously large v_{GS} and v_{DS} biasing it during the transition. If the input transition time is fast compared with the output transition time, the output voltage never reaches equilibrium for any portion of the input transition; the input is successful in completely turning off all devices in off paths before they have time to conduct during a transition. The device that is supposed to be off at the end of the transition never has significant current flowing through it during the transition because it never has a simultaneously large v_{GS} and v_{DS} biasing it during the transition.

21.4   Energy efficiency in digital systems

Equations (21.14), (21.17), and (21.20) are the fundamental equations that describe dynamic, static, and short-circuit power dissipation in digital circuits respectively. Clearly, by reducing f_{clk}, it is possible to reduce dynamic and short-circuit power dissipation. Furthermore, lowering f_{clk} means that a lower value of V_{DD} can be used such that the static power in Equation (21.17) can be reduced as well. However, the decrease of V_{DD} does increase short-circuit power dissipation slightly as per Equation (21.20); nevertheless, short-circuit power dissipation is usually negligible even at relatively low V_{DD}s of $4\phi_t$, the minimum needed to attain nearly rail-to-rail noise margins. Therefore, it is reasonable to expect that one can simply lower V_{DD} to a value larger than the minimum of $4\phi_t$ ($\sim$100 mV) and the value dictated by the robustness considerations of Equation (21.8) ($\sim$ 250 mV for typical values); at this value of V_{DD}, we can then operate at a low value of f_{clk} and achieve low power dissipation. However, achieving low power dissipation without normalizing for f_{clk} is not significant. The important figure of merit for energy-efficient operation is not how small we can make the total power dissipation, P_{TOT}, but how small we can make the energy per cycle of operation, $E_{TOT} = P_{TOT}/f_{clk}$. This energy per cycle of operation, or equivalently, the power-delay product, is a measure of how much power we dissipate to compute at a certain clock rate; equivalently, it is the energy consumed to perform all the computations needed within a clock cycle. We shall now discuss how to compute E_{TOT}. The minimization of E_{TOT} maximizes energy efficiency.

In a complex digital system, $\alpha_{0\to1}$ and C_L will vary at each of the N_{TOT} nodes present in the system. The average fraction, α, of the total system capacitance involved in undergoing an energy-consuming transition is given by

$$\alpha = \frac{\sum_i \alpha_{0\to1}^i C_L^i}{\sum_i C_L^i} \tag{21.21}$$

It is convenient to define an average node capacitance C_L given by

$$C_L = \frac{\sum_i C_L^i}{N_{TOT}}, \tag{21.22}$$

such that the total effective switching capacitance C_{eff} is simply expressed as

$$C_{eff} = \alpha \sum_i C_L^i$$
$$C_{eff} = \alpha N_{TOT} C_L \tag{21.23}$$

The number of devices that are contributing static power within a clock cycle, N_{eff}, is given by

$$N_{eff} = \sum_i \left(\frac{W}{L}\right)_i \qquad (21.24)$$

where $(W/L)_i$ is an abstract representation of the effective W/L that varies from circuit to circuit and with circuit state; e.g., NAND gates may exhibit more leakage with a particular configuration of inputs and exhibit more leakage than a CMOS inverter. Note that the background leakage current contributes all the time, whether we are switching or not. It is convenient to define an average (W/L), denoted by β, and given by

$$\beta = \frac{\sum_i \left(\frac{W}{L}\right)_i}{N_{TOT}} \qquad (21.25)$$

such that the total number of leakage devices is simply expressed as

$$N_{eff} = \beta N_{TOT} \qquad (21.26)$$

Suppose the logic depth in the system is defined to be N_{LD} such that

$$f_{clk} = \frac{1}{N_{LD}T_{inv}} \qquad (21.27)$$

and that the circuit operates at the maximum logic speed for this given logic depth to achieve maximum efficiency. Then, if we ignore the short-circuit contribution to the dynamic energy, the total energy consumed per clock cycle is given by

$$\boxed{E_{TOT} = C_{eff}V_{DD}^2 + N_{eff}\left(I_{OFF}\frac{1}{f_{clk}}\right)V_{DD}} \qquad (21.28)$$

Some simple substitutions for f_{clk} in Equation (21.28) then lead to

$$E_{TOT} = C_{eff}V_{DD}^2 + N_{eff}\left(I_{OFF}\frac{N_{LD}C_LV_{DD}}{I_{ON}}\right)V_{DD}$$

$$E_{TOT} = C_{eff}V_{DD}^2 + N_{eff}N_{LD}C_L\left(\frac{I_{OFF}}{I_{ON}}\right)V_{DD}^2 \qquad (21.29)$$

$$\boxed{E_{TOT} = N_{TOT}C_LV_{DD}^2\left(\alpha + \frac{\beta N_{LD}}{I_{ON}/I_{OFF}}\right)}$$

The boxed expression in Equation (21.28) shows that, for a given V_{DD}, maximizing the I_{ON}/I_{OFF} ratio minimizes E_{TOT} and maximizes energy efficiency. For a given V_{DD}, an exponential relationship in the output current versus input voltage over an input voltage range of $[0, V_{DD}]$ yields a higher I_{ON}/I_{OFF} ratio than a square-law relationship over this same range. Thus, for a given V_{DD}, I_{ON}/I_{OFF} is maximum if purely subthreshold operation is possible over the entire input voltage range. Therefore, in Equation (21.29), energy efficiency is maximized if purely subthreshold operation is possible. Burr and Peterson [2] were the first to explicitly point out this important fact. Equation (21.28) also shows that minimizing logic depth, minimizing W/L ratios, and minimizing α, which is often called the activity factor, minimizes E_{TOT}. Substituting for the dependence of I_{ON}/I_{OFF} on V_{DD} from Equation (21.10), we find in the subthreshold regime that

$$E_{TOT}^{sub} = N_{TOT}\beta N_{LD}C_L V_{DD}^2 \left(\frac{\alpha}{\beta N_{LD}} + e^{-\kappa V_{DD}/\phi_t} \right)$$

$$\alpha_n = \frac{\alpha}{\beta N_{LD}} \quad\quad\quad (21.30)$$

$$\boxed{E_{TOT}^{sub} = (\beta N_{TOT}N_{LD})C_L V_{DD}^2(\alpha_n + e^{-\kappa V_{DD}/\phi_t})}$$

One noteworthy feature of Equations (21.28) and (21.30) is that, in purely subthreshold operation, I_{ON}/I_{OFF} is independent of the threshold voltage V_{T0} such that E_{TOT}^{sub} only depends on V_{DD} and is thus independent of V_{T0}. The factor α_n is a number that measures the relative importance of dynamic power with respect to static power; it is proportional to the activity factor α. A form of Equation (21.30) was first derived in [3].

## 21.5	Optimization of energy efficiency in the subthreshold regime

Figure 21.3 (a) shows a plot of $E_{TOT}^{sub}/(\beta N_{TOT}N_{LD}C_L\phi_t^2)$, a normalized form of E_{TOT}^{sub} in Equation (21.30), vs. V_{DD} for α_n ranging from 10^{-4} to 0.02. We observe that E_{TOT}^{sub} is minimized at an optimal value of V_{DD}, V_{DD}^{opt}, that balances dynamic and static energy dissipation. At this optimum value, V_{DD} is not too high such that dynamic energy dissipation increases but not too low such that the leakage current integrates over a very long period and static energy dissipation increases. Lowering V_{DD} impacts energy efficiency because, at a lower V_{DD}, it takes longer to finish a given operation as the clock speed is lower. Thus, the leakage current is integrated over a longer time, increasing the static energy and consequently the total energy dissipated per cycle. However, lowering V_{DD} does reduce the dynamic energy dissipation. At the optimum V_{DD}, the differential increase in static energy with V_{DD} is just balanced with the differential decrease in dynamic energy with V_{DD}.

Differentiation and geometric analysis of Equation (21.30) reveal that optima only exist for $\alpha_n < \alpha_{max}$ with $\alpha_{max} = 1/(2e^3) \approx 0.0259$. The value of V_{DD}^{opt} cannot be analytically found with well-known computable functions. However, for $\alpha_n < 2\alpha_{max}$, a good empirical approximation for V_{DD}^{opt} can be obtained via numerical regression:

$$\alpha_{max} = \frac{1}{2e^3}$$

$$V_{DD}^{opt} \approx \frac{\phi_t}{\kappa} \left(3 + 1.25\ln\left(\frac{2\alpha_{max}}{\alpha_n}\right) \right) \quad\quad (21.31)$$

Figure 21.3 (b) indicates that the approximation of Equation (21.31) is a good fit for a large range of α_n that occur in practical situations. At $\alpha_n = \alpha_{max}$,

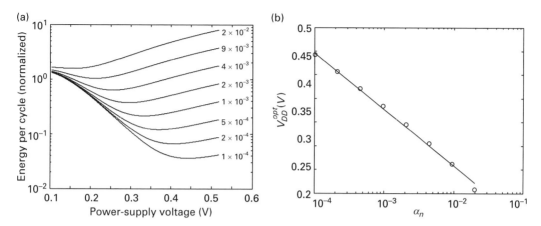

Figure 21.3a, b. (a) Variation of energy efficiency with power-supply voltage in weak inversion for various α_n, a dimensionless measure of the activity factor. (b) Variation of the optimal power-supply voltage in weak inversion with α_n.

$V_{DD}^{opt} = 3\phi_t/\kappa$, its minimum possible value. The inclusion of short-circuit power dissipation of Equation (21.20) has no discernible effect on the curves of Figure 21.3 (a) or those of Figure 21.3 (b).

Substitution of the optimal V_{DD}^{opt} extracted from Equation (21.31) in Equation (21.30) along with simple algebra reveal that the optimal value of E_{TOT}^{sub}, E_{opt}^{sub}, is given by

$$E_{opt}^{sub} \approx \left(\beta N_{TOT} N_{LD}\right)\left(C_L V_{DD}^2\right)\left(\alpha_n + 2\alpha_n^{1.25}\right) \quad (21.32)$$

We notice from Equation (21.32) that the dynamic energy and static energy terms are balanced at the optimum in that both decrease with decreases in α_n, and both terms are of comparable magnitude. We see in Figure 21.3 (a) that the optimum energy is lower at lower values of α_n in accord with Equation (21.32). At low values of α_n, corresponding to small activity factors in Equation (21.29) or Equation (21.30), the relative importance of static energy increases, and V_{DD}^{opt} increases in Equation (21.31) to reduce static energy, consistent with Figure 21.3 (b). Hence, lowering the activity factor of a computation always saves energy and always increases the optimum power-supply voltage.

Duty cycling the input lowers the effective value of f_{clk} in Equation (21.28) and increases the contribution of static energy. Thus, duty cycling the input effectively acts to increase N_{LD} in Equation (21.29), which can also be directly seen from Equation (21.27). The effective increase in N_{LD}, which is a measure of the logic depth of the computation, with the use of duty cycling may appear physically strange at first. The increase in N_{LD} with duty cycling is a mathematical energy equivalence: a large logic depth corresponds to a slow f_{clk} with lots of static devices dissipating energy while a few dynamic devices in the logic path actually switch and compute as the signal propagates through the depth of the logic; duty cycling the clock also leads to lots of static energy dissipation from devices that are mostly inactive while dynamic switching only occurs during a brief part of the active duty cycle. The effective increase in N_{LD} with duty cycling increases the static and overall energy in Equation (21.29), and leads to an increased value of V_{DD}^{opt} and an increased value of E_{opt}^{sub} at this optimum. The increase in V_{DD}^{opt} is consistent with the increase in N_{LD} lowering α_n in Equation (21.30) and thus increasing V_{DD}^{opt} in Equation (21.31). The increase in E_{opt}^{sub} is consistent with the increase in N_{LD} and V_{DD} in Equation (21.32), which dominates over the reduction in α_n. Equation (21.29), which contains no normalizing factors involving N_{LD} that impact α_n, offers the most direct way of seeing that increases in N_{LD} increase static energy and thus overall energy, and therefore that V_{DD} must be increased to increase I_{ON} and combat the increase in static energy. Note that increases in N_{LD} caused by direct increases in N_{LD} rather than by effective increases due to duty cycling have the same impact on energy efficiency.

In subthreshold operation, since I_{ON}/I_{OFF} is independent of V_{T0} in Equation (21.29), the threshold voltage has no effect on E_{TOT}. However, increasing V_{T0} reduces I_{ON} in Equation (21.10), and consequently f_{clk} in Equation (21.27), in an exponential fashion. Hence, we can adjust V_{DD} to an optimal value based on the expected α_n, and adjust the threshold voltage V_{T0} via body biasing to achieve the f_{clk} that is needed for the application:

$$f_{clk} = \frac{I_{T0}}{N_{LD}C_L V_{DD}^{opt}} e^{-\kappa(V_{DD}^{opt}-V_{T0})/\phi_t} \qquad (21.33)$$

The maximum f_{clk} that is possible in subthreshold operation occurs when $V_{T0} = V_{DD}^{opt}$, i.e., at

$$f_{clk,max} = \frac{I_{T0}}{N_{LD}C_L V_{DD}^{opt}} \qquad (21.34)$$

If α_n varies and V_{DD}^{opt} varies to preserve maximum energy efficiency, $(V_{DD}^{opt} - V_{T0})$ must be maintained constant to preserve f_{clk} in Equation (21.33). If V_{T0} is fixed and cannot be varied, then, with V_{DD} set to V_{DD}^{opt}, Equation (21.33) gives us the optimal f_{clk} that maximizes energy efficiency. If V_{DD} is set higher than V_{DD}^{opt}, f_{clk} will correspondingly be higher, but energy efficiency is degraded.

In practice, due to the effects of DIBL discussed in Chapter 6, V_{T0} decreases as the drain-to-source voltage across off transistors increases. For circuits with nearly rail-to-rail operation, I_{OFF} is then increased due to the large $V_{DS} \approx V_{DD}$ across the off leaking devices; the increase in I_{ON} with the decrease in V_{T0} is effectively smaller because, as per the approximations of Figure 21.2, $v_{DS} \approx V_{DD}/2$ over a dynamic transition. Thus, if η is the DIBL coefficient, the net effect is that the I_{ON}/I_{OFF} ratio is degraded to

$$\frac{I_{ON}}{I_{OFF}} \approx e^{(\kappa-\frac{\eta}{2})V_{DD}/\phi_t} \qquad (21.35)$$

from its value on the fourth line of Equation (21.10). The degradation of the I_{ON}/I_{OFF} ratio makes leakage power more significant such that the optimal value of V_{DD} and the optimal value of E_{TOT} in Equation (21.29) increase. Note that, in series or stacked configurations of off transistors, all transistors in the stack will not see a large V_{DS} such that Equation (21.35) tends to underestimate the I_{ON}/I_{OFF} ratio.

21.6   Optimization of energy efficiency in all regimes of operation

If f_{clk} needs to be larger than the maximum in subthreshold operation set by Equation (21.34), moderate-inversion or strong-inversion operation may be necessary. In such cases, assuming that we are not operating in velocity saturation, the empirical EKV expression can be used to modify Equations (21.28) and (21.30) to yield

$$E_{TOT}^{EKV} = (\beta N_{TOT} N_{LD}) C_L V_{DD}^2 \left(\alpha_n + \frac{e^{-\kappa V_{T0}/\phi_t}}{\ln^2(1 + e^{\kappa(V_{DD}-V_{T0})/(2\phi_t)})} \right) \qquad (21.36)$$

Equation (21.36) reduces to Equation (21.30) in subthreshold operation. Figure 21.4 (a) shows a plot of $E_{TOT}^{EKV}/(\beta N_{TOT} N_{LD} C_L \phi_t^2)$ for $\kappa = 0.625$ and $V_{T0} = 0.21\mathrm{V}$ for various α_n. As α_n gets smaller, the relative importance of leakage power becomes more important such that V_{DD} increases to enable larger on currents and faster clock operation; consequently, the leakage current integrates over a shorter time improving energy efficiency. Figure 21.4 (b) shows plots of V_{DD}^{opt} versus α_n for this case.

Velocity saturation can be modeled by using expressions given in Chapter 6 for the on current in Equation (21.29) and Equation (21.30); the subthreshold off current is unchanged. Often an empirical expression $I_{ON} \sim \kappa(V_{DD} - V_{T0})^\gamma$ with $1 < \gamma < 2$ can model operation that is not purely in a velocity-saturated or in a traditional strong-inversion regime. Thus, we can use

$$E_{TOT}^{VS} \approx (\beta N_{TOT} N_{LD}) C_L V_{DD}^2 \left(\alpha_n + \frac{e^{-\kappa V_{T0}/\phi_t}}{\ln^\gamma(1 + e^{\kappa(V_{DD}-V_{T0})/(\gamma\phi_t)})} \right) \qquad (21.37)$$

with $1 < \gamma < 2$ to approximate and smoothly interpolate over all regimes of operation. Figure 21.4 (c) is the plot of Figure 21.4 (a) with identical parameters except that $\gamma = 1.5$ in Equation (21.37) rather than 2 (as is implicit in Equation (21.36)); similarly, Figure 21.4 (d) is a plot of V_{DD}^{opt} vs. α_n for $\gamma = 1.5$ rather than for $\gamma = 2.0$ in Figure 21.4 (b).

For the same V_{DD}, the integration time for the leakage current is longer in the velocity-saturated regime than in traditional above-threshold operation. Therefore, we observe that, for small α_n, the optimal energy is generally higher in Figure 21.4 (c) than in Figure 21.4 (a).

Since increasing V_{DD} does not increase f_{clk} and consequently reduce the leakage integration time or static energy much in a velocity-saturated regime, it is advantageous to lower V_{DD} a little and save more in dynamic energy. Thus, the optimal V_{DD} for the same α_n is lower in Figure 21.4 (d) than in Figure 21.4 (b).

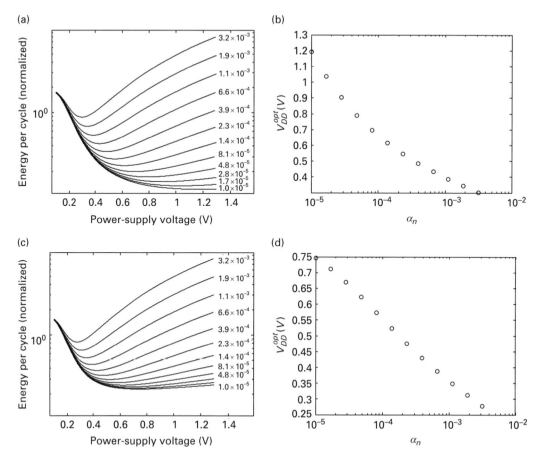

Figure 21.4a, b, c, d. (a) Variation of energy efficiency with power-supply voltage in moderate and strong inversion. (b) Variation of the optimal power-supply voltage versus normalized activity factor in moderate and strong inversion. (c) The same as in (a) but including velocity-saturation effects. (d) The same as in (b) but including velocity saturation effects.

For simplicity, we shall, unless otherwise mentioned, use Equation (21.36) with traditional above-threshold operation in the discussion that follows.

Since V_{T0} does not cancel itself out in Equation (21.36) or in Equation (21.37), in all above-threshold regimes, V_{DD}^{opt} is threshold-voltage dependent. Is there a threshold voltage V_{T0}^{opt} at which the pair $(V_{DD}^{opt}, V_{T0}^{opt})$ leads to a global minimum in the E_{TOT} of Equation (21.36) across all regimes of operation? To minimize E_{TOT} in Equation (21.36), we need to solve a 2D optimization problem with E_{TOT} being a function of (V_{DD}, V_{T0}) and α_n being a parameter that characterizes this function. At first glance, such an optimization may seem trivial: we set $V_{T0} = \infty$ ($\gg V_{DD}$ is fine) in Equation (21.36), which sets I_{OFF} to zero, and operate infinitely slowly in the subthreshold regime with I_{ON} proportional to $e^{\kappa(V_{DD}^{opt} - V_{T0})/\phi_t}$ and V_{DD}^{opt} at its subthreshold value. In an experimental transistor, however, even with $V_{T0} = \infty$, the off current I_{OFF} does not go to zero. At large threshold voltages V_{T0}, the leakage current I_{OFF} is reduced to a sufficiently low value such that other forms of leakage, e.g., diode junction leakage, become important. If we assume that this leakage current is approximately constant at I_{leak} with I_{leak} parametrized by V_{TC} according to

$$I_{leak} \overset{\Delta}{=} I_{T0} e^{-\kappa V_{TC}/\phi_t} \tag{21.38}$$

then we may write

$$I_{OFF} = I_{T0} e^{-\kappa V_{T0}/\phi_t} + I_{leak}$$
$$I_{OFF} = I_{T0} \left(e^{-\kappa V_{T0}/\phi_t} + e^{-\kappa V_{TC}/\phi_t} \right) \tag{21.39}$$

Substitution of Equation (21.39) in Equation (21.28) then leads to a more accurate form of Equation (21.36):

$$\boxed{E_{TOT} = (\beta N_{TOT} N_{LD}) C_L V_{DD}^2 \left(\alpha_n + \frac{e^{-\kappa V_{T0}/\phi_t} + e^{-\kappa V_{TC}/\phi_t}}{\ln^2 (1 + e^{\kappa(V_{DD}-V_{T0})/(2\phi_t)})} \right)} \tag{21.40}$$

In Equation (21.40), when $V_{T0} > V_{TC}$ by a few ϕ_t such that diode-junction leakage dominates and the α_n dynamic-energy term is very small, E_{TOT} is actually minimized in the above-threshold regime with a large V_{DD}. A large f_{clk}, achieved by setting $V_{DD} \gg V_{T0}$ in above-threshold operation, causes integration of the constant diode-junction current over a small $1/f_{clk}$ time period allowing the static energy term to be small; at the optimum, the static-energy term is then comparable to the dynamic-energy term α_n, thus balancing the two energies, and therefore minimizing the total energy E_{TOT}. Without the $e^{-\kappa V_{TC}/\phi_t}$ term in Equation (21.40), we would reach the incorrect conclusion that subthreshold operation with large V_{T0} and large V_{DD} minimizes the overall energy for very small α_n. If α_n is relatively large, Equation (21.40) predicts that the overall energy is minimized when the static-energy term approximately balances the dynamic-energy α_n term; however, the leak current and the integration time in the static-energy term can now be greater such that a lower V_{T0}, lower V_{DD}, and slower subthreshold operation lead to overall energy minimization. In general, for any α_n and V_{TC} pair of parameters, there is an optimal $(V_{DD}^{opt}, V_{T0}^{opt})$ pair.

If we set $\alpha_n = 1.4 \times 10^{-3}$ and $V_{TC} = 0.6$ V in Equation (21.40) with $\kappa = 0.625$, and plot $E_{TOTn} = E_{TOT}/(\beta N_{TOT} N_{LD} C_L \phi_t^2)$ versus V_{DD} and V_{T0}, we get the plot of Figure 21.5 (a). For this particular set of parameters, there is a global minimum at $(V_{DD}^{opt}, V_{T0}^{opt}) = (0.31\,\text{V}, 0.46\,\text{V})$. Figure 21.5 (a) shows that, since this global minimum is in subthreshold operation, the energy is relatively invariant to variations in V_{T0} as Equation (21.30) predicts. The similarity of the cross-sectional shapes in Figure 21.5 (a) to the subthreshold curves of Figure 21.3 (a) and to the above-threshold curves of Figure 21.4 (a) is evident. For $\alpha_n = 1.0 \times 10^{-5}$ and $V_{TC} = 0.47$ V, in a relatively leaky process, the global energy minimum is at $(V_{DD}^{opt}, V_{T0}^{opt}) = (0.67\,\text{V}, 0.46\,\text{V})$ in above-threshold operation. Figure 21.5 (b) shows that the above-threshold energy is more sensitive to the threshold voltage as

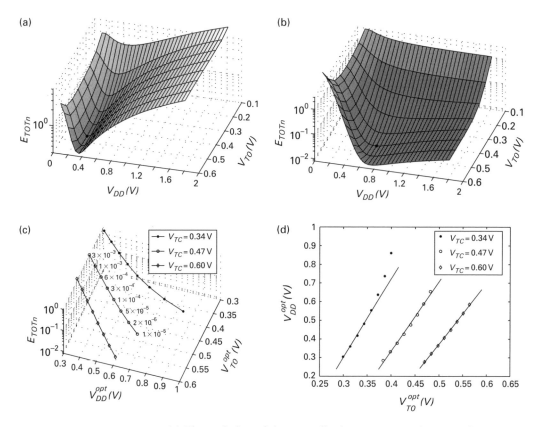

Figure 21.5a, b, c, d. (a) The variation of the normalized energy per cycle versus the power-supply voltage and threshold voltage for relatively high activity factor and relatively low non-subthreshold leakage in all regimes of operation. (b) The variation of energy per cycle versus the power-supply voltage and threshold voltage for relatively low activity factor and high non-subthreshold leakage in all regimes of operation. (c) The normalized energy per cycle versus the optimal power-supply and threshold voltages. (d) The variation of the optimal power-supply voltage versus the optimal threshold voltage at three different levels of non-subthreshold leakage.

expected. Figure 21.5 (c) plots V_{DD}^{opt} and V_{T0}^{opt} for various α_n and three different values of V_{TC} marked in the legend. Not surprisingly, for higher V_{TC}, V_{T0}^{opt} is higher to keep subthreshold leakage comparable to the diode-junction (or other) leakage. Since dynamic and static energy are approximately balanced at the optimum, lower dynamic energy implies lower overall energy. Thus, the optimum energy is lower at lower α_n for all V_{TC} as shown in Figure 21.5 (c).

Figure 21.5 (d) shows that V_{DD}^{opt} and V_{T0}^{opt} track each other for all V_{TC} with higher V_{DD}^{opt} and higher V_{T0}^{opt} corresponding to lower α_n, and lower V_{DD}^{opt} and lower V_{T0}^{opt} corresponding to higher α_n. Such tracking arises because the static-energy term tracks the dynamic-energy α_n term at the optimum, an example of which is seen in subthreshold Equation (21.32). Thus, lower values of α_n, which imply a lower dynamic energy, also lead to lower static energy. Lower static energy is achieved by raising V_{T0} to reduce the leakage current and concomitantly raising V_{DD} to reduce the leakage integration time $1/f_{clk}$; raising V_{DD} ensures that $(V_{DD} - V_{T0})$ is large enough to increase the on current in the denominator of Equation (21.40), thus lowering the leakage-integration time. Some reduction in static energy is achieved by leakage-current reduction via a V_{T0} increase and some reduction in static energy is achieved by integration-time reduction through a V_{DD} increase. The concomitant movement of V_{DD}^{opt} with V_{T0}^{opt} minimizes the energy increase of the common static-and-dynamic V_{DD}^2 prefactor of Equation (21.40) by requiring V_{DD} to not rise as much. Thus, the overall energy is minimized when V_{DD}^{opt} and V_{T0}^{opt} track each other. Figure 21.5 (d) shows that straight-line fits to the V_{DD}^{opt}-versus-V_{T0}^{opt} data are reasonably good, especially for relatively high V_{TC}, the most common practical case.

How do we minimize energy if we need to maintain a certain f_{clk} for our application, which then constrains $V_{DD} - V_{T0}$? In subthreshold operation, we have already provided the answer to this question in Equations (21.33) and (21.34) and the paragraph in which they appear. Therefore, we will focus on answering the question for above-threshold operation. We shall use an above-threshold formulation based on the general EKV model. To preserve analytic tractability and insight, we shall just focus on simple square-law operation. Generalizing to other regimes with equations such as Equation (21.37) or Equation (21.40) is straightforward. Accurate simulations can always provide exact answers if needed. We shall also assume, for simplicity, that $V_{TC} \gg V_{T0}$ in this regime such that subthreshold leakage is the only leakage mechanism.

With f_{clk} fixed, we have a constrained optimization problem rather than an unconstrained one:

Minimize

$$\left(\beta N_{TOT} N_{LD} C_L V_{DD}^2\right)\left(\alpha_n + \frac{I_{OFF}}{I_{ON}}\right)$$

subject to the constraint

$$f_{clk} N_{LD} C_L V_{DD} = I_{ON} = I_{T0}\left(\frac{\kappa}{2}\left(\frac{V_{DD} - V_{T0}}{\phi_t}\right)^2\right) \tag{21.41}$$

with

$$I_{T0} = 2\mu C_{ox} \frac{W}{L} \frac{\phi_t^2}{\kappa}$$

We can first use the constraint in Equation (21.41) to solve for the threshold voltage as a function of V_{DD}:

$$V_{T0} = V_{DD} - \frac{2\phi_t}{\kappa}\sqrt{\left(\frac{f_{clk}}{f_{T0}}\right)\left(\frac{V_{DD}}{\phi_t}\right)}$$

$$f_{T0} = \frac{I_{T0}}{N_{LD} C_L \phi_t} \tag{21.42}$$

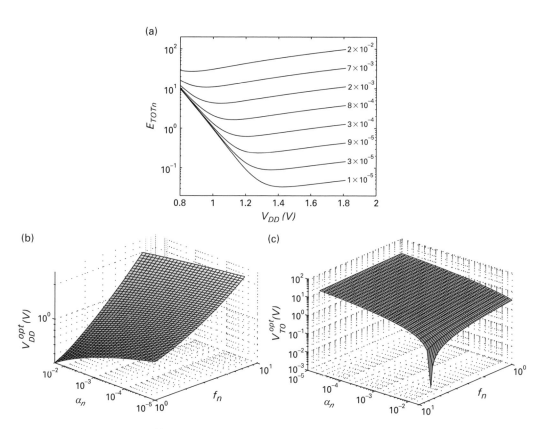

Figure 21.6a, b, c. (a) Energy per cycle versus the power-supply voltage for a fixed constrained frequency of operation f_n at various α_n in above-threshold operation. (b) The variation of the optimal power-supply voltage versus normalized activity factor α_n and normalized frequency of operation f_n. (c) The variation of the optimal threshold voltage versus α_n and f_n.

The normalizing frequency f_{T0} is proportional to the maximum clock frequency possible just at the border between subthreshold and above-threshold operation, i.e., $f_{clk,max}$ in Equation (21.34). Substitution of the value of V_{T0} given by Equation (21.42) to determine I_{ON} in Equation (21.41) and substitution of the constrained value of I_{ON} yields a minimization problem in E_{TOT} with V_{DD} as a variable and α_n and f_{clk} as parameters:

$$E_{TOT} = \left(\beta N_{TOT} N_{LD} C_L \phi_t^2\right) \left(\frac{V_{DD}^2}{\phi_t^2}\right) \left(\alpha_n + \frac{e^{-\kappa V_{DD}/\phi_t} e^{+2\sqrt{\frac{f_{clk}}{f_{T0}}\frac{V_{DD}}{\phi_t}}}}{\frac{f_{clk}}{f_{T0}}\frac{V_{DD}}{\phi_t}}\right) \tag{21.43}$$

While Equation (21.43) cannot be solved in closed form, Figure 21.6 (a) shows a numerical MATLAB plot of $E_{TOTn} = E_{TOT}/\left(\beta N_{TOT} N_{LD} C_L \phi_t^2\right)$ versus V_{DD} for various α_n and $f_n = f_{clk}/f_{T0} = 3$. We observe that V_{DD}^{opt} increases as α_n decreases to maintain the balance between dynamic energy and static energy as in Figure 21.4 (a). Figure 21.6 (b) plots V_{DD}^{opt} for various values of α_n and f_n. We observe that V_{DD}^{opt}

also increases with f_n as expected. Figure 21.6 (c) plots V_{T0}^{opt} computed from Equation (21.42) with $V_{DD} = V_{DD}^{opt}$. We observe that V_{T0}^{opt} must decrease as α_n and f_n increase. To prevent V_{T0}^{opt} from going below 0 V, Equation (21.42) requires that

$$V_{DD} > \frac{4\phi_t f_{clk}}{\kappa^2} \frac{f_{clk}}{f_{T0}} \qquad (21.44)$$

21.7   Varying the power-supply voltage and threshold voltage

From the discussion so far, it must be clear that adaptively increasing V_{DD} to meet the increasing f_{clk} or decreasing activity factor (α) needs of a computation is beneficial; such a strategy is termed dynamic voltage scaling (DVS) [10]. Similarly, adaptively increasing V_{T0} to meet decreasing f_{clk}, decreasing activity factor, or increasing standby requirements is beneficial; such a strategy is termed dynamic threshold voltage scaling (DTVS) [11]. Such strategies are often implemented by having the V_{DD} or V_{T0} of a critical-path replica-biasing circuit at a corner of the chip servo to a desirable input to set V_{DD} or V_{T0} (via body biasing). Biasing the body of both NMOS and PMOS transistors requires a triple-well process and is not always an option. Charge pumps are used to lower body biases below ground or above V_{DD} to increase V_{T0} and reduce leakage. Dc-dc converters are used for efficient conversion between a battery-based V_{DD} and the needed V_{DD}. Feedback schemes with body biasing to match NMOS and PMOS characteristics have been proposed in [8]. Forward body biasing to reduce thresholds is possible as long as the forward body bias is relatively modest and does not significantly increase diode-junction leakage. Similarly, reverse body biasing needs to be done with care to not exacerbate band-to-band-tunneling at the drain junction.

Strategies with dual-threshold voltages, a high V_{T0} for non-critical-path transistors, and a low V_{T0} for critical-path transistors are advantageous even if there is no adaptation. Sleep transistors with higher V_{T0} that are in series with a big computing block accomplish power gating with relatively low standby power and are now ubiquitous. With increasingly more attention paid to matching, increasingly more use of calibration and adaptation loops, increasingly larger propensity for error, digital design is starting to look increasingly analog! In modern 65 nm processes, several energy-efficient systems report optimal V_{DD}s that are near 0.5 V [7], [8] because of the need to keep leakage power in check and the need to be robust to transistor variation. Clock speeds in these designs are $\sim$1 MHz.

21.8   Gated clocks

Gated clocks use control signals that gate or turn them off when they are not needed. For example, if the partial result in a computation, C, ensures that circuitry beyond a certain point will not be needed to compute the final answer of the computation, then we can turn off the clock input to registers that provide inputs to this redundant circuitry. The register inputs to such circuitry then hold their values, do not switch, annulling all switching activity in the circuitry. The gating not only saves dynamic power by making some of the effective switching capacitance due to the clock conditional, but also reduces dynamic power in all the circuits that do not needlessly switch. In logic terms, if C is the control input, ϕ is the ungated clock, and ϕ_g is the gated clock, then

$$\phi_g = \phi.C \tag{21.45}$$

As an example, suppose C is the result of a comparison between the most significant bits (MSBs) in an "$A = B$?" N-bit logic circuit. If $C = 0$, comparisons on all other bits are needless. A gated clock based on C in Equation (21.45) can be generated and fed to registers that hold inputs for comparisons on all the other bits. If A and B are random, half the time there will be no switching activity in any of the other $(N-1)$ bits, and the overall dynamic power is reduced by $(N-1)/(2N)$, i.e., almost by a factor of a half.

21.9   Basics of adiabatic computing

Adiabatic computing attempts to minimize voltage drops across switching devices through very slow operation such that energy dissipation is minimized; the paradigm also recycles and stores switched capacitive node energy such that there is energy recovery. In this section, we shall first discuss the fundamental ideas behind this paradigm. In the next section, we shall provide examples of how it has been applied in practice to generate adiabatic clocks.

Figure 21.7 (a) illustrates the key idea behind adiabatic computing through the operation of a simple RC circuit. The key idea is that if we charge a capacitor through a resistor at an input angular frequency much less than the $1/(RC)$ corner frequency, the phase shift between V_{in} and V_{out} is small and the magnitudes of V_{in} and V_{out} are almost identical; thus, V_{in} and V_{out} track each other very closely; hence, the voltage across and the current through the resistor is always small, and there is little energy dissipation in the resistor. We shall now quantify how low this energy dissipation actually is.

The voltage $V_{out}(s)$ is related to the input $V_{in}(s)$ through the well-known lowpass characteristic

$$V_{out}(s) = \frac{V_{in}(s)}{1 + RCs} \tag{21.46}$$

Thus, the current through the resistor R is given by a high-pass characteristic

$$I_R(s) = \frac{V_{out}(s) - V_{in}(s)}{R}$$
$$I_R(s) = \frac{V_{in}(s)}{R} \frac{RCs}{RCs + 1} \tag{21.47}$$
$$I_R(s) = V_{in}(s) \frac{Cs}{RCs + 1}$$

(a)   (b)

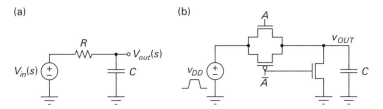

Figure 21.7a, b. (a) An RC circuit. (b) The fundamental canonical circuit of adiabatic computing.

The average power dissipated through the resistor R is given by

$$P_R(j\omega) = \frac{|I_R(j\omega)|^2 R}{2}$$

$$P_R(j\omega) = \left(\frac{|V_{in}(j\omega)|^2}{2R}\right)\left(\frac{\omega^2 (RC)^2}{1 + \omega^2 (RC)^2}\right) \tag{21.48}$$

If $\omega \ll RC$, we can ignore the ω^2 term in the denominator and find that the power dissipated is given by

$$P_R(j\omega) = \left(\frac{|V_{in}(j\omega)|^2}{2R}\right)\omega^2 (RC)^2 \tag{21.49}$$

The energy dissipated per charge-discharge cycle is the average power integrated over a period:

$$E_R(j\omega) = \left(\frac{|V_{in}(j\omega)|^2}{2R}\right)\omega^2 (RC.RC) \times \frac{2\pi}{\omega}$$

$$E_R(j\omega) = \left((2\pi)C\frac{|V_{in}(j\omega)|^2}{2}\right)(\omega RC)$$

$$E_C(j\omega) \equiv C\frac{|V_{in}(j\omega)|^2}{2} \tag{21.50}$$

$$Q \equiv \frac{1}{\omega RC}$$

$$\boxed{E_R(j\omega) = E_C(j\omega)\left(\frac{2\pi}{Q}\right)}$$

Thus, if the input angular frequency $\omega \ll 1/(RC)$, the $E_R(j\omega)$ energy dissipated per charge-discharge cycle is significantly less than the average reactive energy per cycle $E_C(j\omega)$. The parameter $1/(\omega RC) \equiv Q$ is the quality factor of this system, which is proportional to the ratio of the reactive energy to dissipated energy $E_C(j\omega)/E_R(j\omega)$; a high-$Q$ system dissipates little energy per cycle compared with the stored reactive energy. If $\omega \to 0$, $Q \to \infty$, the energy dissipation goes to zero and we have charged the capacitor adiabatically with no heat dissipated in the resistor.

The last line of Equation (21.47) with RCs nearly 0 predicts that the current flowing through the resistor is nearly CdV_{in}/dt and independent of the resistance, input voltage, or output voltage at all times. Intuitively, the current into the capacitor is the current flowing through the resistor, and V_{in} and V_{out} track each other closely such that $CdV_{out}/dt = CdV_{in}/dt$. Thus, in our adiabatic scenario, we have effectively created a time-varying current source that is controlled by the input's rate of change but independent of the input voltage, output voltage, or resistance. If we want to charge V_{out} from 0 to V_{DD} in a fixed amount of time T, we can ask: is the best waveform that dissipates the least energy a slow half sinusoid with a varying current CdV_{in}/dt as we have analyzed? Or is it a more complex waveform? It turns out that a constant current source with CdV_{in}/dt being a constant, i.e., a ramp waveform with $(t, V_{in}(t)) = (t, V_{DD}(t/T))$, $0 < t < T$, is optimally energy efficient as we now explain.

More formally, suppose we have a fixed amount of time T available to charge a capacitance C through a resistance R from its initial condition of 0 to a fixed voltage V. Then, variational calculus shows that the most energy-efficient method for charging C that minimizes dissipation in R is a constant current of value CV/T. The proof of this pleasingly simple result is brilliantly explained in *Feynman Lectures on Physics* in an analogous situation for mechanical systems [12]. We seek to minimize the expected value of a quantity proportional to the square of the current, i.e., the power dissipation I^2R over a time T. The minimum of the expected value of the square of a quantity $\overline{I^2}$ is the square of its mean value $(\overline{I})^2$. This minimum is achieved when there is no variance $\overline{(I - \overline{I})^2}$ about the mean. In our case, zero variance about a fixed $\overline{I} = CV/T$ mean implies that the current must be constant throughout, i.e., we need a current source! Thus, if we want to charge a capacitance from 0 to V_{DD} adiabatically, we must use a gentle ramp with a constant CdV_{in}/dt from 0 to V_{DD}. Once the ramp input reaches V_{DD}, we stay saturated at V_{DD}, at which point there is no voltage across R such that there is no dissipation. The ramp must be gentle enough, i.e., $T \gg RC$, to ensure that there is little voltage drop across R. In fact, the energy dissipation for a ramp input in an RC circuit is $(IR)(CV) = CV^2(RC/T)$. Thus, once again, as in the sinusoidal case of Equation (21.50), we must utilize inputs that are significantly slower than the inherent time constant of the circuit to be energy efficient.

Figure 21.7 (b) shows the charging of a capacitor via a pass-transistor switch with a low on-resistance R_{on}. If the value of $R_{on}C$ is much smaller than the rise time or fall time of the V_{DD} clock input, then v_{OUT} adiabatically follows this input if A is actively high, i.e., if A is held constant at a high logic value. If A is low, then V_{OUT} does not follow the V_{DD} clock input and is discharged to ground via the shunt NMOS transistor. The voltage v_{OUT} is held low at all times if A is low; v_{OUT} follows the v_{DD} clock if A is high. *Thus, v_{OUT} samples the value of A with the v_{DD} clock and is also powered by the clock.* Unlike a traditional static digital circuit, where signals always hold their value irrespective of whether they are sampled by the clock, the value of v_{OUT} can only be established when and if the v_{DD} sampling-and-powering clock arrives. Hence, the term *clock-powered circuits* is often used to describe such adiabatic circuits. Note that the input A serves to gate the v_{DD} clock onto v_{OUT} in an adiabatic fashion exactly implementing Equation (21.45) with $C = A$ and $\phi = v_{DD}$.

The circuit of Figure 21.7 (b) is a foundational circuit in adiabatic computing analogous to the CMOS inverter in traditional digital computing [13]. If the pass-transistor network is changed to a more complex network of switches, we can implement a more complex gated-logic clock. Adiabatic logic typically uses dual pass-transistor networks to generate v_{OUT} and $\overline{v_{OUT}}$ from V_{DD} such that these outputs can themselves be used as the A and $\overline{A}$ control inputs in a following adiabatic circuit. To ensure proper functionality in any adiabatic circuit, A and $\overline{A}$ must be held constant while the v_{DD} clock changes, an issue that significantly complicates timing and pipelining issues in adiabatic logic design. These issues have been solved in reversible adiabatic pipelines [14]. However, they require large overhead such that the improvement in CV^2 switching energy afforded by slow adiabatic operation is compromised by a large increase in C, a large increase in area, and slow performance. Hence adiabatic logic has had relatively little impact in energy-efficient logic design although complete microprocessors have been built [15]. However, adiabatic circuits have had large impact in reducing the switching energy of high-capacitance nodes like clocks in traditional digital circuits, as we now describe.

21.10   Adiabatic clocks

We now discuss how to generate the adiabatic power supply, i.e., the v_{DD} clock, and how to use it in practical digital circuits that themselves may or may not be adiabatic. Figure 21.8 is an example of a gated-clock circuit that implements Equation (21.45) with $\phi = V_{HOTCLK}$, $C = A$, and $\phi_g = V_{OUT}$. If A is high, the low-on-resistance PMOS switch M_1 causes V_{OUT} to adiabatically follow V_{HOTCLK} and $\overline{V_{OUT}}$ to be held at ground. If A is low, V_{OUT} is held at ground, and the low-on-resistance PMOS switch M_2 causes $\overline{V_{OUT}}$ to adiabatically follow V_{HOTCLK}. In either condition, the inductor L, a low-on-resistance PMOS switch with resistance R, and a capacitance C implement a high-Q LCR resonator that creates the sinusoidal waveform V_{HOTCLK}. The input V_{DIGCLK} periodically shorts and fluxes the inductor L to provide energy that compensates for energy lost in the switches and serves to maintain the amplitude of V_{HOTCLK}. If the resonant angular frequency of the LCR resonator and the frequency of V_{HOTCLK} is ω_0, then the Q of the system at $\omega_0 = 2\pi f_0$ is given from Equation (21.50) to be very nearly

$$Q = \frac{1}{\omega_0 RC} = \frac{\sqrt{LC}}{RC} = \frac{\sqrt{L/C}}{R} \tag{21.51}$$

Thus, in order to preserve good energy efficiency such that the $\left(f_0 C V_{DD}^2/2\right)$ reactive energy is $Q/(2\pi)$ times the dissipated energy, we need Q to be sufficiently large. Since C is typically fixed, this requirement implies that L must be sufficiently

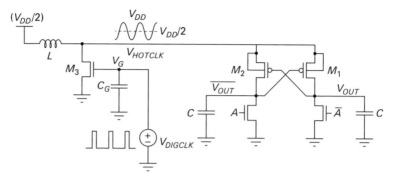

Figure 21.8. Generation of an adiabatic power clock.

large such that $\omega_0 = 1/(\sqrt{(LC)})$ is $\ll 1/(RC)$. So, once again, we find that we must operate slowly to be energy efficient. The inductor provides energy recovery and storage in each charge-discharge cycle of V_{OUT} such that reactive capacitive energy that is not dissipated is converted to inductive energy and then back to capacitive energy for later use. For a fixed R and fixed C, the energy savings per cycle scales like $1/\omega_0$ in Equation (21.51) such that the power savings scale like $1/\omega_0^2$ or linearly with size of the inductance. Equation (21.51) is not exact since the topology of Figure 21.8 functions like an L in parallel with a series RC rather than like a purely series resonator, but for Q greater than 4, the error is less than 7%.

In this particular example, if only V_{OUT} is used as a clock in the system and $\overline{V_{OUT}}$ is a 'dummy output' required to keep the resonator operating at resonance, gating the clock with A costs us $(4\pi/Q)(f_0 C V_{DD}^2/2)$ of energy per cycle. Gating the V_{OUT} clock not only saves switching energy in the clock capacitance but also in all the circuitry served by input registers that are clocked by V_{OUT}. Topologies with adiabatic clocking like that in Figure 21.8 have not only been used in digital systems but also in mixed-signal pattern-recognition systems to achieve excellent energy efficiency [16].

In the analysis of Figure 21.8 so far, we have neglected the switching energy of C_G, which can be significant if M_3 is to have a sufficiently large W/L, operate as a good switch, and have a low on-resistance. The solution is to resonate the capacitance C_G with an inductance as well to lower its dissipative switching energy. Since the drive to C_G is then sinusoidal, it may as well be derived from V_{OUT}. This insight leads to realization that the circuitry of Figure 21.8 is most efficient if *all* the switching and passive devices including the C loads are combined into one differential LC-oscillator topology with C_G sized to be as large as C and forming part of the load. Such a topology is discussed in [17]. In this view, the capacitive loads, the switches, and the energy-recovery inductances are all part of one oscillatory system with no clean separation between what is the drive and what is the load. A disadvantage of such a scheme is that the frequency of oscillation is free running and its phase is not controllable. Controllability of phase is important if several clocks with multiple phases are needed in a digital system.

Circuits and concepts in resonant adiabatic computing bear similarity to circuits and concepts in RF switching-power-amplifier design. For an example of such a design, see the Class-E switching power amplifier described in Chapter 19 in the context of a wireless recharging system for biomedical implants. Adiabatic circuits have been used to reduce power dissipation in other relatively low-speed high-capacitance circuits, e.g., in pad drivers, and in liquid-crystal display drivers [18], [19].

21.11 Architectures and algorithms for improving energy efficiency

We shall now briefly review some key ideas for improving energy efficiency including parallelism, pipelining, symmetry, ordering, and algorithmic improvements.

Parallelism uses a demultiplexer to redirect the high-bandwidth serial input of a computation to many low-throughput identical slow-and-parallel computing units with reduced power-supply voltages. The outputs of these parallel computing units are then sequentially sampled by a fast multiplexer to recreate a high-throughput serial output. Figure 21.9 illustrates the concept, which was proposed in [4]. Dynamic power is saved in a parallel architecture with N computing units because the power of each low-bandwidth parallel unit's power is significantly less than P_{srl}/N; P_{srl} corresponds to the power of a single computational unit operating at the high-throughput rate. However, dynamic power is increased in a parallel architecture with N units because the overhead and capacitance costs of multiplexing and demultiplexing rise with N. There is thus an optimal N at which the power savings are greatest. Parallelism is less useful in the subthreshold regime since the number of parallel units needed to maintain throughput increases exponentially with the reduction in V_{DD} while dynamic power savings are only polynomial in V_{DD} as per Equations (21.10), (21.14), and (21.27); furthermore, the increase in the number of computational units and slower frequency of operation per computational stage increases leakage energy in each computational unit as per Equation (21.28). Nevertheless, for intrinsically parallel power-constrained tasks, e.g., image processing, slow-and-parallel operation is important for low-power design in the digital domain. Slow-and-parallel operation is an extremely important power-saving principle in the analog domain as well. Chapter 19 provides an example for energy-efficient computation with slow-and-parallel operation in a bionic ear processor in the analog domain. Chapter 22 provides a discussion of this general principle in mixed-signal design. Chapters 23 and 24 illustrate the ubiquitous use of the slow-and-parallel principle for energy-efficient computation in mixed-signal biological and bio-inspired systems.

Pipelining allows one to increase throughput by decreasing the logic depth between computational stages that have interposed registers between them. Its operation is similar to the increase in factory productivity obtained by an assembly-line chain of workers with each worker having to do less such that he/she can

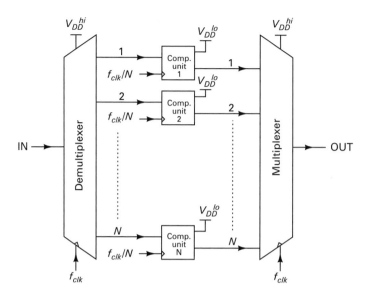

Figure 21.9. An architecture that illustrates the basic ideas behind low-power parallel computing.

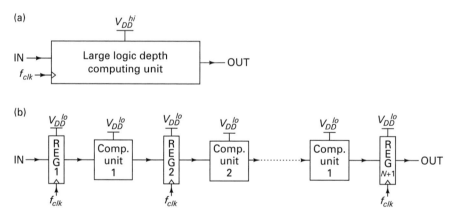

Figure 21.10a, b. (a) A large logic depth computing unit. (b) A pipelined version of the same computing unit that is more energy efficient.

do it faster. Rather than increasing speed, however, if the decrease in logic depth between pipelined stages is used to lower power-supply voltage and maintain throughput, there is more energy-efficient operation. Figures 21.10 (a) and (b) illustrate the concept. There is, however, in a fashion analogous to parallelism, an optimal amount of logic depth between registers. If there is too much pipelining, i.e., there are frequent registers between computational stages of small logic depth, the register power dissipation increases annulling the power savings in the computational stages. Pipelining is more amenable as an architectural low-power technique than parallelism in subthreshold design: the reduced power-supply voltage decreases dynamic power and static power, and the decrease in logic depth also decreases static power, all per Equation (21.29).

The **symmetry** of a computational architecture can significantly affect its energy efficiency. For example, the fast Fourier transform (FFT) exploits symmetry in the basis vectors behind its spectral-analysis computation to create an efficient architecture whose speed and power dissipation scale as $O(N \log_2 N)$ where N is the number of points in the FFT. Symmetry allows factorization and simplification of common operations in the computation such that they may be done more efficiently. Symmetric architectures are easy to lay out, have the best matching, balance critical paths and minimize glitching power, usually have the least wiring capacitance, and thus minimize power and delay. However, as we explained earlier in the example of a 4-input OR gate, they may not optimally exploit the multiplicative reduction in static switching probability that arises in an asymmetric chain-like architecture.

The **order** in which computations are done in an architecture impacts the overall energy efficiency of a system. In general, computations that tend to reduce the information bandwidth of the data to be processed should occur as early as possible, e.g., gain-control computations that reduce dynamic range in an audio input signal, demodulation operations that reduce data bandwidth in a software-defined radio signal, spatial filtering that allows image information to be encoded in a compressed representation etc. This power-saving principle is true in both analog and digital design.

Above all, in low-power digital design, the most important power-saving principle is to compute with an **intelligent-and-efficient algorithm** with an inherently low operation count. For example, algorithms with operations organized in a conditional tree may lead to only a logarithmic number of computing operations. Algorithms that factorize common operations, such as the FFT, are efficient. Clever signal processing in the analog domain is also essential to lowering power. For example, in Chapter 23, we shall see how an ear-inspired RF cochlea does very efficient RF spectrum analysis through the use of an exponentially tapered transmission line to achieve a better tradeoff between analysis time and hardware/power than any other known digital or analog strategy.

References

[1] R. M. Swanson and J. D. Meindl. Ion-implanted complementary MOS transistors in low-voltage circuits. *IEEE Journal of Solid-State Circuits*, 7 (1972), 146–153.

[2] J. Burr and A. M. Peterson, Ultra low power CMOS technology. *Proceedings of the 3rd NASA Symposium on VLSI Design*, Moscow, Idaho, 4.2.11–14.12.13, 1991.

[3] E. Vittoz. Weak Inversion for Ultimate Low-Power Logic. In *Low-power Electronics Design*, ed. C. Piguet. (Boca Raton: CRC Press; 2005), pp. 16–11–16–18.

[4] A. P. Chandrakasan and R. W. Brodersen. Minimizing power consumption in digital CMOS circuits. *Proceedings of the IEEE*, **83** (1995), 498–523.

[5] Christian Piguet. *Low-power Electronics Design* (Boca Raton: CRC Press, 2005).

[6] B. Zhai, L. Nazhandali, J. Olson, A. Reeves, M. Minuth, R. Helfand, Pant Sanjay, D. Blaauw and T. Austin. A 2.60pJ/Inst Subthreshold Sensor Processor for Optimal Energy Efficiency. *Proceedings of the IEEE Symposium on VLSI Circuits*, Honolulu, 154–155, 2006.

[7] J. Kwong, Y. K. Ramadass, N. Verma and A. P. Chandrakasan. A 65 nm Sub-V_t, microcontroller with integrated SRAM and switched capacitor DC-DC converter. *IEEE Journal of Solid-State Circuits*, **44** (2009), 115–126.

[8] Y. Pu, J. P. de Gyvez, H. Corporaal and Y. Ha. An ultra-low-energy/frame multi-standard JPEG co-processor in 65nm CMOS with sub/near-threshold power supply. *Proceedings of the IEEE Solid-State Circuits Conference (ISSCC)*, San Francisco, 146–147,147a, 2009.

[9] V. Ekanayake, C. Kelly IV and R. Manohar. An ultra low-power processor for sensor networks. *Proceedings of the 11th International Conference on Architectural Support for Programming Languages and Operating Systems (ASPLOS)*, Boston, Mass., 27–36, 2004.

[10] S. Lee and T. Sakurai. Run-time voltage hopping for low-power real-time systems. *Proceedings of the 37th Annual IEEE ACM Design Automation Conference*, Los Angeles, CA, 806–809, 2000.

[11] C. H. Kim and K. Roy. Dynamic V_{th} scaling scheme for active leakage power reduction. *Proceedings of the Design, Automation and Test in Europe Conference and Exhibition*, Paris, 163–167, 2002.

[12] R. P. Feynman, M. L. Sands and R. B. Leighton. Lecture 19: The principle of least action. In *The Feynman Lectures on Physics: Commemorative Issue, Vol. 2*, ed. (Reading, MA: Addison Wesley; 1989), pp. 19.11–19.14.

[13] L. Svensson. Adiabatic and Clock-Powered Circuits. In *Low-Power Electronics Design*, ed. C Piguet. (Boca Raton: CRC Press; 2005), pp. 15.11–15.15.

[14] W. C. Athas, L. J. Svensson, J. G. Koller, N. Tzartzanis and E. Y.-C. Chou. Low-power digital systems based on adiabatic-switching principles. *IEEE Transactions on Very Large Scale Integration (VLSI) Systems*, **2** (1994), 398–407.

[15] W. Athas, N. Tzartzanis, W. Mao, L. Peterson, R. Lal, K. Chong, J-S. Moon, L. Svensson and M. Bolotski. The design and implementation of a low-power clock-powered microprocessor. *IEEE Journal of Solid-State Circuits*, **35** (2000), 1561–1570.

[16] R. Karakiewicz, R. Genov and G. Cauwenberghs. 1.1 TMACS/mW Load-Balanced Resonant Charge-Recycling Array Processor. *Proceedings of the IEEE Custom Integrated Circuits Conference (CICC)*, San Jose, California, 603–606, 2007.

[17] W. C. Athas, L. J. Svensson and N. Tzartzanis. A resonant signal driver for two-phase, almost-non-overlapping clocks. *Proceedings of the IEEE International Symposium on Circuits and Systems (ISCAS)* Atlanta, GA, 129–132, 1996.

[18] L. Svensson, W. C. Athas and R. S. C. Wen. A sub-CV^2 pad driver with 10 ns transition time. *Proceedings of the International Symposium on Low Power Electronics and Design*, Monterey, California, 105–108, 1996.

[19] J. Ammer, M. Bolotski, P. Alvelda and T. F. Knight, Jr. A 160 × 120 pixel liquid-crystal-on-silicon microdisplay with an adiabatic DACM. *Proceedings of the IEEE International Solid-State Circuits Conference (ISSCC)*, San Francisco, CA, 212–213, 1999.

22 Principles for ultra-low-power analog and mixed-signal design

If you shut the door to all errors, truth will be shut out.

Rabindranath Tagore

In this chapter, we present ten general principles for ultra-low-power analog and mixed-signal design. We shall begin by comparing the paradigms of analog computation and digital computation intuitively and then quantitatively. The quantitative analysis will be based on fundamental relationships that dictate the reduction of noise and offset with the use of power, area, and time resources in any computation. It shall reveal important tradeoffs in how the power and area resources needed for a computation scale with the precision of computation in analog versus digital systems. From these results, we shall discuss why, from power considerations, there is an optimum amount of analog preprocessing that must be performed before a signal is digitized. If digitization is performed early and at high speed and high precision, as is often the case, the power costs of analog-to-digital conversion and digital processing become large; if digitization is performed too late, the costs of maintaining precision in the analog preprocessing become large; at the optimum, there is a balance between the two forms of processing that minimizes power.

There are detailed similarities between power-saving principles in analog and digital paradigms because they are both concerned with how to represent, process, and transform information with low levels of energy. We shall itemize and discuss several of these similarities. We shall derive from Shannon's theorem on information theory that there is a lower bound on the energy needed to process a bit of information, $kT \ln(2)$, irrespective of whether the information is represented and processed in an analog fashion or in a digital fashion. We shall present five considerations that determine power in all systems, namely task, technology, topology, speed, and precision.

Our analysis suggests that the optimal method for designing the ultra-low-power systems of the future involves collective analog and hybrid computation. Such computation is a hybrid mix of the paradigms of analog and digital computation and is one of the secrets behind the awe-inspiring energy efficiency of biological systems. Chapter 23 dwells on systems involving neuronal computation and Chapter 24 dwells on systems involving cellular computation in some depth.

These systems embody some of the general principles of collective analog and hybrid computation outlined in this chapter. Here, we shall just outline the key ideas and an example that illustrates how such systems can be architected in principle.

A general trend in such bio-inspired systems and in highly energy-efficient systems built by engineers with no knowledge of biology is the increasing presence of feedback interactions between analog and digital portions of a mixed-signal system to improve energy efficiency. Such feedback is already implicitly present in well-known ADC architectures such as sigma-delta and successive-approximation converters, and in ADCs with digital error correction and calibration, all of which represent highly energy-efficient ADC topologies (see Chapter 15). We shall show that such architectures represent a special case of a more general architecture termed a *hybrid state machine* (HSM), a generalization of finite state machines from the digital domain to the hybrid analog-digital domain. Indeed, Chapter 15 on low-power analog-to-digital conversion has already provided an example of such an HSM to create an ADC. Here, we shall show why such mixed-signal feedback architectures are likely to be increasingly important in the future of low-power design and how they may be used to architect general digitally program-mable analog systems and hybrid cellular automata of high energy efficiency in the future.

We discuss how to maintain robustness and flexibility in low-power systems and how robustness and flexibility trade with efficiency. We shall find that feedback and learning provide functions in analog systems analogous to error correction and data compression in digital systems: thus feedback is extremely important in providing a better robustness-efficiency tradeoff in analog systems just as error correction, redundancy, and compression do in digital systems.

We shall outline ten principles for architecting low-power systems, whether analog, digital, or mixed-signal. The digital manifestation of some of these prin-ciples has been discussed in Chapter 21 on ultra-low-power digital design. The analog application of these principles occurs throughout Chapters 11 to 20. We show how general-purpose energy-inefficient systems need to evolve to more energy-efficient special-purpose systems as time and learning allow them to acquire knowledge about their environments.

Applications of many of these principles in concrete contexts can be seen in the biomedical systems described in Chapters 16, 17, 18, 19, and 20. In many highly miniature implantable and noninvasive biomedical systems, the sensing, wireless, power-management, and actuation/stimulation power consumption dominate the power budget of the system. Thus, the costs of analog-to-digital conversion and digital processing can be largely irrelevant. In such cases, the principles for low-power design articulated in this chapter are still applicable to just the analog, RF, and power-management portions of such systems, and the latter chapters have actually applied them in a concrete way. In such systems, ultra-low-power digital processing as discussed in Chapter 21 may still be helpful in improving the energy efficiency of the dominant analog systems via feedback, calibration, learning,

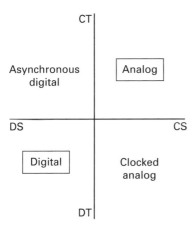

Figure 22.1. The four types of mixed-signal systems. Reprinted with kind permission from [11].

data compression, and mismatch-compensation functions. We discuss how sensors and actuators are also analog processing systems that automatically fit into the framework for low-power mixed-signal design discussed in this chapter. However, their physical state variables are not necessarily electrical.

22.1   Power consumption in analog and digital systems

Figure 22.1 shows that systems can be continuous or discrete in their signal levels (CS or DS), and operate in continuous time or discrete time (CT or DT). Thus, there are four kinds of systems corresponding to each of the four quadrants of Figure 22.1. Examples of systems in the first analog quadrant include continuous-time filters or the retina in the eye. Examples of systems in the second asynchronous digital quadrant include communication systems and pulsatile neuronal systems. Examples of systems in the third digital quadrant include microprocessors and digital signal processors. Examples of systems in the clocked analog quadrant include switched-capacitor filters and synchronous analog-to-digital converters. For simplicity, we will begin by studying the boxed analog and digital systems in the first and third quadrants of Figure 22.1, respectively. Such systems are purely analog, being continuous in both signal levels and time, or purely digital, being discrete in both signal levels and time. Systems in the first and third quadrants represent extremes of the two forms of computation and thus represent a good place to begin; systems in the second and fourth quadrants can then be understood by extrapolation from these extremes. Later, we shall discuss examples of hybrid systems that operate with continuous and discrete signals, and in continuous and discrete time.

We begin with an intuitive comparison of the pros and cons of analog versus digital computation.

Table 22.1 Comparison of analog and digital computation

Analog	Digital
1. Compute on a continuous set, e.g., [0,1], [0, V_{DD}].	1. Compute on a discrete set, e.g., {0,1}, {0, V_{DD}}.
2. The basis functions for computation arise from the *physics* of the computing devices: current-voltage curves of NFETs, PFETs, capacitors, resistors, floating-gate devices, KVL, KCL, etc. The amount of computation squeezed out of a single transistor or device is high.	2. The basis functions for computation arise from the *mathematics* of Boolean logic: logical relations like AND, OR, NOT, NAND, XOR, etc. The transistor is used as a switch, and the amount of computation squeezed out of a single transistor is low.
3. One wire represents many bits of information.	3. One wire represents one bit of information.
4. Computation is offset-prone since it is sensitive to the parameters of the physical devices.	4. Computation is not offset-prone since it is relatively insensitive to the parameters of the physical devices.
5. Noise due to thermal fluctuations in physical devices.	5. Noise due to round-off error and temporal aliasing.
6. Signal not restored at each stage of the computation.	6. Signal restored at each stage of the computation.
7. In a cascade of analog stages, noise starts to accumulate and build up.	7. Round-off error does not accumulate significantly for many computations.
8. Not easily programmable.	8. Easily programmable.
EFFICIENT	**ROBUST**

The intuitive comparison in Table 22.1 suggests that the large number of degrees of freedom exploited by analog computation in each device make it efficient. However, they also make it sensitive to errors in these degrees of freedom. In contrast, the few degrees of freedom exploited by digital computation in each device cause it to be less efficient but the loss of many degrees of freedom is traded for high levels of robustness. Furthermore, there is no general-purpose scalable signal-restoration paradigm in analog computation since we do not know where to restore the signal to in general. Therefore, unlike an arbitrarily complex digital system, an arbitrarily complex purely analog system will always accumulate enough noise such that its function is eventually compromised. To quantify the intuition of the table, we must begin by understanding how noise in devices and noise accumulation affects the precision of analog computation.

22.1.1   Noise in MOS transistors

The input-referred noise voltage $\overline{v_n^2}$ in a transistor with transconductance g_m, width W, length L, oxide capacitance C_{ox} over an operating frequency range of (f_l, f_h) at room temperature T due to thermal noise and $1/f$ noise (or offset) is given by

$$\overline{v_n^2} = \int_{f_l}^{f_h} \left(\frac{4\gamma\,kT}{g_m}\right) df + \left(\frac{B_{imp}}{C_{ox}^2 WL}\right) \frac{df}{f} \tag{22.1}$$

where γ is a noise factor, and B_{imp} is the trap-impurity density, a determinant of the transistor's threshold-voltage offset and its $1/f$ noise. Chapters 7 and 8 provide an in-depth discussion of noise in a transistor. The g_m versus I relationship in a saturated transistor is given by

$$g_m = I^{1.0}\left(\frac{\kappa q}{kT}\right) \longleftarrow \text{subthreshold}$$

$$g_m = I^{0.5}\sqrt{2\mu\kappa C_{ox}\frac{W}{L}} \longleftarrow \text{above-threshold}$$

(22.2)

where I is the dc current flowing through the transistor. Thus, if we define

$$p = \text{exponent of } g_m \text{ versus } I \text{ relation}$$

$$K_w(0.5) = 4kT\left(\frac{\frac{2}{3}}{\sqrt{2\mu\kappa C_{ox}\frac{W}{L}}}\right)$$

$$K_w(1.0) = 4kT\left(\frac{kT}{2q\kappa^2}\right)$$

(22.3)

$$K_f = \frac{B_{imp}}{C_{ox}^2}$$

then we may write

$$\boxed{\overline{v_n^2} = \left(\frac{K_w(p)}{I^p}\right)\Delta f + \left(\frac{K_f}{A}\right)\ln\left(\frac{f_h}{f_l}\right)}$$

(22.4)

where $K_w(p)$ and K_f are technology-dependent constants defined in Equation (22.3), $A = WL$ the transistor's area, and $\Delta f = f_h - f_l$. Equation (22.4) tells us that to reduce input-referred thermal noise, we must increase dc power consumption in the transistor or operate more slowly (reduce Δf); to reduce $1/f$ noise and offset, we must increase the transistor's area. Thus Equation (22.4) is a device noise-resource equation that tells us exactly what the resource consumption of power and area must be in order to attain a given level of input-referred noise in a device. It's a 'no free lunch' equation that states that we must consume resources of time $(1/\Delta f)$ and/or power (I) to average over more numbers of electrons to reduce thermal noise; we must consume resources of area to average over more numbers of traps to reduce $1/f$ noise and offset. Since $K_w(0.5)$ in Equation (22.3) depends on W/L, we can increase transistor area in above-threshold operation to reduce thermal noise. Since $K_w(1.0)$ in Equation (22.3) does not depend on W/L, we cannot increase transistor area in subthreshold operation to reduce thermal noise.

If we compute slowly or adiabatically, $\Delta f \to 0, I \sim (\Delta f)^{1/p}$ to maintain a certain level of thermal noise, and therefore I also goes to zero, which is consistent with Chapter 21. Since $I \sim \left(\Delta f/\overline{v_n^2}\right)^{1/p}$, $p = 1$ in the subthreshold regime and $p = 0.5$ in the above-threshold regime, to attain a given bandwidth Δf at a fixed $\overline{v_n^2}$ or to attain a given $\overline{v_n^2}$ at a fixed Δf, the subthreshold regime consumes less power than the above-threshold regime. It is consequently the most energy-efficient region to operate the transistor in. In contrast, velocity-saturated operation has $0 < p < 0.5$, and is a highly energy-inefficient region to operate the transistor in. Increasing I causes little g_m increase, and consequently little improvement in speed or little reduction in noise; therefore, we shall not operate or analyze the transistor in velocity saturation since it is inefficient in both the analog and digital regimes of operation.

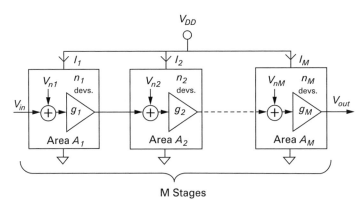

Figure 22.2. Noise accumulation in analog systems. Reprinted with kind permission from [11].

22.1.2   Noise accumulation in analog systems

Figure 22.2 shows an abstraction of a complex analog system with M computing stages. The ith stage has n_i devices, consumes an area A_i, consumes a bias current I_i, provides a gain of g_i to its input, and has an input-referred noise of v_{ni}. The input V_{in} is fed into the first stage and the output V_{out} is taken out of the Mth stage. We would like to minimize the total mean-squared noise at the output, $\overline{v_{no}^2}$, due to all the noise sources, subject to the constraint that the total current (power) and area are constrained to be fixed. That is,

Minimize $\overline{v_{no}^2}$ subject to the constraints

$$
\begin{aligned}
\sum_i I_i &= I_T \\
V_{DD} \sum_i I_i &= P_T \\
\sum_i A_i &= A_T
\end{aligned}
\tag{22.5}
$$

The solution is a simple exercise in Lagrange-multiplier minimization. If we define a set of weights $w_{i,P}$ and $w_{i,A}$ for power and area respectively, given by

$$
\begin{aligned}
G_i &= \prod_{k=i}^{k=M} g_k \\
w_{i,P} &= G_i^{2/(1+p)} n_i \\
w_{i,A} &= G_i n_i
\end{aligned}
\tag{22.6}
$$

then, the power and area distributions of current and area amongst the various computational stages is in proportion to these weights and given by

$$
\begin{aligned}
I_i &= \frac{w_{i,P} I_T}{\sum_i w_{i,P}} \\
A_i &= \frac{w_{i,A} A_T}{\sum_i w_{i,A}}
\end{aligned}
\tag{22.7}
$$

The total output noise at these optimized weights is given by

$$\overline{v_{no}^2} = \frac{\left(\sum_{i=1}^{i=M} w_{i,P}\right)^{(1+p)} K_w(p) \Delta f}{I_T^p} + \frac{\left(\sum_{i=1}^{i=M} w_{i,A}\right)^2 K_f \ln\left(\frac{f_h}{f_l}\right)}{A_T}$$

$$\overline{v_{no}^2} = \frac{V_{DD}^p \left(\sum_{i=1}^{i=M} w_{i,P}\right)^{(1+p)} K_w(p)}{P_T^p} \Delta f + \frac{\left(\sum_{i=1}^{i=M} w_{i,A}\right)^2 K_f}{A_T} \ln\left(\frac{f_h}{f_l}\right) \qquad (22.8)$$

$$\boxed{\overline{v_{no}^2} = \left(\frac{C_w}{P_T^p}\right) \Delta f + \left(\frac{C_f}{A_T}\right) \ln\left(\frac{f_h}{f_l}\right)}$$

Thus, the fundamental device noise-resource equation of Equation (22.4) leads to a system noise-resource equation in Equation (22.8) that is highly analogous to it. We find that the system output noise is reduced if we increase our total power consumption P_T and if we increase our total area consumption A_T. There is no free lunch at the device level and there is no free lunch at the system level either. The technology constants $K_w(p)$ and K_f in Equation (22.4) have now been combined with topological parameters to create system constants C_w and C_f in Equation (22.8) respectively. Interestingly, the optimal weighting of resources given in Equation (22.6) states that, to minimize output noise we should spend significantly more power and area resources in stages that are 1) more complex (n_i is large), or 2) in stages that have large accumulated gain from their noise input to the final output (G_i is large). Therefore, if all g_k are greater than 1 in an amplifying system, the input stage with the largest accumulated gain from input to output should consume the most power and the final stage with the least accumulated gain from input to output should consume the least power. Thus, the front-end preamplifier or first gain stage must always get a generous allocation of power resources in an amplifying system. In an attenuating system where all g_k are less than 1, it is the output stage that is most important in determining the overall noise such that it must get the largest fraction of the power.

22.1.3 The costs of SNR in analog systems

The maximal output signal-to-noise ratio (SNR) in the analog system of Figure 22.2 is given by

$$SNR = \frac{V_{DD}^2}{v_{no}^2} \qquad (22.9)$$

If we substitute for $\overline{v_{no}^2}$ in Equation (22.8) in terms of the SNR of Equation (22.9), and invert Equation (22.8), we can determine that

$$P_T = \left(\frac{C_w(SNR)(\Delta f)}{V_{DD}^2 - SNR(C_f \ln(f_h/f_l)/A_T)}\right)^{1/p}$$

$$A_T = \left(\frac{C_f(SNR)\ln(f_h/f_l)}{V_{DD}^2 - SNR(C_w(\Delta f)/P_T^p)}\right) \qquad (22.10)$$

Equation (22.10) contains a pair of resource-*SNR* equations: the first row of Equation (22.10) tells us how much power we must consume to obtain a given *SNR* for a fixed area consumption; the second row of Equation (22.10) tell us how much area we must consume to obtain a given *SNR* for a fixed power consumption. Thus, for the optimal analog system of Figure 22.2, the power costs for computing increase with increasing output *SNR* and with increasing output bandwidth Δf.

22.1.4   The costs of SNR in digital systems

In digital systems, the costs of computable functions rise polynomially with the number of bits of precision B used to encode the signal: for example, the power and area costs of addition increase linearly with the number of bits of precision; the power and area costs of multiplication increase quadratically with the number of bits of precision. For simplicity, we'll consider systems whose costs scale linearly in the number of bits like addition. Since $B = 1/2(\log_2(1 + SNR))$ from information theory,

$$
\begin{aligned}
P_T &= D_P(\Delta f)(\log_2(1 + SNR)) \\
A_T &= D_A(\log_2(1 + SNR))
\end{aligned}
\tag{22.11}
$$

where D_P and D_A are task, technology, and topology-dependent constants. In Chapter 21, we discuss how V_{DD}, the threshold voltage V_{T0}, the logic depth of the computation N_{LD}, the load capacitance per stage C_L, the average $W/L = \beta$, the total number of switching devices N_{TOT}, and the activity factor α quantitatively determine the total switching energy and leakage energy per cycle, E_{TOT} (see Equation (21.29) in Chapter 21). The power P_T is then simply $E_{TOT}f_{clk}$ or $E_{TOT}(\Delta f)$ in the language used in this chapter. However, we have now explicitly accounted for the fact that N_{TOT} is proportional to $\log_2(1 + SNR)$ if the costs of the computation scale linearly with the bit precision.

22.1.5   The analog-digital SNR-crossover curve

Since analog computation has a rich basis-function set of primitives to compute with, at low values of *SNR*, it is cheap to implement functions: for example, addition can be done by connecting two wires and exploiting Kirchhoff's current law; multiplication can be performed in a handful of transistors with the translinear principle (see Chapter 14) or using $Q = I \times t$ (see Chapter 15), where Q, I, and t are charge, current, and time; filtering is implemented cheaply with transconductors and capacitors (see Chapters 12, 13, and 14); and any polynomially nonlinear dynamical system is efficiently implemented with time-based or current-mode circuits (see section on computational ADCs in Chapter 15 and Chapter 24 on cytomorphic electronics). In contrast, since digital computation only has a logic basis set of primitives to compute with, even a simple one-bit addition with carry needs ~ 20 transistors; a multiplier consumes significant area; filtering needs accumulators, multipliers, and adders; and polynomially nonlinear dynamical systems with feedback, especially if stiff, can be expensive to simulate and implement.

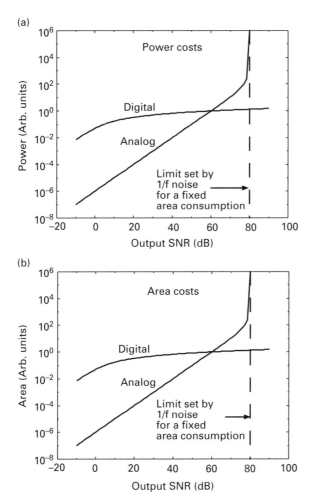

Figure 22.3a, b. Analog-digital crossover curves. Reprinted with kind permission from [11].

Thus, for several computations and typical parameter values corresponding to these computations, a plot of Equations (22.10) and (22.11) in Figures 22.3 (a) and 22.3 (b) reveals that the power and area costs of analog computation are significantly below those of digital computation at low and moderate values of *SNR* respectively.

At large values of *SNR*, maintaining such high *SNR* on a single analog wire requires very low levels of thermal noise and $1/f$ noise, which drastically increases power and area costs for analog computation. In contrast, in digital computation, several 1-bit precise simple logic units can collectively interact with each other to efficiently implement the computation. For example, in the computation of addition, the simple mechanism of a 1-bit carry between adjacent digital adder stages serves to define the overall collective interaction. Thus, the costs for power and area only scale polynomially in the number of bits or logarithmically in the *SNR* as Equation (22.11) reveals. Therefore, Figure 22.3 reveals that at high *SNR* it is significantly more efficient to compute in a digital representation. In fact, Figure 22.3 shows that the analog costs eventually become infinite: at some *SNR*, one is eventually limited by $1/f$ noise such that consuming more power to lower thermal noise does not help in Figure 22.3 (a); at some *SNR*, one is eventually limited by thermal noise such that consuming more area to lower $1/f$ noise does not help in Figure 22.3 (b). As a simple example, 8-bit-precise addition is relatively easy to implement in analog systems and can be significantly cheaper than 8-bit addition in digital systems; 16-bit precise addition in analog systems is extremely challenging but very easy to implement in digital systems; 32-bit precise addition is practically impossible to implement in analog systems but only twice as costly as 16-bit addition in digital systems.

The central take-home lesson from the analog-digital crossover curves of Figures 22.3 (a) and 22.3 (b) is that it pays to compute collectively: several 1-bit precise simple digital units that collectively interact with each other and encode information on several wires/channels will eventually outperform a supercomputing analog unit that tries to compute with and encode all the information on a single channel/wire. Intuitively, a society of digital individuals of 1-bit intelligence who interact with each other in a 1-bit language will collectively outperform an isolated analog genius at any task if the precision needed for the task is sufficiently high or if the complexity of the task is sufficiently high.

Where does complexity enter into the crossover curves of Figure 22.3? The values of C_w and C_f in Equation (22.10) increase more strongly with an increase in complexity, i.e., an increase in M or n_i in Figure 22.2, than do the values of D_P and D_A in Equation (22.11) for an equivalent digital system. The steeper growth in an analog system is due to the lack of signal restoration in each stage such that complex systems require significantly more resources than simple systems within each stage to maintain the output *SNR*. The quantitative scaling of C_w and C_f with M and n_i is given in Equation (22.6), Equation (22.7), and Equation (22.8).

Now that we have learned from digital computation how to compute at high *SNR*, can we use similar ideas in analog computation? Later in this chapter, we will suggest how collective analog systems of the future that function with several moderate-precision analog units, and which communicate with each other through digital or analog variables, can implement a computation at high *SNR* like digital systems. Collective analog systems are ubiquitous in highly energy-efficient biological systems; we shall provide some examples of bio-inspired systems in Chapters 23 and 24 that operate in a collective analog fashion including a radio-frequency silicon cochlea for cognitive radio and cell-inspired biochemical reaction networks for medical applications.

22.1.6   The importance of feedback, calibration, and learning in analog systems

Two factors ensure that analog computation is significantly more power efficient and area efficient than digital computation at low and moderate *SNR*: the rich basis function set and the use of noise-robust feedback topologies. If it were not for feedback, transistor areas would need to be rather large at moderate *SNR* to ensure low $1/f$ noise and offset in Equation (22.4). Such large transistors would increase capacitive circuit loads and increase power consumption to maintain bandwidth. However, a relatively low-cost slow and precise feedback system can attenuate offset and $1/f$ noise via periodic or background calibration (e.g., see the auto-zeroing feedback loops for the micropower log ADC in Chapter 19), such that transistor sizes do not need to be as large and the *SNR* of the overall system is primarily limited by thermal-noise considerations. The feedback system may introduce its own errors such as kT/C sampling errors, and also add its own cost, but these are typically small in good designs. The use of the feedback loop greatly improves the *SNR* of the system and its robustness for a slight increase in cost. In effect feedback helps shift the analog curve in Figure 22.3 to the right such that the analog-digital crossover occurs at higher *SNR*. Similarly, common-mode feedback loops help attenuate common-mode noise sources. Biasing circuits built with positive feedback (e.g., Figure 19.12 in Chapter 19) create references that ensure temperature insensitivity and power-supply-noise immunity. In biological systems, errors are often cancelled by slow learning feedback loops that compensate for these errors. For example, we are constantly adjusting motor parameters in our cerebellums to ensure fine motor movements. If the feedback loop is designed with an integrator in the feedback path, the use of feedback adaptation can serve to differentiate the input. Thus, we have a form of 'delta encoding or delta compression' where only changes in the input are enhanced and reported to future stages of processing, reducing their information bandwidth and power. Such strategies are frequently used in biological systems. Note that error correction does not necessarily have to be analog: for example, in Chapter 19, we show how digital calibration of analog errors optimizes analog system performance in a cochlear-implant (or bionic ear) processor.

22.1.7   Ultra-low-power systems of the future

As we have discussed in detail in Chapter 21, deep submicron digital design with subthreshold operation is somewhat like analog design in that significant attention has to be paid to transistor matching. As in analog design, efficiency is compromised to attain robustness: for example, V_{DD} must be sufficiently high to ensure good noise margins at all process corners in digital design, which degrades the power efficiency. Thus, the stark analog-versus-digital comparison is getting increasingly blurred with time: digital design is incorporating analog aspects and analog design is incorporating digital techniques for calibration and improvement of performance. Later in this chapter, we will see that there are several power-saving principles that are common to analog and digital design.

22.2   The low-power hand

The P_T equation for power in Equation (22.10) shows that there are five determinants of power in analog systems: 1) the task, which in our case is dictated by

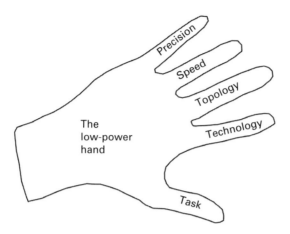

Figure 22.4. The low-power hand.

Figure 22.2; 2) the technology, parametrized by $K_w(p)$ and K_f in Equation (22.3) and V_{DD}, which in turn affect C_w and C_f in Equation (22.8) and thus P_T in Equation (22.10); 3) the *topology*, which is that of Figure 22.2 in our case, but with the optimal weighting of resources given by Equations (22.6) and (22.7), which in turn affect C_W and C_f in Equation (22.8) and thus P_T in Equation (22.10); 4) the speed, which is represented by the bandwidth Δf in Equation (22.10); and 5) the precision which is represented by the *SNR* in Equation (22.10). Every analog system has a power equation of this form with five factors as we first indicated in Chapter 12 (see Equation (12.29)). Now we have re-derived this result in a somewhat more general context. Chapter 13 provided an example in the context of active second-order filters, where the quality factor Q was an additional determinant of the power corresponding to this task, and certain topologies had superior power scaling with Q than others (see Equation (13.32)). Chapter 15 provided many topological examples in the context of the task of ADC conversion with the bandwidth parametrized by the sampling rate f_S and the precision parametrized by 2^{Nbits} in a non-thermal-noise-limited ADC or 2^{2Nbits} in a thermal-noise-limited ADC (see Equation (15.2)). Chapter 18 provided an example in the context of impedance-modulation telemetry where the scaling in power with *SNR* and Δf (called *B* in Chapter 18) followed a gentle one-third power law for strongly coupled RF links (see Equation (18.30)).

In fact the P_T term of Equation (22.11) for digital systems shows that power in digital systems also depends on the same five determinants: the paragraph right after Equation (22.11) show that the task, technology, and topology determine the D_P term, the clock speed determines the Δf term, and the precision determines the $\log_2(1 + SNR)$ term. From the discussion right before Equation (22.11), it must be remembered that the costs will, in general, be a polynomial function of $\log_2(1 + SNR)$ for computable functions.

Figure 22.4 shows a low-power hand with the five determinants of power itemized on it. A low-power designer must constantly work on how to optimize all the fingers within the system hand such that the overall system is low power: For example, she may be asking herself, in the fast technology that just became available, can I redo the task in a new parallel topology where the speed and precision per computing unit may not be so great so I can lower system power? Later, we shall show that the five determinants of power on the hand will naturally lead to some general principles for low-power design.

Table 22.2 summarizes scaling relationships between power, speed, and precision in various tasks, some of which have been implemented in different topologies and discussed in various chapters throughout the book. The technology factors that affect power are presented in the chapters to avoid cluttering the table with details that may distract the reader from the big picture. We have assumed subthreshold operation unless otherwise mentioned.

22.3   The optimum point for digitization in a mixed-signal system

In many biomedical applications, the meaningful information bandwidth at the output is far less than the information bandwidth at the input. For example, in cochlear-implant applications, 16-bit speech information sampled at 44 kHz yields 704 kb/s of raw input information; the 16-electrode output information corresponds to 5-bit current-amplitude information at a 200 Hz sample rate, i.e., 1.6 kb/s of actual discriminable information by a subject. In a paralysis prosthetic, 100 neuronal electrodes in the brain may provide 8-bit, 30 kHz data or 24 Mb/s of raw input information; the output information rate corresponds to 3 output motor-control parameters, updated at 100 Hz and 8-bit precision, i.e., 2.4 kb/s at best. In a pulse-oximetry application, the input photoplethysmographic information is about 16 kb/s while the meaningful output information about oxygen saturation may be a few bits per second. All three of these applications have been discussed in Chapters 19 and 20.

In such biomedical applications it is natural to wonder whether the lowest power method for processing the input information is the ubiquitous ADC-then-DSP architecture shown in Figure 22.5 (a). In this architecture, input from a sensor is transduced by an analog sensor circuit and immediately digitized at maximal bandwidth and maximal precision to yield lots of meaningless numbers that convey little information. The signal is then processed by a high-speed high-precision DSP that processes these numbers to yield meaningful low-bandwidth output information. From our prior discussion, we can conclude that the system is flexible but inefficient because high-bandwidth high-precision ADCs and DSPs are relatively power hungry. In Chapter 15 on low-power ADCs we discussed why high-precision, high-bandwidth ADCs are significantly harder to make energy efficient than low-precision, low-bandwidth ADCs. In Chapter 21, we discussed principles for low-power digital design, which show that high-bandwidth DSPs are significantly harder to make energy efficient than low-bandwidth DSPs.

Table 22.2 Scaling relationships between power, speed, and precision

$$\text{Power} = f\,(\text{Task, Technology, Topology, Speed, Precision})$$

Task	Speed-precision law	Comment
1. First-order active filtering.	$P_{GmC} \sim (f_{crnr})(SNR)$	1. SNR^2 in above-threshold operation. See Chapter 12.
2. Second-order active filtering.	$P_{2Gm} \sim (Q^3 + Q)f_{crnr}(SNR)$ $P_{3Gm} \sim (2Q + 1)f_{crnr}(SNR)$	2. P_{2Gm} and P_{3Gm} represent two topologies. See Chapter 13.
3. Passive filtering or adiabatic operation.	$P_{adbtc} \sim \dfrac{f_{in}SNR}{Q}$	3. The Q can be for a tuned or untuned system. See Chapters 21 and 15.
4. Adaptive Class-A current-mode filtering.	$P_{adpt-clssA} \sim (f_{crnr})(SNR_{max})$ $SNR_{max} \neq$ dynamic range	4. SNR_{max} not being equal to the dynamic range implies that power can be saved. See Chapter 14.
5. Analog-to-digital conversion.	$P_{ADC}^{nthrm} \sim f_s 2^{Nbits}$ $P_{ADC}^{thrm} \sim f_s 2^{2Nbits}$	5. The superscripts *nthrm* and *thrm* correspond to non-thermal-noise-limited and thermal-noise-limited topologies. See Chapter 15.
6. Operational amplification.	$P_{amp} \sim (GBW)\left(\dfrac{1}{\overline{v_n^2}}\right)$ $P_{amp} \sim (GBW)SNR$	6. $GBW =$ Gain-bandwidth product; $\overline{v_n^2} =$ input-referred thermal noise.
7. RF Impedance-modulation communication.	$P_{trnscvr}^{wk} \sim SNR(\Delta f)$ $P_{trnscvr}^{strng} \sim SNR^{1/3}(\Delta f)^{1/3}$	7. The *wk* and *strng* superscripts are for weak and strong coupling. The power includes the transmitting power-amplifier and receiving mixer power. See Chapter 18 for an in-depth discussion of bit-error rate and further details.
8. Sensing.	$P_{snsr} \sim f_{snsr} SNR_{max}$	8. Noise of sensor must be added to electronic noise to determine SNR. SNR_{max} is not necessarily equal to dynamic range in an adaptive sensing system with gain control. See Chapter 19 on imagers and other sensing, Chapter 8, and Chapter 20.
9. Active non-resonant oscillators, e.g., ring, relaxation.	$P_{osc} \sim f_{osc} SNR$	9. SNR corresponds to period jitter.
10. LCR tuned oscillators.	$P_{LCR} \sim \dfrac{f_{osc} SNR}{Q_{osc}}$	10. Since only the dissipated energy is provided by the active system rather than all the energy, watches and high-Q oscillators are energy efficient. See Chapter 13 for a discussion of *LCR* systems.
11. Narrowband LCR tuned amplifiers and sensors.	$P_{tuned-amp} \sim \dfrac{f_{crr}}{Q} SNR$	11. The f_T of the transistors used must be larger than f_{crr} to get significant gain near f_{crr}. Note that signal bandwidth f_{crr}/Q determines power not carrier bandwidth f_{crr}. The SNR degradation of amplifiers is proportional to $(1 + f_{crr}/f_T)$ such that f_{crr} must be significantly less than f_T.
12. Digital systems.	$P_{dgtl} \sim f_{clk} E_{TOT}$ $P_{dgtl} \sim f_{clk} \,\text{Poly}(\log_2(1 + SNR))$	12. Poly( ) is for polynomial function. See Chapter 21 for details on E_{TOT}.

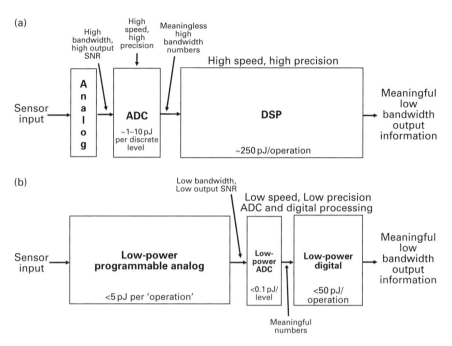

Figure 22.5a, b. (a) A traditional ADC-then-DSP architecture; (b) a more energy-efficient mixed-signal architecture. Analog, ADC, and digital numbers are constantly improving so numbers are only examples.

Figure 22.5 (b) shows a mixed-signal system that uses a low-power analog system to perform significant bandwidth reduction and preprocessing on the sensor input such that low-bandwidth lower-precision information is output for digitization. The techniques and circuits that have been described in this book can be used to design the low-power analog system. The outputs of the low-power analog system are then digitized at high energy efficiency by a low-power ADC with modest requirements; the ADC's outputs can then be processed in a very energy-efficient fashion by a low-bandwidth digital system. The power costs of the analog system in Figure 22.5 (b) increase compared to that in Figure 22.5 (a); the power costs of the ADC and DSP in Figure 22.5 (b) reduce compared to that in Figure 22.5 (a). At a certain optimum level of analog preprocessing, the power costs of the overall system are minimized. The system of Figure 22.5 (b) is not as flexible as that in Figure 22.5 (a). However, if the analog system is made programmable (see Chapter 19 for an in-depth example of a highly programmable analog bionic-ear or cochlear-implant processor for the deaf), the system of Figure 22.5 (b) can be made flexible enough such that the gains in power efficiency more than compensate for some loss in flexibility. The noise and mismatch inherent in analog devices serve to reduce the information content of the computation, but, if sufficient degrees of freedom are allocated to combat these, the output information content, typically significantly less than the raw input information content, can still be robustly preserved. In essence, degrees of freedom in analog processing are not wasted in making every device and every signal robust but in making the overall output robust. In Chapter 19, we discussed how a digitally programmable analog system can be architected to be robust to thermal noise, transistor mismatch, $1/f$ noise, temperature variations, power-supply noise, mixed-signal crosstalk, and yet be highly power efficient.

As we shall discuss later, the architectures of both Figures 22.5 (a) and 22.5 (b) are optimal in different environmental contexts and can possibly co-exist. For now, we shall focus on why the architecture of Figure 22.5 (b) is more power efficient than that of Figure 22.5 (a). To begin, we shall use an example familiar to us from our everyday lives, our cell phone, which will bring home some dramatic lessons in low-power design.

We don't immediately digitize the ~ 1 GHz narrowband analog output from the RF antenna in our cell phones and process all the output information digitally as in the architecture of Figure 22.5 (a). Given that Moore's law has thus far constantly improved digital power efficiency, and that digital systems are highly flexible and robust, it may appear strange that such digitization has not occurred and that completely digital radio or software radio is still very much a dream. The answer is simple: digitizing a 1 GHz signal at 16 bits as is needed even with an incredibly power-efficient 1pJ-per-quantization-level ADC (which has not been built yet at these speeds and precisions) would need at least $4 \times 10^9 \times 2^{16} \times 10^{-12} = 256$ W of power for the ADC alone! The factor of 4 in our calculation is 2 times higher than the Nyquist minimum rate of sampling and is needed to minimize aliasing noise and not require the use of an extremely sharp anti-aliasing filter. The digital post-processing of 4 Gb/s of data is also expensive but it is the ADC that is the bottleneck. The 256 W power consumption is about 3 times larger than the resting power consumption of our own bodies (see Chapter 23) and would certainly cause considerable heating in our pockets. It would increase the size of the battery needed to power a cell phone by about 250-fold.

So what do cell phone radios do? They use analog preprocessing to amplify and convert the high-frequency radio signal to baseband frequencies, often with two mixing stages in a super-heterodyne architecture [1]. The baseband signals can then be digitized by a much-lower-bandwidth ADC and operated on by a much-lower-bandwidth DSP. Thus, the power of the ADC and the power of the DSP are reduced by orders of magnitude due to the ability of the analog preprocessing to extract meaningful information from the input before digitization. The analog circuits are built with power-efficient passive RF filters, power-efficient tuned low-noise amplifiers, power-efficient oscillators, and power-efficient active mixers. The overall cost of the receiver power is in the 10 mW range, about four orders of magnitude more efficient than a brute-force immediate-digitization architecture. The extrapolation from this example is that, even at significantly lower frequencies, analog preprocessing via the architecture of Figure 22.5 (b) can drastically reduce power consumption in an overall system. In fact, in Chapter 19, we showed how the power consumption of a cochlear-implant processor was reduced by more than an order of magnitude due to programmable analog preprocessing; there the bottleneck was the DSP rather than the ADC as in this radio example.

In general, the architecture of Figure 22.5 (a) consumes an amount of power given by

$$P_{trdtnl} = C_A^{frntend}(f_{maxinp})(SNR_{maxinp})$$

$$+ C_{ADClarge}(4f_{maxinp})(SNR_{maxinp}) \tag{22.12}$$

$$+ C_{Dlarge}(4f_{maxinp})(V_{DD\,large}^2)$$

whereas the architecture of Figure 22.5 (b) consumes an amount of power given by

$$P_{lwpr} = C_A^{preprcsng}(f_{maxinp},\ SNR_{maxinp},\ f_{Aoutlow},\ SNR_{Aoutlow})$$

$$+ C_{ADCsmall}(4f_{Aoutlow})(SNR_{Aoutlow}) \tag{22.13}$$

$$+ C_{Dsmall}(4f_{Aoutlow})(V_{DDlow}^2)$$

In Equations (22.12) and (22.13), f_{maxinp} represents the high input bandwidth, SNR_{maxinp} represents the maximal SNR or input dynamic range, $f_{Aoutlow}$ represents the low output bandwidth after analog preprocessing, $SNR_{Aoutlow}$ represents the low output SNR after analog preprocessing, and the constants are labeled in a self-explanatory fashion. The ADC and digital terms in Equations (22.12) and (22.13) are easy to compute from Chapters 15 and 21 respectively. Some portions of the analog preprocessing term depend on f_{maxinp} and SNR_{maxinp} such as those involved with sensing, while others will only depend on $f_{Aoutlow}$ and $SNR_{Aoutlow}$. The sensing term in Equation (22.12) has been combined with the abstract analog-preprocessing function term in Equation (22.13). We have combined these terms because, in general, the analog preprocessing may mix sensing with processing in a smart way to save power. For example, some smart sensors can adjust their output SNR via gain control such that their SNR is relatively constant at all input intensities rather than varying with the input as in a linear system. In such sensors, the sensor's power consumption is not determined by SNR_{maxinp}, the dynamic range of the system, but by a smaller SNR, thus saving power at all input intensities. The sensor's output SNR is made to be relatively constant with input intensity as in a floating-point representation rather than varying as in a fixed-point representation. The biological inner ear or cochlea functions with nearly invariant SNR due to sophisticated gain control as do the photoreceptors in our eye [2], [3], [4]. Chapter 14 provides examples of current-mode filtering circuits that actually implement a constant-SNR scheme for a wide range of input intensities. Note that $C_{ADCsmall}$, C_{Dsmall}, and V_{DDlow} in Equation (22.13) are smaller than $C_{ADClarge}$, C_{Dlarge}, and V_{DDhigh} in Equation (22.12) because the ADC and digital circuitry get significantly more efficient at lower speeds and precisions and can operate at lower V_{DD}'s as well.

An example of the architecture of Figure 22.5 (b) is the bionic-ear or cochlear-implant processor of Figure 19.2. Here, the AFE microphone-preamplifier sensing term is separable from the analog preprocessing term. The AFE microphone preamplifier functions at SNR_{maxinp} ($\sim$77 dB) and f_{maxinp} (20 kHz) since it is linear and operates over the entire dynamic range of the input. The analog preprocessing in Figure 19.2 is composed of all circuitry in the processor after and including the AGC. Circuits after the AGC such as the bandpass filters function at reduced values of $SNR_{Aoutlow}$ (57 dB) and at bandwidths related to their center frequencies. The ADCs function at an overspecified 7-bit precision and 1 kHz, even though 5-bit-precise, 200 Hz sampling would likely suffice for the application. Note that the logarithm function, which is part of the analog preprocessing, is combined with the ADC to make an integrated log ADC in Figure 19.2. Not surprisingly, the highest power circuit is the microphone preamplifier, which has to function over the entire dynamic range at maximum speed: it consumes 106 μW, while all the analog preprocessing consumes 145 μW including ADC digitization power, a power consumption that is about 35 times less than a high-speed ADC-then-DSP processing solution. In this particular instance, a highly energy efficient 16-bit, 44.1 kHz, 50 fJ/(quantization level) ADC alone consumes 145 μW while a DSP with board and chip parasitics consumes nearly 5 mW for 20 MIPS of processing at 250 μW/MIP. High-precision ADCs at the input are needed to cope with the ~80 dB of dynamic range for the input audio signal and to allow for noise budgets in quantization noise, roundoff noise, and temporal aliasing. Hypothetical improvements in digital processing to even 0 μW/MIP at the end of Moore's law, which likely would be unviable given current leakage power consumption scaling, would still lead to a less efficient solution for the ADC-then-DSP approach. The sum of the microphone preamplifier and high-speed ADC alone are at the power level of the entire analog preprocessor of Figure 19.2, so the addition of digital power consumption, however small, will still increase power consumption over that of the analog preprocessing solution. Even a small amount of analog preprocessing such as an analog AGC before digitization reduces power consumption. An interesting question to then ask is where the optimum point for digitization is.

In the architecture of Figure 19.2, digitization can be performed after the microphone, after the AGC, after the bandpass filters, after the energy-extracting envelope detectors, or after the logarithm. The optimum point for digitization in Figure 19.2 was chosen after the logarithm because we were able to design highly energy-efficient AGCs, bandpass filters, envelope detectors and log ADCs in the analog domain (see Chapter 19 for a more detailed description) such that it paid to delay digitization till after the logarithm. We could have chosen to delay digitization to an even further point and go all-analog-all-the way implementing the final current-stimulation drivers that excite the auditory nerve as well. We chose not to do so for three reasons: 1) the use of relatively high-power current-stimulation circuits on an ultra-low-power analog chip significantly degrades the noise performance of the low-power analog circuits through supply-and-substrate coupling; 2) digitization of the log spectral energies allows the chip to report its digital bits across the transcutaneous skin flap of a patient via wireless telemetry (see Chapter 18) thus enabling use in both existent partially implanted systems and in fully implanted systems of the future; 3) digitization of log spectral energies allows flexibility of use with several stimulation circuits and paradigms. Thus, the optimal point for digitization was not picked for pure power-efficiency reasons but also picked to preserve robustness, flexibility, and modularity. The fact that 86 patient parameters could be altered with 373 programmable bits on the chip, and that back-compatibility and modularity were preserved as well imply that a good tradeoff between flexibility and efficiency was made: a 35x power reduction (see Chapter 19) with most of the flexibility intact. Indeed, it was the flexibility and programmability in the analog system that allowed testing of the chip on a deaf subject: She replaced her external digital processor with this chip processor, and was able to understand speech with it on her first try.

22.4 Common themes in low-power analog and digital design

Table 22.3 below summarizes 11 common themes in low-power analog and digital design, many of which have manifested in other chapters of the book.

Table 22.3 Common themes in low-power analog and digital design

Low-power analog	Low-power digital
1. Power $\sim$ Speed $\times$ Poly($2^{\text{precision}}$). So, *slow-and-parallel operation* is a big win. For example, in the bionic-ear processor, many slow-and-parallel moderate-precision log ADCs digitize information more efficiently at its output than one high-speed high-precision linear ADC at its input.	1. Power $\sim$ Speed $\times$ Poly(precision). So, *slow-and-parallel operation* is a big win: $N(f/N)C(V_{DD}^{lo})^2$ is lower than $fC(V_{DD}^{hi})^2$ in above-threshold operation. Parallelism has an overhead cost in the input front-end latches and output multiplexing stages, which need to operate fast to maintain the original speed, so there is an optimum amount of parallelism. See Chapter 21.
2. *Noise and offset management* to maintain speed and precision while lowering power is crucial. Lowering V_{DD} by too much can hurt dynamic range due to the need to operate transistors in saturation. A deep understanding of low-noise design and feedback design is essential.	2. *V_{DD} and threshold-voltage management* to maintain speed and limit static subthreshold leakage power while lowering dynamic and overall power is crucial. Feedback techniques that optimize V_{DD} for energy efficiency or for "just in time" computing are very useful.
3. *Compressive functions* like AGCs that reduce dynamic range at their output save power for later stages, which can operate at lower levels of *SNR* at any instantaneous setting of the AGC, e.g., 80 dB to 60 dB dynamic range reduction by the AGC in the bionic-ear processor.	3. *Compressive functions* like AND that reduce switching activity save power for later stages.
4. *"Pipelined" designs* with many stages of computation that are each doing less are often better than designs where one stage is doing too much. However, there is usually an optimum, e.g., a cascaded-gain amplifier has a better gain-bandwidth product for the same power than a single-stage amplifier but too many stages of gain will hurt its noise performance.	4. *Pipelining* a computation allows one to lower V_{DD} for each simpler stage of the computation. The reduced logic depth reduces the impact of leakage power and improves energy efficiency. However, there is usually an optimum amount of pipelining because latches between pipelined stages consume overhead power.

Table 22.3 (*cont.*)

Low-power analog	Low-power digital
5. *The higher the technology speed, the better the* g_m/C_{par} *ratio, and for a given* g_m/I *ratio, less power is consumed to attain the same speed if speed is limited by* C_{par}. However, there may not be a benefit in an analog design, e.g., a 60 dB envelope detector with dead-zone-limited dynamic range in Chapter 19 improves in power efficiency in a lower-gate-length technology due to lowered parasitics; however, a 75 dB thermal-noise-limited envelope detector experiences no improvement in power efficiency with technology, just area efficiency. In subthreshold, g_m/I does not improve with mobility. The maximum subthreshold current improves with each technology generation.	5. *The higher the technology speed, the faster the inverter switching time for the same* V_{DD}, *the lower are the transistor switching capacitances, and the lower is the switching energy improving power efficiency*. However, the reduction of threshold voltage increases leakage energy due to subthreshold 'leakage' currents in transistors that are off. Chapter 21 discusses an optimal V_{DD} at which switching energy and static subthreshold leakage energy yield the lowest overall optimum energy per operation.
6. *Symmetric designs* have less offset and can be smaller, *saving parasitic power*.	6. *Symmetric logic* designs have better matched delays between logic paths and have *reduced glitching power*.
7. *Gating* of circuits that are not being used saves power. There is usually a startup power cost associated with the turn-on time taken to start up the circuitry. For example, the use of duty-cycled light rather than continuous light in pulse oximetry to reduce LED power consumption; the use of transconductance amplifiers that increase bias current for large differential inputs and that decrease bias current for small differential inputs implements a form of automatic analog gating.	7. *Gating of the clock* for future stages of the computation by outputs from early stages of the computation saves power. See Chapter 21.
8. *Passive systems with high-Q* that only rely on small amounts of restorative energy from active devices to compensate for their loss are highly energy efficient, e.g., watches and tuned oscillators. Such systems trade signal bandwidth around the carrier for energy efficiency. *Adiabatic* multi-level voltage charging saves energy (see Chapters 15 and 19).	8. *High-Q adiabatic circuits* that are resonant or non-resonant operate slowly and trade speed for energy efficiency. See Chapter 21.
9. *Adaptive biasing* such that average performance over the device statistics in an array is significantly better than the worst-case performance is power efficient. For example, adaptive power biasing of neural amplifiers in an array to the required noise floor of a local electrode rather than the lowest noise floor in the array reduces power by an order of magnitude (see Chapter 19).	9. *Redundant* copies of circuits can ensure that outlier device statistics do not compromise power due to worst-case biasing of V_{DD} rather than average-case biasing of V_{DD}. Modest redundancy in circuits improves the robustness-efficiency tradeoff significantly. Error-correcting digital systems are now being explored [5], [6].
10. *Efficient encoding of the computation in well-matched basis functions of a few technological devices saves power*. For example, implementation of a current-mode filter from basis functions of a capacitor and a few	10. *An encoding of the task such that commonly occurring cases are processed efficiently with a few devices saves power*. Such encoding reduces the number of switching operations needed for the computation and thus saves power.

Table 22.3 (*cont.*)

Low-power analog	Low-power digital
exponential devices is very power efficient and area efficient (see Chapter 14).	
11. *Reducing the amount of information that needs to be processed saves power.* The exploitation of knowledge of the signal statistics or knowledge of the error statistics focuses resources on informative regions of the signal space rather than on all regions of the signal space and saves power. For example, adapting a system to the intensity statistics of the signal, e.g., via an AGC, reduces the information content, i.e., the instantaneous input dynamic range and output *SNR* of the signal that needs to be processed, and saves power. The use of learning, feedback, and calibration to remove known environmental, signal, or device errors or to adapt to constant unchanging inputs enhances signal gain over noise gain and yields a better power-*SNR* tradeoff. The use of knowledge can simultaneously improve speed (there is less to do, so it can be done faster) and precision (there is more signal and less noise) at the same power.	11. *An algorithm with a low MIP count reduces the amount of information that must be processed and saves power.* The biggest power-saving principle in digital design is undoubtedly the use of a more efficient algorithm, from which all else follows. For example, FFT processors are energy efficient because the FFT is inherently algorithmically efficient and has a low MIP count. Dynamic switching power is consumed in CMOS only when the signal changes. Thus, constant unchanging signal inputs or inputs with a lot of correlation over time automatically reduce the amount of information that needs to be processed and save switching power.

22.5　The Shannon limit for energy efficiency

Table 22.3 outlined several common themes in low-power digital and analog design. Two common themes in both low-power paradigms are the importance of computing slowly and the importance of reducing the amount of information that needs to be processed. One can ask: is there a limit to the minimum amount of energy needed to process one bit of information, whether that bit is represented in an analog fashion (via an equivalent *SNR*) or in a digital fashion (as an explicit 0 or 1)? Shannon's key theorem from information theory allows one to compute the minimum amount of signal energy needed to encode a bit of information over a noisy communication channel. This minimum is asymptotically approached when one sends the information across the channel very slowly. A computing device that performs a mapping between its input and output, whether discrete or continuous, is a noisy channel that implements a particular transformation to encode the communication between its input and output. Therefore, if we focus on the minimum signal energy at the output of our computing device, Shannon's theorem can provide some insight into the minimum energy needed to compute as well.

Shannon's theorem [7] states that a channel that is communicating with the use of a signal of bandwidth B and a signal-to-noise ratio of *SNR* has a channel capacity C given by

$$C = B \log_2(1 + SNR) \tag{22.14}$$

As long as we attempt to send information at an average bit rate R that is less than C over an asymptotically long limiting time, there can be asymptotically error-free communication. For a given B, if we reduce the SNR, we must reduce R to still satisfy Shannon's theorem. Thus, if we are just at the limit, and we send 1 bit of information in $1/R$ seconds at a net bit rate of R,

$$\frac{R}{B} = \log_2(1 + SNR) \tag{22.15}$$

The SNR in Equation (22.15) can be re-expressed as

$$SNR = \frac{E_b}{N_0} \times \frac{R}{B} \tag{22.16}$$

where N_0 is the noise energy per unit bandwidth, N_0B is the noise power, E_b is the average energy per bit, and E_bR is the signal power. Equation (22.16) is just a re-expression of the usual SNR variable, except that we are now working with actual power units in the numerator and denominator of the SNR rather than with voltage power units as we normally do; that is, if the output capacitor of the computing device is C,

$$SNR = \frac{\overline{V_{sig}^2}}{\overline{v_n^2}}$$
$$\overline{v_n^2} = \frac{kT}{C}$$
$$N_0 = \frac{1}{2}C\overline{v_n^2} = \frac{1}{2}kT \tag{22.17}$$
$$E_b = \frac{1}{2}C\overline{V_{sig}^2}$$

From Equations (22.15) and (22.16), we can compute that at the Shannon limit,

$$E_b = N_0 \frac{2^{R/B} - 1}{R/B} \tag{22.18}$$

If we send the data at a very slow rate, i.e., $R/B \to 0$, we then find from Equations (22.18) and (22.17) that

$$E_b = N_0 \left(\lim_{R/B \to 0} \frac{2^{R/B} - 1}{R/B} \right)$$
$$E_b = N_0 \ln 2 \tag{22.19}$$
$$E_b = \frac{1}{2}kT \ln 2$$

The value of E_b is the minimum signal energy that is present on the output capacitor of the computing system when we charge it up via a charging pathway. From the discussion in Chapter 21, $kT\ln(2)$ is dissipated when we charge the capacitor via a charging pathway; $(1/2)kT\ln(2)$ is stored on the capacitor; and

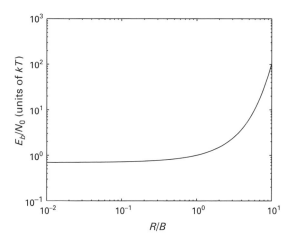

Figure 22.6. The variation of E_b/N_0, the energy per bit, versus R/B, the link spectral efficiency.

$(1/2)kT \ln(2)$ is dissipated when we discharge this capacitor via a separate discharging pathway to terminate our communication signal. Thus, the minimum energy dissipated in the $0 \rightarrow 1$ transition of a computing device that does not recycle its switching energy must at least be

$$E_{\min} = kT \ln(2) \tag{22.20}$$

Figure 22.6 plots E_b/N_0 versus R/B determined from Equation (22.18) and shows that the asymptotic energy minimum is almost reached even for $R/B < 0.1$. Thus, underutilizing the available information bandwidth is highly energy efficient for computation and communication. One reason why Ultra-Wide-Band (UWB) RF systems for short-range communication are quite energy efficient is that they often operate with $R \ll B$ by transmitting high-energy impulses at a periodic frequency that is significantly below their $\sim 7\,\text{GHz}$ bandwidth.

Computing with AND gates or OR gates, that reduce information by transforming a balanced 1-and-0 probability distribution at their input to one weighted more heavily towards 0s or 1s at their output leads to a net reduction in randomness of the states of the computing system, or by definition, a reduction in its entropy. A fundamental law of physics, the second law of thermodynamics postulates that entropy and randomness can only have a net increase in the world; thus the reduction in entropy caused by the AND gate or OR gate must lead to energy dissipated as heat that increases randomness or entropy elsewhere. Typically, the increase in entropy occurs in the environment that the computing system is coupled to, which is called the 'heat reservoir'. Energy is obtained from the power supply and dissipated as heat in the computing device, which then eventually flows to the heat reservoir and increases its entropy. The heat reservoir is modeled as being so large and having so many degrees of freedom that its temperature barely increases in spite of heat being poured into it, one of the reasons that it is called a reservoir. In fact Boltzmann's exponential relation that we have used throughout this book is a consequence of a large heat reservoir and the second law of thermodynamics [8]. On the other hand, for gates such as inverters that are invertible, there is no change in the probability distribution from input to output and no loss in information. Thus, Bennett and Landauer have shown that the minimum energy of Equation (22.20) need NOT be dissipated if we compute infinitely slowly and adiabatically with only invertible functions such as those in inverters [9], [10]. In such cases, we can theoretically compute with zero energy dissipation per bit.

In practice, CV_{DD}^2 switching energies in digital gates today operate at 10,000 to 100,000 kT in a $\sim$100 nm process, so we are quite far from the limit of Equation (22.20). It is unlikely that we will get anywhere close to the limit of Equation (22.20) even at the end of Moore's law because to maintain speed, to reduce the impact of subthreshold leakage, to enable robust operation with device variability, V_{DD} will need to be at least $\sim$0.25 V as we discussed in Chapter 21. Furthermore, we do not have a practical and general paradigm for digital computing today that allows for large-scale computation in the presence of high error rates per device, as would occur if we neared kT and attempted to compute at a finite information rate. As we discuss in Chapter 24, it is remarkable that elementary biochemical molecular operations powered by ATP, the energy-currency molecule of our cells, operate at 20 kT per molecular operation at the nanoscale. Nature is far ahead of man as far as energy efficiency is concerned.

22.6   Collective analog or hybrid computation

Computation in nature appears to follow a paradigm that is not based on either traditional analog computation or on traditional digital computation but on a hybrid combination of both: Figure 22.7 (a) illustrates the idea with an example of an 8-bit precise computation although the concept applies to arbitrary-precision computation. We do not compute with one analog unit exploiting analog basis functions at 8-bit precision on one wire as in traditional analog computation. We do not compute with 8 digital units, each exploiting logic basis functions at 1-bit precision on 8 wires and collective interaction between these 8 units to maintain 8-bit precision as in traditional digital computation. Rather we perform collective analog computation with several moderate-precision analog units, that each exploit the rich basis set of analog functions to compute, and that collectively interact with each other to maintain high-precision information. In Figure 22.7 (a), we show one instantiation of the idea with two 4-bit-precise analog units collectively interacting to maintain 8-bit precision. Since each unit is of modest precision, and encodes the computation in the rich basis function set of technological devices efficiently, each unit is highly energy efficient. The overall computation like traditional digital computation is scalable to arbitrary levels of precision because information and information processing are encoded across several wires and across several computing units respectively in a collective fashion rather than on wire and

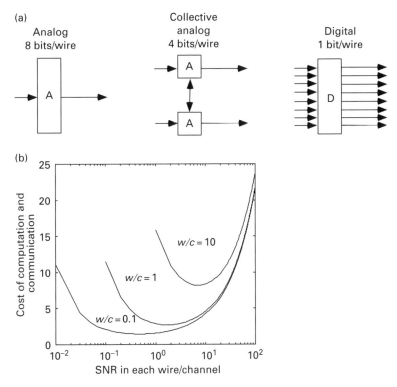

Figure 22.7a, b. (a) The idea behind collective analog computation. (b) The optimal *SNR* per computing channel in collective analog computing. Reprinted with kind permission from [11].

with one computing unit as in traditional analog computation. Thus, the paradigm incorporates the best ideas from the analog and digital worlds, and is likely a very important reason for the energy efficiency of biological computation [11].

One key point to note is that the interactions between analog units in the collective analog topology of Figure 22.7 (a) can be digital. In fact, this is often advantageous because the digital signals then serve the dual role of performing signal restoration within each unit and of being the interaction signals amongst analog units. Periodic signal restoration of analog variables via digitization is critical for enabling scaling of collective analog systems to systems of arbitrary complexity. The digitization must not be done too early such that the costs of signal restoration rise and the power of analog computation is not exploited; the digitization must not be done too late such that the costs of maintaining precision on a single analog channel rise. Thus, just as in the architecture of Figure 22.5 (a), there is an optimal point for digitization [11].

A second key point to note is that collective analog computation is implemented in various embodiments in nature, e.g., computation with cell-cell interactions amongst neurons in the brain, computation with gene-protein and protein-protein interactions amongst molecular state variables within the cell, computation with interactions amongst analog units in the inner ear or cochlea, computation amongst analog units in the retina of eye. Chapters 23 and 24 will delve into some examples of such computation and how we can exploit such ideas for low-power and ultra-fast computations in engineering and medicine. The speed benefits arise primarily from the parallel and efficient nature of the computational encoding.

A third key point to note is that the idea behind collective analog computation is NOT that of multi-valued logic. A principal ingredient of collective analog computation is the use of analog basis functions, not just logic basis functions to compute; multi-level logic, which still uses only logic primitives to compute, has been found to be inferior to traditional logic because 1-bit encoding of digital basis functions in CMOS technology is significantly more efficient than multi-bit encoding of digital basis functions. In essence, analog basis functions implement complex functions efficiently, and if we are just implementing a digital truth table in a slightly different digital way, there is little benefit. Consequently, multi-level schemes have improved spatial density in memories but have had little impact on computational efficiency.

We shall call each analog computing unit and its associated output wire a computational channel. It is natural to ask whether it is more efficient to use four 2-bit-precise analog channels to encode 8 bits of information, two 4-bit-precise analog channels as in Figure 22.7 (a), or 100 0.08-bit-precise channels. Where is the energy-efficient optimum? Computationally speaking, it is always cheaper to compute at low SNR in each channel and have more channels. However, the addition of each channel makes communication, wiring, interaction, and overhead costs, many of which are SNR-independent, more expensive. If we assume that the computational cost per channel is $c(SNR)$ for simplicity and that the SNR-independent wiring/overhead cost per channel is w, if we need to encode N bits of information amongst all computational channels, then the total cost of computation and communication, P_{cllana}, is given by

$$P_{cllana} = (w + cSNR)\frac{N}{(1/2)\log_2(1 + SNR)} \tag{22.21}$$

Figure 22.7 (b) plots $P_{cllana}/(2N)$ versus SNR for several different w/c ratios. We note that there is an optimum SNR per computational channel that balances the overall costs of SNR-dependent and SNR-independent computation and communication. In technologies where w/c is relatively high, we see that the optimum occurs at larger SNR per channel and fewer channels; in technologies where wiring is relatively cheap, we see that the optimum occurs at a lower SNR per channel and more channels.

In the brain's 3D interconnect architecture, the fan-in and fan-out averages to about 6,000, corresponding to relatively cheap wiring costs compared with electronics [12]. Thus, nature is extremely smart in having the collective interaction of relatively noisy low-SNR analog neurons process information such that 22 billion neurons of the brain and its other support cells only consume 14.6 W of power [11]

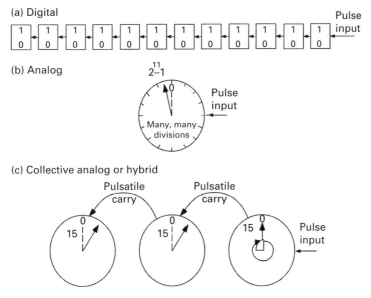

Figure 22.8a, b, c. Pulse counting in (a) digital, (b) analog, and (c) collective analog or hybrid systems. Reprinted with kind permission from [13].

(see Chapter 23). The interactions amongst these analog neurons occur via spiking all-or-none pulsatile digital signals. Spikes have an inherently hybrid analog and digital nature in that each spike is an all-or-none digital event but the time between spikes can be analog making them uniquely suitable for hybrid analog-digital computation [13].

In the body's cells, the *SNR* of individual protein variables in the cell is often near ~10 [14]. Implicit chemical wiring within the intracellular solution of the cell enables a ~30,000-node complex molecular computational network with intense amounts of feedback and hybrid analog-digital computations to be implemented in ~1 pW of power. In Chapter 24, we shall see that there is an astounding similarity between the equations of subthreshold electronics and the equations of a chemical reaction including stochastics. Thus, molecular computations can be efficiently mapped to electronic circuits. Such circuits have applications in the newly emerging fields of systems and synthetic biology, which are poised to revolutionize medicine by enabling an engineering systems-understanding of biology [14], [15].

Thus, nature was very smart in implementing extracellular and intracellular architectures in a highly energy-efficient collective analog fashion as in Figure 22.7 (a). In both cases, analog units interact amongst themselves with both analog and digital signals. It is useful to have a concrete engineering example of how collective analog computation could conceptually work. Figure 22.8 provides such a toy example. Suppose we want to count an incoming number of input pulsatile events, and if they exceed $2^{12} - 1$, output a digital signal that a certain threshold number of pulses have been counted. The digital method of implementing this computation is to cascade 12

1-bit flip flops with some embedded logic to implement a counter as shown in Figure 22.8 (a); we set the counter to 0, and count by having the input pulse serve as a clocking-and-toggle input to the first flip flop; if the final flip flop undergoes a $1 \rightarrow 0$ event, 2^{12} events have been counted. The analog method of implementing this computation is to have each pulse create a small voltage increment ΔV on a capacitor with voltage V; the voltage on the capacitor is initialized at 0 and fed to a comparator; we set the threshold voltage of the comparator $V_T = \Delta V \left(2^{12} + 2^{12} - 1\right)/2$ such that when the 2^{12}th event has caused the voltage on the capacitor to exceed V_T, the comparator fires an output pulse, and resets the capacitor to 0. We have essentially have created an angular state variable $\theta \approx 2\pi(V/V_T)$; when θ exceeds 2π, the phase resets to 0. The angular variable is shown in Figure 22.8 (b).

The digital method of encoding the computation in Figure 22.8 (a) requires 12 1-bit precise units while the analog method of encoding the computation in Figure 22.8 (b) requires one 12-bit precise unit. The collective analog or hybrid method of Figure 22.8 (c) uses three 4-bit precise analog units; each analog unit operates like that in Figure 22.8 (b) except that it is only 4-bit precise rather than 12-bit precise, with each pulse creating voltage increments ΔV such that $\Delta V \left(2^4 + 2^4 - 1\right)/2 = V_T$. The reset or over-ranging spike from one analog unit serves as the input to an adjacent unit on its left, thus serving as a pulse carry much like the binary carry signals passed between flip flops in the digital implementation of Figure 22.8 (a). If the output of the final unit of Figure 22.8 (c) outputs a pulse, 2^{12} events have been counted. In essence, we have created a distributed analog representation of the incoming input information [13]. The architecture of Figure 22.8 (c) encodes intermediate information of the count in analog state variables, but also provides a coarse digital representation of the count, when spikes are carried amongst adjacent units. The architecture of Figure 22.8 (b) maintains all intermediate information of the count in analog state variables. The architecture of Figure 22.8 (a) maintains all intermediate information of the count in digital state variables.

Figure 22.8 is a simple toy example that should not be taken too seriously. Details of the energy costs of (a), (b), and (c) will depend on the energy costs of implementing flip flops versus comparators, whether the comparators are implemented with techniques that have almost no static power and are simple inverter-and-latch-like digital circuits themselves (see Figure 15.17 in Chapter 15), the magnitude of the actual charge increments, whether feedback-calibration schemes have compensated for comparator offset, the amount of subthreshold 'leakage' of the circuits between pulses etc. Counting is a relatively noncomplex computation with good encoding in the digital domain such that one may choose the topology of (a) just for simplicity. What is more important is to understand the scaling of the costs of the three topologies because in more complex computations, where the encoding of the computation in the analog domain, digital domain, and collective analog domain differ significantly, the topologies of (a), (b), and (c) will have significantly different costs. Switching power is reduced by 2^{12} times in the topology of Figure 22.8 (b) compared with that of Figure 22.8 (a) but the energy costs of the comparator are relatively high at 12-bit precision. The topology of Figure 22.8 (c) has approximately 2^4 times lower switching costs than the topology of Figure 22.8 (a), 4 times less static leakage power due to lowered logic depth, and the cost of the 4-bit-precise comparison is $\approx 2^8/3$ times lower than the 12-bit-precise comparison of Figure 22.8 (b). For optimally energy-efficient operation that is not at the mercy of circuit offsets, each of the comparators must be calibrated every so often or constantly in the background with their thresholds set at a value such that 15 input pulses reliably do not trigger a threshold crossing but that 16 input pulses reliably do. As in Figure 22.7 (b) and in Equation (22.21), there is an optimal number of levels to count per analog channel in Figure 22.8 (c) which minimizes energy. This optimum will depend on the technology, circuit details, and topology of the implementations used.

22.7   HSMs: general-purpose mixed-signal systems with feedback

One important reason for the scalability of digital systems is that complex computations can be implemented as a sequence of simple ones. The power of language allows us to program complex computations as an organized combinatorial sequence of simpler ones in a digital computer. Is it possible to do the same for mixed-signal systems that may have digital and analog portions? Would there be any advantage to doing so in terms of energy efficiency?

Figure 22.9 (a) shows the architecture of a finite state machine (FSM), the workhorse of digital computation. Digital computers are simply sophisticated finite state machines operating on a bounded but extremely large memory 'tape'. The digital output is a combinational-logic function of the digital input vector and the current digital state. The state of the system on the clock cycle following the current one is another combinational-logic function of the digital input vector and the current digital state. State transitions only occur after clock edges and the system operates in discrete time.

Figure 22.9 (b) shows a natural generalization of the FSM to the mixed-signal domain termed a *hybrid state machine* (HSM) [13]. The system is composed of a finite state machine with digital state and combinational logic, and an analog dynamical system with analog state and analog basis functions. The analog state is typically always volatile while the digital state could be volatile or nonvolatile. The inputs In_{anlg} and Out_{anlg} represent vector analog inputs and outputs, and the inputs In_{dig} and Out_{dig} represent vector digital inputs and outputs. All inputs and outputs can arise from or act on the external world respectively or from other HSMs that are interacting with this HSM. Digital inputs and outputs can also arise from or act on digital memory as in a traditional computer.

The analog dynamical system (ADS) generates event signals, analogous to clock edges, which trigger state transitions in the FSM. Thus the analog dynamical system affects the digital FSM by causing state changes within it. The events generated by the analog dynamical system can be pulse coded or edge coded.

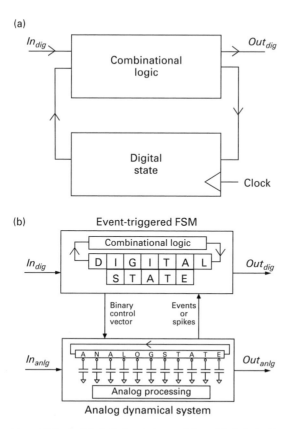

Figure 22.9a, b. (a) A finite state machine. (b) A hybrid state machine (HSM). Reprinted with kind permission from [13].

The events may be periodic synchronous edges as in traditional digital systems but both synchronous and asynchronous edges are allowed. Thus, HSMs operate over all four domains in Figure 22.1. The digital state in the FSM alters parameters and topology in the ADS via the binary control vector in Figure 22.9 (b) that controls the configuration of switches in the ADS. Hence the analog and digital portions of an HSM each affect and control each other, via the feedback loop shown in Figure 22.9 (b). Local analog feedback within the ADS and local digital feedback, present in every FSM, implement feedback loops within the analog and digital portions as well.

The HSM architecture of Figure 22.9 (b) is a generalization of several special-purpose mixed-signal dynamical systems that already exist in the field of analog-to-digital converters (ADCs). We shall provide six examples.

1. The $\Sigma\Delta$ ADC shown in Figure 15.5 has an ADS that is composed of a differential integrator and comparator. The comparator outputs events that are fed back to the integration input of the ADS via a binary control vector. The binary control vector is a simple 1-bit signal that alters DAC parameters that are input to the ADS. The FSM is represented by the digital intelligence that feeds back a 1

in the DAC if the comparator outputs a high event and feeds back a 0 in the DAC if the comparator outputs a low event. The digital LPF that processes the comparator event outputs in Figure 15.5 is also part of the FSM in Figure 22.9 (b).

2. The successive-approximation ADC of Figure 15.4 (b) is also a special case of an HSM. Here, the ADS is a differential comparator that outputs comparator events. These events are fed back to the ADS via a binary control vector that alters parameters in a multi-bit DAC. The FSM is represented by the digital intelligence that sets the parameters of the multi-bit DAC based on comparator event outputs, and by the rest of the circuitry that latch and hold the outputs of the ADC.

3. The neuron-inspired ADC described in 15.8 is also an example of a HSM. The capacitors, charging currents, analog switch network, and pulse-generating comparator form the ADS. The state machine, clock divider, and output registers form the FSM. The events output by ADS cause the FSM to feed back switching signals in a binary control vector that determines which charging current will be used to charge which capacitor, and alters parameters in the comparator.

4. Fast-imprecise ADCs that are calibrated by slow-and-precise ADCs with feedback-calibration signals are also examples of HSMs with mixed-signal feedback.

5. The computational ADC described in Chapter 15 can be represented by an HSM. A digital Taylor-series approximation of the analog input is obtained with an FSM that digitally controls and coordinates what its ADS is doing.

6. Adaptive-precision and adaptive-bandwidth ADCs that alter their precision or bandwidth based on feedback from the higher-level system can save power and better serve the needs of the overall computation rather than simply consume worst-case power. They can be represented by an HSM where the binary control vector alters the number of steps of the ADC conversion or the biasing levels of the analog circuits.

The examples of the previous paragraph show that ADCs represent a class of mixed-signal systems that are implementing a rather simple special-purpose HSM for analog-to-digital conversion. All of the described examples represent highly energy-efficient ADC topologies as we describe in Chapter 15. However, they represent a very narrow class of HSMs that are implementing perhaps 4–5 lines of simple computer code over and over again in their FSMs in a rigid nonflexible way. Can we use programmable HSMs to do more? By simple reconfiguration of portions of their ADS or FSM topologies with switches, HSMs can implement pattern recognition, learning, vector quantization, sequence recognition, arithmetic, any multi-input polynomial function, analog-to-digital conversion, and hybrid control [13]. In general, we may want to build mixed-signal architectures that can implement 1000s of lines of code that sequence such operations. We could imagine using HSM-like architectures to control and coordinate large-scale collective analog computation with many lines of code in the FSM like we control and coordinate combinational-logic operations today. Why would such systems be highly energy efficient? We provide ten reasons.

1. Fine-grained control and integration of 'hardware' and 'software' can be achieved, thus improving energy efficiency. In fact, the boundary between the two becomes very blurred as the ADS determines the FSM states and the FSM states control the parameters of the ADS. Thus, both are equally important and neither is in control of the other. We simply have the hardware and software in an intimate feedback loop.

2. The intelligence can be in both the analog and in the digital domains improving efficiency in both and achieving a more optimized and balanced computational burden in each domain. We can exploit powerful analog basis functions and combinational logic to compute more efficiently rather than always using one stereotyped ADC-then-DSP encoding.

3. Signal-to-symbol conversion is not necessarily through inefficient number representations provided by an ADC but with higher-level symbols. Similarly, signal-to-symbol and symbol-to-signal feedback can be through higher-level representations. Such issues are particularly important in computations that are extremely data intensive, e.g., image processing. Here, the fact that individual numbers carry relatively little information and that an efficient representation and encoding of information really matters is brought home in an obvious way.

4. Networks of interacting HSMs can implement fast energy-efficient information processing for graphics, image-processing, and inherently highly parallel applications as field-programmable-gate-array (FPGA) architectures are increasingly doing. Networks of HSMs can model intracellular and extracellular interactions amongst natural or artificial cells and would constitute an architecture that we may term hybrid cellular automata.

5. Energy-efficient collective analog computation is particularly amenable to mixed-signal HSM computation since calibration, signal-restoring digital operations, and quantization can be interspersed as lines of code amongst regular computational operations as needed. Each code line reconfigures the ADS to perform the required function.

6. We can adapt the ADS and FSM to the signal and device statistics rather than designing for worst-case operation, which significantly improves efficiency.

7. Easy amenability of HSMs to asynchronous operation implies that energy-efficient topologies in digital computation that minimize clocking energy and that are inherently robust to timing errors and device variability can be incorporated.

8. HSMs can represent an architecture for 'analysis by synthesis' where the digital bits represent and model the analog world and feedback-correction signals from the environment constantly update this model to better analyze them. SAR ADCs and $\Sigma\Delta$ ADCs fit particularly well within this framework. We provide other examples of speech locked loops in Chapter 23 that are robust to noise because of inherent model-based feedback tracking of the signal. Such loops are similar to phase locked loops (PLLs) that implement purely analog versions of the concept.

9. Just as clock-gating reduces energy in digital systems, HSMs can enable event-driven analog systems that minimize energy by only turning on high-precision and/or high-speed analog operation in certain states, and by turning off analog devices that are not needed.

10. Both neuronal computation amongst cells in the brain and within molecules in the cell are well described by an HSM framework (see Chapters 23 and 24). Therefore, we have an existence proof from biology of the power of mixed-signal computing with feedback. In summary, the robustness-efficiency and flexibility-efficiency tradeoffs of a system are greatly improved by exploiting the best of the analog and digital worlds.

One challenge in implementing mixed-signal HSM architectures on a large scale is the cross-coupling of undesirable digital signals into analog units, which compromises analog efficiency. A second challenge is that of designing feedback and calibration loops that automatically compensate for device variability in deep submicron processes in a manner that trades robustness for efficiency in an intelligent fashion. Fortunately, such problems are somewhat alleviated in collective analog systems that do not need high levels of local precision, in deep submicron processes where area is a relatively abundant resource, and in fully differential implementations. Furthermore, in energy-efficient subthreshold systems, digital voltage and current levels are modest such that power-supply crosstalk is inherently smaller. Another possible solution is to separate analog and digital systems on different dies and/or technologies that interact via parallel 3D interconnect.

Biology solves such problems very cleverly by using highly localized electro-chemical coupling of signals across synapses in neurons or highly specific chemical binding of molecules in cells. Time will tell whether we will be able to architect such systems on a larger scale successfully as nature has done. Efforts to create algorithmically programmable analog systems in engineering have been explored for image-processing applications [16], [17].

22.8 General principles for low-power mixed-signal system design

We shall now present ten principles for low-power design that apply to digital, analog, and mixed-signal systems. These principles summarize and distill the essence of several concepts and examples discussed throughout the book. Just as in the seven benefits of feedback of Chapter 2, there is redundancy and overlap in these principles: each principle emphasizes a somewhat different aspect of the same underlying truth.

22.8.1 Encode the computation in the technology efficiently

As we have discussed extensively throughout the book, low-power design is inseparable from information-processing design. Given a function $Y(t) = f(X, t)$, basis functions $\{i_{out1} = f_1(\mathbf{v}_{in}), i_{out2} = f_2(d\mathbf{v}_{in}/dt), i_{out3} = f_3(\int \mathbf{v}_{in}), ...\}$ formed by the current-voltage curves of a set of technological devices, and the noise-resource equations of the technological devices (Equation (22.4)), find a topological implementation of the desired function in terms of these devices that maximizes the mutual information between $Y(t)$ and $f(X, t)$ for a fixed power-consumption constraint or per unit system power consumption. Area may or may not be a simultaneous constraint in this optimization problem. A high value of mutual information, measured in units of bits per second, implies that the output encodes a significant amount of desired information about the input, with higher mutual information values typically requiring higher amounts of power consumption. An A/D-DSP-D/A encoding is a general-purpose encoding that is rarely as efficient as a special-purpose encoding customized for the task; general-purpose encodings are a convenient place to begin exploration but special-purpose encodings are essential for efficiency. In the analog domain, efficient encodings enhance signal gain over noise gain. In the digital domain, efficient encodings minimize switching and leakage power. Like all problems that are hard, unsolved, and require artistic human skill, as analog design certainly does and creative aspects of digital design that don't arise from automation do, it is likely that this optimal encoding problem is NP-complete or NP-hard. We have given several examples of encodings that are efficient in particular cases, e.g., Figure 19.2, the bionic-ear or cochlear-implant processor; lock-in amplification in Chapter 8, which encodes the signal such that its processing and amplification are relatively robust to noise in the computing devices. Throughout the book, we have discussed topologies, heuristics, building-block circuits, and principles that are useful in converging to a solution. It is in some ways pleasing that the brilliance and creativity of the human brain, our genius as it were, will likely still be necessary.

22.8.2   Use Subthreshold operation to maximize energy efficiency

Subthreshold operation is advantageous in low-power design for five reasons. 1) The g_m/I ratio is maximum such that speed per watt or precision per watt are maximized. Thus the resource-precision equation, Equation (22.10), shows that we use the least power for a given bandwidth-SNR product in the subthreshold regime. As we have discussed in Chapter 21, it is also the most energy-efficient regime in digital design, since for a given V_{DD}, the on-off current ratio is maximized in this regime. 2) Low values of V_{DSAT} enable V_{DD} minimization and thus save power in both analog and digital design. 3) Effects such as velocity saturation degrade energy efficiency by degrading the g_m/I ratio and by increasing excess thermal noise. Such effects are nonexistent in the subthreshold regime and improve energy efficiency in the analog and digital domains. 4) Since current levels are relatively small in the subthreshold regime, resistive and inductive drops due to parasitics that can degrade g_m are minimized in the subthreshold regime, as is power-supply noise. 5) The subthreshold exponential basis function is a very powerful universal basis function with all order of polynomial terms in its Taylor-series expansion. Thus, polynomially linear and nonlinear current-mode dynamical systems of any order (see Chapters 14 and 24) can be implemented in this regime in a highly energy-efficient fashion; similarly, hyperbolic functions such as tanh, cosh, sinh etc. are easily implemented in this regime. 'Leakage' current as it were can actually profitably be used in the analog domain to compute.

Subthreshold operation has three primary disadvantages: 1) The high g_m/I ratio and exponential sensitivity to voltage and temperature make this regime highly sensitive to transistor mismatch, power-supply noise, and temperature. Thus, biasing circuits such as those of Figure 19.12 are essential for robust operation in analog design as are feedback and calibration architectures. Chapter 19 discusses several examples. Similarly, V_{DD} has to be sufficiently high in digital design (see Equation 21.8) to ensure robust operation across all process corners; certain topologies involving stacking, feedback, and parallel devices must be avoided in digital design (see Chapter 21). 2) The f_T is more subject to parasitics degrading energy efficiency by about a factor of 2. 3) Linearization is more difficult than in other regimes. Chapter 12 discusses how to build linear transconductors in the subthreshold regime.

22.8.3 Exploit Analog preprocessing before digitization in an optimal way

An important principle for efficient encoding is to exploit analog preprocessing to perform information-bandwidth reduction before digitization, e.g., as is essential in low-power radios. There is an optimal amount of analog preprocessing before a signal-restoring or quantizing digitization is performed: too little, and efficiency is degraded due to an increase in power consumption in the ADC and digital processing; too much, and efficiency is degraded because the costs of maintaining precision in the analog preprocessor become too high. The exact optimum depends on the task, technology, and topology. In a practical design, other considerations of flexibility and modularity that are complementary to efficiency will also affect the location of this optimum.

22.8.4 Use Parallel architectures

Parallel architectures utilize many parallel units to implement a complex computation through a divide-and-conquer approach. The speed, complexity, or precision of each parallel unit can be significantly less than that needed in the overall computation. Since the costs of power are typically nonlinear expansive functions of the complexity, speed, or precision, the sum of the power of all N parallel units is significantly less than the power of a complex, high-speed, precise unit doing all the computation:

$$N \times P_{smpl}(f_{lo}(N), SNR_{low}(N)) < P_{cmplx}(f_{hi}, SNR_{hi}) \qquad (22.22)$$

Figure 21.9 illustrates an example in the digital domain and Figure 19.2 illustrates an example in the analog domain. In both domains, if the input is not inherently parallel, e.g., as in an image, then an encoder and/or decoder are needed to perform

1-to-N encoding and N-to-1 decoding. The costs of these encoders and decoders affect the optimum number of parallel units. In Figure 21.9, the encoder is a demultiplexer and the decoder is a multiplexer; power is saved because $Nf_{lo}CV_{DDlo}^2$ is less than $Cf_{hi}V_{DDhi}^2$ in the above-threshold regime. In Figure 19.2, the encoder is a filter bank, and the decoder is a digital multiplexer; power is saved because $N(f_{lo})2^{lo} < f_{hi}2^{hi}$, i.e., bandwidth and precision are reduced in the parallel units. In Chapter 23 we discuss how transmission-line architectures in nature are exploited to share computational hardware amongst many parallel inputs in the retina of the eye, amongst parallel inputs in the dendrites of neurons, and to implement a filter bank in the inner ear or cochlea. The biochemical reaction networks discussed in Chapter 24 are also highly parallel networks. Parallel architectures are capable of high-speed and low-power operation.

22.8.5 Balance Computation and communication costs

A frequent tradeoff in many portable systems is the balance between computation and communication power costs. If one computes too little, there is too much information to transmit, which increases communication costs. If one computes too much, there is little information to transmit, which reduces communication costs, but increases computation costs. The optimum depends on the relative costs of each. Such tradeoffs affect cell-phone/base-station communication, digitization of bits that need to be transmitted wirelessly in brain-machine interfaces and in neural prosthetics (see Chapters 18 and 19), and the general design of complex systems composed of parts that need to interact with each other. Equation (22.21) provides an example of this principle in the context of collective analog computation.

22.8.6 Exploit Collective analog or hybrid computation

Figure 22.7 illustrates the idea behind collective analog or hybrid computation and Figure 22.8 provides an example. This principle is likely the most important principle utilized by nature to save power, and what we believe is an important new frontier in low-power mixed-signal design. Chapter 23 will outline some examples of collective analog computation where the interaction between analog units is purely analog (the RF cochlea for cognitive radio or retinas for image processing). Chapter 24 will outline examples of collective analog computation within cells where the interaction between analog units is both digital and analog. The HSM architecture of Figure 22.9 and our previous associated discussion outline the benefits of mixed-signal systems with feedback.

22.8.7 Reduce the amount of information that needs to be processed

Since energy and information are intimately linked, as per Equation (22.20), reducing the amount of information that needs to be processed saves energy. We have discussed many examples of this principle including the use of AGCs in analog systems, the use of efficient algorithms or clock gating in digital systems, the use of analog preprocessing to reduce information, the use of learning to accumulate knowledge that reduces the amount of information that needs to be processed. The overarching principle of information reduction could encompass all principles in low-power design. We have listed other principles separately because this overarching principle is rather abstract. Its concrete application requires knowledge of algorithms, signal processing, architectures, circuit topologies, and device physics.

22.8.8 Use Feedback and feedforward architectures for improving robustness and energy efficiency

Feedback is important in improving both the efficiency and robustness of a low-power mixed-signal system. The feedback benefits for energy efficiency occur through three primary mechanisms:

a. A lowpass or integrator transfer function in a feedback path provides differentiation, delta encoding, highpass filtering, or 'adaptation' such that unwanted dc information is removed and desirable ac information is propagated to the output. The net reduction in the dynamic range of the input enables one to enhance signal gain over noise gain in the feedforward path without saturation, and thus obtain a better power-*SNR* relationship for the signal. The higher gain also enhances the strength of the signal for future stages of processing, which can then lower their power. Several architectures fall into this category, including the microphone circuit of Chapter 19, the transimpedance amplifier of Chapter 11 with differing ac and dc gain, and the EKG amplifier with common-mode feedback in Chapter 20. This benefit of feedback for energy efficiency could be termed *SNR improvement via feedback attenuation of undesirable signals*. The gain-bandwidth product of the feedback loop is typically small since we only need to reject slowly varying signals in most cases; thus, the energy overhead of the added feedback circuitry is usually small. Clearly, other kinds of filter-rejection transfer functions or other kinds of linear input combinations can be rejected by altering the feedback path. In fact, in Chapter 23, we shall see that the retina in the eye implements highpass spatial filtering by having a lowpass spatial filter in a feedback loop attenuate redundant, correlated, low-spatial-frequency information in its pixels.

b. Nonlinear feedback implements automatic gain control or compression such that characteristics of the architecture such as gain adapt to the current signal statistics. As the signal statistics change, the parameters change such that informative regions of the input space are always mapped to the output with relatively constant *SNR*. The *AGC* circuit described in Chapter 19 and the adaptive Class-A biasing scheme in Chapter 14 fall into this category. In both these cases, we see that the effective overall dynamic range of operation is large while the instantaneous dynamic range and output *SNR* are much smaller.

Thus, for example, we can pay the power costs of an 8-bit computation while having 16-bit dynamic range. We can call this benefit of feedback, *feedback adaptation to signal statistics*. The digital adaptation of V_{DD} as the activity factor changes (see Chapter 21) also falls into this category. The adaptive power biasing of neural amplifier arrays for brain-machine interfaces in Chapter 19 also falls into this category.

c. A very analogous benefit of feedback for energy efficiency to the one just outlined is *feedback attenuation of device noise statistics*. For example, the use of calibration and auto-zeroing feedback loops to attenuate $1/f$ noise and offset in the log ADC of Chapter 19 and the use of digital feedback calibration to improve the performance of analog systems as we have discussed with regard to the HSM architecture of Figure 22.9 fall into this category. The improvement in energy efficiency results because the feedback path is usually either slow-and-precise, simple-and-precise, or infrequent such that it adds little power overhead. Often the same hardware is just reconfigured into a temporary feedback loop with switches such that there is little areal cost except to minimize capacitive charge-injection errors. A fast-and-complex feedforward path can attain precision via a slow and/or simple well-controlled feedback path at cheap cost.

The feedback benefits of robustness are all due to those outlined in Chapter 2, i.e., the reduction of distortion and disturbances at the output due to high feedback loop gain. Feedback provides a significantly better robustness-efficiency tradeoff than can be attained by simply increasing device sizes to improve matching, which is why feedback is critical in moving the analog-digital crossover point in Figure 22.3 to higher *SNR*.

However, all is not rosy with feedback. To exhibit robust performance with good phase margin, a feedback loop often has to operate 2 times to 4 times slower than the maximum operating frequency of the technology. While this may not be limiting in some applications, in others it means a sacrifice in maximal speed.

Closed-loop amplification can also significantly increase the power costs of amplification compared with open-loop schemes. In the case of ADCs, the power costs of getting a gain of 2 are significantly worse with closed-loop feedback amplifiers than with open-loop amplifiers. In such cases, we are giving up the high gain of the feedforward path to attain good closed-loop-gain matching via the feedback path and this is expensive in terms of power. The gain-bandwidth or *SNR*-bandwidth product of the closed-loop topology stays relatively invariant with frequency such that high gain-bandwidth is needed in the feedback loop if we are to be fast and precise at all frequencies of interest. Hence the crossover frequency has to be well above the maximum frequency at which we want to operate at relatively high precision, which increases power consumption. For the same reason, as we outlined in Chapter 12, open-loop-integrator filter topologies are more power efficient than closed-loop-integrator filter topologies. What we are observing here is an example of a robustness-efficiency tradeoff. The feedback is giving us precision in our amplification process and less susceptibility to noise and interference, which the open-loop topologies will almost certainly have, but it is doing so at the cost of reduced efficiency.

When we discussed the benefits of feedback for improving energy efficiency in the three itemized points, we note that feedback improves energy efficiency rather than trading it for robustness as in the paragraph above: there, feedback helps us get rid of slowly changing unwanted signals, helps us operate in information-bearing regions of the signal space, or helps us get rid of slowly changing device errors cheaply. The improvement in all of these cases arises because efficiency improves with the rejection of unwanted disturbances and noise. These disturbances are rejected at relatively low frequencies where the feedback is effective. The signal amplification of the main signal occurs in the feedforward path at higher frequencies where the feedback is ineffective.

Since high gain-bandwidth products in a feedback loop cost power and can lead to instability and ringing, it is advantageous to combine feedforward techniques with feedback techniques such that a feedforward path removes a large fraction of the predictable error or unwanted signal, while a feedback path removes the residual unpredictable error or unwanted signal. Since the error that is needed is then reduced, the loop gain of the feedback loop can be much lower improving stability and energy efficiency. Figure 22.10 illustrates the idea. The figure also indicates that the use of learning can alter parameters in the feedforward compensation, feedback compensation, and main foreground amplification pathways via very slow learning loops over time. Thus, the removal of unwanted signal and undesirable error can get better and better over time as the system gets more familiar with the statistics of the signal and the statistics of the noise. Indeed, that's exactly what biological systems do when they learn to perform a task better.

It is worth noting that the feedforward and feedback portions of Figure 22.10 can be digital in both analog and digital systems. In fact, clock gating that we have discussed in Chapter 21 is an example of a feedforward information-reducing energy-saving principle.

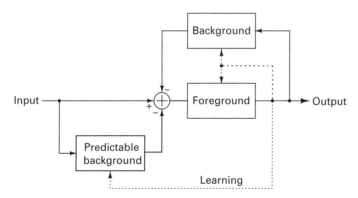

Figure 22.10. Feedforward and feedback architectures in mixed-signal design.

22.8.9   Separate speed and precision in the architecture if possible

It costs power to be simultaneously fast and accurate, whether we are performing analog processing, digital processing, sensing, or actuating. Thus, it is advantageous as far as possible to design architectures where speed and precision are not simultaneously needed. Digital systems in general already incorporate this principle since they can be fast but they are only 1-bit precise. Analog comparators also embody this principle very well, which is why they are significantly more energy efficient than amplifiers; they have also scaled well with technology. Comparator-based ADC topologies are highly energy efficient as we outline in Chapter 15.

A comparator is implemented with a low gain open-loop preamplifier followed by a positive-feedback latch. The preamplifier has a relatively low power cost: the gain-bandwidth product of the open-loop amplifier, which is also proportional to its SNR-bandwidth product, is modest such that the power consumption of the preamplifier is relatively low. The positive-feedback latch achieves high gain through unstable nonlinear amplification. It is therefore highly efficient at having low power consumption for a given speed of operation. In the cascade of the preamplifier and the latch, the preamplifier needs to be somewhat precise as far as its input-referred minimum-detectable signal goes; the relative precision required of the preamplifier decreases with increases in its common-mode voltage. The latch need only be fast. Thus, speed and precision are well separated in the two halves of the comparator since neither circuit, the preamplifier or the latch, need to simultaneously have extreme speed or extreme precision. In Chapter 15, we discuss highly power efficient comparators that have almost no static power consumption through the use of precharging and inverter-like operation. It is interesting that neuronal circuits in nature also use comparators extensively and neurons are highly energy efficient (see Chapter 15).

22.8.10   Operate Slowly and adiabatically if possible

Operating well below the f_T of a technology is power efficient since it allows for adiabatic digital operation by minimizing voltage drops across energy-dissipating resistors. In such operation, the frequency of the clock f_{clk} is well below that of the RC charging times of switches, which leads to a reduction in switching power dissipation by a factor proportional to $f_{clk}RC$. We discussed in Chapter 21 why such energy-efficient adiabatic operation in digital circuits is just a manifestation of high-Q energy-efficient operation with passive devices in analog circuits, e.g., as in tuned LCR oscillators. In analog circuits, operating well below f_T minimizes gate-capacitance switching losses and switch resistance losses, e.g., as in the Class-E switching amplifier discussed in Chapter 16. Operating adiabatically is also advantageous in nerve-stimulation circuits as we discussed in Chapter 19. In RF circuits, operating too close to the maximum f_T increases amplifier input-referred noise due to current and voltage gain reductions, and leads to the need to account for noise

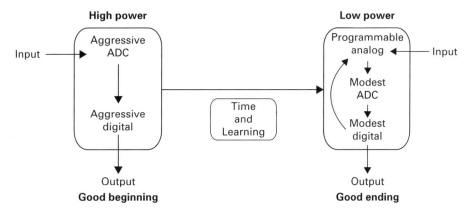

- General purpose
- High ignorance of what is important
- Purely feedforward
- 'Intelligence' only in digital domain

- More special purpose
- High knowledge of what is important
- Adaptation and feedback
- 'Intelligence' in analog and digitial domains

Figure 22.11. The evolution of low-power design.

correlations in transistors via induced gate noise (see Chapters 7 and 8). Thus, in general, in low-power circuits one must try to operate at a factor that is at least 5 times below f_T if possible. Fortunately, f_Ts in subthreshold and moderate inversion are already a few GHz in several commercially available processes. Thus, even for the 10 GHz 0.13 μm RF circuits discussed in Chapter 23 in the context of an RF cochlea, good energy efficiency is obtained in weak and moderate inversion.

Before we end this lengthy section, we would like to point out that robustness and flexibility always trade off with energy efficiency. The extra degrees of freedom necessary to maintain robustness or flexibility invariably hurt efficiency. This principle is true in all systems, whether digital, analog, or mixed-signal.

Finally, a simple *low-power verse* helps us remember all ten principles of low-power mixed-signal design:

> *Energy Saving Angels Play Chess Collectively,*
> *Reducing Future Sins Selectively*

which stands for *Encode Subthreshold Analog Parallel Computation-Communication Collective-analog Reduce Feedback Separate Slowly.* For clarity, the key words in the titles of the various principles have been underlined as well.

22.9   The evolution of low-power design

Figure 22.11 summarizes the pros and cons of the general-purpose ADC-then-DSP architecture versus that of a significantly more energy-efficient architecture that is architected in accord with the low-power principles we have outlined. Both architectures can actually exist in advanced systems, with the general-purpose architecture providing periodic calibration and tuning functions for the more efficient architecture. Low-power design typically evolves from the general-purpose architecture towards the energy-efficient architecture with time and learning. The general-purpose architecture makes sense in initial design iterations when flexibility is important because one's ignorance of the environment is high. When one has acquired knowledge of the environment over time, and complete flexibility is less important, we should evolve towards the more energy-efficient architecture.

22.10   Sensors and actuators

Sensors and actuators have physical state variables that need to be sensed or transformed. They fit within the processing framework of this chapter and therefore need no special treatment: they are simply analog variables in an extended technology that is not purely electrical. In fact, the principles for analog computation that we have outlined imply that we *must* exploit clever signal processing *within* the sensor and actuator whether the technology is mechanical, electrical, or chemical, just like nature does (see the cochlea in Chapter 23 or chemical computations in Chapter 24), to further lower power. The sensor and actuator become part of the analog preprocessing. Adaptive smart sensing may have feedback between the analog front end and the sensor. Sensors and actuators therefore compute! Chapters 8, and 11 to 20 have discussions of low-power sensing and actuating circuits in various biomedical contexts including microphone preamplifiers, imagers, nerve-stimulation circuits, photoreceptors, RF antennas, RF inductive links, and other circuits. Chapter 8 discusses how sensor noise in non-electrical systems affects the overall noise, signal-to-noise ratio, and consequently information encoded by a signal processing system.

References

[1] Thomas H. Lee. *The Design of CMOS Radio-Frequency Integrated Circuits*, 2nd ed. (Cambridge, UK; New York: Cambridge University Press, 2004).

[2] James O. Pickles. *An Introduction to the Physiology of Hearing*, 2nd ed. (London; New York: Academic Press, 1988).

[3] R. Sarpeshkar, R. F. Lyon and C. A. Mead. A low-power wide-dynamic-range analog VLSI cochlea. *Analog Integrated Circuits and Signal Processing*, **16** (1998), 245–274.

[4] Eric R. Kandel, James H. Schwartz and Thomas M. Jessell. *Principles of Neural Science*. 3rd ed. (Norwalk, Conn.: Appleton & Lange, 1991).

[5] S. Das, C. Tokunaga, S. Pant, W. H. Ma, S. Kalaiselvan, K. Lai, D. M. Bull and D. T. Blaauw. RazorII: In situ error detection and correction for PVT and SER tolerance. *IEEE Journal of Solid-State Circuits*, **44** (2009), 32–48.

[6] P. Korkmaz, B. E. S. Akgul and K. V. Palem. Energy, performance, and probability tradeoffs for energy-efficient probabilistic CMOS circuits. *IEEE Transactions on Circuits and Systems I: Regular Papers*, **55** (2008), 2249–2262.

[7] Claude Elwood Shannon and Warren Weaver. *The Mathematical Theory of Communication* (Urbana: University of Illinois Press, 1949).

[8] F. Reif. *Fundamentals of Statistical and Thermal Physics* (New York: McGraw Hill, 1965).

[9] C. H. Bennett and R. Landauer. The fundamental physical limits of computation. *Scientific American*, **253** (1985), 48–56.

[10] R. Landauer. Information is physical. *Workshop on Physics and Computation (PhysComp)*, Dallas, Tex., 1–4, 1992.

[11] R. Sarpeshkar. Analog versus digital: extrapolating from electronics to neurobiology. *Neural Computation*, **10** (1998), 1601–1638.

[12] Christof Koch. *Biophysics of Computation: Information Processing in Single Neurons* (New York: Oxford University Press, 1999).

[13] R. Sarpeshkar and M. O'Halloran. Scalable hybrid computation with spikes. *Neural Computation*, **14** (2002), 2003–2038.

[14] Uri Alon. *An Introduction to Systems Biology: Design Principles of Biological Circuits* (Boca Raton, FL: Chapman & Hall/CRC, 2007).

[15] D. Baker, G. Church, J. Collins, D. Endy, J. Jacobson, J. Keasling, P. Modrich, C. Smolke and R. Weiss. Engineering life: building a fab for biology. *Scientific American*, **294** (2006), 44–51.

[16] P. Dudek and P. J. Hicks. A general-purpose processor-per-pixel analog SIMD vision chip. *IEEE Transactions on Circuits and Systems I: Regular Papers*, **52** (2005), 13–20.

[17] Leon O. Chua and T. Roska. *Cellular Neural Networks and Visual Computing: Foundation and Applications* (Cambridge, UK; New York, NY: Cambridge University Press, 2002).

Section VI

Bio-inspired systems

23 Neuromorphic electronics

It is the unification and simplification of knowledge that gives us hope for the future of our culture. To the extent that we encourage future generations to understand deeply, to see previously unseen connections, and to follow their conviction that such endeavors are noble undertakings of the human spirit, we will have contributed to a brighter future.

Carver Mead

Biological systems are *the* most energy-efficient systems in the world. For example, the ~ 22 billion neuronal cells of the brain dominate the ~ 14.6 W brain power consumption of an average 65 kg male [1], [2]. These numbers imply a power consumption of ~ 0.66 nW per neuron. The hybrid analog-digital brain performs *at least* 6×10^{16} FLOPS (floating-point operations per second) such that its energy efficiency is a staggeringly low 0.24 fJ/FLOP. This energy efficiency is about 5–6 orders of magnitude more efficient than that of even the most energy-efficient digital microprocessor or digital signal processor ever built. The human eye's retina consumes nearly ~ 3.4 mW, making the 135 million photoreceptor array in the eye the lowest power wide-dynamic-range imager and image compressor ever built. The retina in the eye performs sophisticated analog gain control, analog spatial filtering, and analog temporal filtering such that nearly ~ 36 Gb/s of wide-dynamic-range raw image data from its photoreceptor array is compressed to ~ 20 Mb/s of useful optic-nerve spiking output information. The human ear's nano-fluidic, electro mechanical, and electro chemical technology implements more than 1 GFLOPS of analog filtering, analog gain control, and analog spectral-analysis computations in $\sim 14\,\mu$W of power and is a marvel of nanotechnology. The human ear could run on a AAA battery for 15 years. The entire human body only has a basal metabolic power consumption of ~ 81 W [2]. The references and computations that lead to energy estimates for the brain, the body, the eye, and the ear are provided in the Appendix section of this chapter. The Appendix also provides numbers for the power consumption of the other organs of the body. In Chapter 24, we shall discuss the energy consumption of the body's cells.

The overall engineering specifications of the brain, the eye, and the ear, which are listed in Tables 23.1, 23.2, and 23.3 respectively, are awe-inspiring. The brain has a fantastic 3D interconnect technology that enables a fan-in and fan-out of nearly 6,000 connections per neuron compared with ~ 6 in the logic gates of today's microprocessors. Its architecture enables it to vastly outperform even the

Table 23.1 Specifications of the cochlea (and auditory system)

The cochlea and auditory system	
Input dynamic range	$\sim 120\,\mathrm{dB}$
Output dynamic range of nerve firing	$\sim 40\,\mathrm{dB}$
Power dissipation	$\sim 14\,\mu\mathrm{W}$
Power supply voltage	$\sim 150\,\mathrm{mV}$
Detection threshold at 3 kHz	0.05 Angstroms at eardrum
Frequency range	20 Hz – 20 kHz
Number of auditory nerve fibers	$\sim 35{,}000$
Filter bandwidths	$\sim 1/3$ octave
Phase locking threshold	$\sim 5\,\mathrm{kHz}$
Inter-aural time discrimination	$\sim 10\,\mu\mathrm{s}$
Loudness discrimination	1 dB
Frequency discrimination at 1 kHz	2 Hz
Computational rate	$> \sim 1\,\mathrm{GFLOPS}$
Length of uncoiled pea-sized cochlea	35 mm

Table 23.2 Specifications of the retina (and visual system)

The retina and visual system	
Dynamic range	$10^8{:}1$
Power dissipation of retina	$\sim 3.4\,\mathrm{mW}$
Power supply voltage	$\sim 70\,\mathrm{mV}$
Detection threshold of rod photoreceptor	1 photon
Bandwidth of response	12 Hz (rod), 55 Hz (cone)
Number of optic nerve fibers	~ 1.2 million
Spatial acuity at 20/20 vision	~ 2 arc minutes (60 minutes/degree)
Output bit rate of optic nerve	$\sim 20\,\mathrm{Mb/s}$
Vernier hyperacuity	0.2 arc minutes
Number of photoreceptor cells	~ 5.5 million (cones); ~ 130 million (rods)
Number of retinal synapses	~ 1 billion
Computational rate	$> 10\,\mathrm{GFLOPS}$
Area of $\sim 160\,\mu\mathrm{m}$ thick retina	$2500\,\mathrm{mm}^2$

most advanced supercomputers and robots at the tasks it has evolved to be good at, for example, sensorimotor processing, control, pattern recognition, and learning. The eye can detect a single photon and has a spatial acuity of 2 arc minutes that is as good as that determined by diffraction limits of optical physics. The visual system exhibits vernier hyper-acuity that allows it to detect line misalignments that are an order-of-magnitude below that limited by the spatial sampling of the photoreceptors from the eye. At its thermal-noise-limited threshold, the ear can detect a vibration of one-twentieth the diameter of a hydrogen atom at the ear drum. The auditory system can tell the difference in the time of arrival between the left ear and the right ear to within 10 μs even though the time constants of its cells are only 1 ms. The piezoelectric amplifying outer hair cells of the ear have a

Table 23.3 Specifications of the brain

The brain	
Kinds of two-terminal synapses	50–100
Kinds of one-terminal ion channels	50
Power supply voltage	$\sim 125\,\text{mV}$
Number of synapses	240×10^{12}
Number of neurons	~ 22 billion
Number of glial cells	~ 200 billion
Brain power dissipation	$14.6\,\text{W}$
Average firing rate	$\sim 5\,\text{Hz}$
Computational rate	$> 6 \times 10^{16}$ FLOPS
Energy efficiency	< 0.24 fJ/FLOP
Average connections per neuron	~ 6000
Surface area of cortex	$\sim 2500\,\text{cm}^2$
3D synapse grid size of $\sim 2\,\text{mm}$ thick cortex	$1.1\,\mu\text{m}$

piezoelectric coefficient that is a few orders of magnitude larger than that of any artificial piezoelectric known to man. Taken together, these specifications manifest the fantastic technology integration, sophisticated signal processing, and clever architectural design that nature has accomplished over more than a billion years of evolution. Since food, which ultimately provides energy, was a very scarce resource in the history of animal evolution, biological systems have been pushed to get more and more out of less and less by amazingly energy-efficient designs [3], [2]. As Thoreau said, "Nature is full of genius," and we can learn from her. She has figured out how to take many imprecise, noisy parts and put them together to create precise, complex, robust systems that operate in real time with phenomenally high energy efficiency.

Electronics inspired by neurobiology was termed *neuromorphic electronics* by the field's founder, Professor Carver Mead [4], [5]. Mead pointed out that the use of the physical basis functions of analog computation, the intimate integration of logic and memory through mostly local wiring, and the learning capabilities of neurobiological systems were key ingredients to their energy efficiency [4]. It is worth pointing out that *all* biological systems are energy efficient, not just neurobiological ones. For example, in Chapter 24, we shall discuss the amazing energy efficiency of a single cell that performs sophisticated biochemical computations in its gene-protein network. In Chapter 24, we will also discuss how to build *cytomorphic electronics*, a novel form of electronics introduced in this book to describe electronics inspired by cellular and molecular biology. In general biologically inspired engineering systems can be termed *biomorphic systems*. Examples of fluid-mechanical biomorphic systems include airplanes, which take inspiration from the winged flight of a bird. Examples of chemical or material biomorphic systems include the design of coatings for "self-cleaning" windows, which take inspiration from the design of a lotus leaf. Through its clever use of hydrophobicity and hydrophilicity at the leaf surface, the lotus plant's leaves remain remarkably clean in dirty water. They have recently been mimicked in engineering designs [6]. It is worth itemizing where biological designs have or are likely to inspire engineering designs in the future:

- Ultra-low-power and highly energy-efficient sensing, actuating, and information-processing systems, e.g., the brain, the eye, the ear, the muscles, the heart, and molecular sensors and actuators.
- Signal processing and pattern-recognition systems that need to operate in noisy environments and over a wide dynamic range of inputs, e.g., the auditory system excels at speech recognition in noise, cellular biochemical systems operate with noisy molecular inputs and make reliable decisions.
- Robust and efficient computation with noisy and unpredictable devices, e.g., the brain, the ear, and the cell.
- Systems with feedback, adaptation, and learning at multiple spatial and temporal scales, e.g., the brain and the cell.
- Systems that integrate technologies from diverse domains, e.g., the eye, the ear, the hand, and the brain.
- Self-repairing systems, e.g., wound healing in the immune system.
- Self-assembling systems that are robust to error, e.g., the birth of a baby from $\sim$43 cell divisions that eventually all originate from a single fertilized human embryonic cell.
- Energy-harvesting systems, e.g., plants that harvest sunlight for energy and in turn create highly energy-dense molecules from which animal enzyme systems are efficient at extracting energy.
- Robotic systems, e.g., simple houseflies with an ingenious fly-vision system and flight-control system [7], [8]; motor control systems that perform well in spite of large delays in their feedback loops through the use of prediction, e.g., the eye-movement control system [9]; the bio-sonar navigation system of a bat.
- Security systems, e.g., those inspired by the clever positive and negative feedback architectures, specialized and unspecialized architectures of the immune system.
- Materials with biological properties, e.g., a high strength-to-weight ratio of the material in a spider's web.
- Chemical analysis systems, e.g., electronic noses, drug design that mimics enzyme and antibody design to bind targets.

How do biological systems like the ear, the eye, the cell, and the brain achieve good energy efficiency? A few common themes appear to be:

1. The use of *analog* physical basis functions to achieve high computational efficiency rather than just logic basis functions as in traditional digital computation (see Chapter 22 for an extensive discussion of the pros and cons of analog versus digital computing). Biology has a fantastic set of rich technology basis functions to choose from, e.g., the brain can compute with 50–60 different kinds of post-synaptic ion-channel I–V curves rather than just two in electronics

(nFETs and pFETs). It has several different kinds of chemical species that allow extensive signal multiplexing rather than just two in electronics (holes and electrons). The brain also has an adaptive power supply via its mitochondrial cells that automatically increase in number in regions of high power density.

2. The extensive use of *feedback and adaptation* to maintain robustness to device and signal errors and to ensure efficient operation in the most informative regions of the statistical signal space. Feedback and adaptation are so ubiquitous in biology that is hard to find examples where it is not present at any spatial or temporal scale and in any system. Its ubiquity allows for analog signal restoration when negative feedback loops are present and amplification and discretization when positive feedback loops are present.

3. The extensive use of *learning* to allow a system to perfect itself from signal statistics or from explicit examples. Unsupervised learning allows a system to self-organize based on correlation statistics and patterns amongst its inputs. Supervised learning allows a system to alter its parameters based on explicit examples in a supervised fashion. The brain is without doubt the best learning machine that has ever been built, and machine-learning systems are quite far from replicating its speed-of-convergence and generalization abilities.

4. The use of *collective* computation: the use of moderate-precision computing units that collectively interact to perform a more complex computation or a higher-precision computation. We have introduced this idea in Chapter 22 (see Figure 22.7 (a)) and will provide examples of such computation in this chapter.

5. The use of *cellular* architectures with mostly local computation and local communication amongst nearby cells. The intimate integration of processing (logic) and state (memory) in each cell ensures that wiring and communication costs are minimized between cells and within cells unlike in traditional von Neumann architectures. Cells in the body often integrate sensing and actuation within each cellular unit as well. The overall organ system, of which the cells form a part, is then scalable as new cells are added to it and is then robust to the loss of a significant number of cells within it. Examples of cellular architectures range from networks of skin cells in the body to networks of neurons in the brain. Neurons in the brain appear to connect with each other in a stereotypical cellular columnar architecture that is repeated in several areas of the cortex [10]. It is worth remembering that von Neumann himself was researching analog and cellular computation near the end of his life because he was aware of the limitations of his own architecture.

6. The use of 1D or 2D *nonlinear-and-adaptive transmission-line architectures*. Such architectures can process information adiabatically or combine information from many units in a highly slow-and-parallel fashion, both of which are very energy-efficient strategies. The cochlea is well modeled by a pair of coupled active-and-adaptive nonlinear LCR 1D transmission lines with L and C changing with position in the cochlea thus enabling spectral analysis. Sound is input to the beginning of the line. The vocal tract is well modeled by LCR

transmission lines that model the pharyngeal, oral, and nasal airflow tracts with voltage representing pressure and current representing volume velocity. The retina is well modeled by coupled RR 2D transmission lines that enable spatial filtering. Light-dependent sources serve as parallel inputs to one of its transmission lines. The input dendrite of a neuron is often modeled as a passive 1D RC transmission line with several parallel inputs from neurons that connect to the neuron in a distributed fashion throughout the line. It may be more accurate to model these dendrites by active, adaptive, and nonlinear 1D RC transmission lines [1]. The output axons of neurons are well modeled by passive 1D RC transmission lines with periodic active amplifiers for signal restoration. Note that we have used the word 'transmission line' to represent any ladder-like topology with lateral $Z(s)$ impedances and shunt $Y(s)$ admittances while some authors restrict the use of the term to topologies where $Z(s) = Ls$ and $Y(s) = Cs$. Nevertheless, for simplicity we shall avoid the stricter term 'generalized transmission lines'.

7. The use of *hybrid analog-digital* computational architectures that blend analog continuous signal processing and discrete digital symbolic processing. Analog signal processing with basis functions that are matched to the computation provides energy efficiency. Digital symbolic processing allows for decision-making, which improves efficiency by allowing the focused use of resources to only attended parts of the input space, and it allows restoration of noisy analog signals. Neurons have inherently hybrid pulse-based signal representations and are particularly well suited to mixed-signal processing [11]. Gene-protein networks in cells also have inherently hybrid analog-digital signal processing as we shall discuss in Chapter 24. In neuromorphic architectures that blend ideas from biology and engineering, we shall discuss how hybrid analog-digital architectures can enable one to design complex analog computations as a programmable sequence of simpler analog computations.

In this chapter, we shall provide several examples of neuromorphic architectures that illustrate these principles. Chapter 15 has already discussed how a neuron-inspired architecture has led to a very energy-efficient architecture and algorithm for analog-to-digital conversion. That chapter also illustrated how computational ADCs, which digitize computational functions of their analog inputs like neurons do, may be constructed efficiently. Chapter 19 provided an example of an asynchronous interleaved sampling (AIS) algorithm that achieves energy-efficient stimulation in cochlear implants by mimicking the stochastic and asynchronous sampling strategy of auditory neurons. In this chapter, we shall discuss architectures, algorithms, and circuits inspired by those of the inner ear or cochlea, by those of the vocal tract, by those of the retina in the vertebrate eye, by visual motion detectors in flies, and by hybrid analog-digital computation in the brain. Some of these architectures have already had practical impact in medical and engineering applications as we shall discuss. Some promise interesting applications for the future. Medical applications are very natural for bio-inspired architectures since an architecture that mimics biology can help fix it when it does not work. The analog cochlear-implant processor, described in depth in Chapter 19, used several circuits similar to those used in an audio bio-inspired silicon cochlea to perform filtering, energy extraction, and gain control [31], [12], [13], [14].

We shall first summarize two key results from transmission-line theory that are broadly useful in understanding 1D, 2D, and 3D distributed structures in the cochlea, retina, vocal tract, dendrites, axons, and reaction-diffusion networks in cells. This discussion will serve as a short introduction to some background material for those readers unfamiliar with these results. Then, we shall begin with a somewhat unusual and very recent bio-inspired architecture, the *RF cochlea* [12]. The RF cochlea morphs the ear's wave-processing transmission-line architecture to create a wideband, ultra-fast, low-power RF spectrum analyzer. This analyzer operates with RF electromagnetic waves rather than sound waves and with an LCR transmission-line architecture rather than one built with fluid mass, membrane stiffness, and damping as in the biological cochlea. Although the RF cochlea operates at $\sim$10 GHz frequencies rather than $\sim$10 kHz frequencies like the biological cochlea, its principles of operation are very similar. Therefore, it inherits several of the cochlea's benefits including low-power, wide-dynamic-range operation, and the fastest spectrum-analysis algorithm known to man.

Such an RF spectrum analyzer could have applications as a front end in universal, software-defined radio (SDR), and cognitive radios of the future by providing fast approximate estimates of the increasingly crowded radio spectrum. Such estimates may then be used to alter transmitting frequencies, suppress interferers, or improve reception in noisy environments, which can then make much better use of the current poorly utilized radio spectrum in the smart cognitive radio [15]. Then, we shall briefly discuss a bio-inspired companding architecture for spectral-analysis inspired by the cochlea and auditory system that has been shown to improve speech recognition in noise in the deaf and in artificial speech-recognition systems [16], [17]. If we extend ideas behind the spectral analysis from one dimension to two dimensions, companding can sharpen and improve the analysis of noisy images as well. We then discuss another 1D transmission-line chip based on airflow rather than water flow as in the cochlea: a bio-inspired analog integrated-circuit vocal tract, i.e., a chip that talks [18]. The bio-inspired vocal tract is architected to be in a speech-locked loop (SLL) with an audio spectrum analyzer analogous to a phase-locked loop (PLL) in engineering (PLLs are discussed in Chapter 18). By putting speech synthesis and hearing analysis in such an analysis-by-synthesis feedback loop, we show how speech recognition in noise can be potentially more robust: a clean model of how the sound is produced helps understand a noisy utterance of what has just been said. We briefly discuss how such ideas may also be used to construct a heart-locked loop (HLL) that allows estimation of cardiovascular parameters from PPG, EKG, and PCG measurements (see Chapter 20 for a discussion of the heart) [19]. We then discuss how 2D transmission-line architectures have been used to successfully create integrated-circuit models of the retina to perform spatial filtering on the image input. We also discuss how models of image-motion computation in the fly have been used to build visual motion detectors useful in robotic applications.

We discuss the hybrid analog-digital nature of spiking neuronal computation in the brain. We describe how mimicry of the negative and positive feedback in the brain's synaptic networks has led to a bio-inspired circuit that can be simultaneously analog and digital. Such distributed nonlinear feedback architectures break the traditional boundaries between analog and digital computation. Such boundaries are slowly starting to dissolve, and they are likely to continue to dissolve even more in the future.

Hybrid analog-digital architectures inspired by spiking neuronal computation lead to the idea of a hybrid state machine (HSM), a generalization of finite state machines to the hybrid analog-digital domain. Although we have described HSMs in Chapter 22 through the lens of mixed-signal circuits, we shall arrive at them through a different neuron-inspired route in this chapter. Such HSMs allow for general-purpose algorithmically programmable analog computers to be built by composing a complex computation as a sequence of simple ones. There is intimate feedback between the analog and digital portions of an HSM such that the signal-to-symbol mapping of a traditional ADC is not only one way from signal to symbol but also back from symbol to signal. Increasingly, advanced circuits in engineering are using predictive digital compensation of errors in analog systems to enable better robustness and efficiency. Special examples of HSMs include the neuron-inspired ADC described in Chapter 15 and classic $\Sigma\Delta$ ADCs described in Chapter 15. More generally, networks of HSMs serve as architectures for general-purpose programmable hybrid analog-digital cellular computation, like amongst neurons in the brain, or in intracellular or extracellular networks in the body. In fact, visual microprocessors that have been recently built [20] implicitly incorporate a network of HSMs in a special configuration to do ultra-fast, computationally intensive, and highly energy-efficient image processing inspired by the architecture of the retina.

We review how the principle of collective analog or hybrid computation outlined in Figure 22.7 (a) for energy-efficient operation applies to various neurobiological examples that we have discussed. We show that neurobiological architectures appear to obey *all* ten principles for energy-efficient operation that we have outlined in Chapter 22. Biology may continue to be a good place for ultra-low-power engineers to find other such principles in the future.

We conclude by pointing the reader to other work in the field of neuromorphic electronics and to work in biologically inspired systems in general. In biomorphic, neuromorphic, or cytomorphic systems, it is important to keep the insightful "baby" and throw out the cluttering "bath-water" details. Certain architectures in biology may be accidents of evolution, may be more suited to the constraints of a biological organism, and may serve or may have served a purpose that we do not yet understand. Consequently, their relevance to a different engineering context where the constraints are different may be questionable. Birds are not airplanes and airplanes are not birds, although the study of one can shed insight into the study of the other. Hence, it is important to evaluate a biomorphic engineering system by traditional engineering metrics to insightfully understand where value has been added. The big lesson to be learned from biology is that architectural innovations that go beyond the traditional engineering architectures of today are extremely important for energy-efficient and robust architectures of the future. Indeed, the International Roadmap for Semiconductors (ITRS) now recognizes the importance of analog and bio-inspired computation for enabling improvements that go far beyond Moore's law [21]. The field of biologically inspired design suggests that we can mine the intellectual resources of nature to create devices useful to man, just as we have mined her physical resources in the past. Such mining will require us to combine inspiration with perspiration and to understand how nature works with insight.

23.1 Transmission-line theory

In Chapter 17 on antennas, we discussed LC transmission lines composed of an infinite chain of lateral inductances, L, and shunt capacitances, C, to ground. They are a special case of more general transmission lines composed of lateral impedances $Z(s)$ and shunt admittances $Y(s)$ to ground. That is, $Z(s) = Ls$ and $Y(s) = Cs$ correspond to the special case of the transmission line of Chapter 17. In discrete lumped approximations to such lines, $Z(s)$ and $Y(s)$ correspond to actual impedances and admittances. In continuous structures, $Z(s)$ corresponds to the impedance per unit length in the line, and $Y(s)$ corresponds to the admittance per unit length in the line. Two results about these lines are useful in an amazingly broad variety of contexts:

1. At frequency $j\omega$, the line is capable of supporting waves with solutions $\exp(j\omega t - k(\omega)x)$ or equivalently has a propagation transfer function $\exp(-k(\omega)x)$ with

$$k^2(\omega) = Z(j\omega)Y(j\omega)$$
$$\boxed{k(\omega) = \pm\sqrt{Z(j\omega)Y(j\omega)}}$$

(23.1)

With the sign convention that we have used for our wave solution (or our propagation transfer function), the positive root corresponds to forward wave propagation and the negative root corresponds to backward wave propagation.

2. The impedance seen at any point x of the transmission line, termed the characteristic impedance $Z_0(j\omega)$, relates the voltage and current for the forward-going wave $V(j\omega, x) = Z_0(j\omega)I(j\omega, x)$ or the backward going wave $V(j\omega, x) = -Z_0(j\omega)I(j\omega, x)$ and is given by

$$\boxed{Z_0(j\omega) = \sqrt{\frac{Z(j\omega)}{Y(j\omega)}}}$$

(23.2)

Both of these results rise from simple node equations and limit calculus applied to infinite structures, and have been derived in [22]. The equations apply to any generalized $Z(s)$-and-$Y(s)$ transmission line, not just LC transmission lines.

23.2   The cochlea: biology, motivations, theory, and RF-cochlea design

Sound pressure variations in the air are gathered by the sound antenna of our bodies, the externally visible ear lobe or pinna, which forms part of our outer ear (see Figure 19.1). They are then guided through the ear canal and towards the tympanic membrane or ear drum. These sound pressure variations cause the eardrum to vibrate. The vibrations stimulate the bones of the middle ear. At its input end, the middle ear gathers force over a large surface area of vibration of the ear drum; it transmits this force through lever action into a narrower surface area of vibration at its output end, thereby increasing the pressure with which it is able to drive its output. The vibratory output of the middle ear drives a membrane, termed the oval window, in a piston-like fashion. The motions of the oval window create a vibratory input to a fluid-filled coiled box, termed the inner ear or cochlea. The oval window is literally like a window that moves in and out within the bony walls of the fluid-filled cochlear box. If we uncoil the cochlea, and vastly oversimplify its anatomy, the cochlea is a 35 mm long fluid-filled box that is divided into two compartments by a separating membrane termed the basilar membrane. The basilar membrane is stiff at the beginning or basal portion of the cochlea and flexible at the end or apical portion of the cochlea. Fluid motions caused by the to-and-fro motions of the oval window excite basilar-membrane vibrations near the basal end of the cochlea. These vibrations in turn cause fluid motions that excite neighboring apical regions of the basilar membrane and cause them to vibrate. The process repeats in a recursive fashion such that a coupled fluid-membrane traveling wave is created that propagates from the base towards the apex of the cochlea. The wave propagation is similar to that in an LC transmission line with fluid mass analogous to inductance L and membrane compliance (1/stiffness) analogous to capacitance C except that L and C both increase as we move along the line. In reality, the basilar membrane admittance is not just capacitive and has resistive and inductive components that help amplify the wave as it propagates.

A sine-tone sound at a given frequency creates a traveling wave in the transmission-line architecture of the cochlea that propagates down its length. *The wave is amplified as it propagates to a maximal value in a certain best-place region of the cochlea, after which it sharply cuts off and dies.* The wave cuts off because, after a point, the basilar membrane is not stiff enough to keep up with the fast motions of the wave, such that there is a lowpass filtering effect (C and L are too big). Lower-frequency sine tones cause waves that propagate a greater distance along the cochlea and excite maximal vibration of its basilar membrane nearer to its apex or end. Higher-frequency sine tones cause waves that propagate a shorter distance on the cochlea and excite maximal vibration of its basilar membrane nearer to its base or beginning. Thus, the cochlea performs spectrum analysis by implementing a frequency-to-place transform. The membrane vibrations are sensed by rectifying inner hair cells in the cochlea that form part of the organ of Corti. The organ of Corti lies on the basilar membrane throughout its axial length and is consequently able to detect motions anywhere within it. The inner hair cells, which are distributed throughout the organ of Corti, transduce the mechanical vibrations of the basilar membrane into an electrical signal, which is then reported to the brain by the auditory nerve. Piezoelectric outer hair cells are also distributed throughout the length of the cochlea within the organ of Corti and pump energy into the basilar membrane, amplifying its vibrations. Outer hair cells perform nonlinear and dynamic distributed gain control such that soft sounds are amplified more than loud sounds, and a large dynamic range in input sound level is mapped to a small dynamic range in output firing rates of the auditory nerve. An in-depth circuit model of the cochlea may be found in [23]. More biological details may be found in [24].

The ear is similar to an ultra-broadband universal-radio front end for sound waves. The outer ear is analogous to a broadband antenna that transduces energy in a sound wave into vibratory motions within our ears over a 100:1 carrier frequency range, i.e., 100 Hz:10 kHz. The middle ear is analogous to an impedance-matching transformer that trades pressure for reduced velocity of vibration, i.e., it matches the low output impedance of the antenna to the higher input impedance of the cochlea. The inner ear or cochlea is analogous to a broadband distributed transmission-line amplifier that works at multiple RF carrier frequencies separating the broadband RF input into more narrowband cochlear outputs. The rectifying action of the inner hair cell is analogous to the demodulating action of an envelope detector that demodulates the carrier information at each channel of the cochlea down to baseband for eventual transmission to the brain. The piezoelectric amplifying action of the outer hair cells is analogous to amplification with active transistors in an RF amplifier. It is then natural to ask: could we build a radio front end inspired by the ear that, like it, would function over an ultra-wideband 100:1 range of carrier frequencies (say 100 MHz:10 GHz for a radio), would be very low power due to the inherent energy efficiency of distributed amplification, would have a wide dynamic range of operation, and function well with noisy signals like the ear does? We shall show that the ear-inspired RF cochlea that we describe does indeed lead to a very efficient broadband spectrum analyzer that could serve as a universal-radio front end as well. The parallel outputs of the RF cochlea could indeed be used by a back end to create a universal radio with multiple simultaneous channels by using narrowband heterodyning techniques on RF-cochlea outputs. Some details of how such a universal radio could be built with a broadband generalization of heterodyning, i.e., cochlear heterodyning in a *cascaded cochleas architecture*, have been described in [25]. Here, we shall focus on how an RF cochlea can be architected and why its architecture is very efficient for spectrum analysis. In later sections, we shall describe how the noise-robust spectral analysis properties of the cochlea can be exploited in the audio domain, and perhaps in the future in the RF domain as well.

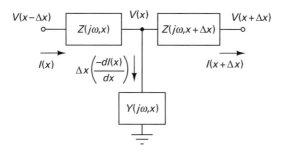

Figure 23.1. A transmission-line model of the cochlea. Reprinted with kind permission from [12] (©2009 IEEE).

23.2.1 Bidirectional transmission-line model of the cochlea

Figure 23.1 reveals a transmission-line model of the biological cochlea where the impedance per unit length, $Z(j\omega, x)$, and admittance per unit length, $Y(j\omega, x)$, vary with position x along the cochlea, and also with frequency $j\omega$ [12]. The voltage $V(x)$ represents the local differential fluid pressure in the cochlea and the current $I(x)$ represents the local fluid volume velocity. The transmission line is created by concatenating several of the T sections shown in Figure 23.1. The spatially varying transmission-line partial differential equation that corresponds to Figure 23.1 is given by

$$\frac{dV}{dx} = -Z(j\omega, x)I(x)$$
$$\frac{dI}{dx} = -Y(j\omega, x)V(x) \tag{23.3}$$

In the biological cochlea, $Z(j\omega, x)$ is well modeled by an inductance whose value increases exponentially as x varies from 0 at the beginning or base of the cochlea to its maximum value at the end or apex of the cochlea. At any x, the admittance $Y(j\omega, x)$ behaves like a capacitance if the frequency at that location is sufficiently low; the magnitude of this capacitance increases exponentially with x; at any x, $Y(j\omega, x)$ changes from being capacitive to resistive to inductive as ω nears and exceeds a characteristic frequency, $\omega_c(x)$, which decreases exponentially with x. We can summarize these characteristics with the following set of relations:

$$\omega_c(x) = \omega_c(0)e^{-x/l}$$
$$L(x) = L(0)e^{+x/l}$$
$$C(x) = C(0)e^{+x/l}$$
$$Z(j\omega, x) = j\omega L(x)$$
$$Y(j\omega, x) \approx j\omega C(x) \text{ iff } \omega \ll \omega_c(x) \tag{23.4}$$
$$Z\left(\frac{j\omega}{\omega_c(0)}, 0\right) = Z\left(\frac{j\omega}{\omega_c(x)}, x\right)$$
$$Y\left(\frac{j\omega}{\omega_c(0)}, 0\right) = Y\left(\frac{j\omega}{\omega_c(x)}, x\right)$$

where l is a length that defines the exponential taper of the cochlea. The identical dependence of the cochlear impedances on a normalized frequency $\omega/\omega_c(x)$ at any x is expressed by the last two rows of Equation (23.4). It is due to scale-invariant properties of an exponentially tapered transmission line. Such scale invariance led Zweig to formulate an ingenious normalized-frequency version of Equation (23.3) [26], [12]: We define the normalized Laplace-transform frequency variable s_n to be

$$s_n = \frac{j\omega}{\omega_c(0)e^{-x/l}} = \frac{j\omega}{\omega_c(0)}e^{+x/l} \tag{23.5}$$

such that

$$\frac{ds_n}{dx} = \frac{s_n}{l} \tag{23.6}$$

Then, Equation (23.3) can be rewritten as

$$\begin{aligned}\frac{dV}{ds_n} \times \frac{ds_n}{dx} &= -Z(\omega,x)I(x) \\ \frac{dI}{ds_n} \times \frac{ds_n}{dx} &= -Y(\omega,x)V(x),\end{aligned} \tag{23.7}$$

which from Equations (23.5) and (23.6) simplifies to

$$\begin{aligned}\frac{dV}{ds_n} &= -\left(\frac{lZ(s_n)}{s_n}\right)I(s_n) \\ \frac{dI}{ds_n} &= -\left(\frac{lY(s_n)}{s_n}\right)V(s_n)\end{aligned} \tag{23.8}$$

Equation (23.8) illustrates that we may write the cochlear partial differential equation (PDE) with s_n rather than x as a variable and with ω being an implicit parameter in the definition of s_n (Equation (23.5)). The advantage of doing so is that it illustrates two very important and beautiful symmetry properties of the cochlea:

1. In normalized s_n coordinates, the solutions $V(s_n)$ and $I(s_n)$ are identical functions at all x in the cochlea. That is, at any location, the frequency-response function is identical to the frequency-response function at any other location, if we normalize frequencies at each location by the corresponding characteristic frequency $\omega_c(x) = \omega_c(0)e^{-x/l}$ at that location.

2. The spatial-response function at a fixed frequency and the frequency-response function at a fixed spatial location have almost identical appearance. The functions look identical if we merely change the X-axis variable from x to $\ln(\omega)$. This interchangeability of frequency and space arises because all functions in the cochlea are functions of $s_n = \omega/\left(\omega_c(0)e^{-x/l}\right)$, a variable that is a nonlinear combination of frequency and space, rather than of space and frequency alone. Thus, it does not matter if s_n changes because x changes and ω is fixed or if s_n changes because ω changes and x is fixed. Since s_n has a linear dependence on ω and an exponential dependence on x, cochlear functions plotted versus $\ln(\omega)$ and x (with the other variable fixed) both change s_n in the same proportion and therefore look identical.

As in Equation (23.1), the PDE of Equation (23.8) has wave solutions of the form $\exp(j\omega t - k_n(s_n)s_n)$ where the wave-number k_n, reciprocally related to the wavelength, now varies with x or equivalently s_n, and is given by

$$
\begin{aligned}
k_n(s_n)^2 &= \left(\frac{lZ(s_n)}{s_n}\right)\left(\frac{lY(s_n)}{s_n}\right) \\
&= \left(\frac{l\omega_c(0)e^{-x/l}j\omega L(0)e^{+x/l}}{j\omega}\right)\left(\frac{lY(s_n)}{s_n}\right) \\
&= (\omega_c(0)L(0)l)\left(\frac{lY(s_n)}{s_n}\right)
\end{aligned}
$$

$$
\boxed{k_n(s_n)^2 = Z_r\left(\frac{lY(s_n)}{s_n}\right)}
$$
(23.9)

with Z_r implicitly defined by the last two rows of Equation (23.9). A lumped approximation to the PDE results if we replace the continuous $Z(s_n)$ and $Y(s_n)$ per unit length by discretized impedances of value

$$
\begin{aligned}
Z_d &= (\Delta x)Z(s_n) \\
Y_d &= (\Delta x)Y(s_n)
\end{aligned}
$$
(23.10)

Then Z_r in Equation (23.9) can be written as

$$
\begin{aligned}
Z_r &= \omega_c(0)L(0)l \\
&= \omega_c(0)L(0)(\Delta x)\frac{l}{\Delta x} \\
&= \omega_c(0)L_0 N_{nat}
\end{aligned}
$$
(23.11)

with $L_0 = L(0)(\Delta x)$ and $N_{nat} = l/(\Delta x)$. The parameter L_0 is the lumped inductance of the first stage of the discrete transmission line. The parameter N_{nat} is the number of discrete transmission-line stages in a length l, i.e., over an *e-fold* of frequency scaling in the exponentially tapered line. The number of discrete transmission-line stages per octave is correspondingly given by $N_{oct} = N_{nat}\ln 2$. The shunt admittance $Y_n(s)$ models the admittance of the organ of Corti and the basilar membrane. Zweig showed from analysis of cochlear measurements that such an admittance must have two pairs of high-Q complex poles, a pair of zeros, capacitive behavior at low frequencies, and inductive behavior at high frequencies [26]. The simplest admittance that satisfies all these requirements and is yet rational is given by

$$
Y(s_n) = \omega_c(0)C(0)s_n\left(\frac{\dfrac{s_n^2}{\mu^2} + \dfrac{s_n}{\mu Q_z} + 1}{\left(\dfrac{s_n^2}{1^2} + \dfrac{s_n}{Q_p} + 1\right)^2}\right)
$$

$$
Y(s_n) = j\omega C(x)\left(\frac{\dfrac{s_n^2}{\mu^2} + \dfrac{s_n}{\mu Q_z} + 1}{\left(\dfrac{s_n^2}{1^2} + \dfrac{s_n}{Q_p} + 1\right)^2}\right)
$$
(23.12)

Allowable values of μ, Q_z, and Q_p at all amplitudes are constrained by the requirement that the zero-crossings in the transient response remain relatively invariant with input amplitude in the biological cochlea [27]. Values of $\mu = 0.76$, $Q_z = 3.8$, and $Q_p = 5.0$ yield good fits to biological data and therefore form good starting points for RF cochlea design as well. The gain, phase, and pole-zero plot of $Y_n(s_n) = Y(s_n)/(\omega(0)C(0))$, a normalized version of the admittance, are shown in Figure 23.2. As in Equation (23.11), we can define

$$Y_r = \omega_c(0)C(0)l$$
$$= \omega_c(0)C(0)\Delta x \frac{N_{nat}}{\Delta x} \qquad (23.13)$$
$$= \omega_c(0)C_0 N_{nat}$$

where $C_0 = C(0)\Delta x$ is the lumped capacitance in the shunt admittance of the first stage. From Equations (23.12) and (23.9), we can show that

$$k_n(s_n) = \sqrt{Z_r\left(\frac{lY(s_n)}{s_n}\right)}$$

$$k_n(s_n) = \sqrt{(l\omega_c(0)L(0))(l\omega_c(0)C(0))}\sqrt{\frac{\left(\frac{s_n^2}{\mu^2} + \frac{s_n}{\mu Q_z} + 1\right)}{\left(\frac{s_n^2}{1^2} + \frac{s_n}{Q_p} + 1\right)^2}}$$

$$P(s_n) = \frac{\sqrt{\frac{s_n^2}{\mu^2} + \frac{s_n}{\mu Q_z} + 1}}{\left(\frac{s_n^2}{1^2} + \frac{s_n}{Q_p} + 1\right)} \qquad (23.14)$$

$$\boxed{k_n(s_n) = \sqrt{Z_r Y_r} P(s_n)}$$

$$k_n(s_n) = \frac{l}{\Delta x}\omega_c(0)\sqrt{L(0)(\Delta x)C(0)(\Delta x)}P(s_n)$$

$$\boxed{k_n(s_n) = N_{nat}(\omega_c(0)\sqrt{L_0 C_0})P(s_n)}$$

Note that $\omega_c(0)$ is the characteristic frequency of the first lumped stage, around which the transition from capacitive to inductive behavior happens, while $(1/\sqrt{(L_0 C_0)})$ is the cutoff frequency of the classic transmission line. This cutoff frequency is set by the lateral inductance of $Z(s_n)$ and the low-frequency admittance properties of $Y(s_n)$.

The transfer function $TF(s_n)$ of the cochlea is defined as the current flowing through the shunt admittance Y, normalized by the input current $I(0)$, and given by

$$TF(s_n) = \frac{\Delta x}{I(0)}\frac{dI}{dx}$$
$$= \frac{\Delta x}{I(0)}\frac{ds_n}{dx} \times \frac{dI}{ds_n} \qquad (23.15)$$
$$= \frac{\Delta x}{I(0)}\frac{s_n}{l}\frac{dI}{ds_n}$$

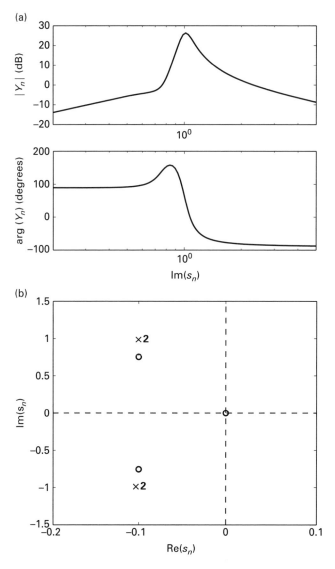

Figure 23.2a, b. The shunt admittance $Y_n(s)$ of the cochlea: (a) gain and phase plots (b) pole-zero plot. Reprinted with kind permission from [12] (©2009 IEEE).

Substituting for dI/ds_n from Equation (23.8), and using $N_{nat} = l/(\Delta x)$ and the boxed expression in Equation (23.9), we get

$$TF(s_n) = \frac{s_n k_n(s_n)^2}{N_{nat}Z_r} \frac{V(s_n)}{I(0)}$$

(23.16)

For a fixed-frequency wave in a transmission line that is not spatially varying, the total phase change in the wave from 0 to x is simply kx where $k = 2\pi/\lambda$ is the wave number with λ being the wavelength; k is constant and independent of x in this case. In a transmission line that is spatially varying, if the spatial variation is sufficiently gentle, the total phase change in the wave from 0 to x may be approximated as $\int k(u)du$, where the lower and upper limits of integration are 0 and x respectively. This approximation is known as the WKB approximation in honor of Wenzel, Kramer, and Brillouin, who used it extensively in quantum mechanical solutions of the Schroedinger wave equation [28]. It may be more appropriate to call it the 'Liouville-Green' (LG) approximation in honor of the authors who actually originated it [29]. If we translate the use of the WKB or LG approximation from the x domain to the s_n domain, we can obtain a solution for $V(s_n)$ in our spatially varying transmission line given by

$$V(s_n) = \frac{\alpha}{\sqrt{k_n(s_n)}} e^{\pm \int\limits_{s_0}^{s_n} k_n(s)ds} \tag{23.17}$$

where α is some constant and $s_0 = j\omega/\omega_c(0)$. In Equation (23.17), the $\pm$ symbol allows for forward-going and backward-going waves on the transmission line and the $\sqrt{(k_n(s_n))}$ term arises from ensuring that the energy of the wave over one period is conserved as $k_n(s_n)$ varies. Substituting, this value of $V(s_n)$ in Equation (23.16), we obtain

$$TF(s_n) = \frac{s_n k_n(s_n)^{3/2}}{N_{nat}Z_r} \left[c_1 e^{+\int\limits_{s_0}^{s_n} k_n(s)ds} + c_2 e^{-\int\limits_{s_0}^{s_n} k_n(s)ds} \right] \tag{23.18}$$

where c_1 and c_2 are constants defined by boundary conditions. The forward-going wave corresponds to losing phase if $e^{j\omega t}$ is input at the beginning of the line, i.e., to the term with a negative exponent in Equation (23.18), and the backward-going wave corresponds to gaining phase, i.e., to the term with a positive exponent in Equation (23.18). In the cochlea, we would like to avoid the backward-propagating c_1 term in Equation (23.18) as much as possible. Backward-going waves interfere with and potentially cancel the forward-going wave, prolong the response to an impulse input due to multiple reflections, and lead to a complex standing-wave encoding of the cochlear input that is hard to decode. In the biological cochlea, the transmission line is terminated at its end by the effective lumped impedance of the 'helicotrema' [24]. This termination ensures that the line impedance of the cochlea at its end is matched to that of the termination impedance. Therefore, reflection of the forward-going wave that leads to a backward-going wave is minimal, and only leads to tiny oto-acoustic emissions. An RF cochlea needs termination as well. To compute the needed termination impedance, we can as in Equation (23.2) evaluate the position-dependent input impedance of the cochlear transmission line to be

$$Z_{in}(s_n) = \sqrt{\frac{lZ(s_n)/s_n}{lY(s_n)/s_n}}$$

$$= \sqrt{\frac{j\omega L(0)e^{+x/l}}{j\omega C(0)e^{+x/l}}} \sqrt{\frac{\left(\frac{s_n^2}{1^2} + \frac{s_n}{Q_p} + 1\right)^2}{\frac{s_n^2}{\mu^2} + \frac{s_n}{\mu Q_z} + 1}}$$

(23.19)

$$\boxed{Z_{in}(s_n) = \sqrt{\frac{L(0)}{C(0)} \frac{\left(\frac{s_n^2}{1^2} + \frac{s_n}{Q_p} + 1\right)}{\sqrt{\frac{s_n^2}{\mu^2} + \frac{s_n}{\mu Q_z} + 1}}}}$$

$$Z_{in}(s_n) = \sqrt{\frac{L(0)}{C(0)} \frac{1}{P(s_n)}}$$

The boxed expression in Equation (23.19) implies that

$$Z_{in}(s_n)|_{s_n=0} = \sqrt{\frac{L(0)}{C(0)}} = Z_0$$

(23.20)

$$Z_{in}(s_n)|_{s_n=\infty} = \sqrt{\frac{L(0)}{C(0)}}(\mu s_n) = j\frac{\omega}{\omega_c(0)e^{-x/l}}(\mu Z_0)$$

Thus, the input impedance of the cochlea at any location changes from being resistive to inductive with the transition occurring near the characteristic frequency $\omega_c(x) = \omega_c(0)e^{-x/l}$ of the location. Note that the input impedance is always at Z_0 at frequencies well below the characteristic frequency at any location such that very-low-frequency inputs travel along the entire length of the cochlea as though it were a uniform transmission-line with constant impedance Z_0 until they reach its end. Intermediate and high frequencies are amplified and then severely attenuated before they reach the end of the cochlea. If L_c is the length of the cochlea, the very low frequencies with values near and below $\omega_c(0)e^{-L_c/l}$ can potentially create significant reflected energy. To prevent such frequencies from being reflected back into the transmission line, the end of the cochlea must be terminated with an impedance identical to that given in the boxed expression of Equation (23.19) with $s_n = j\omega/(\omega_c(0)e^{-L_c/l})$. While an exact creation of such an irrational impedance requires an infinite transmission line itself, Equation (23.20) suggests a good approximation: a resistance of value Z_0 in series with an inductance of value $(\mu Z_0)e^{L_c/l}/\omega_c(0)$. The biological cochlea intelligently prevents very low frequencies from being effectively coupled into the cochlea due to stiffness-limited effects in the middle ear, thus reducing the probability of low-frequency reflections. It also prevents very high frequencies from being effectively coupled into the cochlea due to mass-limited effects in the middle ear. Very high frequencies can lead to reflections at the input of the cochlea. In the RF cochlea, we simply design

$$Z_0 = \sqrt{\frac{L(0)}{C(0)}} = \sqrt{\frac{L(0)\Delta x}{C(0)\Delta x}} = \sqrt{\frac{L_0}{C_0}} = 50\,\Omega, \tag{23.21}$$

i.e., we ratio discrete-lumped-stage inductances and capacitances in the RF cochlea to yield 50 Ω, an easily implemented and common value in RF circuits and RF antennas near 1 GHz. Although this impedance value is nearly the same throughout the cochlea, its value in the first few stages is the primary determinant of the input impedance of the cochlea.

The RF cochlea approximates $Z(s)$ and $Y(s)$ in the continuous PDE with lumped-circuit stages that effectively provide discrete sampling of continuous functions. If this sampling is too coarse, the signal sees abrupt impedance changes between adjacent stages. The changes are especially abrupt in regions near $s_n = 1$, where $k_n(s_n)$ has a large magnitude and the wave has a small wavelength. The impedance changes lead to undesirable inter-stage reflections *within* the cochlea not just at its ends and parasitic resonances. To minimize such reflections and to ensure adequate sampling, we can require that the phase change in the wave between two adjacent sampling points is less than π. This requirement ensures that we have more than two sampling points per spatial wavelength, the bare minimum required by the Nyquist sampling theorem. Thus,

$$k_n(s_n^{i+1} - s_n^i) < \pi$$

$$\left| k_n \frac{ds_n}{dx} \Delta x \right| < \pi$$

$$\left| k_n \frac{s_n}{l} \Delta x \right| < \pi$$

$$\left| k_n \frac{s_n}{N_{nat}} \right| < \pi \tag{23.22}$$

$$\omega_c(0)\sqrt{L_0 C_0}\,|P(s_n)|\frac{\omega}{\omega_c(0)e^{-x/l}} < \pi$$

$$\omega < \frac{1}{(\omega_c(0)\sqrt{L_0 C_0})}\frac{\pi\omega_c(x)}{|P(j\omega/\omega_c(x))|}$$

$$\boxed{\eta_{cf} = \omega_c(0)\sqrt{L_0 C_0}}$$

$$\boxed{\omega < \frac{1}{\eta_{cf}}\frac{\pi\omega_c(x)}{|P(j\omega/\omega_c(x))|}}$$

In Equation (23.22), the parameter η_{cf} is a dimensionless measure of the characteristic frequency of cochlear stages around which the transition from capacitive to inductive behavior happens ($\omega_c(0)$), measured with respect to the cutoff frequency of the classic LC transmission line, $\sqrt{(L_0 C_0)}$. Although η_{cf} is measured at the beginning of the line ($x = 0$), it is the same throughout the cochlea since all inductances, capacitances, and characteristic frequencies scale in a similar exponential fashion. Thus, high-Q $P(s_n)$ functions in Equation (23.14), which lead to large magnitudes in $P(s_n)$, require a small value of η_{cf} to ensure that signals near the characteristic frequency, $\omega_c(x)$, do not cause reflections. The scale-invariant and frequency behavior of the cochlea is such that, if we ensure that Equation (23.22) is valid at $x = 0$, it will automatically be valid at all x: the cochlea keeps reducing the signal bandwidth ω in proportion to $\omega_c(x)$, and η_{cf} and $P(s_n)$ are invariant with x.

23.2.2   Unidirectional cochlear model

Due to the care that the cochlea takes in minimizing reflections within its architecture, backward-going waves typically have small amplitude. They are only important for measuring oto-acoustic emissions, a likely side effect of imperfections in its design [27]. Thus, most of the important aspects of cochlear function may be understood by focusing exclusively on the WKB solution for forward-going waves. Lyon and Mead showed that mapping this solution to a circuit architecture was simply and elegantly implemented with a filter cascade [30]. The first audio silicon cochlea was constructed with an all-pole filter cascade and G_m–C filters. Various improvements that exhibit good performance have also used an all-pole filter cascade [31], [29]. The choice of the filters in the cascade have been somewhat arbitrary. The all-pole filter choice has led to problems of excessive group delay. We shall now show that it is possible to derive these filters in a first-principles way from the known characteristics of the biological cochlea.

The WKB solution of Equation (23.18) with $c_1 = 0$ allows us to focus only on forward-going waves. Since the pre-exponential term typically varies significantly more slowly than the exponential term, it can be assumed constant. The integration can then be broken up into small parts such that

$$\exp\left(-\int_{s_0}^{s_n} k_n(s)ds\right) = \prod_i \exp\left(-\int_{s_{i-1}}^{s_i} k_n(s)ds\right) = \prod_i H_i \qquad (23.23)$$

with the H_i being filter transfer functions such that the product of these transfer functions leads to a filter cascade. Since $\omega_c(x) = \omega_c(0)e^{-x/l}$, the discretized version of s, s_i, is given by $s_i = (j\omega/\omega_c(0))e^{i/N_{nat}}$ with N_{nat} being the number of filters for every e-fold in frequency. The parameter ds is discretized to $\Delta s_i = s_i - s_{i-1}$, which for large N_{nat} evaluates to s_i/N_{nat}. If N_{nat} is large enough such that k_n remains approximately constant between s_i and s_{i-1}, H_i can be approximated as

$$H_i \approx \exp(-(k_n(s_i) \times \Delta s_i)) \qquad (23.24)$$

Furthermore, if $N_{nat} \gg 1$, $\exp(-x) \approx 1/(1+x)$. Therefore, we may write

$$H_i \approx \frac{1}{1 + \dfrac{k_n(s_i)s_i}{N_{nat}}} \qquad (23.25)$$

The transfer functions will vary with i as $k_n(s_i)$ and s_i vary with i as we move along the filter cascade from beginning to end. If we set $\omega_c(0)\sqrt{(L_0 C_0)} = \eta_{cf}$ in Equation (23.14) and substitute for $k_n(s_i)$ in Equation (23.25) we get

$$H_i = \frac{1}{1 + \eta_{cf} P(s_i)s_i} \qquad (23.26)$$

To make $P(s_i)$ rational, we set $Q_z = 0.5$ in Equation (23.14). This value of Q_z enables the square-root term in $P(s_i)$ to have a perfect square as its argument and consequently leads to its easy evaluation. The net rational solution of $P(s_i)$ may then be implemented with a finite number of components. Substituting for $P(s_i)$ in Equation (23.26) with $Q_z = 0.5$ then leads to an explicit rational formulation of H_i given by

$$H(s_n) = \frac{s_n^2 + \dfrac{s_n}{Q_p} + 1}{s_n^2\left(1 + \dfrac{\eta_{cf}}{\mu}\right) + s_n\left(\dfrac{1}{Q_p} + \eta_{cf}\right) + 1} \qquad (23.27)$$

We have used a normalized frequency variable, s_n, for all transfer functions as usual with $s_n = j\omega/\omega_c(i)$ in the discrete unidirectional cochlea rather than $s_n = j\omega/\omega_c(x)$ in the continuous biological cochlea, with $\omega_c(i)$ proportional to $\exp(-i/N_{nat})$. For the unidirectional cochlea, we use $Q_p = 5$, $\eta_{cf} = 0.5$, $\mu = 0.2$. Figure 23.3 (a) shows the frequency characteristics of $H(s_n)$ and Figure 23.3 (b) shows the pole-zero plot of $H(s_n)$ in the s_n plane.

The parameter N_{nat} can be different between the unidirectional and bidirectional cochleas. The two-pole-two-zero transfer function used in Equation (23.27) to create a filter cascade has two advantages over all-pole transfer functions that have been previously used: 1) the group delay is greatly reduced because delays due to poles are cancelled by delays due to zeros; 2) the lowpass rolloff slope of the cochlea is increased by the collective action of the complex poles and zeros after the cutoff frequency. As we shall see in Equation (23.28), the sharpening of the cochlear rolloff slope improves its frequency resolution.

From Equation (23.14), a high value N_{nat} leads to a proportionately larger value of $k_n(s_n)$, and consequently an exponentially larger gain and exponentially larger high-frequency rolloff slope in the $\exp\left(-\int k_n(s)ds\right)$ cochlear transfer function. Note that the rolloff slope is usually measured in logarithmic units, e.g., 240 dB/decade for a twelve-pole filter rolloff. Thus, increasing N_{nat} increases the slope measured in such units proportionately with N_{nat}. A good approximation to cochlear transfer functions at any location is to assume that N_{nat} identical filters with $H(s_n)$ as in Equation (23.27) create it, i.e., that it is well described by $H(s_n)^{N_{nat}}$.

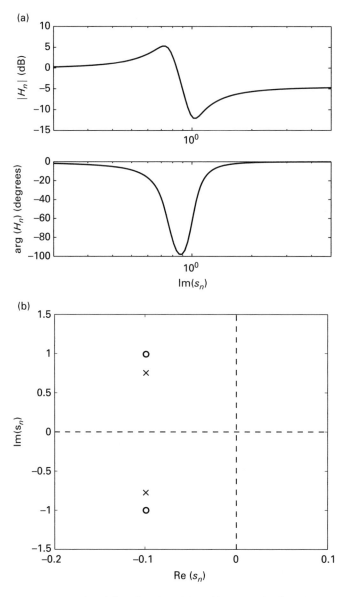

Figure 23.3a, b. A unidirectional cochlear filter transfer function. Reprinted with kind permission from [12] (©2009 IEEE).

Increasing N_{nat} is thus beneficial for increasing filter order, increasing rolloff slope, and improving the minimum detectable signal: the gain of the cochlea is such that signal is amplified more than noise when gain increases, either because N_{nat} increases or because the resonant gain per stage increases; thus, the minimum detectable signal decreases if gain is increased [31]. However, increasing N_{nat} also leads to higher total output noise, less output dynamic range, more compression, more stages, more area, and more power consumption [31]. A cochlear designer must choose N_{nat} based on all of these tradeoffs.

22.2.3 The genius of the cochlear spectral-analysis algorithm

Figure 23.4 (c) illustrates that, because of the exponentially tapered structure of the cochlea, its transfer function (TF) at any position is produced by a "sliding window" of the approximately N_{nat} stages basal of that position. Thus any cochlear *TF* is well approximated as a cascade of N_{nat} identical stages, each of which contributes an effective multiple-pole rolloff to the overall cochlear rolloff slope [31]. Therefore, if the high-frequency rolloff slope of an individual cochlear stage's transfer function compared with a one-pole rolloff is R_{mult}, the overall cochlear rolloff slope is $R_{mlt}N_{nat}$ dB/dB in a log-log amplitude-versus-frequency plot. If the maximal SNR of a cochlear output is *SNR*, then the effective frequency resolution of the cochlear spectrum analyzer $\Delta f/f$ is approximately given by

$$\frac{\Delta f}{f} \approx \frac{1}{\sqrt{SNR}} \frac{1}{R_{mult}N_{nat}} \tag{23.28}$$

Equation (23.28) can be derived by comparing two adjacent cochlear-stage outputs for a sine-tone input, one of which has a signal that is just larger than that of its neighbors by a noise floor. The actual frequency resolution may be improved by averaging across many cochlear outputs and by using phase cues to obtain frequency information such that Equation (23.28) should only be viewed as an approximation of the final spectral resolution possible in the auditory system. Effects such as gain control [31] and two-tone suppression in the cochlea [32] interact with amplitude and phase cues such that the frequency resolution has an *SNR* dependence that is more complex than that given in Equation (23.28). The resonant gain in the cochlea helps improve the *SNR* but the ultimate frequency resolution is determined by the sharp rolloff slopes in the cochlea, and not only by the resonant Q. The parameter N_{nat} also affects the *SNR*. In spite of all these complexities, Equation (23.28) illustrates that increasing R_{mult} with a two-pole-two-zero filter, increasing N_{nat}, and increasing *SNR* improve frequency resolution in the cochlea.

The analysis time τ_i to resolve the i_{th} frequency bin is limited by the sum of the settling times of stages basal of i, and is approximately $N_{nat}/\omega_c(i)$. Therefore, the analysis time for the whole spectrum is on the order of N_{nat} cycles of the lowest frequency of interest. When β, the ratio of maximum and minimum analyzed frequencies, is fixed, the total number of stages $N = N_{nat}(\ln(\beta) + 1)$, such that $\tau_i \propto N$. The 1 comes about because the very first output bin of a cochlear spectrum analyzer needs an extra N_{nat} number of stages basal to it. Since each stage in the cochlea is a frequency-scaled copy of a canonical prototype, its hardware and power requirements also scale as the total number of filters, i.e., N. In contrast, in a bank of constant-Q independent filters that also analyze the spectrum in a scale-invariant cochlear-like fashion, to create rolloff slopes proportional to N_{nat}, we

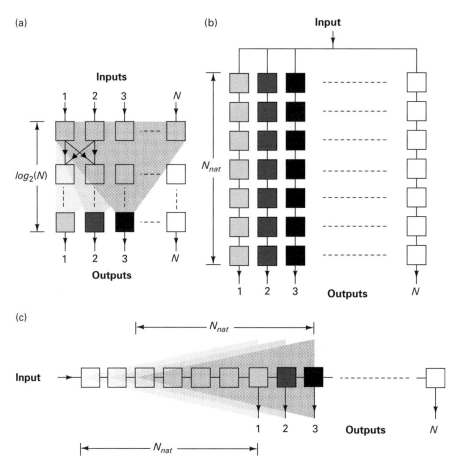

(a)

Inputs

1 2 3   N

$log_2(N)$

1 2 3   N

Outputs

(b)   Input

N_{nat}

1 2 3   Outputs   N

(c)

N_{nat}

Input

N_{nat}

1 2 3   Outputs   N

Figure 23.4a, b, c. Comparison of the FFT (a), analog filter bank (b), and cochlear architecture (c) for spectrum analysis. Reprinted with kind permission from [12] (©2009 IEEE).

need N_{nat} filtering blocks per stage. Since there are N such stages, which, unlike in the cochlea, are not shared in an exponentially tapered transmission line, the hardware costs, as measured by the total number of blocks, scales like $N \times N_{nat} \propto N^2$ if β is fixed. Figure 23.4 (b) illustrates the architectural scaling in this case. The time taken for spectrum analysis in the filter bank is given by the sum of the settling times of the N_{nat} filtering blocks in each stage, which still scales like N as in the cochlea. The fast Fourier transform (FFT), which, unlike the cochlea and constant-Q filter banks has fixed-bandwidth frequency bins, takes $O(N \ln(N))$ time (measured by the number of multiply-and-add operations) and $O(N \ln(N))$ hardware (measured by the number of multipliers and adders) to perform spectrum analysis as shown in Figure 23.4 (a). The output bins of the cochlea and filter banks are available and updated in parallel, which allows them to continuously monitor the whole spectrum. This behavior is in contrast to that of most commercial RF spectrum analyzers, which are of the swept-sine or super-heterodyne type. In this type of analyzer a simple frequency bin is sampled and updated at a given time, causing aliasing of nonstationary spectra. Due to the fundamental time-vs.-frequency-resolution uncertainty principle of Fourier transforms, the analysis time per higher-resolution frequency scales as $1/(1/N)$ *and* the number of frequency bins scales as N, such that the analysis time of the swept-sine analyzer scales as N^2 [33], [34]. Thus, the cochlear architecture, which achieves linear scaling in time, hardware, and power, is the most efficient architecture for spectrum analysis known to man, outperforming the FFT, a filter-bank architecture, and swept-sine spectrum analyzers. It ingeniously exploits the scale-invariant nature of an exponential to achieve this efficiency as illustrated by Figure 23.4.

The cochlea does floating-point scale-invariant spectrum analysis unlike the FFT, which performs fixed-point spectrum analysis with higher frequency resolution at higher center frequencies. If fixed-bandwidth resolution is needed, say to listen to a particular radio channel, an *RF fovea* can operate on cochlear outputs to obtain such resolution via narrowband mixing and filtering techniques [25]. The RF cochlea gets the big picture, quickly performing fast broadband spectrum analysis over the entire spectrum. The rest of the 'RF auditory system' then zooms in on particular cochlear outputs to obtain more detailed information about the small picture. This idea is analogous to one employed by the eye. The central foveal region in the eye obtains detailed information about interesting regions of an image and has 2 arc minute angular resolution; the spatial resolution of the eye constantly degrades as one moves towards the periphery of the eye worsening to more than a degree. The overall eye quickly gets the big picture and we zoom in to interesting regions of the image, afterward, by moving the foveal region of the eye to analyze them further. The combination of an RF cochlea and an RF fovea can then function like an eye-and-ear inspired radio performing broadband and high-resolution RF signal detection and analysis in a highly energy-efficient fashion in multiple parallel channels simultaneously. Some of these ideas of signal processing may be important for universal, software, and cognitive radios of the future and could be implemented in software rather than hardware [25]. Further details of the RF cochlea have been described in a recent paper [12].

23.3   Integrated-circuit unidirectional and bidirectional RF cochleas

Figure 23.5 shows a circuit schematic of a bidirectional RF cochlea. To allow parasitic shunt capacitances to ground to be absorbed into the cochlear architecture, this architecture maps pressure to current and volume velocity to voltage. The ZY lateral-shunt transmission-line architecture of the cochlea that we have described thus far then transforms to a $Y'Z'$ architecture with $Z \rightarrow Y'$ and $Y \rightarrow Z'$; current and voltage variables get interchanged with each other; inductances map to capacitances and vice versa; series configurations map to parallel configurations; the output is a voltage derivative instead of a current derivative as shown;

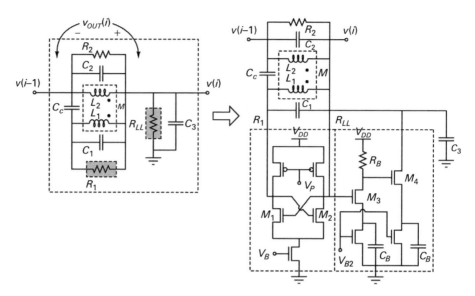

Figure 23.5. A bidirectional cochlear stage. Reprinted with kind permission from [12] (©2009 IEEE).

in essence, we create a dual architecture that flips currents to voltages and vice versa but the overall equations are otherwise identical. Another advantage of this design is that unwanted common-mode signals on ground are automatically subtracted out.

The L_1C_1 and L_2C_2 resonators are coupled capacitively via C_c and via mutual-inductance transformer action to create the needed pole-zero $Z'(s)$ equivalent to the $Y_n(s_n)$ of Figure 23.2. The coupled-resonator frequency response seen in Figure 16.7 (a) in Chapter 16 is also due to similar pole-zero pairs. The resistance R_1 is a negative resistance needed to pump energy into the traveling wave basal to the peak with R_2 being positive for overall stability. The negative resistance is implemented via a classic cross-coupled M_1–M_2 pair whose bias current set by V_B, sets R_1 at nearly $-2/g_m^{M1}$. A parasitic resistance in L_2 can form a resistive divider with Z_{in} such that low-frequency signals in the RF cochlea are slightly attenuated per stage, and this attenuation can compound to create a large overall attenuation at the end of the cochlea: e.g. $(0.99)^{40} \approx 0.6$. The purpose of R_{LL}, a negative resistance, is to help create a negative resistance that counters this line loss. The resistance R_{LL} is implemented by the cascaded negative-gain stages of M_3 and M_4 with C_B ensuring that the negative resistance of $-1/\left(\left(g_m^{M3}R_B\right)g_m^{M4}\right)$ is primarily of importance only at RF frequencies. Nominally, g_m^{M3} and g_m^{M4} are nearly equal. The output voltages $v_{OUT}(n)$ are preamplified by a gain of approximately 5.5 before their envelopes are detected and read out by a simple passive diode-capacitor circuit. The preamplification helps reduce the dead zone of the envelope detector. Higher performance envelope detectors like those described in Chapter 19 may also be used if they are suitably altered to work at higher RF frequencies.

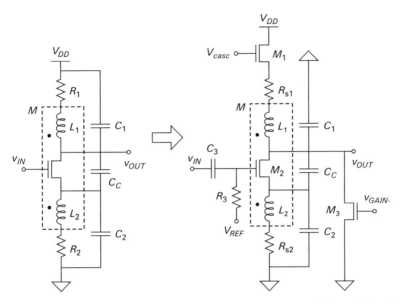

Figure 23.6. A unidirectional cochlear stage. Reprinted with kind permission from [12] (©2009 IEEE).

The voltage V_P is set by a low-frequency negative-feedback loop that senses the dc voltage of the line, compares it to a reference and increases or decreases V_P appropriately.

Figure 23.6 shows a circuit schematic of a unidirectional RF cochlea. Once again, the $L_1 C_1$ and $L_2 C_2$ resonators are capacitively coupled via C_c and via mutual-inductance transformer action to create the needed pole-zero pair of $H_n(s)$ in Figure 23.3. Gain at $s_n \approx 1$ is provided by the transistor M_2 which has a parallel resonant load and parallel resonant source degeneration. The low-frequency gain $TF(s_n \ll 1)$ is nearly

$$TF(s_n \ll 1) = (R_{s1} + 1/g_{m1})\frac{g_{m2}}{g_{m2}R_{s2} + 1} \tag{23.29}$$

and is set by a source-degenerated g_{m2} looking into a series load. The use of M_1 isolates each stage from parasitic inductances in the supply. The capacitance C_3 and R_3 form a high-pass filter that enables ac coupling between stages. The transistor M_3 allows M_1 to have more dc current than M_2 thus attenuating the g_{m2}/g_{m1} ratio and reducing the gain of Equation (23.29). A low-frequency 22 MHz signal is injected into the cochlea, and a feedback loop attempts to equalize the envelope at the input and the output of the cochlea by negative-feedback biasing of the gate of M_3 (v_{GAIN-} in Figure 23.6) such that the overall cochlea has a net low-frequency gain of 1. In a practical implementation of the cochlea, it was found that this low-frequency gain was below 1 even with M_3 completely turned off due to a larger-than-expected parasitic R_{s2} [12]. A simple improvement to fix such a

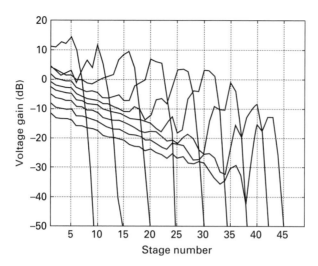

Figure 23.7. Outputs from the RF Cochlea. Reprinted with kind permission from [12] (©2009 IEEE).

problem is to add a pFET transistor M_4 to the output that can subtract from M_1's current, thus increasing the g_{m2}/g_{m1} ratio and consequently the gain.

The bias voltage V_{REF} sets the dc bias current through M_1 and M_2 and thus determines g_{m1} and g_{m2}. The amount of resonant peak gain in each cochlear stage is approximately $g_{m2}(\sqrt{(L_1/C_1)})$. Thus, by changing g_{m2}, we can alter this gain and control the overall peak gain in the cochlea. A gain-setting feedback loop on a matched version of transistor M_2 with R_{s2} source degeneration uses a DAC-determined R_L to set V_{REF} via negative-feedback biasing such that, at 22 MHz, $g_{m2} \approx 1/R_L$ in the replica circuit. This replica bias value of V_{REF} is then used to bias all M_2 transistors on the RF cochlea chip. The chip also contained a single-transistor common-gate broadband low-noise amplifier (LNA) with its bias current set such that $1/g_s \approx 50\,\Omega$ could match RF-source inputs. A gain of 3 in the LNA was achieved via resistive loading at its drain.

Figure 23.7 shows spatial responses of the bidirectional RF cochlea to logarithmically spaced input frequencies between 1 GHz and 8 GHz at a fixed input power level of −10 dBm obtained from a chip built in 0.13 μm technology [12]. The line loss due to parasitics is not completely canceled leading to low-frequency attenuation. The resonant gain, sharp rolloff, and logarithmically spaced filter peaks are very similar to those in the biological cochlea. Further details of integrated-circuit RF cochleas are provided in [12]. Figure 1.7 in Chapter 1 shows a layout picture of the bidirectional RF cochlea chip. Both unidirectional and bidirectional versions described in that article have similar input-referred 70 dB dynamic range and operate with 180–300 mW of power. They span a 600 MHz to 8 GHz frequency range in the implementation described in [12]. Direct digitization of the RF bandwidth for a software or cognitive radio would consume nearly 100 times as much power. An analog constant-Q filter bank with equivalent performance would need approximately 20 times the hardware. The bidirectional cochlea operated with $N_{nat} = 20$ while the unidirectional cochlea operated with $N_{nat} = 16$. The cochleas had 50 (or 51) stages.

23.4   Audio cochleas and bio-inspired noise-robust spectral analysis

Audio silicon cochleas have been the subject of much research [30], [29], [31], [36], [37]. Unlike RF silicon cochleas, inductors are created via implicit or explicit gyration with second-order $G_m - C$ transconductor or current-mode filtering circuits as described in Chapters 12, 13, and 14. For a given Q, an active inductor has Q^2 times more noise than a passive inductor. Thus, the power consumption needed to maintain SNR for a given bandwidth is higher in an audio silicon cochlea compared with an RF cochlea where true inductors are used. It is also worse than in biology where masses, noiseless, lossless reactive elements, can effectively implement practical small-size inductors at audio frequencies. Models that explicitly represent the dimension of the cochlea orthogonal to its length-to-length axis lead to 2D cochleas. The main advantage of 2D cochleas is that they can potentially yield steeper rolloff slopes. Due to fluid flow changing from long-wave whole-fluid motion throughout the cochlear duct to short-wave local fluid motion [29] only near the basilar membrane, the axial inductance in the 1D transmission line effectively increases from several inductances in parallel at low frequencies to few inductances in parallel at high frequencies. Consequently, the rolloff slope steepens due to this additional inductance-decreasing mechanism. In practice, however, resistive and active elements have been used to mimic these 2D effects since non-lossy inductors are not available at audio [29], [37]. Such elements contribute significant excess noise, and dynamic range and power consumption are consequently severely compromised. Therefore, in practical systems, altering 1D models from using all-pole filters to using the two-pole-two-zero filters shown in Figure 23.3 can have more immediate engineering impact. For example, a recent audio silicon cochlea implemented such filters in a current-mode topology to achieve 79 dB dynamic range at 41 μW power consumption with a constant output SNR of 40 dB at most signal levels versus a 60 dB, 0.5 mW, 23 dB SNR former state-of-the-art cochlea with automatic gain control [38], [31]. Several other performance metrics in the cochlea such as rolloff slope, group delay, higher quality factor (Q), and less stages needed for gain build up were improved as well. A key circuit innovation in this cochlea was the use of a novel technique, known as adaptive balanced biasing, that greatly improved the power efficiency of current-mode filters for a given SNR (by Q), and dynamic range (by Q^4) respectively [38]. Several details in the cochlea can be mimicked in silicon as well, but will not be discussed here, e.g., the exponential taper in the biological cochlea alters to a more linear taper at frequencies lower than 1 kHz etc.

Figure 23.8 shows a spectral-analysis architecture inspired by the operation of gain-control mechanisms and the phenomenon of two-tone suppression in the cochlea and auditory system. This bio-inspired companding (compressing and

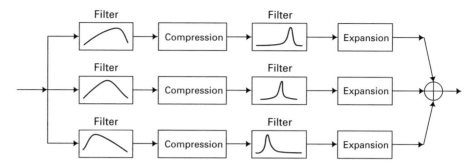

Figure 23.8. The bio-inspired companding architecture. Reprinted with kind permission from [39] ©2007 Acoustical Society of America.

expanding) architecture has led to improved spectral analysis of noisy signals, improving the performance of deaf patients and artificial speech-recognition systems [39], [16], [17]. An input signal is analyzed by a bank of spectral channels. Each spectral channel is composed of a relatively broadband bandpass filter whose output sets the gain of a compressor with instantaneous or dynamic gain control. The compressor amplifies soft signals more than loud signals such that the output dynamic range of the compressor is less than the input dynamic range of the compressor. The biological cochlea performs such gain control and compression to ensure that nearly 120 dB of dynamic range in input sound levels is compressed to about 40 dB of output dynamic range in auditory-nerve firing rates. The output of the compressor is then filtered by a relatively narrowband bandpass filter that removes distortion and out-of-band spectral components in its channel. The output of the narrowband filter determines the gain of an expander, typically via dynamic gain control. The output dynamic range of the narrowband filter is expanded back to original by the expander or maintains some compression. For example, if $1/n_1$ is the compression power law of the compressor and n_2 is the expansion power law of the expander, the overall output of a spectral channel sees a compression power law of n_2/n_1 if $n_2 < n_1$. There is no overall compression if $n_2 = n_1$. If the goal is to get a noise-suppressed audio output, the outputs of the spectral channels can be added back to recreate the original signal as shown in Figure 23.8. If the goal is to get a noise-suppressed spectrum, the outputs of the spectral channel are left and are not added back.

In the companding architecture, in a given spectral channel, out-of-band signals can affect the gain of an in-band signal since the broadband filter sets the gain of the compressor. If the input signal is dominated by a simple strong tone, then in the spectral channel(s) where this tone lies within the bandwidth of the narrowband filter, the tone will simply get compressed, expanded, and emerge unscathed as it was at the input. In adjacent and nearby channels, however, where this tone is within the bandwidth of the broadband filter but not within that of the narrowband filter, the tone will suppress the gain of these spectral channels to their preferred narrowband signals: it reduces the gain of the compressor to a value appropriate for a louder level than a softer level; the narrowband filter of these spectral channels then removes the dominant tone via filtering; after the expansion, the weak in-band signals emerge at a suppressed level because they are not sufficiently preamplified by the compressor whose gain is set too low by the dominant out-of-band tone. Thus, strong tones effectively inhibit and suppress the strength of signals that are near them in frequency. Consequently, the spectral contrast of these tones with respect to their neighbors improves, sharpening the spectral peaks in the signal. However, the audibility of signals with spectra that are distant from the dominant tone is not much affected: their spectral energy is distant from the dominant tone in frequency and does not lie within the bandwidth of the broadband filter, such that they are not suppressed. The local winner-take-all effect thus improves local spectral contrast without degrading audibility. Noise acts like a weak signal that is also suppressed and the spectrogram appears to be sharper and cleaned up. Examples of signals and spectrograms that have been processed by the companding architecture may be found in the paper where companding was first proposed [32]. Compressor and expander parameters and filter bandwidths can be tuned to optimize the level of suppression and the locality of the suppression.

There are several areas where companding can be useful since it is a general spectral-analysis technique. The most obvious lies in the audio domain itself. Companding improves the hearing of vowels in noise. Vowels have spectral peaks known as formants, which are enhanced via companding. Companding is neutral towards consonants but somewhat helpful since consonants provide a context surrounding the vowel that affects the exact location of the vowel's spectral peak, and companding can help enhance detection of these peaks. Overall there is a significant improvement in hearing in noise for the deaf and in artificial speech-recognition front ends [39], [16], [17]. Versions of companding suitable to digital FFT-based schemes have also been built [40]. Since the suppression effects arise primarily from the compressor, it is worth noting that the benefits of companding are present even if the output spectral information remains encoded in the compressed space and is not decoded back into the original input space. In the cochlea, the broadband compression is performed by the effects of sounds on the nonlinearity and dynamic gain control of the outer hair cells that they affect, while the narrowband filtering is performed partly by resonant filtering within the cochlea and partly by higher auditory centers. Such centers may perform expansion on the compressed outputs from the cochlea to yield an overall linear input-output response, but an overall linear response is not essential. Note that the spectral companding architecture of Figure 23.8 is different from the companding current-mode architectures of Chapter 14, since those architectures do not incorporate broadband and narrowband filters, which are essential. If 1D temporal filters are replaced by 2D spatial filters, a straightforward application of the ideas from the time domain to the spatial domain leads to image enhancement via companding rather than sound enhancement. Finally, if the outputs of companding are used in a feedforward topology, they can be used to detect dominant interferers in the signal, subtract these out from the signal, and amplify the subtracted residue to then detect weaker signals. The latter strategy may be useful in RF, where a dominant interferer can make it hard to detect desired weaker signals.

We shall end our tour of the cochlea by noting that, while biology has inspired several interesting architectures like the RF cochlea and companding architecture in engineering, engineering has in turn been useful to biology. A simple feedback and circuit model of the cochlea showed how fast 100 kHz amplification with slow 1 ms outer hair cells is possible in the cochlea, a two-decade-long mystery in hearing [23]. The essence is to realize that negative feedback in the cochlea serves to speed up time constants and increase Q via a simple root-locus plot as it does in several feedback systems (see discussions of feedback in Chapters 2, 9, and 10). Thus, the open-loop time constant of the outer hair cell is not predictive of its closed-loop time constant in the cochlea in much the same manner that the open-loop 10 Hz time constant of an operational amplifier is not predictive of its 1 MHz closed-loop performance in a negative-feedback buffer. In fact, the work in [23] showed that the gain-bandwidth product of the outer hair cells is large enough such that the use of negative feedback and sufficient gain bandwidth easily enables fast amplification and provides a good fit to experimentally observed cochlear data. Circuit models of biology can undoubtedly exploit the intuitive power of circuits and feedback to shed insight into the partial differential equations of biology in several other systems in the future, insight that is not easily attained by highly computationally intensive brute-force computer simulations. We shall return to this theme in the next chapter, where we discuss circuit models of gene-protein and protein-protein networks within the cell.

23.5   A bio-inspired analog vocal tract

Figure 23.9 shows a conceptual diagram of how speech synthesis and hearing analysis can be put together in a feedback loop analogous to the phase locked loop (PLL) discussed in Chapter 18. The feature comparator compares a spectral representation of the output synthesized by a vocal tract with a spectral representation of what has just been heard. The spectral representations are created by auditory processors that may be sophisticated and mimic the cochlea, e.g., a companding spectral representation; or they may be more straightforward spectral representations, e.g, that output by the bionic-ear processor described in Chapter 19 or an FFT. The results of the comparisons are used to drive a vocal-tract controller that outputs articulatory motor-control information that drives the vocal tract to produce sound. Such motor information could include configurations for articulators like the tongue, lips, jaw, velum, and larynx position. The feedback loop is architected such that the speech synthesized is both a good match to what has just been heard and can be produced by relatively smooth articulator dynamics that minimize articulator state changes and consequently muscle effort in the vocal tract. The auditory processors and feature comparator

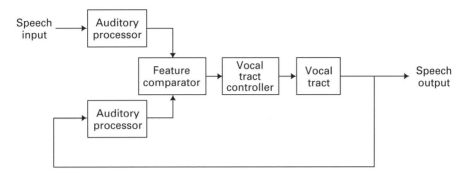

Figure 23.9. A speech locked loop (SLL). Reprinted with kind permission from [18] (©2008 IEEE).

are analogous to the phase detector in a PLL, the vocal-tract controller is analogous to the charge pump and loop filter in the PLL, and the vocal tract itself is analogous to the voltage-controlled oscillator (VCO) in the PLL. Hence, the feedback architecture of Figure 23.9 is termed a speech locked loop (SLL) and this method of recognition is an example of an analysis-by-synthesis approach: the speech is analyzed by extracting the information needed to synthesize it. The synthesis information (output of the vocal-tract controller) provides articulator and consequently phoneme information that then represents and recognizes the speech.

The potential benefits of such an approach are three-fold. 1) Since we have a model of how the speech is being produced, i.e., the vocal tract, noisy speech is recognized via a best-match scenario that minimizes the effects of the noise. The process is analogous to the fitting of noisy data by a straight-line model of it to minimize regression error. More sophisticated schemes could incorporate models of the noise and of the environment to further improve performance. 2) The speech can be coded into a very compressed articulator representation at a low bit rate that minimizes the needed storage or communication channel capacity, for example, in cellular phones. 3) Information such as speaker identification and emotional information that are hard to obtain from an auditory representation alone can be more transparently extracted. The SLL of Figure 23.9 is inspired by the fact that there are intriguing connections between motor and auditory representations of speech in the brain via the arcuate fasiculus [10]. It is not surprising that in order to analyze something, it is useful to know how to synthesize it.

The essence of the algorithm used by the SLL is simple. 1) A stored babble codebook allows look up of vocal-tract configurations such that any heard 10 ms spectrogram segment can be mapped to, say, 9 vocal-tract configurations that best reproduce it. Such a codebook can be learned and stored by babbling, i.e., randomly changing vocal-tract configurations and storing the association between vocal-tract configurations and the resulting spectrograms. 2) In a sequence of N spectrogram snippets, dynamic programming is used to find a lowest-total-cost path through the 9^N possible sequence of articulator states [41]. The costs for each sequence include matching-fidelity costs that quantify the discrepancy between what is heard and what is produced for each articulator state and travel costs between articulator states that quantify the effort needed to transition between successive states. In analogy with the spectrogram, which is a sequence of spectral vectors, we term the sequence of articulator vectors a *vocalogram*. Remarkably, the overall process has similarities to how hidden Markov model (HMM) speech recognition works except that there is a physical meaning to the state representations, far fewer states are needed, and learning and generalization in noise become possible because the data are not overfit by a mere statistical model. We shall now briefly describe how to create an analog bio-inspired vocal tract. The SLL provides context and motivation for why such analog vocal tracts can be very useful, even if they are imperfect. It also leads to related ideas of other analysis-by-synthesis architectures like a heart locked loop (HLL) that combines sensory information about the heart (ECG, PPG, PCG, . . . see Chapter 20) and a model of the heart (see Chapter 20) to extract information about cardiac state that could potentially be more robust in noise [19].

We speak by modulating the flow of air expired by our lungs. This airflow is modulated as it makes its way through a tube or tract known as the vocal tract. The airflow emerges out of the lungs, through a narrow constriction known as the glottis, through the intra-oral tract in our neck and mouth, through a supra-glottal constriction formed between the tongue and mouth roof, through the oral tract in our mouth, and finally out of our lips as radiated sound. The modulations occur all along the vocal tract, periodically at the glottal constriction when the vocal folds oscillate and vibrate for voiced sounds, at the velum, a trapdoor that closes off airflow to the nasal pathway (or not), and by the tongue, lips and jaw, which constrict or increase the cross-sectional area of the pathway to reduce or increase airflow, respectively. The lips can also extend the length of the vocal tract a little. Sound is primarily radiated out the lips but also slightly at the nose for certain sounds. Vowels correspond to a relatively open vocal tract with relatively high volume velocity of flow and relatively high radiated sound pressure. Consonants are typically associated with a narrow supra-glottal constriction, a relatively low volume velocity of flow, and relatively low radiated sound pressure [42]. Vowels and consonants frequently alternate with each other in speech.

The vocal tract can be approximated as a non-uniform acoustic tube, with time-varying cross-sectional areas, that is terminated by the glottis at one end, and the lips and/or nose at the other. Since the cross-sectional dimensions of the tube are small compared to the wavelength of sound, the waves that propagate along the tube may be approximated as being planar. The wave equation for one-dimensional planar sound propagation in a lossless uniform tube of circular cross-section is given by

$$-\frac{\partial P}{\partial x} = \frac{\rho}{A}\frac{\partial U}{\partial t}$$

$$-\frac{\partial U}{\partial x} = \frac{A}{\rho c^2}\frac{\partial P}{\partial t},$$

(23.30)

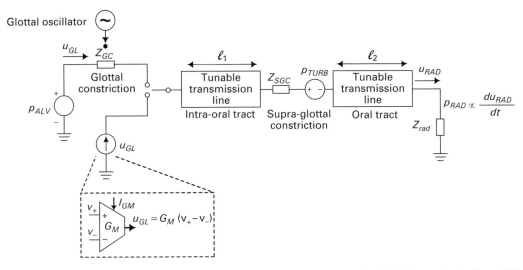

Figure 23.10. Architecture of the analog vocal tract. Reprinted with kind permission from [18] (©2008 IEEE).

where P is the sound pressure, U is the volume velocity, ρ is the density of air, c is the speed of sound, and A is the area of cross-section. The pressure P is analogous to voltage V, and U is analogous to current. The volume of air in a tube exhibits an acoustic inertance ρ/A due to its mass which opposes acceleration. The electrical analog of such an acoustic inertance is an inductance L. The air also exhibits an acoustic compliance $A/(\rho c^2)$ due to its compressibility, which opposes changes in volume. The electrical analog of this acoustic compliance is a capacitance C. Thus, the electrical analog of Equation (23.30) is

$$
\begin{aligned}
-\frac{\partial V}{\partial x} &= L \frac{\partial I}{\partial t} \\
-\frac{\partial I}{\partial x} &= C \frac{\partial V}{\partial t}
\end{aligned}
\tag{23.31}
$$

where L and C are the inductance and capacitance per unit length.

Figure 23.10 shows an integrated-circuit analog vocal tract (AVT). The AVT represents the intra-oral and oral tracts as transmission lines of length l_1 and l_2 respectively. Each line comprises a cascade of tunable two-port elements, corresponding to a concatenation of short cylindrical acoustic tubes of length l with varying cross-sections $A(x)$ that vary as x varies from the beginning to the end of the tract. The error introduced by spatial quantization is kept small by making l small compared with the wavelength of sound corresponding to the maximum frequency of interest (10 to 20 kHz). Each two-port is an electrical LCR π-circuit element where the series inductance L and the shunt capacitance C may be controlled by physiological parameters corresponding to articulatory movement such as the movement of the tongue, jaw, or lips. The resistance is needed to model viscous flow losses. Losses at the walls of the vocal tract are modeled by shunt conductances G [18].

Speech is produced by controlled variations of the cross-sectional areas along the tube by the articulators in conjunction with the application of one or two sources of excitation: 1) a periodic source at the glottis, and/or 2) a turbulent noise source p_{TURB} at some point along the tract. Vocal fold vibration produces a periodic interruption of the air flow from the lungs to supra-glottal vocal tract. At most frequencies of interest, the glottal source has a high acoustic impedance compared to the driving point impedance of the vocal tract. Consequently, a current source u_{GL} may be used as the electrical analog that approximates the volume-velocity source at the glottis. Alternatively, the constriction at the glottis may be represented by a variable impedance modulated by a glottal oscillator to simulate the opening and closing of the vocal folds. The switches in Figure 23.10 indicate that either u_{GL} or the combination of Z_{GC} and p_{ALV} can be used. The u_{GL} current source can be implemented with voltage inputs and a transconductor as shown. The pressure source of the lungs is represented by p_{ALV} and a glottal-oscillator source modulates Z_{GC}, an impedance that models losses occurring at the glottis due to laminar and turbulent flow. Laminar flow is characterized by a linear relationship between volume velocity and the differential pressure across the glottal constriction and is easily mimicked by a linear conductance with a linear I–V curve [42]. Turbulent flow is similarly mimicked by a square-root I–V curve [42], [18]. Both linear and square-root I–V curves can be implemented with a novel electronically tunable linear or nonlinear resistor [43]. This resistor uses trans-linear negative-feedback biasing of the gate of a MOSFET to achieve any desired linear or nonlinear I–V characteristic with zero dc offset and is described in detail in [43]. The impedance Z_{GC} is comprised as a series combination of two such electronically tunable resistances, one of which is architected to be linear, and the other of which is architected to have a square-root conductance. The location of p_{TURB} is downstream of the supra-glottal constriction location. We approximate the turbulent noise generated by p_{TURB} with a signal produced by modulating the cross-sectional area of a two-port section downstream of the constriction in a noisy fashion. At the lips, the oral-tract transmission line is terminated by a radiation impedance Z_{rad} that is primarily inductive [42]. The radiated sound pressure at the mouth, p_{RAD}, is proportional to the derivative of the current flowing in Z_{rad}, i.e., the voltage across Z_{rad} if Z_{rad} is an inductance. A transmission line to model the nasal tract can also be incorporated into an enhanced version of Figure 23.10 but the work in [18] did not implement such a line for simplicity.

The value of L in a line section increases if the local area of cross-section of the corresponding region of vocal tract decreases and may be electronically tuned by changing a bias current. The value of C in a line section decreases if the local area of cross-section of the corresponding region of vocal tract decreases and may also be electronically tuned. The product of L and C in any section is invariant to ensure that the speed of sound propagation in the transmission line is invariant with cross-sectional area. The lines are implemented with gyrators and judicious

(a)

(b)

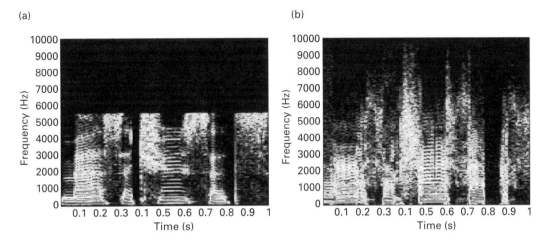

Figure 23.11a, b. (a) Spectrogram with intentional lowpass filtering above 5 kHz.
(b) Spectrogram output by the analog vocal tract that reintroduces missing components
in (a). Reprinted with kind permission from [18] (©2008 IEEE).

use of transconductor-capacitor circuits [18]. Feedback circuits ensure that dc
offset does not greatly accumulate and degrade performance in the transmission
line. More details of the analog vocal tract, including a description of how cross-
sectional areas for various sounds are determined and related to articulator
movements to create a babble code book are described in [18].

Figure 23.11 (a) shows a spectrogram of the word "Massachusetts" with intentional lowpass filtering at 5.5 kHz. Figure 23.11 (b) shows a re-synthesized version
of this spectrogram using the vocal tract in a SLL configuration such that we try
to match the sound that we have just heard. The spectrogram of the original
recording (Figure 23.11 (a)) and the synthesized sound (Figure 23.11 (b)) show
good matching between trajectories and formants indicating good performance by
the analysis-by-synthesis technique. Furthermore, high-frequency speech components that were missing in Figure 23.11 (a) have been re-introduced by the AVT in
Figure 23.11 (b). The AVT synthesizes all and only speech signals and thus provides
a measure of signal restoration. Such signal-restorative properties are important
in robust speech recognition in noise. Indeed, tests of the recognition of vowel-
consonant syllables with the AVT in an SLL show improved performance in noise.

23.6   Bio-inspired vision architectures

Two examples of bio-inspired vision chips include the silicon retina and fly-vision-
inspired motion-detection chips. The silicon retina, an electronic model of the
retina, a thin sheet of tissue in the eye, performs spatial bandpass filtering and
compression and was first built by Carver Mead [44]. Examples of silicon retinas
may be found in [45], [46]. Chips for detection of image motion, inspired by
Reichardt motion-detector circuits in the fly [7], may be found in [48], [49].

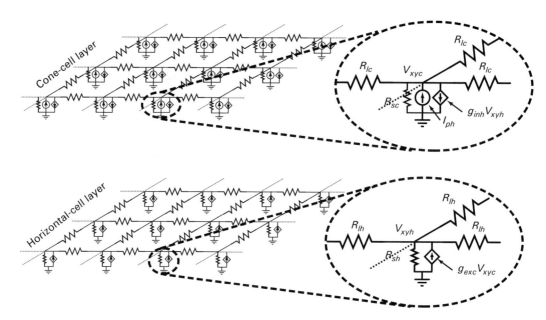

Figure 23.12. A model for the outer plexiform layer of the retina.

We shall only focus on the key architectural ideas and insights since implementation details may be found in the cited papers.

Figure 23.12 reveals a circuit model of the outer-plexiform layer of the retina [47]. It consists of two coupled resistive grids that implement a spatial bandpass filter. The top-layer resistive grid with lateral resistances R_{lc} in Figure 23.12 represents the coupling between the light-sensitive input photoreceptor cone cells of the retina. The bottom-layer resistive grid in Figure 23.12 with lateral resistances R_{lh} represents the coupling between the horizontal cells of the retina. The resistances R_{sc} and R_{sh} represent shunt resistances in the respective layers. Together with the lateral resistances, the shunt resistances serve to create a 2D coupled ZY-transmission-line-like topology in each layer with $Z = R_{lc}$ or R_{lh}, and $Y = 1/R_{sc}$ or $1/R_{sh}$, respectively. The input photocurrents are represented by I_{ph} in the cone-cell layer. The cone-cell and horizontal-cell layer form local negative-feedback coupling. The transconductance g_{exc} represents excitatory coupling that serves to increase horizontal-cell-layer voltage V_{xyh} when the cone-cell-layer voltage V_{xyc} rises at image pixel location (x, y). The transconductance g_{inh} represents inhibitory coupling that serves to decrease cone-cell-layer voltage V_{xyc} when the horizontal-cell-layer voltage V_{xyh} rises at pixel location (x, y). This circuit shown in the breakout boxes of Figure 23.12 is repeated at each pixel location (x, y) throughout the length of the ZY 2D transmission lines to create a highly parallel circuit architecture that performs spatial bandpass filtering.

The bandpass filtering arises because R_{lc}/R_{sc} is much larger than R_{lh}/R_{sh} such that the spatial impulse response in the cone-cell layer is tall and narrow while the spatial impulse response in the horizontal-cell layer is shallow and broad.

To understand the overall spatial impulse response of Figure 23.12, let's begin by assuming there is only one impulsive photocurrent input I_{ph} at the pixel (0, 0). At small distances from the pixel (0, 0) in the cone-cell layer, the lateral excitation from the cone input overwhelms the feedback inhibition from the horizontal-cell layer. Thus cone cells serve to locally excite each other at small distances from their pixel locations. At larger distances from (0, 0) in the cone-cell layer, the feedback inhibition from the horizontal-cell layer overwhelms the lateral excitation from the cone input. Thus, cone cells effectively serve to inhibit each other at large distances from their pixel locations. The net effect is to create a center-surround spatial-impulse response in the cone-cell layer with a local excitatory center that is tall, positive, and narrow, and a broader inhibitory surround that is shallow, negative, and broad. The center-surround impulse response manifests as a spatial bandpass filter in the spatial-frequency domain: when the impulse response is convolved with the image input, very-low-spatial-frequency inputs $i_{ph}(x, y)$ (almost uniform) cancel out in the V_{xyc} output due to balanced contributions from the excitation and inhibition; very-high-spatial-frequency image inputs $i_{ph}(x, y)$ cancel out because their fine spatial variations average out within the excitatory center or inhibitory surround during convolution; intermediate-spatial-frequency inputs $i_{ph}(x, y)$ are most effective at leading to a value of V_{xyc} with good gain with respect to their input amplitudes.

In many electronic retinas, e.g., [44], the width of the excitatory center region is significantly smaller than that of the surround region such that the excitatory region may be approximated by an impulse. In such a case, the retina serves as a highpass spatial filter that accentuates spatial derivatives, i.e., it amplifies edges in the image. Note that the architecture of Figure 23.12 is then the spatial equivalent of the classic highpass temporal filter created by having a lowpass filter in a negative-feedback loop. In general, the overall retina is the 2D spatial equivalent of a 1D temporal bandpass filter created by having a feedforward lowpass filter (that in the cone-cell layer) and a feedback lowpass filter (that in the horizontal cell layer) coupled together in a negative-feedback loop. A simple root-locus analysis then reveals that the overall output is bandpass. This intuition is borne out by a more formal analysis. The equations that describe Figure 23.12 are given by

$$
\nabla^2 \begin{pmatrix} V_{xyc} \\ V_{xyh} \end{pmatrix} = \begin{pmatrix} \dfrac{R_{lc}}{R_{sc}} & -g_{inh}R_{lc} \\ g_{exc}R_{lh} & \dfrac{R_{lh}}{R_{sh}} \end{pmatrix} \begin{pmatrix} V_{xyc} \\ V_{xyh} \end{pmatrix} + \begin{pmatrix} I_{ph}R_{lc} \\ 0 \end{pmatrix} \tag{23.32}
$$

As is customary in linear analysis, we set $I_{ph} = 0$ and determine the eigensolutions $V_{xyc} = c_1 \exp(k_x x + k_y y)$ and $V_{xyh} = c_2 \exp(k_x x + k_y y)$ into Equation (23.32), we find that

$$
\left(k_x^2 + k_y^2\right) \begin{pmatrix} V_{xyc} \\ V_{xyh} \end{pmatrix} = \begin{pmatrix} \dfrac{R_{lc}}{R_{sc}} & -g_{inh}R_{lc} \\ g_{exc}R_{lh} & \dfrac{R_{lh}}{R_{sh}} \end{pmatrix} \begin{pmatrix} V_{xyc} \\ V_{xyh} \end{pmatrix} \tag{23.33}
$$

With some algebra and by setting determinants to zero, Equation (23.33) leads to a quadratic equation whose solution is given by

$$k_r^2 = k_x^2 + k_y^2 = \frac{\left(\dfrac{R_{lh}}{R_{sh}} + \dfrac{R_{lc}}{R_{sc}}\right) \pm \sqrt{\left(\dfrac{R_{lh}}{R_{sh}} - \dfrac{R_{lc}}{R_{sc}}\right)^2 - 4(g_{inh}R_{lc})(g_{exc}R_{lh})}}{2} \tag{23.34}$$

Equation (23.34) yields roots for k_r^2 that vary like that in a classic two-real-pole root locus with root-locus gain parameter corresponding to $(g_{inh}R_{lc})(g_{exc}R_{lh})$. The two different roots for k_r^2 that are obtained from Equation (23.34) lead to two different values for k_r, which correspond to the reciprocal of the center and the surround exponential-decay lengths respectively. Equation (23.34) is a slightly more complex version of the classic 1D transmission-line of Equation (23.1) that applies to the coupled-transmission-line Equation (23.33) rather than to a single transmission line. In fact, the alert reader will recall that the bidirectional RF (and biological) cochlea is actually architected with coupled 1D resonator transmission lines (see Figure 23.5 and [23]); the feedback coupling between these lines leads to an effective 1D line with a more complex coupled-resonator Y that we analyzed (Figures 23.1, 23.2, and 23.3). In the RF cochlea, we began with the analysis of a 1D transmission line (Figure 23.1) and found that it was most easily synthesized by coupled 1D transmission lines (Figure 23.5). In the retina, we began with a coupled-line architecture and analyzed it to obtain an equivalent 1D k_r corresponding to a radial coordinate. Thus, there are deep similarities between the 1D and 2D ZY-transmission-line architectures in the cochlea and in the retina respectively. Both structures can be well modeled by coupled transmission-line equations with coupled LCR 1D architectures in the cochlea and coupled $R_l R_s$ 2D architectures in the retina. The similarities between audition and vision extend beyond the cochlea and retina and are deep and many. They can often be exploited to port insight or inspiration from one domain into another. For example, Chapter 19 suggests how inspiration from cochlear implants for the deaf may lead to useful architectures in brain implants for the blind.

A compact-and-elegant current-mode implementation of the model of Figure 23.12 is described in [45] with subthreshold transistors. Unfortunately, the resulting spatial bandpass filter is rather nonlinear with input intensity due to the body effect in subthreshold transistors and the time constants of the response are relatively slow. In spite of the compact implementation, pixel spatial resolution is compromised compared with that in commercial imagers. Nevertheless, with some engineering improvements and modifications, such circuit architectures could be used in conjunction with inputs derived from another imager to implement highly parallel efficient spatial filtering. The biological retina has a marvelous 3D technology with image processing from the photoreceptor layer to the final optic-nerve output done in a layer-by-layer fashion such that spatial resolution is not compromised.

Figure 23.13 reveals a cellular-neural-network (CNN) architecture that has been used by Roska [20] to model more complex processing in the retina besides

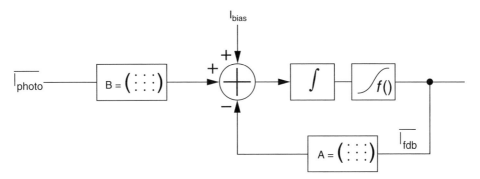

Figure 23.13. A cellular-neural-network architecture for vision processing.

that in its outer-plexiform layer (the processing in this layer is well modeled by Figure 23.12). This architecture is capable of modeling dynamic responses as well. The A and B 'template matrices' model nearest-neighbor and next-nearest-neighbor interactions between outputs of pixels and inputs of pixels respectively. The resistive grids of Figure 23.12 easily fit into this formulation. The integration in the feedforward path of Figure 23.13 allows for dynamic vision processing, which is important in the extraction of motion. The saturation nonlinearity and bias input in Figure 23.13 allow for thresholding, signal restorative, and pattern-recognition functions. Vision chips with impressive computational capabilities have been built that exploit the highly parallel analog processing of this architecture [20]. Most importantly, the A and B templates can be digitally programmed such that a vision processor can execute a complex visual program through sequential instructions as in a microprocessor. We shall return to the theme of algorithmically programmable analog processing in a more general context in the next section.

A fly performs amazing feats of navigation with inputs from its simple and highly ingenious visual system. The fly's visual system controls its winged flight in a sensorimotor feedback loop [7], [8]. An important function of its visual system is its ability to detect image and ego (self) motion. Figure 23.14 (a) reveals a motion-processing facilitate-and-sample (FS) architecture inspired by the correlation-based motion-processing architecture of the common house fly [49]. In response to fast brightness transients, generated, for example, by a moving edge, temporal-edge detectors (E) output brief current impulses and correspondingly brief v_F voltages in each half of the circuit, i.e., v_{F1} and v_{F2} in Figure 23.14 (a). The v_F pulses are termed fast pulses. The current impulses are fed to pulse-shaping circuits (P) that convert the current impulses into slowly decaying voltage pulses, v_S, in each half, i.e., v_{S1} and v_{S2} in Figure 23.14 (a). The slow pulses are created by a parallel diode-capacitor circuit whose initial condition is established by the fast input current pulse, and whose subsequent dynamics after the input current pulse has terminated is established by a diode-capacitor decay. The slow v_S pulses are termed facilitation pulses, while the fast v_F pulses are

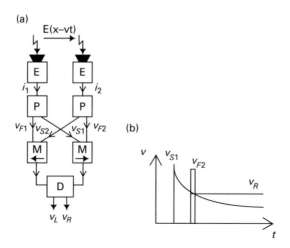

Figure 23.14a, b. (a) A bio-inspired facilitate-and-sample motion-processing architecture. (b) Waveforms in the architecture. Reproduced with kind permission from [49] (©1996 IEEE).

termed sampling pulses, which is why the overall architecture is termed a facilitate-and-sample architecture.

Voltage pulses from adjacent pixels are fed into two motion circuits (M) computing velocity for opposite directions (v_L and v_R) along one dimension. The rightward direction of motion is shown at the top of Figure 23.14 (a). Figure 23.14 (b) reveals the analog output voltage v_R of the right M circuit for a rightward direction of motion. The voltage v_R equals the voltage of the broad facilitating v_{S1} pulse at the time of arrival of the narrow sampling pulse v_{F2}. Slower velocities lead to a late time of arrival of v_{F2} such that the sampled value of v_{S1} at v_R is small. Thus, speed is encoded in the sampled value of v_R. For leftward motion, the fast sampling pulse v_{F2} precedes the slow facilitation pulse v_{S1} such that the output voltage v_R is sampled to be at the low starting value of v_{S1}, encoded as 0. Thus, the analog output voltage v_R encodes speed for its preferred rightward direction of motion only and outputs zero for its opposite null leftward direction of motion. Similarly, the analog output voltage v_L encodes speed for leftward motion only and is also direction selective. In addition, the direction-selection circuit (D) suppresses undesirable responses in the null directions for each half to prevent temporal aliasing.

In a diode-capacitor circuit an impulsive-current input can establish an initial-condition current of $I(0)$ in the diode and a corresponding voltage of $v_S(0)$ on the capacitor C across the diode. The diode current $i(t)$ and voltage $v_S(t)$ at a time t after this initial condition may be shown to be given by

$$i(t) = \frac{I(0)}{1 + \frac{I(0)t}{C(\phi_t/\kappa)}}$$

$$v_S(t) = V_S(0) + \frac{\phi_t}{\kappa} \ln\left(\frac{i(t)}{I(0)}\right)$$

(23.35)

where ϕ_t/κ is the characteristic exponential voltage of the diode. An intriguing property of Equation (23.35) is that if $I(0)t$ is sufficiently large, the value of $i(t)$ becomes independent of the initial condition $I(0)$ and only dependent on $C(\phi_t/\kappa)/t$. Thus, if we output and sample the current $i(t)$ from the pulse-shaping circuit P rather than the voltage $v_S(t)$, the speed is directly encoded in the value of the current (since speed = distance/t). Its value is also relatively invariant with the strength of the impulse current input from E. The initial-condition invariance and the $1/t$ property of diode-capacitor dynamics make it uniquely useful for wide-dynamic-range velocity sensing. The circuit of Figure 23.14 shares many of the features of the correlation-based motion circuits of the fly but the FS architecture and the use of diode-capacitor dynamics represent engineering improvisations on the basic motion-processing architecture of the fly. Further details of motion-sensing architectures are provided in [49]. The use of fly-vision and other bio-inspired circuits for controlling robots is described in [48], [50].

23.7   Hybrid analog-digital computation in the brain

We constantly make discrete digital decisions that alter our continuous analog behaviors. For example, we may decide to change our mode of transportation from one analog form, walking, where we can walk with graded speeds, to another analog form, running, where we can also run with graded speeds. We understand visual scenes as compositions of discrete symbolic objects with graded analog properties. Speech recognition involves a constant interplay between parsing analog auditory signals into discrete digital symbols which in turn can change the interpretation of past, current, or future analog signals. Thus the inherent nature of sensory and motor tasks has a hybrid analog-digital aspect. Not surprisingly, architectures in the brain have evolved hybrid analog-digital architectures to process such information.

Figure 23.15 shows a cartoon drawing of a neuron with its input tree-like dendrites, output axon, and somatic spike-generation region. Spikes are all-or-none digital pulse events generated predominantly near the somatic region of a neuron when the neuronal voltage on a capacitance in that region crosses a threshold. The spikes are conveyed by the output axon wire to other neurons that the neuron connects to. The connections between neurons occur at synapses. Synapses are effectively spike-dependent electrochemical g_m generators. They convert the input digital spike impulse arriving from a presynaptic transmitting neuronal axon into an exponential analog impulse-response current on the receiving dendrite of the postsynaptic neuron. The dendrite of the receiving neuron processes all of the input currents from all of its synapses in a highly parallel ($\sim$ 6,000-input) adaptive-and-nonlinear ZY transmission-like network [1]. The net result of all this processing is the generation of new spikes in the receiving neuron, which are then conveyed to other neurons that it connects to. The architecture of the neuron is inherently suited to spatiotemporal filtering, pattern recognition and

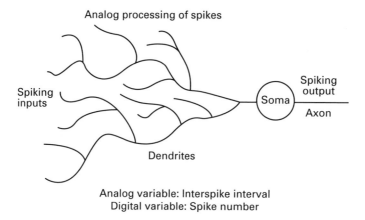

Figure 23.15. A cartoon drawing of a neuron.

signal reconstruction [51]. Adaptation of the synaptic g_m through ingenious automatic learning rules at the synapse leads to powerful computational networks that learn [52]. There is a constant alternation between digital and analog representations amongst neurons in a network as we move from digital output spikes in one neuron to analog input currents in the neuron that it connects to. However, it is worth noting that interspike intervals are analog while spike number is digital, such that there is always an inherently hybrid nature to spike-based signal representations in neurons.

Given the hybrid nature of sensory and sensorimotor tasks that we perform, and the hybrid analog-digital nature of neuronal signal processing in the cortex, it is natural to imagine that our brains compute in a hybrid fashion. An analysis in [53] suggested that one of the key reasons for the efficiency of the human brain, which only consumes 14.6 W of power and performs complex computations in real time, is the collective analog or hybrid nature of its architecture. Neurons are individually imprecise and noisy but interact with each other collectively to perform precise and/or complex computations as in the middle portion of Figure 22.7 (a). The interactions amongst neurons occur via spike signals just as in the example of Figure 22.8 (c). Furthermore neurons can exhibit simultaneous analog and digital behavior unlike electronic circuits that are purely digital or purely analog.

Figure 14.22 in Chapter 14 illustrates a current-mode circuit that models a pattern of synaptic connectivity often seen amongst neurons in cortex, the amazing sheet of tissue that envelopes our brains. We have already discussed this circuit in Chapter 14 in the context of current-mode circuits, so we shall not discuss it in depth here. We just remind the reader that excitatory positive-feedback connections in the brain are mostly local while inhibitory negative-feedback connections in the brain tend to have a broader range of connectivity. The net result of such distributed positive and negative feedback connectivity, whether in the brain or in electronics, leads to behavior that is the hybrid of that exhibited by a many-input digital latch and a classic analog amplifier. Such feedback leads to the simultaneous digital selection of certain inputs and analog amplification of these inputs while all other inputs are suppressed. The brain can amplify certain digital inputs that it 'attends to' in an analog fashion while ignoring others. The complex nonlinear feedback dynamics automatically determines which inputs are chosen and amplified. Further details are described in [54].

23.8   Spike-based hybrid computers

We have discussed the energy-efficient benefits of mixed-signal or hybrid computation in Chapter 22. How do we build a hybrid computer that can implement a complex computation as a digitally programmable sequence of simpler analog computations, each of moderate precision? Or, to put it differently, can we create a digital microprocessor with an instruction set that controls a sequence of analog computations rather than just a sequence of digital ones? How do we build hybrid analog-digital cellular automata that function like interacting neurons or interacting cells in biology to implement powerful network-based computations?

Pulsatile spikes or equivalent event-based signals are natural for hybrid computation because spike number is discrete while inter-spike intervals are analog. Spikes can serve as overranging-and-reset signals that allow one to create a distributed representation of an analog number over many moderate-precision analog channels rather than one precise analog channel as we outlined in Chapter 22 (Figure 22.8 (c)). They can be used as asynchronous triggering events that cause state changes. Their finite pulse width can be used in the same manner that a clock is used in traditional digital systems to allow circuitry to settle. They can serve to create pulse-to-charge synapses as in a neuron. Inter-spike intervals between spikes can be used as start and stop signals for reconfiguring currents in an analog circuit. Inter-spike intervals can be quantized to finite precision by a clock to perform signal restoration or analog-to-digital conversion (see Chapter 15 for the discussion of a neuron-inspired spike-based ADC). Spike signals can be quantized to 1-bit precision by comparing their time of occurrence with respect to other events to make decisions. Rounding up or rounding down of spike count can be used to implement signal restoration in analog systems as well. Thus, the use of spikes can serve as a powerful signal-to-symbol and symbol-to-signal interface between continuous dynamical systems and discrete dynamical systems. Spikes appear to be natural for hybrid computation.

Hybrid state machines (HSMs), which we introduced in Chapter 22, and which are shown in Figure 22.9 (b), exploit these unique features of spikes. The HSM generalizes the idea of a finite state machine (FSM) to the hybrid domain and is a conceptually important block in a hybrid computer. HSMs may be built with comparator events rather than pulsatile spikes, but for all of the reasons that we have outlined, spikes appear to be more natural. Although we have discussed HSMs from the perspective of mixed-signal circuit design in Chapter 22, we shall discuss them again in this chapter in a slightly different neuron-inspired context. HSMs were first conceived in such a context [11].

The HSM, shown in Figure 22.9 (b) on the right, is composed of a spike-generating analog dynamical system with volatile analog state, analog inputs, and analog outputs; it also has a digital finite state machine with combinational logic, nonvolatile digital state, digital inputs and outputs. The spikes from the analog system trigger state transitions in the digital finite state machine. Hereafter, we shall refer to the digital portion of the HSM as a spike-triggered finite state machine, abbreviated by SFSM, and denote the analog dynamical system as ADS. The states and inputs of the SFSM reconfigure the parameters or inputs of the analog dynamical system via a binary control vector. This vector may explicitly create an analog input via a DAC (digital-to-analog converter), or switch an analog input on or off, or reconfigure the topology of the analog system by switching certain connections on or off. Thus, in an HSM, the digital portion has control over the analog portion, and the analog portion has control over the digital portion. If we desire storage of the volatile analog state on a particular capacitor, we can perform an on-the-fly spike-based A/D quantization of the charge via the neuron-inspired ADC described in Chapter 15, a special-purpose subroutine HSM with two capacitors, and store the result in the SFSM. In keeping with the philosophy of collective analog computing in Figure 22.7 the quantization is only performed to a moderate precision that is characteristic of the analog channel. Analog signal restoration can occur through quantization or because some capacitors are reset at the beginning of each state.

It is useful to understand examples of architectures that already exist that represent special cases of the more general HSM architecture. As we have discussed in Chapter 22, several ADC architectures described in Chapter 15 such as those based on successive approximation, $\Sigma\Delta$, neuron-inspired, and algorithmic conversion may be viewed as being special HSMs. Computational ADCs described in Chapter 15 are also examples of HSMs. Hidden Markov models (HMMs) used in speech recognition can be modeled by HSMs with parameters that change from state to state such that the noise and topology in each state are different. Aircraft controllers that transition between different smooth analog controllers depending on whether the aircraft is cruising, landing, or taking off are HSMs. In general, hybrid controllers that are used in robots and industrial control are also examples of HSMs. Gene-protein networks within cells may be approximated by models with different transcription-factor configurations that adapt to the analog protein environment or analog protein inputs to the cell; these transcription factors may be approximated as switching on or switching off certain genes within the cell such that protein processing in the cell is state dependent.

Networks of HSMs that interact via their digital bits or via analog signals can be potentially used to create powerful hybrid analog-digital cellular automata that are useful for modeling highly parallel computations, e.g., for image processing or for modeling biochemical reaction networks within or between cells (described in

Chapter 24, the next chapter). Networks of HSMs that interact via digital or analog signals are examples of collective analog or hybrid systems outlined in Figure 22.7 (a).

23.9   Collective analog or hybrid systems

Examples in this chapter have shown that many analog components can interact via a shared 1D or 2D transmission line architecture to implement powerful parallel, e.g., the silicon retina, or efficient scale-invariant bio-inspired computation, e.g., the exponentially tapered RF cochlea. They are examples of collective analog computation in Figure 22.7 (a) where the interactions amongst components are all analog. The RF cochlea and biological cochlea illustrate that noiseless inductors (masses) and capacitors (stiffnesses) with some active amplification and gain control can interact with each other to perform highly energy-efficient and ultra-fast filtering, amplification, compression, and spectral analysis that is robust to noise in its signals. In the case of the RF cochlea the exponential nature of the transmission line leads to scale-invariant convergence of noise from its devices as well [31]. Retinas and RF cochleas may be viewed as examples of analog preprocessing architectures that encode information about their analog inputs in an intelligent and distributed fashion without merely reporting raw digitized numbers as in Figure 22.5 (b). After such analog preprocessing, the outputs of the retina and cochlea are sampled and reported as spikes in the optic nerve and auditory nerve respectively.

Interacting spiking neuronal systems are examples of collective analog computation in Figure 22.7 (a) where the interaction amongst components is via spikes. Since spikes represent an all-or-none variable we may view such interactions as digital. Since the time of arrival of a spike is analog the interactions amongst neurons also have an analog aspect to them.

Collective analog computation is one of the ten principles for energy-efficient computation that we itemized in Chapter 22. Does biology obey other energy-efficient principles that we itemized in Chapter 22? It appears to obey *all* of them.

23.10   Energy efficiency in neurobiological systems

We shall now discuss examples from neurobiology that suggest that it is phenomenally energy efficient, partly because it obeys all ten principles for low power computation that we discussed in Chapter 22. Clearly there are others that are likely still unknown to us. Our ignorance of neurobiology is at least 10 times larger than our knowledge in all fields of engineering. So, the discussion below should just be viewed as a first step in understanding the reasons for energy efficiency in neurobiological systems.

23.10.1   Encode the computation in the technology efficiently

The traveling wave architecture in the ear exploits fluid masses, spring stiffnesses, piezoelectric outer-hair-cell amplifiers, motion-sensing inner hair cells, and electro-chemical synapses to fashion an elegant low-noise spectrum analyzer and compression system. It is an example of advanced BioMEMS, nanoscale engineering, microfluidics, mechatronics, all rolled into one pea-sized cochlea. Each of these technologies lead to basis-function primitives that are useful for the distributed filtering and gain-control computations, which are performed with 14 μW of power dissipation.

The rich technology of the brain, summarized in Table 23.3, allow it to utilize a rich array of devices and 3D interconnect to construct the most amazing nonlinear feedback system that has ever been built, which is created by multiple connections amongst its neuronal cells. Fast impulsive spiking events from its outputs cause slower adaptive analog synaptic events that are combined in parallel in the non-linear, adaptive, and active RC transmission line that forms the dendrite of the neuron. The basis functions implemented by this architecture are ideally suited for vector pattern recognition, learning, sequential and parallel computation, sampling, signal reconstruction, signal restoration, and hybrid analog-digital computation.

23.10.2   Use subthreshold operation to maximize energy efficiency

The founder of neuromorphic electronics, Carver Mead, noticed the use of Boltzmann exponential I–V relationships in both the ion channels (holes or 'channels' in the nerve membrane that allow Na^+, K^+, Cl^- and other ions to travel through them and that form its elementary conductance units) and the synapses of neuro-biology [5]. It was in fact this observation that lead to the founding of the new field of neuromorphic electronics. The g_m/I_{DS} relationship, which is defined by the slope of the exponential, is in fact 6 times higher for neurobiological ion channels compared with that of an electronic diode but the effective power supply voltage in neurobiology is only about 125 mV.

23.10.3   Exploit analog preprocessing before digitization in an optimal way

The cochlea, retina, and the dendrite in a neuron are examples of the use of analog preprocessing before spike-based sampling and quantization. We do not know enough to know whether nature has optimized the amount of analog processing within its dendrite, retina, or cochlea before such digitization. However, it certainly does a lot of analog preprocessing.

23.10.4   Use parallel architectures

The architectures of the brain, the retina, and the cochlea are highly parallel. In the brain, neurons process information in a highly parallel hybrid analog-digital fashion. In the retina, rod and cone cells, horizontal cells, bipolar cells, amacrine cells, and ganglion cells are hooked to each other in a highly parallel 2.5D architecture to do parallel processing of the retinal image. In the cochlea, 3,500 inner hair cells and 10,000 outer hair cells work in parallel with each other, coupled to one another via the motions of the basilar membrane, the tectorial membrane, and the cochlear fluid.

23.10.5   Balance computation and communication costs

Estimates of energy consumption by neurons in the brain show that they spend about as much energy in communicating with action potentials as in computing with them [55].

23.10.6   Exploit collective analog or hybrid computation

We have already discussed this point in Section 23.9 and partly with reference to Figure 22.7 in Chapter 22.

23.10.7   Reduce the amount of information that needs to be processed

If the retina did not preprocess and compress the information that it sends to the brain, we would be processing about $\sim$36 Gb/s of wide-dynamic-range raw image data from its photoreceptor array rather than $\sim$20 Mb/s of useful optic-nerve spiking output information. If we assume that the energy consumption per bit does not change, the power consumption of the brain would then need to be $\sim$22 kW, more than that of 20 furnaces! The auditory nerve is architected to fire spikes more often in response to sound envelope onsets than to steady-state sound envelopes [24]. Throughout the nervous system and in the brain, there are integrators or lowpass filters that form parts of an adaptive feedback loop such that neurons only fire in response to changes in their inputs rather than to constant inputs [10]. They are constantly learning to adapt away useless information that they can predict. This principle is so ubiquitous in the organization of the nervous system, that we refer readers to [10] for at least 100 such examples.

23.10.8   Use feedback and feedforward architectures for improving robustness and energy efficiency

A good example of feedforward processing in the nervous system is the *vestibular ocular reflex* (VOR) [10]. The motion of the head to the right triggers a compensatory reflexive motion of the eyes to the left such that the image that the eye is watching does not blur when the head is moved. The use of feedforward processing allows for quick fast responses but requires constant learning and calibration to ensure that the eye-motion to head-motion gain is just right. It avoids the need for complex feedback loops that would need to model the dynamics of the head and the eye in detail to prevent overshoots and instability while maintaining fast response times.

The use of integrators and lowpass filters in feedback loops serve to improve energy efficiency by reducing the amount of information that needs to be processed, while simultaneously removing dc errors and device mismatch that prevent robust operation. Feedback loops for gain control in the cochlea and photo-receptors ensure relatively constant SNR operation over a very wide dynamic range of input signals, which would otherwise compromise information fidelity and energy efficiency. Feedback and feedforward gain control is ubiquitous amongst neurons in the brain at multiple temporal and spatial scales [1].

Chapter 2 provides several examples of feedback loops in biology.

23.10.9   Separate speed and precision in the architecture if possible

The somatic regions of neurons shown in Figure 23.15 function as a simple thresholding comparator that generates spikes. Comparators, as we explain in Chapter 22, use preamplification with modest gain to obtain precision and positive feedback to obtain speed.

The retina in the eye has a central foveal region made up of cone cells that are relatively slow but that have precise spatial and contrast resolution. The peripheral region of the eye is made up of rod cells that are relatively fast but that have imprecise spatial and contrast resolution.

The right brain is, in general, better at getting the 'big picture' quickly but is relatively imprecise. It is important for recognizing novel threats and responding to predators quickly without getting bogged down in irrelevant detail that could lead to death. In contrast, the left brain is, in general, better at getting the 'details' slowly and is relatively precise [56]. Interestingly, some remarkable experiments [56] have shown that patients with left-brain damage can reproduce the overall big-picture of a visual scene from memory but are unable to reproduce its details. In contrast, patients with right-brain damage can reproduce minute details in a visual scene from memory but are unable to tie them together in a big picture.

23.10.10   Operate slowly and adiabatically if possible

Since the biological cochlea can be described as a high-Q coupled-resonator transmission-line structure, it is inherently adiabatic as per the discussion in Chapter 21. That is one of the key reasons for its incredibly low power dissipation of $\sim 14\,\mu$W.

Dendrites in neurons in the brain are architected with distributed RC transmission lines, which because of their distributed nature have an effectively smaller time constant for their inputs. For example, a well-known result in RF electronics is that the effective RC time constant associated with a distributed polysilicon gate input to a transistor with gate length L is $(RL)(CL)/3$ rather than $(RL)(CL)$, where R and C are the resistance and capacitance per unit length of the polysilicon wire respectively. The 1/3 factor arises because the entire R and the entire C do not filter the incoming gate input at any point in the distributed RC line such that different points in the channel see different amounts of incremental filtering. The effective speedup in time constant due to a distributed structure rather than a lumped structure implies that, for a given input frequency of operation, the operation is more adiabatic: the cutoff frequencies of the structure are now larger relative to the input frequency. The brain's average rate of spike production is near 5 Hz and the time constants of its neuronal structures vary from 100 ms to 0.1 ms. Thus, the brain operates in a relatively slow fashion in some of its structures.

23.11   Other work

Neuromorphic electronics is still in its infancy because understanding nature with insight has taken time and will continue to take time. Nevertheless, we can pick increasingly high-hanging fruit as engineering helps understand biology better and biology in turn inspires better engineering, leading to a powerful positive-feedback loop. Besides some of the RF, auditory, speech, vision, and hybrid analog-digital examples presented in this chapter, we mention a few examples for the reader interested in exploring further. Spike-based communication circuits for interfacing chips termed *address event representation* have been proposed to connect chips, e.g., [57], [58]. Biosonar systems inspired by the bat are being actively researched and applied to artificial sonar systems [59]. Bio-inspired robotic systems have been important in creating ingeniously simple-but-clever robots [60], [50]. Learning and pattern-recognition circuits have been built [61], [62], [63]. A learning architecture inspired by neural encoding-decoding architectures has been proposed for decoding of neural signals in paralysis prosthetics [51], [64]. A table of some of the labs working in the general area of neuromorphic electronics may be found in [13]. Finally, the bio-inspired revolution is expanding beyond neurons and beyond electrical engineering to other biomorphic domains: as just one example in a large space, extremely large single-crystal structures inspired by biological material formation have led to very innovative materials-science designs [65].

In the next chapter, we shall study computation and circuits within cells, which can inspire cell-inspired or cytomorphic architectures in the future.

23.12   Appendix: Power and computation in the brain, eye, ear, and body

The brain's neuronal cells output $\sim$1 ms pulses (spikes) at an average rate of 5 Hz [55]. The 240 trillion synaptic connections [1] amongst the brain's neurons thus lead to a computational rate of *at least* 10^{15} synaptic operations per second. A synapse implements multiplication and filtering operations on every spike and sophisticated learning operations over multiple spikes. If we assume that synaptic multiplication is at least one floating-point operation (FLOP), the 20 ms second-order filter impulse response due to each synapse is 40 FLOPS, and that synaptic learning requires at least 10 FLOPS per spike, a synapse implements at least 50 FLOPS of computation per spike. The nonlinear adaptation-and-thresholding computations in the somatic regions of a neuron implement almost 1200 floating-point operations (FLOPS) per spike [66]. Thus, the brain is performing at least $50\,\text{FLOPS} \times 5\,\text{Hz} \times 240 \times 10^{12} + 1200\,\text{FLOPS} \times 5\,\text{Hz} \times 22 \times 10^{9} \approx 6 \times 10^{16}$ FLOPS per second. The energy efficiency of the brain is a staggering $14.6/\left(6 \times 10^{16}\right) = 0.24\,\text{fJ/FLOP}$. In computing this estimate, we have ignored several complex spatiotemporal transmission-line-like computations in the dendritic regions of a neuron, which may significantly increase its computational rate. Thus, these numbers should be viewed as a lower bound on the computational rate of the brain. The power dissipation of the brain is estimated in [2]: the brain of a 65 kg male consumes nearly 14.6 W.

No one has measured or estimated the actual number of synapses in the retina. However, based on a suggestion of Professor Simon Laughlin (personal communication) and Professor Peter Sterling, both expert neurobiologists, a good guesstimate is to multiply the number of cone cells by 20 and the number of rod cells by 1, which yields an estimate for the initial retinal synapses near its photoreceptor layers, and then multiply this overall number by 5 to get an estimate for synapses over all of its layers. Such a computation yields (5.5 million $\times$ 20 + 130 million $\times$ 1) $\times$ 5 = 1.2 billion synapses. We have listed $\sim$1 billion in Table 23.2.

Recent measurements show that a single mouse rod cell consumes 10^{8} ATP/s reducing to 2.5×10^{7} ATP/s in the dark [67]. Cone cells need to have lots of ion channels open to register fluctuations in photon catch (usual $\sqrt{N}$ arguments as in the rest of this book) and be assumed to consume ATP at the same rate as rod cells. The inner retina in the monkey (and therefore, likely human) seems to consume ATP at the same rate per gram as that of the outer and inner retina in the rat. Given all these considerations, a reasonable assumption to make (personal communication from Simon Laughlin) is that human rods account for 40% of the whole retina's maximal power consumption in the dark. Thus, the power consumption of the retina can be estimated to be 2.5×10^{8} ATP/s $\times$ 20 kT/ATP $\times$ 130 million rods = 2.6 mW in the dark. Interestingly, another estimate obtained in a different way yields numbers that are close to this number: The dark consumption for pure rods in the mouse corresponds to 13 μmol ATP/g/min, similar to values measured for outer retina in the rat [67]. Rods are 50% of the total wet weight of mouse retina.

The weight of the human retina is 2500 mm^2 (area) $\times$ 160 μm (avg. thickness) $\times$ 1000 kg/m^3 (density in SI units) = 0.4 grams. Thus, the power consumption of human rods in the dark may be estimated to be 0.2 grams $\times$ 13 μmol ATP/g/min $\times$ 20 kT/ATP = 2.1 mW. If we assume that outer retina power consumption is dominated by the rods, and that the inner and outer retina consume at the same rate in humans, then the total power consumption of the retina in the dark may be estimated to be 2.1 mW $\times$ 2 = 4.2 mW. We list the average of $(2.6 + 4.2)/2 = 3.4\,\text{mW}$ as our estimate for the total power consumption of the retina in Table 23.2. We thank Simon Laughlin for his generous assistance in helping us estimate the number of synapses in the retina and the power consumption of the eye.

The eye operates over a 10^{8} : 1 dynamic range in light levels ($\sim$27 bits in a linear system) with 135 million pixels at about a sample rate of 12 Hz, i.e., the raw input bit rate is 36 Gb/s while the output bit rate of the optic nerve is nearly 20 Mb/s [68]. If we assume that each synapse in the retina does the equivalent of a floating-point multiplication operation, the 1 billion synapses in the retina operating at 12 Hz yield a computational rate of at least 12 GFLOPS.

The rolloff slopes of the ear's filters have been measured to be at least 240 dB/octave indicating that they are the equivalent of at least a 40th order filter [24]. Since there are 3,500 inner hair cells in the cochlea, each of which leads to a spectral output (about 10 auditory nerves represent each output in a redundant fashion), we have at least $3,500 \times 40 = 144,000$ multiplies per cochlear output over a 120 dB dynamic range of inputs. The maximal sampling rate is near the 5 kHz phase-locking rate for the first few stages of the cochlea, which dominate its computational rate of $144,000 \times 5\,\text{kHz} = 0.72$ GFLOPS. From impedance and energy measurements of the cochlea we can estimate the biological cochlea's power consumption to be $0.4\,\mu\text{W/mm} \times 35\,\text{mm} = 14\,\mu\text{W}$ [69]. A 1 nA bias current (upper bound) associated with each of the cochlea's 10,000 hair cells (inner and outer hair cells) with a 125 mV power supply yields $1.25\,\mu\text{W}$. If ATP chemical energy is converted to electrical work with an efficiency of 0.5, the static electrical energy consumption of the $14\,\mu\text{W}$ is at least $2.5\,\mu\text{W}$.

The power consumption of other bodily organs is given in [2]. The brain consumes 14.6 W; the heart 9.7 W, the kidneys 7.0 W, the liver 17.1 W, the gastro-intestinal tract 13.4 W, the skeletal muscles 13.5 W, the lungs 4.0 W, and the skin 1.5 W for a staggeringly low grand total of 80.8 W for the basal metabolic rate of the whole body!

References

[1] Christof Koch. *Biophysics of Computation: Information Processing in Single Neurons* (New York: Oxford University Press, 1999).

[2] L. C. Aiello. Brains and guts in human evolution: The expensive tissue hypothesis. *Brazilian Journal of Genetics*, **20** (1997).

[3] John Morgan Allman. *Evolving Brains* (New York: Scientifc American Library: Distributed by W.H. Freeman and Co., 1999).

[4] C. A. Mead. Neuromorphic electronic systems. *Proceedings of the IEEE*, **78** (1990), 1629–1636.

[5] Carver Mead. *Analog VLSI and Neural Systems* (Reading, Mass.: Addison-Wesley, 1989).

[6] A. Solga, Z. Cerman, B. F. Striffler, M. Spaeth and W. Barthlott. The dream of staying clean: Lotus and biomimetic surfaces. *Bioinspiration and Biomimetics*, **2** (2007), 126.

[7] B. Hassenstein and W. Reichardt. Systemtheoretische analyse der zeit-, reihenfolgen-und vorzeichenauswertung bei der bewegungsperzeption des rüsselkäfers chlorophanus. *Zeitschrift für Naturforschung*, **11** (1956), 513–524.

[8] M. H. Dickinson, F. O. Lehmann and S. P. Sane. Wing rotation and the aerodynamic basis of insect flight. *Science*, **284** (1999), 1954–1960.

[9] D. A. Robinson, J. L. Gordon and S. E. Gordon. A model of the smooth pursuit eye movement system. *Biological Cybernetics*, **55** (1986), 43–57.

[10] Eric R. Kandel, James H. Schwartz and Thomas M. Jessell. *Principles of Neural Science*. 3rd ed. (Norwalk, Conn.: Appleton & Lange, 1991).

[11] R. Sarpeshkar and M. O'Halloran. Scalable hybrid computation with spikes. *Neural Computation*, **14** (2002), 2003–2038.

[12] S. Mandal, S. M. Zhak and R. Sarpeshkar. A bio-inspired active radio-frequency silicon cochlea. *IEEE Journal of Solid-State Circuits*, **44** (2009), 1814–1828.

[13] R. Sarpeshkar. Brain power: Borrowing from biology makes for low power computing. *IEEE Spectrum*, **43** (2006), 24–29.

[14] R. Sarpeshkar, C. D. Salthouse, J. J. Sit, M. W. Baker, S. M. Zhak, T. K. T. Lu, L. Turicchia and S. Balster. An ultra-low-power programmable analog bionic ear processor. *IEEE Transactions on Biomedical Engineering*, **52** (2005), 711–727.

[15] J. Mitola III and G. Q. Maguire Jr. Cognitive radio: making software radios more personal. *IEEE Personal Communications*, **6** (1999), 13–18.

[16] A. Bhattacharya and F. G. Zeng. Companding to improve cochlear-implant speech recognition in speech-shaped noise. *The Journal of the Acoustical Society of America*, **122** (2007), 1079.

[17] J. Guinness, B. Raj, B. Schmidt-Nielsen, L. Turicchia and R. Sarpeshkar. A companding front end for noise-robust automatic speech recognition. *Proceedings of the IEEE International Conference on Acoustics, Speech, and Signal Processing (ICASSP)*, Philadelphia, Penn., 2005.

[18] K. H. Wee, L. Turicchia and R. Sarpeshkar. An Analog Integrated-Circuit Vocal Tract. *IEEE Transactions on Biomedical Circuits and Systems*, **2** (2008), 316–327.

[19] J. Bohorquez, W. Sanchez, L. Turicchia and R. Sarpeshkar, An integrated-circuit switched-capacitor model and implementation of the heart. *Proceedings of the First International Symposium on Applied Sciences on Biomedical and Communication Technologies (ISABEL)*, Aalborg, Denmark, 1–5, 2008.

[20] Leon O. Chua and T. Roska. *Cellular Neural Networks and Visual Computing: Foundation and Applications* (Cambridge, UK; New York, NY: Cambridge University Press, 2002).

[21] International Technology Roadmap for Semiconductors Emerging Research Devices Report, 2007.

[22] Thomas H. Lee. *The Design of CMOS Radio-Frequency Integrated Circuits*, 2nd ed. (Cambridge, UK; New York: Cambridge University Press, 2004).

[23] T. K. Lu, S. Zhak, P. Dallos and R. Sarpeshkar. Fast cochlear amplification with slow outer hair cells. *Hearing Research*, **214** (2006), 45–67.

[24] James O. Pickles. *An Introduction to the Physiology of Hearing*. 2nd ed. (London; New York: Academic Press, 1988).

[25] S. Mandal, S. Zhak and R. Sarpeshkar, inventors. *Architectures for Universal or Software Radio*, U.S. Provisional Patent 60/870,719, filed December 19, 2006; Utility Patent 11/958,990, filed December 18, 2007.

[26] G. Zweig. Finding the impedance of the organ of Corti. *The Journal of the Acoustical Society of America*, **89** (1991), 1229.

[27] C. A. Shera. Mammalian spontaneous otoacoustic emissions are amplitude-stabilized cochlear standing waves. *The Journal of the Acoustical Society of America*, **114** (2003), 244.

[28] Leonard I. Schiff. *Quantum Mechanics*. 3rd ed. (New York,: McGraw-Hill, 1968).

[29] L. Watts. *Cochlear mechanics: Analysis and analog VLSI*. Ph.D. Thesis, Electrical Engineering, California Institute of Technology (1992).

[30] R. F. Lyon and C. A. Mead. An analog electronic cochlea. *IEEE Transactions on Acoustics, Speech and Signal Processing*, **36** (1988), 1119–1134.

[31] R. Sarpeshkar, R. F. Lyon and C. A. Mead. A low-power wide-dynamic-range analog VLSI cochlea. *Analog Integrated Circuits and Signal Processing*, **16** (1998), 245–274.

[32] L. Turicchia and R. Sarpeshkar. A bio-inspired companding strategy for spectral enhancement. *IEEE Transactions on Speech and Audio Processing*, **13** (2005), 243–253.

[33] William McC Siebert. *Circuits, Signals, and Systems* (Cambridge, Mass.; New York: MIT Press; McGraw-Hill, 1986).

[34] E. M. Williams. Radio-Frequency Spectrum Analyzers. *Proceedings of the IRE*, **34** (1946), 18p–22p.

[35] S. Mandal, S. Zhak and R. Sarpeshkar, inventors. *Architectures for Universal or Software Radio*.

[36] W. Liu, A. G. Andreou and M. H. Goldstein. Voiced-speech representation by an analog silicon model of the auditory periphery. *IEEE Transactions on Neural Networks*, **3** (1992), 477–487.

[37] T. J. Hamilton, C. Jin, A. van Schaik and J. Tapson, A 2-D silicon cochlea with an improved automatic quality factor control-loop. *Proceedings of the IEEE International Symposium on Circuits and Systems (ISCAS)*, Seattle, Wash., 1772–1775, 2008.

[38] S. M. Zhak. *Modeling and design of an active silicon cochlea*. Ph.D. Thesis, Electrical Engineering, Massachusetts Institute of Technology (2008).

[39] A. J. Oxenham, A. M. Simonson, L. Turicchia and R. Sarpeshkar. Evaluation of companding-based spectral enhancement using simulated cochlear-implant processing. *The Journal of the Acoustical Society of America*, **121** (2007), 1709–1716.

[40] B. Raj, L. Turicchia, B. Schmidt-Nielsen and R. Sarpeshkar. An FFT-based companding front end for noise-robust automatic speech recognition. *EURASIP Journal on Audio, Speech, and Music Processing*, **Article ID 65420** (2007), 13.

[41] R. Bellman. Dynamic programming. *Science*, **153** (1966), 34–37.

[42] Kenneth N. Stevens. *Acoustic Phonetics* (Cambridge, MA: MIT Press, 2000).

[43] K. H. Wee and R. Sarpeshkar. An electronically tunable linear or nonlinear MOS resistor. *IEEE Transactions on Circuits and Systems I: Regular Papers*, **55** (2008), 2573–2583.

[44] M. A. Mahowald and C. A. Mead. Silicon Retina. In *Addison-Wesley VLSI Systems Series*, ed. C A Mead. (Reading, Mass.: Addison-Wesley; 1989), pp. 257–278.

[45] K. A. Boahen and A. G. Andreou, A contrast sensitive silicon retina with reciprocal synapses. *Proceedings of the IEEE Neural Information Processing Systems (NIPS)*, Denver, Colorado, 764–772, 1992.

[46] T. Delbruck. Silicon retina with correlation-based, velocity-tuned pixels. *IEEE Transactions on Neural Networks*, **4** (1993), 529–541.

[47] T. Yagi, Y. Funahashi and F. Ariki, Dynamic model of dual layer neural network for vertebrate retina. *Proceedings of the International Joint Conference on Neural Networks (IJCNN)*, Washington, DC, 787–789, 1989.

[48] R. R. Harrison and C. Koch. A silicon implementation of the fly's optomotor control system. *Neural Computation*, **12** (2000), 2291–2304.

[49] R. Sarpeshkar, J. È. Kramer, G. Indiveri and C. Koch. Analog VLSI architectures for motion processing: From fundamental limits to system applications. *Proceedings of the IEEE*, **84** (1996), 969–987.

[50] G. Indiveri. Neuromorphic analog VLSI sensor for visual tracking: circuits and application examples. *IEEE Transactions on Circuits and Systems II: Analog and Digital Signal Processing*, **46** (1999), 1337–1347.

[51] Chris Eliasmith and C. H. Anderson. *Neural Engineering: Computation, Representation, and Dynamics in Neurobiological Systems* (Cambridge, Mass.: MIT Press, 2003).

[52] J. Herz, A. Krogh and R. G. Palmer. *Introduction to the Theory of Neural Computation* (Reading, Mass.: Addison Wesley, 1991).

[53] R. Sarpeshkar. Analog versus digital: extrapolating from electronics to neurobiology. *Neural Computation*, **10** (1998), 1601–1638.

[54] R. H. R. Hahnloser, R. Sarpeshkar, M. A. Mahowald, R. J. Douglas and H. S. Seung. Digital selection and analogue amplification coexist in a cortex-inspired silicon circuit. *Nature*, **405** (2000), 947–951.

[55] D. Attwell and S. B. Laughlin. An energy budget for signaling in the grey matter of the brain. *Journal of Cerebral Blood Flow and Metabolism*, **21** (2001), 1133–1145.

[56] P. F. MacNeilage, L. J. Rogers and G. Vallortigara. Origins of the left & right brain. *Scientific American*, **301** (2009), 60–67.

[57] S-C. Liu, J. È. Kramer, G. Indiveri, T. Delbruck, T. Burg and R. Douglas. Orientation-selective a VLSI spiking neurons. *Neural Networks*, **14** (2001), 629–643.

[58] K. A. Boahen. Point-to-point connectivity between neuromorphic chips using address events. *IEEE Transactions on Circuits and Systems II: Analog and Digital Signal Processing*, **47** (2000), 416–434.

[59] T. K. Horiuchi. "Seeing" in the dark: Neuromorphic VLSI modeling of bat echolocation. *IEEE Signal Processing Magazine*, **22** (2005), 134–139.

[60] M. A. Lewis, R. Etienne-Cummings, A. H. Cohen and M. Hartmann. Toward biomorphic control using custom aVLSI CPG chips. *Proceedings of the International Conference on Robotics and Automation (ICRA)*, 494–500, 2000.

[61] C. Diorio, P. Hasler, A. Minch and C. A. Mead. A single-transistor silicon synapse. *IEEE Transactions on Electron Devices*, **43** (1996), 1972–1980.

[62] Gert Cauwenberghs and Magdy A. Bayoumi. *Learning on Silicon: Adaptive VLSI Neural Systems* (Boston: Kluwer Academic, 1999).

[63] Y. Berg, R. L. Sigvartsen, T. S. Lande and A. Abusland. An analog feed-forward neural network with on-chip learning. *Analog Integrated Circuits and Signal Processing*, **9** (1996), 65–75.

[64] B. I. Rapoport, W. Wattanapanitch, J. L. Penagos, S. Musallam, R. Andersen and R. Sarpeshkar, A biomimetic adaptive algorithm and low-power architecture for implantable neural decoders. *Proceedings of the 31st Annual International Conference of the IEEE Engineering in Medicine and Biology Society (EMBC)*, Minneapolis, MN, 2009.

[65] J. Aizenberg. Crystallization in Patterns: A Bio-Inspired Approach. *Advanced Materials*, **16** (2004), 1295–1302.

[66] E. M. Izhikevich. Which model to use for cortical spiking neurons? *IEEE Transactions on Neural Networks*, **15** (2004), 1063–1070.

[67] H. Okawa, A. P. Sampath, S. B. Laughlin and G. L. Fain. ATP consumption by mammalian rod photoreceptors in darkness and in light. *Current Biology*, **18** (2008), 1917–1921.

[68] K. Koch, J. McLean, M. Berry, P. Sterling, V. Balasubramanian and M. A. Freed. Efficiency of information transmission by retinal ganglion cells. *Current Biology*, **14** (2004), 1523–1530.

[69] B. M. Johnstone. Genesis of the cochlear endolymphatic potential. *Current Topics in Bioenergetics*, **2** (1967), 335–352.

24 Cytomorphic electronics: cell-inspired electronics for systems and synthetic biology

Any living cell carries with it the experience of a billion years of experimentation by its ancestors.

Max Delbrück

The cells in the human body provide examples of phenomenally energy-efficient sensing, actuation, and processing. An average $\sim$10 μm-size human cell hydrolyzes several energy-carrying adenosine-tri-phosphate (ATP) molecules within it to perform nearly $\sim$10^7 ATP-dependent biochemical operations per second [1]. Since, under the conditions in the body, the hydrolysis of each ATP molecule provides about 20 kT (8×10^{-20} J) of metabolic energy, the net power consumption of a single human cell is an astoundingly low 0.8 pW! The $\sim$100 trillion cells of the human body thus have an average resting power consumption of $\sim$80 W, consistent with numbers derived from the Appendix of Chapter 23.

The cell processes its mechanical and chemical input signals with highly noisy and imprecise parts. Nevertheless, it performs complex, highly sensitive, and collectively precise hybrid analog-digital signal processing on its inputs such that reliable outputs are produced. Such signal processing enables the cell to sense and amplify minute changes in the concentrations of specific molecules amidst a background of confoundingly similar molecules, to harvest and metabolize energy contained in molecules in its environment, to detoxify and/or transport poisonous molecules out of it, to sense if it has been infected by a virus, to communicate with other cells in its neighborhood, to move, to maintain its structure, to regulate its growth in response to signals in its surround, to speed up chemical reactions via sophisticated enzymes, and to replicate itself when it is appropriate to do so. The $\sim$30,000-node gene-protein and protein-protein molecular interaction networks within a cell that implement and regulate these functions are a true marvel of nanotechnology. The nanotechnology of man appears crude and primitive when contrasted with that in nature's cells.

In this chapter, we show that the equations that describe subthreshold transistor operation and the equations that describe chemical reactions have strikingly detailed similarity, including stochastic properties. Therefore, any chemical reaction can be efficiently and programmably represented with a handful of subthreshold (or bipolar) transistors that comprise an analog circuit. Intracellular protein-protein biochemical reaction networks can then potentially be modeled by hooking such circuits to each other. DNA-protein networks can also be efficiently modeled with such analog circuits. DNA-protein networks can be modeled even more efficiently with hybrid analog-digital circuits that approximate nonlinear analog characteristics with more approximate digital ones. Since extracellular cell-cell networks also rely on molecular binding and chemical reactions, networks such as hormonal networks or neuronal networks can also be efficiently modeled. Thus, in the future, we can potentially attempt to simulate cells, organs, and tissues with ultra-fast highly parallel analog and hybrid analog-digital circuits including molecular stochastics and cell-to-cell variability on large-scale electronic chips. Such molecular-dynamics simulations are extremely computationally inten-sive especially when the effects of noise, nonlinearity, network-feedback effects, and cell-to-cell variability are included. Stochastics and cell-to-cell variabililty are highly important factors for predicting a cell's response to drug treatment, e.g., the response of tumor cells to chemotherapy treatments [2]. We will show in this chapter that circuit, feedback, and noise-analysis techniques described in the rest of this book can shed insight into the systems biology of the cell [3]. For example, flux balance analysis is frequently used to reduce the search space of parameters in a cell [4]. It is automatically implemented as Kirchhoff's current law in circuits since molecular fluxes map to circuit currents. Similarly, Kirchhoff's voltage law automatically implements the laws of thermodynamic energy balance in chemical-reaction loops. We shall provide other examples throughout the chapter. An excellent introduction to systems biology may be found in [5]. Robustness analysis of the circuit using return-ratio techniques can shed insight in the future into which genes, when mutated, will lead to disease in a network, and which will not. Circuit-design techniques can also be mapped to create synthetic-biology circuits that perform useful functions in the future [6].

Circuits in biology and circuits in electronics may be viewed as being highly similar with biology using molecules, ions, proteins, and DNA rather than electrons and transistors. Just as neural circuits have led to biologically inspired neuro-morphic electronics, cellular circuits can lead to a novel biologically inspired field that we introduce in this chapter and term *cytomorphic electronics*. In fact, we will show that there are many similarities between spiking-neuron computation and cellular computation. The hybrid state machines (HSMs) described in Chapters 22 and 23 (see Figure 22.9 (b)) provide a useful framework for thinking about cellular circuits with active DNA genes in the cell being represented by digital variables and protein concentrations represented by analog variables. DNA-protein inter-actions are represented by interactions between the analog and digital parts of an HSM. Networks of hybrid state machines can model networks of cells. The ZY transmission-line architectures that we discussed in Chapter 23 for describing computation in the retina, the cochlea, the vocal tract, and neuronal dendrites are also useful for understanding nonlinear reaction-diffusion partial differential equations in cells. Ingenious nonlinear transmission lines are exploited by the cell to create spatially decaying molecular-concentration profiles that are robustly invariant to changes in protein-production rates during development.

Leonardo da Vinci, perhaps the first bio-inspired engineer in mankind, said: 'Human subtlety will never devise an invention more beautiful, more simple or more direct than does nature because in her inventions nothing is lacking, and nothing is superfluous.' One is indeed awed and humbled as one learns more and more about the ingenious operation of even a single cell. We have much to learn from nature in architecting clever electronics, algorithms, and nanostructures. This chapter will only attempt to scratch the surface.

24.1   Electronic analogies of chemical reactions

Figure 24.1 illustrates that there are striking similarities between chemical reaction dynamics (Figure 24.1 (a)) and electronic current flow in the subthreshold regime of transistor operation (Figure 24.1 (b)): electron concentration at the source is analogous to reactant concentration; electron concentration at the drain is analogous to product concentration; forward and reverse current flows in the transistor are analogous to forward and reverse reaction rates in a chemical reaction; the forward and reverse currents in a transistor are exponential in voltage differences at its terminals analogous to reaction rates being exponential in the free energy differences in a chemical reaction; increases in gate voltage lower energy barriers in a transistor increasing current flow analogous to the effects of enzymes or catalysts in chemical reactions that increase reaction rates; and the stochastics of Poisson shot noise in subthreshold transistors are analogous to the stochastics of molecular shot noise in reactions. These analogies suggest that one can mimic and model large-scale chemical-processing systems in biological and artificial networks very efficiently on an electronic chip at time scales that could be a million to billion times faster. No one, thus far, appears to have exploited the detailed similarity behind the equations of chemistry and the equations of electronics to build such networks. The single-transistor analogy of Figure 24.1 is already an exact representation of the chemical reaction $A \rightleftharpoons B$ including

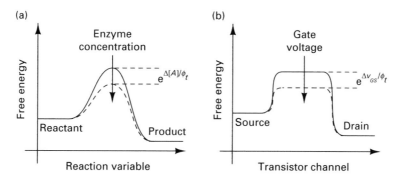

Figure 24.1a, b. Similarities between chemical reaction dynamics (a) and subthreshold transistor electronic flow (b). Reprinted with permission from [9] (©2009 IEEE).

stochastics, with forward electron flow from source to drain corresponding to the $A \rightarrow B$ molecular flow and backward electron flow from drain to source corresponding to the $B \rightarrow A$ molecular flow. In this chapter, we shall build on the key idea of Figure 24.1 to show how to create current-mode subthreshold transistor circuits for modeling arbitrary chemical reactions. We can then create large-scale biochemical reaction networks from such circuits for modeling computation within and amongst cells. Before we do so, we provide another example of a voltage-mode resistor circuit that can efficiently represent the differential equations that describe association and dissociation chemical reactions.

An association chemical reaction, for example, that between an enzyme E and a substrate S, is described by

$$E + S \underset{k_r}{\overset{k_f}{\rightleftharpoons}} ES \tag{24.1}$$

In Equation (24.1), the enzyme binds to the substrate to create a bound enzyme ES or E_b via a forward reaction with reaction-rate constant k_f. The product E_b also dissociates via a backward reaction with reaction-rate constant k_r to recreate E and S. The total amount of enzyme $E_t = E + E_b$ is a constant invariant with time and composed of enzyme in free (E) or bound (E_b) form. If the concentrations of each variable are denoted by square brackets around the variable and the substrate concentration is assumed constant at $[S]$, the equations that describe the mass-action kinetics of the chemical reaction are given by

$$\frac{d[E_b]}{dt} = k_f[E][S] - k_r[E_b] \tag{24.2}$$
$$[E] = [E_t] - [E_b]$$

If $C = 1$, the equations of Equation (24.2) are represented by the resistive-divider circuit of Figure 24.2 (a) with resistances of value $1/(k_f[S])$ and $1/k_r$; the currents through these resistances represent the forward and backward fluxes of the chemical reaction in the first line of Equation (24.2). At equilibrium, when the capacitor voltage has reached a constant value, these two currents are equal and the circuit functions like a resistive divider. From simple resistive-divider analysis, some algebra reveals that, under these conditions,

$$[E_b] = [E_t]\left(\frac{[S]/K_d}{([S]/K_d) + 1}\right)$$

$$K_d = \frac{k_r}{k_f} \tag{24.3}$$

$$K_{eq} = \frac{k_f}{k_r} = \frac{1}{K_d}$$

Thus, when $[S]$ is significantly greater than K_d, the enzyme substrate binding is said to exhibit saturation, since the fraction of bound enzyme is limited by the total enzyme concentration $[E_t]$ rather than by the substrate concentration $[S]$. The circuit of Figure 24.2 (a) reveals that circuit analogies allow us to rapidly

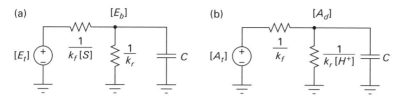

Figure 24.2a, b. Michaelis-Menten association reaction and acid dissociation reaction.

understand Michaelis-Menten kinetics of enzyme-substrate saturation or of another association reaction as resistive-divider saturation.

In many biological systems, typically $[E_b]$ leads to significantly slower production of a product P at a flux rate of $v[E_b]$ via a further unidirectional reaction, accompanied by unbinding of the enzyme and substrate. This reaction is easily represented by including a resistance of value $1/v$ in parallel with the resistance $1/k_r$ in Figure 24.2 (a); in addition, a voltage-dependent transconductance whose output current depends on $[E_b]$ as $v[E_b]$ charges a capacitance $C_2 = 1$; the capacitor's voltage then represents the concentration of P. Thus, simple circuit blocks can also represent Michaelis-Menten kinetics more exactly if needed. However, the essential dynamics of Michaelis-Menten kinetics are well represented by Equations (24.2), (24.3), and Figure 24.2 (a).

A dissociation reaction, e.g., that due to an acid, is given by

$$HA \underset{k_r}{\overset{k_f}{\rightleftharpoons}} H^+ + A^- \tag{24.4}$$

The reader should be able to show that the circuit of Figure 24.2 (b) represents this reaction with $[A_t]$ being the amount of acid in total form (undissociated as HA) or dissociated as $[A^-] = [A_d]$, i.e., $[A_t] = [HA] + [A_d]$.

Every chemical reaction, e.g., the enzyme-substrate binding illustrated in Figure 24.3, implicitly has two negative-feedback loops embedded within it. The first feedback loop arises because the concentration of reactants falls if the concentration of products builds, thus slowing the forward reaction; this constitutes the major loop of Figure 24.3. The second feedback loop arises because the backward reaction speeds up when the concentration of products builds; this constitutes the minor loop in Figure 24.3. At steady state, the $1/s$ integrator has infinite gain such that the feedback path of the minor loop determines the closed-loop transfer function of the minor loop. Thus, the feedback loop of Figure 24.3 also leads to Michaelis-Menten saturation except that the saturation is now viewed as a high-loop-gain effect with loop gain $[S]/K_d$. The major loop in Figure 24.3 corresponds to the restorative current through resistance $1/k_f$ in Figure 24.2 (a). The minor loop in Figure 24.3 corresponds to the restorative current through $1/k_r$ in Figure 24.2 (a).

One difference between the chemical resistive-divider circuit of Figure 24.2 (a) and actual electronic resistive-divider circuits lies in their noise properties. The current in real resistors flows by drift while their $4kTR$ current noise per unit

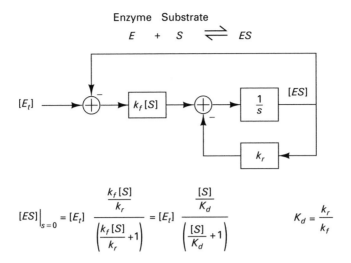

Figure 24.3. The two feedback loops embedded in any chemical reaction.

bandwidth is due to the shot noise of internal diffusion currents (see Chapter 7 on noise). In contrast, in Figure 24.2 (a), the resistor current *and* resistor noise currents are both due to diffusion currents that exhibit Poisson noise statistics. Thus, in Figure 24.2 (a) the resistor's noise is simply the shot noise of the current flowing through it and varies as the current through the resistor varies. Hence, the subthreshold transistor analogy of Figure 24.1 is more accurate than that of the resistor analogy of Figure 24.2 (a) since it gets both the current and the noise exactly correct (for the $A \rightleftharpoons B$ reaction). Nevertheless, if we use the correct shot-noise formulas that pertain to the resistors of Figure 24.2 (a), we can still exploit the techniques of Chapters 7 and 8 to compute the total concentration noise $\overline{v_n^2}$ on the capacitor C: the shot noise of the currents flowing through the $1/(k_f[S])$ and $1/(k_r)$ resistors is converted to voltage noise on the capacitor C as in a simple RC circuit with two resistors in parallel. Since we are interested in mass currents rather than charge currents, we set

$$q = 1$$
$$C = \text{Volume} \tag{24.5}$$

If we assume that $[S]$ is nearly constant and exhibits no noise, for example, because $[S]$ is large, we get

$$k_f' = k_f[S]$$

$$\overline{v_n^2} = 2q \left(E_t \left(1 - \frac{k_f'}{k_f' + k_r} \right) k_f' + E_t \left(\frac{k_f'}{k_f' + k_r} \right) k_r \right) \times \frac{1}{\left(k_f' + k_r \right)^2} \times \frac{k_f' + k_r}{2\pi C} \times \frac{\pi}{2}$$

$$\overline{v_n^2} = 4q E_t \left(\frac{k_f' k_r}{k_f' + k_r} \right) \frac{1}{\left(k_f' + k_r \right)} \times \frac{1}{4C}$$

$$= \frac{E_t}{C} \frac{k_f' k_r}{\left(k_f' + k_r\right)^2}$$

$$\boxed{\overline{v_n^2} = \frac{E_t}{C} \left(\frac{k_f'}{k_f' + k_r}\right) \left(\frac{k_r}{k_f' + k_r}\right)}$$

(24.6)

The result of Equation (24.6) has the $1/C$ scaling expected from the usual kT/C relationship for the noise on a capacitor. If we compute the total fluctuation in the number of bound enzyme molecules, σ_N, on the capacitor C (i.e., the chemical analog of the charge fluctuation on capacitor C rather than the voltage fluctuation), we get

$$\sqrt{\sigma_N^2} = \left(\sqrt{\frac{E_t}{C} \left(\frac{k_f'}{k_f' + k_r}\right) \left(\frac{k_r}{k_f' + k_r}\right)}\right) C$$

$$\sigma_N = \left(\sqrt{E_t C \left(\frac{k_f'}{k_f' + k_r}\right) \left(\frac{k_r}{k_f' + k_r}\right)}\right)$$

(24.7)

$$\boxed{\sigma_N = \sqrt{N_t p(1 - p)}}$$

where $N_t = E_t C$ is the total number of enzyme molecules within the volume compartment C of the reaction, whether bound or unbound, and p is the probability that an enzyme molecule is bound. Thus, we find that the noise is maximized when the probability of an enzyme molecule being bound is ½ and minimized when it is near 1 or 0. Intuitively, a chemical reaction that has a forward flux greatly in excess of the reverse flux ($p = 1$) or vice versa ($p = 0$) will exhibit little noise since almost all molecules will be bound or unbound respectively. Similarly, a transport channel in a cell membrane exhibits the least noise when the probability for its opening is near 1 or 0 [7].

24.2   Log-domain current-mode models of chemical reactions and protein-protein networks

The dynamics of a chemical species i with concentration x_i in a chemical reaction can be described by a differential equation of the form

$$\frac{dx_i}{dt} = \sum_j c_j + \sum_l k_{il} u_l + \sum_m k_{im} x_m + \sum_{lm} k_{ilm} x_l x_m + \sum_{np} k_{inp} x_n u_p + \cdots \quad (24.8)$$

where various externally controlled input species $u(t)$ and state-variable chemical species $x(t)$ interact via zeroth order (the c_j terms), first order (the k_{il} and k_{im} terms), second order (the k_{ilm} and k_{inp} terms), or higher-order interactions to create fluxes that increase x_i (positive kinetic rate constants k) or decrease x_i (negative kinetic rate constants k). In almost all chemical reactions, reaction mechanisms at practical temperatures, pressures, and concentrations are such that only terms up to second order in Equation (24.8) are sufficient for describing the dynamics of all species that are involved. Higher-order terms due to the simultaneous association or dissociation of more than two chemical species have vanishingly low probability: most reactions that involve the association or dissociation of more than two species usually occur via chemical intermediates with at most two species associating or dissociating at any given time. Thus, truncating Equation (24.8) after the second-order terms is usually an excellent description of chemical dynamics for all species. Note that an equation such as Equation (24.8) must be written for every state-variable species i involved in the reaction. Furthermore, the outputs $y_i(t)$ can be a linear combination of the input species

$$y_i = \sum_j r_{ij}x_j + \sum_j s_{ij}u_j \qquad (24.9)$$

Thus, chemical reaction dynamics are typically described by polynomially nonlinear differential equations with an order of two. Linear systems are an example of polynomial differential equations with an order of one. Note that we have assumed mass-action kinetics and a well-mixed spatially homogeneous medium in formulating Equation (24.8).

We know from Chapter 14 on current-mode circuits that translinear circuits are capable of creating static nonlinear systems and linear dynamical systems through the use of exponential nonlinear devices. Given the similarities between chemical computation and electronic computation illustrated in Figure 24.1, both of which involve exponential nonlinearities, we may wonder if log-domain bipolar and subthreshold circuits with exponential nonlinearities can mimic the dynamics of chemical reaction networks. We now show that it is indeed possible to do so, in fact, for polynomially nonlinear dynamical systems of any order. For the most part, we shall focus on second-order systems only since they are most relevant for modeling chemical reactions within cells.

To map chemical reaction networks to current-mode circuits, we must have an appropriate amplitude and time-constant scaling from the chemical to the electrical domain. Amplitude scaling is easily accomplished by setting

$$\frac{x_i}{X_0} = \frac{i_i}{I_0} \qquad (24.10)$$

Equation (24.10) states that the normalized concentration of the chemical concentration x_i with respect to a global reference concentration, X_0, is equal to the normalized current, i_i, with respect to a global reference current I_0. To make the normalized electrical state variable i_i/I_0 have a time derivative that is α times faster than the normalized chemical state variable x_i/X_0, the kinetic rate constants in the chemical equation need to be replaced by equivalent electrical rate constants. Thus, a kinetic term in a chemical differential equation will scale according to

$$
\begin{aligned}
\frac{dx_i}{dt} &= \cdots + \sum_{lm} k_{ilm} x_l x_m + \cdots \\
X_0 \frac{d(x_i/X_0)}{dt} &= \cdots + X_0^2 \sum_{lm} k_{ilm} \frac{x_l}{X_0} \frac{x_m}{X_0} + \cdots \\
X_0 \left(\frac{1}{\alpha}\right) \frac{d(i_i/I_0)}{dt} &= \cdots + X_0^2 \sum_{lm} k_{ilm} \frac{i_l}{I_0} \frac{i_m}{I_0} + \cdots \\
\frac{d(i_i/I_0)}{dt} &= \cdots + \sum_{lm} (\alpha X_0^1 k_{ilm}) \frac{i_l}{I_0} \frac{i_m}{I_0} + \cdots
\end{aligned}
\tag{24.11}
$$

That is a kinetic rate constant transforms according to

$$
k \rightarrow \alpha X_0^{(S-1)} k
\tag{24.12}
$$

when we map from the chemical domain to the electrical domain; here, S is the order of the term in the chemical equation (zeroth, first-order, second-order, ...). Note that input u terms in the chemical equation are also normalized by X_0, so the scaling of Equation (24.12) also applies to them.

The last row of Equation (24.11) can be transformed into a form suitable for a log-domain dynamical system, if we multiply the left hand and right hand sides by the reciprocal of i_i/I_0 (see Chapter 14), perform some further multiplications on both sides, and do some algebra:

$$
\begin{aligned}
\frac{1}{(i_i/I_0)} \frac{d(i_i/I_0)}{dt} &= \cdots + \sum_{lm} (\alpha X_0^1 k_{ilm}) \frac{\frac{i_l}{I_0} \frac{i_m}{I_0}}{(i_i/I_0)} + \cdots \\
\frac{C_i \frac{\phi_t}{\kappa}}{i_i} \frac{di_i}{dt} &= \cdots + I_0 \sum_{lm} (\alpha \frac{C_i \frac{\phi_t}{\kappa}}{I_0} X_0^1 k_{ilm}) \frac{\frac{i_l}{I_0} \frac{i_m}{I_0}}{(i_i/I_0)} + \cdots \\
C_i \frac{dv_{Ci}}{dt} &= \cdots + I_0 \sum_{lm} (\beta_{ilm}) \frac{\frac{i_l}{I_0} \frac{i_m}{I_0}}{(i_i/I_0)} + \cdots \\
\beta_{ilm} &= \alpha \frac{C_i \frac{\phi_t}{\kappa}}{I_0} X_0^1 k_{ilm} \\
\beta_{ilm} &= \alpha \tau_{0i} X_0^1 k_{ilm}
\end{aligned}
\tag{24.13}
$$

where v_{Ci} is the log-domain voltage on capacitor C_i that represents $\log(x_i/X_0)$ as $(\phi_t/\kappa)\ln(i_i/I_0)$ with (ϕ_t/κ) being the characteristic subthreshold exponential voltage ($\kappa = 1$ in the case of a bipolar), and $\tau_{0i} = C_i(\phi_t/\kappa)/I_0$ being a characteristic electrical time constant. All terms undergo similar transformation like the second-order term in Equation (24.13): that is, Equation (24.8) is described by the following transformation

$$
\frac{dx_i}{dt} = \cdots + \sum_m k_{im} x_m + \sum_{lm} k_{ilm} x_l x_m + \sum_{np} k_{inp} x_n u_p + \cdots
$$

$$
C_i \frac{dv_{Ci}}{dt} = I_0 \left(\cdots + \sum_m \beta_{im} \frac{i_m/I_0}{i_i/I_0} + \sum_{lm} \beta_{ilm} \frac{(i_l/i_0)(i_m/I_0)}{i_i/I_0} + \sum_{np} \beta_{inp} \frac{(i_n/I_0)(u_p/I_0)}{i_i/I_0} + \cdots \right)
\tag{24.14}
$$

where the dimensionless β_i kinetic rate constants in the electrical system are related to their corresponding chemical ones according to

$$\beta_{i..} = \left(\alpha\tau_{0i}X_0^{S-1}\right)k_i \qquad (24.15)$$

where $S = [0, 1, 2]$ depending on whether the term is zeroth, first, or second order. Note that Equations (24.14) and (24.15) describe transformations for a particular chemical species i but an identical transformation applies to all other species involved in the dynamics with one capacitor C_i per state variable x_i. That is, if the original chemical dynamical system had N state variables, M inputs, and P outputs, the transformed electrical dynamical system will have N capacitors, M current inputs, and P current outputs. More formally, consider a reaction system consisting of N species $\mathbf{x}$, M inputs $\mathbf{u}$, and P outputs $\mathbf{y}$. If we assume zeroth, first, and second order mass-action kinetics only, and model forward and backward reactions through separate unidirectional differential equations, we may describe such a system by

$$\frac{d\mathbf{x}}{dt} = \mathbf{C} + \mathbf{Dx} + \mathbf{E}(\mathbf{x} \otimes \mathbf{x}) + \mathbf{Fu} + \mathbf{G}(\mathbf{x} \otimes \mathbf{u})$$
$$\mathbf{y} = \mathbf{Hx} + \mathbf{Ku} \qquad (24.16)$$

where $\otimes$ is the outer product of two vectors. The similarity of Equation (24.16) to the standard state-space equations of linear control theory should be evident [8]. The $(\mathbf{x} \otimes \mathbf{x})$ terms add polynomial nonlinearities of order 2 such that Equation (24.16) differs from the standard linear equations of control theory.

As an example, to implement the differential equations that model the second-order unidirectional association reaction $A + B \rightarrow C$, the current-mode integrator of Chapter 14 is modified to create the log-domain circuit of Figure 24.4. This circuit implements the differential equations

$$\beta = \beta_1\beta_2$$
$$C\frac{dv_C}{dt} = -\beta\frac{i_A i_B}{i_C}$$
$$C\frac{dv_B}{dt} = +\beta\frac{i_A i_B}{i_B} = +\beta i_A \qquad (24.17)$$
$$C\frac{dv_A}{dt} = +\beta\frac{i_A i_B}{i_A} = +\beta i_B$$

Note that the signs of the currents have been reversed because the state variable v_i in Figure 24.4 is referenced to V_{DD}, i.e., given by $V_{DD}-v_i$. The parameters β_1 and β_2 are altered via DAC-programmable bias currents with $\beta = \beta_1\beta_2$; thus, a large dynamic range of programmability in β is achieved with two lower dynamic range DACs.

In Figure 24.4, we can keep the concentrations of i_A constant and i_B constant (hold the voltage of nodes v_A and v_B) and implement the backward reaction $(C \rightarrow A + B)$ via a similar circuit and then vary $K_d = I_0\beta_r/\beta_f$ by varying the backward-reaction β_r compared with the forward-reaction β_f. In normal non-log-voltage units, the overall current-mode circuit then simulates a constant

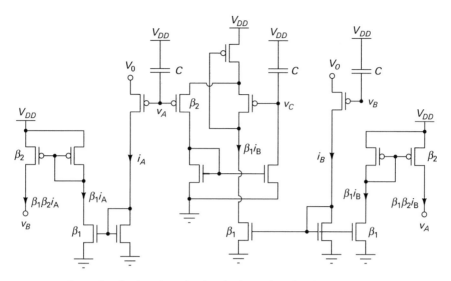

Figure 24.4. Second-order log-domain chemical reaction circuit. Reprinted with permission from [9] (©2009 IEEE).

forward-reaction current source (that exhibits Poisson noise statistics) hooked to a conductance proportional to K_d, which is in parallel with a capacitance (and the current source). The voltage on the capacitance represents the concentration of C. Due to the reverse-reaction current flowing through it, the conductance proportional to K_d exhibits Poisson statistics as well. In effect, we have a parallel RC circuit fed by a current source with shot noise from the current source and shot noise from the current flowing through the R. Note that the power spectral density of the current through R is NOT $4kTG$ but $2qI$ in such chemical resistances as we discussed when we derived Equation (24.7). If the forward-reaction current source has a value I, the voltage v_C will equilibrate at a value such that the backward-reaction current through R balances it. Thus, we may expect the noise voltage on the capacitor C to be given by

$$\overline{v_n^2} = (2qI + 2qI)\left(R^2\right)\left(\frac{\pi}{2}\right)\left(\frac{1}{(2\pi)RC}\right)$$
$$= \frac{q(IR)}{C}$$

(24.18)

The signal-voltage power on the capacitor is given by

$$v_C^2 = (IR)^2$$

(24.19)

Thus, the net *SNR* of the circuit in this scenario is given by

$$SNR = \frac{v_C^2}{\overline{v_n^2}}$$
$$= \frac{C(IR)}{q}$$

(24.20)

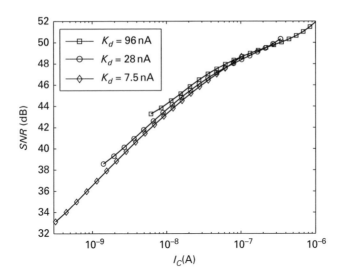

Figure 24.5. The *SNR* variation versus I_C when I_A and I_B are maintained constant in a bidirectional association reaction. Reprinted with permission from [9] (©2009 IEEE).

In the actual current-mode circuit, *IR* is represented by an equivalent output current I_C such that Equation (24.20) predicts that the *SNR* of the current-mode circuit increases as the mean value of I_C increases and is only dependent on the net value of I_C (or *IR*), not the value of *R* or K_d. Indeed, Figure 24.5 confirms these predictions when noise simulations are performed on the bidirectional current-mode equivalent circuit. Figure 24.5 indicates that the stochastics of the current-mode circuits that we have designed do indeed automatically represent those seen in chemical reactions. In Figure 24.5, the saturation at large *SNR* is primarily due to flicker noise. The value of *C* can be used to scale the *SNR* as was also confirmed by simulations (but not shown). At very low *SNR* (say below 20 dB) parasitic capacitances and leakage currents limit the reliable control of *SNR* in the circuit. In these situations, the circuit described in Figure 24.12 allows us to reliably add artificial noise in a quiet electronic circuit with the right Poisson properties. The circuit of Figure 24.12 is described later in the context of DNA-protein dynamics but it can be adapted for use in any chemical reaction.

A 1.5 mm × 1.5 mm proof-of-concept protein-protein network chip with 81 second-order reaction blocks, 40 first-order reaction blocks, 40 zeroth-order reaction blocks, 32 state variables, 16 inputs and 8 outputs in a 0.18 μm process has been reported in [9]. The reaction rates, initial conditions, volume compartments of reactions (capacitances), and which molecular species are involved in which reaction are completely programmable via reconfigurable digital bits. The chip has the potential for fast highly parallel stimulation with stochastics and cell-to-cell variability automatically included. Data from this chip for a simple reaction network given by $A + A \rightarrow B$; $B \rightarrow C$ is shown in Figure 24.6 and compared with a MATLAB simulation of the same system. In both cases, we

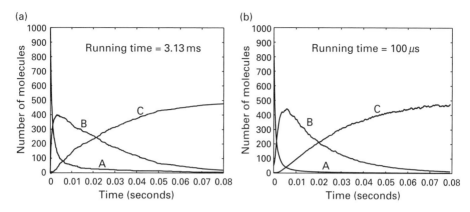

Figure 24.6a, b. MATLAB (a) versus chip data (b). Reprinted with permission from [9] (©2009 IEEE).

see that as A is consumed, B rises, and then as B is consumed and transformed to C, B falls, while C rises. The MATLAB simulation runs in 3.13 ms while the chip's simulation runs in $100\,\mu s$. The performance speedup of 30 times in this first-iteration chip was limited by data-acquisition hardware not being able to acquire data from the chip fast enough. Due to the fast highly parallel nature of the analog implementations, speedups of 10^3 times should be possible in the near future and speedups greater than 10^6 times should eventually be possible.

The automatic inclusion of stochastics in the analog chips that we have described is highly beneficial for simulating large-scale protein networks in the cell in an ultra-fast fashion. In spite of efficient stochastic algorithms being available [10], the computation time of molecular-dynamics simulators increases precipitously when stochastics are included. Currently, the real-time simulation of just 30 state variables with stochastics is quite challenging. However, large-scale protein networks in the cell can have up to 30,000 state variables in human cells. Our knowledge of the topology and parameters of such biochemical networks is growing every day although our ability to predict what they will do, given this knowledge, is still very limited. Thus, programmable analog chips that rapidly emulate stochastic biochemical reaction networks in cells may be very important in the future in understanding and simulating cellular function on a large scale. The parallel and compact nature of the implementation allows scalability of large-scale networks onto a single chip or a few chips. Furthermore, they can be used to provide plausible ranges for unknown parameters in reaction networks via rapid exploration of such networks subject to certain known experimental facts and constraints. Elegant stimulus-design criteria for distinguishing between models for biochemical reaction mechanisms based on control theory can be used with such chips in the feedback loop [11]. Machine-learning techniques such as stochastic gradient descent, regression, and gradient descent can be implemented on a digital computer to measure and configure analog protein-network chips such that they optimize fits to experimental data, are consistent with known constraints, and/or maximize objective functions known to be of biological importance like cell growth. In essence, analog chips that model chemical reaction networks can serve as 'special-purpose' ALUs that implement nonlinear dynamical systems optimized for simulating biochemical reaction networks.

24.3   Analog circuit models of gene-protein dynamics

Figure 24.7 illustrates the basics of gene-protein interactions in cells. An *inducer* molecule, e.g., glucose or S_X, may enter a cell and cause biochemical reaction events that eventually, e.g., via a protein-protein network, lead to the activation of a particular protein called a *transcription factor X*. When activated, $X \rightarrow X^*$, with X^* being the active form of the transcription factor. The activation of the transcription factor most often occurs because a molecule, typically the inducer, binds to the transcription factor and changes its shape. The activated transcription factor X^* can then bind to DNA near or within specific *promoter* binding sites on the DNA that have a particular sequence of A, T, C, or G nucleotides within them. These sites are effective in binding the particular transcription factor. *Transcription* is the process whereby the enzyme RNA polymerase (RNAp in Figure 24.7) converts the DNA sequence of a gene into a corresponding messenger RNA (mRNA) transcript sequence that is eventually translated into a protein. The binding of the transcription factor causes the transcription rate of a gene near the promoter to be increased if the transcription factor is an *activator* or decreased if the transcription factor is a *repressor*. In Figure 24.7, a repressor transcription factor Y that is activated to Y^* by an inducer S_Y is shown. Ribosome molecules in

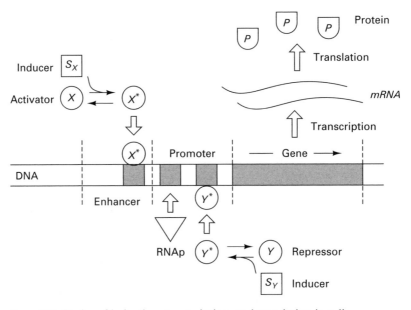

Figure 24.7. Basics of induction, transcription, and translation in cells.

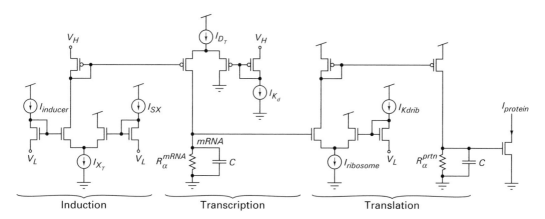

Figure 24.8. Basic induction, transcription, and translation circuit.

the cell *translate* the mRNA transcript into a sequence of corresponding amino acids to form a protein. The final translated protein can act as a transcription factor for other genes in a gene-protein network or affect other proteins in a protein-protein network or both. The translated protein can also serve as an activator or repressor for its own gene. Readers interested in further details of molecular biology should consult [1].

Figure 24.8 reveals a circuit model of induction, transcription, and translation. If the inducer concentration, $I_{inducer}$, is significantly greater than I_{SX}, the K_d for inducer-transcription-factor binding, most of the transcription-factor molecules X will be transformed into an active state X^*. The current output of the subthreshold differential pair that represents induction quantitatively models this process:

$$I_{X^*} = I_{X_T} \left(\frac{\dfrac{I_{inducer}}{I_{SX}}}{\dfrac{I_{inducer}}{I_{SX}} + 1} \right) \tag{24.21}$$

Equation (24.21) is an exact model of Michaelis-Menten binding as per Equation (24.3) with I_{X_T} representing the total concentration of transcription factor, X_T, whether activated or not. Similarly, if the activated transcription-factor concentration I_{X^*} is significantly greater than K_d, the dissociation constant for DNA-transcription-factor binding, the rate of production of mRNA transcripts by the enzyme RNA polymerase will be near its maximal value, assuming, without loss of generality, that the transcription factor in question is an activator. The current output, I_m, of the subthreshold differential pair that represents transcription in Figure 24.8 quantitatively models this process:

$$I_m = I_{D_T} \left(\frac{\dfrac{I_{X*}}{I_{Kd}}}{\dfrac{I_{X*}}{I_{Kd}} + 1} \right) \tag{24.22}$$

In Equation (24.22), I_m represents the rate of mRNA production and I_{D_T} is the maximal rate of mRNA production. Equation (24.22) suggests that the absence of activator shuts off mRNA production to 0. In practice, one can add a constant-current term to Equation (24.22) to model the fact that there is a basal mRNA production rate even when there is no activator present. Equation (24.22) can be converted into a repressor equation if we use current from the other arm of the transcription differential-pair in Figure 24.8. The current I_m leads to a final mRNA concentration described by a Laplace-transform lowpass filter equation

$$M_{RNA}(s) = I_m(s) \frac{R_\alpha^{mRNA}}{1 + R_\alpha^{mRNA} Cs} \tag{24.23}$$

In Figure 24.8 and Equation (24.23), $mRNA.R_\alpha^{mRNA}$ is the rate of degradation of mRNA due to enzymes that actively degrade mRNA in a linear fashion and C is the volume of the reaction compartment of the cell. The time course of transcription is typically determined by the $R^{mRNA}C$ time constant since inducer binding is relatively rapid. Although Figure 24.8 represents mRNA as a voltage variable for simplicity, the $R^{mRNA}C$ circuit can be replaced by a current-input current-output lowpass filter that outputs an I_{mRNA} current instead of an mRNA output voltage. If I_{mRNA} represents the mRNA concentration, we can represent the process of translation by the overall equation

$$I_{protein}(s) = I_{ribosome} \left(\frac{\dfrac{I_{mRNA}}{I_{Kdrib}}}{\dfrac{I_{mRNA}}{I_{Kdrib}} + 1} \right) \left(\frac{R_\alpha^{prtn}}{1 + R_\alpha^{prtn} Cs} \right) \tag{24.24}$$

with parameters as shown in Figure 24.8. Typically, the protein degradation time constant $R_\alpha^{prtn}C$ is the largest time constant in the overall process, with a value ranging from 30 minutes (in bacteria in high-growth medium) to over 20 hours (in human cells). Certain proteins, for example, those in certain cells in the eye, may not be degraded at all.

Equations (24.21) through (24.24) represent the basics of gene-protein dynamics when chemical binding occurs between two molecules that are each monomers. Frequently, the transcription factor has maximal binding efficacy to DNA when it is a dimer (two identical molecules bound to each other) or is even a quadrimer (four identical molecules bound to each other). In such situations, Equation (24.22) can be well approximated by an equation of the form

$$I_m = I_{D_T} \left(\frac{\left(\dfrac{I_{X^*}}{I_{Kd}} \right)^n}{\left(\dfrac{I_{X^*}}{I_{Kd}} \right)^n + 1} \right) \tag{24.25}$$

where $n = 2$ in the case of dimer binding or 4 in the case of quadrimer binding [5]. The parameter n is often termed the 'Hill coefficient' and is implicitly 1 in Equations (24.21) through (24.24). At Hill coefficients that are large, Equation (24.25) is often described by a digital approximation: if the transcription factor is an activator, we assume that mRNA transcript production is at its maximal rate I_{D_T} when $I_{X^*} > K_d$ and 0 otherwise; if the transcription factor is a repressor, we assume that mRNA transcript production is at 0 when $I_{X^*} > K_d$ and I_{D_T} otherwise.

Circuits to create Hill coefficients with any analog value from, say, 1 to 4 can be designed by exploiting a strategy similar to that used in Chapter 19 to create power-law coefficients in AGC circuits. The input diode-connected transistors in the differential pairs of Figure 24.8 are replaced by a transistor with a buffered WLR (Chapter 12) of transconductance G_1 from its drain to gate and a buffered WLR of transconductance G_2 from its gate to a reference. Programming of G_1 and G_2 with DAC currents then ensures that a power-law of approximately $(G_1 + G_2)/G_1$ is obtained.

Note that, for simplicity, we have ignored the body effect in Figure 24.8 by setting $\kappa = 1$. In an actual implementation, as we have discussed in Chapter 14 on current-mode circuits, source-tied-to-well pFET transistors should be used to implement all differential pairs and lowpass filters in Figure 24.8. In processes where the body of nFETs is fixed at a global substrate voltage, nFETs should only be used in fixed-ratio current mirrors.

24.4   Logic-like operations in gene-protein circuits

When *E. coli* are cultured in a medium that lacks glucose but has lactose, these bacteria increase the transcription rate of certain genes, which are normally expressed at a very low level. These genes produce proteins that help metabolize lactose to obtain energy, help transport lactose into the cell, and help detoxify toxic metabolites caused by lactose metabolism. If glucose is present, these genes are not expressed: it is significantly cheaper for the cell to metabolize glucose rather than lactose if glucose is present. Therefore, the bacteria do not bother making proteins useful for lactose metabolism if glucose is present. Effectively, the cell behaves as though the expression of these genes is based on the logical expression NOT (Glucose) AND Lactose.

Experiments have shown that the absence of glucose induces an activator transcription factor called CRP to become active and bind DNA in a certain enhancer region of the DNA. Thus, in Figure 24.7, activated CRP would correspond to X^*. The binding of CRP enhances the transcription rate of RNA polymerase in a catalytic fashion by helping recruit RNA polymerase to bind near the enhancer and thus begin transcription of genes near the enhancer region. The presence of lactose induces a repressor transcription factor called LacI to become inactive and unbind from DNA in a promoter region of the DNA that is adjacent to the enhancer region where CRP binds. Thus, in Figure 24.7, activated LacI would correspond to Y^*, and the inactivation of Y^* to Y by the lactose inducer would correspond to LacI unbinding from the DNA. RNA polymerase must bind to DNA in the promoter region for transcription of nearby genes to begin. The unbinding of LacI serves to unblock the binding of RNA polymerase on the promoter and consequently enables transcription to proceed. Thus, the interactions between the transcription factors on DNA is

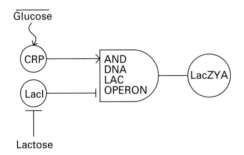

Figure 24.9. LacZ circuit. Reprinted with permission from [14] (©2009 IEEE).

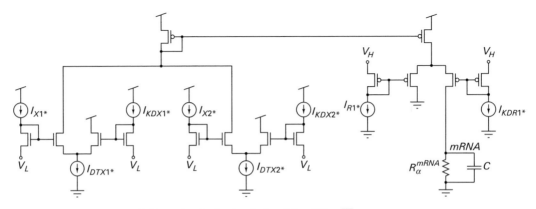

Figure 24.10. DNA analog logic circuit for $\left(X_1^* + X_2^*\right) * \overline{R_1^*}$.

analogous to the manner in which multiple inputs affect the output of a logic gate as shown in Figure 24.9. The AND DNA Lac operon increases the transcription of the sequentially adjacent LacZ, LacY, and LacA genes, if the CRP activator is bound and the LacI repressor is unbound. These sequentially adjacent genes are said to be on a common *operon* whose transcription is controlled by the binding/unbinding of the CRP and LacI transcription factors to the enhancer and promoter DNA regions respectively, which are both near each other and near the Lac genes. In general, interactions between transcription factors can lead to multiple-input logic functions. The true interactions are, of course, not digital, but analog. Further details can be found in [12].

Figure 24.10 shows an example of an analog logic circuit that implements the function

$$M_{mRNA}(s) = \left(I_{DTX_1^*}\left(\frac{\frac{I_{X1^*}}{I_{KDX1^*}}}{\frac{I_{X1^*}}{I_{KDX1^*}}+1}\right) + I_{DTX_2^*}\left(\frac{\frac{I_{X2^*}}{I_{KDX2^*}}}{\frac{I_{X2^*}}{I_{KDX2^*}}+1}\right)\right)\left(\frac{1}{1+\frac{I_{R1^*}}{I_{KDR1^*}}}\right)\left(\frac{R_\alpha^{mRNA}}{1+R_\alpha^{mRNA}Cs}\right)(24.26)$$

This function is the analog version of the digital logic function $\left(X_1^* + X_2^*\right) \cdot \overline{R_1^*}$ and models the effect of two activator transcription factors X_1^* and X_2^* that

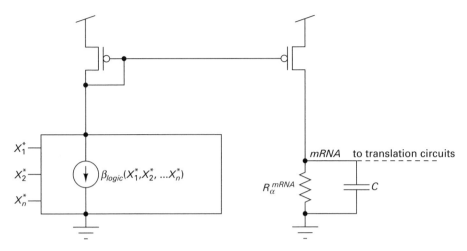

Figure 24.11. Simplified DNA logic circuit.

interact in an OR fashion and share a common repressor R_1^*. Equation (24.26) is a consequence of applying subthreshold differential-pair equations to each of the three differential pairs of Figure 24.10. These differential pairs model classic monomer Michaelis-Menten binding.

Figure 24.11 shows a hybrid analog-digital circuit that approximates gene-protein interactions with digital logic functions rather than with graded analog current-mode circuits as in Equation (24.26) or in Figure 24.10. In this circuit, the rate of production of mRNA varies with the value of a digital vector input D_n defined by

$$u(x) = 1 \text{ iff } x > 0 \text{ and } 0 \text{ otherwise}$$

$$D_n = \left(u\left(\frac{X_1^*}{K_{DX_1^*}} - 1\right), u\left(\frac{X_2^*}{K_{DX_2^*}} - 1\right), ..., u\left(\frac{X_n^*}{K_{DX_n^*}} - 1\right) \right) \quad (24.27)$$

with the rate being a function of D_n given by

$$\beta_{mRNA} = \beta_{\text{logic}}(D_n) \quad (24.28)$$

That is for every discrete value of D_n, e.g., 001010, there is a corresponding analog rate of mRNA production, β_{001010} that can be listed in a lookup table. In essence, we have a logic DAC that converts D_n to analog production rates rather than a number-based DAC.

The presence of R^{mRNA} in Figure 24.11 models mRNA degradation. Translation circuits are implemented as shown in Figure 24.8 and only indicated with dotted lines in Figure 24.11. While Figure 24.11 is an approximate representation of DNA-protein interactions, and a decent approximation only at high Hill coefficients, it can still provide insight in several cases. In fact, an insightful systems-biology text is largely based on differential equations that are identical to those created by the circuit of Figure 24.11 [5].

24.5   Stochastics in DNA-protein circuits

Poisson noise in mRNA-production flux can be mimicked by Poisson electronic current noise in a manner analogous to that discussed previously for protein-production flux. However, the noise levels in mRNA production for some genes can be high enough such that extremely low currents and extremely small capacitances become necessary in electronics to mimic the same low signal-to-noise ratio (SNR) in biology. The resultant noise in electronics is then not well controlled or predictable. Therefore, it is sometimes advantageous to artificially introduce a controlled level of noise in a relatively quiet electronic circuit to mimic high-noise signals in biology. Figure 24.12 illustrates a circuit for doing so.

In Figure 24.12, the current-mode integrator with output capacitor C and $I_{R\alpha}$ implement a current-mode version of the $R^{mRNA}C$ lowpass filter in Figure 24.11. That is, $2I_{R\alpha}$ and C correspond to I_A and C in Figure 14.9 in Chapter 14 on current-mode circuits. Similarly, v_{mRNA} and i_{mRNA} correspond to v_{OUT} and i_{OUT} in Figure 14.9 respectively. Instead of a constant $2I_{R\alpha}$ in a traditional current-mode lowpass filter, however, the leak current $2I_{R\alpha}$ is pseudo-randomly switched on and off with a duty cycle of 0.5. Thus, the average value of the leak current that sets the lowpass filter time constant is $I_{R\alpha}$ as in a traditional circuit but the random switching introduces a stochasticity in this leak current. The log voltage on the current-mode capacitor is exponentiated and converted to a current i_{mRNA} that encodes the level of mRNA as in any current-mode circuit. The current i_{mRNA} is gained up by β_{SNR} and used to control the frequency, f_{CCO}, of a current-controlled oscillator (CCO). The output switching frequency of the oscillator is proportional to i_{mRNA} according to $f_{CCO} = \beta_{SNR}i_{mRNA}/q_{CCO}$, where q_{CCO} depends on design parameters internal to the CCO. Thus, as mRNA levels rise, the control current and switching frequency of the CCO rise in proportion. The linear feedback shift register (LFSR) converts the digital output of the CCO to a random switching signal via a classic pseudo-random-number generator technique [13]. Thus, the output of the LFSR randomly switches the $I_{R\alpha}$ current on and off with a switching frequency f_{CCO} that is proportional to the mRNA level encoded by i_{mRNA}. Consequently, as mRNA levels rise, a consequence of a higher mRNA production rate, the arrival

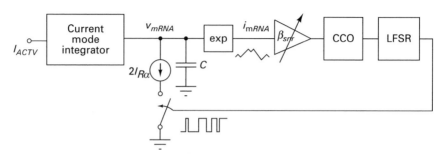

Figure 24.12. Artificial noise generation circuit for low SNR. Reprinted with permission from [14] (©2009 IEEE).

rate, f_{CCO}, of charge packets with value $2I_{R\alpha}/f_{CCO}$ increases even though the mean value of current stays at $0.5f_{CCO}(2I_{R\alpha}/f_{CCO}) = I_{R\alpha}$. The noise power spectral density of the switching current at the log-voltage node is given by the usual shot-noise formula with an effective $q = I_{R\alpha}/f_{CCO}$ and the mean current equals $I_{R\alpha}$ (see Chapters 7 and 8):

$$\Delta I^2_{psd} = 2\left(\frac{I_{R\alpha}}{f_{CCO}}\right)(I_{R\alpha}) \tag{24.29}$$

If the conductance at this node is $G(f)$, the output noise power spectral density is given by

$$\begin{aligned}
\Delta I^2_{psdout}(f) &= 2\left(\frac{I_{R\alpha}}{f_{CCO}}\right)(I_{R\alpha})\frac{g_m^2}{G(f)^2} \\
&= 2\left(\frac{I_{R\alpha}}{(\beta_{SNR}I_{mRNA}/q_{CCO})}\right)(I_{R\alpha})\frac{I^2_{mRNA}/(\phi_t/\kappa)^2}{G(f)^2} \\
&= 2\frac{q_{CCO}}{\beta_{SNR}}\frac{(I_{R\alpha})^2}{G(f)^2(\phi_t/\kappa)^2}I_{mRNA}
\end{aligned} \tag{24.30}$$

while the output signal power is given by

$$I^2_{out} = \frac{(I_{R\alpha})^2}{G(f)^2}\frac{I^2_{mRNA}}{(\phi_t/\kappa)^2} \tag{24.31}$$

since the dc input from the current mode integrator must be $I_{R\alpha}$ as well. Comparing Equations (24.30) and (24.31), we see that the ratio of the output signal power and output noise power spectral density behave exactly as one would expect for a classic Poisson shot-noise current source except that the charge on the electron has been replaced by q_{CCO}/β_{SNR}. Thus, the value of q_{CCO}/β_{SNR} serves effectively like the charge on the electron, which we can increase if we want more noise (by decreasing β_{SNR}) and which we can decrease if we want less noise (by increasing β_{SNR}).

The input labeled I_{ACTV} in Figure 24.12 is a logical signal that is activated when a switching transition in D_n leads to mRNA production. The current I_{ACTV} represents $\beta_{logic}(D_n)$ in Equation (24.28). Transients in I_{ACTV} will lead to $R^{mRNA}C$-like dynamics in the mRNA level. As mRNA levels change, the output noise of the circuit of Figure 24.12 will change as well ensuring that noise dynamics and signal dynamics are correlated as in any real biological or artificial system.

Figure 24.13 shows experimental data obtained from a chip containing this circuit. Relatively low values of SNR are set and measured. Note that the SNR is not in dB but in regular squared units. The protein SNR is higher than the mRNA SNR because translation of mRNA to protein involves additional lowpass filtering on the mRNA output (see Figure 24.8). The additional lowpass filtering reduces the output protein noise and improves SNR. Other measurements from the chip confirm that the probability distributions of currents are Poisson [14] as they should be.

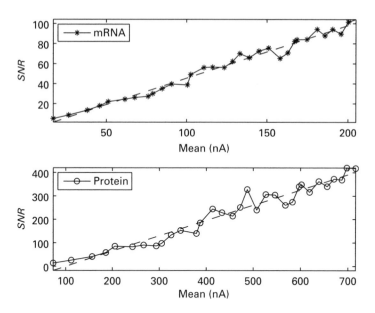

Figure 24.13. SNR measurements from the circuit of Figure 24.12. Reprinted with permission from [14] (©2009 IEEE).

24.6   An example of a simple DNA-protein circuit

Figure 24.14 shows an example feed-forward logic (FFL) circuit that is extremely common in *E. coli* bacteria and that has been extensively studied [5]. When the output genes of the circuit are activated, they produce proteins that are useful for metabolizing and processing arabinose, a sugar. The lactose circuit of Figure 24.9 implements NOT (glucose) AND lactose while the circuit of Figure 24.14 implements NOT (glucose) AND (arabinose). Hence, the circuits of Figures 24.9 and 24.14 have similar functions but there are significant differences in their topologies that lead to differences in how they process their inputs.

The CRP and AraC transcription factors that are activated by glucose and arabinose signals respectively, are in series in Figure 24.14. That is, AraC production begins only if the CRP protein, which is activated by the absence of glucose, activates transcription of genes that produce AraC. If arabinose is present, it induces AraC to become activated, and the combination of an active CRP and an active AraC lead to production of the AraBAD proteins via the AND DNA Ara operon. The logic of Figure 24.14 is redundant: it implements an AND function via the series combination of CRP and AraC and also via the AND DNA operon. In contrast, the logic of Figure 24.9 implements the AND function only via the AND DNA operon. Why does nature have such redundancy? Data in Figures 24.15 (a) and 24.15 (b) from an integrated-circuit chip that models the gene-protein circuit of Figure 24.14 illustrate the purpose of the redundancy. The data are in accord with work described in [5]. In this data, the arabinose inducer

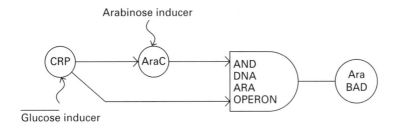

Arabinose inducer

Glucose inducer

Figure 24.14. The arabinose circuit. Reprinted with permission from [14] (©2009 IEEE).

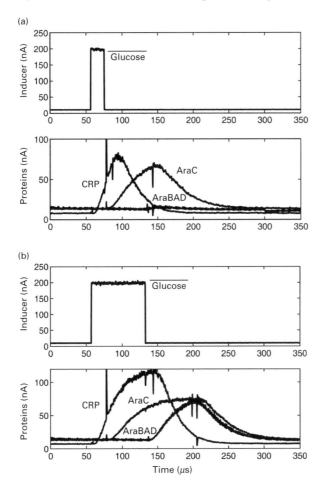

Figure 24.15a, b. Arabinose circuit responses for short (a) and long (b) pulse inputs of the inducer. The glitches are due to computer data-acquisition artifacts. Reprinted with permission from [14] (©2009 IEEE).

sugar concentration is large and constant such that AraC is instantly converted into its active form when produced. Figure 24.15 (a) shows that short absences of glucose inducer do not allow enough time for AraC protein production to build up and reach equilibrium such that sufficiently large amounts of active CRP and

AraC are not simultaneously present; thus, AraBAD protein production is never triggered if glucose is only briefly absent. Figure 24.15 (b) shows that long absences of glucose inducer allow enough time for AraC protein production to build up and reach equilibrium such that sufficiently large amounts of active CRP and AraC are simultaneously present; thus, AraBAD protein production is triggered if glucose is absent for a consistently long duration. Hence, the FFL circuit of Figure 24.14 ensures that the relatively expensive production of proteins needed to process arabinose, a sugar that has a higher cost/benefit ratio than glucose, is only initiated if glucose has been absent for a sufficiently long time.

In many molecular circuits in bacteria and yeast, delays in transcription and translation are relatively negligible compared to mRNA or protein degradation time constants, respectively. Thus, they can often be approximated by increasing these time constants a little. In mammalian cells, however, the delay in the transcription of relatively long genes, which is $\sim$30 minutes, can exceed the mRNA degradation time, which can range from 10 minutes to 10 hours. In certain feedback circuits, the representation of these delays is crucial. Such delays can be programmably represented by conventional clock-counting or one-shot techniques that activate β_{logic} in Figure 24.11 only after a delay. Such delays can be programmably incorporated into the chip from which the data of Figures 24.13 and 24.15 were gathered as described in [14].

24.7   Circuits-and-feedback techniques for systems and synthetic biology

Circuits can often shed insight into biology that is harder to obtain in a different language. For example, work described in [15] has shown how slow outer hair cells in the ear with a time constant of 1 ms can amplify sounds at frequencies over 100 kHz, a two-decade-old mystery. This work shows that while the open-loop time constant of the piezoelectric outer hair cell is 1 ms, the presence of negative feedback in the cochlea with a gain-bandwidth product that is large enough results in a closed-loop system capable of amplification at higher frequencies: the process is analogous to how an operational amplifier with an open-loop time constant of 10 Hz can exploit negative feedback to build an amplifier with a gain of 10^3 at 10 kHz because its gain-bandwidth product is 10^6. In the cochlea, the situation is slightly more complex since the overall negative-feedback loop in the cochlea has a resonance as well, but a simple negative-feedback and root-locus analysis shows that the resonant gain is also improved by negative feedback, thus further enabling high-frequency amplification [15]. Thus, circuits-and-feedback concepts can shed light into how biological systems work. This book has presented circuit models of the heart in Chapter 20, of the neuron in Chapter 15, and of the ear, vocal tract, retina, and neurons in Chapter 23, which allow us to rapidly understand how these systems work. Circuit models are efficient because they are inherently designed to graphically represent interactions between devices in a system in a meaningful way that lead to functional sub-blocks. Such sub-blocks can parse complex systems

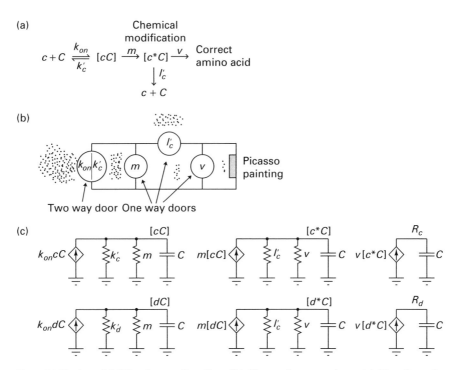

Figure 24.16a, b, c. (a) Kinetic proofreading. (b) Picasso-lover analogy. (c) Circuit analogy.

into meaningful network motifs such as the FFL sub-block that we have discussed [5]. In contrast, representing complex systems via a set of random connections amongst network nodes does not yield much insight into the function of the complex system although it may be mathematically equivalent to a circuit.

As an example of how circuits can shed insight into systems biology, we illustrate how the language of circuits can provide a useful interpretation of kinetic proofreading circuits in cells, which are ubiquitously present in many of its subsystems. Kinetic proofreading allows the cell to reduce the discrimination error rate between highly similar molecules without spending too much time or designing highly specific recognition molecules to do so [16]. Figure 24.16 (a) illustrates the biochemical process; Figure 24.16 (b) provides a Picasso-lover analogy of the process that is adapted from a textual description of this analogy in [5]; Figure 24.16 (c) provides a circuit model of the process.

Figure 24.16 (a) illustrates the chemical-reaction cascade involved in the binding of a specific tRNA codon molecule, which contains a three-letter snippet of RNA, to a complementary three-letter snippet in an mRNA transcript to create a bound species. The tRNA codon is denoted by c, the corresponding mRNA snippet is denoted by C, and the bound species is denoted by cC. Such binding triggers events within the ribosome protein translating machinery in the cell. These events translate a three-letter codon word formed from an alphabet of 4 molecular letters to a particular amino acid. Each of the 64 possible codons is translated by the ribosome machinery to a corresponding amino acid that it codes for. Some amino acids are coded for by more than one codon such that there is some redundancy in the many-to-one translation mapping. The net result is that the ribosome machinery translates the sequence of letters in mRNA to an equivalent sequence of amino acids. Certain codons are not translated into amino acids but function as 'start' and 'stop' codons for the ribosome machinery. Further details are described in any standard molecular-biology text, e.g., [1].

To prevent a translation error, only the correct complementary tRNA codon must bind to the mRNA, such that a wrong amino acid is not incorporated into the protein. Since two tRNA codons that differ by only one letter have similar binding $K'_d s$, however, there is a relatively high probability of error ($\sim 10^{-2}$). One solution is to simply design the binding to be more specific, but this solution costs time: since the association rate, k_{on}, is diffusion-limited and similar for all tRNA codons, the specificity of binding is decided by k'_c, the rate of dissociation or unbinding. High specificity implies that each codon is bound for a long time because of a low dissociation rate k'_c. Since the ribosome can't move to translating the next codon until the current bound tRNA drops off the mRNA, a low k'_c improves specificity but lowers the overall rate of translation. Even with this undesirable tradeoff, designing highly specific binding with a low k'_c is not easy.

To reduce the error rate of translation, the cell implements the methylation scheme shown in Figure 24.16 (a). The bound tRNA species cC is methylated by enzymes at a rate m to create the species $c*C$ via a unidirectional reaction; only the methylated species leads to translation at a unidirectional rate v. The affinity of the methylated tRNA codon for the mRNA is similar to that of the unmethylated tRNA. Thus, the rate of dissociation l'_c of the methylated species from the mRNA is proportional to k'_c and high-affinity tRNA codons bind longer to the mRNA, whether methylated or unmethylated. Unlike the reversible binding of the k_c/k'_c process, however, if the tRNA codon unbinds from the mRNA, it dissociates into an un-methylated form and cannot mount back onto the mRNA directly until it reforms cC as shown in Figure 24.16 (a). High-affinity tRNA codons are more likely to remain bound through the process of methylation and not fall off the mRNA; wrong tRNA codons are more likely to fall off and have a significantly lower probability of translation in a given time: these tRNA codons need to 'get back in line' to the beginning of the chemical cascade, form cC, and then form c^*C to get translated, which delays their translation. The unidirectionality and delay of methylation thus serve to improve the probability of correct translation.

The Picasso-lover analogy of Figure 24.16 (b) is useful in understanding kinetic proofreading. In a particular museum of fine arts, we would like to separate true Picasso lovers from fake ones who are not as sincerely passionate about his paintings. Given their higher affinity for Picasso, true Picasso lovers are more likely to linger for a longer time near one of his paintings or wait longer to see it than fake lovers. The k_{on}/k'_c two-way door in Figure 24.16 (b) represents the rate at which Picasso lovers enter (k_{on}) and leave (k'_c) a cC chamber with higher-affinity lovers having a lower leaving rate k'_c. The one-way 'm' door allows those in the cC

chamber to enter the c^*C chamber at a rate m. Those who enter this chamber can leave it via the one-way l'_c door at a rate l'_c that is proportional to their k'_c. If they do so, because they've been too impatient to wait to see the Picasso painting, they need to get back in line and re-enter via the two way k_{on}/k'_c door to see the Picasso painting again. Thus, in the c^*C chamber, those with low k'_c and consequently low l'_c have a higher probability of being present than fake lovers who've gotten impatient and left. The one-way 'v' door finally rewards those left in the c^*C chamber to enter the chamber where the Picasso painting is actually displayed at a rate v. Thus, only true Picasso lovers will likely see the painting and have their desire to see it translated into reality while others will likely have left since they had two chances to do so.

Figure 24.16 (c) is the circuit representation of Figures 24.16 (a) and 24.16 (b). The same identical circuit is shown for a tRNA codon c and a tRNA codon d with differing rates of unbinding in methylated and unmethylated states. The tRNA codon c represents a true codon while the tRNA codon d represents a codon that can falsely bind to the mRNA. We can see from this circuit immediately that the ratio of wrongly translated d, R_d, to correctly translated c, R_c, is given by simple analysis of a two-stage 'cascaded amplifier' topology:

$$\frac{R_d}{R_c} = \frac{d}{c} \left(\frac{k'_c + m}{k'_d + m} \right) \left(\frac{l'_c + v}{l'_d + v} \right)$$

$$\approx \frac{d}{c} \left(\frac{k'_c}{k'_d} \right) \left(\frac{l'_c}{l'_d} \right) \tag{24.32}$$

$$= \frac{d}{c} \left(\frac{k'_c}{k'_d} \right)^2$$

where the approximation holds if $m \ll k'_c$ or k'_d and $v \ll l'_c$ or l'_d. Typically, c and d are equally abundant such that $d/c \approx 1$ in Equation (24.32). By having an additional stage of methylated amplification, we improve the R_d/R_c error rate from a linear dependence on k'_c/k'_d to a square dependence on k'_c/k'_d in Equation (24.32). Thus, if $k'_c/k'_d = 10^{-2}$ the mistranslation error rate is reduced to 10^{-4} because of the single methylation stage of amplification. If there are n stages of methylation, we improve the R_d/R_c ratio by an $(n+1)$th power law due to $(n+1)$ gain stages of amplification. Thus, kinetic proofreading is seen to be analogous to improving the discriminability of one set of proportional resistances (the c set) from another set of proportional resistances (the d set) by having them be part of a multiple-gain-stage topology rather than a single-gain-stage topology. Since delays or time constants from multiple gain stages only add while gains multiply, kinetic proofreading exploits the energy-efficient principle of distributed-gain amplification, a principle common in photoreceptors [17], in the cochlea [18], in the dendrites of neurons, and in transmission-line circuits in general. Chapter 23 discusses several distributed-gain transmission-line topologies in some depth. Furthermore, amplification requires one-way propagation of signals to prevent loads in late stages of amplification from compromising performance in early stages. The unidirectionality of the methylation process ensures such one-way propagation.

Robustness-efficiency, analog-digital, and other feedback-circuits-and-systems tradeoffs occur in cells just as in ultra-low-power analog and mixed-signal circuit design. Ultra-low-power analog electronic circuits face very similar tradeoffs like cells in biology because of their need to operate quickly, accurately and robustly despite mismatched and noisy components and signals, a necessary consequence of having very low levels of available power and space. Thus, analog circuits can shed insight into cellular systems biology just as they have in systems neurobiology in the past [19], [20]. They can also lead to the creation of new synthetic-biology circuits that are borrowed from circuit design. Simple topologies like ring oscillators have already been created [21] but there is still a lot of room for the creation of interesting circuits in cells that may truly have important medical applications. For example, next-generation antibiotic design involves the clever use of genetic-circuit engineering with bacteriophages [22]. Cellular biology, in turn, which performs 10 million biochemical operations at 1 pW of power, can inspire new circuit and system designs as several neurobiological architectures in the ear, eye, and brain have previously done (see Chapter 23). It is the author's firm belief that the field of cytomorphic electronics that we have suggested in this chapter is a field waiting to be born. This chapter has just scratched its surface. We shall now briefly highlight two broad areas that we believe to be particularly promising for the future.

24.7.1   Circuits-and-feedback analysis of stochastics and system dynamics in a cell

In many chapters of this book, we have shown how a thorough analysis of the noise performance of an ultra-low-power circuit allows quantitative predictions of its stochastic performance that are confirmed by experiment (for example, see Chapters 8, 11, 12, 13, 15, and 19). We have shown how noise sources that occur at different points in a circuit or equivalently at different points in a feedback block diagram have differing noise transfer functions (for example, see Chapters 11, 12, 13, 14, 15, 19, and 20) that make their variability more or less important. The same process can be applied to cells. The classic two-stage transcription-translation process of a cell, along with activation of the transcription factor, and binding of the transcription factor to DNA, is shown in Figure 24.17 (a). The feedback block diagram of Figure 24.17 (b) represents a simplified version of Figure 24.17 (a) that represents typical cellular operation. The square wave A in Figure 24.17 (b) represents rapid binding (k_a) and unbinding (γ_a) of the transcription factor to DNA in Figure 24.17 (a), which typically occurs at a fast enough rate such that it can be represented by a mean bound $\bar{A}$ in the lowpass circuits that follow it as in Figure 24.17 (c). Simplification of the block diagram of Figure 24.17 (b) and representation of the bound transcription factor concentration by its mean results in Figure 24.17 (c). Note that A is itself the result of a fast inducer binding-and-unbinding process to the transcription factor, which activates the transcription factor from X to X^*. The block diagram of Figure 24.17 (b) is also a simplified representation of the circuit of Figure 24.8.

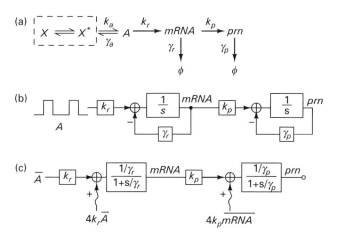

Figure 24.17a, b, c. (a) Induction, transcription and translation processes within a cell. (b) A feedback block diagram representation of Figure 24.16 (a). (c) Simplification of the feedback block diagram of Figure 24.16 (b). Reprinted with permission from [14] (©2009 IEEE).

The input shot-noise sources in Figure 24.17 (b) are due to transcription/translation molecular production (k) and degradation (γ) rates, which are balanced and equal to each other [14]. Analysis of Figure 24.17 (c) reveals that the total mRNA and protein power spectral densities are then given by

$$\sigma_{mRNA}^2 = \frac{k_r \bar{A}}{\gamma_r} = \overline{mRNA}$$

$$\beta = \frac{k_p}{\gamma_p} \tag{24.33}$$

$$\overline{prn} = \beta(\overline{mRNA})$$

$$\sigma_{prn}^2 = \overline{prn}(\beta + 1)$$

Equation (24.33) is a good model of the noise if the dynamics of protein degradation are much slower than that of mRNA degradation, i.e., $\gamma_p \ll \gamma_r$, such that the effective noise bandwidth is set by the protein degradation time constant, which is typically the case.

If we represent the mRNA concentration by the base current in a bipolar and the protein concentration by a collector current in a bipolar transistor, then Equation (24.33) exhibits the same mean and stochastic properties as the base and collector current in the bipolar:

$$\sigma_{base}^2 = 2q\overline{i_B}(\Delta f)$$

$$\overline{i_C} = \beta\overline{i_B} \tag{24.34}$$

$$\sigma_{coll}^2 = 2q\overline{i_C}(\beta + 1)(\Delta f)$$

The 'burst factor' β in Equation (24.33) is the protein/mRNA amplification ratio just like the current gain in bipolar transistors. Not surprisingly, there is excess noise of $(1 + \beta)$ because the amplified base-current (mRNA) noise dominates over the intrinsic collector (protein) noise. Experimental measurements of noise in cells confirm Equation (24.33) exactly and use a full-blown stochastic analysis to derive it [23]. The simple circuits insight afforded by the similarity of Equations (24.34) and (24.33) makes the stochastic analysis transparent. Therefore, we can leverage off 50 years of knowledge in circuit design to shed insight into what biology is doing. The mathematical soul of an architecture can be very similar across different fields although its physical embodiment in each field appears to be very different. Therefore, the wheel is often reinvented in each field.

While biochemical circuits have quadratic nonlinearities due to association and dissociation reactions, making purely linear-systems analyses inadequate, one can still perform small-signal and stochastic analyses about equilibrium molecular concentrations in the cell that shed insight. Furthermore, large-signal analyses that are performed by holding one molecular concentration constant while varying that of another in an association or dissociation reaction are still useful (e.g. see the Michaelis-Menten resistive-divider analysis of Figure 24.2 in this chapter). Indeed, engineering analyses have been done in cells for small-scale systems and have lead to results in good accord with experiment [24]. In the future, techniques for noise analysis of circuits that have been expounded in this book could be ported over to analyze stochastics in cells on a larger scale than the basic Equation (24.33). Stochastic variation of protein concentrations within cells is highly significant and important for predicting a cell's response to its environment or to drug treatments. For example, a recent paper has shown that variations in protein concentration successfully account for variability in the response of tumor cells to chemotherapy treatment for cancer [2].

24.7.2   Return-ratio robustness analysis of parameter-and-topology variations in a cell

In Chapter 10, we showed how return-ratio techniques allow us to predict the variation in the closed-loop transfer function of a circuit when one of its parameters varies. Specifically, we derived that the relative change in the transfer function of a circuit, $d(TF_{gm})/TF_{gm}$, is related to the relative variation in the parameters of one of its elements, dg_m/g_m, according to

$$\boxed{\frac{d(TF_{gm})}{TF_{gm}} = \left(1 - \frac{TF_0}{TF_{gm}}\right)\left(\frac{dg_m/g_m}{1 + R_{inputnulled}}\right)} \qquad (24.35)$$

where TF_0 is the transfer function when $g_m = 0$ and $R_{inputnulled}$ is the return ratio of the element. From Figures 24.4, 24.8, 24.10, or 24.11, variations in gene or protein parameters can be mapped to equivalent variations in circuit parameters. Equation (24.35) can then be used to predict how robust the circuit is to variation in a particular circuit parameter. Such an analysis can then shed insight into which gene/protein mutations and alterations sensitively impact the performance of a network and which do not. Gene knockouts represent an extreme version of this paradigm with g_m changed from a finite value to 0. Such analyses can be considerably more subtle and robust to model error than digital-circuit techniques that have attempted to do the same by treating all variations as 'on' or 'off' and modeled all DNA circuits as being purely digital.

24.8 Hybrid analog-digital computation in cells and neurons

Computation in a neuronal network and computation in a gene-protein network share intriguing similarities at the signal, circuit, and system level. Table 24.1 below highlights some of these similarities.

These analogies suggest that neuronal networks in the brain have similarities to ultra-fast, highly plastic, large-scale gene-protein networks with lots of connections per node. The speed of neuronal responses is matched to the time scales

Table 24.1 Similarities between neuronal and cellular computation

Property	Neuronal equivalent	Cellular equivalent
1. Basic computational unit device	A neuron	A gene
2. Discrete symbolic digital output of device	A spike	An mRNA transcript
3. Translation of symbol to signal	Post-synaptic potential (PSP)	Translation of mRNA transcript to protein
4. Processing of signals to create symbols	Analog and digital dendritic processing of multiple inputs to neuron	Analog logic circuits on DNA from multiple inputs
5. Connection weighting	Synaptic weight	K_d and transcription-factor binding
6. Kinds of connections	Excitatory and inhibitory	Activatory and repressory
7. Dale's rule	Output connection signs of a single neuron are correlated	Output connection signs of a single gene are correlated
8. Topological similarities	Shunting inhibition near soma with many excitatory inputs at other dendritic locations	Common repressor on DNA promoter with many activator transcription factor inputs.
9. Adaptation	Learning	Evolution
10. Connections per node	$\sim$6000	$\sim$12
11. Number of nodes	$\sim$22 billion	$\sim$30,000
12. Distributed ZY transmission-line processing	In dendrites	Reaction-diffusion networks within and outside cells.
13. Power consumption	$\sim$0.66 nW per neuron 22 billion neurons in the brain	$\sim$1 pW per cell $\sim$100 trillion cells in the body

at which the brain needs to control the body to react to fast changes in its environment. The speed of cellular responses is matched to the time scales at which the gene-protein network needs to control the cell to react to slow changes in its environment. Therefore, neuromorphic and cytomorphic circuits may share several similarities. In fact, the hybrid state machine (HSM) that was discussed in Chapter 22 (see Figure 22.9 (b)) and Chapter 23 on neuromorphic electronics functions as a useful description of computation within a cell as well: the analog state of the HSM is represented by protein concentrations and analog processing is accomplished via protein-protein interactions; the digital state of the HSM is represented by a subset of active/inactive genes; the configuration of these genes alters a binary control vector that affects protein production rates; protein concentrations that cross certain thresholds create logical comparator transitions (spikes are derivatives of these) that serve as inputs to the analog logic DNA gates (as opposed to a hard finite state machine in the HSM). Networks of cells can communicate and compute as networks of HSMs. Finally, we should mention that protein-protein biochemical networks perform computations that are similar to those of neuronal networks as well: similarities in pattern recognition in kinase protein networks and in *C. elegans* neuronal networks are discussed in [5].

The similarities between neuronal computation and cellular computation imply that the principles for energy-efficient operation that we discussed in Chapter 22 and that we showed are obeyed by neurobiology in Chapter 23 are also obeyed by the cell. In fact, since cells arose much earlier than brains in evolution, it may be more accurate to say that the cell is the pioneer of low-power biological computation. Neurons in the brain may have inherited many of the nice energy-efficient computational principles in the cell for free, and reapplied them on faster time scales and in networks with more connectivity and relatively fast plasticity.

In Chapter 23, we discussed ZY transmission lines in the context of neurobiological computation in dendrites, the cochlea, the retina, and vocal tract. Distributed ZY transmission lines are used in cells as well: ingenious ZY transmission lines with diffusion-based resistive Z and quadratic/polynomial shunt Y conductances implement spatially decaying molecular profiles [5]. These profiles are robust to variations in conditions at the beginning of the line: if the local concentration at a point on the line increases, the polynomially increasing Y causes a faster decay, restoring the profile to its invariant shape; if the local concentration at a point on the line decreases, the polynomially increasing Y causes a slower decay, once again restoring the profile to an invariant shape. The use of nonlinear negative feedback in the ZY transmission line to create invariant spatial profiles is analogous to the use of nonlinear negative feedback in the diode-capacitor circuit of Chapter 23 (see Figures 23.14 (a) and (b) and Equation (23.35)). There, we described how to create invariant temporal profiles that are robust to initial conditions on the capacitor through the use of an exponential nonlinearity (an infinitely high-order polynomial); here we create invariant spatial profiles that are robust to conditions at the beginning of the line.

Such ZY transmission lines are exploited by cells during development to ensure robustness in morphogen gradients independent of variations in protein-production rates. The concentrations of morphogen molecules can decide the fate of a cell during development, morphing it from being a universal stem cell (say) to a neuron depending on the concentration of the morphogen. Therefore, the robust self-assembly of a human being is in part determined by ingenious analog non-linear partial differential equations within our mothers! As the Nobel Prize winning physicist, Richard Feynman, said: "The imagination of nature is far, far greater than the imagination of man" [25].

References

[1] Harvey F. Lodish. *Molecular Cell Biology*, 6th ed. (New York: W.H. Freeman, 2008).

[2] S. L. Spencer, S. Gaudet, J. G. Albeck, J. M. Burke and P. K. Sorger. Non-genetic origins of cell-to-cell variability in TRAIL-induced apoptosis. *Nature*, **459** (2009), 428–432.

[3] K. A. Janes, J. G. Albeck, S. Gaudet, P. K. Sorger, D. A. Lauffenburger and M. B. Yaffe. A systems model of signaling identifies a molecular basis set for cytokine-induced apoptosis. *Science*, **310** (2005), 1646–1653.

[4] Bernhard Palsson. *Systems Biology: Properties of Reconstructed Networks* (New York: Cambridge University Press, 2006).

[5] Uri Alon. *An Introduction to Systems Biology: Design Principles of Biological Circuits* (Boca Raton, FL: Chapman & Hall/CRC, 2007).

[6] D. Baker, G. Church, J. Collins, D. Endy, J. Jacobson, J. Keasling, P. Modrich, C. Smolke and R. Weiss. Engineering life: building a fab for biology. *Scientific American*, **294** (2006), 44–51.

[7] Thomas Fischer Weiss. *Cellular Biophysics* (Cambridge, Mass.: MIT Press, 1996).

[8] Gene F. Franklin, J. David Powell and Abbas Emami-Naeini. *Feedback Control of Dynamic Systems*, 5th ed. (Upper Saddle River, N.J.: Pearson Prentice Hall, 2006).

[9] S. Mandal and R. Sarpeshkar, Log-domain Circuit Models of Chemical Reactions. *Proceedings of the IEEE Symposium on Circuits and Systems (ISCAS)*, Taipei, Taiwan, 2009.

[10] D. T. Gillespie. A general method for numerically simulating the stochastic time evolution of coupled chemical reactions. *Journal of Computational Physics*, **22** (1976), 403–434.

[11] J. F. Apgar, J. E. Toettcher, D.Endy, F. M. White and B. Tidor. Stimulus design for model selection and validation in cell signaling. *Public Library of Science Computational Biology*, **4** (2008), e30.

[12] M. Ptashne and A. Gann. *Genes and Signals* (Cold Spring Harbour, New York: CSHL Press, 2002).

[13] R. N. Mutagi. Pseudo noise sequences for engineers. *Electronics and Communication Engineering Journal*, **8** (1996), 79–87.

[14] S. Mandal and R. Sarpeshkar. Circuit models of stochastic genetic networks. *Proceedings of the IEEE Biological Circuits and Systems Conference*, Beijing, China, 2009.

[15] T. K. Lu, S. Zhak, P. Dallos and R. Sarpeshkar. Fast cochlear amplification with slow outer hair cells. *Hearing Research*, **214** (2006), 45–67.

[16] J. J. Hopfield. Kinetic proofreading: a new mechanism for reducing errors in biosynthetic processes requiring high specificity. *Proceedings of the National Academy of Sciences*, **71** (1974), 4135–4139.

[17] M. Tavakoli, L. Turicchia and R. Sarpeshkar. An ultra-low-power pulse oximeter implemented with an energy efficient transimpedance amplifier. *IEEE Transactions on Biomedical Circuits and Systems*, (2009).

[18] R. Sarpeshkar, R. F. Lyon and C. A. Mead. A low-power wide-dynamic-range analog VLSI cochlea. *Analog Integrated Circuits and Signal Processing*, **16** (1998), 245–274.

[19] Carver Mead. *Analog VLSI and Neural Systems* (Reading, Mass.: Addison-Wesley, 1989).

[20] D. W. Tank and J. J. Hopfield. Collective computation in neuron like circuits. *Scientific American*, **257** (1987), 104–114.

[21] M. B. Elowitz and S. Leibler. A synthetic oscillatory network of transcriptional regulators. *Nature*, **403** (2000), 335–338.

[22] T. K. Lu and J. J. Collins. Engineered bacteriophage targeting gene networks as adjuvants for antibiotic therapy. *Proceedings of the National Academy of Sciences*, **106** (2009), 4629.

[23] E. M. Ozbudak, M. Thattai, I. Kurtser, A. D. Grossman and A. van Oudenaarden. Regulation of noise in the expression of a single gene. *Nature Genetics*, **31** (2002), 69–73.

[24] J. T. Mettetal, D. Muzzey, C. Gomez-Uribe and A. van Oudenaarden. The frequency dependence of osmo-adaptation in Saccharomyces cerevisiae. *Science*, **319** (2008), 482.

[25] R. P. Feynman. *The Pleasure of Finding Things Out: The Best Short Works of Richard P. Feynman* (Cambridge, Mass.: Perseus Books, 1999).

Section VII

Energy sources

25 Batteries and electrochemistry

I have been so electrically occupied of late that I feel as if hungry for a little chemistry: but then the conviction crosses my mind that these things hang together under one law.

Michael Faraday

Ultra-low-power electronic systems often utilize batteries. The battery is frequently the heaviest and most critical portion of a portable system. The battery's size, weight, energy density, power density, form factor, cycling properties and lifetime can dictate constraints under which a portable electronic system must operate. Such constraints include the overall power dissipation of the system, the number of recharges that are possible in the system, the expected lifetime of the system, the overall size of the system, and the overall cost of the system. In this chapter, we shall provide a brief introduction to batteries with an emphasis on intuition and knowledge that is useful for providing insight into battery operation and ultra-low-power system design.

First, we shall begin with an exposition of the basic principles by which chemical energy is converted to electrical energy in a battery. Then, we shall discuss an equivalent circuit for a battery that intuitively describes several effects that are observed in practical batteries. We shall discuss the basics of lithium-ion and zinc-air batteries, two of the most common and most efficient batteries in existence in ultra-low-power electronic systems, particularly biomedical ones. We shall also briefly discuss the benefits of fuel cells and ultra capacitors, devices that are capable of being used in conjunction with or instead of batteries, and that are likely to be important in the future. Throughout our discussion, we shall focus on fundamental tradeoffs and practical characteristics that are seen in all batteries, irrespective of differences in their detailed chemistry.

25.1 Basic operation of a battery

A battery is a device that converts chemical energy to electrical energy by transforming the energy of a chemical reaction to electrical energy in a controlled fashion. The energy conversion and chemical reaction ideally only occur when the battery is used, i.e., when it discharges. The conversion is performed directly without the burning of chemical fuel in a battery. Thus, batteries have a

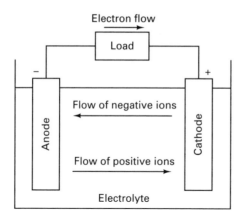

Figure 25.1. Basic battery with anode and cathode.

significantly higher conversion efficiency than heat engines that burn fuel to extract mechanical or electrical energy. Heat engines such as combustion engines are subject to certain thermodynamic Carnot limits on their energy-conversion efficiency whereas batteries are not subject to these same limits.

Batteries exploit a special kind of chemical reaction wherein electrons are transferred from a reducing-agent reactant at the anode, or negative terminal, of the battery to an oxidizing-agent reactant at the cathode, or positive terminal, of the battery in what is termed a redox reaction. The transfer of electrons ideally only occurs via an electrical circuit that is external to the battery such that the electron transfer and the rate at which the chemical reaction proceeds is determined by the electrical circuit that the battery is connected to. Note that the naming convention for the anode and cathode or negative and positive terminals of a battery respectively are opposite to that used in vacuum-tube or in some solid-state devices.

A substance is said to be reduced when it gains electrons and is said to be oxidized when it loses electrons. Reducing agents give up their electrons to other substances relatively easily such that they are oxidized and the substance that they give their electrons to is reduced. Oxidizing agents extract electrons from other substances such that they are reduced and the substance that they extract electrons from is oxidized. Metals and hydrogen are good reducing agents and form good battery anode reactants while metal oxides and oxygen are good oxidizing agents and form good battery cathode reactants. Batteries convert the energy released by the redox chemical reaction between the reducing-agent reactant at the anode and the oxidizing-agent reactant at the cathode to electrical energy. The energetically favorable chemical reaction proceeds downhill, and the downhill flow of electrons is exploited by the battery to do work in the external electric circuit. If the external electric circuit is a resistor, the chemical energy of the reaction is ultimately dissipated as heat in the resistor.

Figure 25.1 illustrates the basic configuration of a battery which is made up of two electrodes, an anode and a cathode, and an electrolyte solution. Electrons are conducted via current-collecting metallic wires that are connected to the anode and cathode and flow from the high-electron-energy negative anode to the low-electron-energy positive cathode in the external electric circuit. Ions are conducted in the electrolyte solution with positively charged cations flowing away from the anode and towards the cathode and negatively charged anions flowing away from the cathode and towards the anode. In steady-state operation, the net conventional ionic current in the electrolyte from, say, the anode to the cathode matches the conventional current in the external circuit from the cathode to the anode such that current flows in a loop in the series circuit of Figure 25.1.

25.2 Example mechanism for battery operation

We shall begin by describing a symmetric and conceptually simple mechanism of battery operation that is meant to illustrate key principles. The actual mechanism of operation in real batteries varies depending on their chemistry. Therefore, the description below should not be taken to be representative of any specific battery but viewed more as a tutorial. Later in the chapter, we shall discuss the operation of lithium-ion and zinc-air batteries and discuss their specific chemistries and mechanisms of operation.

The anode electrode, made up of a reducing-agent species R_A like a metal or metal composite, is oxidized via the anode half reaction

$$R_A \rightleftarrows R_A^{n+} + ne^- \tag{25.1}$$

to create R_A^{n+} that goes into solution and n electrons that remain in the anodic material. The electrons remaining in the anodic material charge it to a negative potential. If the battery is supplying current to an external electrical circuit, then some of these electrons are conducted away from the anode to the external circuit. The cation R_A^{n+} enters the solution or electrolyte and diffuses away from the anode towards the main bulk of the electrolyte.

The anodic forward reaction rate (left-to-right) in Equation (25.1) is exponentially sped up if the voltage at the anode becomes less negative since electrons are more likely to remain in the anode rather than recombine if the anode is less negative. The anodic backward reaction rate (right-to-left) in Equation (25.1) is exponentially sped up if the voltage at the anode becomes more negative since the R_A^{n+} ions prefer to recombine with the electrons at the anode if the anode voltage is more negative. The anodic reaction is energetically favorable in the forward direction such that there is a net tendency at the anode for R_A to ionize and release electrons at the anode until R_A^{n+} and ne^- accumulate in large enough concentrations near the anode and the anode voltage gets sufficiently negative such that the forward and backward reaction rates balance at equilibrium (zero current from battery) or at steady-state conditions (constant current draw from battery). In steady-state conditions, the anode's negative voltage is less negative than during open-circuit conditions to ensure that the forward reaction rate exceeds the backward reaction rate by just the right amount such that the net rate of production of R_A^{n+} is $(1/n)$ the electron-current production rate in accord with Equation (25.1).

The cathode electrode, made up of an oxidizing-agent species O_C like a metal oxide or oxygen, is reduced by the cathode half reaction

$$O_C + ne^- \rightleftarrows O_C^{n-} \tag{25.2}$$

to create O_C^{n-} that goes into solution and a paucity of n electrons in the cathodic material. This paucity of electrons in the cathodic material charges it to a positive potential. If the battery is supplying current to an external electrical circuit, then some of these electrons are provided by the external circuit to the cathode. The anion O_C^{n-} enters the solution or electrolyte and diffuses away from the cathode towards the main bulk of the electrolyte.

The cathodic forward reaction rate (left-to-right) in Equation (25.2) is exponentially sped up if the voltage at the cathode becomes less positive since electrons are more likely to leave the cathode if it is less positive. The cathodic backward reaction rate (right-to-left) in Equation (25.2) is exponentially sped up if the voltage at the cathode becomes more positive since the O_C^{n-} ions prefer to give up their electrons to the cathode if the cathode voltage is more positive. The cathodic reaction is energetically favorable in the forward direction such that there is a net tendency at the cathode for O_C to ionize and extract electrons from the cathode until O_C^{n-} accumulates in large enough concentrations near the cathode and the cathode voltage gets sufficiently positive such that the forward and backward reaction rates balance at equilibrium (zero current from battery) or at steady-state conditions (constant current draw from battery). In steady-state conditions, the cathode's positive voltage is less positive than at open-circuit conditions to ensure that the forward reaction rate exceeds the backward reaction rate by just the right amount such that the net rate of production of O_C^{n-} is $(1/n)$ the electron-current consumption rate in accord with Equation (25.2).

The anodic and cathodic reactions described by Equation (25.1) and Equation (25.2) along with the explanation of battery operation based on them constitutes the simplest and most symmetric example of battery operation. The cationic diffusion current of R_A^{n+} that flows away from the anode towards the electrolyte bulk is equal to the external electronic current that flows away from the anode such that the anode voltage has no net charge gain or loss and maintains its steady-state value. The anionic diffusion current of O_C^{n-} that flows away from the cathode towards the bulk is equal to the external electronic current that flows towards the cathode such that the cathode voltage has no net charge gain or loss and maintains its steady-state value. Both ionic currents are equal to their respective electronic currents in magnitude and flow towards the common 'ground' in the center of the electrolyte. The two grounded currents may be combined to create one net floating ionic current that flows from anode to cathode, the direction of flow of conventional ionic current. This conventional ionic current matches the external conventional current that flows from cathode to anode such that the current flow is continuous throughout the series loop of the battery circuit.

Real batteries can have only one ionic current that serves to ensure equality between electronic current and ionic current at both the anode and cathode by flowing directly from one electrode towards the other in the electrolyte. For example, in a lithium-ion battery, the lithium ion flows from the anode to the cathode and ensures equality of electronic current and ionic current at both electrodes and equality of conventional current in the electrolyte and conventional current in the external circuit [1]. The lithium ion ensures such equality by participating in half reactions at both the anode and the cathode, as we discuss later, such that stochiometric chemical balance at equilibrium or at steady state in each half reaction leads to current balance throughout the circuit. Equivalently, the lithium-ion battery has a true floating ionic current from the anode to the cathode rather than two grounded ionic currents that can be combined to form one floating ionic current as in the example described by Equation (25.1) and Equation (25.2). Similarly, in a hydrogen-oxygen fuel cell that we discuss later, in an acid electrolyte, H^+ ions diffuse away from the anode and participate in the cathodic half reaction to create one floating ionic current. In a hydrogen-oxygen fuel cell, in an alkaline electrolyte, OH^- ions diffuse away from the cathode and participate in the anodic half reaction to create one floating ionic current.

Our description thus far has neglected the influence of mass transport, i.e., the influence of the ionic diffusion currents in limiting the maximum rate of current production by the battery. It has also neglected the influence of the finite conductivity of the electrolyte in causing ohmic source resistance in the battery. Finally, we would like a more quantitative but still intuitive understanding of battery operation that incorporates chemical energetics, mass transport, and ohmic resistance into a simple current-voltage relationship that approximates the behavior of the battery. Therefore, we will now discuss how to create a large-signal dc equivalent circuit of the battery that provides an intuitive and economical understanding of these effects in battery operation. We first discuss how chemical reaction kinetics affect the current measured at a battery electrode.

25.3   Chemical reaction kinetics and electrode current

Our discussion in this section presents a simplified view of reaction kinetics at actual electrodes in batteries. It is a good description of certain kinds of redox flow batteries and of batteries where the surface morphology of the electrode may be assumed to be relatively intact during charging and discharging. Parts of our description will also apply to a simplified view of kinetics in a lithium-ion battery. The kinetics of an actual lithium-ion battery are considerably more complicated than we describe due to electrode porosity and solid-state diffusion within the electrode. Therefore, the reader should view the description below as a first step in

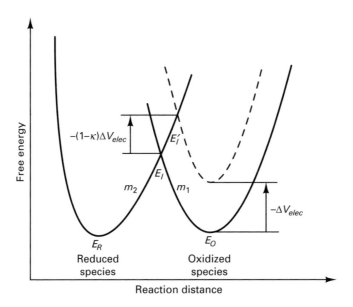

Figure 25.2. Potential energy curve for oxidized and reduced species inside a battery.

understanding the fundamentals of battery operation. These fundamentals lay a foundation for understanding more complex effects.

To begin, consider the energetics of a hypothetical redox reaction where a positively charged oxidized species in the electrolyte reacts with electrons at an electrode to generate a neutral reduced species. That is,

$$Z^{n+} + ne^- \rightleftharpoons Z \qquad (25.3)$$

where Z^{n+} is the oxidized species and Z is a neutral reduced species. Such a reaction does indeed occur at the cathode terminal of a lithium-ion battery with Z being Li and $n = 1$. The energetics of such a reaction can be represented by two intersecting potential-energy curves in Figure 25.2 that represent the potential energy of the oxidized or reduced species respectively, each with a stable minimum [1]. When the forward reaction occurs, the oxidized species changes state from the rightward minimum to the leftward minimum. When the backward reaction occurs, the reduced species changes state from the leftward minimum to the rightward minimum. The forward reaction rate decreases exponentially as the energy difference between the point at the intersection of the two curves and the rightward minimum increases, i.e., the forward reaction rate is proportional to $\exp(-(E_I - E_O))/kT$. The backward reaction rate decreases exponentially as the energy difference between the point at the intersection of the two curves and the leftward minimum increases, i.e., the backward reaction rate is proportional to $\exp(-(E_I - E_R))/kT$. The forward reaction rate also corresponds stochiometrically to a forward electronic current flowing from the electrode that is consumed by the Z^{n+} ions. The backward reaction rate also corresponds stochiometrically to a backward electronic current flowing away from the electrode that is generated by the decomposition of Z. The forward reaction current is analogous to a subthreshold source-to-drain forward current and the backward reaction current is analogous to a subthreshold drain-to-source backward current. The similarities between Figure 24.1 in Chapter 24 and Figure 25.2 in this chapter reflect the fact that all figures represent energy-dependent flows, whether in chemical reactions, in subthreshold electronic conduction, or in electrochemical reactions. The electronic currents in Figure 25.2 that we have discussed are termed Faradaic currents. Faradaic currents are computed by establishing stochiometric balance of chemical reactions at the electrode.

At some equilibrium potential between the electrode and electrolyte, there is a balance between the forward reaction rate and the backward reaction rate under open-circuit conditions resulting in no net current entering or departing at the electrode and no net forward or backward chemical reaction. If the potential of the electrode is forced to be more negative than this equilibrium potential by ΔV_{elec}, i.e., the change in potential of the electrode is $-\Delta V_{elec}$, the potential energy of the oxidized species Z^{n+} in the electrolyte increases since it becomes energetically less favorable for Z^{n+} to remain in solution rather than combine with electrons from the electrode via the forward reaction of Equation (25.3). The potential-energy curve in Figure 25.2 for the oxidized species Z^{n+} therefore shifts upward by $-\Delta V_{elec}$ while the potential-energy curve for the neutral reduced species Z remains unchanged. The forward reaction rate and electron consumption in Equation (25.3) then increase as the energy barrier for the oxidized species to transform into the reduced species becomes lower in Figure 25.2, i.e., the forward reaction rate is proportional to $\exp(-(E_I' - E_O)/kT)$. The backward reaction rate falls as the energy barrier for the reduced species to transform into the oxidized species increases in Figure 25.2, i.e., the backward reaction rate is proportional to $\exp(-(E_I' - E_R)/kT)$.

It has been empirically found that the slopes of the potential-energy curves for the oxidized and reduced species in their region of overlap can be approximated by straight lines with slopes of m_1 and m_2 respectively as shown in Figure 25.2 [2]. Then, if we define

$$\kappa = \frac{m_1}{m_1 + m_2} \tag{25.4}$$

we can show via simple algebra that the energy barriers for the forward and backward reaction rates, as measured in voltage units $(\Delta V = \Delta E/(nq))$, alter by

$$\Delta V_{bck} = -(1 - \kappa)\Delta V_{elec}$$
$$\Delta V_{frd} = +\kappa\Delta V_{elec} \tag{25.5}$$

Note that, in Equation (25.5), if ΔV_{elec} is negative as we have assumed, ΔV_{bck} is positive such that the backward-reaction ionic rate falls exponentially and ΔV_{frd} is negative such that the forward reaction ionic rate increases exponentially. In general, the net electronic current is dependent on the difference between the forward and backward reaction rates through the stoichiometry of Equation (25.3). Therefore, for the electronic current we may write

$$i = i_0 \left(e^{-\kappa q n (\Delta V_{elec})/(kT)} - e^{+(1-\kappa) q n (\Delta V_{elec})/(kT)} \right)$$ (25.6)

which is historically known as the Butler-Volmer equation [2]. The parameter i_0, the exchange current, is proportional to the area of the electrode and the rate of the forward or backward reaction at equilibrium, when the electrode potential with regard to the electrolyte is at its open-circuit value. Note that the total electronic current i at the electrode, which is the difference between the forward and backward currents, is 0 at equilibrium as confirmed by setting $\Delta V_{elec} = 0$ in Equation (25.6). However, it is composed of a forward and backward electronic exchange current, each of value i_0, which cancel, analogous to the I_S or equilibrium current of a pn junction.

When ΔV_{elec} is nonzero by more than a few (kT/qn) either the forward or backward current dominate in Equation (25.6) such that we may assume that only one term is dominant. Thus, the first approximation to an electrode-electrolyte interface is a voltage source of value equal to the open-circuit voltage in series with a diode-like exponential conductance described by one of the terms of Equation (25.6). Since all electrodes are subject to similar redox reactions, extending such an analysis to a battery electrode reveals that a battery electrode is well approximated by a voltage source in series with a diode-like exponential conductance in steady-state behavior. The voltage source is negative at the anode and positive at the cathode: at equilibrium, the anode's negative voltage slows its forward reaction that tends to provides electrons to the anode, and the cathode's positive voltage slows its forward reaction that tends to consume electrons from the cathode. At equilibrium, where there is no steady-state current, we must retain both exponential terms in Equation (25.6) to faithfully describe electrode operation, i.e., the electrode is then described by an open-circuit voltage in series with two back-to-back diode-like exponential elements but with different exponential I–V slopes, given by κ and $(1 - \kappa)$, respectively. The parameter κ is often called the charge-transfer coefficient [2].

25.4   Mass-transport limitations

The analysis thus far suggests that the external current provided by the battery is limited by the rate of the chemical reaction in the battery. Usually, the external current is limited by the rate at which diffusion in the electrolyte can take away ions from the electrode towards the bulk of the electrolyte or bring ions from the bulk of the electrolyte towards the electrode to be consumed. The voltage drop across the electrode-electrolyte junction adjusts from its open-circuit value to establish a steady-state concentration ratio of the oxidized and reduced species at the electrode that yields a forward-backward reaction-rate imbalance that is equal to that set by the ionic diffusion current. At steady-state, this ionic diffusion current equals the forward-backward rate imbalance, which in turn equals the external current leaving or entering the electrode such that all concentrations are constant and all currents in series are identical. Since diffusion is governed by the value of spatial concentration gradients of chemical species between the electrode and the electrolyte bulk, the maximal possible value of these gradients ultimately determines the maximum current that the battery can provide.

In the case of Equation (25.3), which is representative of the cathodic reaction of a lithium-ion battery, diffusion limitations imply that the maximal value of the spatial gradient occurs when the concentration of Z^{n+} approaches zero at the electrode because ions are consumed at the cathode as fast as the diffusion current from the bulk gets them there, i.e.,

$$I_{batt}^{max} \equiv I_{LC}^{disch} = \frac{qnD_C A_C}{\delta_C}\left(C_{Znp}^{blk} - 0\right) \tag{25.7}$$

where the diffusion constant D, diffusion length δ, and electrode area A have subscripts of C to remind us that these limitations apply to the cathode but not necessarily the anode in this particular battery. We have only focused on the first term of Equation (25.6) for now (the positive κ term). The value of C_{Znp}^{blk} is the value of the concentration of Z^{n+} in the bulk electrolyte, often established by the relatively invariant electrolyte concentration, which can be composed of a large reservoir of Z^{n+} ions. Typically only $D/\delta = m$, the mass-transport factor, is measurable and known. If the battery's external current I_{batt} is less than this maximal value,

$$I_{batt} = \frac{qnD_C A_C}{\delta_C}\left(C_{Znp}^{blk} - C_{Znp}^{elec}\right) \tag{25.8}$$

where C_{Znp}^{elec} is now a non-zero concentration value of the Z^{n+} ions right near the electrode. From Equations (25.7) and (25.8), some algebra reveals that

$$\left(1 - \frac{I_{batt}}{I_{LC}^{disch}}\right) = \frac{C_{Znp}^{elec}}{C_{Znp}^{blk}} \tag{25.9}$$

When $I_{batt} = 0$, $C_{Znp}^{elec} = C_{Znp}^{blk}$ since no current flow implies a zero spatial gradient. The first term in Equation (25.6) was computed under equilibrium conditions with an implicit value for $C_{Znp}^{elec} = C_{Znp}^{blk}$. When current is flowing, the concentration of Z^{n+} ions is attenuated by a factor of $C_{Znp}^{elec}/C_{Znp}^{blk}$ where $C_{Znp}^{elec}/C_{Znp}^{blk}$ is given by Equation (25.9). That is,

$$i_0 e^{-\kappa qn(\Delta V_{elec})/(kT)} \longrightarrow i_0 e^{-\kappa qn(\Delta V_{elec})/(kT)} \frac{C_{Znp}^{elec}}{C_{Znp}^{blk}}$$

$$i_0 \longrightarrow i_0 \frac{C_{Znp}^{elec}}{C_{Znp}^{blk}} \tag{25.10}$$

$$i_0 \longrightarrow i_0 \left(1 - \frac{I_{batt}}{I_{LC}^{disch}}\right)$$

Thus, the net effect of mass-transport limitations is to lower the effective value of i_0 for the first term in Equation (25.6) due to the reduced value of the concentration of the Z^{n+} ions near the electrode when current is flowing. Similarly, due to mass-transport limitations having to do with the diffusion of Z^{n+} ions away from the electrode, we can define a limiting current I_{LC}^{ch} such that the second term in Equation (25.6) (the second negative $(1 - \kappa)$ term) is altered such that

$$i_0 \longrightarrow i_0 \left(1 - \frac{I_{batt}}{I_{LC}^{ch}}\right) \tag{25.11}$$

The maximal value of I_{LC}^{ch} is determined by the maximum concentration of Z^{n+} at the electrode, which in turn is determined by the maximal surface concentration of active Z at the electrode: Z^{n+} is limited to this maximal surface concentration since all of Z has to be converted to Z^{n+} such that the surface concentration of Z goes to zero at the electrode. Hence the net effect of mass-transport limitations is to transform Equation (25.6) to

$$i_C = i_{0C}\left(1 - \frac{I_{batt}}{I_{LC}^{dsch}}\right)e^{-\kappa_C qn(\Delta V_{elec}^C)/(kT)} - i_{0C}\left(1 - \frac{I_{batt}}{I_{LC}^{ch}}\right)e^{+(1-\kappa_C)qn(\Delta V_{elec}^C)/(kT)} \tag{25.12}$$

We have used superscripts of ch and $dsch$ because of the importance of each of the terms of Equation (25.12) to the discharging (forward reaction in Equation (25.3)) or charging (backward reaction in Equation (25.3)) of a battery, in this example at the cathode of the battery, indicated by the subscript C. A similar analysis for the anode electrode of the battery yields

$$i_A = i_{0A}\left(1 - \frac{I_{batt}}{I_{LA}^{ch}}\right)e^{-(1-\kappa_A)qn(\Delta V_{elec}^A)/(kT)} - i_{0A}\left(1 - \frac{I_{batt}}{I_{LA}^{disch}}\right)e^{+\kappa_A qn(\Delta V_{elec}^A)/(kT)} \tag{25.13}$$

It must be noted that the values of i_0, κ, etc. are different for the anode and the cathode corresponding to the different chemical energetics and mass transport at each electrode, which is why we have used subscripts of C or A in Equation (25.12) or Equation (25.13) respectively. During discharge, only the first term in Equation (25.12) or the second term in Equation (25.13) dominate, ΔV_{elec}^C is negative, ΔV_{elec}^A is positive, conventional current at the cathode is positive, and conventional current at the anode is negative. During charge, only the second term in Equation (25.12) or the first term in Equation (25.13) dominate, ΔV_{elec}^C is positive, ΔV_{elec}^A is negative, conventional current at the cathode is negative, and conventional current at the anode is positive. For Equation (25.12) and Equation (25.13) to be valid, i_A and i_C have to both equal I_{batt} by dropping the appropriate ΔV_{elec}^C or ΔV_{elec}^A across the cathode or anode respectively. We shall discuss how these voltage drops become precipitous when I_{batt} nears I_L in later sections.

25.5   Large-signal equivalent circuit of a battery

Figure 25.3 shows a large-signal equivalent circuit of a battery that represents Equations (25.12) and (25.13) in circuit form with exponential elements (κ or $(1 - \kappa)$ elements) shown as diodes. The pre-exponential I_S or equilibrium current of the diode is marked on the Figure 25.3 for each diode in terms of the respective i_0 and the mass-transport I_L parameters. The open-circuit voltages for the cathode and anode half cells are represented with voltage sources with the total open-circuit battery voltage being given by the difference between the cathodic voltage source (positive) and the anodic voltage source (negative). The dark-face diodes correspond to the exponential terms that are dominant during battery discharge while the light-face diodes correspond to the exponential terms that are dominant during battery charge. The conventional current I_{batt} flows through an external circuit impedance that the battery performs electrical work on from the cathode to the anode while electron current flows from the anode to the cathode. The circuit also models the electrolyte ohmic resistance due to its finite ionic conductivity via the R_{elec} resistances. The electrolyte resistance can be improved by increasing concentrations of its ionic species and by geometric design: increasing cross-sectional area between the anode and cathode and decreasing the distance between anode and cathode. The ionic conductivity of a battery usually increases with temperature such that R_{elec} falls at high temperatures. The significance of Q_A and Q_C in Figure 25.3 with respect to the state of charge (SOC) and usage of the battery are explained later. For now, we assume that all the parameters in Figure 25.3 such as i_{0A} and I_L remain constant in the battery independent of how long the battery has been discharged. Unless otherwise noted, we shall focus primarily on the discharging properties of the battery when it is used in a circuit.

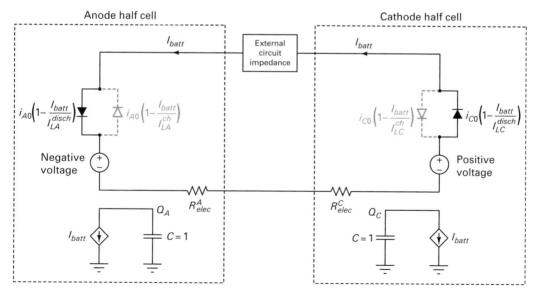

Figure 25.3. Large-signal dc equivalent circuit of a battery.

Figure 25.3 illustrates that any current draw from the battery will reduce its potential from the open-circuit voltage due to voltage drops across the forward-reaction diodes termed *activation polarization*. Furthermore, voltage drops in the electrolyte caused by the ohmic losses $I_{batt}(R_{elec}^A + R_{elec}^C)$ in Figure 25.3 also reduce the open-circuit voltage of the battery when current is drawn from it due to *ohmic polarization*. Finally, voltage drops due to mass-transport limitations, i.e., the I_L-dependent pre-exponential I_S terms of the diodes in Figure 25.3, also reduce the open-circuit voltage of the battery when current is drawn from it due to *concentration polarization*. To find the battery voltage for a given current I_{batt}, we must subtract ohmic polarization losses from the battery's open-circuit voltage and invert Equations (25.12) and (25.13) to solve for the ΔV_{elec} losses due to activation and concentration polarization at each electrode. An exact closed-form inversion of Equation (25.12) and Equation (25.13) is difficult. Can we find an approximate solution that still provides insight?

For small battery currents, the small-signal conductance of each of the two parallel back-to-back diodes at both the anode and cathode cause the resistances at both these electrodes to appear linear. For large battery currents, we can safely ignore the charging diode at both electrodes and assume that only the forward discharging diode is active at both of them. An approximation that straddles both this small-current linear regime and the large-current regime can be found by assuming that the I_L parameters for charging and discharging are nearly equal, by assuming that κ is near 0.5, and by inverting Equations (25.12) and (25.13) to solve for the ΔV_{elec} drops across both electrodes. Some algebra then reveals that the battery voltage is given by

$$
\begin{aligned}
V_{batt} \approx & \left(V_{oc}^{cathode} - V_{oc}^{anode} \right) \\
& - I_{batt}\left(R_{elec}^C + R_{elec}^A \right) \\
& - \left(\frac{kT}{qn\kappa_C} \sinh^{-1}\left(\frac{\kappa_C \dfrac{I_{batt}}{i_{0C}}}{1 - \dfrac{I_{batt}}{I_{LC}^{disch}}} \right) + \frac{kT}{qn\kappa_A} \sinh^{-1}\left(\frac{\kappa_A \dfrac{I_{batt}}{i_{0A}}}{1 - \dfrac{I_{batt}}{I_{LA}^{disch}}} \right) \right)
\end{aligned}
\tag{25.14}
$$

The second row of Equation (25.14) from top to bottom characterizes ohmic losses while the third row characterizes a combination of activation and concentration polarization losses. Since the I_{LC} terms in Equation (25.14) can be neglected for small currents, $\sinh^{-1}(x) \approx x$ for small x, and $\sinh^{-1}(x) \approx \ln(2x)$ for large x, we can create a small-current and large-current version of Equation (25.14) that are useful in interpreting experimental V–I battery curves. The small-current version of Equation (25.14) leads to

$$
\begin{aligned}
V_{batt}^{sml} = & \left(V_{oc}^{cathode} - V_{oc}^{anode} \right) \\
& - I_{batt}\left(R_{elec}^C + R_{elec}^A \right) \\
& - I_{batt}\left(\frac{kT}{qni_{0C}} + \frac{kT}{qni_{0A}} \right)
\end{aligned}
\tag{25.15}
$$

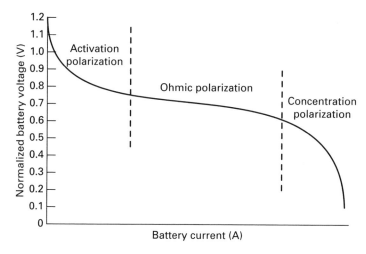

Figure 25.4. Three types of current-dependent battery polarization.

We note that Equation (25.15) is not dependent on κ and that the effects of activation polarization are well represented by a small-signal *charge-transfer resistance* $kT/(qni_0)$ or *Faradaic resistance* for each electrode. The large-current version of Equation (25.14) is given by

$$
\begin{aligned}
V_{batt}^{lrg} &= \left(V_{oc}^{cathode} - V_{oc}^{anode}\right) \\
&\quad - I_{batt}\left(R_{elec}^{C} + R_{elec}^{A}\right) \\
&\quad - \left(\frac{kT}{qn\kappa_C}\ln\left(2\kappa_C\frac{I_{batt}}{i_{0C}}\right) + \frac{kT}{qn\kappa_A}\ln\left(2\kappa_A\frac{I_{batt}}{i_{0A}}\right)\right) \\
&\quad - \left(\frac{kT}{qn\kappa_C}\ln\left(\frac{1}{1-\dfrac{I_{batt}}{I_{LC}^{disch}}}\right) + \frac{kT}{qn\kappa_A}\ln\left(\frac{1}{1-\dfrac{I_{batt}}{I_{LA}^{disch}}}\right)\right)
\end{aligned}
\tag{25.16}
$$

The ohmic, activation, and concentration polarization losses are represented by the second, third, and fourth rows of Equation (25.16) respectively. The activation and concentration losses are now separable unlike in Equation (25.14).

Figure 25.4 plots the typical *V–I* relationship for a battery showing regions where losses due to activation polarization, ohmic polarization, and concentration polarization dominate. The steep drop in voltage of the battery occurs when I_{batt} is slightly less than I_{LA}^{disch} or I_{LC}^{disch}, whichever is lower. Several commercial lithium-ion and zinc-air batteries have curves that are similar to the one drawn [1]. We shall later present experimental measurements from a commercial lithium-ion battery that are very similar in appearance to that shown in Figure 25.4. The power drawn from the battery is maximized in more central regions of the curve rather than at very low currents or at very high currents.

25.6   Battery voltage degradation with decreasing state of charge

When a battery has been used to supply current for a while there is a loss of active material from the electrodes, the concentrations of product species build in the bulk electrolyte, thin films may deposit on the electrode degrading diffusion coefficients, and electrolyte resistance typically increases. Experimentally, one observes that the battery voltage starts to fall, and in fact falls in a fashion that looks remarkably similar to the curve of Figure 25.4 except that the X-axis is time rather than current. In a lithium-ion battery, for example, there is a gradual loss in capacity as the lithium ions move from being present at the anode to being present at the cathode. The battery is characterized as losing charge, and as it loses charge, the battery's output voltage degrades. We say that the battery's output voltage is a function of its *state of charge* (SOC). Freshly charged batteries at the beginning of their discharge cycle that have barely been used have a maximum state of charge and output relatively large voltages. Batteries that have been heavily used and that are near the end of their discharge cycle have a low state of charge and output relatively small voltages.

While very complex models have been formulated to characterize battery voltage degradation as a function of their state of charge, we shall begin by approximating the overall process as one that causes the maximum electrode currents at the anode I_{LA}^{disch} and cathode I_{LC}^{disch} to fall with time. This abstraction is useful because, to a large extent, the increase in bulk electrolyte species, the loss of active material from the electrodes, or diffusion-coefficient degradation will degrade the maximal spatial gradient in the battery, thus degrading I_L. Hence, if we assume that there is an SOC variable Q_A at the anode, and an SOC variable Q_C at the cathode that represents the coulombic charge-capacity state at these electrodes, we may describe the battery SOC by the following simple and empirical equations:

$$Q_A(t) = Q_A^{max} - \int_0^t I_{batt}(x)dx$$

$$Q_C(t) = Q_C^{max} - \int_0^t I_{batt}(x)dx \tag{25.17}$$

$$I_{LA}^{disch}(t) = \frac{Q_A(t)}{\tau_A}$$

$$I_{LC}^{disch}(t) = \frac{Q_C(t)}{\tau_C}$$

From Equation (25.7), the values of τ_A and τ_C are related to relatively constant mass-transport factors, (D_A/δ_A) or (D_C/δ_C), and geometric areal cross-sections, A_A or A_C, in the battery. Equation (25.17) allows us to estimate how the battery discharge time, i.e., the time until its voltage reaches the steep-fall-off portion of Figure 25.4, changes as a function of the external current draw in the battery. If we assume a constant current of I_{batt}, then this time t_{dsch} will occur approximately when either $Q_A(t)$ or $Q_C(t)$ reaches a value such that I_L at the corresponding limiting electrode equals I_{batt}. Thus,

$$\frac{\left(Q_{ah}^{max} - I_{batt}t_{dsch}\right)}{\tau} = I_{batt}$$

$$\boxed{t_{dsch} = \frac{Q_{ah}^{max}}{I_{batt}} - \tau}$$

(25.18)

Here Q_{ah}^{max} and τ correspond to the limiting battery-electrode parameters in Equation (25.17), i.e., Q_A^{max} and τ_A if the anode limits battery discharge time, or Q_C^{max} and τ_C if the cathode limits battery discharge time. The parameter Q_{ah}^{max} may be viewed as the maximum ampere-hour charge capacity of the battery. If Q_{ah}^{max}/I_{batt} is much larger than τ in Equation (25.18), i.e., the battery current I_{batt} is small, the latter equation predicts that the expected battery discharge time is inversely proportional to its current consumption I_{batt}. However, at large values of I_{batt}, Equation (25.18) predicts that the expected battery discharge time Q_{ah}^{max}/I_{batt} overestimates the actual battery discharge time by τ.

The degradation in battery discharge time with increasing I_{batt} is often described as a degradation in the ampere-hour rating of a battery with increasing current. The discharge current required to discharge the battery in one hour based on the manufacturer's ampere-hour rating is known as C. Thus, a 1 A h battery has a 1 C discharge current of 1 A and a discharge lifetime of 1 hour with this current. The same 1 A h battery when discharged at $C/30$, i.e., at 1 A/30 = 33 mA, will have a discharge lifetime greater than 30 hours, with an apparent improvement in its ampere-hour capacity at low discharge currents. In fact, it is not uncommon to have a two-fold improvement in the ampere-hour capacity for such a 30-fold decrease in the discharge current. The relationship between the battery discharge time t_{dsch} and I_{batt} is often described by an equation known as Peukert's law, which uses an empirical $(1/I_{batt}^z)$ term with $z > 1$ to describe the worsening in discharge time with increasing current. However, we can predict such battery degradation based on Equation (25.18) without the need for an empirical power law: Equation (25.18) can be expressed in an equivalent form by expressing $Q_{ah} = I_{batt}t_{dsch}$, the actual ampere-hour rating of the battery, as a function of I_{batt}. To do so, we multiply both sides of Equation (25.18) by I_{batt}, and find that

$$\boxed{Q_{ah} = Q_{ah}^{max} - I_{batt}\tau}$$

(25.19)

Does Equation (25.19) describe real batteries?

Figure 25.5 (a) shows battery voltage-versus-time discharge curves obtained from a AA-size rechargeable 750 mA h lithium-ion battery for various different values of I_{batt} ranging from 75 mA to 750 mA. The battery was fully charged to an open-circuit voltage of 4.2 V and allowed to rest for 3 hours in an open-circuit condition before the initiation of each discharge curve. After this rest period, the battery was then discharged at constant I_{batt} until its voltage reached 3 V, after which the discharge current was shut off to prevent damage to the battery.

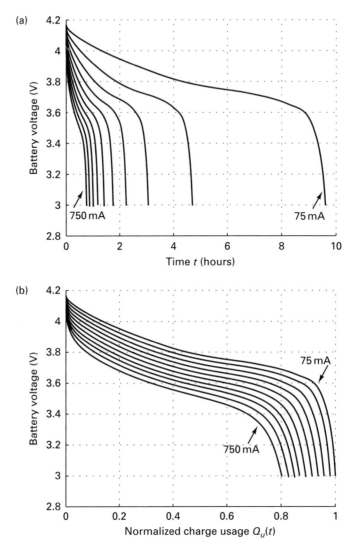

Figure 25.5a, b. Battery voltage versus time in (a) or versus normalized charge usage (b).

The current I_{batt} was maintained constant using a current source throughout the battery discharge. If we define t_{dsch} as the time taken by the battery to reach 3 V, a voltage in its precipitous region as shown in Figure 25.5 (a), we can compute the battery capacity $Q_{ah} = I_{batt}t_{dsch}$ for each discharge curve. We can estimate Q_{ah}^{max} as the value of Q_{ah} corresponding to the minimal current of 75 mA. If we replot the data of Figure 25.5 (a) versus $Q_u(t) = I_{batt}t/Q_{ah}^{max}$ rather than versus time, i.e., plot the data versus the normalized charge usage $Q_u(t)$, we see that Q_{ah} is indeed lower for batteries used with large values of I_{batt} rather than for those used with small values of I_{batt}. In Figure 25.5 (b) for instance, $Q_{ah} = 0.8\, Q_{ah}^{max}$ for the $I_{batt} = 750$ mA case versus Q_{ah}^{max} for the $I_{batt} = 75$ mA case.

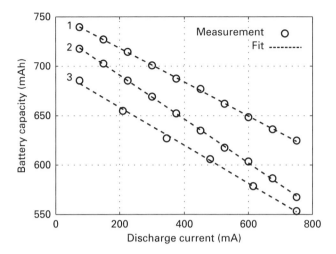

Figure 25.6. Loss in battery capacity with increasing discharge current.

Figure 25.6 plots Q_{ah} versus I_{batt} for three different lithium-ion batteries. We see that Equation (25.19) is a good fit to all three of them. For battery 1, we found $Q_{ah}^{max} = 753\,\mathrm{mA\,h}$ and $\tau = 0.17\,\mathrm{h}$; for battery 2, we found $Q_{ah}^{max} = 735\,\mathrm{mA\,h}$ and $\tau = 0.22\,\mathrm{h}$; for battery 3, we found $Q_{ah}^{max} = 697\,\mathrm{mA\,h}$ and $\tau = 0.19\,\mathrm{h}$. If Q_{ah} were not a function of I_{batt}, the curves in Figure 25.6 would all be flat horizontal lines with $\tau = 0$. Instead, they are lines with a finite slope of $-\tau$ as predicted by Equation (25.19).

From the discussions leading to Equations (25.17), (25.18), and (25.19), we can define an effective normalized SOC variable for the battery $Q_n(t)$ that is given by

$$Q_n(t) = \left(1 - \frac{\int_0^t I_{batt}(\eta)d\eta}{Q_{ah}^{max}}\right)$$

$$Q_n(t) = \frac{\int_0^t I_{batt}(\eta)d\eta}{Q_{ah}^{max}}$$

$$Q_n(t) = 1 - Q_u(t)$$

(25.20)

From Equation (25.17), as the battery is discharged, $Q_n(t)$ falls such that internal variables within it like I_L change according to

$$I_L(t) = I_L^{max}Q_n(t)$$

(25.21)

Note that $I_L(t)$ and I_L^{max} are based on combining parameters at the anode and the cathode, which may actually be different, into one effective parameter in Equation (25.21). Similarly, as the battery discharges, internal concentrations of reactants such as, for example, the number of lithium ions in the anode falls, or concentrations of reactants and products get more equalized. Thus as the battery reaction proceeds downhill, the battery's stored free energy and consequently its open-circuit voltage V_{oc} decreases as Q_n decreases. This drop in free energy as a reaction proceeds downhill is true in all chemical reactions and batteries are no exception. The electrolyte resistance R_{elec} also frequently increases as Q_n decreases.

Figure 25.3 incorporates the first two rows of Equation (25.17) into the large-signal dc equivalent circuit of the battery. Note that the recharging current of the battery, which is obtained by reversing the direction of flow of the discharging current in Figure 25.3, flows through different exponential diode elements in Figure 25.3 compared with the discharging current. Recharging the battery also serves to restore the values of Q_A and Q_C after they have drooped during battery discharge to Q_A^{max} and Q_C^{max} by changing the currents on the Q_A and Q_C capacitors from a discharging current to a charging current in Figure 25.3. The restoration of Q_A and Q_C then serve to restore the I_L, V_{oc} and R_{elec} parameters. In practice, the restoration is not perfect on each charge-discharge cycle such that there are only a finite number of cycle lifetimes in the battery. Miniscule losses in capacity, e.g., 0.03% loss in Q_A^{max} and/or Q_C^{max} on each charge-discharge cycle, compound to eventually cause a 50% capacity degradation in 1200 cycles. This degradation phenomenon is similar to the "Problem of the 9's" in charge-coupled-device (CCD) cameras: the 0.9999 efficiency of charge transfer in a single analog-shift-register stage of a CCD can compound to create a 50% loss of charge after 3600 stages of shifting.

Finally, the self-discharge properties of a battery can be modeled by including large resistances from the Q_A and Q_C capacitors in Figure 25.3 to ground. Self-discharge often occurs due to parasitic reactions in a battery on a very slow time scale. It is responsible for a loss in capacity of the battery even when it is not being used and is simply being stored on the shelf.

25.7   Small-signal equivalent circuit of a battery and of electrodes

To derive a small-signal equivalent circuit for the battery, we must model capacitive effects and conductive effects in the battery. We shall first focus exclusively on capacitive effects before combining conductive and capacitive effects into a complete small-signal model.

25.7.1   Capacitive analysis

Figure 25.7 reveals that a charged conductive battery electrode attracts water-dipole molecules and ions of opposite charge near it. The example shown is for that of a negatively charged metal electrode and positive ions. The water dipoles are composed of slightly negatively charged oxygen and slightly positively charged hydrogen atoms. The ions are typically surrounded by water dipoles to form hydrated ions. The configuration shown in Figure 25.7 illustrates that the configuration of ions and water dipoles near an electrode forms what is termed as a

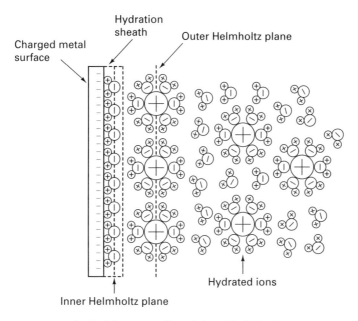

Figure 25.7. Physical interpretation of the Helmholtz layer at the electrode-electrolyte interface. (Figure courtesy of Luke Theogarajan.)

Helmholtz double layer, a double-layer coating of the electrode [2]. The double layer is composed of a layer of water dipoles nearest the electrode that form a hydration sheath followed by a layer of hydrated ions. The two layers lead to the definition of an inner Helmholtz plane and an outer Helmholtz plane as shown. Beyond the double layer, water dipoles and hydrated ions are present in a less structured and more loosely organized fashion with positive ion concentrations falling in a Boltzmann exponential fashion as we move away from the electrode. Thus the capacitance between the electrode and the bulk electrolyte may be modeled as being composed of two capacitances in series, one due to the Helmholtz double layer that is invariant with potential and one that exhibits a cosh-like dependence on the electrode potential due to the exponential distribution of ionic species near the electrode. It may then be shown that the net capacitance is given by

$$\frac{1}{C_{Stern}} = \frac{1}{C_{Hlm}} + \frac{1}{C_{GC}}$$

$$C_{Hlm} = \frac{\varepsilon_r \varepsilon_0 A}{d_{OHP}}$$

$$C_{GC} = \frac{\varepsilon_o \varepsilon_r A \cosh\left(\frac{nqV_{elec}}{2kT}\right)}{L_D} \tag{25.22}$$

$$L_D = \frac{1}{qn} \sqrt{\frac{\varepsilon_o \varepsilon_r (kT)}{2N_{elec}}}$$

where ε_r is the relative permittivity of water, N_{elec} is the concentration of the positive ions in the bulk electrolyte, L_D, termed the Debye length, is the scaling length over which the diffuse ionic charge concentration beyond the double layer falls by an e-fold, and d_{OHP}, the distance from the electrode to the outer Helmholtz plane, is 0.5–1 nm [2]. The subscripts on the capacitances stand for the double-layer Helmholtz capacitance, the Gouy-Chapman cosh-like capacitance, and the total Stern capacitance, in honor of the pioneers who were important in formulating their values. The cosh-like capacitance arises because ionic charges pack together more tightly near the low-energy electrode when the magnitude of the electrode potential V_{elec} is higher.

The presence of the Helmholtz double layer often complicates electrochemical analyses, since the relevant electrode potential for determining ionic concentrations is that at the edge of the Helmholtz double layer, which is always lower in magnitude than that at the electrode. However, we have not and shall not be too careful in our use of potentials in this regard: it is difficult to always estimate such potentials correctly and the general nature of equations is usually unchanged except for a scale factor.

25.7.2   Conductive analysis

We shall begin by analyzing charge-transfer effects due to reaction kinetics, ignoring mass-transfer effects due to diffusion. Then we shall treat mass-transfer effects and finally combine them with charge-transfer effects and the electrolyte resistance to compute the overall small-signal model of the battery.

25.7.2.a   Charge-transfer analysis

In Figure 25.3, when there is no current flowing through the battery, an exchange current of i_{0A} flows through both diodes at the anode. Thus, from the usual small-signal resistance for an exponential element, we derive that the small-signal open-circuit resistance r_{ct}^A at the anode is given by

$$\frac{1}{r_{ct}^A} = \frac{qn}{kT}(\kappa_A i_{0A} + (1-\kappa_A)i_{0A})$$

$$= \frac{qn}{kT}i_{0A} \tag{25.23}$$

$$r_{ct}^A = \frac{kT}{qni_{0A}}$$

Similarly, the small-signal open-circuit resistance at the cathode is given by

$$r_{ct}^C = \frac{kT}{qi_{0C}} \tag{25.24}$$

The total small-signal open-circuit resistance of the battery is then given by

$$r_{ct} = \frac{kT}{qni_{0A}} + \frac{kT}{qni_{0C}} \tag{25.25}$$

This resistance is termed the charge-transfer resistance in electrochemistry [2] and is consistent with our alternative derivation of it in Equation (25.15).

If the battery is discharging a current of I_{batt} such that, to a good approximation, only the discharging exponential elements in Figure 25.3 are significant, then the small-signal charge-transfer resistance of the battery at this operating point is given by the sum of the small-signal resistances of the discharging diodes at the anode and cathode respectively:

$$r_{ct}^{ibatt} = \frac{kT}{qn\kappa_A I_{batt}} + \frac{kT}{qn\kappa_C I_{batt}} \tag{25.26}$$

25.7.2.b   Mass-transfer analysis

When the potential of an electrode changes, there is a change in the concentration of reducing and/or oxidizing species at the electrode, which in turn leads to a change in the diffusion current. For example, in the cathodic forward reaction of a lithium-ion battery, which is well represented by Equation (25.3) with $n = 1$, the diffusion current is given by Equation (25.8) with

$$I_{batt} = \frac{qnD_C A_C}{\delta_C}\left(C_{Znp}^{blk} - C_{Znp}^{elec}\right)$$

$$C_{Znp}^{elec} = C_{Znp}^{blk} e^{-qn\kappa_C \Delta V_{elec}/kT} \tag{25.27}$$

such that when ΔV_{elec} varies at the cathode, the diffusion current in Equation (25.27) varies. However, the diffusion current described by the first line of Equation (25.27) is really a steady-state solution to a partial differential equation given by

$$D_C \frac{\partial^2 C_{Znp}(x)}{\partial x^2} = \frac{\partial C_{Znp}}{\partial t}$$

$$C_{Znp}\big|_{x=0} = C_{Znp}^{blk} e^{-qn\kappa_C \Delta V_{elec}/kT} \tag{25.28}$$

$$C_{Znp}\big|_{x=\infty} = C_{Znp}^{blk}$$

where D_C is the diffusion constant for the ionic species Z^{n+} near the environment of the cathode. If we represent concentration by a voltage and diffusive concentration flows by mass currents (rather than charge currents), a circuit model of Equation (25.28) is an infinite one-dimensional transmission line composed of a lateral resistance per unit length of $1/(D_C A)$ and a shunt capacitance per unit length of A. From standard transmission-line theory (see Equation (23.2) in Chapter 23 and Chapter 17), such a transmission line has an equivalent characteristic mass impedance looking in at the electrode of

$$Z_M(\omega) = \sqrt{\frac{1}{(D_C A_C)(j\omega A_C)}} \tag{25.29}$$

The diffusive flow of the Z^{n+} ions causes an actual electrical current flow as well since Z^{n+} is a charged species. The electrical admittance G_W seen at the electrode due to the diffusive flow is then given by

$$G_W(\omega) = -\left(\frac{\partial C_{Znp}^{elec}}{\partial(\Delta V_{elec})}\right)\left(\frac{1}{Z_M(\omega)}\right)(nq) \tag{25.30}$$

The first bracketed term in Equation (25.30) yields the change in boundary condition at the electrode due to the electrode-potential change, the second bracketed term yields the mass-current flow through the transmission line caused by this change in boundary condition at the beginning of the line, and the third bracketed term converts a mass current to an electrical charge current. By differentiating the second line of (25.27) and substituting the value of Z_M given by Equation (25.29) into Equation (25.30), we get

$$G_W^C(\omega) = \left(C_{Znp}^{elec}\frac{(qn)\kappa_C}{kT}\right)\left(A_C\sqrt{\frac{D_C\omega}{2}}(1+j)(nq)\right) \tag{25.31}$$

The *Warburg impedance* is defined to be $1/G_w^C$ and is the diffusion impedance seen at the electrode. From Equation (25.31), it is then given by

$$\boxed{Z_W^C(\omega) = \left(\frac{kT}{nq\kappa_C}\right)\left(\frac{1}{nqC_{Znp}^{elec}A_C\sqrt{2D_C\omega}}\right)(1-j)} \tag{25.32}$$

First, we note that Equation (25.32) reveals that the capacitive reactance and resistance components of the Warburg impedance are always *identical* since the coefficient of the 1 and the $-j$ terms are identical and that both terms have a $(1/\sqrt{\omega})$ dependence on the frequency. If we substitute for C_{Znp}^{elec} from Equation (25.9), we can rewrite Equation (25.32) as

$$\boxed{Z_W^C(\omega) = \left(\frac{kT}{nq\kappa_C}\right)\left(\frac{1}{nqC_{Znp}^{blk}\left(1-\frac{I_{batt}}{I_{LC}}\right)A_C\sqrt{2D_C\omega}}\right)(1-j)} \tag{25.33}$$

Equation (25.33) reveals that Z_W^C becomes infinite when I_{batt} is at the limiting diffusion current I_{LC}. This finding is consistent with the infinite slope in Figure 25.4 when I_{batt} nears i_{LC} and with our intuition that the diffusion current barely changes with voltage once it has saturated. Another interesting point to note in Equation (25.32) is that the denominator of the second bracketed term is the current obtained at the electrode if C_{Znp}^{elec} diffused with a linear spatial gradient and fell to zero over a length $\delta_C = \sqrt{(D_C/(2\omega))}$. Since $kT/(nq\kappa_C)$ is the characteristic thermal voltage, Equations (25.32) and (25.33) represent the ratio of a voltage and a current and thus yield an impedance.

25.7.3   Overall small-signal circuit

Figure 25.8 configures the charge-transfer resistance r_{ct}, the Warburg impedance $Z_w(\omega)$, the Stern capacitance C_{stern}, and the electrolyte resistance R_{elec} in both the

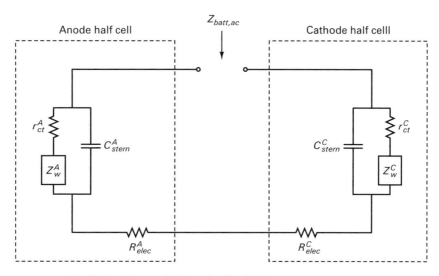

Figure 25.8. Small-signal ac equivalent circuit of a battery.

anodic and cathodic half circuits to create a complete small-signal equivalent for the battery, Z_{batt}^{ac} [2]. That is

$$Z_{batt}^{ac}(\omega) = \left(r_{ct}^{C} + Z_{W}^{C}(\omega)\right) \| \left(\frac{1}{j\omega C_{stern}^{C}}\right)$$

$$+ \left(r_{ct}^{A} + Z_{W}^{A}(\omega)\right) \| \left(\frac{1}{j\omega C_{stern}^{A}}\right) + \left(R_{elec}^{C} + R_{elec}^{A}\right)$$

(25.34)

The anodic side has impedance formulas analogous to ones that we have explicitly discussed for the cathodic side.

The electrolyte resistance is determined by evaluating the resistance of all three-dimensional electrolytic conduction paths from anode to cathode. In the case of two nearby rectangular electrodes of equal surface area $A = A_C = A_A$, separated by a length L that is much smaller than any dimension of the electrode, to a good approximation,

$$R_{elec} = R_{elec}^{C} + R_{elec}^{A}$$

$$= \frac{\rho_{elec}(L/2)}{A} + \frac{\rho_{elec}(L/2)}{A}$$

(25.35)

In Equation (25.35), ρ_{elec} is the specific resistivity of the electrolyte, which is a function of the ionic concentrations and ion mobilities. The specific resistivity of an electrolyte varies from 2–$10\,\Omega$ cm for aqueous electrolytes, 10–$50\,\Omega$ cm for inorganic electrolytes, 100–$1000\,\Omega$ cm for organic electrolytes, 1000 to $10^{7}\,\Omega$ cm for polymer electrolytes, and 10^{5}–$10^{8}\,\Omega$ cm for inorganic solid electrolytes [1].

For electrodes that are widely separated, or in more complex three-dimensional geometries, the exact evaluation of R_{elec} may involve a three-dimensional integral. As a simple example, consider a spherical anodic electrode of radius R sourcing current into an electrolyte symmetrically in all dimensions with the current traveling away to infinity. In practice the return cathode electrode would pick up this current but we shall only focus only on the anode for now. The net half resistance R_{elec}^A as measured at the anode can then be approximated by

$$R_{elec}^A = \int_R^\infty \frac{\rho}{4\pi r^2} dr = \frac{\rho}{4\pi R} \tag{25.36}$$

Equation (25.36) illustrates that having electrolytes with low resistivities and electrodes with large surface area will minimize R_{elec} in the battery as Equation (25.35) also predicts. If one electrode is of much larger surface area than another, then the electrode of smaller surface area will be dominant in determining the total R_{elec} of the battery.

25.7.4   Small-signal circuits of electrodes

Figure 25.8 is also a good model for electrodes used in implantable electronics, e.g., in extracellular saline solution in cochlear, retinal, brain, and cardiac implants. Implantable electronics is discussed in Chapter 19. In recording applications, R_{elec} and C_{stern} capacitances typically dominate at most frequencies of interest with C_{Hlm} being the limiting small series capacitance rather than C_{GC} in Equation (25.22). In stimulation applications care must be taken to not excite Faradaic reactions such that the influence of r_{ct} in Figure 25.8 or the influence of the diodes in Figure 25.3 is small. The emphasis on charge-balanced ac-current-only stimulation in these applications attempts to ensure that any electronic charge that is put into the electrode on one phase is removed in the next phase. Thus, potentially toxic redox reactions, which would stoichiometrically require net electron consumption or donation, are minimized. Large series blocking capacitors for coupling the current stimulation to the electrode are often used. Transiently large voltage levels, e.g., ± 10 V, across an electrode are common with charge-balanced, short-pulse, mA-level stimulation in applications like cochlear implants, and appear to be tolerated well by patients as long as there is little net dc current error. Chapter 19 discusses how to create blocking-capacitor-free miniature circuits with little dc current error (see Figure 19.15). The Warburg impedance Z_W can and frequently does play a part: the electrode impedance versus frequency often has a $1/\sqrt{f}$ character to it in its real and imaginary part as expected from Equation (25.33). The terms anode and cathode are no longer relevant since one electrode may be a ground electrode in a *monopolar* configuration and the other electrode could be one of a differential pair of electrodes in a *bipolar* differential configuration.

25.8   Operation of a lithium-ion battery

Lithium-ion batteries have rapidly become the rechargeable battery of choice in batteries that range from very small (< 1 mA h) capacities in some implantable biological applications to batteries with very large capacities in mobile-vehicle applications. They are actively being researched for use in hybrid gasoline-electric and electric vehicles that may need 50 kW power outputs. Their popularity stems from their good energy density ($\sim$620 J/g and $\sim$1440 J/cm^3), good power density ($\sim$0.5 W/g average and $\sim$1.8 W/g peak), ability to operate at high discharge rates of 5 C and at pulsatile discharge rates of 25 C, their ability to operate from $-40\,^\circ$C to $+60\,^\circ$C, high operating voltages (2.5–4.2 V), their relatively low fade capacity on repeated charge-discharge cycles ($\sim$0.03% capacity fade per cycle and $\sim$0.3 mV loss in open-circuit voltage per cycle), relatively low self-discharge rate ($\sim$5% per month if unused and not refrigerated), and advances in their ease of manufacture [1].

Unlike NiMH batteries or other batteries, lithium-ion batteries do not have a memory effect. Batteries with a memory effect need to be deeply discharged and then recharged in order to achieve good capacities. Lithium-ion batteries, on the other hand, usually have better performance and longer cycle life if they are not deeply discharged. For example, their cycle life at 100% depth of discharge (DOD) (Q_A or Q_C are discharged almost to zero in Figure 25.3) is usually only 1000–3000 cycles. At 20%–40% DOD (Q_A or Q_C are only discharged to 0.8 or 0.6 of their initial fully-charged value), some measurements have shown that lithium-ion batteries are capable of 20,000 cycles of charging and recharging [1]. Thus, lithium-ion batteries are optimally suited for ultra-low-power systems that operate for a long time between recharges with only a modest DOD. Hence, ultra-low-power dissipation also prolongs battery life and thus product life and system cost.

The greatest drawback of lithium-ion batteries is that they are relatively sensitive to overcharging and undercharging and need precision battery-management circuitry to ensure that they operate safely and efficiently. For example, they need protective circuitry to ensure that they are not overcharged beyond 4.2 V, which can cause thermal runaway and venting. Deep discharges to voltages below 2 V cause capacity degradation. Thus, Li-ion batteries typically employ mechanical disconnect devices that open on overheating and battery-management circuits that provide protection from over-charging, over-discharging, or over-temperature conditions. Undercharging a lithium-ion battery to less than the optimal 4.2 V by even a few percent can lead to a significant loss of capacity (e.g., a -1.2% undercharge causes a 9% loss in capacity in some batteries) [3]. Thus, lithium-ion batteries often incorporate battery-management circuits that ensure good efficiency and safety.

A typical battery-management circuit for a lithium-ion battery is charged in three different regimes. If the battery has been deeply discharged, a small preconditioning current is used to charge the battery until the battery enters a regime of modest DOD. Then, it is charged with constant current, typically $1C$ or less until its voltage nears 4.2 V. After that, it is charged with a constant voltage of 4.2 V until the charging current falls below a minimum threshold value, typically smaller than $C/16$. Batteries that are charged with a constant current in excess of $1C$ show degradation in cycle lifetimes and, in general, slower charging improves battery lifetimes. Lead-acid batteries are also charged with a CCCV (constant-current-constant-voltage) scheme while NiMH and NiCd batteries are typically charged with a constant-current scheme. Simple and efficient analog circuits for charging lithium-ion batteries are described in [4].

Lithium-ion batteries operate with a rocking-chair chemistry. During discharge, lithium ions and electrons leave the anode material [1]. The lithium ions travel through the electrolyte, the electrons travel through the external circuit, and both meet up at the cathode to deposit intercalated lithium into it via a redox reaction. During charge, lithium ions and electrons leave the cathode material. The lithium ions travel through the electrolyte, the electrons travel through the external circuit, and both meet up at the anode to deposit intercalated lithium into it via a redox reaction. Thus, lithium ions rock back and forth between the anode and cathode depending on whether the battery is being discharged or recharged respectively. The anode material is most commonly lithiated graphite, i.e., Li_xC. The cathode material is most commonly lithium cobalt oxide, i.e., $Li_{1-x}CoO_2$. The anode and cathode half reactions are given by

$$Li_xC \rightleftharpoons xLi^+ + xe^- + C$$
$$Li_{1-x}CoO_2 + xLi^+ + xe^- \rightleftharpoons LiCoO_2$$

(25.37)

The left-to-right arrow corresponds to discharging the battery while the right-to-left arrow corresponds to charging the battery.

The electrolyte in lithium-ion batteries is typically a salt of lithium, $LiPF_6$, dissolved in a solvent composed of a mixture of ethylene carbonate and propylene carbonate at a relatively high 1 M concentration [1]. The resistivities of such electrolytes are near $100\,\Omega\,cm$. Ethylene carbonate is associated with a low fade capacity and good cycle life but the presence of other carbonates in the solvent lowers viscosity and the freezing point of the mixture. Separators between the anode and cathode are made of 10–30 μm microporous polyolefin materials made of polyethylene, polypropylene, or laminates of both. Such separators with pore sizes less than 1 μm are easily wetted by the electrolyte. They are compatible and stable in contact with the electrode, electrolyte, and other materials. The current collector for the anode is typically a copper foil and that for the cathode is an aluminium foil. The current collectors adhere to the electrodes with a binder, typically made of polyvinylidene fluoride (PVDF) and a high-surface-area carbon-black or graphite conductive diluent.

Lithium ion batteries are made in cylindrical and prismatic geometries. In flat-plate prismatic designs, the current from several parallel electrodes is often collected to increase the effective electrode surface area in a given volume and thus increase power density. Snake-like windings of electrode materials with separators in between squeeze high-surface-area electrodes into a small volume in wound prismatic cells. In an 18 mm × 65.0 mm prismatic battery, the energy density is limited by the fact that the electrode materials that yield energy only account for about 60% of the volume, while the rest is consumed by the separator, the electrolyte, the can, and a cell header incorporating mechanical breaker and vent mechanisms [1]. Larger batteries in general have better energy densities because

active materials that yield energy form a larger part of the battery in its core while
the costs of the housing and parasitics are minimized at the battery's periphery:
the surface-area-to-volume ratio decreases with the size of the battery such that
peripheral portions of the battery form a smaller fraction of the overall weight of
the battery.

25.9   Operation of a zinc-air battery

Zinc-air batteries are primary batteries that are non-rechargeable. In contrast, a
lithium-ion battery is termed a secondary battery because it is rechargeable.
Zinc-air button batteries are currently the highest energy-density batteries that
are readily commercially available ($\sim$1050 J/g). They are dominant in the hearing-
aid market and are important in all ultra-low-power biomedical applications
where a replaceable battery outside the body may be used. The high energy density
of zinc-air batteries arises because of the relatively energy-rich redox chemical
reactions that form the basis of the battery, and because they have an implicit
oxygen cathode. Since oxygen can be harvested from air near the battery, the
weight of the cathodic half cell in a zinc-air battery is minimized, leading to
superior energy density. Their chief disadvantage is that their exposure to the
atmosphere increases their sensitivity to the environment. Zinc-air batteries
operate best at 60% relative humidity: lower humidities cause drying out of the
water in the battery, while high humidities flood the battery and limit its power
output [1].

The cathodic half reaction is given by

$$\frac{1}{2}O_2 + H_2O + 2e^- \longrightarrow 2OH^- \tag{25.38}$$

The anodic half reaction is given by

$$Zn \longrightarrow Zn^{2+} + 2e^-$$

$$Zn^{2+} + 2OH^- \longrightarrow Zn(OH)_2 \tag{25.39}$$

$$Zn(OH)_2 \longrightarrow ZnO + H_2O$$

Water, electrons, and OH^- act as intermediates in the overall reaction, which is
simply the oxidation of zinc by oxygen:

$$Zn + \frac{1}{2}O_2 \longrightarrow ZnO \tag{25.40}$$

A nickel-mesh screen serves as the cathodic current collector and also supports
a mixture of carbon blended with MnO_2 to catalyze the cathodic reaction.
A hydrophobic Teflon layer contacts the nickel mesh and minimizes water-vapor
entry into the cell. The entry of air into the cathode is regulated by air holes or a
gas diffusion membrane and an air distribution layer to ensure even distribution of

oxygen on the cathode. The nickel mesh contacts an external cathode can, which serves as one terminal of the battery, at its circumferential periphery. The cathode can-like container also doubles as a mechanical housing for the various cathodic layers.

A powdered zinc amalgam, gelling agent, and potassium-hydroxide electrolyte contact an external upside-down top anodic can that forms the other terminal of the battery. The anodic can is engulfed by the cathodic can which surrounds it concentrically at its sides but not at its top. The cans are insulated from each other by a plastic gasket. A separator layer separates the anodic contents from the cathodic wire mesh and catalyst layers.

Zinc-air batteries have good energy densities because the air cathode can harvest oxygen from its environment and thus the weight due to the normal cathodic half of a battery is approximately zero. However, their power density is usually relatively poor ($\sim 0.1\,\mathrm{W/g}$): the power density is limited by the diffusion rate of oxygen into the cell, regulated by the size of air holes or micro-porous diffusion layers. The rate of diffusion must be kept low to avoid environmental degradation and a short shelf life.

The components of zinc-air batteries are similar to that of the common $1.5\,\mathrm{V}$ $\mathrm{Zn\text{-}KOH\text{-}MnO_2}$ (anode-electrolyte-cathode) alkaline batteries that dominate the primary-battery market in cylindrical AAAA, N, AAA, AA, C, D, and F sizes. In alkaline batteries, the oxidant is $\mathrm{MnO_2}$ and is stored in the battery rather than being oxygen that is harvested from the air in zinc-air batteries.

25.10   Basic operation of fuel cells

Fuel cells are close cousins of batteries. Like batteries, they convert the energy from a chemical reaction into electrical energy. However, unlike batteries, which carry their reactant sources of energy in their electrodes, the source of energy in fuel cells arises from an external reactant fuel and oxidant that are supplied to the anode and cathode of the fuel cell, respectively. The electrodes in the fuel cell serve as catalysts and electron carriers for the chemical reactions between the fuels but do not participate directly in the energy-producing chemical reactions. Fuel cells thus separate their power-generation capability from their energy-storage capability unlike batteries where both capabilities are intimately integrated. The replenishment of fuel (and/or oxidant if it is not harvested from the air) is analogous to the recharging of a conventional battery. The fuels used in fuel cells typically have significantly higher energy densities than the active materials used in batteries, potentially increasing their attractiveness when compared with batteries. However, fuel cells have significantly higher overhead costs than batteries that degrade their overall energy density and make them more costly and complex to build: they require a fuel-supply system that regulates fuel flows, may require thermal and fluid-management subsystems, may need stacking to yield a useful output voltage, and may need an oxidant subsystem.

A basic hydrogen-oxygen fuel cell functions as follows [1]. Hydrogen gas that is supplied to the anodic half of the fuel cell creates electrons and protons via the anodic half reaction

$$H_2 \longrightarrow 2H^+ + 2e^- \tag{25.41}$$

The electrons are deposited at the anode while the protons go into the electrolyte of the fuel cell [1]. In small fuel cells, the electrolyte is usually a solid-polymer proton-conducting membrane electrolyte such as Nafion, a product made by the Dupont Corporation. The protons travel through the electrolyte and react with oxygen gas that is supplied to the cathodic half of the fuel cell via the redox half reaction

$$2H^+ + \frac{1}{2}O_2 + 2e^- \longrightarrow H_2O \tag{25.42}$$

The open-circuit voltage that corresponds to the free energy of the overall chemical reaction is 1.23 V but most fuel cells operate near 0.6 V due to activation, ohmic, and concentration polarization losses. Thus, several cells are often stacked in series if a higher voltage is needed. The Nafion membrane also partly serves to separate gases at the anode and cathode and prevents them from interacting directly with each other and thus short-circuiting the energy-production mechanism of the battery. It must therefore have a minimum thickness, which is generally in the $20\,\mu m$–$100\,\mu m$ range.

25.11   Energy density, power density, and system cost

In a battery, energy density and power density trade off with each other because high power density arises from large electrode surface areas and high energy density arises from large electrode volumes. To get good power density, one needs a high surface area of the active electrode material to allow for a good rate of consumption of the active material. To get high energy density, one needs a high volume of the active electrode material such that there is plenty of active material in the battery's overall volume and consequently high energy density. Thus, thick electrodes that fill up a given battery volume lead to high energy density but low power density. Thin electrodes that are snaked or spiraled to fill up a given battery volume lead to high power density but lower energy density. High power-density batteries have a higher volume-and-weight fraction of ancillary non-energy contributing components in the battery like current substrates, which lowers the overall energy density of the battery. A well-known curve known as the Ragone curve expresses the energy-density versus power-density tradeoff in a battery [1], [5]. An example of a Ragone curve for a battery is shown in Figure 25.9.

Energy and power densities can be measured per unit volume (volumetric density) or per unit weight (gravimetric density). Both forms are important since energy and power per unit volume or per unit weight are key in most applications.

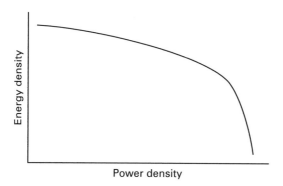

Figure 25.9. Example Ragone curve.

Often, the terms energy density and power density are used to describe volumetric numbers while the adjective 'specific' refers to gravimetric numbers. For simplicity, we shall use energy density and power density to describe both volumetric and gravimetric numbers.

When the power drawn from a battery is near its absolute maximum, $I_{batt}\tau \approx Q_{ah}^{max}$, such that Q_{ah} in Equation (25.19) is 0, and consequently the energy capacity or energy density of the battery is near 0; the open-circuit voltage of the battery is also severely degraded under such extreme conditions. When I_{batt} is almost 0, Q_{ah} in Equation (25.19) is near Q_{ah}^{max}, the battery's open-circuit value is near its maximum with almost no activation, ohmic, or concentration polarization losses, and consequently the energy capacity or energy density of the battery is at its maximum. These extremes of operation determine the limits of maximum energy density and maximum power density on the Ragone curve. Increasing the surface area/volume of the electrodes by having thin electrodes increases the maximum value of I_L, thus reducing τ and improving the power density but also compromises Q_{ah}^{max}, thus reducing the energy density.

Primary non-rechargeable batteries are built to have good energy densities such that they last long, and therefore tend to compromise power density. Secondary rechargeable batteries do not need to run as long as primary batteries since they can be recharged. Therefore, they usually have higher power densities and lower energy densities.

An example that illustrates the degradation in energy density with small size due to surface-area-versus-volume considerations is provided by the common cylindrical alkaline 1.5 V cells that dominate the primary battery market: Typical average ampere-hour capacities for AAAA, AAA, and AA batteries are 0.56 A h, 1.15 A h, and 2.67 A h, respectively as cylindrical radii and cell diameters increase from the smallest-size AAAA to the largest-size AA. These data were averaged from various manufacturers and are taken from [1]. The corresponding energy densities adapted from [1] are then given by 465 J/g, 564 J/g, and 627 J/g, respectively and improve as a larger fraction of the volume is devoted to energy storage in the battery.

Fuel cells can optimize power-generation by scaling up to high fuel-supply rates and large-area electrodes if high power densities are required and independently optimize energy storage by increasing the amount of fuel stored based on the desired operation lifetime. However, they are also subject to the energy density versus power density tradeoff of batteries. If more volume is devoted to energy-storing fuel, less is available for the structures needed for power-generation such that energy density rises while power density falls. If more volume is devoted to structures needed for power generation, less is available for energy-storing fuel such that power density rises while energy density falls.

Fuel cells are significantly more energy efficient than batteries at low power densities where the infrastructure for power generation is a small fraction of the overall volume, most of which can then be devoted to the storage of highly-energy-dense fuels. For example, the energy density of methanol is approximately 19,500 J/g while that of liquid hydrogen is 143,000 J/g. It is not surprising, therefore, that space applications have used fuel-cell-based technology.

Hybrid systems that combine fuel cells with batteries are being researched to get the best of both worlds: batteries are used for transiently high-power discharges that the fuel cell is not capable of handling, while fuel cells provide the average low-power steady discharge at excellent energy density most of the time, and also serve to recharge the battery. Such schemes are useful in applications that have a high peak-to-average ratio of power consumption, for example, in cell phones where peak-transmission-power and average-standby-power requirements are significantly different from each other.

Battery–super-capacitor hybrid systems represent a variation on the same battery–fuel-cell hybrid theme. Readily available ultra-capacitors have low energy densities of approximately ~ 10 J/g but they can be recharged for 10^5–10^6 cycles, and, due to their extremely high surface areas, are capable of power densities that can be 1000 times higher than those of typical batteries [6]. Thus, ultra-capacitors or super-capacitors can be used for transient high-power-discharges while the battery is used for the steady-state average-low-power discharge. The battery also serves to recharge the ultra-capacitor, which can be done very quickly.

Batteries and fuel cells already operate within a factor of 2–3 of their theoretical energy density limits as can be computed from the free-energy-change (voltage limit) and coulombic stoichiometry (current limit) of their overall chemical reaction [1]. Thus, advances in battery efficiencies can theoretically yield a factor of 2 or 3 performance improvement at most; in practice, improvements of a few tens of percent are extremely hard to achieve. Hence, advances in ultra-low-power electronics are essential to attain order-of-magnitude or several-order-of-magnitude improvements in the lifetimes of portable systems.

Ultra-low-power electronics allows portable-product lifetime to increase in a nonlinear expansive fashion due to four benefits:

1. The lifetime is increased because the power drain on the battery is lower.
2. Low power densities can be traded for higher energy densities in a battery such that a battery that lasts for a significantly longer time can be made by trading surface area for volume.

3. Batteries that do not need to be discharged as deeply in an ultra-low-power electronic system fade more slowly and can be recharged several more times.
4. Low-power operation minimizes all polarization losses in the battery (activation, ohmic, and concentration) and thus maximizes the battery's voltage. The lack of losses improves the efficiency of energy extraction and thus prolongs the time between recharges.

An ultra-low-power electronic system reduces battery weight and size, consequently system weight and size, and consequently system cost. The materials cost of packaging is reduced because less material is needed in a smaller system, batteries with lower performance specifications are significantly cheaper to make, costs due to heating concerns are reduced, increased system lifetimes amortize costs over a longer time, a small lightweight product significantly increases the value of the product to the consumer, and enables scaling to larger more-complex systems to become possible. A small, portable low-power system can make solutions in tightly constrained spaces of the body with low tissue heating possible in implantable medical applications that could not have been possible otherwise. Indeed, implantable batteries in cardiac pacemakers have already revolutionized patients' lives. Ultra-low-power electronics will undoubtedly enable several more improvements in medical implants for patient treatment as we have discussed in Chapter 19.

We have discussed the harvesting of energy from RF sources in Chapters 16 and 17 for implantable (see Chapter 19) and for noninvasive medical applications (see Chapter 20), respectively. In the succeeding Chapter 26, we shall discuss other sources of energy that can be harvested such as solar energy, vibratory energy from the body, thermal energy from body heat, and chemical energy from organic or bio-molecules. We shall see that insights from ultra-low-power system design, which we have focused on mostly for small-scale medical applications, and mostly in electronics, are also important at larger scales and in non-electronic applications like transportation. To meet the 15 TW power budget of the world with renewable energy sources alone, a necessity after oil is highly scarce in 3–4 decades, is challenging. Just as low-power electronic design and battery design can undergo joint optimization to create a win-win scenario for both, low-power transportation design and renewable energy generation can undergo joint optimization to create a win-win solution for both.

References

[1] David Linden and Thomas B. Reddy. *Handbook of Batteries*, 3rd ed. (New York: McGraw-Hill, 2002).
[2] Allen J. Bard and Larry R. Faulkner. *Electrochemical Methods: Fundamentals and Applications*, 2nd ed. (New York: Wiley, 2001).
[3] S. Dearborn. Charging Li-ion batteries for maximum run times. *Power Electronics Technology Magazine*, (2005), 40–49.

[4] M. Chen and G. A. Rincon-Mora. Accurate, Compact, and Power-Efficient Li-Ion Battery Charger Circuit. *IEEE Transactions on Circuits and Systems II: Express Briefs*, **53** (2006), 1180–1184.

[5] D. V. Ragone. Review of Battery Systems for Electrically Powered Vehicles. *Mid-Year Meeting of the Society of Automotive Engineers*, Detroit, Michigan, 1968.

[6] Lund Instrument Engineering Inc. Available from: www.powerstream.com.

26 Energy harvesting and the future of energy

Nature uses only the longest threads to weave her patterns, so that each small piece of her fabric reveals the organization of the entire tapestry.

Richard P. Feynman

Energy surrounds us, is within us, and is created by us. In this chapter, we shall discuss how systems can harvest energy in their environments and thus function without needing to constantly carry their own energy source. The potential benefits of an energy-harvesting strategy are that the lifetime of the low-power system is then not limited by the finite lifetime of its energy source, and that the weight and volume of the system can be reduced if the size of the energy-harvester is itself small. The challenges of an energy-harvesting strategy are that many energy sources are intermittent, can be hard to efficiently harvest, and provide relatively low power per unit area. Thus, energy-harvesting systems are usually practical only if the system that they power operates with relatively low power consumption.

We shall begin by discussing energy-harvesting strategies that have been explored for low-power biomedical and portable applications. First, we discuss the use of strategies that function by converting mechanical body motions into electricity. A circuit model developed for describing energy transfer in inductive links in Chapter 16 is extremely similar to a circuit model that accurately characterizes how such mechanical energy harvesters function. Thus, tradeoffs on maximizing energy efficiency or energy transfer are also similar. Energy harvesting with RF energy is discussed extensively in Chapters 16 and 17, so we shall not discuss it in this chapter. Then, we discuss the use of thermoelectric strategies that function by converting body heat into electricity. A fundamental thermodynamic principle limits the energy efficiency of a 'heat engine', whether in an internal combustion engine in a car, in a refrigerator, or in a thermoelectric device powered by body heat. The limiting efficiency is called the *Carnot efficiency*. The Carnot efficiency and models of heat flow from the body will help us understand the limits of operation of thermoelectric energy harvesting.

This book has largely discussed ultra-low-power systems at relatively small spatial scales in biomedical and in bio-inspired systems, mostly in the 10^{-12} W to 10^{-2} W range. In this final chapter, we shall see that principles of low-power design are also relevant to systems at large spatial scales with gigantic power consumption, e.g., a 40 kW gasoline-powered car moving at 30 mph, which if operated for 1 hour each day leads to an average power consumption of 1.67 kW.

The average human being on Earth consumes 2.5 kW of power such that our planet's current aggregate power consumption is roughly 15 TW. The average power consumption of people in richer countries is higher than that in poorer countries. For example, the average person in the United States consumes 10.4 kW [1].[1] We have been able to sustain such power consumption thus far largely because the 46,400 J/g energy density of gasoline, the 53,600 J/g energy density of natural gas, and the 32,500 J/g energy density of coal, and their relative abundance, have enabled us to burn energy at a profligate rate. In comparison, a well-optimized lithium-ion battery for portable applications operates at 650 J/g. Gasoline is currently cheaper per liter than bottled water in the United States.

For every kW h of oil, natural gas, or coal that is consumed, 250 g, 190 g, and 300 g, respectively, of CO_2 is dumped into our atmosphere [2]. This means that 5.5 tons of CO_2 is generated on average per person per year, increasing CO_2 levels by ~2.5 ppm (parts per million) per year today [3]. The accumulation of CO_2 has increased the atmospheric concentration from 280 ppm in pre-industrial times to ~390 ppm today [4]. The pace of CO_2 emissions is expected to increase significantly as India, China, and other developing nations output more CO_2. For every ppm increase in CO_2, the average Earth temperature appears to rise due to a greenhouse effect [3]. Many climatologists believe that there will be serious and irreversible consequences to world climate, partly due to positive-feedback loops, if the CO_2 concentrations increase significantly beyond 550 ppm.

The profligate burning of fossil fuels will lead to their inevitable extinction, which is not only catastrophic for energy and climate reasons, but also because they are quite useful for making several materials like plastics cheaply. Due to the need for minimizing fossil-fuel CO_2 emissions that impact climate change and due to the exhaustion of these fossil-fuel energy sources, our planet will need to function increasingly on renewable energy sources. These sources include solar power, wind power, hydroelectric power, wave power, tidal power, geothermal power, and biofuels. Since the areal power densities of these sources are relatively small, it is imperative that our power consumption be reduced. Most of our power consumption arises from transportation, heating, electricity usage, and material-synthesis costs.

We discuss how electric cars, powered by batteries driving motors, enable improvements in transport energy efficiency, i.e., energy consumed per person-km, over those of gasoline-powered cars. We shall discuss an equivalent circuit for a car, which will allow us to draw on principles of low-power design in electronics to understand how power consumption in cars can and is being reduced. We shall compare the energy efficiency of advanced electric cars versus cheetahs, the fastest land animals on earth. Even though legged locomotion is significantly less

---

[1] Interestingly, the average national per-capita income of a person in K\$ divided by 4 is a good predictor of that nation's average per-person power consumption in kW.

efficient than wheels on flat terrains, we shall see that animals have impressively good transport energy efficiency when compared with even highly energy-efficient electric cars.

We will focus on two renewable sources that are likely to be very important in our future, namely, solar photovoltaics and biofuels. The basic principles of phototransduction described in Chapter 11 will be useful for understanding how solar photovoltaic cells function. We shall delve deeper into phototransduction in this chapter to understand the limits of solar-cell efficiency. Solar photovoltaic sources are important at small scales, e.g., for solar photovoltaic cells that power portable and biomedical applications, and also at large scales, e.g., for 300 MW electric generators. Solar energy is widely viewed as the most important renewable energy source because of its relatively high power density and ubiquitous presence [5]. We shall discuss some challenges in making solar electricity generation cost effective. We conclude by discussing biofuels, which are created by plants storing the energy of sunlight in chemical bonds through the process of photosynthesis. Biofuels represent an energy-dense method for the storage and distribution of solar energy. Such biofuels could be useful in cars and in implantable biomedical systems in the future.

26.1   Sources of energy

Figure 26.1 shows six common sources of energy that we can harvest. We have discussed RF energy harvesting in near-field systems in Chapter 16 for biomedical implants and in far-field systems for cardiac monitoring in Chapters 17 and 20. In general, ambient RF energy from cell phones and wireless devices in the environment may be harvested. Implantable biomedical systems can potentially harness the energy of blood flow or the energy of airflow during respiration to function; work in this area is just beginning. Ultra-low-power outdoor monitoring applications can exploit potential differences between two points on a tree trunk, which can vary by a few hundreds of mV, to operate [6]. In this chapter, we shall primarily focus on inertial-motion, heat, and solar energy harvesting.

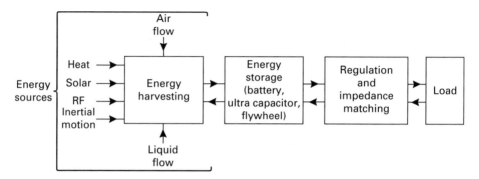

Figure 26.1. A typical energy-harvesting architecture.

At first, we shall only focus on harvesting at small scales for low-power biomedical applications.

Noninvasive and implanted biomedical systems can harvest the energy in the inertial motions of the bodily limbs or the head to which they are attached to function. Small ultra-low-power implanted systems that are attached to the heart or to the lungs can harness the mechanical energy of heart or lung motions to operate. The body is maintained at nearly 37 °C while the environment is usually below this temperature. Therefore, the flow of heat energy from the body towards its surround can be harnessed in a thermoelectric device to operate electronics attached to the body. For example, the micropower EKG or PPG amplifiers discussed in Chapter 20 can be powered in such a fashion. Solar cells attached to the body can power noninvasive electronics attached near them as they now power watches. In general, several energy sources are intermittent, e.g., solar energy is only available during the day, mechanical energy is only available during motions of the body, and RF energy may only be available when there is a wireless device in the environment. The intermittency of the energy implies that there is need for storage of the harvested energy, e.g., in a battery or in a large capacitor as shown in Figure 26.1. The energy-storage system serves to smooth energy fluctuations such that power is always reliably available to the load. To prevent residual power-supply fluctuations output by the energy-storage system from affecting the electronics that it powers, and to ensure that there is good impedance matching for maximum or energy-efficient transfer of power to the load, a regulation and impedance-matching stage is usually necessary. For example, the load may need high voltage and low current while the energy harvester inherently provides low voltage and high current. Thus, a dc-to-dc up-converter from the output of the energy-storage element to the load may be necessary. In the RF antenna-based energy-harvesting system that we discussed in Chapter 17, the energy harvester is an antenna, and the energy-storage and impedance-matching functions are combined in the charge pump and in the capacitors of the pump. In general, to reduce the variability in available energy and to gather more energy, several energy sources can be simultaneously harvested and stored.

Before we begin with our discussion of inertial-motion mechanical-energy harvesting, we shall digress briefly to explain how electrical circuit models of mechanical systems are constructed.

26.2   Electrical circuit models of mechanical systems

The electrical equivalents of Newton's three laws of motion in mechanical systems are as follows:

1. Newton's first law:
   Every body continues in a state of rest or in its state of motion unless it is acted on by a force.

Electrical equivalent:

Every capacitor holds its charge unless it is charged or discharged by an electrical current.

2. Newton's second law:

$$F = mdv/dt \tag{26.1}$$

F is the force, m is the mass, and v is the velocity of the moving or stationary mass.

Electrical equivalent:

$$I = CdV/dt \tag{26.2}$$

I is the current, C is the capacitance, and V is the voltage on the capacitor.

3. Newton's third law:

For every action, there is an equal and opposite reaction.

Electrical equivalent:

In any two-terminal electrical element, whether active or passive, dependent or independent, linear or nonlinear, the current flowing into one terminal on the element is equal to the current flowing out of the other terminal of the element.

In the formulation above, current is analogous to a force, capacitance is analogous to a mass, and voltage is analogous to a velocity. The electrical equivalent of Newton's third law is such that it is automatically satisfied and represented in any circuit. Mutual interactions between two bodies are represented as a floating current between two nodes such that one of the currents through the two-terminal element creates a sink current on the node that it is attached to while its paired current creates a source current on the node that it is attached to. Thus, Newton's third law is nothing more or less than stating that a floating current source between two nodes may always be represented as a grounded sink current at one node and a grounded source current at the other node. The automatic and natural representation of Newton's third law by a circuit makes electrical representations of mechanical systems powerful because one is relieved from the burden of having to constantly keep track of symmetric pushing and pulling between bodies. Furthermore, force balancing is also automatic. Since the voltage on a capacitor stops changing when all the currents flowing towards (or away from) it sum to zero, Kirchhoff's current law is the law of force balance. Vector forces require 3D electrical circuits because the electrical analogies of mechanical systems hold separately for each of the x, y, and z components of force and velocity. For example, Figure 17.1 shows how circuit descriptions of Maxwell's equations conceptually represent vectors.

In the formulation above, capacitance is a mass. If

$$F = k \int vdt$$

$$I = \frac{1}{L} \int Vdt, \tag{26.3}$$

then the reciprocal of an inductance represents a spring stiffness, or equivalently inductance represents a compliance. Mechanical damping is represented by a conductance:

$$F = \eta v$$
$$I = GV$$

(26.4)

Thus, a resonant mechanical mass-spring-damper system acted on by a force is represented by a parallel *LCR* resonator sourced by a current.

Frequently, a dual version of Equations (26.1), (26.2), (26.3), and (26.4) is used to represent mechanical systems by an electrical equivalent: force is represented by a voltage, velocity is represented by a current, mass is represented by an inductance, damping is represented by a resistance, and compliance is represented by a capacitance. In this analogy, a resonant mechanical-spring-damper system acted on by a force is represented by a series *LCR* resonator sourced by a voltage. Both forms are mathematically equivalent. However, one form is often more intuitive than the other and one should always work with a form that is the most intuitive. For example, in purely mechanical systems composed of interacting solids, if the equivalence described by Equations (26.1), (26.2), (26.3), and (26.4) is used, a parallel mechanical geometry maps to a parallel electrical topology, and a series mechanical geometry maps to a series electrical topology; the dual analogy flips parallel mechanical geometries to series electrical topologies and vice versa and is less intuitive. In contrast, in mechanical systems involving fluids, if we represent pressure by voltage and volume velocity by current, parallel fluid geometries map to parallel electrical circuits and series fluid geometries map to series electrical circuits; thus, the dual analogy is more intuitive for fluids. In piezoelectric electromechanical devices, forces cause charge displacements and voltages cause mechanical displacements. Thus, for reasons of symmetry, in piezoelectric devices, it is more natural to represent force by a voltage and velocity by a current.

26.3   Energy harvesting of body motion

Mechanical energy harvesting has been performed with three kinds of devices, namely, electromagnetic, electrostatic, and piezoelectric. An electromagnetic device converts flux changes induced by mechanical motion into an electrical voltage as in hydroelectric generators. If the voltage across a sensing capacitance is fixed, an electrostatic device, e.g., like the MEMS capacitance discussed in Chapter 8, converts capacitance changes due to mechanical displacements into charge changes. Electrostatic devices also convert capacitance changes into voltage changes if the charge on the sensing capacitance is fixed. A force imposed on a piezoelectric device causes mechanical deformation and charge changes within it. The charge changes manifest as a voltage across the piezoelectric device's electrical capacitance. An exhaustive review of energy harvesting with all three kinds of

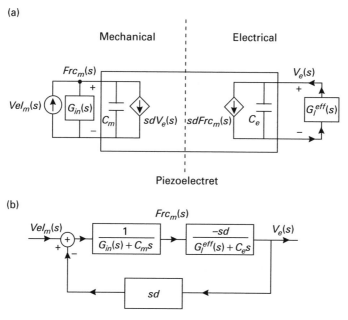

Figure 26.2. A circuit description of a piezoelectret in (a) and a feedback block diagram that represents this circuit in (b).

devices may be found in [7]. Work in [8] has shown that models for electromechanical energy harvesters are mathematically identical across all three classes of devices. Therefore, for reasons of brevity, we shall focus primarily on piezoelectric energy harvesters.

In all such passive devices, the presence of mechanical-to-electrical transduction implies that there is also correspondingly electrical-to-mechanical transduction in the reverse direction. The presence of transduction in both directions, each of which affects the other, leads to a feedback loop in the device. For example, electrical generators or electromagnetic energy harvesters don't just convert mechanical motion to an electric voltage. Their operation causes a 'back torque' in addition to the mechanical torque driving the generator because the electric voltage that is generated also causes the generator to behave like a motor. In electrostatic devices, mechanical displacements lead to voltage or charge changes and also changes in the attractive electrostatic force between the capacitor plates. Piezoelectric devices are no exception. Applied force induces charge motions within the piezoelectric, which manifest as a voltage across their electrical capacitance. Applied voltage induces mechanical displacement changes, which manifest as a 'back force' across their mechanical compliance. Figure 26.2 (a) reveals a two-port electromechanical circuit that represents the functioning of a piezoelectric device. Figure 26.2 (b) reveals the feedback loop that represents this two port.

In Figure 26.2 (a), for convenience, we operate with current, which is the derivative of charge, such that all device characterizations can be done in terms of voltage and current rather than in terms of voltage and charge. The mechanical compliance of the piezoelectret is represented by C_m. The electrical capacitance is represented by C_e. The mechanical force is represented by the voltage $Frc_m(s)$ and the electrical voltage by $V_e(s)$. The piezoelectric forward-and-back coefficients from the mechanical to the electrical domain and vice versa are typically symmetric and represented by d. The inertial input to the piezoelectric device is represented by a Norton velocity source of value $Vel_m(s)$ in parallel with an output admittance $G_{in}(s)$. The output admittance is often due to an inertial mass M with impedance Ms, i.e., $G_{in}(s) = 1/(Ms)$. The output voltage of the piezoelectric device drives an electrical load with effective admittance $G_l^{eff}(s)$. Some algebra maps the two-port of Figure 26.2 (a) to the feedback loop of Figure 26.2 (b). For maximum efficiency, or maximum power transfer, we should configure $G_{in}(s)$ such that it is resonant with $C_m s$, and arrange $G_l^{eff}(s)$ such that it is resonant with $C_e s$. The alert reader will then immediately notice that Figure 26.2 (a) is exactly the dual circuit of the resonant mutual-impedance link that we described in Chapter 16. The similarity of Figure 26.2 (a) to its dual circuit in Figure 16.2 and the similarity of Figure 26.2 (b) to the feedback loop in Figure 16.3 are striking: in mapping Figure 16.2 to its dual version in Figure 26.2 (a), we simply exchange voltage for current, impedance for conductance, and a series two-port circuit for a parallel two-port circuit. The piezoelectric coefficient d is analogous to the mutual inductance M, C_m is analogous to L_1, C_e is analogous to L_2, and we can define a coupling coefficient k given by

$$k^2 = \frac{d^2}{C_m C_e} \tag{26.5}$$

identical to that defined in other treatments [9]. Therefore, we can exploit the analysis discussed in Chapter 16 to analyze piezoelectrets since the mathematics is virtually identical. We can define a reflected admittance $G_{refl}(s)$ analogous to the reflected impedance of Chapter 16 that is given by

$$G_{refl}(s) = -\frac{d^2 s^2}{C_e s + G_l^{eff}(s)} \tag{26.6}$$

This admittance is reflected from the electrical domain to the mechanical domain and appears in parallel with $G_{in}(s)$. From Chapter 16, for maximal energy efficiency, the resonance in the electrical and mechanical domain must both occur at the same optimal $\omega = \omega_n$. For maximal energy efficiency in the 'primary' mechanical domain, the reflected conductance must be much greater than the conductive portion of $G_{in}(s)$. For maximal energy efficiency in the 'secondary' electrical domain, the conductive portion of $G_l^{eff}(s)$ must be much greater than an effective parasitic conductance, G_e, that is in parallel with C_e, and which represents electrical losses. Note that G_e is not shown in Figure 26.2 (a) since it is a parasitic. To determine the maximum overall energy efficiency, we can define an effective quality factor Q_1 for the resonator in the mechanical domain and an unloaded quality factor Q_2 for the resonator in the electrical domain.

Then, Equations (16.35) and (16.36) in Chapter 16 apply exactly to piezoelectric energy harvesters and describe their energy efficiency. From [8], since the mathematics of through-and-across variables is very similar for electromagnetic and electrostatic energy harvesters, we can analyze other motion-energy harvesters through the equations of Chapter 16 as well. Through variables are analogous to generalized current variables while across variables are analogous to generalized voltage variables. Later in this chapter, we shall discuss the operation of electric motors, which are electromagnetic energy generators (harvesters) that operate in reverse. This discussion will further illustrate the similarity between different kinds of electromechanical devices.

In Chapter 16, since we had a required load power consumption in the secondary and we wanted to ensure that the reflected power consumption in the primary due to this load was minimal, we focused on optimizing energy efficiency. In many energy-harvesting situations, energy efficiency may not be as important as maximizing energy transfer, i.e., getting as much absolute energy out of the harvester as possible, even if it means that a large fraction of energy is wasted. For example, in a resistive-divider circuit composed of a voltage source with a source impedance R_S driving a load impedance R_L, energy efficiency is maximized when $R_L \gg R_S$; maximum energy is transferred from the source to the load when $R_L = R_S$, where the energy efficiency is only 50%. In this case, in the terminology of Chapter 16, it can be shown that for maximal power transfer

$$Q_L^{opt} = \frac{Q_2}{1 + k^2 Q_1 Q_2} \tag{26.7}$$

The power dissipated in the electrical load P_e at this optimal value is related to the power dissipated at the mechanical input with no reflected load (d or $k = 0$), P_m, according to

$$\frac{P_e}{P_m} = \frac{1}{4}\left(\frac{k^2 Q_1 Q_2}{k^2 Q_1 Q_2 + 1}\right) \tag{26.8}$$

Piezoelectric harvesters for wireless sensor networks are described in [10] and have generated 180–335 μW in $1\,\text{cm}^3$ of volume. They can be adapted for use in the noninvasive medical-monitoring systems described in Chapter 20. A piezoelectric energy harvester that scavenges energy from compression of the shoe sole has been able to generate 0.8 W of electrical power [11], [12]. Attempts to generate large amounts of electrical power from body motions, however, create a significant reflected electrical load on the mechanical side such that the metabolic effort needed to generate electrical power is consciously felt by the user. Since only 25% of the chemical oxidative energy of glucose is output as useful mechanical work by the body, even a highly efficient energy harvester at 31% can lead to a metabolic load to the body that is 12 times greater than the energy being harvested. One innovative effort to reduce such metabolic loading on the body uses an electromagnetic energy harvester placed on a knee brace slightly above the knee that harvests energy only during leg decelerations. It helps the leg to decelerate by serving as an effective energy-harvester 'brake' to slow the leg at the end of a leg-extension movement. The metabolic load on the muscle is then reduced on average compared with the condition when an energy-harvesting brake is not present [13]. In the proof-of-concept design, the weight of the device, however, increased the mean metabolic load of walking by 20%. The device successfully generated 5 W of power. An electrostatic generator meant to harness ventricular wall motions of the heart generated $36 \, \mu$W with simulated heart motions, sufficient to power a cardiac pacemaker. However, it was too big to implant and test directly on the heart [14]. In general, devices less than $1 \, cm^3$ in volume are unlikely to generate more than 1 mW of power from body motions [7], but for many low-power applications like we have discussed in Chapter 20 $100 \, \mu$W is more than adequate. One challenge in the field is that it is easy to make small devices that have high resonant frequencies but most of the power spectrum of motion energy is below 100 Hz. Furthermore, if the body motion is far in excess of the maximal motions possible in a small device, resonant amplification is not necessarily an advantage. Non-resonant conversion strategies are being investigated [15].

26.4   Energy harvesting of body heat

When heat flows from a hot body at temperature T_{high} to a cold body at temperature T_{low}, some of the heat energy can be harnessed to perform useful work. For example, the internal combustion engine in a car burns gasoline fuel in a controlled fashion, which releases energy primarily as heat. A fraction, i.e. ~25%, of this heat energy is exploited to perform useful mechanical work. A large fraction, i.e., ~75%, of it is wasted as heat from the radiator to the surround. In the body, energy in glucose molecules is first converted to energy in many smaller energy-carrying molecules called adenosine tri-phosphate (ATP) at nearly 50% efficiency within our cells. The ATP molecules serve as universal energy currency throughout the cell and power various activities in the cell that perform useful work. For example, ATP powers electricity generation across all cell membranes in all cells of the body and also powers the contractions of muscle cells. The efficiency of energy conversion from ATP to useful work is nearly 50%. Thus, the overall efficiency from fuel to useful work in the body is also $50\% \times 50\% = 25\%$ as in a gasoline engine. Hence 75% of the energy in the food that we eat is converted to heat energy. This heat energy is used to maintain the body at an internal $37 \, °$C temperature significantly higher than the external temperature, at say $22 \, °$C, and compensates for heat lost from the body to the environment.

With temperature analogous to voltage, and heat flow analogous to current, a circuit for thermoelectric generation is as shown in Figure 26.3 [16]. Each resistance in Figure 26.3 is described by an Ohm's law equation of the form

$$\Delta T_x = I_{heat} R_x \tag{26.9}$$

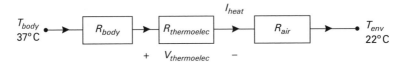

Figure 26.3. A circuit model for body heat loss.

where ΔT_x is the temperature drop across the resistance R_x, and is measured in units of °C. The heat flow, I_{heat}, is measured in units of W/m^2. From Equation (26.9), the thermal resistance is then measured in units of m^{2}K/W. The $R_{thermoelec}$ element is built with a cascade of several BiTe Seebeck-effect thermopiles that each provide about 0.2 mV/°C of output voltage. Such thermopiles are built by bringing two dissimilar metals together at two junctions, one at the hot side and one at the cold side. A series stack of several of these thermopiles is necessary to develop voltages of ~1 V. For example, the recent design described in [16] used $158 \times 8 \times 4 = 5056$ of these devices in series to develop nearly 0.7 V. What determines the values of R_{body}, R_{air}, and $R_{thermoelec}$, and thus the value of I_{heat}?

Since 0.75 of the body's resting power dissipation of 81 W is dissipated over ~2 m^2 surface area, the net average heat flow out of the body may be expected to be ~30 W/m^2 under resting or sleeping conditions. Under normal conditions, where the power dissipation averages to 125 W – 150 W, it has been measured to be ~60 W/m^2 or ~6 mW/cm^2 when the ambient temperature is 28 °C [17]. At steady state, the heat flow out of the body must be matched by the heat that it generates to ensure that the body does not heat up or cool down. Not surprisingly, the value of the body's effective thermal resistance, R_{body}, is altered via blood-vessel dilation and constriction and other feedback mechanisms to ensure that the body's temperature is maintained. There is variance in the heat flow at different positions in the body. For example, the relatively hot blood in the radial artery on the underside of the forearm, where watches are worn, is only separated from the ambient air by a ~7 mm layer of skin without any heat-insulating muscle. Thus, the heat flow in this region of the body is around 100 W/m^2 or 10 mW/cm^2. The thermal resistance of the body, which is ~500 cm^{2}K/W, is reduced in this region to ~100 cm^{2}K/W.

The thermopile resistance, $R_{thermopile}$, is determined by the heat-conduction properties of the thermopile material, the cross-sectional area of the legs that join together to create its junctions, and its length. Larger cross-sectional areas and smaller lengths lead to lower resistances. The dependence of thermopile heat resistance on geometry is similar to that of electrical resistances except that the heat conductivity κ plays the role of the electrical conductivity σ. In Figure 26.3, $R_{thermoelec}$ must have its geometry designed such that it is comparable to $R_{body} + R_{air}$. It is hard to make it significantly larger than this value without making devices too long or cross-sectional areas too thin, since the value of R_{air} is typically quite high. For example, commercially available thermopiles have $R_{thermopile}$ at ~50 cm^{2}K/W while R_{air} is ~1000 cm^{2}K/W. The value of R_{air} is

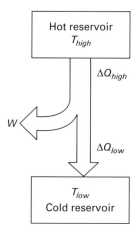

Figure 26.4. A heat engine.

determined by radiative and convective losses in air. The presence of wind lowers R_{air} by promoting heat exchange and increasing heat flow. A value of $R_{thermopile}$ that is significantly smaller than $R_{body} + R_{air}$ will not have much temperature dropped across it, and lead to a loss in sensitivity.

The device described in [16] achieved $\sim$250 μW of power extraction with a heat flow of 20 mW/cm^2 across a 6 cm^2 wristwatch-sized device with an ambient air temperature of 22 °C. The output voltage into a matched load was 0.93 V. Given that we have 120 mW of heat flowing into our device, why are we only able to extract 250 μW? A fundamental limit known as the Carnot efficiency limits the amount of power that can be extracted from a thermoelectric device.

Figure 26.4 shows what is termed as a 'heat engine', i.e., a system that generates useful work W as heat flows from a 'hot reservoir' at temperature T_{high} to a 'cold reservoir' at temperature T_{low}. An amount of heat, ΔQ_{high}, flows out of the hot reservoir, some of it generates useful work W, and the rest, ΔQ_{low}, flows into the cold reservoir. By energy conservation, it is necessary that

$$\Delta Q_{high} = \Delta Q_{low} + W \qquad (26.10)$$

The reservoirs are assumed to have so many degrees of freedom in which to absorb heat energy, i.e., a high heat capacity, such that their temperature barely changes with the modest amount of heat drawn out of or poured into them. But what determines how much of ΔQ_{high} ends up as useful work W, rather than wasted heat ΔQ_{low}?

The fundamental second law of thermodynamics in physics states that the amount of disorder in the world, measured by 'entropy', can only have a net increase or remain the same. It is based on the fact that disordered and highly random system states where energy is distributed equally amongst many degrees of freedom are statistically significantly more likely than ordered system states where energy is concentrated amongst a few degrees of freedom. In fact, temperature is just a measure of the average random thermal energy per degree of freedom in a system. A degree of freedom represents the voltage on a capacitor, the current in an inductor, the position of a spring, or the velocity of a mass in a physical system. (Degrees of freedom are discussed in Chapter 7 in the context of the equipartition theorem.) Heat flows from a hot high-temperature body to a cold low-temperature body because the random thermal energy tries to redistribute itself equally amongst all degrees of freedom in both the hot body and the cold body. Since the hot body has more average thermal energy per degree of freedom than the cold body, there is a net thermal energy flow from the hot body to the cold body as the energy redistribution occurs. When a small amount of heat ΔQ_{low} flows into a heat reservoir of temperature T_{low}, its entropy is defined to increase by $\Delta Q_{low}/T_{low}$ since the number of accessible states in the reservoir increases as more energy is poured into it, leading to more uncertainty about its state, and more disorder. When a small amount of heat ΔQ_{high} flows out of a heat reservoir of temperature T_{high}, its entropy is defined to decrease by $\Delta Q_{high}/T_{high}$ since the number of accessible states in the reservoir decreases as energy leaves it, leading to less uncertainty about its state, and less disorder. Since the net entropy change must be nonzero by the second law

$$-\frac{\Delta Q_{high}}{T_{high}} + \frac{\Delta Q_{low}}{T_{low}} \geq 0 \qquad (26.11)$$

Some algebra on Equations (26.10) and (26.11) then reveals that

$$W \leq \Delta Q_{high}\left(1 - \frac{T_{low}}{T_{high}}\right) \qquad (26.12)$$

Thus, the maximum efficiency of the heat engine, that is the fraction of heat ΔQ_{high} that is converted to useful work W, is limited to a maximum value known as the Carnot efficiency, C,

$$C = 1 - \frac{T_{low}}{T_{high}} \qquad (26.13)$$

The Carnot efficiency sets limits on the efficiencies of steam engines, plane engines, car engines, and on our thermoelectric harvester as well. If the body is at $T_{high} = 37\,°C$, and the ambient temperature is at $T_{low} = 22\,°C$, the maximum possible efficiency for the thermoelectric harvester is given by

$$C_{thrmhrv} = 1 - \frac{273 + 22}{273 + 37}$$
$$= 0.0484 \qquad (26.14)$$

Thus, if 120 mW flows into a thermoelectric energy harvester, the best we can hope to do is extract 5.8 mW of power. It is not atypical for an experimental system to operate at 10% of the limiting possible Carnot efficiency, which is only achievable at infinitely slow operation. The system described in [16] achieves nearly 4% of the Carnot limit but it is one of the best systems reported thus far. Its delivered power density of $\sim 0.2\,\mathrm{W/m}^2$ is in excess of what an average 10%-efficient solar cell might deliver in indoor environments. An efficient charge pump for such thermal harvesters is described in [18].

26.5 Power consumption of the world

Mackay estimates the average power consumption of an affluent British citizen today in his book [1]. If we adapt his units of $1\,\mathrm{kW\,h/day}$ to simple kW units with the conversion factor $1\,\mathrm{kW\,h/day} = 41.67\,\mathrm{W}$, we find that this consumption may be broken down as shown in Table 26.1.

The costs of Table 26.1 are estimated for an affluent British citizen. An average British citizen actually consumes $5.2\,\mathrm{kW}$, an average European citizen consumes $5.46\,\mathrm{kW}$, while an average American citizen consumes nearly $10.4\,\mathrm{kW}$. The world average is $2.5\,\mathrm{kW}$ with great variance across nations. Since there are

Table 26.1 Power consumption of the world

Item	Power consumption	Comment
1. Car usage	1.67 kW	30 mph at 30 mpg for ~ 1 hour at $10\,\mathrm{kW\,h/}$ liter for gas with 3.8 liters $= 1$ gallon. Or equivalently, the cost of an average $42\,\mathrm{kW}$ car driven for ~ 1 hour each day.
2. One transatlantic flight per year on a Boeing 747	1.25 kW	Such planes operate at 0.14 mpg but amortize this cost over ~ 400 passengers such that they effectively operate at $\sim 60\,\mathrm{mpg}$ per person. The power consumption of a Boeing 747 is $\sim 150\,\mathrm{MW}$.
3. Heating	1.540 kW	Not important in some geographical areas.
4. Material synthesis energy costs	2.08 kW	It costs energy to manufacture appliances.
5. Electric lighting	0.167 kW	Estimated for an average home.
6. Electric gadgets	0.208 kW	Washers, dryers, cell phones, etc.
7. Material transport	0.500 kW	Trucking and transportation costs to move materials.
8. Food	0.625 kW	This energy cost in food only tracks industrial energy flows associated with food, not the natural embedded energy in food. For example, it costs energy to transport food, and to maintain animals to be used later as food.
9. Defense	0.167 kW	These national costs are amortized per person.
Total	**8.207 kW**	Does not include the cost of imported goods, which bear their own energy costs, at 1.667 kW.

approximately 6 billion people on our planet today, the power consumption of the world is 15 TW. The electricity consumption of the world is 2 TW. However, since typical generating stations burn fossil fuels like coal to generate electricity and are only 40% efficient, the actual power consumption due to electricity use is 5 TW. We notice that a large fraction of the power consumption of the world revolves around transportation, heating, and electricity costs. This book has already discussed principles for lowering power in electrical systems. Now, we shall discuss some principles for the design of low-power transportation systems of the future.

26.6   A circuit model for car power consumption

Figure 26.5 shows a circuit model of a car that is useful for understanding factors that affect its power consumption. We shall use current to represent force and voltage to represent velocity in accord with Equations (26.1), (26.2), (26.3), and (26.4). Thus, mass is represented by a capacitance, mechanical damping by a conductance, and mechanical compliance by an inductance. A chemomechanical dependent force i_{ENG} due to the burning of fuel along with a Norton-equivalent mechanical admittance G_{ENG} represents the characteristics of the engine power source. The fuel-to-mechanical work efficiency is typically 25% such that the $i_{ENG}v_{ENG}$ power output by the engine requires $4i_{ENG}v_{ENG}$ power to be extracted from the chemical fuel. The motions of the engine are periodic. The engine force is conveyed via gears to provide force to the car wheels. The transformer in Figure 26.5 represents a lossless gearbox (and transmission) that performs an impedance transformation. The reflected admittance of the secondary wheel-and-road side to the primary engine side must be such that most of the power output by the engine is dissipated in the reflected admittance, not in G_{ENG}, which is usually the case. As the characteristics of the impedance in the secondary change with flat, uphill, or downhill road conditions, the gear ratios are changed such that this efficiency is preserved.

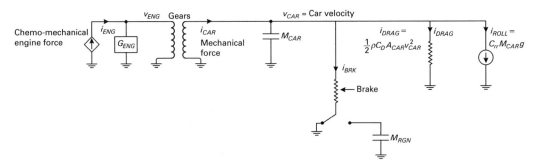

Figure 26.5. Equivalent circuit of a car showing losses due to air drag, rolling friction, braking, and chemical-to-mechanical energy conversion.

The mechanical current i_{CAR} charges the capacitance M_{CAR} to a voltage v_{CAR}, which represents the car's velocity. When M_{CAR} is charged, the car accelerates in accord with Newton's second law. The car has three force currents i_{DRAG}, i_{ROLL}, and i_{BRK} respectively that oppose i_{CAR} and attempt to decelerate the car. When no brake is applied, i_{BRK} is 0. The brake current is shown as a switched resistance in Figure 26.5. When the car is moving at a steady velocity and no brake is applied, i_{CAR} must balance i_{DRAG} and i_{ROLL} such that there is no charging or discharging current on the capacitor and the velocity of the car is maintained at a constant value.

The drag force i_{DRAG} is due to viscous air resistance caused by the fluid moving past the car. It can semi-empirically be represented by the current through a quadratic conductance [19]:

$$i_{DRAG} = \frac{1}{2}\rho C_D A_{CAR} v_{CAR}^2 \qquad (26.15)$$

The parameter A_{CAR} is the effective cross-sectional area of the car, which must be kept small to reduce air drag. That is why most natural creatures and artificial transport mechanisms that move efficiently, e.g., trains, are architected to be long and thin such that A_{CAR} is small within a given volume constraint. We can estimate A_{CAR} as $3\,\mathrm{m}^2$ for a 1.5 m high and 2 m wide car. The parameter C_D is called the coefficient of drag and is typically near 0.3 in most cars and lower in streamlined racing cars. The parameter ρ is the density of air, which is $1.3\,\mathrm{kg/m}^3$. The mass of the car, M_{CAR}, is typically one tonne, i.e., 1000 kg.

The force i_{ROLL} is the force due to 'rolling friction' in the car. It is due to the fact that the tires slightly distort and recover shape when they move, and some of the energy in the tires and tire bearings is dissipated. The force i_{ROLL} is semi-empirically represented by [19], [20]

$$i_{ROLL} = C_{rr}M_{CAR}g \qquad (26.16)$$

with $C_{rr} = 0.013$ for a relatively smooth road and average car tires. The parameter C_{rr} is relatively invariant with speed.

The braking force i_{BRK} dissipates the car's kinetic energy as heat when the brake is applied and the braking resistance is switched to 'ground', i.e., to zero velocity as shown in Figure 26.5. Just as the discharge current in digital CMOS design dissipates the $(1/2)CV^2$ capacitive energy stored in a voltage node as heat, the braking force dissipates the $(1/2)M_{CAR}v_{CAR}^2$ kinetic energy stored in the mass of the car as heat. As in adiabatic CMOS design, discussed in Chapter 21, the switched braking energy can be partially recovered and stored on an ultra-capacitor, flywheel, or battery and then used to provide energy back to the car when it is time to accelerate. Hybrid cars recover 50% of the switching energy through such 'regenerative braking' strategies. Equation (21.50) in Chapter 21 showed that, in a high-quality-factor system, the switching energy can be reduced by as much as $2\pi/Q$ where Q is the quality factor of the system. In Figure 26.5, we have represented the regenerative storage abstractly by a mass M_{RGN} although any form of storage may be used.

Lots of braking corresponds to a high activity factor or lots of switching in electronic systems. A high maximum velocity of the car corresponds to a large V_{DD} in electronic systems. The i_{DRAG} and i_{ROLL} forces correspond to static 'leakage' currents in electronic systems. One leak current increases quadratically with voltage (the drag current) and one leak current is a constant voltage-independent current like subthreshold leakage current in digital systems (the roll current). We can therefore draw upon principles learned in low-power digital design to reduce static energy and dynamic switching energy for a given distance of transport. But what is a good metric for energy-efficient transportation?

As we discussed in Chapter 21, the energy per cycle of operation, $E_{TOT} = P_{TOT}(1/f_{clk})$, is used to characterize the energy efficiency of digital systems. Transport is rarely periodic such that 'a cycle of operation' does not make sense in transportation systems. However, the analogy to speed of operation, f_{clk}, is the average velocity of travel, v_{av}. Thus, the average power consumption divided by the average velocity of travel might be a good metric. If the total time of travel is denoted by t_{trv},

$$I_{mtrc} = \frac{P_{av}}{v_{av}} = \frac{P_{av}t_{trv}}{v_{av}t_{trv}} = \frac{E_{TOT}}{d_{TOT}} \tag{26.17}$$

Thus, our metric inspired from electronics is mathematically equivalent to the metric that is actually used to characterize energy efficiency by transportation engineers: the energy consumed, E_{TOT}, divided by the total distance traveled, d_{TOT}. It is pleasing that metrics used in completely different fields are intuitively similar.

Suppose the average distance between braking stops is d_{stp} and that we can recover a fraction α_{rgn} of the kinetic energy wasted during braking. Also suppose that the time spent to accelerate to cruising velocity V_{MX} or to brake to 0 is negligible compared with the time spent traveling at this cruising velocity, i.e., we have a square-wave-like profile in our velocity. Then, from Equations (26.15), (26.16), (26.17), and Figure 26.5, the metric for transport energy efficiency is given by

$$I_{mtrc} = \left(\frac{(1 - \alpha_{rgn})M_{CAR}V_{MX}^2}{2d_{stp}} + \frac{1}{2}\rho C_D A_{CAR} V_{MX}^2 + C_{rr}M_{CAR}g \right) \tag{26.18}$$

In Equation (26.18), I_{mtrc} is a current, which implies that we are energy efficient for a given distance of travel if we can minimize the average force or 'thrust' required during this travel. Equation (26.18) also reveals that, to be energy efficient, d_{stp} and α_{rgn} should be high such that we do not brake often and that we recover most of the energy when we do, respectively; that M_{CAR} should be low to minimize rolling friction and energy dissipated during braking; that the frontal area of the car, A_{CAR}, and the drag coefficient, C_D, should be minimized by having the car created in a tear drop shape; and that V_{MX} should be low to minimize drag and braking forces. If the car were an electrical system, we would state that, for low-power operation, the activity factor should be small and recycling efficiency should be near 1 such that switching energy is minimized; that capacitances should

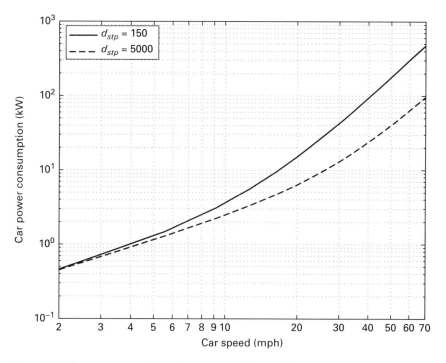

Figure 26.6. Power consumption of a car versus its speed.

be small; that leakage currents should be optimized to be small; and that the power-supply voltage, V_{DD}, of the system should be small.

Equation (26.18) reflects the energy efficiency of the secondary wheel side. If we reflect this back to the primary side, the actual energy efficiency of transport, which is related to the fuel consumed, is given by

$$I_{mtrc}^{act} = 4 \times \frac{1}{e(V_{MX})} \times \left(\frac{(1 - \alpha_{rgn})M_{CAR}V_{MX}^2}{2d_{stp}} + \frac{1}{2}\rho C_D A_{CAR} V_{MX}^2 + C_{rr} M_{CAR} g \right), \quad (26.19)$$

where the factor of 4 arises from the 25% fuel-to-mechanical-work efficiency, and the $e(V_{MX})$ function represents the fact that the efficiency of the engine changes with speed V_{MX}. It is usually maximum at an optimal value of V_{MX} due to the characteristics of G_{ENG} and the nature of the fuel-to-mechanical-energy transfer. The power consumption of a car on the secondary side is just V_{MX} times I_{mtrc} since I_{mtrc} represents the force thrust of the car. Reflecting this power consumption back to the secondary side, and assuming that $e(V_{MX})$ is 1, for simplicity, we can compute the power consumption of the car to be

$$P_{CAR} = 4\left(\frac{(1 - \alpha_{rgn})M_{CAR}V_{MX}^3}{2d_{stp}} + \frac{1}{2}\rho C_D A_{CAR} V_{MX}^3 + C_{rr} M_{CAR} g V_{MX} \right) \quad (26.20)$$

Figure 26.6 plots the power consumption, P_{CAR}, versus its velocity V_{MX} for $\alpha_{rgn} = 0$, $M_{CAR} = 1000$ kg, $\rho = 1.3$ kg/m^3, $C_D = 0.3$, $A_{CAR} = 3$ m^2, $C_{rr} = 0.013$, and $g = 9.8 \, \text{m/s}^2$ for two different values of $d_{stp} = 150 \, \text{m}$ (city driving) and $d_{stp} = 5000 \, \text{m}$ (highway driving). The transition of a car from a linear V_{MX} power consumption to a cubic V_{MX}^3 power consumption occurs when drag and/or braking forces eventually overwhelm the forces of rolling friction. While Equation (26.20) is only an approximation since $e(V_{MX})$ has simply been set to 1, and it ignores several other details of a complex system like a car, it does approximate actual car power consumptions, especially if V_{MX} is not too small. For example, the power consumption of the car at 30 mph with the parameters that we have chosen is nearly 40 kW. At lower speeds, the higher curve with a smaller d_{stp} more accurately reflects car power consumption. At higher speeds, the lower curve with a larger d_{stp} more accurately reflects actual car power consumption. A real car's power consumption will transition between curves like those in Figure 26.6 as d_{stp} increases with speed.

The transport energy efficiency of a gasoline car given by Equation (26.19) for the parameters used in Figure 26.6 at $d_{tsp} = 150 \, \text{m}$ at 30 mph is nearly 3000 N. The 'N' stands for newtons, a unit of force. The car consumes 3000 J of energy to go 1 m or, equivalently, the car burns $\sim$40 kW of power to go 13 m/s. The transport energy efficiency of a car is usually optimal at a moderate speed where $e(V_{MX})$ is high in Equation (26.19) and the V_{MX} drag and braking terms are not too large. The transport energy efficiency that we have computed is indeed near to that observed in real cars.

Part of the staggering transport inefficiency of a car arises from the fact that it has to transport its own weight (1000 kg), which is significantly in excess of the weight of its cargo ($\sim$65 kg for one person). The presence of just four persons in the car improves its transport efficiency per person by a factor of almost four. Trains take this idea to an extreme and also incorporate other ideas making them highly energy efficient. Trains are significantly more energy efficient per person than cars for six reasons. First, since they rarely stop and brake and have a dedicated track to run on, the braking term in Equation (26.18) is nearly 0, except when the train stops at the end of its journey, so we can set it to 0. Second, they employ a 'collective' or 'parallel' low-power principle (see Chapters 21 and 22) to amortize transportation costs per person to a small value: if N persons occupy a train, the effective transport energy efficiency *per person* is given from Equation (26.18) by

$$I_{mtrc}^{perprsn} = \left(\frac{1}{2} \rho C_D \frac{A_{TRN}}{N} V_{MX}^2 + C_{rr} \left(\frac{M_{TRN} + N M_{prsn}}{N} \right) g \right)$$

$$(26.21)$$

$$I_{mtrc}^{perprsn} = \left(\frac{1}{2} \rho C_D \frac{A_{TRN}}{N} V_{MX}^2 + C_{rr} \left(\frac{M_{TRN}}{N} + M_{prsn} \right) g \right)$$

Hence, the fuller a train is, the lower its drag and rolling-friction terms per person, and the more efficient it becomes. Third, their long and lean design with relatively low frontal area reduces drag. Fourth, the coefficient of rolling friction, C_{rr}, for steel on steel is 0.002 rather than 0.013 for a car. Fifth, they can run at an optimal speed where their engine has optimal efficiency. Sixth, they can be run on electricity and thus have engine efficiencies of over 80%. The overall result is that trains can operate at high speeds with a transport energy efficiency of 58 N per person, about 50 times less than that of a car. In practice, trains are rarely full in developed countries. Nevertheless, in Japan, rail transport operates at 216 N-per-person efficiency, i.e., it has a 14 times better transport efficiency than that of an average car today.

26.7   Electric cars versus gasoline cars

Given the high energy efficiency of electric engines in trains, can we build purely electric cars that are more efficient than gasoline cars? Purely electric cars function by using a battery to power a motor, which, after gearing, provides torque to turn the wheels of the car. The battery takes the place of the fuel as the energy source, and the motor takes the place of the car engine. The primary side of Figure 26.5 in a purely electrical car is different from that in a gasoline car. The secondary side of Figure 26.5 is identical in a purely electrical car to that in a gasoline car. From now on, for brevity, we shall refer to purely electrical cars as electric cars. Before we compare the efficiency of electric cars versus gasoline cars, it is useful to understand how a motor works.

Figure 26.7 (a) reveals the equivalent electromechanical circuit of a motor and Figure 26.7 (b) reveals the feedback loop that describes Figure 26.7 (a). In Figure 26.7 (a), $V_{in}(s)$ represents the input voltage source to the motor with source impedance $Z_{in}(s)$. The motor has an electrical impedance $Z_e(s)$ and a 'back emf', $K\Omega_m(s)$, which is proportional to its angular velocity $\Omega_m(s)$. The back emf arises, because, just as in a piezoelectret, mechanical-to-electrical transduction occurs simultaneously with electrical-to-mechanical transduction. Equivalently, all motors generate a back emf because they are also electrical generators. The net current in the motor, $I_{in}(s)$, generates a torque, $\Gamma(s)$, which drives the mechanical admittance of the motor, $G_m(s)$, and the effective mechanical admittance of the load, $G_l^{eff}(s)$, to create the motor's angular velocity, $\Omega_m(s)$. The admittance of the motor, $G_m(s)$, is typically inertial/capacitive with a little loss. To find $G_l^{eff}(s)$, the effective load on the motor, we need to model an electric car. We can use the model shown in Figure 26.5 to model an electric car. The load $G_l^{eff}(s)$ is the reflected load of the secondary side of Figure 26.5, which appears, after gearing, as a load to the motor on the primary side. If the torque and angular velocity of the motor are represented by τ and ω, respectively, and if the transformer, which represents the gears, is lossless,

$$\tau_m\omega_m = i_{CAR}v_{CAR} \tag{26.22}$$

As an aside, note that, if R is the wheel radius on the secondary side of Figure 26.5, we can also choose to parametrize the mechanical output variables of the car by the torque on the wheel and the angular velocity of the wheel, i.e.,

$$\tau_{CAR} = i_{CAR}R$$
$$\omega_{CAR} = v_{CAR}/R \tag{26.23}$$

(a)

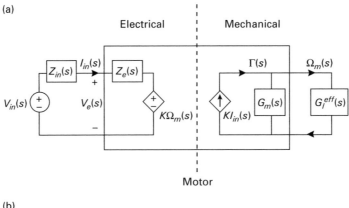

(b)

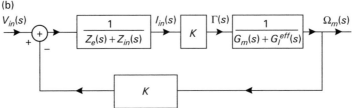

Figure 26.7. Equivalent circuit of an electric motor in (a) and a feedback equivalent in (b).

In Figure 26.7 (a), $KI_{in}(s)$ is analogous to $I_{ENG}(s)$ in Figure 26.5 and similarly $G_m(s)$ is analogous to $G_{ENG}(s)$. Thus, the overall circuit of an electric car is formed by having the circuit of Figure 26.7 (a) replace the primary side of Figure 26.5 *and* with $G_l^{eff}(s)$ replaced by the gears and secondary side of Figure 26.5. Hence, there are two 2-port circuits in an electric car, which are cascaded with one another: The first occurs due to the electromechanical motor two-port circuit of Figure 26.7 (a) and the second occurs due to the simple gearing transformer of Figure 26.5.

The mechanical load of the motor is reflected to the electrical side in Figure 26.7 (a) as an equivalent electrical impedance of value

$$Z_{refl}^{mtr}(s) = \frac{K^2}{G_l^{eff}(s) + G_m(s)} \tag{26.24}$$

which appears in series with $Z_{in}(s)$ and $Z_e(s)$ in Figure 26.7 (a), i.e., the reflected impedance $Z_{refl}^{mtr}(s)$ replaces the $K\Omega_m(s)$ dependent generator in Figure 26.7 (a). A physical interpretation of Equation (26.24) leads to the realization that admittance on the secondary side in Figure 26.5 simply transforms to a scaled identical electrical admittance $G_{refl}^{mtr}(s) = 1/Z_{refl}^{mtr}(s)$ that is an exact mimic of the mechanical admittance. If g is the gear ratio, greater than 1, that represents the up-conversion of force from the primary to the secondary side in Figure 26.5, and $G_{CAR}(s) = I_{CAR}(s)/V_{CAR}(s)$ is the car admittance, then, from Equations (26.24) and Figures 26.5 and 26.7(a), we get

$$G_{refl}^{mtr}(s) = \frac{G_m(s)}{K^2} + \frac{1}{g^2 K^2} G_{CAR}(s) \tag{26.25}$$

Note that Equation (26.25) is only a small-signal frequency-domain characterization of the reflected impedance of the car. However, the feedback loop of Figure 26.7 (b), suggests that even if the $\Gamma \to \Omega$ or equivalently $i_{CAR} \to v_{CAR}$ transformation is nonlinear due to the nonlinear load characteristics of the car (quadratic drag conductance and fixed rolling-resistance current source in Figure 26.5), we can still represent this relationship as a nonlinear block in an equivalent time-domain feedback loop version of Figure 26.7 (b). We can then reflect this nonlinear block into the electrical domain of the motor by replacing the back emf by a nonlinear $I\text{–}V$ block that characterizes the car. That is, in Equation (26.25), we simply use the scaling constants $1/K^2$ and $1/(g^2K^2)$ on the $I\text{–}V$ curves that characterize the motor admittance g_m or the car admittance g_{CAR} in the time domain respectively, and reflect these summed-and-scaled $I\text{–}V$ curves into the electrical domain as a nonlinear $I\text{–}V$ element. A small-signal frequency analysis about each large-signal operating point as in transistor circuits, however, is still useful for providing intuition.

The factors that affect the efficiency of electric cars are the same ones that affect the efficiency of mutual-inductance links that we discussed in Chapter 16. Most of the torque current of the motor in Figure 26.7 (a) should flow through $G_l^{eff}(s)$ rather than through $G_m(s)$ to preserve good efficiency in the mechanical output domain. The motor mass is significantly below the car's mass and the motor damping losses are usually significantly less than the drag and other car losses. Thus, good efficiency in the mechanical domain can be achieved with modest gear ratios in Equation (26.25). Most of the voltage drop of the motor's input voltage should be dropped across the reflected impedance $Z_{refl}^{mtr}(s)$ rather than across $Z_{in}(s)$ or $Z_e(s)$ to preserve good efficiency in the electrical input domain, that is $G_{refl}^{mtr}(s)$ should be sufficiently small. Thick wiring in the motor windings reduces electrical losses R_e, large batteries have low output source impedance, and typically the motor's electrical inductance L_e is such that the electrical time constant of the motor, L_e/R_e, is much less than the mechanical time constant caused by the reflected impedance. Hence, if K, which is primarily determined by the amount of flux in the motor, is sufficiently large, $G_{refl}^{mtr}(s)$ can be made small enough in Equation (26.25) such that efficiency in the electrical domain is excellent. The overall efficiency of the motor is the product of the efficiency in the electrical domain times the efficiency in the mechanical domain. Motors can operate with excellent energy efficiency. Indeed, the Tesla Roadster electric car has achieved efficiencies that average 92%, which are significantly higher than the 25% efficiency of a gasoline car engine [21].

The Tesla Roadster electric car is an impressive engineering feat since it achieves the specifications of a high-performance sports car with excellent transport energy efficiency and zero emissions. Its source of power is a 450 kg lithium-ion battery with 53 kW h capacity (424 J/g) capable of 200 kW output. The car itself weighs 1222 kg, has a peak mechanical power output of 189 kW, accelerates from 0 to 60 mph in 3.9 seconds, has a top speed of 125 mph, and can go 244 miles on a single battery charge. Its battery lifetime is limited to about ~ 500 charge-recharge cycles ($\sim$ 100,000 miles), and it takes 3.5 hours for a full battery recharge although a full recharge may rarely be necessary. It implements regenerative braking. The lithium-ion battery has several short-circuit protection features including in-built fuses that disconnect it in situations of high temperature and pressure. The battery is architected to be safe even during collisions.

Most importantly, the Tesla Roadster's transport energy efficiency is $\sim$500 N, about 6 times better than that of an average car. The transport efficiencies of several lighter, lower-range, and low-speed electric cars are not significantly different from that of the Roadster and some are much worse [1]. To be fair to the average gasoline car, though, the Tesla Roadster uses high-grade electric energy, while the average car needs to extract its energy from fossil fuel. Most electricity generating fossil-fuel plants are 30%–40% efficient such that one could argue that the real improvement of a Roadster is a factor of $2\times$ to $2.4\times$. Nevertheless, the Tesla Roadster does illustrate that direct conversion between high-grade forms of energy, e.g., electrical to mechanical rather than from chemical-to-heat-to-mechanical as in a gasoline car, is efficient. In the future, if electricity is generated in solar plants, such a car could indeed have a zero-emission footprint, especially if it is manufactured in plants using solar electricity as well. Even though the energy density of the lithium-ion battery that was used is 11 times less than that of gasoline, the weight of the car is manageable because the heavy gasoline engine is replaced with an electric motor.

26.8   Cars versus animals

Another impressive example in transportation engineering is the cheetah (*Acinonyx jubatus*), the fastest land animal. Its top sprint speed has been measured to be 30 m/s, i.e., 68 mph [22]. It can accelerate to 68 mph in 3 seconds, faster than a Tesla Roadster, which gets to 60 mph in 3.9 seconds, and faster than most high-performance cars [23], [24]. Since the average cheetah weighs nearly 50 kg, we can estimate that its mechanical power output during this acceleration is $(1/2)50(30^2)/3 = 7.5$ kW. Its rudder-like tail enables it to make incredibly quick turns during its chase of a prey animal. The cheetah uses its spring-like backbone to partly store and regenerate energy in each stride and is airborne for more than half its stride. Its transportation efficiency for aerobic speeds, which can typically be maintained for long distances only if they are less than half the top speed [25], has been measured to be equivalent to 0.14 ml of oxygen consumption per g.km [26]. From the energetics of a glucose or carbohydrate reaction, and from the weight of an average cheetah, these numbers work out to an energy efficiency of 132 N. The cheetah's transport energy efficiency is 4 times better than that of a highly energy-efficient electric car. What is more impressive is that this transport energy efficiency is achieved even though the cheetah has to make do with a 25% efficient engine (fuel-to-mechanical-work) and that it runs with legs, not as optimal as wheels on flat terrain. For example, the energy efficiency of humans walking at

2.5 mph can be calculated from measurements in [27] to be close to that of a cheetah, 130 N, but humans riding average unoptimized bicycles with wheels can achieve 81 N even though the drag coefficient of such bicycles is extremely poor (0.9 vs. 0.3 in a car).[2] Gazelles and goats have a transport energy efficiency similar to that of a cheetah at aerobic speeds [26]. Thus, many animals including the cheetah achieve a 4 times improvement in transport efficiency over a highly energy-efficient electric car, while operating with at least a $(0.94/0.25) \times (130/81) = 6$ times disadvantage compared to it (inefficient engine, no wheels). Pronghorn antelopes, the prey of cheetahs, have top speeds that are nearly as fast as cheetahs since they need to outrun the cheetah to live [29]. However, unlike sprinting cheetahs, they can maintain a 45 mph speed (20 m/s) for long sustained periods of time with an energy efficiency comparable to cheetahs, to which they are similar in weight. The power consumption of an antelope running at 30 mph is 1.9 kW (130 N $\times$ 13 m/s), about 20 times less than an average gasoline car at the same speed.

Airplanes and birds both operate near the fundamental limits set by the laws of aerodynamics and fly at an optimal speed needed to support their weight and minimize air drag [30]. A bar-tailed godwit bird flying nonstop from Alaska to New Zealand (7,008 miles) at 36 mph on its stored-fat fuel [31] has nearly the same range as the maximal range of a Boeing 747–300 (7,440 miles) flying at 555 mph on its stored jet fuel. The bar-tailed godwit takes 8.1 days for its flight over the central Pacific Ocean and maintains a 9 times increase in its basal metabolic rate. Somehow, it endures sleep deprivation and potential dehydration to complete its heroic journey [31]. There is significantly more room for improvement in land-transport energy efficiency than in air-transport energy efficiency [1]. However, new designs for micro-air vehicles are exploring the use of flapping and flexible wings that are aerodynamically important in small birds but less important for large airplanes and in big birds that glide [32].

Could animal transport inspire the design of cars? It is possible that it will, but it will require deep and insightful knowledge of *both* animal transport and car engineering to pluck this high-hanging fruit. Clearly, many of the constraints and goals of human transport are different from those of animal transport and we can't incorporate our energy for high-speed transport within our own bodies like fast animals do. Therefore, it is likely that insights and principles will be useful, not details, just as has occurred in bird-versus-airplane design since the days of the Wright brothers. However, paradoxically, to get these insights, one will need to understand a lot of details in both fields. This principle is true in all of bio-inspired design. For example, in the RF-cochlea section in Chapter 23, we described how the architecture of a biological cochlea inspired the design of an efficient RF

---

[2] In contrast, on recumbent bicycles where the supine rider is enclosed in a carbon-fiber-Kevlar-composite shell, the drag coefficient of the overall highly streamlined bicycle can be quite low. Consequently, Sam Whittingham has achieved a top speed > 82 mph on such a bicycle and an energy efficiency near ~41 N [28]. P. Grogan. Sam Whittingham tops 80 mph – on a push-bike. *The Sunday Times.* September 20, 2009.

spectrum analyzer for advanced radio applications. This example required a good knowledge of both cochlear models and of traditional RF design, failing which the algorithmic insight of a cochlear model would have been missed or the experimental performance of the RF design would have been poor.

26.9   Principles of low-power design in transportation

Why do animals have excellent transportation efficiency? One key is that they are *light*, the analog of having small capacitance to reduce power in electronic design. A car weighs 1000 kg but transports a 65 kg human. If four people use a car, the car's per-person transport energy efficiency improves by a factor of almost 4. So, we are observing the classic flexibility-efficiency tradeoff of low-power design. If the car is to be flexible enough to handle situations involving transport of up to 4 persons, its efficiency for transporting a single person is degraded. Degrees of freedom needed to maintain flexibility hurt energy efficiency as in electronic systems. In fact, animals pay a price for flexibility as well. In order to have a universal energy currency molecule, ATP, available to power various activities flexibly, they have to suffer the inefficiency of two energy-conversion steps, one from food to ATP, and another from ATP to useful work, rather than one direct step that converts from fuel to electricity, as some bacteria accomplish. All energy-efficient transport vehicles from trains to cheetahs have encoded in the shape of their bodies a long-and-lean structure that minimizes drag. Trains exploit parallelism to achieve better transport efficiency than cheetahs when full (see Equation (26.21)). Animals recycle spring energy in their tendons to improve their transport efficiency just as cars with regenerative braking or adiabatic digital circuits do [29]. The car is also subject to a robustness-efficiency tradeoff as in other low-power systems: heavier cars do better in accidents but are highly energy inefficient. So, there are clear connections between the principles of low-power transportation and the principles of low-power electronic design which we discussed in Chapters 21 and 22. But what is the connection between information and energy in transportation, a connection that led to several power-saving principles in prior chapters?

The physical variables that change state when we move are our position and velocity. Transportation may be described as an information-processing problem where our state needs to change from x_0, our initial position, to x_f, our final desired position. To change this state variable from x_0 to x_f, we need to change the state of another of our variables, our velocity, such that its integral over time is $x_f - x_0$. And, as is true for all systems, it costs energy to maintain state (maintain velocity in spite of drag and rolling friction) and to transform state (the costs of increasing car kinetic energy). We exploit the technology of transport, which can be designed in various topologies (gasoline cars, electric cars, trains), to solve this task. Higher average speeds at which the task is solved lead to higher power consumption. The feedback loop implemented by the visual sensing system of the driver and his control strategy of the actuating system, i.e., the car, ensures that the car stops and starts at needed positions along the way, with what is usually adequate precision. Thus, the task, technology, topology, speed, and precision costs of a low-power system illustrated in the low-power hand of Figure 1.1 also apply to cars. The car already implements one good principle of low-power design through its use of a feedback loop, i.e., it separates the costs of speed and precision by having an accurate sensor (the driver's eyes) and control system (the driver's brain) determine the precision of transport while the actuator determines its speed. The mutual information that is of relevance in a transportation task is that between a desired smooth, relatively fast transport trajectory in the head of the driver and the actual transport trajectory that is achieved. In the future, a trajectory that weighs the costs of carbon emissions will also be important.[3]

One principle of low-power design that cars can exploit in the future lies in improving the balance between computation costs and communication costs. Cars can wirelessly communicate with traffic lights and with each other such that the transportation of several drivers is more optimal, therefore saving energy. For example, traffic lights could automatically adapt their timing within a reasonable range such that the directions and locations of high flux have lower waiting times than the directions and locations of lower flux. Traffic lights can be coordinated and synchronized like interacting phase-locked loops that receive correction inputs based on traffic flux counts. Traffic lights could also adapt to patterns that are automatically recognized as being due to an accident scenario. The power costs of wireless transmission for relatively short ranges is extremely cheap, especially when compared with the phenomenal power costs of transportation (1 to 10 W versus tens of kW). Furthermore, energy harvesting from LEDs in traffic lights or RF transmissions from incoming cars can provide constant recharging boosts to such systems such that they can be self-powered (see Chapter 17 for a discussion of far-field wireless recharging systems). Car-to-car hopping can be used for longer-range communication, which is significantly more power efficient per unit distance than a non-hopping strategy ($N(R/N)^2 > R^2$ for $N > 1$ in an N-hop network). Needless to say, the benefits of such sensor-network schemes will have to be weighed against their costs and ease of implementation within an existing infrastructure. Adaptive traffic lights and car-to-car communication are being researched [33].

We shall now shift gears from discussing how to minimize power consumption to discussing how to generate power. We shall begin with what is likely to be the most important source of the power in our future, solar electricity.

---

[3] Accidents result because of conflicting control algorithms in the heads of different drivers, the imprecision and/or slow reaction times of a drunk-driver's control algorithm, or the disobedience of traffic rules. Thus, driving precision is strongly determined by feedback loops. The power costs of precision are largely borne by the driver and are relatively small. From [27], D. J. Morton and D. D. Fuller. *Human Locomotion and Body Form* (New York, NY: Waverly Press, Inc., 1952), they are estimated to be an additional 83 W over the basal 81 W metabolic rate of the driver.

26.10   Solar electricity generation

David Goodstein and others have pointed out that the only renewable energy source capable of solely powering our planet at its expected and future power consumption without an incredible use of land area is the sun [5].[4] The sun transmits $1366\,\mathrm{W/m^2}$ of power to the Earth when its rays are orthogonal to a location on it [34]. However, 30% of this radiation is reflected back into space, partly by clouds, and about 19% is absorbed directly by the atmosphere [35]. Attenuation factors that vary as the angle of the incident radiation varies throughout the day, the variation in latitude of various places on Earth, the variation in cloud cover in different regions, the complete absence of the sun at night, and the variation in sunshine with seasons cause the power density of the sun to fluctuate over spatial location and over time. The power density of the sun at a particular region on Earth, termed its *insolation,* is often integrated over the span of a day and expressed in units of $\mathrm{kW\,h/m^2}$ per day. Multiplication of the number quoted in $\mathrm{kW\,h/m^2}$ per day by $1000/24$ yields the average daily insolation in units of $\mathrm{W/m^2}$. The average annual insolation can range from $100\,\mathrm{W/m^2}$ in Helsinki, Finland, to $320\,\mathrm{W/m^2}$ in Inyokern, California, USA. Not surprisingly, there is more variability across seasons at extreme latitudes than at equatorial latitudes. To find the insolation at any latitude and longitude on earth, or at the location where you live, visit [36]. Insolation for various major US cities is available at [37].

Solar *photovoltaic* cells or *photovoltaics* that convert solar energy to electricity have efficiency limits that are determined by laws of physics that govern the interaction of light with matter. We shall discuss some of these laws. Losses due to shadowing, due to reflection at the cell surface, due to a loss in light collection area (a loss in 'fill factor' as in the imagers described in Chapter 19) can further reduce efficiency. Low-efficiency cells are typically cheap to manufacture while high-efficiency cells are typically expensive to manufacture. The efficiency of solar cells can range from 2% to 40%. Most commercial systems that can be mass manufactured are in the 10%–20% range today. We shall first discuss how solar photovoltaics function and then discuss fundamental limits on their energy efficiency.

In Chapter 11, we discussed how light creates electrons in pn junctions, and how we could exploit such phototransduction to create a photoreceptor. In Chapter 19, we discussed how to build low-power imagers using pn junctions. The basic principles of phototransduction discussed for these applications also apply to solar cells. Solar radiation is largely composed of energy in the visible light and near-infrared regions. Figure 11.2 (a) reveals a pn junction formed by abutting a semiconductor of n-type material with a semiconductor of p-type material. Figure 11.2 (b) reveals the energy diagram that describes such junctions with E_C and E_V representing the minimal and maximal energy of the conduction band and valence band respectively. The depletion region created by the equilibration of

---

[4] Nuclear energy from fission sources will eventually also be non-renewable and there has not yet been a breakthrough in nuclear fusion, which could potentially be renewable.

drift and diffusion currents at the border between the n-type and p-type regions causes the bending of the band energies. The electric field in the depletion region is such that positive depletion charge in the n-type region, created when electrons diffuse away from the n-type region, raises the potential of the region or equivalently lowers the electron energy in the n-type region. Similarly, negative depletion charge in the p-type region, created when holes diffuse away from the p-type region, lowers the potential of the region or equivalently raises the electron energy in the p-type region. In a junction with zero voltage across it, at thermal equilibrium, the Fermi level E_F is the same at all spatial locations such that the average energy of an electron is the same at all spatial locations. If a photon with energy hv greater than the band-gap energy $E_G = E_C - E_V$ is absorbed by the junction, the energy can be used to promote an electron from a low-energy state in the valence band to a higher energy state in the conduction band. The absorption of energy creates a hole in the valence band in the energy state that the electron has come from and an electron in the conduction band in the energy state that the electron has gone to.

The created electrons travel 'home' to their native majority-carrier n-type region and the created holes travel home to their native majority-carrier p-type region because these regions represent attractive regions of low energy. Figure 11.2 (b) shows that electrons created in an n-type region do not travel since they are already in a region of low energy. Similarly, holes created in a p-type region do not travel since they are already in a region of low energy. In any case, the absorption of the photon results in the net arrival of an electron in the n-type region and the net arrival of a hole in the p-type region. The arrival of electrons in the n-type region and the arrival of holes in the p-type region effectively results in a floating current source across the junction. Figure 11.2 (c) in Chapter 11 shows that light effectively shifts the I–V curve of the junction due to the presence of the floating current within it. If the junction is in a short-circuit configuration, the current will appear as an enhanced reverse-bias current. If the junction is in an open-circuit mode, the floating current forward-biases the junction to a voltage such that the forward-bias current balances the light-dependent reverse-bias current. Consequently, the open-circuit voltage is at a steady-state value. In Figure 11.2 (c), the short-circuit current corresponds to the value at $V = 0$ while the open-circuit voltage corresponds to the value at $I = 0$.

Suppose the junction can be characterized by the usual equation for a pn junction,

$$I = I_S(e^{qV/kT} - 1) \tag{26.26}$$

where I is the forward-bias current through the junction and V is the forward-bias voltage. The parameter I_S is determined by minority-carrier concentrations, the area of cross-section, and carrier diffusion lengths within the junction [38]. Suppose initially that all photons are of frequency v and have an energy hv. Then, if the probability that a photon of energy, hv, creates an electron hole pair is $\alpha(v)$ and there are I_l photons per second arriving over the collection area of the junction, the current in the junction including the reverse-bias photocurrent is given by

$$I = I_S(e^{qV/kT} - 1) - \alpha(v)qI_l \tag{26.27}$$

The short-circuit current is obtained by setting $V = 0$ in Equation (26.27). Thus,

$$I_{sc} = -\alpha(v)qI_l \tag{26.28}$$

The open-circuit voltage, V_{oc}, of the junction is given by setting $I = 0$ in Equation (26.27). We find that

$$V_{oc} = \frac{kT}{q}\ln\left(\frac{\alpha(v)qI_l}{I_S} + 1\right) \tag{26.29}$$

The incoming radiation has a power of hvI_l while the power of the solar cell cannot be greater than $V_{oc}I_{sc}$. Therefore, the ratio $(V_{oc}I_{sc})/(hvI_l)$ establishes a crude upper bound on the solar-cell efficiency. From Figure 11.2 (c), since I_{sc} and V_{oc} cannot simultaneously be maximal, the actual power output of the solar cell IV is maximized when $I < I_{sc}$ and $V < V_{oc}$. For a given I_l set by solar insolation, it is clear that the efficiency of the solar cells is maximized when $\alpha(v)$ is maximum, which improves both the open-circuit voltage and short-circuit current, and when I_S is minimum, which improves the open-circuit voltage. What determines $\alpha(v)$, what is the distribution of photons of a given frequency v in solar radiation, and what is the exact bound on the limit of solar-cell efficiency?

Shockley and Queisser provided insight into the limits of solar-cell efficiency in a landmark paper [39]. The analysis in their paper indicates that the limits of solar-cell efficiency in a single-bandgap pn junction receiving 1 sun's worth of radiation is 31%. We shall summarize the key ideas in an intuitive fashion here. Readers interested in further details should consult [40]. Figure 26.8 (a) shows that, when a photon with energy $hv > E_G$ creates an electron-hole pair, energy in excess of E_G is quickly lost as heat such that only an energy equal to E_G is available as electricity. Thus, while high-energy photons have a high probability of creating an electron hole pair since many possible states are available for their creation, a good fraction of their energy is lost as heat. Lower-energy photons with an energy hv just greater than E_G have better efficiencies of energy extraction. Shockley and Quiesser assumed that any photon with energy $hv > E_G$ that created an electron-hole pair would do so with effective energy E_G and that any photon with energy $hv < E_G$ would create no electron-hole pair.

Figure 26.8 (b) shows the known 6000 K black-body spectrum of solar radiation, i.e., the solar photon probability distribution for photons of various frequencies v. The shaded area to the right of the minimal $hv_g = E_G$ frequency yields the net fraction of photons in solar radiation that contribute to electric energy generation. Each of the high-energy photons that are represented in this region contributes an energy of E_G to electric generation and wastes $hv - E_G$ as heat energy. From the entire probability distribution of Figure 26.8 (b), and the energy hv of single photons, we can compute the total incoming energy in solar radiation.

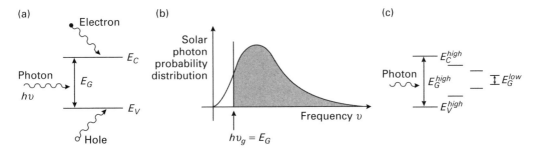

Figure 26.8. Generation of an electron-hole pair by a photon in a solar cell in (a). The only photons that can generate such electron-hole pairs must have energy greater than the bandgap energy such that only a fraction of the incoming solar black-body spectrum at 6000 K can be converted to electricity as shown in (b). Solar photovoltaics that attempt to improve energy efficiency use materials with multiple bandgaps as shown in (c) to increase their photon collection efficiency.

From the shaded area in Figure 26.8 (b), we can compute the electric energy generated by the high-energy photons. The ratio of the electric energy to the total incoming radiation energy then yields an ultimate limit for the solar-cell efficiency.

Shockley and Quiesser showed that their ultimate limit could be attained if the only method for electron-hole pair destruction at 300 K is radiative, i.e., incoming thermal energy at 300 K from the environment creates electron-hole pairs, which then recombine to generate outgoing 300 K blackbody radiation that balances such generation. In this limit, the value of I_S is as low as it can possibly be, and the open-circuit voltage of the junction asymptotes to the bandgap voltage, $V_G = E_G/q$, of the semiconductor. Any other forms of recombination, e.g., due to impurities in the semiconductor, decrease the minority-carrier lifetime τ, and consequently decrease the minority-carrier diffusion length and increase I_S. Hence, the open-circuit voltage given in Equation (26.29) is reduced. The use of pure semiconductors is thus important for achieving high efficiency, but making pure materials is expensive. The solar cell must be thick enough such that the probability of absorbing a photon is high. Figure 11.6 (a) and Figure 11.6 (b) in Chapter 11 show that bluer photons are absorbed at shallower depths while redder photons are absorbed at deeper depths. The solar cell should not be too thick since the chance for electron-hole recombination is then increased. Hence, there is an optimal thickness in solar cells that maximizes efficiency. However, maximum power transfer is not attained at this optimum. Topologies are now being explored in which electron-hole pairs are created by photons along one spatial dimension while electrons and holes travel a short distance in an orthogonal direction to create an electrical voltage; thus, the electron-hole pairs are given little opportunity to recombine [41].

Figure 26.8 (c) illustrates an idea for increasing the fraction of photons that contribute to electrical energy in a solar cell. If we have multiple pn junctions made of materials with progressively smaller bandgaps, we can first extract the

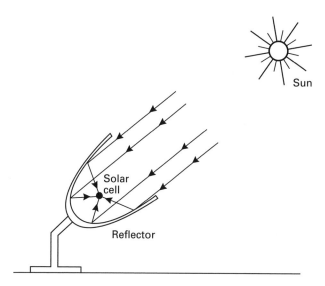

Figure 26.9. A solar concentrator to improve solar-cell efficiency.

energy in the highest-energy photons efficiently, then in those with moderate energy, then in those with the lowest energy, and so on. Such an hierarchical spectral-energy extraction scheme is very much like that in a biological cochlea or in an RF cochlea (see Chapter 23), i.e., it is like a 'cochlear solar cell'. The overall scheme extracts incoming solar photon energy in all spectral bands such that the area of the shaded region in Figure 26.8 is maximized. It also extracts this energy in a fashion such that little of it is wasted as heat. In theory, such schemes have been shown to be capable of nearly 70% efficiency [41].

Figure 26.9 illustrates an idea for improving solar-cell efficiency in a 'solar concentrator'. We gather radiation over a large area and focus it into a small area such that the effective intensity of the sun per unit area is increased from '1 sun' to, say, '400 suns'. The I_l/I_S ratio in Equation (26.29) then increases, improving open-circuit voltage and thus efficiency. One advantage of the concentrator is that smaller active areas of pure material are needed for the same power, which can minimize cost. However, such schemes have to track the sun to ensure that its radiation does not move off the focal point where the solar cell is located. Solar concentrators have indeed improved efficiencies over simple flat-panel solar cells.

The ultimate limit on solar-cell efficiency occurs when a graded-bandgap or multiple-junction material like that in Figure 26.8 (b) is combined with a concentrator like that in Figure 26.9 to create a structure where it appears that the solar cell is 'surrounded' by 6000 K black-body radiation from the sun, on all sides. It absorbs all of this radiation, and then reradiates a small fraction of it as black-body radiation at 300 K. In this limit, the solar cell is just a 'heat engine' as shown in Figure 26.4: it absorbs heat from the 6000 K 'hot reservoir' of the sun, converts most of it to useful electrical work, and loses some of it as radiated heat to the surrounding 300 K 'cold reservoir'. Thus, from the Carnot efficiency limit of Equation (26.13), the best possible efficiency of a solar cell is given by

$$C_{slr} = 1 - \frac{300}{6000} = 95\% \qquad (26.30)$$

The best solar cells that have been built today operate very near 40%, with every small percent improvement that is eked out requiring ingenuity and relatively expensive fabrication. The key bottleneck to solar-electricity generation on a large scale in the world today is the implementation of cost-effective strategies that are also efficient.

If S is the solar insolation in watts per square meter, e is the efficiency of the solar cell expressed as a fraction, $C(e)$ is the cost in dollars per square meter of installation of a solar plant, N is the desired number of years to recoup the installation investment, and i is the average rate of inflation over N years expressed as a percentage, then S_{cost}, the cost in cents per kW h of solar electricity, can be shown to be

$$S_{cost} = 11.4 \left(\frac{C(e)}{eSN} \right) \left(1 + \frac{i}{100} \right)^N \qquad (26.31)$$

Hence, if $e = 0.1$, $C(0.1) = \$600/m^2$, $S = 190\,W/m^2$, $N = 30$ years, and $i = 3\%$, then $S_{cost} = 29$ cents per kWh. Hence, competing with the cost of generating fossil-fuel electricity at 4 cents per kW h is difficult. However, with increasing research and with increasing economies of scale, $C(e)$ has been constantly reducing. An important win-win situation can occur if we lower power consumption such that the net change to the user is cost neutral or only results in a modest increase in cost: the electricity costs more but we use less of it such that our overall cost is relatively unchanged.

The intermittency of solar energy implies that we must draw on stores when it is not available, e.g., at night, and replenish these stores during the day. Various storage options are being explored including compressed air, flywheels, and batteries. One promising option may be the use of electrochemical capacitors, sometimes called ultra-capacitors or super-capacitors, which are capable of many cycles of rapid charge and discharge and that have relatively high power densities [42]. Such capacitors are essentially high-surface-area double-layer capacitors, i.e., the Helmholtz capacitors described in Chapter 25.

Ultra-capacitors are complementary to batteries that have fewer cycles of charge and discharge but relatively high energy density. They have already been used for regenerative braking applications in electric cars and hybrid cars. Ultra-capacitors implement short-term storage of energy, and batteries implement medium-term storage of energy. The ultimate in long-term storage of energy is to convert solar energy to a highly energy-dense chemical fuel. Biology has accomplished such storage via the process of photosynthesis in plants for hundreds of millions of years. We shall now briefly discuss biofuels and their importance.

26.11   Biofuels

Figure 26.10 illustrates an essential feedback loop between plants and animals. Plants harvest the energy in sunlight to split water into hydrogen and oxygen as part of the process of photosynthesis. The hydrogen is bound with CO_2 to create energy-rich molecules such as glucose ($C_6H_{12}O_6$) which in turn is often bound up in polymers like starch and cellulose. Plants also generate the oxygen that we breathe. Animals eat plant foods (or eat other animals that eat plant foods) and oxidize $C_6H_{12}O_6$ molecules to water (H_2O) and carbon dioxide (CO_2). The energy derived from the process of oxidation generates ATP, which is used to power various energy-consuming processes in animals. Thus, plants are the solar cells and fuel generators for animals. Animals provide raw materials useful to plants.

Research is under way to create biofuels that can power cars in the future by converting grasses and non-edible plants that contain cellulose to biofuels, i.e., to create 'grassoline' [43]. Such fast-growing plants can grow in land areas where normal food crops cannot, use relatively little water, and do not encroach on valuable farm land. Since plants absorb CO_2 in the atmosphere, which is then returned to the atmosphere when the fuel is burned, biofuels are net carbon neutral. The process of converting cellulose in plants to fuel in an economical fashion is technically very challenging and is an active area of research. Scientists are attempting to take inspiration from bacteria and fungi in the guts of cows and in termites respectively to understand how to digest these recalcitrant plants and thus create economical biofuels [44], [45].

If electric cars operate with sufficiently low power consumption in the future, biofuels could potentially power fuel-cell-based batteries in these cars, i.e, directly convert the chemical energy in a fuel to electricity, rather than burning it via a heat engine as in a conventional gasoline-powered car. Chapter 25 contains a discussion of how fuel cells operate. The high energy density of biofuels implies that car batteries can then be lighter, and the use of an electric motor rather than a heavy engine can further lighten the car. It is worth noting that biofuel-based fuel cells have been explored for implantable applications for a long time [46]. A microfluidic fuel cell suitable for implantable applications has recently been described [47]. One challenge in the operation of such biofuel cells has been that

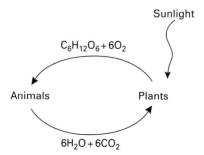

Figure 26.10. Photochemical energy flows in nature.

the enzymes that are used to oxidize the fuel lose their efficacy after some time, making them unattractive in long-term implants or in car applications that may need years of battery operation. Cells solve these problems by constantly degrading and regenerating enzymes needed for various biological processes such that they always maintain their efficacy.

26.12   Energy use and energy generation

Low-power systems can enable sources of energy that would normally be impractical for powering an application to become practical. In inductive links (Chapter 16), in piezoelectric harvesters, in electric motors, and in electric cars, we have seen that a low-power system can improve the energy efficiency of an overall system by altering the effective reflected load seen by the energy source. A system is most energy efficient when there is minimal power transfer but only achieves 50% efficiency at maximal power transfer. In the chapter on batteries (Chapter 25), we saw that a low-power system does not increase battery lifetime merely because of a low-power draw but also because it enables higher energy density, higher efficiency, and lower fade capacity in the battery. In electric cars, the use of a relatively light and efficient electric motor rather than a heavy engine enabled an energy source with significantly lower energy density, i.e., a battery, to become practical for powering a car. The cost effectiveness of solar electricity is improved if electricity consumption can be reduced, thus enabling green electricity rather than 'red' electricity. The principles of adiabatic design in Chapter 21, the Shannon limit on the minimum energy needed to compute in Chapter 22, and the Ragone-curve tradeoff between energy density and power density in Chapter 25 all reveal that if you can pull energy out of a source slowly, you can waste less of it and create a higher capacity to store it. The central take-home lesson from these numerous examples is that energy use and energy generation are deeply linked. We must try to optimize them jointly rather than treat them as two separate problems.

References

[1] David J. C. MacKay. *Sustainable Energy – Without the Hot Air* (Cambridge, UK: UIT Cambridge, Ltd., 2009).

[2] Department for Environment Food and Rural Affairs (DEFRA), *Guidelines for Company Reporting on Greenhouse Gas Emissions*, London (2005); available from: http://www.defra.gov.uk/environment/business/reporting/pdf/envrpgas-annexes.pdf.

[3] D. L. Green and Scientific American. *Oil and the Future of Energy: Climate Repair, Hydrogen, Nuclear Fuel, Renewable and Green Sources, Energy Efficiency* (Guilford, Conn.: The Lyons Press, 2007).

[4] T. R. Karl and K. E. Trenberth. Modern global climate change. *Science*, **302** (2003), 1719–1723.

[5] D. Goodstein. *Out of Gas: The End of the Age of Oil* (New York, NY: W.W. Norton and Company, 2004).

[6] A. Q. Ansari and D. J. F. Bowling. Measurement of the Trans-Root Electrical Potential of Plants Grown in Soil. *New Phytologist*, **71** (1972), 111–117.

[7] P. D. Mitcheson, E. M. Yeatman, G. K. Rao, A. S. Holmes and T. C. Green. Energy harvesting from human and machine motion for wireless electronic devices. *Proceedings of the IEEE*, **96** (2008), 1457–1486.

[8] S. Roundy. On the effectiveness of vibration-based energy harvesting. *Journal of Intelligent Material Systems and Structures*, **16** (2005), 809.

[9] C. Bright. Energy Coupling and Efficiency. *Sensors*, **18** (2001), in special section on Motion Control.

[10] S. Roundy, P. K. Wright and J. M. Rabaey. *Energy Scavaging for Wireless Sensor Networks*. 1st ed. (Boston, Mass.: Kluwer Academic Publishers, 2003).

[11] Thad E. Starner and Joseph A. Paradiso. Human generated power for mobile electronics. In *Low Power Electronics Design*, ed. C Piguet. (Boca Raton, Flor.: CRC Press; 2005).

[12] J. A. Paradiso and T. Starner. Energy scavenging for mobile and wireless electronics. *IEEE Pervasive Computing*, **4** (2005), 18–27.

[13] J. M. Donelan, Q. Li, V. Naing, J. A. Hoffer, D. J. Weber and A. D. Kuo. Biomechanical energy harvesting: Generating electricity during walking with minimal user effort. *Science*, **319** (2008), 807.

[14] R. Tashiro, N. Kabei, K. Katayama, E. Tsuboi and K. Tsuchiya. Development of an electrostatic generator for a cardiac pacemaker that harnesses the ventricular wall motion. *Journal of Artificial Organs*, **5** (2002), 239–245.

[15] D. Spreemann, Y. Manoli, B. Folkmer and D. Mintenbeck. Non-resonant vibration conversion. *Journal of Micromechanics and Microengineering*, **16** (2006), 169.

[16] V. Leonov, T. Torfs, P. Fiorini, C. Van Hoof and L. IMEC. Thermoelectric converters of human warmth for self-powered wireless sensor nodes. *IEEE Sensors Journal*, **7** (2007), 650–657.

[17] J. L. Montieth and Lawrence Mount, editors. *Heat Loss from Animals and Man* (London: Butterworth & Co. Publishers Ltd.; 1974).

[18] I. Doms, P. Merken, R. Mertens and C. Van Hoof. Integrated capacitive power-management circuit for thermal harvesters with output power 10 to 1000 μW. *Proceedings of the IEEE International Solid-State Circuits Conference (ISSCC)*, San Francisco, Calif., 300–301, 301a, 2009.

[19] J. Y. Wong. *Theory of Ground Vehicles*. 4th ed. (Hoboken, New Jersey: John Wiley & Sons, 2008).

[20] J. C. Dixon. *Tires, Suspension and Handling*. 2nd ed. (Warrendale, Penn.: Society of Automotive Engineers, 2004).

[21] Tesla Motors. Available from: www.teslamotors.com.

[22] N. C. C. Sharp. Timed running speed of a cheetah (*Acinonyx jubatus*). *Journal of Zoology*, **241** (1997), 493–494.

[23] M. Hildebrand. Motions of the running cheetah and horse. *Journal of Mammalogy*, **40** (1959), 481–495.

[24] L. L. Marker and A. J. Dickman. Morphology, physical condition, and growth of the cheetah (*Acinonyx jubatus jubatus*). *Journal of Mammalogy*, **84** (2003), 840–850.

[25] T. Garland. The relation between maximal running speed and body mass in terrestrial mammals. *Journal of Zoology*, **199** (1983), 70.

[26] C. R. Taylor, A. Shkolnik, R. Dmi'el, D. Baharav and A. Borut. Running in cheetahs, gazelles, and goats: energy cost and limb configuration. *American Journal of Physiology*, **227** (1974), 848–850.

[27] D. J. Morton and D. D. Fuller. *Human Locomotion and Body Form* (New York, NY: Waverly Press, Inc., 1952).

[28] P. Grogan. Sam Whittingham tops 80 mph – on a push-bike. *The Sunday Times*. September 20, 2009.

[29] R. McNeill Alexander. *Principles of Animal Locomotion* (Woodstock, Oxfordshire, UK: Princeton University Press, 2003).

[30] H. Tennekes. *The Simple Science of Flight: From Insects to Jumbo Jets* (Cambridge, Mass.: MIT Press, 1997).

[31] R. E. Gill, T. L. Tibbitts, D. C. Douglas, C. M. Handel, D. M. Mulcahy, J. C. Gottschalck, N. Warnock, B. J. McCaffery, P. F. Battley and T. Piersma. Extreme endurance flights by landbirds crossing the Pacific Ocean: ecological corridor rather than barrier? *Proceedings of the Royal Society, B.*, **276** (2009), 447.

[32] W. Shyy, M. Berg and D. Ljungqvist. Flapping and flexible wings for biological and micro air vehicles. *Progress in Aerospace Sciences*, **35** (1999), 455–505.

[33] V. Gradinescu, C. Gorgorin, R. Diaconescu, V. Cristea and L. Iftode, Adaptive Traffic Lights Using Car-to-Car Communication. *Proceedings of the Vehicular Technology Conference*, Dublin, Ireland, 21–25, 2007.

[34] F. S. Johnson. The solar constant. *Journal of the Atmospheric Sciences*, **11** (1954), 431–439.

[35] M. Pidwirny and D. Budikova. Earth's energy balance. In *Encyclopedia of Earth*, ed. C J Cleveland. (2008), Available from http://www.eoearth.org/article/Earth's_energy_balance

[36] Atmospheric Science Data Center. NASA Surface Meteorology and Solar Energy Calculator. Available from: http://eosweb.larc.nasa.gov/cgi-bin/sse/grid.cgi?uid = 3030.

[37] Advanced Energy Group. Solar Insolation for US Major Cities. Available from: http://www.solar4power.com/solar-power-insolation-window.html.

[38] Yannis Tsividis. *Operation and Modeling of the MOS Transistor*, 3rd ed. (New York: Oxford University Press, 2008).

[39] W. Shockley and H. J. Queisser. Detailed balance limit of efficiency of p-n junction solar cells. *Journal of Applied Physics*, **32** (1961), 510–519.

[40] M. A. Green. *Third Generation Photovoltaics: Advanced Solar Energy Conversion* (Berlin: Spinger-Verlag, 2004).

[41] N. S. Lewis. Toward cost-effective solar energy use. *Science*, **315** (2007), 798–801.

[42] J. R. Miller and P. Simon. Electrochemical Capacitors for Energy Management. *Science*, **321** (2008), 651.

[43] G. W. Huber and G. Dale. Grassoline at the Pump. *Scientific American*, **301** (2009), 52–59.

[44] E. M. Rubin. Genomics of cellulosic biofuels. *Nature*, **454** (2008), 841–845.

[45] P. Heinzelman, C. D. Snow, I. Wu, C. Nguyen, A. Villalobos, S. Govindarajan, J. Minshull and F. H. Arnold. A family of thermostable fungal cellulases created by

structure-guided recombination. *Proceedings of the National Academy of Sciences*, **106** (2009), 5610.

[46] H. H. Kim, N. Mano, Y. Zhang and A. Heller. A miniature membrane-less biofuel cell operating under physiological conditions at 0.5 V. *Journal of The Electrochemical Society*, **150** (2003), A209.

[47] Y. A. Song, C. Batista, R. Sarpeshkar and J. Han. Rapid fabrication of microfluidic polymer electrolyte membrane fuel cell in PDMS by surface patterning of perfluorinated ion-exchange resin. *Journal of Power Sources*, **183** (2008), 674–677.

Epilogue

Information is represented by the states of physical devices. It costs energy to transform or maintain the states of these physical devices. Thus, energy and information are deeply linked. It is this deep link that allows us to articulate information-based principles for ultra-low-power design that apply to biology or to electronics, to analog or to digital systems, to electrical or to non-electrical systems, at small scales or at large scales. The graphical languages of circuits and feedback serve as powerful unifying tools to understand or to design low-power systems that range from molecular networks in cells to biomedical implants in the brain to energy-efficient cars.

A vision that this book has attempted to paint in the context of the fields of ultra-low-power electronics and bioelectronics is shown in the figure below. Engineering can aid biology through analysis, instrumentation, design, and repair (medicine). Biology can aid engineering through bio-inspired design. The positive-feedback loop created by this two-way interaction can amplify and speed progress in both disciplines and shed insight into both. It is my hope that this book will bring appreciation to the beauty, art, and practicality of such synergy and that it will inspire the building of more connections in one or both directions in the future.

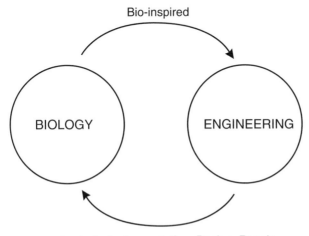

Epilogue The two-way flow between biology and engineering.

Bibliography

Advanced Energy Group. Solar Insolation for US Major Cities. Available from: http://www.solar4power.com/solar-power-insolation-window.html.

Affymetrix. Available from: http://www.affymetrix.com/index.affx.

L. C. Aiello. Brains and guts in human evolution: The expensive tissue hypothesis. *Brazilian Journal of Genetics*, **20** (1997).

J. Aizenberg. Crystallization in Patterns: A Bio-Inspired Approach. *Advanced Materials*, **16** (2004), 1295–1302.

E. V. Aksenov, Y. M. Ljashenko, A. V. Plotnikov, D. A. Prilutskiy, S. V. Selishchev and E. V. Vetvetskiy. Biomedical data acquisition systems based on sigma-delta analogue-to-digital converters. *Proceedings of the 23rd Annual International IEEE Engineering in Medicine and Biology Conference (EMBS)*, Istanbul, Turkey, 3336–3337, 2001.

R. McNeill Alexander. *Principles of Animal Locomotion* (Woodstock, Oxfordshire, UK: Princeton University Press, 2003).

John Morgan Allman. *Evolving Brains* (New York: Scientifc American Library: Distributed by W. H. Freeman and Co., 1999).

Uri Alon. *An Introduction to Systems Biology: Design Principles of Biological Circuits* (Boca Raton, FL: Chapman & Hall/CRC, 2007).

J. Ammer, M. Bolotski, P. Alvelda and T. F. Knight, Jr. A 160×120 pixel liquid-crystal-on-silicon microdisplay with an adiabatic DACM. *Proceedings of the IEEE International Solid-State Circuits Conference (ISSCC)*, San Francisco, CA, 212–213, 1999.

H. Andersson and A. Van den Berg. Microfluidic devices for cellomics: a review. *Sensors and Actuators: B. Chemical*, **92** (2003), 315–325.

A. G. Andreou, K. A. Boahen, P. O. Pouliquen, A. Pavasovic, R. E. Jenkins and K. Strohbehn. Current-mode subthreshold MOS circuits for analog VLSI neural systems. *IEEE Transactions on Neural Networks*, **2** (1991), 205–213.

A. Q. Ansari and D. J. F. Bowling. Measurement of the Trans-Root Electrical Potential of Plants Grown in Soil. *New Phytologist*, **71** (1972), 111–117.

J. F. Apgar, J. E. Toettcher, D. Endy, F. M. White and B. Tidor. Stimulus design for model selection and validation in cell signaling. *Public Library of Science Computational Biology*, **4** (2008), e30.

S. K. Arfin, M. A. Long, M. S. Fee and R. Sarpeshkar. Wireless Neural Stimulation in Freely Behaving Small Animals. *Journal of Neurophysiology*, (2009), 598–605.

W. C. Athas, L. J. Svensson, J. G. Koller, N. Tzartzanis and E. Y.-C. Chou. Low-power digital systems based on adiabatic-switching principles. *IEEE Transactions on Very Large Scale Integration (VLSI) Systems*, **2** (1994), 398–407.

W. C. Athas, L. J. Svensson and N. Tzartzanis. A resonant signal driver for two-phase, almost-non-overlapping clocks. *Proceedings of the IEEE International Symposium on Circuits and Systems (ISCAS)* Atlanta, GA, 129–132 1996.

W. Athas, N. Tzartzanis, W. Mao, L. Peterson, R. Lal, K. Chong, J-S. Moon, L. Svensson and M. Bolotski. The design and implementation of a low-power clock-powered microprocessor. *IEEE Journal of Solid-State Circuits*, **35** (2000), 1561–1570.

Atmospheric Science Data Center. NASA Surface Meteorology and Solar Energy Calculator. Available from: http://eosweb.larc.nasa.gov/cgi-bin/sse/grid.cgi?uid = 3030.

D. Attwell and S. B. Laughlin. An energy budget for signaling in the grey matter of the brain. *Journal of Cerebral Blood Flow and Metabolism*, **21** (2001), 1133–1145.

A. T. Avestruz, W. Santa, D. Carlson, R. Jensen, S. Stanslaski, A. Helfenstine and T. Denison. A 5 μW/channel Spectral Analysis IC for Chronic Bidirectional Brain-Machine Interfaces. *IEEE Journal of Solid-State Circuits*, **43** (2008), 3006–3024.

D. Baker, G. Church, J. Collins, D. Endy, J. Jacobson, J. Keasling, P. Modrich, C. Smolke and R. Weiss. Engineering life: building a fab for biology. *Scientific American*, **294** (2006), 44–51.

M. W. Baker and R. Sarpeshkar. A low-power high-PSRR current-mode microphone preamplifier. *IEEE Journal of Solid-State Circuits*, **38** (2003), 1671–1678.

M. W. Baker and R. Sarpeshkar. Low-Power Single-Loop and Dual-Loop AGCs for Bionic Ears. *IEEE Journal of Solid-State Circuits*, **41** (2006), 1983–1996.

M. W. Baker and R. Sarpeshkar. Feedback analysis and design of RF power links for low-power bionic systems. *IEEE Transactions on Biomedical Circuits and Systems*, **1** (2007), 28–38.

Constantine A. Balanis. *Antenna Theory: Analysis and Design*. 3rd ed. (Hoboken, NJ: John Wiley & Sons, Inc., 2005).

Allen J. Bard and Larry R. Faulkner. *Electrochemical Methods: Fundamentals and Applications*. 2nd ed. (New York: Wiley, 2001).

R. Bashir. BioMEMS: State-of-the-art in detection, opportunities and prospects. *Advanced Drug Delivery Reviews*, **56** (2004), 1565–1586.

D. J. Beebe, G. A. Mensing and G. M. Walker. Physics and applications of microfluidics in biology. *Annual Review of Biomedical Engineering*, **4** (2002), 261–286.

R. Bellman. Dynamic programming. *Science*, **153** (1966), 34–37.

C. H. Bennett and R. Landauer. The fundamental physical limits of computation. *Scientific American*, **253** (1985), 48–56.

Y. Berg, R. L. Sigvartsen, T. S. Lande and A. Abusland. An analog feed-forward neural network with on-chip learning. *Analog Integrated Circuits and Signal Processing*, **9** (1996), 65–75.

S. N. Bhatia and C. S. Chen. Tissue engineering at the micro-scale. *Biomedical Microdevices*, **2** (1999), 131–144.

A. Bhattacharya and F. G. Zeng. Companding to improve cochlear-implant speech recognition in speech-shaped noise. *The Journal of the Acoustical Society of America*, **122** (2007), 1079.

Harold S. Black, inventor. *Wave Translation System*, US Patent Number 2,102,671.

R. B. Blackman. Effect of feedback on impedance. *Bell Systems Technical Journal*, **22** (1943).

K. A. Boahen. Point-to-point connectivity between neuromorphic chips using address events. *IEEE Transactions on Circuits and Systems II: Analog and Digital Signal Processing*, **47** (2000), 416–434.

K. A. Boahen and A. G. Andreou. A contrast sensitive silicon retina with reciprocal synapses. *Proceedings of the IEEE Neural Information Processing Systems (NIPS)*, Denver, Colorado, 764–772, 1992.

Hendrik W. Bode. *Network Analysis and Feedback Amplifier Design* (New York, NY: Van Nostrand, 1945).

J. Bohorquez, W. Sanchez, L. Turicchia and R. Sarpeshkar. An integrated-circuit switched-capacitor model and implementation of the heart. *Proceedings of the First International Symposium on Applied Sciences on Biomedical and Communication Technologies (ISABEL)*, Aalborg, Denmark, 1–5, 2008.

P. D. Bradley, An ultra low power, high performance medical implant communication system (MICS) transceiver for implantable devices. *IEEE Biomedical Circuits and Systems Conference (BioCAS)*, London, UK, 158–161, 2006.

J. C. Bramwell and A. V. Hill. The velocity of the pulse wave in man. *Proceedings of the Royal Society of London. Series B, Containing Papers of a Biological Character*, **93** (1922), 298–306.

C. Bright. Energy Coupling and Efficiency. *Sensors*, **18** (2001), in special section on Motion Control.

T. Buchegger, G. Oßberger, A. Reisenzahn, E. Hochmair, A. Stelzer and A. Springer. Ultra-wideband transceivers for cochlear implants. *EURASIP Journal on Applied Signal Processing*, **2005** (2005), 3069–3075.

K. Bult. Analog design in deep sub-micron CMOS. *Proceedings of the 26th European Solid-State Circuits Conference (ESSCIRC)*, Stockholm, Sweden, 126–132, 2000.

T. P. Burg, M. Godin, S. M. Knudsen, W. Shen, G. Carlson, J. S. Foster, K. Babcock and S. R. Manalis. Weighing of biomolecules, single cells and single nanoparticles in fluid. *Nature*, **446** (2007), 1066–1069.

M. A. Burns, B. N. Johnson, S. N. Brahmasandra, K. Handique, J. R. Webster, M. Krishnan, T. S. Sammarco, P. M. Man, D. Jones and D. Heldsinger. An integrated nanoliter DNA analysis device. *Science*, **282** (1998), 484.

J. Burr and A. M. Peterson. Ultra low power CMOS technology. *Proceedings of the 3rd NASA Symposium on VLSI Design*, Moscow, Idaho, 4.2.11–14.12.13, 1991.

H. B. Callen and T. A. Welton. Irreversibility and generalized noise. *Physical Review*, **83** (1951), 34–40.

Gert Cauwenberghs and Magdy A. Bayoumi. *Learning on Silicon: Adaptive VLSI Neural Systems* (Boston: Kluwer Academic, 1999).

A. P. Chandrakasan and R. W. Brodersen. Minimizing power consumption in digital CMOS circuits. *Proceedings of the IEEE*, **83** (1995), 498–523.

S. Chatterjee, Y. Tsividis and P. Kinget. 0.5-V analog circuit techniques and their application in OTA and filter design. *IEEE Journal of Solid-State Circuits*, **40** (2005), 2373–2387.

C. H. Chen and M. J. Deen. Channel noise modeling of deep submicron MOSFETs. *IEEE Transactions on Electron Devices*, **49** (2002), 1484–1487.

M. Chen and G. A. Rincon-Mora. Accurate, Compact, and Power-Efficient Li-Ion Battery Charger Circuit. *IEEE Transactions on Circuits and Systems II: Express Briefs*, **53** (2006), 1180–1184.

L. J. Chu. Physical limitations of omnidirectional antennas. *Technical Report (Research Laboratory of Electronics, Massachusetts Institute of Technology)*, **64** (1948).

Leon O. Chua and T. Roska. *Cellular Neural Networks and Visual Computing: Foundation and Applications* (Cambridge, UK; New York, NY: Cambridge University Press, 2002).

S. Das, C. Tokunaga, S. Pant, W. H. Ma, S. Kalaiselvan, K. Lai, D. M. Bull and D. T. Blaauw. RazorII: In situ error detection and correction for PVT and SER tolerance. *IEEE Journal of Solid-State Circuits*, **44** (2009), 32–48.

W. C. Dash and R. Newman. Intrinsic optical absorption in single-crystal germanium and silicon at 77 K and 300 K. *Physical Review*, **99** (1955), 1151–1155.

S. Dearborn. Charging Li-ion batteries for maximum run times. *Power Electronics Technology Magazine*, (2005), 40–49.

T. Delbruck. Bump circuits for computing similarity and dissimilarity of analog voltages. *Proceedings of the International Joint Conference on Neural Networks*, Seattle, WA, 475–479, 1991.

T. Delbruck. Silicon retina with correlation-based, velocity-tuned pixels. *IEEE Transactions on Neural Networks*, **4** (1993), 529–541.

T. Delbruck and C. A. Mead. Analog VLSI phototransduction by continuous-time, adaptive, logarithmic photoreceptor circuits. *CalTech CNS Memo*, **30** (1994).

T. Delbruck and C. A. Mead. Adaptive photoreceptor with wide dynamic range. *Proceedings of the IEEE International Symposium on Circuits and Systems (ISCAS)*, London, UK, 339–342, 1994.

T. Denison, K. Consoer, A. Kelly, A. Hachenburg, W. Santa, M. N. Technol and C. Heights. A 2.2 µW 94nV/√Hz, Chopper-Stabilized Instrumentation Amplifier for EEG Detection in Chronic Implants. *Digest of Technical Papers of the IEEE International Solid-State Circuits Conference (ISSCC 2007)*, San Francisco, CA, 162–594, 2007.

Department for Environment Food and Rural Affairs (DEFRA). *Guidelines for Company Reporting on Greenhouse Gas Emissions*, London (2005); available from: http://www.defra.gov.uk/environment/business/reporting/pdf/envrpgas-annexes.pdf.

M. H. Dickinson, F. O. Lehmann and S. P. Sane. Wing rotation and the aerodynamic basis of insect flight. *Science*, **284** (1999), 1954–1960.

C. Diorio, P. Hasler, A. Minch and C. A. Mead. A single-transistor silicon synapse. *IEEE Transactions on Electron Devices*, **43** (1996), 1972–1980.

J. C. Dixon. *Tires, Suspension and Handling*. 2nd ed. (Warrendale, Penn.: Society of Automotive Engineers, 2004).

I. Doms, P. Merken, R. Mertens and C. Van Hoof. Integrated capacitive power-management circuit for thermal harvesters with output power 10 to 1000 µW. *Proceedings of the IEEE International Solid-State Circuits Conference (ISSCC)*, San Francisco, Calif., 300–301, 301a, 2009.

J. M. Donelan, Q. Li, V. Naing, J. A. Hoffer, D. J. Weber and A. D. Kuo. Biomechanical energy harvesting: Generating electricity during walking with minimal user effort. *Science*, **319** (2008), 807.

T. G. Drummond, M. G. Hill and J. K. Barton. Electrochemical DNA sensors. *Nature Biotechnology*, **21** (2003), 1192–1199.

P. Dudek and P. J. Hicks. A CMOS general-purpose sampled-data analogue microprocessor. *Proceedings of the IEEE International Symposium on Circuits and Systems (ISCAS)*, Geneva, Switzerland, 2000.

P. Dudek and P. J. Hicks. A general-purpose processor-per-pixel analog SIMD vision chip. *IEEE Transactions on Circuits and Systems I: Regular Papers*, **52** (2005), 13–20.

V. Ekanayake, C. Kelly IV and R. Manohar. An ultra low-power processor for sensor networks. *Proceedings of the 11th International Conference on Architectural Support for Programming Languages and Operating Systems (ASPLOS)*, Boston, Mass., 27–36, 2004.

K. L. Ekinci, X. M. H. Huang and M. L. Roukes. Ultrasensitive nanoelectromechanical mass detection. *Applied Physics Letters*, **84** (2004), 4469–4471.

Chris Eliasmith and C. H. Anderson. *Neural Engineering: Computation, Representation, and Dynamics in Neurobiological Systems* (Cambridge, Mass.: MIT Press, 2003).

M. B. Elowitz and S. Leibler. A synthetic oscillatory network of transcriptional regulators. *Nature*, **403** (2000), 335–338.

Emerging Research Devices Report: International Technology Roadmap for Semiconductors, 2007.

C. C. Enz and G. C. Temes. Circuit techniques for reducing the effects of op-amp imperfections: autozeroing, correlated double sampling, and chopper stabilization. *Proceedings of the IEEE*, **84** (1996), 1584–1614.

Christian Enz and Eric A. Vittoz. *Charge-based MOS Transistor Modeling: The EKV Model for Low-Power and RF IC Design* (Chichester, England; Hoboken, NJ: John Wiley, 2006).

R. G. H. Eschauzier and J. H. Huijsing. *Frequency Compensation Techniques for Low-Power Operational Amplifiers* (Dordrecht: Springer, 1995).

Walter R. Evans. *Control-System Dynamics* (New York, NY: McGraw-Hill, 1954).

R. M. Fano. Theoretical limitations on the broadband matching of arbitrary impedances. *Technical Report (Research Laboratory of Electronics, Massachusetts Institute of Technology)*, **41** (1948).

L. Fay, V. Misra and R. Sarpeshkar. A Micropower Electrocardiogram Recording Amplifier. *IEEE Transactions on Biomedical Circuits and Systems*, **vol. 3**, No. 5, pp. 312–320, October 2009.

R. P. Feynman, M. L. Sands and R. B. Leighton. Lecture 19: The principle of least action. In *The Feynman Lectures on Physics: Commemorative Issue*, **Vol. 2**, ed. (Reading, MA: Addison Wesley; 1989), pp. 19.11–19.14.

R. P. Feynman. *The Pleasure of Finding Things Out: The Best Short Works of Richard P. Feynman*. (Cambridge, Mass.: Perseus Books, 1999).

J. K. Fiorenza, T. Sepke, P. Holloway, C. G. Sodini and L. Hae-Seung. Comparator-Based Switched-Capacitor Circuits for Scaled CMOS Technologies. *IEEE Journal of Solid-State Circuits*, **41** (2006), 2658–2668.

Gene F. Franklin, J. David Powell and Abbas Emami-Naeini. *Feedback Control of Dynamic Systems*. 5th ed. (Upper Saddle River, N.J.: Pearson Prentice Hall, 2006).

D. R. Frey. State-space synthesis and analysis of log-domain filters. *IEEE Transactions on Circuits and Systems II: Analog and Digital Signal Processing*, **45** (1998), 1205–1211.

L. M. Friesen, R. V. Shannon, D. Baskent and X. Wang. Speech recognition in noise as a function of the number of spectral channels: comparison of acoustic hearing and cochlear implants. *The Journal of the Acoustical Society of America*, **110** (2001), 1150.

J. Fritz, E. B. Cooper, S. Gaudet, P. K. Sorger and S. R. Manalis. Electronic detection of DNA by its intrinsic molecular charge. *Proceedings of the National Academy of Sciences*, **99** (2002), 14142–14146.

S. Gabriel, R. W. Lau and C. Gabriel. The dielectric properties of biological tissues: III. Parametric models for the dielectric spectrum of tissues. *Physics in Medicine and Biology*, **41** (1996), 2271–2293.

T. Garland. The relation between maximal running speed and body mass in terrestrial mammals. *Journal of Zoology*, **199** (1983), 70.

A. Gerosa, A. Maniero and A. Neviani. A fully integrated two-channel A/D interface for the acquisition of cardiac signals in implantable pacemakers. *IEEE Journal of Solid-State Circuits*, **39** (2004), 1083–1093.

E. Ghafar-Zadeh and M. Sawan. A hybrid microfluidic/CMOS capacitive sensor dedicated to lab-on-chip applications. *IEEE Transactions on Biomedical Circuits and Systems*, **1** (2007), 270–277.

M. Ghovanloo and S. Atluri. A Wide-Band Power-Efficient Inductive Wireless Link for Implantable Microelectronic Devices Using Multiple Carriers. *IEEE Transactions on Circuits and Systems I: Regular Papers*, **54** (2007), 2211–2221.

V. Giannini, P. Nuzzo, V. Chironi, A. Baschirotto, G. Van der Plas and J. Craninckx. An 820 μW 9b 40MS/s Noise-Tolerant Dynamic-SAR ADC in 90 nm Digital CMOS. *Proceedings of the IEEE International Solid-State Circuits Conference (ISSCC)*, San Francisco, CA, 238–239, 2008.

B. Gilbert. Translinear circuits: An historical overview. *Analog Integrated Circuits and Signal Processing*, **9** (1996), 95–118.

R. E. Gill, T. L. Tibbitts, D. C. Douglas, C. M. Handel, D. M. Mulcahy, J. C. Gottschalck, N. Warnock, B. J. McCaffery, P. F. Battley and T. Piersma. Extreme endurance flights by landbirds crossing the Pacific Ocean: ecological corridor rather than barrier? *Proceedings of the Royal Society, B.*, **276** (2009), 447.

D. T. Gillespie. A general method for numerically simulating the stochastic time evolution of coupled chemical reactions. *Journal of Computational Physics*, **22** (1976), 403–434.

D. Goodstein. *Out of Gas: The End of the Age of Oil* (New York, NY: W.W. Norton and Company, 2004).

W. Gosling. *Radio Antennas and Propagation* (Oxford, UK: Newnes, 2004).

V. Gradinescu, C. Gorgorin, R. Diaconescu, V. Cristea and L. Iftode. Adaptive Traffic Lights Using Car-to-Car Communication. *Proceedings of the Vehicular Technology Conference*, Dublin, Ireland, 21–25, 2007.

P. Gray, P. Hurst, S. Lewis and R. Meyer. *Analysis and Design of Analog Integrated Circuits.* 4th ed. (New York: Wiley, 2001).

A. C. R. Grayson, R. S. Shawgo, A. M. Johnson, N. T. Flynn, Y. Li, M. J. Cima and R. Langer. A bioMEMS review: MEMS technology for physiologically integrated devices. *Proceedings of the IEEE*, **92** (2004), 6–21.

D. L. Green and Scientific American. *Oil and the Future of Energy: Climate Repair, Hydrogen, Nuclear Fuel, Renewable and Green Sources, Energy Efficiency* (Guilford, Conn.: The Lyons Press, 2007).

M. A. Green. *Third Generation Photovoltaics: Advanced Solar Energy Conversion.* (Berlin: Spinger-Verlag, 2004).

P. Grogan. Sam Whittingham tops 80 mph – on a push-bike. *The Sunday Times.* September 20, 2009.

J. Guinness, B. Raj, B. Schmidt-Nielsen, L. Turicchia and R. Sarpeshkar. A companding front end for noise-robust automatic speech recognition. *Proceedings of the IEEE International Conference on Acoustics, Speech, and Signal Processing (ICASSP)*, Philadelphia, Penn., 2005.

A. Gupta, D. Akin and R. Bashir. Single virus particle mass detection using microresonators with nanoscale thickness. *Applied Physics Letters*, **84** (2004), 1976–1978.

B. B. Haab, M. J. Dunham and P. O. Brown. Protein microarrays for highly parallel detection and quantitation of specific proteins and antibodies in complex solutions. *Genome Biology*, **2** (2001), 1–13.

R. H. R. Hahnloser, R. Sarpeshkar, M. A. Mahowald, R. J. Douglas and H. S. Seung. Digital selection and analogue amplification coexist in a cortex-inspired silicon circuit. *Nature*, **405** (2000), 947–951.

H. Hamanaka, H. Torikai and T. Saito. Spike position map with quantized state and its application to algorithmic A/D converter. *Proceedings of the International Symposium on Circuits and Systems (ISCAS)*, Vancouver, BC, 673–676, 2004.

T. J. Hamilton, C. Jin, A. van Schaik and J. Tapson. A 2-D silicon cochlea with an improved automatic quality factor control-loop. *Proceedings of the IEEE International Symposium on Circuits and Systems (ISCAS)*, Seattle, Wash., 1772–1775, 2008.

T. Handa, S. Shoji, S. Ike, S. Takeda and T. Sekiguchi. A very low-power consumption wireless ECG monitoring system using body as a signal transmission medium. *Proceedings of the International Conference Solid-State Sensors and Actuators*, Chicago, IL, 1003–1007, 1997.

R. R. Harrison and C. Charles. A low-power low-noise CMOS amplifier for neural recording applications. *IEEE Journal of Solid-State Circuits*, **38** (2003), 958–965.

R. R. Harrison and C. Koch. A silicon implementation of the fly's optomotor control system. *Neural Computation*, **12** (2000), 2291–2304.

R. R. Harrison, P. T. Watkins, R. J. Kier, R. O. Lovejoy, D. J. Black, B. Greger and F. Solzbacher. A low-power integrated circuit for a wireless 100-electrode neural recording system. *IEEE Journal of Solid-State Circuits*, **42** (2007), 123–133.

B. Hassenstein and W. Reichardt. Systemtheoretische analyse der zeit-, reihenfolgen-und vorzeichenauswertung bei der bewegungsperzeption des rüsselkäfers chlorophanus. *Zeitschrift für Naturforschung*, **11** (1956), 513–524.

P. Heinzelman, C. D. Snow, I. Wu, C. Nguyen, A. Villalobos, S. Govindarajan, J. Minshull and F. H. Arnold. A family of thermostable fungal cellulases created by structure-guided recombination. *Proceedings of the National Academy of Sciences*, **106** (2009), 5610.

J. Herz, A. Krogh and R. G. Palmer. *Introduction to the Theory of Neural Computation* (Reading, Mass.: Addison Wesley, 1991).

M. Hildebrand. Motions of the running cheetah and horse. *Journal of Mammalogy*, **40** (1959), 481–495.

R. Hintsche, C. Kruse, A. Uhlig, M. Paeschke, T. Lisec, U. Schnakenberg and B. Wagner. Chemical microsensor systems for medical applications in catheters. *Sensors & Actuators B.: Chemical*, **27** (1995), 471–473.

M. Ho, P. Georgiou, S. Singhal, N. Oliver and C. Toumazou, A bio-inspired closed-loop insulin delivery based on the silicon pancreatic beta-cell. *Proceedings of the IEEE International Symposium on Circuits and Systems (ISCAS)*, Seattle, Wash., 1052–1055, 2008.

L. R. Hochberg, M. D. Serruya, G. M. Friehs, J. A. Mukand, M. Saleh, A. H. Caplan, A. Branner, D. Chen, R. D. Penn and J. P. Donoghue. Neuronal ensemble control of prosthetic devices by a human with tetraplegia. *Nature*, **442** (2006), 164–171.

F. N. Hooge. $1/f$ noise. *Physica*, **83** (1976), 14–23.

J. J. Hopfield. Kinetic proofreading: a new mechanism for reducing errors in biosynthetic processes requiring high specificity. *Proceedings of the National Academy of Sciences*, **71** (1974), 4135–4139.

T. K. Horiuchi. "Seeing" in the dark: Neuromorphic VLSI modeling of bat echolocation. *IEEE Signal Processing Magazine*, **22** (2005), 134–139.

G. W. Huber and G. Dale. Grassoline at the Pump. *Scientific American*, **301** (2009), 52–59.

G. Indiveri. Neuromorphic analog VLSI sensor for visual tracking: circuits and application examples. *IEEE Transactions on Circuits and Systems II: Analog and Digital Signal Processing*, **46** (1999), 1337–1347.

Inductance Calculator. Available from: http://www.technick.net/public/code/cp_dpage.php?aiocp_dp = util_inductance_calculator.

International Technology Roadmap for Semiconductors. Available from: http://public.itrs.net/.

T. Ishikawa, T. S. Aytur and B. E. Boser. A wireless integrated immunosensor. *Complex Medical Engineering*, (2007), 555.

Italian National Research Council and Institute for Applied Physics. Available from: http://niremf.ifac.cnr.it/tissprop/.

E. M. Izhikevich. Which model to use for cortical spiking neurons? *IEEE Transactions on Neural Networks*, **15** (2004), 1063–1070.

K. A. Janes, J. G. Albeck, S. Gaudet, P. K. Sorger, D. A. Lauffenburger and M. B. Yaffe. A systems model of signaling identifies a molecular basis set for cytokine-induced apoptosis. *Science*, **310** (2005), 1646–1653.

A. Javey, R. Tu, D. B. Farmer, J. Guo, R. G. Gordon and H. Dai. High Performance n-Type Carbon Nanotube Field-Effect Transistors with Chemically Doped Contacts. *Nano Letters*, **5** (2005), 345–348.

K. Jensen, K. Kim and A. Zettl. An atomic-resolution nanomechanical mass sensor. *Nature Nanotechnology*, **3** (2008), 533–537.

F. S. Johnson. The solar constant. *Journal of the Atmospheric Sciences*, **11** (1954), 431–439.

B. M. Johnstone. Genesis of the cochlear endolymphatic potential. *Current Topics in Bioenergetics*, **2** (1967), 335–352.

Eric R. Kandel, James H. Schwartz and Thomas M. Jessell. *Principles of Neural Science*. 3rd ed. (Norwalk, Conn.: Appleton & Lange, 1991).

R. Karakiewicz, R. Genov and G. Cauwenberghs. 1.1 TMACS/mW Load-Balanced Resonant Charge-Recycling Array Processor. *Proceedings of the IEEE Custom Integrated Circuits Conference (CICC)*, San Jose, California, 603–606, 2007.

T. R. Karl and K. E. Trenberth. Modern global climate change. *Science*, **302** (2003), 1719–1723.

S. K. Kelly and J. Wyatt. A power-efficient voltage-based neural tissue stimulator with energy recovery. *Proceedings of the IEEE International Solid-State Circuits Conference (ISSCC)*, San Francisco, CA, 228–524, 2004.

C. H. Kim and K. Roy. Dynamic V_{th} scaling scheme for active leakage power reduction. *Proceedings of the Design, Automation and Test in Europe Conference and Exhibition*, Paris, 163–167, 2002.

H. H. Kim, N. Mano, Y. Zhang and A. Heller. A miniature membrane-less biofuel cell operating under physiological conditions at 0.5 V. *Journal of The Electrochemical Society*, **150** (2003), A209.

R. Kline. Harold Black and the negative-feedback amplifier. *IEEE Control Systems Magazine*, **13** (1993), 82–85.

E. A. M. Klumperink, S. L. J. Gierkink, A. P. Van der Wel and B. Nauta. Reducing MOSFET $1/f$ noise and power consumption by switched biasing. *IEEE Journal of Solid-State Circuits*, **35** (2000), 994–1001.

P. K. Ko. Approaches to Scaling. In *Advanced MOS Device Physics*, ed. N. G. Einspruch, and G. S. Gildenblat (San Diego: Academic Press; 1989), pp. 1–37.

Christof Koch. *Biophysics of Computation: Information Processing in Single Neurons* (New York: Oxford University Press, 1999).

K. Koch, J. McLean, M. Berry, P. Sterling, V. Balasubramanian and M. A. Freed. Efficiency of information transmission by retinal ganglion cells. *Current Biology*, **14** (2004), 1523–1530.

R. Kolarova, T. Skotnicki and J. A. Chroboczek. Low frequency noise in thin gate oxide MOSFETs. *Microelectronics Reliability*, **41** (2001), 579–585.

P. Korkmaz, B. E. S. Akgul and K. V. Palem. Energy, performance, and probability tradeoffs for energy-efficient probabilistic CMOS circuits. *IEEE Transactions on Circuits and Systems I: Regular Papers*, **55** (2008), 2249–2262.

J. D. Kraus and R. J. Marhefka. *Antennas for All Applications* (New York: McGraw-Hill, 2002).

M. L. Kringelbach, N. Jenkinson, S. L. F. Owen and T. Z. Aziz. Translational principles of deep brain stimulation. *Nature Reviews Neuroscience*, **8** (2007), 623–635.

G. Kron. Equivalent circuit of the field equations of Maxwell-I. *Proceedings of the IRE*, **32** (1944), 289–299.

J. Kwong, Y. K. Ramadass, N. Verma and A. P. Chandrakasan. A 65 nm Sub-V_t, microcontroller with integrated SRAM and switched capacitor DC-DC converter. *IEEE Journal of Solid-State Circuits*, **44** (2009), 115–126.

M. F. Land. Movements of the retinae of jumping spiders (Salticidae: Dendryphantinae) in response to visual stimuli. *Journal of Experimental Biology*, **51** (1969), 471–493.

R. Landauer. Information is physical. *Workshop on Physics and Computation (PhysComp)*, Dallas, Tex., 1–4, 1992.

J. Lazzaro, S. Ryckebusch, M. A. Mahowald and C. A. Mead. Winner-take-all networks of O (n) complexity. *Advances in Neural Information Processing Systems*, **1** (1989), 703–711.

S. Lee and T. Sakurai. Run-time voltage hopping for low-power real-time systems. *Proceedings of the 37th Annual IEEE ACM Design Automation Conference*, Los Angeles, CA, 806–809, 2000.

Thomas H. Lee. *The Design of CMOS Radio-Frequency Integrated Circuits*. 2nd ed. (Cambridge, UK; New York: Cambridge University Press, 2004).

M. Lehmann, W. Baumann, M. Brischwein, H. J. Gahle, I. Freund, R. Ehret, S. Drechsler, H. Palzer, M. Kleintges and U. Sieben. Simultaneous measurement of cellular respiration and acidification with a single CMOS ISFET. *Biosensors and Bioelectronics*, **16** (2001), 195–203.

V. Leonov, T. Torfs, P. Fiorini, C. Van Hoof and L. Imec. Thermoelectric converters of human warmth for self-powered wireless sensor nodes. *IEEE Sensors Journal*, **7** (2007), 650–657.

M. A. Lewis, R. Etienne-Cummings, A. H. Cohen and M. Hartmann. Toward biomorphic control using custom aVLSI CPG chips. *Proceedings of the International Conference on Robotics and Automation (ICRA)*, 494–500, 2000.

N. S. Lewis. Toward cost-effective solar energy use. *Science*, **315** (2007), 798–801.

K. K. Likharev. Single-electron devices and their applications. *Proceedings of the IEEE*, **87** (1999), 606–632.

David Linden and Thomas B. Reddy. *Handbook of Batteries*. 3rd ed. (New York: McGraw-Hill, 2002).

S-C. Liu, J. È. Kramer, G. Indiveri, T. Delbruck, T. Burg and R. Douglas. Orientation-selective aVLSI spiking neurons. *Neural Networks*, **14** (2001), 629–643.

W. Liu, A. G. Andreou and M. H. Goldstein. Voiced-speech representation by an analog silicon model of theauditory periphery. *IEEE Transactions on Neural Networks*, **3** (1992), 477–487.

Harvey F. Lodish. *Molecular Cell Biology*. 6th ed. (New York: W.H. Freeman, 2008).

P. C. Loizou. Mimicking the human ear. *IEEE Signal Processing Magazine*, **15** (1998), 101–130.

T. K. Lu and J. J. Collins. Engineered bacteriophage targeting gene networks as adjuvants for antibiotic therapy. *Proceedings of the National Academy of Sciences*, **106** (2009), 4629.

T. K. Lu, S. Zhak, P. Dallos and R. Sarpeshkar. Fast cochlear amplification with slow outer hair cells. *Hearing Research*, **214** (2006), 45–67.

M. Lundstrom. Elementary scattering theory of the Si MOSFET. *IEEE Electron Device Letters*, **18** (1997), 361–363.

Mark Lundstrom and Jing Guo. *Nanoscale Transistors: Device Physics, Modeling and Simulation* (New York: Springer, 2006).

R. F. Lyon and C. A. Mead. An analog electronic cochlea. *IEEE Transactions on Acoustics, Speech and Signal Processing*, **36** (1988), 1119–1134.

D. J. C. MacKay. *Sustainable Energy: Without the Hot Air* (Cambridge, UK: UIT Cambridge Ltd., 2009).

P. F. MacNeilage, L. J. Rogers and G. Vallortigara. Origins of the left & right brain. *Scientific American*, **301** (2009), 60–67.

M. A. Maher and C. A. Mead. A physical charge-controlled model for the MOS transistor. *Proceedings of the Advanced Research in VLSI Conference*, Stanford, CA, 1987.

M. A. Maher and C. A. Mead. Fine Points of Transistor Physics. In *Analog VLSI and Neural Systems*, ed. C Mead (Reading, MA: Addison-Wesley; 1989), pp. 319–338.

M. A. Mahowald and C. A. Mead. Silicon Retina. In *Addison-Wesley VLSI Systems Series*, ed. C A Mead (Reading, Mass.: Addison-Wesley; 1989), pp. 257–278.

S. Mandal and R. Sarpeshkar. Low-power CMOS rectifier design for RFID applications. *IEEE Transactions on Circuits and Systems I: Regular Papers*, **54** (2007), 1177–1188.

S. Mandal and R. Sarpeshkar. Power-Efficient Impedance-Modulation Wireless Data Links for Biomedical Implants. *IEEE Transactions on Biomedical Circuits and Systems*, **2** (2008), 301–315.

S. Mandal and R. Sarpeshkar. Circuit models of stochastic genetic networks. *Proceedings of the IEEE Biological Circuits and Systems Conference*, Beijing, China, 2009.

S. Mandal and R. Sarpeshkar. Log-domain Circuit Models of Chemical Reactions. *Proceedings of the IEEE Symposium on Circuits and Systems (ISCAS)*, Taipei, Taiwan, 2009.

S. Mandal, S. K. Arfin and R. Sarpeshkar. Sub-μHz MOSFET $1/f$ noise measurements. *Electronics Letters*, **45** (2009).

S. Mandal, L. Turicchia and R. Sarpeshkar. A Low-Power Battery-Free Tag for Body Sensor Networks. *IEEE Pervasive Computing*, **in press** (2009).

S. Mandal, S. Zhak and R. Sarpeshkar, inventors. *Architectures for Universal or Software Radio*. U.S. Provisional Patent 60/870,719, filed December 19, 2006; Utility Patent 11/958,990, filed December 18, 2007.

S. Mandal, S. M. Zhak and R. Sarpeshkar. A bio-inspired active radio-frequency silicon cochlea. *IEEE Journal of Solid-State Circuits*, **44** (2009), 1814–1828.

L. L. Marker and A. J. Dickman. Morphology, physical condition, and growth of the cheetah (Acinonyx jubatus jubatus). *Journal of Mammalogy*, **84** (2003), 840–850.

A. L. McWorther. $1/f$ noise and germanium surface properties. In *Semiconductor Surface Physics*, ed. R. H. Kingston (Philadelphia, Penn.: University of Pennsylvania Press; 1957), pp. xvi, 413 p.

C. A. Mead. Neuromorphic electronic systems. *Proceedings of the IEEE*, **78** (1990), 1629–1636.

C. A. Mead. Scaling of MOS technology to submicrometer feature sizes. *The Journal of VLSI Signal Processing*, **8** (1994), 9–25.

Carver Mead. *Analog VLSI and Neural Systems* (Reading, Mass.: Addison-Wesley, 1989).

Carver Mead. *Collective Electrodynamics: Quantum Foundations of Electromagnetism* (Cambridge, MA: MIT Press, 2000).

J. T. Mettetal, D. Muzzey, C. Gomez-Uribe and A. van Oudenaarden. The frequency dependence of osmo-adaptation in Saccharomyces cerevisiae. *Science*, **319** (2008), 482.

R. A. C. Metting van Rijn, A. Peper and C. A. Grimbergen. High-quality recording of bioelectric events. *Medical and Biological Engineering and Computing*, **28** (1990), 389–397.

A. C. H. MeVay and R. Sarpeshkar. Predictive comparators with adaptive control. *IEEE Transactions on Circuits and Systems II: Analog and Digital Signal Processing*, **50** (2003), 579–588.

R. D. Middlebrook. Null double injection and the extra element theorem. *IEEE Transactions on Education*, **32** (1989), 167–180.

D. Miklavcic, N. Pavselj and F. X. Hart. Electric properties of tissues. In *Wiley Encyclopedia of Biomedical Engineering*, ed. M. Akay (New York: John Wiley & Sons; 2006).

J. R. Miller and P. Simon. Electrochemical Capacitors for Energy Management. *Science*, **321** (2008), 651.

B. A. Minch, Synthesis of multiple-input translinear element log-domain filters. *Proceedings of the IEEE International Symposium on Circuits and Systems (ISCAS)*, Orlando, FL, 697–700, 1999.

P. D. Mitcheson, E. M. Yeatman, G. K. Rao, A. S. Holmes and T. C. Green. Energy harvesting from human and machine motion for wireless electronic devices. *Proceedings of the IEEE*, **96** (2008), 1457–1486.

J. Mitola III and G. Q. Maguire Jr. Cognitive radio: making software radios more personal. *IEEE Personal Communications*, **6** (1999), 13–18.

J. L. Montieth and Lawrence Mount, editors. *Heat Loss from Animals and Man* (London: Butterworth & Co. Publishers Ltd.; 1974).

G. E. Moore. Cramming more components onto integrated circuits. *Electronics*, **38** (1965), 114–117.

D. J. Morton and D. D. Fuller. *Human Locomotion and Body Form* (New York, NY: Waverly Press, Inc., 1952).

Kary B. Mullis, François Ferré and Richard Gibbs. *The Polymerase Chain Reaction* (Boston: Birkhäuser, 1994).

B. Murmann and B. E. Boser. A 12-bit 75-MS/s pipelined ADC using open-loop residue amplification. *IEEE Journal of Solid-State Circuits*, **38** (2003), 2040–2050.

R. N. Mutagi. Pseudo noise sequences for engineers. *Electronics and Communication Engineering Journal*, **8** (1996), 79–87.

K. Natori. Ballistic metal-oxide-semiconductor field effect transistor. *Journal of Applied Physics*, **76** (1994), 4879–4890.

K. Natori. Scaling Limit of the MOS Transistor–A Ballistic MOSFET. *IEICE Transactions on Electronics*, **E84-C** (2001), 1029–1036.

Nam Trung Nguyen and Steven T. Wereley. *Fundamentals and Applications of Microfluidics* (Boston, MA: Artech House, 2002).

H. Nyquist. Thermal agitation of electric charge in conductors. *Physical Review*, **32** (1928), 110–113.

H. Nyquist. Regeneration theory. *Bell Systems Technical Journal*, **11** (1932), 126–147.

S. O'Driscoll, A. Poon and T. H. Meng. A mm-sized implantable power receiver with adaptive link compensation. *Digest of Technical Papers of the IEEE International Solid-State Circuits Conference (ISSCC)*, San Francisco, California, 294–295, 295a, 2009.

M. O'Halloran and R. Sarpeshkar. A 10-nW 12-bit accurate analog storage cell with 10-aA leakage. *IEEE Journal of Solid-State Circuits*, **39** (2004), 1985–1996.

M. O'Halloran and R. Sarpeshkar, An analog storage cell with 5 electron/sec leakage. *Proceedings of the IEEE International Symposium on Circuits and Systems (ISCAS)*, Kos, Greece, 557–560, 2006.

H. Okawa, A. P. Sampath, S. B. Laughlin and G. L. Fain. ATP consumption by mammalian rod photoreceptors in darkness and in light. *Current Biology*, **18** (2008), 1917–1921.

S. J. Osterfeld, H. Yu, R. S. Gaster, S. Caramuta, L. Xu, S. J. Han, D. A. Hall, R. J. Wilson, S. Sun and R. L. White. Multiplex protein assays based on real-time magnetic nanotag sensing. *Proceedings of the National Academy of Sciences*, **105** (2008), 20637.

A. J. Oxenham, A. M. Simonson, L. Turicchia and R. Sarpeshkar. Evaluation of companding-based spectral enhancement using simulated cochlear-implant processing. *The Journal of the Acoustical Society of America*, **121** (2007), 1709–1716.

E. M. Ozbudak, M. Thattai, I. Kurtser, A. D. Grossman and A. van Oudenaarden. Regulation of noise in the expression of a single gene. *Nature Genetics*, **31** (2002), 69–73.

Bernhard Palsson. *Systems Biology: Properties of Reconstructed Networks* (New York: Cambridge University Press, 2006).

J. A. Paradiso and T. Starner. Energy scavenging for mobile and wireless electronics. *IEEE Pervasive Computing*, **4** (2005), 18–27.

S. J. Park, T. A. Taton and C. A. Mirkin. Array-based electrical detection of DNA with nanoparticle probes. *Science*, **295** (2002), 1503–1506.

S. Pavan, N. Krishnapura, R. Pandarinathan and P. Sankar. A Power-Optimized Continuous-Time Delta Sigma ADC for Audio Applications. *IEEE Journal of Solid-State Circuits*, **43** (2008), 351–360.

G. L. Pearson. Shot Effect and Thermal Agitation in an Electron Current Limited by Space Charge. *Physics*, **6** (1935).

J. S. Pezaris and R. C. Reid. Demonstration of artificial visual percepts generated through thalamic microstimulation. *Proceedings of the National Academy of Sciences*, **104** (2007), 7670–7675.

James O. Pickles. *An Introduction to the Physiology of Hearing*. 2nd ed. (London; New York: Academic Press, 1988).

M. Pidwirny and D. Budikova. Earth's energy balance. In *Encyclopedia of Earth*, ed. C. J. Cleveland. (2008), Available from http://www.eoearth.org/article/Earth's_energy_balance.

Christian Piguet. *Low-power Electronics Design* (Boca Raton: CRC Press, 2005).

A. S. Porret, J. M. Sallese and C. C. Enz. A compact non-quasi-static extension of a charge-based MOS model. *IEEE Transactions on Electron Devices*, **48** (2001), 1647–1654.

PowerStream Technology, Lund Instrument Engineering, Inc. Available from: www. powerstream.com.

M. Ptashne and A. Gann. *Genes and Signals* (Cold Spring Harbour, New York: CSHL Press, 2002).

Y. Pu, J. P. de Gyvez, H. Corporaal and Y. Ha. An ultra-low-energy/frame multi-standard JPEG co-processor in 65 nm CMOS with sub/near-threshold power supply. *Proceedings of the IEEE Solid-State Circuits Conference (ISSCC)*, San Francisco, 146–147,147a, 2009.

M. Puurtinen, J. Hyttinen and J. Malmivuo. Optimizing bipolar electrode location for wireless ECG measurement–analysis of ECG signal strength and deviation between individuals. *International Journal of Bioelectromagnetism*, **7** (2005), 236–239.

F. Raab. Idealized operation of the class E tuned power amplifier. *IEEE Transactions on Circuits and Systems*, **24** (1977), 725–735.

D. V. Ragone. Review of Battery Systems for Electrically Powered Vehicles. *Mid-Year Meeting of the Society of Automotive Engineers*, Detroit, Michigan, 1968.

B. Raj, L. Turicchia, B. Schmidt-Nielsen and R. Sarpeshkar. An FFT-based companding front end for noise-robust automatic speech recognition. *EURASIP Journal on Audio, Speech, and Music Processing*, **Article ID 65420** (2007), 13.

B. I. Rapoport, W. Wattanapanitch, J. L. Penagos, S. Musallam, R. Andersen and R. Sarpeshkar. A biomimetic adaptive algorithm and low-power architecture for implantable neural decoders. *Proceedings of the 31st Annual International Conference of the IEEE Engineering in Medicine and Biology Society (EMBC)*, Minneapolis, MN, 2009.

Behzad Razavi. *RF Microelectronics* (Upper Saddle River, NJ: Prentice Hall, 1998).

F. Reif. *Fundamentals of Statistical and Thermal Physics* (New York: McGraw Hill, 1965).

James K. Roberge. *Operational Amplifiers: Theory and Practice* (New York: Wiley, 1975).

D. A. Robinson, J. L. Gordon and S. E. Gordon. A model of the smooth pursuit eye movement system. *Biological Cybernetics*, **55** (1986), 43–57.

M. Roham, J. M. Halpern, H. B. Martin, H. J. Chiel and P. Mohseni. Wireless amperometric neurochemical monitoring using an integrated telemetry circuit. *IEEE Transactions on Biomedical Engineering*, **55** (2008), 2628–2634.

S. Roundy. On the effectiveness of vibration-based energy harvesting. *Journal of Intelligent Material Systems and Structures*, **16** (2005), 809.

S. Roundy, P. K. Wright and J. M. Rabaey. *Energy Scavaging for Wireless Sensor Networks*. 1st ed. (Boston, Mass.: Kluwer Academic Publishers, 2003).

E. M. Rubin. Genomics of cellulosic biofuels. *Nature*, **454** (2008), 841–845.

J. M. Sallese, M. Bucher, F. Krummenacher and P. Fazan. Inversion charge linearization in MOSFET modeling and rigorous derivation of the EKV compact model. *Solid State Electronics*, **47** (2003), 677–683.

C. D. Salthouse and R. Sarpeshkar. A practical micropower programmable bandpass filter for use in bionic ears. *IEEE Journal of Solid-State Circuits*, **38** (2003), 63–70.

R. Sarpeshkar. *Efficient precise computation with noisy components: Extrapolating from an electronic cochlea to the brain*. PhD Thesis, Computation and Neural Systems, California Institute of Technology (1997).

R. Sarpeshkar. Analog versus digital: extrapolating from electronics to neurobiology. *Neural Computation*, **10** (1998), 1601–1638.

R. Sarpeshkar. Brain power: Borrowing from biology makes for low power computing. *IEEE Spectrum*, **43** (2006), 24–29.

R. Sarpeshkar and M. O'Halloran. Scalable hybrid computation with spikes. *Neural Computation*, **14** (2002), 2003–2038.

R. Sarpeshkar and M. Tavakoli, inventors. *An Ultra-Low-Power Pulse Oximeter Implemented with an Energy-Efficient Transimpedance Amplifier*, U.S. Provisional Patent 60/847/034, filed September 25, 2006; Utility Patent 11/903,571, filed September 24, 2007.

R. Sarpeshkar, M. W. Baker, C. D. Salthouse, J. J. Sit, L. Turicchia and S. M. Zhak, An analog bionic ear processor with zero-crossing detection. *Proceedings of the IEEE International Solid State Circuits Conference (ISSCC)*, San Francisco, CA, 78–79, 2005.

R. Sarpeshkar, T. Delbruck and C. A. Mead. White noise in MOS transistors and resistors. *IEEE Circuits and Devices Magazine*, **9** (1993), 23–29.

R. Sarpeshkar, J. È. Kramer, G. Indiveri and C. Koch. Analog VLSI architectures for motion processing: From fundamental limits to system applications. *Proceedings of the IEEE*, **84** (1996), 969–987.

R. Sarpeshkar, R. F. Lyon and C. A. Mead. A low-power wide-linear-range transconductance amplifier. *Analog Integrated Circuits and Signal Processing*, **13** (1997), 123–151.

R. Sarpeshkar, R. F. Lyon and C. A. Mead. A low-power wide-dynamic-range analog VLSI cochlea. *Analog Integrated Circuits and Signal Processing*, **16** (1998), 245–274.

R. Sarpeshkar, C. D. Salthouse, J. J. Sit, M. W. Baker, S. M. Zhak, T. K. T. Lu, L. Turicchia and S. Balster. An ultra-low-power programmable analog bionic ear processor. *IEEE Transactions on Biomedical Engineering*, **52** (2005), 711–727.

R. Sarpeshkar, W. Wattanapanitch, S. K. Arfin, B. I. Rapoport, S. Mandal, M. W. Baker, M. S. Fee, S. Musallam and R. A. Andersen. Low-Power Circuits for Brain-Machine Interfaces. *IEEE Transactions on Biomedical Circuits and Systems*, **2** (2008), 173–183.

R. Sarpeshkar, L. Watts and C. A. Mead. Refractory neuron circuits. *Computation and Neural Systems Memo CNS TR-92–08*, (1992).

Leonard I. Schiff. *Quantum Mechanics*. 3d ed. (New York,: McGraw-Hill, 1968).

E. M. Schmidt, M. J. Bak, F. T. Hambrecht, C. V. Kufta, D. K. O'Rourke and P. Vallabhanath. Feasibility of a visual prosthesis for the blind based on intracortical micro stimulation of the visual cortex. *Brain*, **119** (1996), 507.

M. D. Scott, B. E. Boser and K. S. J. Pister. An ultra-low-energy ADC for smart dust. *IEEE Journal of Solid-State Circuits*, **38** (2003), 1123–1129.

Second Sight Completes US Phase I Enrollment and Commences European Clinical Trial for the Argus II Retinal Implant. [Press Release] Available from: http://www.2-sight.com/press-release2-15-final.html.

E. Seevinck. Companding current-mode integrator: A new circuit principle for continuous-time monolithic filters. *Electronics Letters*, **26** (1990), 2046–2047.

E. Seevinck, E. A. Vittoz, M. du Plessi, T. H. Joubert and W. Beetge. CMOS translinear circuits for minimum supply voltage. *IEEE Transactions on Circuits and Systems II: Analog and Digital Signal Processing*, **47** (2000), 1560–1564.

Self-Resonant Frequency of Single-Layer Solenoid Coils. Available from: http://www.smeter.net/feeding/transmission-line-choke-coils.php#Circuit.

Claude Elwood Shannon and Warren Weaver. *The Mathematical Theory of Communication* (Urbana: University of Illinois Press, 1949).

N. C. C. Sharp. Timed running speed of a cheetah (*Acinonyx jubatus*). *Journal of Zoology*, **241** (1997), 493–494.

R. K. Shepherd, N. Linahan, J. Xu, G. M. Clark and S. Araki. Chronic electrical stimulation of the auditory nerve using non-charge-balanced stimuli. *Acta Oto-Laryngologica*, **119** (1999), 674–684.

C. A. Shera. Intensity-invariance of fine time structure in basilar-membrane click responses: Implications for cochlear mechanics. *The Journal of the Acoustical Society of America*, **110** (2001), 332.

C. A. Shera. Mammalian spontaneous otoacoustic emissions are amplitude-stabilized cochlear standing waves. *The Journal of the Acoustical Society of America*, **114** (2003), 244.

W. Shockley and H. J. Queisser. Detailed Balance Limit of Efficiency of p-n Junction Solar Cells. *Journal of Applied Physics*, **32** (1961), 510–519.

W. Shyy, M. Berg and D. Ljungqvist. Flapping and flexible wings for biological and micro air vehicles. *Progress in Aerospace Sciences*, **35** (1999), 455–505.

William McC Siebert. *Circuits, Signals, and Systems* (Cambridge, Mass.; New York: MIT Press; McGraw-Hill, 1986).

J. J. Sit and R. Sarpeshkar. A micropower logarithmic A/D with offset and temperature compensation. *IEEE Journal of Solid-State Circuits*, **39** (2004), 308–319.

J. J. Sit and R. Sarpeshkar. A low-power, blocking-capacitor-free, charge-balanced electrode-stimulator chip with less than 6 nA DC error for 1mA full-scale stimulation. *IEEE Transactions on Biomedical Circuits and Systems*, **1** (2007), 172–183.

J. J. Sit and R. Sarpeshkar. A Cochlear-Implant Processor for Encoding Music and Lowering Stimulation Power. *IEEE Pervasive Computing*, **1** (2008), 40–48.

J. J. Sit, A. M. Simonson, A. J. Oxenham, M. A. Faltys, R. Sarpeshkar and C. MIT. A low-power asynchronous interleaved sampling algorithm for cochlear implants that encodes envelope and phase information. *IEEE Transactions on Biomedical Engineering*, **54** (2007), 138–149.

N. O. Sokal and A. D. Sokal. Class E – A new class of high-efficiency tuned single-ended switching power amplifiers. *IEEE Journal of Solid-State Circuits*, **10** (1975), 168–176.

A. Solga, Z. Cerman, B. F. Striffler, M. Spaeth and W. Barthlott. The dream of staying clean: Lotus and biomimetic surfaces. *Bioinspiration and Biomimetics*, **2** (2007), 126.

Y. A. Song, C. Batista, R. Sarpeshkar and J. Han. Rapid fabrication of microfluidic polymer electrolyte membrane fuel cell in PDMS by surface patterning of perfluorinated ion-exchange resin. *Journal of Power Sources*, **183** (2008), 674–677.

S. L. Spencer, S. Gaudet, J. G. Albeck, J. M. Burke and P. K. Sorger. Non-genetic origins of cell-to-cell variability in TRAIL-induced apoptosis. *Nature*, **459** (2009), 428–432.

D. Spreemann, Y. Manoli, B. Folkmer and D. Mintenbeck. Non-resonant vibration conversion. *Journal of Micromechanics and Microengineering*, **16** (2006), 169.

T. M. Squires and S. R. Quake. Microfluidics: fluid physics at the nanoliter scale. *Reviews of Modern Physics*, **77** (2005), 977–1026.

T. M. Squires, R. J. Messinger and S. R. Manalis. Making it stick: convection, reaction and diffusion in surface-based biosensors. *Nature Biotechnology*, **26** (2008), 417–426.

C. Stagni, D. Esposti, C. Guiducci, C. Paulus, M. Schienle, M. Augustyniak, G. Zuccheri, B. Samor, L. Benini and B. Ricco. Fully electronic CMOS DNA detection array based on capacitance measurement with on-chip analog-to-digital conversion. *Proceedings of the*

IEEE International Solid-State Circuits Conference (ISSCC), San Francisco, CA, 69–78, 2006.

Thad E. Starner and Joseph A. Paradiso. Human generated power for mobile electronics. In *Low Power Electronics Design*, ed. C. Piguet (Boca Raton, Flor.: CRC Press; 2005).

E. Stern, J. F. Klemic, D. A. Routenberg, P. N. Wyrembak, D. B. Turner-Evans, A. D. Hamilton, D. A. LaVan, T. M. Fahmy and M. A. Reed. Label-free immunodetection with CMOS-compatible semiconducting nanowires. *Nature*, **445** (2007), 519–522.

Kenneth N. Stevens. *Acoustic Phonetics* (Cambridge, MA: MIT Press, 2000).

L. Svensson. Adiabatic and Clock-Powered Circuits. In *Low-Power Electronics Design*, ed. C. Piguet. (Boca Raton: CRC Press; 2005), pp. 15.11–15.15.

L. Svensson, W. C. Athas and R. S. C. Wen, A sub-CV^2 pad driver with 10 ns transition time. *Proceedings of the International Symposium on Low Power Electronics and Design*, Monterey, California, 105–108, 1996.

R. M. Swanson and J. D. Meindl. Ion-implanted complementary MOS transistors in low-voltage circuits. *IEEE Journal of Solid-State Circuits*, **7** (1972), 146–153.

T. G. Tang, Q. M. Tieng and M. W. Gunn. Equivalent circuit of a dipole antenna using frequency-independent lumped elements. *IEEE Transactions on Antennas and Propagation*, **41** (1993), 100–103.

D. Tank and J. Hopfield. Simple 'neural' optimization networks: An A/D converter, signal decision circuit, and a linear programming circuit. *IEEE Transactions on Circuits and Systems*, **33** (1986), 533–541.

D. W. Tank and J. J. Hopfield. Collective computation in neuronlike circuits. *Scientific American*, **257** (1987), 104–114.

R. Tashiro, N. Kabei, K. Katayama, E. Tsuboi and K. Tsuchiya. Development of an electrostatic generator for a cardiac pacemaker that harnesses the ventricular wall motion. *Journal of Artificial Organs*, **5** (2002), 239–245.

M. Tavakoli and R. Sarpeshkar. An offset-canceling low-noise lock-in architecture for capacitive sensing. *IEEE Journal of Solid-State Circuits*, **38** (2003), 244–253.

M. Tavakoli, L. Turicchia and R. Sarpeshkar. An ultra-low-power pulse oximeter implemented with an energy efficient transimpedance amplifier. *IEEE Transactions on Biomedical Circuits and Systems*, **in press** (2009).

C. R. Taylor, A. Shkolnik, R. Dmi'el, D. Baharav and A. Borut. Running in cheetahs, gazelles, and goats: energy cost and limb configuration. *American Journal of Physiology*, **227** (1974), 848–850.

H. Tennekes. *The Simple Science of Flight: From Insects to Jumbo Jets* (Cambridge, Mass.: MIT Press, 1997).

F. E. Terman. *Radio Engineers Handbook* (New York: McGraw-Hill, 1943).

Tesla Motors. Available from: www.teslamotors.com.

Texas Instruments. TMS320C55x Technical Overview, Literature Number SPRU393. Technical Report: Texas Instruments Incorporated, February 2000.

Texas Instruments BGA Mechanical Data, Document MPBG021C. Technical Report: Texas Instruments Incorporated, May 2002.

Texas Instruments. TMS320VC5510 Power Consumption Summary, Literature Number SPRA972. Application Report: Texas Instruments Incorporated, November 2003.

L. Theogarajan, J. Wyatt, J. Rizzo, B. Drohan, M. Markova, S. Kelly, G. Swider, M. Raj, D. Shire and M. Gingerich. Minimally invasive retinal prosthesis. *Proceedings of the IEEE International Solid-State Circuits Conference (ISSCC)*, San Francisco, CA, 99–108, 2006.

B. J. Thompson, D. O. North and W. A. Harris. Fluctuations in space-charge-limited currents at moderately high frequencies. *RCA Review*, **4** (1940).

J. Thoné, S. Radiom, D. Turgis, R. Carta, G. Gielen and R. Puers. Design of a 2Mbps FSK near-field transmitter for wireless capsule endoscopy. *Sensors & Actuators: A. Physical* (2008).

C. Toumazou, F. J. Lidgey and D. Haigh. *Analogue IC design: The Current-Mode Approach* (Herts, United Kingdom: Institution of Engineering and Technology, 1990).

Yannis Tsividis. *Mixed Analog-Digital VLSI Devices and Technology* (Singapore: World Scientific Publishing Company, 2002).

Yannis Tsividis. *Operation and Modeling of the MOS Transistor*. 3rd ed. (New York: Oxford University Press, 2008).

Y. P. Tsividis, V. Gopinathan and L. Toth. Companding in signal processing. *Electronics Letters*, **26** (1990), 1331–1332.

Y. Tsividis, N. Krishnapura, Y. Palaskas and L. Toth. Internally varying analog circuits minimize power dissipation. *IEEE Circuits and Devices Magazine*, **19** (2003), 63–72.

L. Turicchia and R. Sarpeshkar. A bio-inspired companding strategy for spectral enhancement. *IEEE Transactions on Speech and Audio Processing*, **13** (2005), 243–253.

L. Turicchia, S. Mandal, M. Tavakoli, L. Fay, V. Misra, J. Bohoruez, W. Sanchez and R. Sarpeshkar. Ultra-low-power Electronics for Non-invasive Medical Monitoring. *Proceedings of the IEEE Custom Integrated Circuits Conference (CICC)*, Invited Paper, 6–5, San Jose, CA, 2009.

L. Turicchia, M. O'Halloran, D. P. Kumar and R. Sarpeshkar. A low-power imager and compression algorithms for a brain-machine visual prosthesis for the blind. *Proceedings of the SPIE*, San Diego, CA, 7035101–7035113, 2008.

Aldert Van der Ziel. *Noise: Sources, Characterization, Measurement* (Englewood Cliffs, N.J.: Prentice-Hall, 1970).

M. van Elzakker, E. van Tuijl, P. Geraedts, D. Schinkel, E. Klumperink and B. Nauta. A 1.9 μW 4.4 fJ/Conversion-step 10 b 1MS/s Charge-Redistribution ADC. *Digest of Technical Papers IEEE International Solid-State Circuits Conference (ISSCC)*, San Francisco, CA, 244–610, 2008.

M. E. van Valkenburg. *Analog Filter Design* (New York: Oxford University Press, Inc., 1982).

E. Vittoz. Weak Inversion for Ultimate Low-Power Logic. In *Low-power Electronics Design*, ed. C. Piguet (Boca Raton: CRC Press; 2005), pp. 16–11–16–18.

Vatché Vorpérian. *Fast Analytical Techniques for Electrical and Electronic Circuits* (Cambridge; New York: Cambridge University Press, 2002).

G. S. Wagner. *Marriott's Practical Electrocardiography*. 10th ed. (Philadelphia: Lippincott, Williams & Wilkins, 2001).

R. H. Walden. Analog-to-digital converter survey and analysis. *IEEE Journal on Selected Areas in Communications*, **17** (1999), 539–550.

H. Wang, Y. Chen, A. Hassibi, A. Scherer and A. Hajimiri. A frequency-shift CMOS magnetic biosensor array with single-bead sensitivity and no external magnet. *Digest of Technical Papers IEEE International Solid-State Circuits Conference (ISSCC)* San Francisco, CA, 438–439, 2009.

L. Wang, M. Kondo and A. Bill. Glucose metabolism in cat outer retina. Effects of light and hyperoxia. *Investigative Ophthalmology & Visual Science*, **38** (1997), 48–55.

X. Wang, P. J. Hurst and S. H. Lewis. A 12-bit 20-Msample/s pipelined analog-to-digital converter with nested digital background calibration. *IEEE Journal of Solid-State Circuits*, **39** (2004), 1799–1808.

W. Wattanapanitch, M. Fee and R. Sarpeshkar. An energy-efficient micropower neural recording amplifier. *IEEE Transactions on Biomedical Circuits and Systems*, **1** (2007), 136–147.

L. Watts. *Cochlear mechanics: Analysis and analog VLSI*. Ph.D. Thesis, Electrical Engineering, California Institute of Technology (1992).

K. H. Wee and R. Sarpeshkar. An electronically tunable linear or nonlinear MOS resistor. *IEEE Transactions on Circuits and Systems I: Regular Papers*, **55** (2008), 2573–2583.

K. H. Wee, L. Turicchia and R. Sarpeshkar. An Analog Integrated-Circuit Vocal Tract. *IEEE Transactions on Biomedical Circuits and Systems*, **2** (2008), 316–327.

M. S. Wegmueller, A. Kuhn, J. Froehlich, M. Oberle, N. Felber, N. Kuster and W. Fichtner. An Attempt to Model the Human Body as a Communication Channel. *IEEE Transactions on Biomedical Engineering*, **54** (2007), 1851–1857.

J. D. Weiland and M. S. Humayun. Visual prosthesis. *Proceedings of the IEEE*, **96** (2008), 1076–1084.

Thomas Fischer Weiss. *Cellular Biophysics* (Cambridge, Mass.: MIT Press, 1996).

C. J. Wiggers. The heart. *Scientific American*, **195** (1957).

E. M. Williams. Radio-Frequency Spectrum Analyzers. *Proceedings of the IRE*, **34** (1946), 18p–22p.

B. B. Winter and J. G. Webster. Reduction of interference due to common mode voltage in biopotential amplifiers. *IEEE Transactions on Biomedical Engineering*, **30** (1983), 58–62.

K. D. Wise. Silicon microsystems for neuroscience and neural prostheses. *IEEE Engineering in Medicine and Biology Magazine*, **24** (2005), 22–29.

U. Wismar, D. Wisland and P. Andreani. A 0.2V 0.44/spl mu W 20 kHz Analog to Digital $\Sigma\Delta$ Modulator with 57 fJ/conversion FoM. *Proceedings of the 32nd European Solid-State Circuits Conference (ESSCIRC)*, 187–190, 2006.

L. S. Y. Wong, S. Hossain, A. Ta, J. Edvinsson, D. H. Rivas, H. Naas, C. R. M. Div and C. A. Sunnyvale. A very low-power CMOS mixed-signal IC for implantable pacemaker applications. *IEEE Journal of Solid-State Circuits*, **39** (2004), 2446–2456.

J. Y. Wong. *Theory of Ground Vehicles*. 4th ed. (Hoboken, New Jersey: John Wiley & Sons, 2008).

J. Wyatt and J. Rizzo. Ocular implants for the blind. *IEEE Spectrum*, **33** (1996), 47–53.

Y. Xia and G. M. Whitesides. Soft lithography. *Annual Review of Materials Science*, **28** (1998), 153–184.

T. Yagi, Y. Funahashi and F. Ariki. Dynamic model of dual layer neural network for vertebrate retina. *Proceedings of the International Joint Conference on Neural Networks (IJCNN)*, Washington, D.C., 787–789, 1989.

Guang Zong Yang. *Body Sensor Networks* (London: Springer-Verlag, 2006).

H. Y. Yang. *A time-based energy-efficient analog-to-digital converter*. Ph.D. Thesis, Electrical Engineering and Computer Science, Massachusetts Institute of Technology (2006).

H. Y. Yang and R. Sarpeshkar. A time-based energy-efficient analog-to-digital converter. *IEEE Journal of Solid-State Circuits*, **40** (2005), 1590–1601.

H. Y. Yang and R. Sarpeshkar. A Bio-Inspired Ultra-Energy-Efficient Analog-to-Digital Converter for Biomedical Applications. *IEEE Transactions on Circuits and Systems I: Regular Papers*, **53** (2006), 2349–2356.

R. F. Yazicioglu, P. Merken, R. Puers and C. Van Hoof. A 200 μW Eight-Channel EEG Acquisition ASIC for Ambulatory EEG Systems. *IEEE Journal of Solid-State Circuits*, **43** (2008), 3025–3038.

C. J. Yu, Y. Wan, H. Yowanto, J. Li, C. Tao, M. D. James, C. L. Tan, G. F. Blackburn and T. J. Meade. Electronic detection of single-base mismatches in DNA with ferrocene-modified probes. *Journal of the American Chemical Society*, **123** (2001), 11155–11161.

M. R. Yuce, T. Dissanayake and H. C. Keong. Wireless telemetry for electronic pill technology. *Proceedings of the IEEE Conference on Sensors*, Christchurch, New Zealand, 2009.

B. Zhai, L. Nazhandali, J. Olson, A. Reeves, M. Minuth, R. Helfand, Pant Sanjay, D. Blaauw and T. Austin. A 2.60pJ/Inst Subthreshold Sensor Processor for Optimal Energy Efficiency. *Proceedings of the IEEE Symposium on VLSI Circuits*, Honolulu, 154–155, 2006.

S. M. Zhak. *Modeling and design of an active silicon cochlea*. Ph.D. Thesis, Electrical Engineering, Massachusetts Institute of Technology (2008).

S. M. Zhak, M. W. Baker and R. Sarpeshkar. A low-power wide dynamic range envelope detector. *IEEE Journal of Solid-State Circuits*, **38** (2003), 1750–1753.

B. Zhao, J. S. Moore and D. J. Beebe. Surface-directed liquid flow inside microchannels. *Science*, **291** (2001), 1023–1026.

V. V. Zhirnov, R. K. Cavin III, J. A. Hutchby and G. I. Bourianoff. Limits to binary logic switch scaling–a gedanken model. *Proceedings of the IEEE*, **91** (2003), 1934–1939.

G. Zweig. Finding the impedance of the organ of Corti. *The Journal of the Acoustical Society of America*, **89** (1991), 1229.

G. Zweig, R. Lipes and J. R. Pierce. The cochlear compromise. *The Journal of the Acoustical Society of America*, **59** (1976), 975.

Index

in birds, 845
in cheetahs, 844
in planes, 845
of antelopes, 845
of walking, 845
wheels, 845
tricuspid valve, 583
tunneling, 151
Tsividis, 68, 354
Tzu, 103

ultra-capacitors, 536, 819, 853
ultra-low-leakage switch, 563
ultra-low-noise amplifier, 208
ultra wide band (UWB), 489, 525, 673
uplink, 491

ventricle, 581
visual prosthesis
    cochlear-implant inspired, 570

compression and coding in, 571
Vittoz, 129, 354
vocal tract, 730
    integrated-circuit implementation, 731
    noise robustness, 733

Warburg impedance, 810
wearable electronics, 579
well-input amplifier, 304
wide linear range amplifier (WLR), 303
winner-take-all circuit, 373
    large-signal operation, 379
    small-signal operation, 375
WKB approximation, 713
world electricity consumption, 836
world power consumption, 835

zinc-air battery
    basic numbers, 815
    chemistry, 815